The Biology of Paramyxoviruses

Edited by

Siba K. Samal

VA-MD Regional College of Veterinary Medicine
University of Maryland
College Park, MD
USA

Caister Academic Press

Caister Academic Press
Norfolk, UK

www.caister.com

British Library Cataloguing-in-Publication Data
A catalogue record for this book is available from the British Library

ISBN: 978-1-904455-85-1

Cover designed by Sachin Kumar, University of Maryland. The background image, provided by Dr Griffith D. Parks, Wake Forest University School of Medicine, is a single transmission electron microscopic image of SV5 virus. The nucleocapsid that has emerged from the particle has a 'herringbone-like' structure that represents turns of the helical nucleocapsid.

Printed and bound in Great Britain

Contents

Contributors

Danielle E. Anderson
INRS-Institut Armand-Frappier
University of Quebec
Laval, QC
Canada

danielle.anderson@duke-nus.edu.sg

Amiya K. Banerjee
Department of Molecular Genetics
Section of Virology
Lerner Research Institute
Cleveland Clinic
Cleveland, OH
USA

banerja@ccf.org

Michael D. Baron
Institute for Animal Health
Pirbright, Surrey
UK

michael.baron@bbsrc.ac.uk

Santanu Chattopadhyay
Department of Molecular Genetics
Section of Virology
Lerner Research Institute
Cleveland Clinic
Cleveland, OH
USA

santanuchattopadhyay@yahoo.com

Peter L. Collins
Laboratory of Infectious Diseases
National Institute of Allergy and Infectious Diseases
National Institutes of Health
Bethesda, MD
USA

pcollins@niaid.nih.gov

Melissa M. Coughlin
Division of Viral Diseases
Measles, Mumps, Rubella and Herpes Viruses Laboratory Branch
Centers for Disease Control and Prevention
Atlanta, GA
USA

mcoughlin@cdc.gov

James E. Crowe, Jr.
Vanderbilt Vaccine Center
Vanderbilt University Medical Center, T-2220 Medical Center North
Nashville, TN
USA

james.crowe@vanderbilt.edu

Sébastien Delpeut
INRS-Institut Armand-Frappier
University of Quebec
Laval, QC
Canada

sebastien.delpeut@iaf.inrs.ca

W. Paul Duprex
Department of Microbiology
Boston University
School of Medicine and National Emerging Infectious Diseases Laboratories
Boston, MA
USA

pduprex@bu.edu

Frank Esper
Department of Molecular Genetics
Section of Virology
Lerner Research Institute
Cleveland Clinic
Cleveland, OH
USA

frank.esper@case.edu

Bin Gotoh
Department of Pathology
Division of Microbiology and Infectious Diseases
Shiga University of Medical Science
Seta, Otsu
Shiga
Japan

bing@belle.shiga-med.ac.jp

Takashi Irie
Department of Virology
Graduate School of Biomedical Sciences
Hiroshima University
Kasumi Minami-ku
Hiroshima
Japan

tirie@hiroshima-u.ac.jp

John B. Johnson
Department of Microbiology and Immunology
Wake Forest University School of Medicine
Winston-Salem, NC
USA

jojohnson@wfubmc.edu

Michael K. Lo
Division of Viral Diseases
Measles, Mumps, Rubella and Herpes Viruses Laboratory Branch
Centers for Disease Control and Prevention
Atlanta, GA
USA

mlo@cdc.gov

Brian W.J. Mahy
Centers for Disease Control and Prevention, MS61
Atlanta, GA
USA

virusjournals@gmail.com

Mary J. Manuse
Department of Microbiology and Immunology
Wake Forest University School of Medicine
Winston-Salem, NC
USA

mmanuse@wfubmc.edu

Veronika von Messling
INRS-Institut Armand-Frappier
University of Quebec
Laval, QC
Canada

veronika.vonmessling@iaf.inrs.ca

Yoshiyuki Nagai
RIKEN Center of Research Network for Infectious Diseases
Chiyoda-ku
Tokyo
Japan

yoshi-nagai@riken.jp

Griffith D. Parks
Department of Microbiology and Immunology
Wake Forest University School of Medicine
Winston-Salem, NC
USA

gparks@wfubmc.edu

Paul A. Rota
Division of Viral Diseases
Measles, Mumps, Rubella and Herpes Viruses Laboratory Branch
Centers for Disease Control and Prevention
Atlanta, GA
USA

prota@cdc.gov

Steven Rubin
Division of Viral Products
Center for Biologics Evaluation and Research
U.S. Food and Drug Administration
Bethesda, MD
USA

Steven.Rubin@fda.hhs.gov

Siba K. Samal
VA-MD Regional College of Veterinary Medicine
University of Maryland
College Park, MD
USA

ssamal@umd.edu

Bevan Sawatsky
INRS-Institut Armand-Frappier
University of Quebec
Laval, QC
Canada

bevan.sawatsky@iaf.inrs.ca

Sibylle Schneider-Schaulies
Institute for Virology and Immunobiology
University of Wuerzburg
Wuerzburg
Germany

s-s-s@vim.uni-wuerzburg.de

Akira Takakura
Central Institute for Experimental Animals
Nogawa
Miyamae, Kawasaki
Japan

takakura@ciea.or.jp

Corinne Vandermeulen
Centre for Youth Health Care - K.U. Leuven
Kapucijnenvoer
Leuven
Belgium

corinne.vandermeulen@uzleuven.be

Lin-Fa Wang
CSIRO Livestock Industries
Australian Animal Health Laboratory
Geelong, Victoria
Australia

linfa.wang@csiro.au

John V. Williams
Department of Pediatrics
School of Medicine
Division of Infectious Diseases
Vanderbilt University
Nashville, TN
USA

john.williams@vanderbilt.edu

Yoshikazu Yonemitsu
Graduate School of Pharmaceutical Sciences
R&D Laboratory for Innovative Biotherapeutics
Kyushu University
Maidashi
Higashi-ku
Fukuoka
Japan

yonemitsu@med.kyushu-u.ac.jp

I dedicate this book to my late assistant, Ireen Dryburgh-Barry, who inspired me to undertake this project. She envisioned this project with me and assisted me in the beginning of preparing this book. Ireen passed away on 26 December 2009 after losing her battle to lung cancer. She was an outstanding person and a constant source of encouragement and inspiration during the 16 years that I was privileged to work with her at the University of Maryland.

Thank you, Ireen.

Preface

Paramyxoviruses are one of the most challenging families of viruses that infect a diverse range of hosts including mammals, birds, reptiles and fish. This diverse family encompasses many important human pathogens such as measles, mumps, parainfluenza, respiratory syncytial, and metapneumoviruses; animal pathogens such as rinderpest, canine distemper and Newcastle disease viruses; and zoonotic pathogens such as Nipah and Hendra viruses. Paramyxoviruses, such as SV-5 and Sendai viruses, have been major subjects for basic research by molecular biologists, and the knowledge gained has greatly improved our understanding of the structure and replication of viruses and their interactions with animal cells. The list of paramyxoviruses is already long and new paramyxoviruses are being isolated on a continuous basis from a wide variety of animals. The natural history and disease potential of many paramyxoviruses are poorly understood. There are many paramyxovirus diseases of human and animals for which effective vaccines are currently not available.

Previously, genetic manipulation of paramyxoviruses was not possible because the genome is not infectious alone and RNA recombination is essentially non-existent. This posed an obstacle to the molecular and biological characterization of these viruses. However, during the last fifteen years, methods of producing infectious paramyxoviruses from cDNA clones (reverse genetics) have been developed. This has greatly improved our understanding of the functions of each gene in replication and pathogenesis of these viruses. Another important aspect of reverse genetics is that vaccines can now be designed for some of the paramyxovirus diseases for which either vaccines are not currently available or the available vaccines are not satisfactory.

The purpose of this book is to incorporate the recent advances made in the biology of all paramyxoviruses in one place. The chapters in this book not only cover the genome structure and replication of paramyxoviruses, but also viral pathogenesis, control of virus infections and virus evolution. This book represents the most comprehensive and up-to-date information on this important group of human and animal viruses. Since the book is designed primarily as a reference, the authors have attempted to make each chapter comprehensible when read by itself, even though this involves repetition of information in other chapters. This book is a valuable source of timely information for virologists, microbiologists, molecular biologists, biochemists, immunologists, physicians, veterinarians, and scientists in related fields.

The broad coverage of viruses in this book would not have been possible without the contributions from many leading scientists, all experts in their individual fields. I express my most sincere gratitude to all of the authors for having joined in this effort. I would also like to thank LaShae Green, who provided assistance in proofreading and formatting the chapters in addition to corresponding with authors. My special thanks to Sachin Kumar, Baibaswata Nayak and Anandan Paldurai in addition to my graduate students and post-doctoral fellows for their help on this project as well.

Siba K. Samal

Introduction

Brian W. J. Mahy

The chapters in this excellent book provide a truly comprehensive account of all known paramyxoviruses, a group whose members include highly pathogenic viruses affecting the human population, as well as animals and birds.

In the early days of virology, the word 'myxovirus' was coined for a group of viruses that had common features, namely an affinity for mucoproteins, and an enzyme activity that attacks the mucoprotein substrate (Andrewes *et al.*, 1955). This group of viruses included fowl plague, first discovered in 1902 (Centanni, 1902), and later shown to be a highly virulent form of avian influenza virus (Davenport *et al.*, 1960), as well as the human viruses, influenza, parainfluenza, mumps, and avian Newcastle disease virus. Soon after this group was named, however, a number of biological differences as well as structural differences were noted between influenza and fowl plague viruses on the one hand, and parainfluenza, mumps, and Newcastle disease viruses on the other (Andrewes and Worthington, 1959; Franklin and Wecker, 1959). This led to the concept of two kinds of myxovirus (Waterson, 1962), now known as the families Orthomyxoviridae and *Paramyxoviridae*. With hindsight, we now know that orthomyxoviruses are totally different in their structure and replication from paramyxoviruses, and their only features in common are the properties of binding to mucoproteins by a virion envelope protein and release by a receptor destroying enzyme, the neuraminidase, as originally recognized (Andrewes *et al.*, 1955). The orthomyxovirus genome consists of negative-stranded RNA which is segmented, and transcription of this genome occurs within the nucleus and requires capped oligonucleotide primer RNAs (10–13 nucleotides long) that are derived from newly synthesized host cell mRNAs by endonuclease activity of the virus PB2 protein. Because of this intimate involvement with host cell transcription, orthomyxovirus replication is blocked by substances such as actinomycin D or α-amanitin, which do not affect paramyxovirus replication.

The paramyxoviruses are characterized by a genome of negative-sense single-stranded ribonucleic acid in one continuous strand (15,156 to 18,246 nucleotides in length). It is contained within a helical nucleocapsid, 18 nm in diameter and up to a 1000 nm in length. In the virion this nucleocapsid is contained within an approximately spherical envelope 150–200 nm in diameter which is derived from the host cell membrane during budding. Larger virions of 500–600 nm in diameter and filamentous forms are occasionally observed. There are two surface glycoproteins, 8–12 nm in length: the fusion (F) and the attachment protein (G, or H, or HN). The HN has both haemagglutinating and neuraminidase activity. There is also a non-glycosylated envelope protein, M. There are three internal nucleocapsid-associated proteins, an RNA-binding protein (NP), a phosphoprotein (P), and a large polymerase protein (L). Virus replication occurs in the cytoplasm, where the RNA genome is transcribed into 6–10 separate mRNAs. An interesting feature of paramyxoviruses is that, in general, their genome RNA contains a total number of nucleotides that is a multiple of six (Calain and Roux, 1993), apparently because six nucleotides bind to each subunit of the nucleocapsid protein. As new paramyxoviruses have been discovered, (such as the henipaviruses) they have all been found to follow this rule.

We now recognize two subfamilies of the family *Paramyxoviridae*. These are the *Paramyxovirinae*, which include five genera (*Avulovirus, Henipavirus, Morbillivirus, Respirovirus* and *Rubulavirus*) and the *Pneumovirinae*, which include two genera (*Pneumovirus* and *Metapneumovirus*). The members of the *Pneumovirinae* differ from members of the *Paramyxovirinae* in having 10 separate transcriptional units instead of six or seven, extensive O-linked glycosylation of the G protein, a smaller nucleocapsid diameter (17 nm), and longer surface glycoproteins (13–17 nm).

Paramyxoviruses include some of the most important causes of serious human disease, such as respiratory syncytial disease virus and measles virus, as well as devastating diseases of birds (Newcastle disease virus and turkey rhinotracheitis virus), carnivores (canine distemper virus), and cattle (rinderpest virus). The huge losses caused to cattle herds by rinderpest virus were so great in the eighteenth century that it led to the establishment of the first schools of veterinary medicine in Europe (in France in 1762 and in Great Britain in 1791). Following the development and use of the Plowright vaccine (Plowright, 1984), this disease appears to have been globally eradicated in 2010 (see Chapter 9), so becoming only the second virus disease that has been eradicated since the eradication of variola (smallpox) in 1977.

Considering the remarkably wide spectrum of diseases caused by paramyxoviruses, it is surprising that a number of serious diseases remained unrecognized until relatively recently. These include Hendra virus, first recognized in 1994, which causes fatal disease in horses and humans; Nipah virus, first recognized in 1997, which causes fatal disease in pigs and humans; the human metapneumovirus, a ubiquitous pathogen of infants, children and the elderly, first recognized in 2001, and several new aquatic paramyxoviruses first reported in 2003 (see Chapter 12). Comparative genome nucleotide sequence analysis showed that Hendra virus and Nipah virus are closely related, and are now both members of the genus *Henipavirus* (see Chapter 6). They both appear to be enzootic in fruit-eating bats in Australia and South-East Asia; an earlier study of a fruit-eating bat in Brazil led to the discovery of Mapuera virus in 1979, though this virus has not so far been shown to be pathogenic in any species. On the basis of genome sequence analysis, Mapuera virus has now been classified as a member of the genus *Rubulavirus*. More recently other new rubulaviruses of pigs have been found in Asia and Australia, namely Menangle virus and Tioman virus, though their pathogenic potential is not clear (see Chapter 12). There seems little doubt that more paramyxoviruses will be discovered in the future as genomic approaches are applied more widely.

It is clear from this extremely comprehensive book on paramyxoviruses that enormous progress has been made in our understanding and treatment of the most serious human diseases which they cause, namely measles, mumps and respiratory syncytial viruses. However it is also clear from Chapter 4 that a great deal of basic research to understand paramyxovirus replication and pathogenesis has been carried out using the rodent pathogen, Sendai virus, which on its first isolation in 1953 was erroneously thought to cause fatal pneumonia in newborn children. It is reasonable to speculate that this harmless virus has contributed more to our basic understanding of paramyxoviruses than any other member of the family.

References

Andrewes, C.H., Bang, F.B., and Burnet, F.M. (1955). A short description of the Myxovirus group (influenza and related viruses). Virology *1*, 176–184.

Andrewes, C.H and Worthington, G. (1959). Some new or little-known respiratory viruses. Bull. World Health Organ. *20*, 435–443.

Calain, P., and Roux, L. (1993). The rule of six, a basic feature for efficient replication of Sendai virus defective interfering RNA. J. Virol. *67*, 4822–4830.

Centanni, E. (1902). Die Vogelpest, Zentbl. Bakt. Paraskitkde, Abt. I, *31*, 145–152, 182–201.

Davenport, F.M., Rott, R., and Schafer, W. (1960). Physical and biological properties of influenza virus components obtained after ether treatment. J. Exp. Med. *112*, 765–782.

Franklin, R.M., and Wecker, E. (1959). Inactivation of some animal viruses by hydroxylamine and the structure of ribonucleic acid. Nature *184,* 343–345.

Plowright, W. (1984). The duration of immunity in cattle following inoculation of rfinderpest cell culture vaccine. J. Hygiene (London) *92,* 285–296.

Waterson, A.P. (1962). Two kinds of myxovirus. Nature *24,* 1163–1164.

Mumps Virus

1

Steven Rubin and Corinne Vandermeulen*

Abstract

Mumps virus, a member of the *Rubulavirus* genus in the *Paramyxoviridae* family, causes a communicable illness characterized by a wide array of acute inflammatory reactions. Mumps outbreaks were common in the pre-vaccine era, with nearly everyone having been exposed by adolescence based on the presence of virus-specific antibody. The hallmark of infection is the painful enlargement of the parotid gland, and, although perceived as a benign childhood disease, mumps was the leading cause of aseptic meningitis and sensorineural deafness prior to the adoption of widespread vaccine use. The changing epidemiology of mumps in the post-vaccine era has renewed interest in the virus, specifically in regards to the biology of the virus, development of new diagnostic tools, investigation of immune responses following vaccination and the epidemiological effects of mumps vaccination in large populations. This chapter reviews the current knowledge of mumps virus as an infectious agent, the immune response to infection and its epidemiology, prevention and control measures.

History

Mumps is one of the earliest recorded diseases of man, affecting populations worldwide for thousands of years. The disease was first recorded in 400 BC by Hippocrates in his seminal text '*Of the Epidemics*' wherein a contagious illness characterized by inflammation of the salivary glands and of the testes was described (T.E.C., Jr., 1967). The first detailed scientific description of mumps was made in 1773 by the British physician Robert Hamilton who described, in addition to parotitis and orchitis, 'involvement of the brain attended by a train of terrible symptoms, with death sometimes ending the conflict', describing for the first time several cases of fulminant mumps encephalitis (Hamilton, 1790). Mumps has been of particular importance in military mobilization, being one of the most significant causes of disability among military troops during the First and Second World Wars, surpassed only by influenza and gonococcal infection (Gordon and Heeren, 1940; Gordon and Kilham, 1949).

A viral aetiology for the disease was suggested as early as 1908 by the Italian physician Saverio Granata, who examined the agent in rabbits (Granata, 1908); however, it was not proven until 1933 when, in a remarkable series of experiments, Johnson and Goodpasture transferred the disease from experimentally infected monkeys to neighbourhood children. Of four children with a prior history of mumps or mumps exposure, none showed reactions to inoculation of a bacteria-free, filter-sterilized emulsion of infected monkey parotid gland material. In contrast, of 13 children believed to be mumps naive, nine (70%) developed clinically apparent parotitis or complained of symptoms consistent with mumps. Saliva from one of the cases was inoculated back into naive monkeys via the intra-parotid route, producing parotitis,

*Corresponding author

thereby fulfilling Koch's postulates (Johnson and Goodpasture, 1934, 1935).

Habel and Enders separately demonstrated that mumps virus could be isolated and propagated in embryonated eggs and the subsequent introduction of tissue culture as a more efficient means of virus propagation greatly facilitated studies on the physical and biological properties of the virus (Habel, 1945; Enders, 1946; Henle and Deinhardt, 1955). These systems led to the development of an inactivated vaccine in 1946 (Habel, 1946, 1951a,b) and to the first live attenuated vaccine in 1954 (Smorodintsev and Klyatchko, 1954, 1958; Klyatchko and Smorodintsev, 1958). Following implementation of live, attenuated mumps vaccines in national immunization programmes, disease incidence declined sharply in all countries where widely used. In the USA, for example, mumps incidence declined from >100 cases per 100,000 population in the pre-vaccine era (pre-1967) to less than 1 case per 100,000 population following the institution of a two-dose measles–mumps–rubella (MMR) vaccine schedule (McNabb *et al.*, 2007). The economic impact of vaccine use has proven remarkable. Based on a study of cases in the USA, it has been estimated that over $1.5 billion in costs are averted annually by vaccinating against mumps, equivalent to a savings of $3614 per case prevented per year (Zhou *et al.*, 2004). Similar success in the control of mumps has been achieved in other countries through high vaccine coverage (Centers for Disease Control and Prevention, 1978; Peltola *et al.*, 2000; Sartorius *et al.*, 2005).

Infectious agent

Classification and genome organization

Mumps virus is a member of the *Paramyxoviridae* family, genus *Rubulavirus* (Carbone and Rubin, 2007). Other members of this genus include human parainfluenza virus types 2 and 4, parainfluenza virus 5 (formerly SV5), Mapuera virus, porcine rubulavirus (formerly La-Piedad-Michoacan-Mexico virus), and simian virus 41 (Lamb and Parks, 2007). Viruses tentatively assigned to this genus include Menangle virus, Tioman virus, and Tuhoko virus types 1, 2, and 3. Mumps virions are enveloped, pleomorphic particles with a diameter of about 200 nm (Fig. 1.1).

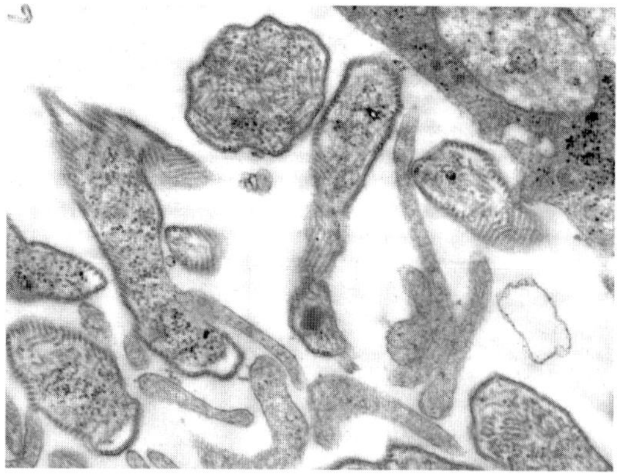

Figure 1.1 Transmission electron micrograph showing pleomorphic mumps virions produced in Vero cells (cell surface shown in upper right). Reproduced with permission from CDC/A. Harrison and F.A. Murphy (http://phil.cdc.gov/phil/details.asp).

The virus genome is a non-segmented, single-stranded RNA macromolecule of negative polarity containing 15,384 nucleotides encoding the following genes in 3′–5′ genomic order: nucleoprotein (N), V/P/I protein, matrix protein (M), fusion glycoprotein (F), small hydrophobic protein (SH), haemagglutinin–neuraminidase glycoprotein (HN) and large protein (L). The genes are separated by untranslated regions (UTRs) of variable length consisting of gene end (GE) sequences containing transcriptional and translational termination signals, followed by a short intergenic region of one to seven nucleotides and a stretch of gene start (GS) sequences containing transcription and translation initiation signals for the next gene (Elango *et al.*, 1988; Lamb and Parks, 2007). The genome termini consist of a 3′ extracistronic leader of 55 bases and a 5′ extracistronic trailer of 24 bases, both containing sequences responsible for regulating transcription and replication levels (Elango, 1988; Okazaki, 1992; Tanabayashi, 1990).

Viral genes and proteins

N gene

The N gene encodes a 61–73 kDa, 549 amino acid protein consisting of a highly conserved amino-terminal domain and a hypervariable carboxyl-terminal domain. The protein complexes with the genomic RNA to form the ribonucleoprotein (RNP), a helical structure that acts as the template for viral RNA synthesis. The RNP is linked to the virion membrane through interactions between the viral N and M proteins (see M gene below). In contrast to most paramyxoviruses, the N-terminal domain of the mumps virus N protein (not the hypervariable carboxyl-terminal domain) binds the P protein (Schmitt, 2010), which along with the L protein forms the RNA dependent RNA polymerase (RdRp; Lamb and Parks, 2007).

V/P/I gene

The V/P/I gene encodes three distinct proteins. A faithful transcript of the V/P/I gene produces the non-structural V protein, whereas co-transcriptional addition of two non-templated guanine (G) residues between nucleotide positions 461–466 produces the P protein and addition of one or four non-templated G residues produces the I protein (Takeuchi *et al.*, 1988; Elliott *et al.*, 1990; Paterson and Lamb, 1990). The accuracy of insertion of the non-templated G residues may vary with the mumps virus strain (Elliott *et al.*, 1990). All three proteins share a common amino terminus of 155 amino acids.

The V gene open reading frame encodes a 25–28 kDa, 224 amino acid protein involved in circumventing the innate cellular antiviral response to infection. The C-terminal domain of the mumps virus V protein has been demonstrated to bind to the melanoma differentiation-associated gene 5 (MDA-5) product, an activator of the interferon (IFN) b promoter, resulting in inhibition of IFN production by infected cells (Andrejeva *et al.*, 2004). The C-terminal domain of the V protein can also bind to several of cellular signal transducer and activator of transcription (STAT) proteins, leading to their ubiquitination and subsequent degradation through the proteosome pathway. Key among these are STAT-1 and STAT-2, which are mediators of the IFN signal transduction pathway that produces an antiviral state (Yokosawa *et al.*, 1998; Kubota *et al.*, 2001; Nishio *et al.*, 2002; Ulane *et al.*, 2003; Rosas-Murrieta, 2010). The mumps virus V protein also targets STAT3, another protein involved in cellular antiviral responses (Nishio *et al.*, 2002; Ulane, 2003; Puri *et al.*, 2009). As a non-structural protein, the V protein is not an integral part of the virion.

The P gene open reading frame encodes a 41–47 kDa, 391 amino acid protein. Studies of Sendai virus (a related member of the *Paramyxoviridae* family) indicate that the P protein binds to the L protein forming the RdRp, which is responsible for transcription, replication, methylation, capping and polyadenylation (Horikami *et al.*, 1992; Curran *et al.*, 1994; Holmes and Moyer, 2002; Lamb and Parks, 2007). Carboxyl-terminal domains of the P protein are believed to facilitate the binding of the RdRp complex to the RNP, whereas the amino terminus is thought to be essential for viral replication (Curran *et al.*, 1994; Smallwood *et al.*, 1994).

The I gene open reading frame encodes a 19-kDa, 170 amino acid protein, all but 15 of which are common to the P and V proteins (Elliot *et al.*, 1990; Paterson and Lamb, 1990). The protein (sometimes referred to as the NS2) has been identified in virus infected cells, but has not specifically been identified in the virion (Paterson and Lamb, 1990) and its function has not been established.

M gene

The M gene encodes a 41–42 kDa, 375 amino acid protein which plays a major role in virus budding. Indeed, the M protein alone is capable of inducing virus-like particle (VLP) formation (Li *et al.*, 2009). No other virus protein singly expressed is capable of inducing VLPs. M proteins reside immediately below the plasma membrane of the infected cell, and based on studies of related viruses, the M protein is involved in physically associating the RNP with areas of host plasma membrane containing the viral HN and F glycoproteins, presumably by forming a bridge between the N proteins of the RNP and the cytoplasmic tails of the glycoproteins (Matsumoto, 1982; Ali and Nayak, 2000; Takimoto and Portner, 2004). There is conflicting evidence whether or not the M protein plays a role in regulating the transition between transcription and replication. Using a rabies virus reverse genetics system, Finke *et al.* showed that virus transcription was markedly inhibited in cells overexpressing the M protein, but despite reduced levels of mRNA, genome RNA levels were relatively unchanged and in the absence of M protein, levels of genome RNA were decreased. These results suggest that, at least for a rhabdovirus, expression of the M protein is associated with inhibition of transcription and promotion of replication (Finke *et al.*, 2003). In contrast, for Sendai virus, Mottet-Osman *et al.* (2007) found no effect of the M protein on levels of transcription or replication.

F gene

The F gene encodes a 64–74kDa, 538 amino acid precursor glycoprotein, which by inference from studies of Sendai virus is proteolytically cleaved in the host cell by the cellular protease furin after transport from the endoplasmic reticulum through the trans-Golgi network (Gardner *et al.*, 2007), to produce disulfide linked glycopeptides F_1 (58–61kDa, 436 amino acids) and F_2 (10–16kDa, 102 amino acids), which are expressed on the outer surface of the virion (Naruse *et al.*, 1981; Merz *et al.*, 1983). The ectodomain of the F_1 subunit contains a hydrophobic region at its amino terminus consisting of 10–15 amino acids referred to as the fusion peptide, which triggers virus-to-cell and cell-to-cell fusion, but only in the presence of the viral HN glycoprotein (Malik *et al.*, 2007b). Early studies on antibodies directed against the F protein, which is expressed on the virion surface and is essential to virus infectivity and spread, indicated that these antibodies were possibly involved in neutralizing virus infectivity. While sera from hamsters infected with vaccinia virus recombinants expressing the mumps virus F protein were found to neutralize mumps virus infectivity *in vitro* (Houard *et al.*, 1995), several other studies failed to demonstrate the viral F protein as a target of neutralizing antibody (Orvell *et al.*, 1984; Wolinsky *et al.*, 1985; Love *et al.*, 1986; Tecle *et al.*, 2000). Nonetheless, *in vivo*, anti-mumps virus F protein antibodies were demonstrated to provide protection against virus-induced encephalitis and neuronal necrosis (Love *et al.*, 1986; Houard *et al.*, 1995), but it is not clear whether this was owing to virus neutralization or antibody-mediated prevention of F protein-mediated cytopathology.

SH gene

The SH gene encodes a 6kDa, 57 amino acid protein. The SH protein has not been detected in the mumps virion, however, an analogous protein has been identified in the human parainfluenza virus 5 (hPIV-5) virion, suggesting that the SH protein may be an integral component of the mumps virus particle (Hiebert *et al.*, 1988). The Enders strain of mumps virus contains a mutation in the F gene polyadenylation site (termination signal), resulting in a bicistronic F-SH mRNA, of which only the F cistron is translated, indicating that the SH gene product is not required for mumps virus replication (Afzal *et al.*, 1990; Takeuchi *et al.*, 1996). Studies of related viruses indicate that the SH protein plays a role in evasion of the host anti-viral response via blocking apoptosis (He *et al.*, 1998, 2001; Wilson *et al.*, 2006; Bao *et al.*, 2008). The SH gene sequence, being the most variable of all the mumps virus genes, is used in virus genotyping for molecular epidemiological purposes.

HN gene

The HN gene encodes a 74–80kDa, 582 amino acid disulfide-bonded oligomer expressed on the surface of the virion and is responsible for virus binding and release (Orvell, 1978). The haemagglutinin portion of the protein binds the virus to its cellular receptor, sialic acid (N-acetyl neuraminic acid), a sugar group common to many glycosylated molecules, while the neuraminidase portion of the protein cleaves sialic acid from sugar side chains, thereby releasing progeny virions from the cell surface (Jensik and Silver, 1976; Orvell, 1978). The neuraminidase also functions to disengage virus from inappropriately bound cells or debris and to remove sialic acid from nascent virions to prevent self-association (Merz and Wolinsky, 1983). The haemagglutination and neuraminidase activities can be separately inhibited using different monoclonal antibodies, initially suggesting that these activities reside within two separate sites within the protein (Iorio and Bratt, 1984; Orvell, 1984). Subsequent structural studies refute this notion, indicating a single, switchable site for neuraminidase and sialic acid binding activities (Crennell *et al.*, 2000; Connaris *et al.*, 2002). In addition to receptor binding and neuraminidase activities, the mumps virus HN protein also possesses fusion promotion activity (Malik *et al.*, 2007b). Studies of related viruses indicate that the HN and F glycoproteins physically associate prior to fusion and, following HN binding to its cellular receptor, the F protein is released permitting the necessary conformational changes required for fusion activity (Connaris *et al.*, 2002; Takimoto *et al.*, 2002; Gravel and Morrison, 2003; Morrison, 2003). The HN protein is the major target of neutralizing antibodies (Somboonthum *et al.*, 2007; Liang *et al.*, 2008). Neutralizing epitopes have been mapped to three regions, amino acids 265–288, 329–340 and 352–360 (Server *et al.*, 1982; Kovamees *et al.*, 1990; Orvell *et al.*, 1997; Cusi *et al.*, 2001), all present on the surface of the protein's predicted 3D structure derived by homology modelling using coordinates of HN of Newcastle disease virus (NDV; Crennell *et al.*, 2000). Using a structure-based conformational epitope prediction programme, Kulkarni-Kale *et al.* (2007) identified a total of 16 potential conformational epitopes on the HN protein, five of which overlap with the three experimentally proven neutralizing epitopes. The validity of the remaining predicted epitopes has yet to be demonstrated.

The large gene

The L gene, the last transcriptional unit in the genome, is the largest of the mumps virus genes encoding a 180–200 kDa, 2261 amino acid protein. The L protein is often referred to as the polymerase protein, highlighting its functional role as part of the RdRp complex, which is responsible for transcription, replication, methylation, capping and polyadenylation (Lamb and Parks, 2007). The mumps virus L protein shares significant homology with those of most other paramyxoviruses with alignments demonstrating six highly conserved domains designated I–VI (Poch *et al.*, 1990; Parks *et al.*, 1992; Sidhu *et al.*, 1993). By inference from studies of related paramyxoviruses, domains I and V appear to regulate the transition from transcription to replication (Chandrika *et al.*, 1995; Cortese *et al.*, 2000) and domains II and III appear to constitute the active site for viral RNA polymerase activity (Jablonski *et al.*, 1991; Muller *et al.*, 1994). Mutational studies have also shown changes in domains IV and VI to dramatically affect RNA synthesis (Feller *et al.*, 2000). Alignments among paramyxoviruses subfamilies have identified additional conserved sequence motifs outside of these domains, which have also been found to play key roles in RNA synthesis (McIlhatton *et al.*, 1997; Chattopadhyay *et al.*, 2004).

Virus transcription and replication

Upon adsorption of the virus to its receptor (sialoglycoconjugates) the viral membrane fuses with the cellular plasma membrane releasing the RNP into the cell cytoplasm. Primary transcription is mediated by the viral RdRp, which is associated with the incoming RNP. The RdRp recognizes a single promoter at the 3′ end of the genomic RNA synthesizes the (+) leader RNA and terminates at the end of the leader region before reinitiating at the beginning of the first gene (N). There is some evidence that the RdRp can bind directly to its promoter in the UTR at the first GS sequence immediately downstream the (+) leader (Banerjee *et al.*, 1991; Vulliemoz *et al.*, 2005; Cordey and Roux, 2006). Genome transcription terminates upon reaching a GE signal in the UTR that follows the gene. The GE contains a polyuridine tract of 4–7 nucleotides. The RdRp of non-segmented negative-strand RNA viruses is unique in that is does not disengage from the template upon release of the mature mRNA, but instead remains bound to the template and scans for the next GS site in the UTR (Banerjee *et al.*, 1991; Stillman and Whitt, 1998; Fearns and Collins, 1999; Cordey and Roux, 2006). In this fashion, genome transcription is said to occur by a stop–start mechanism.

The long polyadenylated mRNA produced by transcription is thought to be the result of the RdRp 'stuttering' at the polyuridine tracts at the end of each gene, producing long stretches of non-templated adenine residues. The failure of the RdRp to terminate at GE polyuridine tracts results in 'read-through' transcripts, consisting of polycistronic mRNA species, of which only the upstream cistron is translated (Rassa and Parks, 1998). Such read-through transcription events are prevalent among members of the *Mononegavirales*, suggesting a role in the regulation of gene expression (Masters and Samuel, 1984; Elliott *et al.*, 1989; Bousse *et al.*, 2002; Poenisch *et al.*, 2008).

Occasionally, during polyadenylation when the association between the RdRp and the RNP is least stable, the RdRp disassociates from the RNP. Since the RdRp can only initiate at the 3′ end of the genome, a decreasing gradient of monocistronic mRNAs are produced for genes located further from the 3′-end promoter. Thus, the concentration of N protein mRNA is greatest and that for the subsequent transcripts are reduced in proportion to their distance from the 3′ genome promoter (Rassa and Parks, 1998; Hausmann *et al.*, 1999). This property of negative strand RNA viruses does not appear to be arbitrary in that genes whose products are required in large amounts (e.g. the N protein) are located more proximal to the promoter and those needed in catalytic amounts (e.g. the L protein) are more distal.

In contrast to transcription, replication to produce antigenomic and genomic RNA requires that the RdRp ignore the GS and GE signals and produce full-length complementary RNA instead of discrete monocistronic mRNAs (Lamb and Parks, 2007). The anti-genomic RNA serves as a template for the synthesis of full-length progeny genomes, which can be used for synthesis of viral mRNAs or templates for further replication, or can be assembled into infectious virions (Lamb and Parks, 2007). Whether the RdRp acts to transcribe or replicate the genome is believed to be a consequence of the amount of available soluble N protein (Patton *et al.*, 1984; Howard and Wertz, 1989; Curran and Kolakofsky, 1999). In this model, when the concentration of N is high (e.g. after long periods of transcription), the leader RNA becomes encapsidated at its 5′ end resulting in the inability of the RdRp to recognize the GE signal and instead continues along the template, reading through the intergenic sequences, resulting in replication of the entire genomic RNA. At some point, the concentration of soluble N in the cytoplasm will decrease (owing to high replication activity). However, this does not, as one might expect, lead to a switch back to transcription as suggested by studies of respiratory syncytial virus where increasing levels of N did not influence the ratio of transcription to replication (Fearns *et al.*, 1997).

Genomic RNAs are needed for incorporation into newly formed virus particles, to serve as templates for mRNA synthesis and to serve as templates for antigenome synthesis, whereas antigenomic RNAs is only needed to serve as a template to amplify genomic RNA. Thus significantly higher levels of genomic RNA (produced by the antigenomic promoter) are required compared with antigenomic RNA (produced by the genomic promoter; Mottet and Roux, 1989; Lamb and Kolakofsky, 2001; Le Mercier *et al.*, 2002). Based on studies of PIV-5, the amounts of RNA replicated by the genome promoter are several times lower than amounts of RNA replicated from the antigenomic promoter (Keller and Parks, 2003). This differential replication activity was shown to be driven by 3′-*cis* acting elements in both promoters (Keller and Parks, 2003).

In addition to exerting control over replication, *cis*-acting signals may also control transcription, ultimately affecting levels of translated protein. For most members of the *Paramyxoviridae* family, including mumps virus, there is a high degree of UTR variability (in length as well as sequence), even within a virus genome (Lamb and Parks, 2007), suggestive of their role in differentially regulating polymerase functions (Parks *et al.*, 2001). This is supported by studies demonstrating the effect of replacement of nucleotides in the portion of the UTR containing transcription and translation initiation signals for the downstream gene. Such changes have been shown to result either in significant attenuation of downstream gene transcription or overexpression of the gene and altered virus phenotype, both *in vitro* and *in vivo* (He and Lamb, 1999; Kato *et al.*, 1999; Finke *et al.*, 2000; Anderson and von Messling, 2008; Yan and Samal, 2008). The effect of the GE UTR sequence stretch on transcriptional and translational activity for the upstream gene has been less well described.

Virus assembly, budding and release

Following assembly of the virus components at the cellular membranes, mumps virus virions, like all enveloped virus particles, are formed as they emerge from the infected cell by a process referred to as budding. Although not specifically demonstrated for mumps virus, based on inference from studies with other paramyxoviruses, virus assembly probably takes place along lipid raft microdomains within the cellular membrane (Harrison *et al.*, 2010). The mechanism by which viral glycoproteins are preferentially attracted to these regions is poorly understood, but appears to involve specific transmembrane domains of the viral glycoproteins (Schmitt and Lamb, 2004). Budding from other cellular regions, as has been documented for vesicular stomatitis virus and rabies virus, has not been ruled out.

The key protein involved in orchestrating virus component assembly appears to be the M protein, which as discussed above, is believed to physically 'connect' the viral RNP with the cytoplasmic tails of the cell surface expressed viral glycoproteins. Paramyxoviruses, including mumps virus, genetically engineered with partial or complete M gene deletions were found to be severely limited or completely unable to bud from the surfaces of infected cells (Cathomen *et al.*, 1998; Schmitt *et al.*, 2005; Patch *et al.*, 2008; Li *et al.*, 2009;). Although mumps virus particle formation is dependent on the M protein, efficient budding also requires the N and F proteins (Li *et al.*, 2009). The mechanism of mumps virus budding and particle release remains poorly understood. Studies by Li *et al.* (2009) have linked mumps virus budding with recruitment of cellular Class E proteins which comprise part of the endosomal sorting complexes required for transport (ESCORT). Interestingly, recruitment of cellular factors is believed to involve virus matrix protein late domain sequences, as implicated for many negative-strand RNA viruses, including vesicular stomatitis virus, rabies, Ebola virus, lymphocytic choriomeningitis virus and Lassa fever virus; however, most paramyxoviruses, including mumps virus, lack late domains (Harrison *et al.*, 2010). Whether or not other cellular proteins are involved in budding and release, or even if the ESCORT complex is required, are unknown.

Reverse genetics

The establishment in 1994 by Schnell *et al.* (1994) of a system to recover rabies virus entirely from complementary DNA (cDNA-encoded RNA) was a breakthrough in the field of genetics and has since been adapted to most other *Mononegavirales* (Collins and Crowe, 2007; Lamb and Parks, 2007). The first reverse genetic system for mumps virus was described by Clarke *et al.* in 2000 and was based on the highly neuroattenuated Jeryl Lynn vaccine strain. This vaccine strain, produced by Merck and Co, is composed of two distinct, but homologous mumps virus strains termed Jeryl Lynn-5 (JL-5) and Jeryl Lynn-2 (JL-2) existing in a 5:1 ratio (Afzal *et al.*, 1993). The full-length mumps virus cDNA clone was prepared from a single JL-5 (major component) plaque of the vaccine preparation. The cDNA clone was assembled 5′ end to 3′ end by successive cloning of contiguous cDNA fragments into the Stratagene plasmid vector pBluescript II SK(+) between a T7 RNA polymerase promoter and the hepatitis delta ribozyme followed by a T7 RNA polymerase transcription termination sequence. Like all paramyxoviruses, replication of the viral RNA requires the presence of the viral N, P and L proteins which form the RdRp. Thus, rescue of the full-length virus requires co-transfection with separate plasmids containing the mumps virus N, P and L genes. Once in RNP form, the transcript corresponding to the viral antigenome (positive strand) can be replicated by the cDNA encoded support proteins to form the negative strand viral genome thereby initiating the infectious cycle. This is the most common approach used in the rescue of *Mononegavirales*. An alternative approach is to construct the full-length plasmid so that instead of the transcript encoding the antigenome, the transcript corresponds to the viral genome (negative strand). This approach has been used with limited success to rescue Sendai virus and human parainfluenza-3 virus, but was not a viable means of rescuing most other *Mononegavirales* attempted, most probably because of hybridization of the mRNAs encoding the N, P and L support proteins to the negative strand transcript (Conzelmann, 2004).

Because the plasmids used in the rescue of mumps virus are under the control of a T7 RNA polymerase promoter, the host cells used in the rescue system must express the bacteriophage T7 RNA polymerase. This is typically accomplished either by co-infection of the cells with a recombinant vaccinia virus expressing the T7 polymerase (e.g. vTF7–3 or MVA-T7), co-transfection of cells with a separate plasmid expressing T7 polymerase, or by use of cells that constitutively express the T7 polymerase (Fuerst *et al.*, 1986; Radecke *et al.*, 1995; Sutter and Moss, 1995; Buchholz *et al.*, 1999). Use of alternative, potentially more efficient rescue systems, such as use of plasmids that are under the control of an RNA polymerase II promoter (Martin *et al.*, 2006) has not been attempted for mumps virus.

The use of reverse genetics coupled with experimental mutagenesis to generate infectious viruses from cloned cDNA is a powerful technique allowing for the study of the biology of the virus and molecular basis for pathogenicity and attenuation. For mumps viruses, such studies have only recently commenced and have resulted in a better understanding of the role of specific genes in virus phenotype; however, these appear to be virus strain specific (Lemon *et al.*, 2007; Malik *et al.*, 2007a, 2009; Ninomiya *et al.*, 2009; Sauder *et al.*, 2009). To date, no genetic markers have been identified that universally predict the virulence or attenuation profile of mumps viruses. Further work in this area is of particular importance for vaccine development, given that with the exception of the Jeryl Lynn and Rubini vaccine strains, nearly all other vaccine strains have been causally linked with aseptic meningitis owing to inadequate attenuation (Plotkin and Rubin, 2008).

Pathogenesis

Host range, entry, spread and tissue tropism

Although humans are the only natural host and there is no known animal reservoir, several animal species can be experimentally infected with mumps virus (Carbone and Rubin, 2007). Based on observations that susceptible individuals in close contact are a prerequisite for an outbreak and that transmission has been demonstrated following nasal or buccal mucosal inoculation of virus suggests that infection is spread by oropharyngeal secretions via direct contact, respiratory droplet or fomites (Johnson and Goodpasture, 1935; McGuiness and Gall, 1944; Henle *et al.*, 1948a; Gordon and Kilham, 1949).

As a respiratory infection, it is presumed that primary replication occurs in the upper respiratory tract epithelium and regional lymph nodes, but this has not been specifically demonstrated for mumps virus (Arguedas *et al.*, 2004; Leinikki, 2009). Although viraemia has only rarely been demonstrated (Kilham, 1948; Overman, 1958), the frequency by which this occurs appears high based on the regularity of a wide array of acute inflammatory responses related to infection of glandular tissues and the central nervous system (CNS; Arguedas *et al.*, 2004; Leinikki, 2009). Based on limited data, viraemia develops early after the primary infection and lasts for 3–5 days (Fig. 1.2), coinciding with the development of virus-specific IgA. Based on studies in hamsters injected with virus intraperitoneally, viraemia is thought to be T cell-associated (Wolinsky *et al.*, 1976), which may explain the ability of the virus to spread systemically despite the rapid development of a humoral immune response (Wolinsky *et al.*, 1976; Ukkonen *et al.*, 1981; Fleischer and Kreth, 1982). However, virus has been isolated from serum in humans, suggesting that mumps virus may also be cell-free during viraemia (Overman, 1958).

CNS infection in mumps is common. Studies in laboratory animals indicate that virus infected mononuclear cells or cell-free virus either crosses the blood brain barrier and enters the CSF or infects the endothelial cells which then become the source of virus entering the CSF (Wolinsky *et al.*, 1976). Once within the CSF, virus (either cell-associated or cell-free) infects the cells of the choroid plexus allowing for virus dissemination throughout the ventricular system, including the subarachnoid space. This is probably the path by which meningitis results. The ependymal cells that line the ventricular cavities of the brain are a major target of the virus, providing a means of contiguous spread of the virus deeper into the brain parenchyma and resulting in encephalitis (Levenbuk *et al.*, 1979). While primary encephalitis is typically a response to direct viral infection of neural cells, cases of post-infectious encephalitis also occur, but as a result of an autoimmune attack on CNS myelin sheaths triggered by the neural

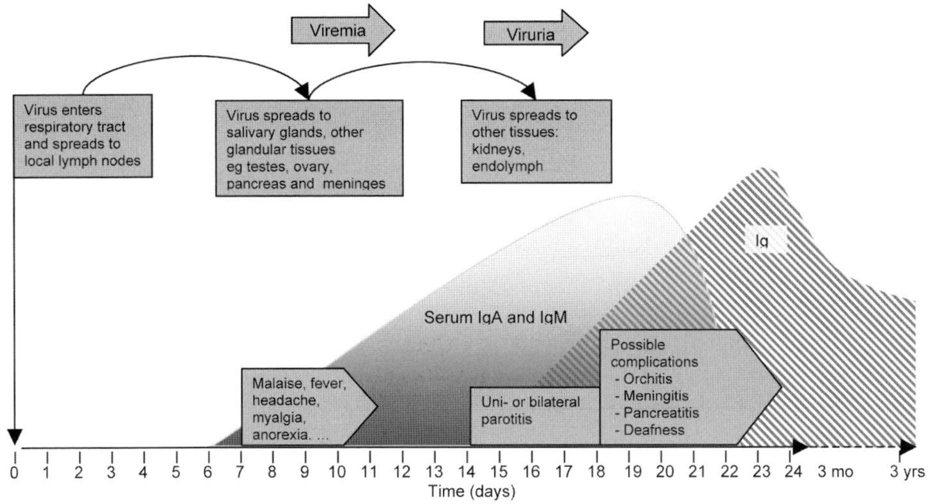

Figure 1.2 Pathogenesis of mumps (Kilham, 1948; Overman, 1958; Black and Houghton, 1967; Brown *et al.*, 1970; Bjorvatn, 1974; Friedman, 1981; Morgan-Capner, 2004; Leinikki, 2009).

breakdown products of the primary infection (Litman and Baum, 2010).

Animal models and virulence factors

Experimental mumps infection has been mostly limited to the study of virus neurovirulence and neuroattenuation. Most of what is known about the pathogenesis of mumps comes from studies in hamsters and Old World monkeys (*Cercopithecidae*). Intracerebral or intraperitoneal inoculation of mumps virus into the suckling hamster results in a massive inflammatory response including meningitis, encephalitis, cerebellitis, ventriculitis and hydrocephalus (Overman *et al.*, 1953; Ennis *et al.*, 1969; Kilham and Margolis, 1975; Wolinsky *et al.*, 1976, 1978; Takano *et al.*, 1994; Takikita *et al.*, 2001). Similar findings have also been reported in monkeys (Findlay and Clarke, 1934; Chu *et al.*, 1951; St Geme *et al.*, 1972; Yuzepchuk *et al.*, 1975; London *et al.*, 1979; Rozina and Hilgenfeldt, 1985) and are consistent with outcomes of mumps virus infections in man. Although such parallels support the applicability of these animal models for the study of mumps pathogenesis, the neuropathology induced by attenuated versus virulent mumps virus strains cannot be reliably distinguished in these models, limiting their value in the study of virus virulence factors (Buynak and Hilleman, 1966; Ennis *et al.*, 1969; Kilham and Margolis, 1975; Wolinsky and Stroop, 1978; McCarthy *et al.*, 1980; Rozina and Hilgenfeldt, 1985; Afzal *et al.*, 1999; Rubin *et al.*, 1999). Therefore, development and validation of other animal models has been a major focus of recent efforts towards an understanding of virus attenuation. The two most promising new animal models are the marmoset (*Callithrix jacchus*) and the Lewis rat. Very little, if any, work has been accomplished in mice since mumps virus infection in this species tends to be abortive, thereby limiting the value of a mouse model (Hayashi *et al.*, 1976; Kristensson *et al.*, 1984; Yamada *et al.*, 1984; Tsurudome *et al.*, 1987).

In marmosets, the frequency and severity of meningitis and encephalitis following intravenous, intracerebral or intraspinal inoculation was found to track with the known attenuation state of the virus for humans (Saika *et al.*, 2002, 2004, 2006). In rats inoculated intracerebrally as newborns, hydrocephalus is the major neuropathological finding, the severity of which was also found to track with the known human clinical virulence profile of the assayed virus (Rubin *et al.*, 1998, 2000, 2005). Thus, these models not only have the potential to facilitate studies of the molecular basis of virus neurotoxicity and neuroattenuation, but, pending further validation, might also serve as preclinical neurovirulence safety tests to support the development and licensure of safe mumps vaccines.

Using the rat model, investigations have identified numerous genetic changes throughout the mumps virus genome associated with altered virulence, indicating that the genetic basis of mumps virus neurovirulence is polygenic (Cusi *et al.*, 1998; Rubin *et al.*, 2003; Santos-Lopez *et al.*, 2006; Sauder *et al.*, 2006; Lemon *et al.*, 2007; Malik *et al.*, 2007a,b, 2009; Shah *et al.*, 2009). Thus, it is perhaps not surprising that no simple pattern of genomic mutations capable of discriminating virulent from attenuated mumps virus strains has been identified (Cusi *et al.*, 1998; Amexis *et al.*, 2003; Ivancic *et al.*, 2005; Sauder *et al.*, 2006).

Immune responses to virus infection

Humoral immunity

Early humoral immune response to natural infection by mumps virus follows a course typical of that to respiratory viruses, i.e. local secretion of salivary IgA, followed by production of virus-specific serum IgA and IgM. Mumps virus-specific serum IgA and IgM are detectable by the onset of disease, peak 1–2 weeks later, and then gradually decline over a 2-month period (Fig. 1.2; Brown *et al.*, 1970; Bjorvatn, 1974; Friedman, 1981; Ukkonen *et al.*, 1981; van der Logt *et al.*, 1982; Gut *et al.*, 1995; Arguedas *et al.*, 2004). IgG antibody development lags behind IgM by 1–2 weeks, but unlike IgM, peak levels can persist without decline for several months and can be detected indefinitely in most cases (Brown *et al.*, 1970). During primary infection, IgG3-specific antibodies are predominant, whereas in later stages isotypes IgG1 and IgG2 predominate. Neutralizing antibodies appear during convalescence and are believed to afford lifelong protection; however, as discussed later, this view is changing (Philip *et al.*, 1959; Meyer, 1962; Black and Houghton, 1967; Brown *et al.*, 1970; Biedel, 1978; Gut *et al.*, 1995; Yoshida *et al.*, 2008). Upon a second exposure to mumps virus, IgM and IgA antibodies may not be detectable and the IgG1 subclass predominates and is typically of high avidity compared with the relatively low-avidity IgG antibodies induced during primary infection (Gut *et al.*, 1995). Mumps virus-specific memory B cells can be detected decades after mumps disease (Amanna *et al.*, 2007).

Cell-mediated immunity

Mumps virus-induced cell-mediated immune responses have been demonstrated by lymphocytotoxicity assay and by the identification of interferon gamma in serum, saliva and cerebrospinal fluid following infection with mumps virus (Morishima *et al.*, 1980; Martin *et al.*, 1991;Ichiyama *et al.*, 2005; Jokinen *et al.*, 2007; Hanna-Wakim *et al.*, 2008). The magnitude of the lymphocytotoxicity response does not appear to correlate with levels of serum antibody (Chiba *et al.*, 1976), and interferon levels, which decrease steadily from the first day of symptom onset, do not correlate with illness severity (Larke, 1967; Waddell *et al.*, 1968). During mumps meningitis, interferon and CD4 and CD8 T cells can be detected in the CSF, suggesting that mumps virus activates the cell-mediated immune system directly in the CNS (Nagai *et al.*, 1983). Lymphoproliferative responses and IFN-γ can be induced following *in vitro* stimulation with mumps antigen many years following mumps disease, even in seronegative persons (Jokinen *et al.*, 2007; Hanna-Wakim *et al.*, 2008), however, it remains unclear whether cell-mediated immunity plays a role in disease recovery or protection against complications, as has been indicated with other viruses (van Els and Nanan, 2002).

Clinical features

In the pre-vaccine era, the majority of cases occurred in children aged 5–9 years; disease in people 15 years of age or older was uncommon (Litman and Baum, 2010). In more recent times, the majority of cases have occurred in those aged 15–24 years (Dayan and Rubin, 2008; Barskey *et al.*, 2009; Jick *et al.*, 2009). This age incidence shift is probably a consequence of an increased susceptibility of the adolescent population owing to waning of vaccine-induced immunity (discussed later) coupled with increased opportunities for transmission given that the majority of recent cases have occurred in high-density, close contact settings, such as university and college campuses (Anderson and Seward, 2008).

Mumps has a relatively long incubation period with a mean of 17 days (range 12–25 days). The virus can be isolated from saliva up to 7 days prior to and after the onset of clinical symptoms (Utz *et al.*, 1957), a time period during which the patient is infectious. Notably, in approximately one-third of infections only non-specific or primarily respiratory symptoms occur, although these individuals can readily transmit the virus as well (Henle *et al.*, 1948b; Philip *et al.*, 1959; World Health Organization, 2007a). Mumps virus reinfections can occur; however, symptomatology associated with a second attack tends to be milder (Philip *et al.*, 1959; Meyer, 1962; Biedel, 1978; Gut *et al.*, 1995; Yoshida *et al.*, 2008).

Mumps is generally a mild self-limiting disease, although serious complications may occur. Mumps is typically characterized by prodromal non-specific symptoms such as fever, headache, anorexia, myalgia and malaise lasting for several days and is typically followed by enlargement of one or more of the salivary glands, usually the parotid (Fig. 1.3). Parotitis is often bilateral (Leinikki, 2009) and is associated with ear pain and tenderness on palpitation of the angle of the jaw. Although the standard clinical case definition of mumps is based on the acute onset of swelling of the parotid or other salivary glands (Centers for Disease Control and Prevention, 2006a), other symptoms, such as orchitis and meningitis can occur in the absence of salivary gland swelling (Philip *et al.*, 1959). Salivary gland swelling usually subsides after 4–7 days, but can persist for 10 days or longer.

Epididymitis and orchitis occur in approximately 20–30% of post-pubertal males. The onset of orchitis usually occurs about 10 days after the onset of parotitis and is associated with constitutional symptoms such as high-grade fever, vomiting, headache and malaise (Davis *et al.*, 2010). Although orchitis can be bilateral in 30% of cases, it seldom leads to sterility. Owing to the recent shift in the peak age incidence of mumps from the five to nine year old group to the 15- to 24-year-old group, there has been a corresponding increase in cases of mumps orchitis (Singh *et al.*, 2006; Davis *et al.*, 2010).

The virus frequently invades the CNS as was shown by the high frequency of pleocytosis in the CSF in mumps patients (Finkelstein, 1938). Nevertheless, symptomatic infection, typically aseptic meningitis, is seen in less than 10% of clinical cases. Encephalitis and other more serious complications of virus infection of the CNS are rare, but can lead to permanent sequelae (Russell and Donald, 1958; Johnstone *et al.*, 1972). Although the overall case fatality ratio for mumps is low (1 per 10,000), that for patients presenting with encephalitis is approximately 14 per 1000 (Centers for Disease Control and Prevention, 1972, 1978). Symptoms of CNS involvement is usually the result of direct virus invasion of the CNS, although a post-infectious autoimmune aetiology has been described (Russell and Donald,

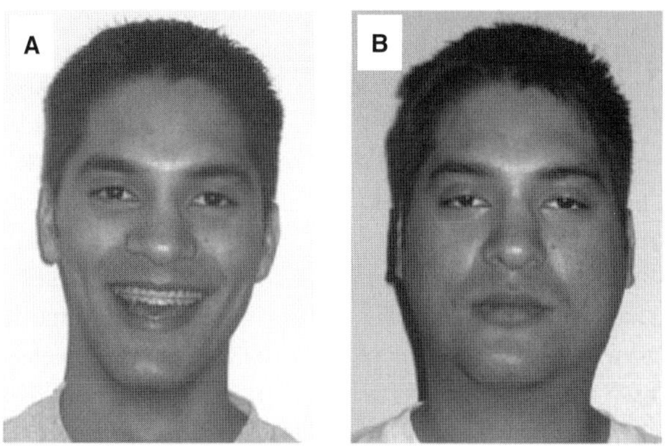

Figure 1.3 Mumps parotitis in an adolescent prior to mumps infection (A) and on day 3 (B) of acute bilateral parotitis. Courtesy of patient C.M. and the Cleveland Clinic, with permission.

1958; Johnstone *et al.*, 1972; Litman and Baum, 2010). For reasons that are unclear, symptomatic CNS infection is more common in adults than children, and occurs with a 3:1 male:female ratio (Koskiniemi *et al.*, 1983; Leboreiro-Fernandez *et al.*, 1997).

Sensorineural deafness, which can be transient or permanent, is a major complication of mumps virus infection with incidence rates estimated to be as high as 1:1000 cases (Kawashima *et al.*, 2005; Hashimoto *et al.*, 2009). Deafness is usually unilateral, sudden in onset and profound (Unal *et al.*, 1998). Hearing loss can occur before, after, at the same time or in the complete absence of parotid swelling (Kanra *et al.*, 2004). Deafness is believed to be the result of virus infection of the endolymph of the ear, damaging the organ of Corti, tectorial membrane or the myelin sheath of the eighth cranial nerve (McLean *et al.*, 1960; Kanra *et al.*, 2004).

Pancreatitis is diagnosed in approximately 4% of the cases (Plotkin and Rubin, 2008), although it may be more prevalent as suggested by gastro-intestinal symptoms such as nausea, vomiting, epigastric and left upper quadrant pain, which can occur in up to 50% of cases (Philip *et al.*, 1959). Considering possible virus mediated damage to pancreatic islet cells it has been postulated that mumps virus could be a cause of diabetes; however, the association is tenuous. Some studies (Gamble, 1980; Hyoty *et al.*, 1988; Goto *et al.*, 2008) have reported an association between mumps and juvenile-onset or type I insulin-dependent diabetes mellitus (IDDM), but other studies have not (Ratzmann *et al.*, 1985; Schulz *et al.*, 1987; Goldacre *et al.*, 2005). There is no association between mumps and type II diabetes (Lohr and Oldstone, 1990).

Virus has been detected in the urine in more than half of hosptial-admitted cases (Utz *et al.*, 1958, 1964). Thus, if viruria is used as an indication of kidney infection, then kidney involvement is not uncommon; however, symptomatic nephritis is rarely diagnosed. Other infrequent complications of mumps include oophoritis, mastitis, myocarditis, arthralgia and thrombocytopenia (Philip *et al.*, 1959; Roberts and Fox, 1965; Graham *et al.*, 1974; Ninomiya *et al.*, 1977). Mumps virus can infect the placenta and the fetus and has been linked with an increased rate of spontaneous abortion and intrauterine fetal death; however, the incidence of such events has not been examined in controlled studies (Ornoy and Tenenbaum, 2006). Fetal malformations following infection with mumps virus during pregnancy have not been reported (World Health Organization, 2005).

Diagnosis

Clinical

The standard clinical case definition of mumps is the acute onset of unilateral or bilateral swelling of the parotid or other salivary glands lasting more than 2 days without other apparent cause (Centers for Disease Control and Prevention, 2006a). While parotitis is the hallmark of mumps, in many cases, salivary gland swelling may not be apparent. This is especially true in individuals with mumps meningitis, who frequently present in the absence of detectable salivary gland enlargement (Ritter, 1958; Johnstone *et al.*, 1972). Notably, salivary gland swelling is also caused by a number of other infectious agents, drugs and conditions (see Differential diagnosis) and when disease incidence is low, the impact of such alternative aetiologies dramatically reduces the positive predictive value of a clinical diagnosis (Pelosi *et al.*, 2001; Guy *et al.*, 2004). Thus, not only should many of the other widely varying constitutional mumps symptoms, as reviewed earlier, be considered in the clinical diagnosis, but laboratory testing is essential for case confirmation.

Laboratory

A laboratory diagnosis of mumps is currently based on isolation of the virus in culture, detection of viral RNA by reverse-transcription polymerase chain reaction (RT-PCR) or serological confirmation. The highest success rates for virus detection are obtained using buccal or throat swabs, although virus can also be detected in saliva, urine and CSF (Litman and Baum, 2010). Despite the apparent frequency of viraemia (based on the frequency of systemic infection), virus has only rarely been isolated from blood (Kilham, 1948; Overman,

1958), possibly related to the presence of anti-mumps virus antibodies, which begin to appear at the time of symptoms.

The window of opportunity for successful virus isolation or detection once symptoms are evident is limited. Recent studies indicate successful detection in only 14–35% of cases when the clinical specimen is collected within 4 days of symptom onset (Bitsko *et al.*, 2008; Reid *et al.*, 2008; Rota *et al.*, 2009). Virus isolation/detection success rates dramatically decline beyond this time frame. These data are from studies mostly including previous vaccinees who later contracted disease, and, thus, the low rates of detection may be a consequence of low levels of virus shedding, which might not be the case in unvaccinated individuals.

Historically, detection of virus has been made by incubation of a clinical specimen on a susceptible cell line (typically Vero cells) followed by immunohistochemical staining; however, this method has, for good reason, been supplanted by the more rapid technique of RT-PCR (Poggio *et al.*, 2000; Uchida *et al.*, 2005; Rota *et al.*, 2009). The mumps virus SH gene is the most common target for RT-PCR assays, as its detection not only confirms infection, but its nucleotide sequence can be used to identify the specific mumps virus strain, which is helpful in establishing epidemiological trends.

Owing to limited abilities to conduct PCR-based laboratory investigations in-house, most laboratories use commercially available enzyme-linked immunosorbent assay (ELISA) kits for detection of virus-specific IgM and IgG antibodies to aid in diagnosis. One major advantage of serological testing over RT-PCR testing is that the window for detection of viral RNA is a matter of a few days post-symptom onset levels of IgM remain elevated for several weeks or months (Ukkonen *et al.*, 1981; Benito *et al.*, 1987) and the IgG response can persist indefinitely. Nonetheless, the timing of sera collection for detection of IgM and IgG is important. False-negative IgM results have been reported using serum collected during the first few days of clinical presentation (Cunningham *et al.*, 2006; Warrener and Samuel, 2006; Krause *et al.*, 2007), and in previously infected or immunized individuals, IgM responses may not occur. In countries with high vaccine coverage, testing of acute and convalescent serum pairs for an increase in IgG titre is often more informative, but such pairs are often not available.

Owing to the ease in which IgG and IgM can be measured and quantitated by ELISA, this technique has for the most part supplanted use of the more labour intensive tests of complement fixation (CF), haemagglutination inhibition (HI) or virus neutralization (NT). Notably, comparison of different tests in the same individual have shown discrepancies, indicating that different tests measure different types of antibodies or immunity. For example, unlike virus NT tests, which measure the amount of functional neutralizing antibody, the ELISA detects all virus-specific antibodies, including those that are not protective. Thus, 'seroconversion' can be demonstrated by ELISA even in the absence of neutralizing antibody (Christenson and Bottiger, 1990), raising doubt whether a demonstration of seroconversion by ELISA is necessarily clinically meaningful. Despite such limitations, the simplicity and speed by which samples can be assessed by ELISA makes this assay format best suited for large-scale serosurveys, an endeavour that could not be practically accomplished by other means. CF and HI tests are no longer recommended for assessing mumps immunity or for screening purposes owing to issues with assay sensitivity and specificity (Leinikki and Passila, 1977; Zimmerman *et al.*, 2002).

Although some studies have reported a good correlation between ELISA and neutralization test data (Tsuji *et al.*, 1983; Nigro *et al.*, 1986; Usonis *et al.*, 2001; Mauldin *et al.*, 2005), given that there is no specific level of antibody detected by ELISA that has been correlated with protection, ELISA data cannot be used to assess functional immunity. In contrast, there is a strong association between the presence of neutralizing antibody and protection from disease (Vicens *et al.*, 1966; Ennis, 1969); thus, the NT test is considered the gold standard in measuring protective immunity. Unfortunately, the NT test has not been standardized, an international reference serum is lacking and there is no established neutralizing antibody titre cut-off for protection. Studies conducted during the pre-vaccine era indicate that a neutralizing antibody titre of ≥1:8 is protective (Bashe, 1953; Mayer, 1966; Brunell, 1968; Ennis, 1969), however, the relevance of these data to the present day situation where antibody is induced by vaccination, rather than natural infection, is not known.

Differential diagnosis

During a mumps outbreak or epidemic, when parotitis is present, the clinical diagnosis of mumps is straightforward. However, when disease incidence is low, other causes of parotitis should be considered and laboratory testing is required for case confirmation. Infectious causes of parotitis include other viruses, e.g. Epstein–Barr virus (EBV), human parainfluenza virus 3 (hPIV-3), adenovirus, HIV-1, Coxsackievirus, influenza A virus, parvovirus B19, lymphocytic choriomeningitis virus, human herpesvirus 6 (HHV-6) or Gram-positive bacteria, e.g. atypical mycobacteria and *Bartonella* sp. bacteria (Litman and Baum, 2010). These agents do not produce parotitis on an epidemic scale and all can be easily differentiated from mumps virus by serology, culture or RT-PCR. Parotitis can also be seen in Sarcoidosis, Sjögren's syndrome, Mikulicz's syndrome, Parinaud's syndrome, uraemia, diabetes mellitus, starch ingestion, malnutrition, cirrhosis and associated with various drugs (Litman and Baum, 2010). Other entities should be considered when manifestations consistent with mumps appear in organs other than the parotid.

Prevention and control

Vaccination is the only practical control measure for the prevention of mumps. Isolation of individuals with mumps to limit person-to-person virus spread is of limited utility considering that virus is shed prior to the appearance of clinical symptoms and a significant proportion of infected individuals are asymptomatic.

Passive immunization

Post-exposure prophylactic administration of immune globulin is of limited effectiveness in the prevention of mumps and its complication and is no longer practised. Nevertheless, maternal antibody transferred across the placenta, appears to protect infants from developing mumps during the first year of life (Plotkin and Rubin, 2008). Vaccination after exposure is not harmful and may possibly avert later disease (Centers for Disease Control and Prevention, 2009).

Active immunization

Vaccine development

The first vaccines against mumps, developed in the 1950s, were killed vaccines, mainly used in the armed forces in the USA and Finland (Peltola, 1993). Although killed vaccines showed some efficacy, the duration of immunity was short, and their use was abandoned (Minor, 1997). Live, attenuated mumps vaccines were first developed in Russia in the mid-1950s (Smorodintsev *et al.*, 1954, 1958, 1965; Klyatchko and Smorodintsev, 1958). All mumps vaccines in current use are live, attenuated viruses and are mostly available in combination with other vaccines such as the MMR or the MMR–varicella (MMRV) combination.

The Jeryl Lynn, Urabe-Am9, and Rubini vaccine strains have been the most widely used strains. The Jeryl Lynn vaccine, produced in the 1960s by Merck and Company, is composed of two distinct, but homologous mumps virus strains, JL-2 and JL-5, existing in a 1:5 ratio (Afzal *et al.*, 1993). In 1998, GlaxoSmithKline (then SmithKline Beecham Biologicals) clonally isolated the major component of the Jeryl Lynn strain (JL-5) by limit dilution and expanded the clone in chick embryo cultures to obtain a seed virus to produce a 'new' vaccine strain called RIT-4385, which was incorporated with measles and rubella vaccine strains to produce the MMR vaccine Priorix (Usonis *et al.*, 1998). RIT-4385 appears to be equally immunogenic (based on seroconversion rates) as Jeryl Lynn, and like Jeryl Lynn, has not been causally associated with aseptic meningitis (Plotkin and Rubin, 2008).

The Urabe-Am9 strain was developed by the Biken Institute in Japan and obtained after isolation from the saliva of a mumps patient followed by serial passage in the amniotic cavity of chick embryos and in quail embryo fibroblasts. The vaccine was also produced in Europe by GlaxoSmithKline, Chiron and Aventis Pasteur. The Urabe-Am9 vaccine, like the Jeryl Lynn vaccine, is composed of a mixture of virus variants. Reports of vaccine-associated meningitis began appearing in 1991 and by 1993 several countries discontinued its use (Miller *et al.*, 1993). Nevertheless, it is still used today in several countries based on the benefit of protection outweighing the small risk of non-life threatening adverse events (Bonnet *et al.*, 2006). Initial molecular studies of the vaccine indicated that a variant population possessing a lysine (K) at amino acid position 335 in the haemagglutinin–neuraminidase protein was associated with cases of post-vaccination meningitis (Brown *et al.*, 1996; Brown and Wright, 1998). Notably, K335 is present in the Jeryl Lynn vaccine strain and in other lots of Urabe Am9 vaccines, which have not been causally associated with aseptic meningitis, and, thus, this genetic locus itself is probably not responsible for the adverse safety profile of the Urabe-Am9 vaccine (Katayama *et al.*, 1993; Colville *et al.*, 1994; Afzal *et al.*, 1998; Amexis *et al.*, 2001; da Cunha *et al.*, 2002; Furesz, 2002; Ki *et al.*, 2003; Sauder *et al.*, 2006; Shah *et al.*, 2009).

The Rubini mumps vaccine virus was derived from an isolate obtained from the urine of a child in Switzerland in 1974 and was then passaged into the amniotic or allantoic cavity of embryonated hens' eggs before adaptation to the human diploid cell line MRC-5, which is the substrate for production. Sequence analysis of the SH gene demonstrates that the vaccine strain is 99% identical at the nucleotide level to the tissue culture-adapted Enders strain prepared in the USA in 1945. As described later, this vaccine was later found to offer virtually no protective efficacy and is no longer in use.

Dosage, route and adverse reactions

Mumps vaccination is primarily given in the form of the trivalent MMR vaccine. The vaccine is administered subcutaneously. As with other live virus vaccines, mumps vaccines should not be given to immunocompromised patients; however, HIV positive subjects who are asymptomatic or have only mild symptoms can be vaccinated since they do not generally experience complications (World Health Organization, 2007a).

Fever is the most common systemic adverse reaction following MMR vaccination, with 5–15% of recipients developing a temperature of 39.4°C (102.92°F) or higher, usually 7–12 days after vaccination and generally lasting 1–2 days. Most people with fever are otherwise asymptomatic. The most common mumps vaccine-specific side-effect is parotitis and/or submaxillary gland swelling, occurring in approximately 1% of vaccinees (Popow-Kraupp *et al.*, 1986; Christenson *et al.*, 1983; Dos Santos *et al.*, 2002). Aseptic meningitis, which is usually mild, has been reported in association with several of the vaccine strains, including the Urabe, Leningrad–Zagreb, and Leningrad-3 strains (Brown *et al.*, 1991, 1996; Afzal *et al.*, 1998; Miller *et al.*, 2007). Rates of this adverse event range from less than 1 case per 100,000 doses to 1 case per 400 doses and vary according to manufacturer, the index of clinical suspicion, and the intensity of surveillance (World Health Organization, 2007a). Aseptic meningitis has not been demonstrated to occur following vaccination with the Jeryl Lynn strain. Other, more rare, vaccine-associated complications include orchitis, pancreatitis, and sensorineural deafness (Plotkin and Rubin, 2008).

Anaphylaxis has been reported following vaccination, but is exceedingly rare. Such severe allergic reactions are believed to be in response to egg proteins, neomycin, and gelatin contained in some vaccines. The current opinion is to immunize children with mild egg allergy, but a cautious approach is recommended for persons with a history of anaphylaxis or serious immediate reactions after ingestion of egg products or after previous vaccinations (Khakoo and Lack, 2000; Patja *et al.*, 2001). Persons who have a history of anaphylactic reactions to neomycin or gelatin should not receive vaccines containing these ingredients.

Immune responses

Mumps vaccines (live) are able to elicit humoral and cell-mediated immune responses, which are similar to the immune responses following natural disease, although, levels of neutralizing antibodies after vaccination have been reported to be lower than after natural mumps infection (Weibel *et al.*, 1967; Fedova *et al.*, 1987). The clinical significance of this, if any, is not known.

Seroconversion rates following immunization with a single dose of most mumps vaccines, including the Jeryl Lynn and Urabe-Am9 mumps vaccines are similarly high, exceeding 90% (Stokes *et al.*, 1967; Weibel *et al.*, 1967; Young *et al.*, 1967; Ehrengut *et al.*, 1983; Beck *et al.*, 1989; Usonis *et al.*, 2001; Wellington and Goa, 2003). In contrast, significantly lower seroconversion rates have been measured following immunization with a single dose of the Rubini vaccine strain (Schwarzer *et al.*, 1998; Tischer and Gerike, 2000), consistent with its extremely poor efficacy (Dayan and Rubin, 2008). As of 2001, the WHO no longer recommends the use of the Rubini vaccine strain for national immunization programmes (World Health Organization, 2001). The first dose of mumps vaccine is typically administered at 12–18 months of age. Administration at six months of age or earlier may result in lower humoral immune responses owing to interference by circulating maternal antibodies and/or an intrinsic limitation of the infant's immune system (Gans *et al.*, 2003). Nevertheless, despite lower immune responses at this age, recipients seem to be well primed, since they respond well to a second immunization given at 12 months of age (Gans *et al.*, 2001, 2003). Administration of vaccine at nine months of age may also be a feasible strategy since vaccination at this age was not found to be associated with lower levels of antibody production (Klinge *et al.*, 2000; Gans *et al.*, 2003). Although immunity to mumps after natural infection or vaccination was historically believed to be lifelong, as discussed later, there is a wealth of information indicating waning immunity and susceptibility to reinfection, even in two-dose vaccines (Davidkin *et al.*, 2008; Rubin *et al.*, 2008; LeBaron *et al.*, 2009).

As is the case for natural infection, cell mediated immune responses, as indicated by T-cell proliferation and IFN-γ production assays, are induced by immunization. In one study, cellular immunity was measured 21 years after a single dose of mumps-containing vaccine. Of 50 persons evaluated, 49 (98.0%) possessed strong lymphoproliferative responses irrespective of the antibody titres measured (Jokinen *et al.*, 2007). Similar results were reported by others (Dhiman *et al.*, 2005; Hanna-Wakim *et al.*, 2008; Vandermeulen *et al.*, 2009). Whether or not cellular immunity plays a large role in protection against mumps infection or complications is unclear.

Only one study examined the presence of mumps-specific memory B cells following immunization (Vandermeulen *et al.*, 2010). In that study, peripheral blood mononuclear cells (PBMCs) from mumps antibody seronegative and seropositive vaccines were injected intrasplenically into immunoincompetent mice. Mumps virus specific antibodies were only induced in mice inoculated with PBMCs from mumps antibody-seropositive vaccines, indicating an absence of circulating memory B cells in seronegative individuals (Vandermeulen *et al.*, 2010). Notably, T cells from the seronegative subjects were capable of mounting a lymphoproliferative response to stimulation, indicating that the vaccinated seronegative individuals were not examples of primary vaccine failures.

Vaccine efficacy

True vaccine efficacy is assessed in randomized, double-blind placebo-controlled clinical trials, typically prelicensure. Few such studies have been undertaken following the licensure of the Jeryl Lynn vaccine strain in 1967, as withholding of a vaccine from a control group of children once its efficacy has been established is regarded as unethical. Therefore, efficacy is often estimated from vaccine effectiveness studies performed in the field after the vaccine has been in use. Although vaccine effectiveness studies potentially suffer from inaccurate case definition, incomplete surveillance, limited case ascertainment, and inaccurate determination of vaccination status, in comparison to controlled efficacy studies, vaccine effectiveness studies better reflect the performance of the vaccine in routine clinical practice.

All vaccines in current use are effective, although the duration of immunity may not be as long as once believed (discussed below). The effectiveness of one dose of mumps vaccine ranges from approximately 70% to 90% and that for two doses ranges from approximately 80% to 95% (Dayan and Rubin, 2008; Plotkin and Rubin, 2008). Although the effectiveness of two doses of vaccine tend to be greater than one dose, the differences have often been found to not be statistically significant (Sartorius *et al.*, 2005; Cohen *et al.*, 2007; Schaffzin *et al.*, 2007; Marin *et al.*, 2008).

Epidemiology

Mumps contagiousness is similar to that of influenza and rubella, but less than that of measles or varicella (Gordon and Heeren, 1940; Anderson and May, 1990). The number of secondary cases of infection expected to result from an index case of mumps in a fully susceptible population (R_0 or basic reproductive number) is 10–12 (Philip *et al.*, 1959; Anderson and May, 1990). Approximately one third of mumps infections are subclinical, although these patients still contribute to the circulation of the virus (Henle *et al.*, 1948a; Meyer, 1962). Mumps is endemic worldwide, occurring epidemically every 2–5 years in unvaccinated populations (Centers for Disease Control and Prevention, 1974, 1991). Development and widespread sustained use of live, attenuated mumps vaccines has significantly reduced disease incidence by limiting virus circulation. Despite these successes, only one country (Finland) has claimed elimination of indigenous disease (Peltola *et al.*, 1994), but it is unlikely that such a state can be maintained long-term owing to an increasing population of susceptibles as a result of growth of the unvaccinated population and waning immunity (Peltola *et al.*, 2008).

Pre-vaccination era

In the pre-vaccine era, mumps was characterized by wavelike progression through a population with seasonal rise in the winter and spring months (Gordon and Heeren, 1940; Gordon and Kilham, 1949; World Health Organization, 2007a). The estimated annual global incidence was 100 to 1000 per 100,000 population, an incidence that continues to be reported in countries not using mumps vaccines (World Health Organization, 2007a,b). The highest incidence occurred in children 5–9 of age and, by age 15, 92% of a given population had evidence of a previous mumps infection (Philip *et al.*, 1959; Meyer, 1962; Mortimer, 1978). Mumps was, and continues to be, rarely seen in children younger than 1 year of age, most probably owing to acquisition of maternal antibody (Plotkin and Rubin, 2008). As with nearly all infectious diseases spread by the respiratory route, intensity of exposure, such as crowding in families, military or college settings, is a major risk factor (Gordon and Heeren, 1940; Gordon and Kilham, 1949; Philip *et al.*, 1959; Cooney *et al.*, 1975).

Post-vaccination era

The majority of industrialized countries and many countries in economic transition have incorporated mumps vaccination into their national immunization programmes, in most cases by using the combined MMR vaccine. As of July 2010, a total of 117 (61%) of the 193 member states of the World Health Organization included mumps vaccine in their routine national immunization programmes, nearly all using the trivalent MMR vaccine. Mumps immunization has yet to be implemented in the routine immunization schedules of most developing countries.

The herd immunity threshold has been estimated to be 90–92% (Anderson, 1992); however, recent evidence of waning immunity raises doubts whether herd immunity can be maintained. Based on a logistic regression model, the odds ratio for developing mumps increases by 27% with each year since vaccination (Vandermeulen *et al.*, 2004), consistent with findings of time since vaccination as a risk factor (Cohen *et al.*, 2007; Muhsen *et al.*, 2008; Castilla *et al.*, 2009; LeBaron *et al.*, 2009). The significance of this has been highlighted by outbreaks in highly vaccinated populations globally. For example, in 2006 the USA experienced its largest mumps outbreak in 19 years with 6584 cases, compared with approximately 300 cases annually between 1999 and 2005 (Rubin *et al.*, 2008). A number of outbreak investigations revealed that nearly all cases had received the recommended two doses of MMR vaccine (Cortese *et al.*, 2008; Marin *et al.*, 2008; Rota *et al.*, 2009). Similarly, between 2009 and 2010, a large mumps outbreak occurred in New York involving over 1500 cases. Among the patients for whom vaccination status was reported, 75% had received two doses (Centers for Disease Control and Prevention, 2010). Outbreaks attributable to two-dose failure were also reported in the Czech Republic in 2005–2006, where 70% of cases had received two doses of vaccine (Boxall *et al.*, 2008), in Australia in 2007–2008 where 76% of the most affected age group had received two doses of vaccine (Bangor-Jones *et al.*, 2009), and in the North Wales region of the UK, in 2008–2009, where 87% of case-patients had received two doses of MMR vaccine (Roberts *et al.*, 2009). Notably, the overwhelming majority of cases in these recent outbreaks occurred among individuals 15–24 years of age, a cohort immunized more than 10 years earlier.

A second feature common to recent outbreaks in vaccinated populations is the genotype of the virus isolated from cases. Genotyping of mumps viruses is performed for epidemiological purposes and is based on the hypervariable nucleotide sequence of the small hydrophobic (SH) gene (Yeo *et al.*, 1993; Tecle *et al.*, 2001). Currently, 13 different genotypes have been reported, identified as genotypes A-M (Fig. 1.4; Jin *et al.*, 2005; Santos *et al.*, 2008). In recent mumps outbreaks in the Western hemisphere, nearly all clinical virus isolates were found to be members of the genotype G group of viruses (Savage *et al.*, 2005; Watson-Creed *et al.*, 2006; Centers for Disease Control and Prevention, 2006a,b,c; Peltola *et al.*, 2007; Public Health Agency of Canada, 2007, 2009; Kubinyiova *et al.*, 2008) whereas the predominant vaccines used in these countries (Jeryl Lynn and Jeryl Lynn-based) are genotype A viruses. In the Asia-Pacific region, also using the Jeryl Lynn vaccines in Australia and China, caused by genotype J and F strains were isolated (Bangor-Jones *et al.*, 2009; Chen *et al.*, 2009). This raises the possibility that immune responses raised against one mumps virus genotype may not be sufficiently potent against heterologous mumps virus genotypes. This notion is supported by reports of cases of mumps re-infection by heterologous mumps virus strains and by laboratory data demonstrating virus strain-specific neutralizing epitopes (Yates *et al.*, 1996; Tecle *et al.*, 1998; Nojd *et al.*, 2001; Montes *et al.*, 2002; Orvell *et al.*, 2002; Rubin *et al.*, 2006). In one recent study, it was demonstrated that although serum from Jeryl Lynn vaccines possessed neutralizing activity against a genotype G outbreak strain, neutralizing antibody titres were significantly lower against the genotype G strain compared with the homologous vaccine strain (Rubin *et al.*, 2008). Of note, outbreaks in vaccines are not a problem limited to persons vaccinated with the Jeryl Lynn strain, as outbreaks have been reported among recipients of the Leningrad-Zagreb and Urabe-Am9 vaccine strains, which are non-genotype A viruses (Paccaud *et al.*, 1995; Schlegel *et al.*, 1999; Ong *et al.*, 2005; Hindiyeh *et al.*, 2009). In addition, there is practical evidence against the notion of genotype-specific immunity. For example, in the US, mumps incidence declined by 99.9% (Centers for Disease Control and Prevention, 1978; McNabb *et al.*, 2007) in association with the use of the Jeryl Lynn vaccine, the only live, attenuated vaccine strain ever used in the U.S. That this was accomplished in the face of wild-type circulating mumps virus strains belonging to multiple heterologous genotypes (Rota *et al.*, 2007) supports the general idea of sufficient effectiveness of vaccine-induced antibodies against heterologous mumps virus strains. However, this does not rule out a possible conflict between the vaccine genotype and that of currently circulating mumps virus strains. Nevertheless, mumps vaccines are not 100% effective, thus, in a highly vaccinated population, mumps cases can be expected in vaccinees.

Perspectives and future directions in research

There are numerous unresolved issues related to the biology of mumps virus and the host response to infection, creating a gap in our understanding of the pathogenesis and epidemiology of the disease. This has been made quite evident when considering historic and current issues with mumps vaccine safety and efficacy.

Our current understanding of the pathogenesis of mumps is derived from experimental infection of animals by unnatural routes, precluding the identification of the true primary and secondary sites of virus replication and the mechanisms and kinetics of spread between these sites. This impinges on our ability to rationally design attenuated vaccines, a practical concern made evident by the unfortunate clinical experience with many mumps vaccines. The recent development of relevant animal models of disease coupled with

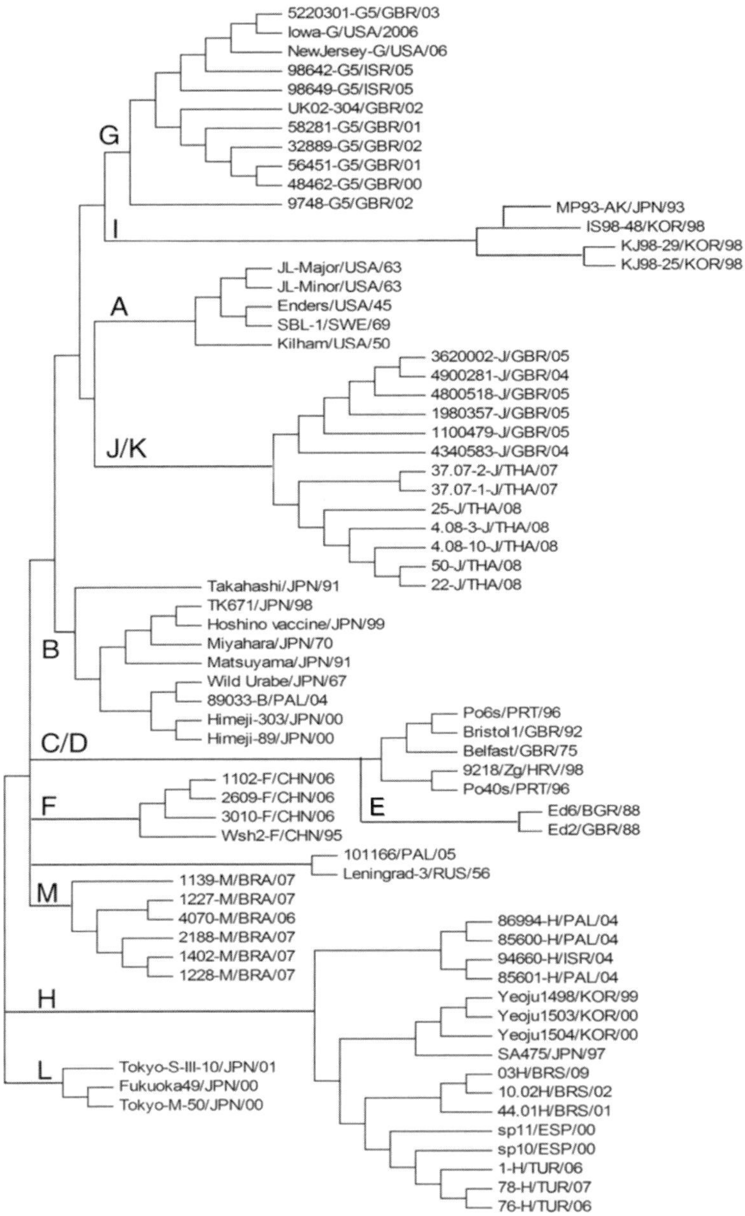

Figure 1.4 Phylogenetic analysis of the mumps virus SH gene sequences obtained from the EMBL/GenBank database showing groupings of genotypes A–M. Tree constructed using the neighbour-joining method.

the availability of reverse genetic systems for the manipulation and rescue of mumps viruses of differing phenotype affords, for the first time, a means to identify virulence factors. Not only will this advance the field of mumps virology, but will probably benefit the understanding of the molecular basis of virulence for other paramyxoviruses as well.

Another important area for future research is the study of the genetic stability of the virus. In comparison to measles virus for which low mutation rates have been identified among field isolates (Rima *et al.*, 1997; Peltola *et al.*, 2007) and relatively low sequence variation has been observed between strains (Rota *et al.*, 1996), mumps viruses appears to be much less genetically stable

(Amexis *et al.*, 2001; Muhlemann, 2004; Sauder *et al.*, 2006). This is also indicated by laboratory data showing few, if any, genomic changes with measles virus following serial passage *in vitro* in contrast to numerous genomic changes with mumps viruses passaged under similar conditions (Amexis *et al.*, 2002; Rubin *et al.*, 2003; Sauder *et al.*, 2006; Borges *et al.*, 2008). Interestingly, in stark contrast to recent mumps outbreaks where there is very little evidence of sparing of those previously vaccinated (Cortese *et al.*, 2008; Rota *et al.*, 2009), measles outbreaks are characterized by virus spread almost exclusively through the small numbers of unvaccinated individuals within a largely vaccinated population (Parker *et al.*, 2006; Dallaire *et al.*, 2009). Whether or not the relative genetic instability of mumps virus facilitates virus spread and is responsible for the changing epidemiology of the disease or the effectiveness of the vaccine remains to be explored.

In addition to outstanding questions related to the biology of the virus itself, numerous issues of importance remain unanswered regarding the host immune response to infection. Although neutralizing antibody is a correlate of protection, the level of antibody required has not been established. Further, little is known about the role of cell mediated immune responses in protection against infection and complications.

Lastly, as the bulk of available data indicates waning immunity as a factor in recent outbreaks and the impact of different vaccination schedules should be explored. Ironically, waning immunity may be a consequence of the remarkable success of the vaccination programmes themselves. As vaccine coverage increases, natural transmission pathways become interrupted, dramatically reducing levels of endemically circulating wild-type viruses, which in the past served to periodically boost the vaccine-induced immune response and combat waning immunity. Numerous studies have demonstrated that levels of vaccine-induced antibodies decline over time and, not surprisingly, that the risk of contracting mumps increases with time since immunization (Davidkin *et al.*, 1995, 2008; Vandermeulen *et al.*, 2004; Cohen *et al.*, 2007; Muhsen *et al.*, 2008; Castilla *et al.*, 2009; LeBaron *et al.*, 2009). Recent current outbreaks therefore indicate that the protective immunogenicity of mumps vaccine strains is less sustainable than initially believed based on studies conducted decades ago when mumps virus was endemic. This poses a somewhat difficult situation, where despite sustained two-dose vaccine coverage levels approaching 100% in some populations, herd immunity cannot be maintained. Thus, either more effective vaccines are needed, or an additional dose of vaccine administered during adolescence may be called for. Based on evidence of waning immunity being a factor for nearly all vaccine strains (Boulianne *et al.*, 1995; Miller *et al.*, 1995; Usonis *et al.*, 2001; Raut *et al.*, 2007), development of a new vaccine strain that will not suffer from this problem seems unlikely. Further, better matching of the mumps virus vaccine strain genotype with contemporary virus genotypes does not appear to be a sound approach given that antibodies produced by any new vaccine will probably wane at the same or similar rate as seen for current vaccines. Further, it is not clear if lower antibody titres to heterologous virus strains is clinically meaningful. Thus, changing to a three-dose mumps-containing vaccine schedule, with the third dose administered in late adolescence, appears to be the most promising approach to stemming the emergence of mumps outbreaks. In the absence of addressing the issue of waning immunity, more and probably larger mumps outbreaks can be expected in the future.

References

Afzal, M.A., Elliott, G.D., Rima, B.K., and Orvell, C. (1990). Virus and host cell-dependent variation in transcription of the mumps virus genome. J. Gen. Virol. *71*, 615–619.

Afzal, M.A., Pickford, A.R., Forsey, T., Heath, A.B., and Minor, P.D. (1993). The Jeryl Lynn vaccine strain of mumps virus is a mixture of two distinct isolates. J. Gen. Virol. *74*, 917–920.

Afzal, M.A., Yates, P.J., and Minor, P.D. (1998). Nucleotide sequence at position 1081 of the hemagglutinin-neuraminidase gene in the mumps Urabe vaccine strain. J. Infect. Dis. *177*, 265–266.

Afzal, M.A., Marsden, S.A., Hull, R.M., Pipkin, P.A., Bentley, M.L., and Minor, P.D. (1999). Evaluation of the neurovirulence test for mumps vaccines. Biologicals *27*, 43–49.

Ali, A., and Nayak, D.P. (2000). Assembly of Sendai virus: M protein interacts with F and HN proteins and with the cytoplasmic tail and transmembrane domain of F protein. Virology *276*, 289–303.

Amanna, I.J., Carlson, N.E., and Slifka, M.K. (2007). Duration of humoral immunity to common viral and vaccine antigens. N. Engl. J. Med. *357,* 1903–1915.

Amexis, G., Fineschi, N., and Chumakov, K. (2001). Correlation of genetic variability with safety of mumps vaccine Urabe AM9 strain. Virology *287,* 234–241.

Amexis, G., Rubin, S., Chizhikov, V., Pelloquin, F., Carbone, K., and Chumakov, K. (2002). Sequence diversity of Jeryl Lynn strain of mumps virus: quantitative mutant analysis for vaccine quality control. Virology *300,* 171–179.

Amexis, G., Rubin, S., Chatterjee, N., Carbone, K., and Chumakov, K. (2003). Identification of a new genotype H wild-type mumps virus strain and its molecular relatedness to other virulent and attenuated strains. J. Med. Virol. *70,* 284–286.

Anderson, R.M. (1992). The concept of herd community and the design of community-based immunization programmes. Vaccine *10,* 928–935.

Anderson, R.M., and May, R.M. (1990). Immunisation and herd immunity. Lancet *335,* 641–645.

Anderson, D.E., and von Messling, V. (2008). Region between the canine distemper virus M and F genes modulates virulence by controlling fusion protein expression. J. Virol. *82,* 10510–10518.

Anderson, L.J., and Seward, J.F. (2008). Mumps epidemiology and immunity: the anatomy of a modern epidemic. Pediatr. Infect. Dis. J. *27,* S75–79.

Andrejeva, J., Childs, K.S., Young, D.F., Carlos, T.S., Stock, N., Goodbourn, S., and Randall, R.E. (2004). The V proteins of paramyxoviruses bind the IFN-inducible RNA helicase, mda-5, and inhibit its activation of the IFN-beta promoter. Proc. Natl. Acad. Sci. U.S.A. *101,* 17264–17269.

Arguedas, A., Janai, H., and Marks, M. (2004). Mumps. In Infectious Diseases, Gorbach, S.L., Bartlett, J.G., Blackow, N., eds. (Lippincot Williams & Wilkins: Philadelphia), pp. 1210–1214.

Banerjee, A.K., Barik, S., and De, B.P. (1991). Gene expression of nonsegmented negative strand RNA viruses. Pharmacol. Ther. *51,* 47–70.

Bangor-Jones, R.D., Dowse, G.K., Giele, C.M., van Buynder, P.G., Hodge, M.M., and Whitty, M.M. (2009). A prolonged mumps outbreak among highly vaccinated Aboriginal people in the Kimberley region of Western Australia. Med. J. Aust. *191,* 398–401.

Bao, X., Kolli, D., Liu, T., Shan, Y., Garofalo, R.P., and Casola, A. (2008). Human metapneumovirus small hydrophobic protein inhibits NF-kappaB transcriptional activity. J. Virol. *82,* 8224–8229.

Barskey, A.E., Glasser, J.W., and LeBaron, C.W. (2009). Mumps resurgences in the United States: A historical perspective on unexpected elements. Vaccine *27,* 6186–6195.

Basche, W.J., Jr., Gottlieb, T., Henle, G., Henle, W. (1953). Studies on the prevention of mumps. VI. The relationship of neutralizing antibodies to the determination of susceptability and to the evaluation of immunization procedures. J. Immunol. *71,* 76–85.

Beck, M., Welsz-Malecek, R., Mesko-Prejac, M., Radman, V., Juzbasic, M., Rajninger-Miholic, M., Prislin-Musklic, M., Dobrovsak-Sourek, V., Smerdel, S., and Stainer, D.W. (1989). Mumps vaccine L-Zagreb, prepared in chick fibroblasts. I. Production and field trials. J. Biol. Stand. *17,* 85–90.

Benito, R.J., Larrad, L., Lasierra, M.P., Benito, J.F., and Erdociain, F. (1987). Persistence of specific IgM antibodies after natural mumps infection. J. Infect. Dis. *155,* 156–157.

Biedel, C.W. (1978). Recurrent mumps parotitis following natural infection and immunization. Am. J. Dis. Child. *132,* 678–680.

Bitsko, R.H., Cortese, M.M., Dayan, G.H., Rota, P.A., Lowe, L., Iversen, S.C., and Bellini, W.J. (2008). Detection of RNA of mumps virus during an outbreak in a population with a high level of measles, mumps, and rubella vaccine coverage. J. Clin. Microbiol. *46,* 1101–1103.

Bjorvatn, B. (1974). Incidence and persistence of mumps-specific IgM and IgA in the sera of mumps patients. Scand. J. Infect. Dis. *6,* 125–129.

Black, F.L., and Houghton, W.J. (1967). The significance of mumps hemagglutinin inhibition titers in normal populations. Am. J. Epidemiol. *85,* 101–107.

Bonnet, M.C., Dutta, A., Weinberger, C., and Plotkin, S.A. (2006). Mumps vaccine virus strains and aseptic meningitis. Vaccine *24,* 7037–7045.

Borges, M.B., Caride, E., Jabor, A.V., Malachias, J.M., Freire, M.S., Homma, A., and Galler, R. (2008). Study of the genetic stability of measles virus CAM-70 vaccine strain after serial passages in chicken embryo fibroblasts primary cultures. Virus Genes *36,* 35–44.

Boulianne, N., De Serres, G., Ratnam, S., Ward, B.J., Joly, J.R., and Duval, B. (1995). Measles, mumps, and rubella antibodies in children 5–6 years after immunization: effect of vaccine type and age at vaccination. Vaccine *13,* 1611–1616.

Bousse, T., Matrosovich, T., Portner, A., Kato, A., Nagai, Y., and Takimoto, T. (2002). The long noncoding region of the human parainfluenza virus type 1 f gene contributes to the read-through transcription at the m-f gene junction. J. Virol. *76,* 8244–8251.

Boxall, N., Kubinyiova, M., Prikazsky, V., Benes, C., and Castkova, J. (2008). An increase in the number of mumps cases in the Czech Republic, 2005–2006. Euro. Surveill. *16,* 18842.

Brown, G.C., Baublis, J.V., and O'Leary, T.P. (1970). Development and duration of mumps fluorescent antibodies in various immunoglobulin fractions of human serum. J. Immunol. *104,* 86–94.

Brown, E.G., Furesz, J., Dimock, K., Yarosh, W., and Contreras, G. (1991). Nucleotide sequence analysis of Urabe mumps vaccine strain that caused meningitis in vaccine recipients. Vaccine *9,* 840–842.

Brown, E.G., Dimock, K., and Wright, K.E. (1996). The Urabe AM9 mumps vaccine is a mixture of viruses differing at amino acid 335 of the hemagglutinin-neuraminidase gene with one form associated with disease. J. Infect. Dis. *174,* 619–622.

Brown, E.G., and Wright, K.E. (1998). Genetic studies on a mumps vaccine strain associated with meningitis. Rev. Med. Virol. *8,* 129–142.

Brunell, P.A., Brickmann, A., D'Hose, D., Steinberg, S. (1968). Ineffectiveness of isolation of patients as a method of preventing the spread of mumps. Failure of the mumps skin test antigen to predict immune status. N. Engl. J. Med. 279, 1357–1361

Buchholz, U.J., Finke, S., and Conzelmann, K.K. (1999). Generation of bovine respiratory syncytial virus (BRSV) from cDNA: BRSV NS2 is not essential for virus replication in tissue culture, and the human RSV leader region acts as a functional BRSV genome promoter. J. Virol. *73,* 251–259.

Buynak, E.B., and Hilleman, M.R. (1966). Live attenuated mumps virus vaccine. 1. Vaccine development. Proc. Soc. Exp. Biol. Med. *123,* 768–775.

Carbone, K., and Rubin, S. (2007). Mumps Virus. In Fields Virology, Knipe, D.M., Howley, P.M., eds. (Wolters Kluwer, Lippinocott Williams & Wilkins: Philadelphia), pp. 1527–1542

Castilla, J., Garcia Cenoz, M., Arriazu, M., Fernandez-Alonso, M., Martinez-Artola, V., Etxeberria, J., Irisarri, F., and Barricarte, A. (2009). Effectiveness of Jeryl Lynn-containing vaccine in Spanish children. Vaccine *27,* 2089–2093.

Cathomen, T., Mrkic, B., Spehner, D., Drillien, R., Naef, R., Pavlovic, J., Aguzzi, A., Billeter, M.A., and Cattaneo, R. (1998). A matrix-less measles virus is infectious and elicits extensive cell fusion: consequences for propagation in the brain. EMBO J. *17,* 3899–3908

Centers for Disease Control and Prevention. (1972). Mumps surveillance report. MMWR Morb. Mortal. Wkly Rep. *21,* 1–12.

Centers for Disease Control and Prevention. (1974). Mumps Surveillance 1973. MMWR Morb. Mortal. Wkly Rep. *23,* 431.

Centers for Disease Control and Prevention. (1978). Mumps- United States. MMWR Morb. Mortal. Wkly Rep. *27,* 379–381.

Centers for Disease Control and Prevention. (1991). Summary of notifiable diseases, United States, 1991. MMWR Morb. Mortal. Wkly Rep. *40,* 3.

Centers for Disease Control and Prevention. (2006a). Brief report: update: mumps activity – United States, January 1–October 7, 2006. MMWR Morb. Mortal. Wkly Rep. *55,* 1152–1153.

Centers for Disease Control and Prevention. (2006b). Mumps epidemic – United kingdom, 2004–2005. MMWR Morb. Mortal. Wkly Rep. *55,* 173–175.

Centers for Disease Control and Prevention. (2006c). Update: multistate outbreak of mumps – United States, January 1–May 2, 2006. MMWR Morb. Mortal. Wkly Rep. *55,* 559–563.

Centers for Disease Control and Prevention. (2009). Mumps. In Epidemiology and Prevention of Vaccine-preventable Diseases, Atkinson, W., Wolfe, C., Hamborsky, J., McIntyre, L., eds. (Public Health Foundation: Washington DC), pp. 189–197.

Centers for Disease Control and Prevention. (2010). Update: mumps outbreak – New York and New Jersey, June 2009–January 2010. MMWR Morb. Mortal. Wkly Rep. *59,* 125–129.

Chandrika, R., Horikami, S.M., Smallwood, S., and Moyer, S.A. (1995). Mutations in conserved domain I of the Sendai virus L polymerase protein uncouple transcription and replication. Virology *213,* 352–363.

Chattopadhyay, A., Raha, T., and Shaila, M.S. (2004). Effect of single amino acid mutations in the conserved GDNQ motif of L protein of Rinderpest virus on RNA synthesis *in vitro* and *in vivo*. Virus Res. *99,* 139–145.

Chen, M., Zhang, T.G., Chen, L.J., Wu, J., Yang, J., and Zhang, W. (2009). [Comparison of genotype characteristics between the circulating mumps virus strain in Beijing area and the vaccine strain.]. Zhonghua Liu Xing Bing Xue Za Zhi *30,* 1184–1188.

Chiba, Y., Dzierba, J.L., Morag, A., and Ogra, P.L. (1976). Cell-mediated immune response to mumps virus infection in man. J. Immunol. *116,* 12–15.

Christenson, B., and Bottiger, M. (1990). Methods for screening the naturally acquired and vaccine-induced immunity to the measles virus. Biologicals *18,* 207–211.

Christenson, B., Heller, L., and Bottiger, M. (1983). The immunizing effect and reactogenicity of two live attenuated mumps virus vaccines in Swedish schoolchildren. J. Biol. Stand. *11,* 323–331.

Chu, T.H., Cheever, F.S., Coons, A.H., and Daniels, J.B. (1951). Distribution of mumps virus in the experimentally infected monkey. Proc. Soc. Exp. Biol. Med. *76,* 571–574.

Clarke, D.K., Sidhu, M.S., Johnson, J.E., and Udem, S.A. (2000). Rescue of mumps virus from cDNA. J. Virol. *74,* 4831–4838.

Cohen, C., White, J.M., Savage, E.J., Glynn, J.R., Choi, Y., Andrews, N., Brown, D., and Ramsay, M.E. (2007). Vaccine effectiveness estimates, 2004–2005 mumps outbreak, England. Emerg. Infect. Dis. 13, 12–17.

Collins, P., and Crowe J.E.J. (2007). Respiratory syncytial virus and metapneumovirus. In Fields Virology, Knipe, D.M., Howley, P.M., eds. (Wolters Kluwer, Lippinocott Williams & Wilkins: Philadelphia), pp. 1617–1618.

Colville, A., Pugh, S., and Miller, E. (1994). Withdrawal of a mumps vaccine. Eur. J. Pediatr. *153,* 467–468.

Connaris, H., Takimoto, T., Russell, R., Crennell, S., Moustafa, I., Portner, A., and Taylor, G. (2002). Probing the sialic acid binding site of the hemagglutinin-neuraminidase of Newcastle disease virus: identification of key amino acids involved in cell binding, catalysis, and fusion. J. Virol. *76,* 1816–1824.

Conzelmann, K.K. (2004). Reverse genetics of *mononegavirales*. Curr. Top. Microbiol. Immunol. *283,* 1–41.

Cooney, M.K., Fox, J.P., and Hall, C.E. (1975). The Seattle virus watch. VI. Observations of infections with and illness due to parainfluenza, mumps and respiratory syncytial viruses and *Mycoplasma pneumoniae*. Am. J. Epidemiol. *101,* 532–551.

Cordey, S., and Roux, L. (2006). Transcribing paramyxovirus RNA polymerase engages the template at its 3′ extremity. J. Gen. Virol. *87,* 665–672.

Cortese, C.K., Feller, J.A., and Moyer, S.A. (2000). Mutations in domain V of the Sendai virus L polymerase protein uncouple transcription and replication and differentially affect replication *in vitro* and *in vivo*. Virology *277,* 387–396.

Cortese, M.M., Jordan, H.T., Curns, A.T., Quinlan, P.A., Ens, K.A., Denning, P.M., and Dayan, G.H. (2008). Mumps vaccine performance among university students during a mumps outbreak. Clin. Infect. Dis. *46,* 1172–1180.

Crennell, S., Takimoto, T., Portner, A., and Taylor, G. (2000). Crystal structure of the multifunctional paramyxovirus hemagglutinin-neuraminidase. Nat. Struct. Biol. *7,* 1068–1074.

Cunningham, C., Faherty, C., Cormican, M., and Murphy, A.W. (2006). Importance of clinical features in diagnosis of mumps during a community outbreak. Ir. Med. J. *99,* 171–173.

Curran, J., and Kolakofsky, D. (1999). Replication of paramyxoviruses. Adv. Virus Res. *54,* 403–422.

Curran, J., Pelet, T., and Kolakofsky, D. (1994). An acidic activation-like domain of the Sendai virus P protein is required for RNA synthesis and encapsidation. Virology *202,* 875–884.

Cusi, M.G., Santini, L., Bianchi, S., Valassina, M., and Valensin, P.E. (1998). Nucleotide sequence at position 1081 of the hemagglutinin-neuraminidasegene in wild-type strains of mumps virus is the most relevant marker of virulence. J. Clin. Microbiol. *36,* 3743–3744.

Cusi, M.G., Fischer, S., Sedlmeier, R., Valassina, M., Valensin, P.E., Donati, M., and Neubert, W.J. (2001). Localization of a new neutralizing epitope on the mumps virus hemagglutinin-neuraminidase protein. Virus Res. *74,* 133–137.

da Cunha, S.S., Rodrigues, L.C., Barreto, M.L., and Dourado, I. (2002). Outbreak of aseptic meningitis and mumps after mass vaccination with MMR vaccine using the Leningrad-Zagreb mumps strain. Vaccine *20,* 1106–1112.

Dallaire, F., De Serres, G., Tremblay, F.W., Markowski, F., and Tipples, G. (2009). Long-lasting measles outbreak affecting several unrelated networks of unvaccinated persons. J. Infect. Dis. *200,* 1602–1605.

Davidkin, I., Valle, M., and Julkunen, I. (1995). Persistence of anti-mumps virus antibodies after a two-dose MMR vaccination. A nine-year follow-up. Vaccine *13,* 1617–1622.

Davidkin, I., Jokinen, S., Broman, M., Leinikki, P., and Peltola, H. (2008). Persistence of measles, mumps, and rubella antibodies in an MMR-vaccinated cohort: a 20-year follow-up. J. Infect. Dis. *197,* 950–956.

Davis, N.F., McGuire, B.B., Mahon, J.A., Smyth, A.E., O'Malley, K.J., and Fitzpatrick, J.M. (2010). The increasing incidence of mumps orchitis: a comprehensive review. BJU Int. *105,* 1060–1065.

Dayan, G.H., and Rubin, S. (2008). Mumps outbreaks in vaccinated populations: are available mumps vaccines effective enough to prevent outbreaks? Clin. Infect. Dis. *47,* 1458–1467.

Dhiman, N., Ovsyannikova, I.G., Jacobson, R.M., Vierkant, R.A., Pankratz, V.S., Jacobsen, S.J., and Poland, G.A. (2005). Correlates of lymphoproliferative responses to measles, mumps, and rubella (MMR) virus vaccines following MMR-II vaccination in healthy children. *115,* 154–161.

Dos Santos, B.A., Ranieri, T.S., Bercini, M., Schermann, M.T., Famer, S., Mohrdieck, R., Maraskin, T., and Wagner, M.B. (2002). An evaluation of the adverse reaction potential of three measles–mumps-rubella combination vaccines. Rev. Panam. Salud Publica. *12,* 240–246.

Ehrengut, W., Georges, A.M., and Andre, F.E. (1983). The reactogenicity and immunogenicity of the Urabe Am. 9 live mumps vaccine and persistence of vaccine induced antibodies in healthy young children. J. Biol. Stand. *11,* 105–113.

Elango, N., Varsanyi, T.M., Kovamees, J., and Norrby, E. (1988). Molecular cloning and characterization of six genes, determination of gene order and intergenic sequences and leader sequence of mumps virus. J. Gen. Virol. *69,* 2893–2900.

Elliott, G.D., Afzal, M.A., Martin, S.J., and Rima, B.K. (1989). Nucleotide sequence of the matrix, fusion and putative SH protein genes of mumps virus and their deduced amino acid sequences. Virus Res. *12,* 61–75.

Elliott, G.D., Yeo, R.P., Afzal, M.A., Simpson, E.J., Curran, J.A., and Rima, B.K. (1990). Strain-variable editing during transcription of the P gene of mumps virus may lead to the generation of non-structural proteins NS1 (V) and NS2. J. Gen. Virol. *71,* 1555–1560.

van Els, C.A., and Nanan, R. (2002). T-cell responses in acute measles. Viral Immunol. *15,* 435–450.

Enders, J. (1946). Mumps:techniques of laboratory diagnosis, tests for susceptibility, and experiments on specific prophylaxis. J. Pediatr. *29,* 129–142.

Ennis, F.A. (1969). Immunity to mumps in an institutional epidemic. Correlation of insusceptibility to mumps with serum plaque neutralizing and haemagglutination-inhibiting antibodies. J. Infect. Dis. *119,* 654–657.

Ennis, F.A., Hopps, H.E., Douglas, R.D., and Meyer, H.M., Jr. (1969). Hydrocephalus in hamsters: induction by natural and attenuated mumps viruses. J. Infect. Dis. *119,* 75–79.

Fearns, R., and Collins, P.L. (1999). Model for polymerase access to the overlapped L gene of respiratory syncytial virus. J. Virol. *73,* 388–397.

Fearns, R., Peeples, M.E., and Collins, P.L. (1997). Increased expression of the N protein of respiratory syncytial virus stimulates minigenome replication but does not alter the balance between the synthesis of mRNA and antigenome. Virology *236,* 188–201.

Fedova, D., Bruckova, M., Plesnik, V., Slonim, D., Sejda, J., Svandova, E., and Kubinova, I. (1987). Detection of postvaccination mumps virus antibody by neutralization test, enzyme-linked immunosorbent assay and sensitive haemagglutination inhibition test. J. Hyg. Epidemiol. Microbiol. Immunol. *31,* 409–422.

Feller, J.A., Smallwood, S., Horikami, S.M., and Moyer, S.A. (2000). Mutations in conserved domains IV and VI of the large (L) subunit of the sendai virus RNA polymerase give a spectrum of defective RNA synthesis phenotypes. Virology *269,* 426–439.

Findlay, G., and Clarke, L. (1934). The experimental production of mumps in monkeys. Br. J. Exp. Pathol. *15,* 309–313.

Finke, S., Cox, J.H., and Conzelmann, K.K. (2000). Differential transcription attenuation of rabies virus genes by intergenic regions: generation of recombinant viruses overexpressing the polymerase gene. J. Virol. *74*, 7261–7269.

Finke, S., Mueller-Waldeck, R., and Conzelmann, K.K. (2003). Rabies virus matrix protein regulates the balance of virus transcription and replication. J. Gen. Virol. *84*, 1613–1621.

Finkelstein, H. (1938). Meningo-encephalitis in mumps. *JAMA 3*, 17–19.

Fleischer, B., and Kreth, H.W. (1982). Mumps virus replication in human lymphoid cell lines and in peripheral blood lymphocytes: preference for T cells. Infect. Immun. *35*, 25–31.

Friedman, M.G. (1981). Salivary IgA antibodies to mumps virus during and after mumps. J. Infect. Dis. *143*, 617.

Fuerst, T.R., Niles, e.g. Studier, F.W., and Moss, B. (1986). Eukaryotic transient-expression system based on recombinant vaccinia virus that synthesizes bacteriophage T7 RNA polymerase. Proc. Natl. Acad. Sci. U.S.A. *83*, 8122–8126.

Furesz, J. (2002). Safety of live mumps virus vaccines. J. Med. Virol. *67*, 299–300.

Gamble, D.R. (1980). Relation of antecedent illness to development of diabetes in children. Br. Med. J. *281*, 99–101.

Gans, H., Yasukawa, L., Rinki, M., DeHovitz, R., Forghani, B., Beeler, J., Audet, S., Maldonado, Y., and Arvin, A.M. (2001). Immune responses to measles and mumps vaccination of infants at *6, 9,* and 12 months. J. Infect. Dis. *184*, 817–826.

Gans, H., DeHovitz, R., Forghani, B., Beeler, J., Maldonado, Y., and Arvin, A.M. (2003). Measles and mumps vaccination as a model to investigate the developing immune system: passive and active immunity during the first year of life. Vaccine *21*, 3398–3405.

Gardner, A.E., Martin, K.L., and Dutch, R.E. (2007). A conserved region between the heptad repeats of paramyxovirus fusion proteins is critical for proper F protein folding. Biochemistry *46*, 5094–5105.

Goldacre, M.J., Wotton, C.J., Yeates, D., Seagroatt, V., and Neil, A. (2005). Hospital admission for selected single virus infections prior to diabetes mellitus. Diabetes Res. Clin. Pract. *69*, 256–261.

Gordon, J.E., and Heeren, R.H. (1940). The epidemiology of mumps. Am. J. Med. Sci. *200*, 338–359.

Gordon, J.E., and Kilham, L. (1949). Ten years in the epidemiology of mumps. Am. J. Med. Sci. *218*, 338–359.

Goto, A., Takahashi, Y., Kishimoto, M., Nakajima, Y., Nakanishi, K., Kajio, H., and Noda, M. (2008). A case of fulminant type 1 diabetes associated with significant elevation of mumps titers. Endocr. J. *55*, 561–564.

Graham, D.Y., Brown, C.H., 3rd, Benrey, J., and Butel, J.S. (1974). Thrombocytopenia. A complication of mumps. JAMA *227*, 1162–1164.

Granata, S. (1908). Sulla etiologia degli orecchioni da virus filtrable. Med. Ital. *6*, 647–672.

Gravel, K.A., and Morrison, T.G. (2003). Interacting domains of the HN and F proteins of newcastle disease virus. J. Virol. *77, 11*,040–11049.

Gut, J.P., Lablache, C., Behr, S., and Kirn, A. (1995). Symptomatic mumps virus reinfections. J. Med. Virol. *45*, 17–23.

Guy, R.J., Andrews, R.M., Kelly, H.A., Leydon, J.A., Riddell, M.A., Lambert, S.B., and Catton, M.G. (2004). Mumps and rubella: a year of enhanced surveillance and laboratory testing. Epidemiol. Infect. *132*, 391–398.

Habel, K. (1945). Cultivation of mumps virus in the developing chick embryo and its application to studies of immunity to mumps in man. Public Health Rep. *60*, 201–212.

Habel, K. (1946). Preparation of mumps vaccines and immunization of monkeys against experimental mumps infection. Public Health Rep. *61*, 1655–1664.

Habel, K. (1951a). Vaccination of human beings against mumps; vaccine administered at the start of an epidemic. I. Incidence and severity of mumps in vaccinated and control groups. Am. J. Hyg. *54*, 295–311.

Habel, K. (1951b). Vaccination of human beings against mumps; vaccine administered at the start of an epidemic. II. Effect of vaccination upon the epidemic. Am. J. Hyg. *54*, 312–318.

Hamilton, R. (1790). An account of a distemper by the common people of England vulgarly called the mumps. Trans. Royal Soc. Ed. *2*, 59–72.

Hanna-Wakim, R., Yasukawa, L.L., Sung, P., Arvin, A.M., and Gans, H.A. (2008). Immune responses to mumps vaccine in adults who were vaccinated in childhood. J. Infect. Dis. *197*, 1669–1675.

Harrison, M.S., Sakaguchi, T., and Schmitt, A.P. (2010). Paramyxovirus assembly and budding: building particles that transmit infections. Int. J. Biochem. Cell Biol. *42*, 1416–1429

Hashimoto, H., Fujioka, M., and Kinumaki, H. (2009). An office-based prospective study of deafness in mumps. Pediatr. Infect. Dis. J. *28*, 173–175.

Hausmann, S., Garcin, D., Delenda, C., and Kolakofsky, D. (1999). The versatility of paramyxovirus RNA polymerase stuttering. J. Virol. *73*, 5568–5576.

Hayashi, K., Ross, M.E., and Notkins, A.L. (1976). Persistence of mumps viral antigens in mouse brain. Jpn J. Exp. Med. *46*, 197–200.

He, B., Leser, G.P., Paterson, R.G., and Lamb, R.A. (1998). The paramyxovirus SV5 small hydrophobic (SH) protein is not essential for virus growth in tissue culture cells. Virology *250*, 30–40.

He, B., and Lamb, R.A. (1999). Effect of inserting paramyxovirus simian virus 5 gene junctions at the HN/L gene junction: analysis of accumulation of mRNAs transcribed from rescued viable viruses. J. Virol. *73*, 6228–6234.

He, B., Lin, G.Y., Durbin, J.E., Durbin, R.K., and Lamb, R.A. (2001). The SH integral membrane protein of the paramyxovirus simian virus 5 is required to block apoptosis in MDBK cells. J. Virol. *75*, 4068–4079.

Henle, G., and Deinhardt, F. (1955). Propagation and primary isolation of mumps virus in tissue culture. Proc. Soc. Exp. Biol. Med. *89,* 556–560.

Henle, G., Henle, W., Wendell, K.K., and Rosenberg, R. (1948a). Isolation of mumps virus from human beings with induced apparent or inapparent infections. J. Exp. Med. *88,* 223–232.

Henle, G., Henle, W., and Harris, S. (1948b). The use of complement-fixation technic in the analysis of two institutional outbreaks of mumps. Pediatrics *1,* 593–600.

Hiebert, S.W., Richardson, C.D., and Lamb, R.A. (1988). Cell surface expression and orientation in membranes of the 44-amino-acid SH protein of simian virus 5. J. Virol. *62,* 2347–2357.

Hindiyeh, M.Y., Aboudy, Y., Wohoush, M., Shulman, L.M., Ram, D., Levin, T., Frank, T., Riccardo, F., Khalili, M., *et al.* (2009). Characterization of large mumps outbreak among vaccinated Palestinian refugees. J. Clin. Microbiol. *47,* 560–565.

Holmes, D.E., and Moyer, S.A. (2002). The phosphoprotein (P) binding site resides in the N terminus of the L polymerase subunit of sendai virus. J. Virol. *76,* 3078–3083.

Horikami, S.M., Curran, J., Kolakofsky, D., and Moyer, S.A. (1992). Complexes of Sendai virus NP-P and P-L proteins are required for defective interfering particle genome replication *in vitro*. J. Virol. *66,* 4901–4908.

Houard, S., Varsanyi, T.M., Milican, F., Norrby, E., and Bollen, A. (1995). Protection of hamsters against experimental mumps virus (MuV) infection by antibodies raised against the MuV surface glycoproteins expressed from recombinant vaccinia virus vectors. J. Gen. Virol. 76, 421–423.

Howard, M., and Wertz, G. (1989). Vesicular stomatitis virus RNA replication: a role for the NS protein. J. Gen. Virol. *70,* 2683–2694.

Hyoty, H., Leinikki, P., Reunanen, A., Ilonen, J., Surcel, H.M., Rilva, A., Kaar, M.L., Huupponen, T., Hakulinen, A., Makela, A.L., *et al.* (1988). Mumps infections in the etiology of type 1 (insulin-dependent) diabetes. Diabetes Res. *9,* 111–116.

Ichiyama, T., Maeba, S., Suenaga, N., Saito, K., Matsubara, T., and Furukawa, S. (2005). Analysis of cytokine levels in cerebrospinal fluid in mumps meningitis: comparison with echovirus type 30 meningitis. Cytokine *30,* 243–247.

Iorio, R.M., and Bratt, M.A. (1984). Monoclonal antibodies as functional probes of the HN glycoprotein of Newcastle disease virus: antigenic separation of the hemagglutinating and neuraminidase sites. J. Immunol. *133,* 2215–2219.

Ivancic, J., Gulija, T.K., Forcic, D., Baricevic, M., Jug, R., Mesko-Prejac, M., and Mazuran, R. (2005). Genetic characterization of L-Zagreb mumps vaccine strain. Virus Res. *109,* 95–105.

Jablonski, S.A., Luo, M., and Morrow, C.D. (1991). Enzymatic activity of poliovirus RNA polymerase mutants with single amino acid changes in the conserved YGDD amino acid motif. J. Virol. *65,* 4565–4572.

Jensik, S.C., and Silver, S. (1976). Polypeptides of mumps virus. J. Virol. *17,* 363–373.

Jick, H., Chamberlin, D.P., and Hagberg, K.W. (2009). The origin and spread of a mumps epidemic: United kingdom, 2003–2006. Epidemiology *20,* 656–661.

Jin, L., Rima, B., Brown, D., Orvell, C., Tecle, T., Afzal, M., Uchida, K., Nakayama, T., Song, J.W., Kang, C., *et al.* (2005). Proposal for genetic characterisation of wild-type mumps strains: preliminary standardisation of the nomenclature. Arch. Virol. *150,* 1903–1909.

Johnson, C.D., and Goodpasture, E.W. (1934). An investigation of the etiology of mumps. J. Exp. Med. *59,* 1–19.

Johnson, C.D., and Goodpasture, E.W. (1935). The etiology of mumps. Am. J. Hyg. *21,* 46–57.

Johnstone, J.A., Ross, C.A., and Dunn, M. (1972). Meningitis and encephalitis associated with mumps infection. A 10-year survey. Arch. Dis. Child. *47,* 647–651.

Jokinen, S., Osterlund, P., Julkunen, I., and Davidkin, I. (2007). Cellular immunity to mumps virus in young adults 21 years after measles-mumps-rubella vaccination. J. Infect. Dis. *196,* 861–867.

Kanra, G., Isik, P., Kara, A., Cengiz, A.B., Secmeer, G., and Ceyhan, M. (2004). Complementary findings in clinical and epidemiologic features of mumps and mumps meningoencephalitis in children without mumps vaccination. Pediatr. Int. *46,* 663–668.

Katayama, K., Oya, A., Tanabayashi, K., Okazaki, K., Hishiyama, M., Yamazaki, S., and Yamada, A. (1993). Differentiation of mumps vaccine strains from wild viruses by single-strand conformation polymorphism of the P gene. Vaccine *11,* 621–623.

Kato, A., Kiyotani, K., Hasan, M.K., Shioda, T., Sakai, Y., Yoshida, T., and Nagai, Y. (1999). Sendai virus gene start signals are not equivalent in reinitiation capacity: moderation at the fusion protein gene. J. Virol. *73,* 9237–9246.

Kawashima, Y., Ihara, K., Nakamura, M., Nakashima, T., Fukuda, S., and Kitamura, K. (2005). Epidemiological study of mumps deafness in Japan. Auris Nasus Larynx *32,* 125–128.

Keller, M.A., and Parks, G.D. (2003). Positive- and negative-acting signals combine to determine differential RNA replication from the paramyxovirus simian virus 5 genomic and antigenomic promoters. Virology *306,* 347–358.

Khakoo, G.A., and Lack, G. (2000). Recommendations for using MMR vaccine in children allergic to eggs. BMJ *320,* 929–932.

Ki, M., Park, T., Yi, S.G., Oh, J.K., and Choi, B. (2003). Risk analysis of aseptic meningitis after measles–mumps–rubella vaccination in Korean children by using a case-crossover design. Am. J. Epidemiol. *157,* 158–165.

Kilham, L. (1948). Isolation of mumps virus from the blood of a patient. Proc. Soc. Exp. Biol. Med. *69,* 99.

Kilham, L. (1951). Mumps virus in human milk and in milk of infected monkey. J. Am. Med. Assoc. *146,* 1231–1232.

Kilham, L., and Margolis, G. (1975). Induction of congenital hydrocephalus in hamsters with attenuated and natural strains of mumps virus. J. Infect. Dis. *132,* 462–466.

Klinge, J., Lugauer, S., Korn, K., Heininger, U., and Stehr, K. (2000). Comparison of immunogenicity and reactogenicity of a measles, mumps and rubella (MMR) vaccine in German children vaccinated at 9–11, 12–14 or 15–17 months of age. Vaccine *18,* 3134–3140.

Klyatchko, N.S., and Smorodintsev, A.A. (1958). Live anti-mumps vaccine. II. Epidemiological effectiveness of the immunization of children with a single intradermal injection of live anti-mumps vaccine. Acta Virol. *2,* 145–151.

Koskiniemi, M., Donner, M., and Pettay, O. (1983). Clinical appearance and outcome in mumps encephalitis in children. Acta Paediatr. Scand. *72,* 603–609.

Kovamees, J., Rydbeck, R., Orvell, C., and Norrby, E. (1990). Hemagglutinin-neuraminidase (HN) amino acid alterations in neutralization escape mutants of Kilham mumps virus. Virus Res. *17,* 119–129.

Krause, C.H., Molyneaux, P.J., Ho-Yen, D.O., McIntyre, P., Carman, W.F., and Templeton, K.E. (2007). Comparison of mumps-IgM ELISAs in acute infection. J. Clin. Virol. *38,* 153–156.

Kristensson, K., Orvell, C., Malm, G., and Norrby, E. (1984). Mumps virus infection of the developing mouse brain–appearance of structural virus proteins demonstrated with monoclonal antibodies. J. Neuropathol. Exp. Neurol. *43,* 131–140.

Kubinyiova, M., Benes, C., Prikazsky, V., Roubalova, K., and Castkova, J. (2008). Mumps vaccination in the Czech Republic. Euro. Surveill. *13,* 18920.

Kubota, T., Yokosawa, N., Yokota, S., and Fujii, N. (2001). C terminal CYS-RICH region of mumps virus structural V protein correlates with block of interferon alpha and gamma signal transduction pathway through decrease of STAT 1-alpha. Biochem. Biophys. Res. Commun. *283,* 255–259.

Kulkarni-Kale, U., Ojha, J., Manjari, G.S., Deobagkar, D.D., Mallya, A.D., Dhere, R.M., and Kapre, S.V. (2007). Mapping antigenic diversity and strain specificity of mumps virus: a bioinformatics approach. Virology *359,* 436–446.

Kunkel, U., Driesel, G., Henning, U., Gerike, E., Willers, H., and Schreier, E. (1995). Differentiation of vaccine and wild mumps viruses by polymerase chain reaction and nucleotide sequencing of the SH gene: brief report. J. Med. Virol. *45,* 121–126.

Lamb, R.A., and Kolakofsky, D. (2001). *Paramyxoviridae*: the viruses and their replication. In Fields Virology, Knipe, D.M., and Howley, P.M., eds. (Lippincott Williams & Wilkins: Philadelphia), pp. 1305–1340.

Lamb, R.A., and Parks, G.D. (2007). *Paramyxoviridae*: The viruses and their replication. In Fields Virology, Knipe, D.M., and Howley, P.M., eds. (Lippincott Williams & Wilkins: Philadelphia), pp. 1449–1496.

Larke, R.P. (1967). Interferon in the cerebrospinal fluid of children with central nervous system disorders. Can. Med. Assoc. J. *96,* 21–32.

Le Mercier, P., Garcin, D., Hausmann, S., and Kolakofsky, D. (2002). Ambisense sendai viruses are inherently unstable but are useful to study viral RNA synthesis. J. Virol. *76,* 5492–5502.

LeBaron, C.W., Forghani, B., Beck, C., Brown, C., Bi, D., Cossen, C., and Sullivan, B.J. (2009). Persistence of mumps antibodies after 2 doses of measles–mumps–rubella vaccine. J. Infect. Dis. *199,* 552–560.

Leboreiro-Fernandez, A., Moura-Ribeiro, M.V., Leboreiro, I.E., Sawan, F.M., Ventura, A.R., and Barbosa, K.C. (1997). Mumps meningoencephalitis. An epidemiological approach. Arq. Neuropsiquiatr. *55,* 12–15.

Leinikki, P. (2009). Mumps. In Principles and Practices of Clinical Virology, Zuckerman, A.J., Schoub, B.D., Griffiths, P.D., Mortimer, P., eds. (Wiley-Blackwell: West-Sussex), pp. 593–600.

Leinikki, P.O., and Passila, S. (1977). Quantitative, semiautomated, enzyme-linked immunosorbent assay for viral antibodies. J. Infect. Dis. *136 (Suppl),* S294–299.

Lemon, K., Rima, B.K., McQuaid, S., Allen, I.V., and Duprex, W.P. (2007). The F gene of rodent brain-adapted mumps virus is a major determinant of neurovirulence. J. Virol. *81,* 8293–8302.

Levenbuk, I.S., Nikolayeva, M.A., Chigirinsky, A.E., Ralf, N.M., Kozlov, V.G., Vardanyan, N.V., Sliopushkina, V.G., Kolomiets, O.L., Rukhamina, M.L., and Grigoryeva, L.V. (1979). On the morphological evaluation of the neurovirulence safety of attenuated mumps virus strains in monkeys. J. Biol. Stand. *7,* 9–19.

Li, M., Schmitt, P.T., Li, Z., McCrory, T.S., He, B., and Schmitt, A.P. (2009). Mumps virus matrix, fusion, and nucleocapsid proteins cooperate for efficient production of virus-like particles. J. Virol. *83,* 7261–7272.

Liang, Y., Ma, S., Yang, Z., Liu, L., Wang, L., Wang, J., Jiang, L., Shi, C., Dong, C., and Li, Q. (2008). Immunogenicity and safety of a novel formalin-inactivated and alum-adjuvanted candidate subunit vaccine for mumps. Vaccine *26,* 4276–4283.

Litman, N., and Baum, S.G. (2010). Mumps Virus. In Principles and Practice of Infectious Diseases, Mandell, G.L., Donlin, R., eds. (Churchill Livingstone: Philadelphia), pp. 2201–2206.

van der Logt, J.T., Heessen, F.W., van Loon, A.M., and van der Veen, J. (1982). Hemadsorption immunosorbent technique for determination of mumps immunoglobulin M antibody. J. Clin. Microbiol. *15,* 82–86.

Lohr, J.M., and Oldstone, M.B. (1990). Detection of cytomegalovirus nucleic acid sequences in pancreas in type 2 diabetes. Lancet *336,* 644–648.

London, W.T., Kent, S.G., Palmer, A.E., Fucillo, D.A., Houff, S.A., Saini, N., and Sever, J.L. (1979). Induction of congenital hydrocephalus with mumps virus in rhesus monkeys. J. Infect. Dis. *139,* 324–328.

Love, A., Rydbeck, R., Utter, G., Orvell, C., Kristensson, K., and Norrby, E. (1986). Monoclonal antibodies against the fusion protein are protective in necrotizing mumps meningoencephalitis. J. Virol. *58,* 220–222.

McCarthy, M., Jubelt, B., Fay, D.B., and Johnson, R.T. (1980). Comparative studies of five strains of mumps virus *in vitro* and in neonatal hamsters: evaluation of growth, cytopathogenicity, and neurovirulence. J. Med. Virol. *5*, 1–15.

McGuiness, A.C., and Gall, E.A. (1944). Mumps at army camps in 1943. War. Med. *5*, 95.

McIlhatton, M.A., Curran, M.D., and Rima, B.K. (1997). Nucleotide sequence analysis of the large (L) genes of phocine distemper virus and canine distemper virus (corrected sequence). J. Gen. Virol. *78*, 571–576.

McLean, D.M., Walker, S.J., and McNaughton, G.A. (1960). Mumps meningoencephalitis: a virological and clinical study. Can. Med. Assoc. J. *83*, 148–151.

McNabb, S.J., Jajosky, R.A., Hall-Baker, P.A., Adams, D.A., Sharp, P., Anderson, W.J., Javier, A.J., Jones, G.J., Nitschke, D.A., Worshams, C.A., *et al.* (2007). Summary of notifiable diseases – United States, 2005. MMWR Morb. Mortal. Wkly Rep. *54*, 1–92.

Malik, T., Sauder, C., Wolbert, C., Zhang, C., Carbone, K.M., and Rubin, S. (2007a). A single nucleotide change in the mumps virus F gene affects virus fusogenicity *in vitro* and virulence *in vivo*. J. Neurovirol. *13*, 513–521.

Malik, T., Wolbert, C., Mauldin, J., Sauder, C., Carbone, K.M., and Rubin, S.A. (2007b). Functional consequences of attenuating mutations in the haemagglutinin neuraminidase, fusion and polymerase proteins of a wild-type mumps virus strain. J. Gen. Virol. *88*, 2533–2541.

Malik, T.H., Wolbert, C., Nerret, L., Sauder, C., and Rubin, S. (2009). Single amino acid changes in the mumps virus haemagglutinin–neuraminidase and polymerase proteins are associated with neuroattenuation. J. Gen. Virol. *90*, 1741–1747.

Marin, M., Quinlisk, P., Shimabukuro, T., Sawhney, C., Brown, C., and Lebaron, C.W. (2008). Mumps vaccination coverage and vaccine effectiveness in a large outbreak among college students – Iowa, 2006. Vaccine *26*, 3601–3607.

Martin, A., Staeheli, P., and Schneider, U. (2006). RNA polymerase II-controlled expression of antigenomic RNA enhances the rescue efficacies of two different members of the *Mononegavirales* independently of the site of viral genome replication. J. Virol. *80*, 5708–5715.

Martin, R., Vallbracht, A., and Kreth, H.W. (1991). Interferon-gamma secretion by *in vivo* activated cytotoxic T lymphocytes from the blood and cerebrospinal fluid during mumps meningitis. J. Neuroimmunol. *33*, 191–198.

Masters, P.S., and Samuel, C.E. (1984). Detection of *in vivo* synthesis of polycistronic mRNAs of vesicular stomatitis virus. Virology *134*, 277–286.

Matsumoto, T. (1982). Assembly of paramyxoviruses. Microbiol. Immunol. *26*, 285–320.

Mauldin, J., Carbone, K., Hsu, H., Yolken, R., and Rubin, S. (2005). Mumps virus-specific antibody titers from pre-vaccine era sera: comparison of the plaque reduction neutralization assay and enzyme immunoassays. J. Clin. Microbiol. *43*, 4847–4851.

Merz, D.C., and Wolinsky, J.S. (1983). Conversion of nonfusing mumps virus infections to fusing infections by selective proteolysis of the HN glycoprotein. Virology *131*, 328–340.

Merz, D.C., Server, A.C., Waxham, M.N., and Wolinsky, J.S. (1983). Biosynthesis of mumps virus F glycoprotein: non-fusing strains efficiently cleave the F glycoprotein precursor. J. Gen. Virol. *64*, 1457–1467.

Meyer, M.B. (1962). An epidemiologic study of mumps; its spread in schools and families. Am. J. Hyg. *75*, 259–281.

Mayer, M.B., Stifler, W.C. and Joseph, J.M. (1966). Evaluation of mumps vaccine given after exposure to mumps, with special reference to the exposed adult. Pediatrics *37*, 304–315.

Miller, E., Goldacre, M., Pugh, S., Colville, A., Farrington, P., Flower, A., Nash, J., MacFarlane, L., and Tettmar, R. (1993). Risk of aseptic meningitis after measles, mumps, and rubella vaccine in UK children. Lancet *341*, 979–982.

Miller, E., Hill, A., Morgan-Capner, P., Forsey, T., and Rush, M. (1995). Antibodies to measles, mumps and rubella in UK children 4 years after vaccination with different MMR vaccines. Vaccine *13*, 799–802.

Miller, E., Andrews, N., Stowe, J., Grant, A., Waight, P., and Taylor, B. (2007). Risks of convulsion and aseptic meningitis following measles-mumps-rubella vaccination in the United Kingdom. Am. J. Epidemiol. *165*, 704–709.

Minor, P.D. (1997). Laboratory tests for mumps vaccines. Biologicals *25*, 35–40.

Montes, M., Cilla, G., Artieda, J., Vicente, D., and Basterretxea, M. (2002). Mumps outbreak in vaccinated children in Gipuzkoa (Basque Country), Spain. Epidemiol. Infect. *129*, 551–556.

Morgan-Capner, P. (2004). Measles, Mumps and Rubella, vol 212 (Mosby: Philadelphia).

Morishima, T., Miyazu, M., Ozaki, T., Isomura, S., and Suzuki, S. (1980). Local immunity in mumps meningitis. Am. J. Dis. Child. *134*, 1060–1064.

Morrison, T.G. (2003). Structure and function of a paramyxovirus fusion protein. Biochim. Biophys. Acta *1614*, 73–84.

Mortimer, P.P. (1978). Mumps prophylaxis in the light of a new test for antibody. Br. Med. J. *2*, 1523–1524.

Mottet, G., and Roux, L. (1989). Budding efficiency of Sendai virus nucleocapsids: influence of size and ends of the RNA. Virus Res. *14*, 175–187.

Mottet-Osman, G., Iseni, F., Pelet, T., Wiznerowicz, M., Garcin, D., and Roux, L. (2007). Suppression of the Sendai virus M protein through a novel short interfering RNA approach inhibits viral particle production but does not affect viral RNA synthesis. J. Virol. *81*, 2861–2868.

Muhlemann, K. (2004). The molecular epidemiology of mumps virus. Infect. Genet. Evol. *4*, 215–219.

Muhsen, K., Aboudy, Y., Mendelson, E., Green, M.S., and Cohen, D. (2008). Prevalence of mumps antibodies in the Israeli population in relation to mumps vaccination policy and incidence of disease. Epidemiol. Infect. *136*, 688–693.

Muller, R., Poch, O., Delarue, M., Bishop, D.H., and Bouloy, M. (1994). Rift Valley fever virus L segment: correction of the sequence and possible functional role of newly identified regions conserved in RNA-dependent polymerases. J. Gen. Virol. *75,* 1345–1352.

Nagai, H., Morishima, T., Morishima, Y., Isomura, S., and Suzuki, S. (1983). Local T cell subsets in mumps meningitis. Arch. Dis. Child. *58,* 927–928.

Naruse, H., Nagai, Y., Yoshida, T., Hamaguchi, M., Matsumoto, T., Isomura, S., and Suzuki, S. (1981). The polypeptides of mumps virus and their synthesis in infected chick embryo cells. Virology *112,* 119–130.

Nigro, G., Nanni, F., and Midulla, M. (1986). Determination of vaccine-induced and naturally acquired class-specific mumps antibodies by two indirect enzyme-linked immunosorbent assays. J. Virol. Methods *13,* 91–106.

Ninomiya, N., Maeda, T., and Matsuda, I. (1977). Thrombocytopenic purpura occurring during the early phase of a mumps infection. Helv. Paediatr. Acta *32,* 87–89.

Ninomiya, K., Kanayama, T., Fujieda, N., Nakayama, T., Komase, K., Nagata, K., and Takeuchi, K. (2009). Amino acid substitution at position 464 in the haemagglutinin-neuraminidase protein of a mumps virus Urabe strain enhanced the virus growth in neuroblastoma SH-SY5Y cells. Vaccine *27,* 6160–6165.

Nishio, M., Garcin, D., Simonet, V., and Kolakofsky, D. (2002). The carboxyl segment of the mumps virus V protein associates with Stat proteins *in vitro* via a tryptophan-rich motif. Virology *300,* 92–99.

Nojd, J., Tecle, T., Samuelsson, A., and Orvell, C. (2001). Mumps virus neutralizing antibodies do not protect against reinfection with a heterologous mumps virus genotype. Vaccine *19,* 1727–1731.

Okazaki, K., Tanabayashi, K., Takeuchi, K., Hishiyama, M., and Yamada, A. (1992). Molecular cloning and sequence analysis of the mumps virus gene encoding the L protein and the trailer sequence. Virology *188,* 926–930.

Ong, G., Goh, K.T., Ma, S., and Chew, S.K. (2005). Comparative efficacy of Rubini, Jeryl-Lynn and Urabe mumps vaccine in an Asian population. J. Infect. *51,* 294–298.

Ornoy, A., and Tenenbaum, A. (2006). Pregnancy outcome following infections by coxsackie, echo, measles, mumps, hepatitis, polio and encephalitis viruses. Reprod. Toxicol. *21,* 446–457.

Orvell, C. (1978). Immunological properties of purified mumps virus glycoproteins. J. Gen. Virol. *41,* 517–526.

Orvell, C. (1984). The reactions of monoclonal antibodies with structural proteins of mumps virus. J. Immunol. *132,* 2622–2629.

Orvell, C., Alsheikhly, A.R., Kalantari, M., and Johansson, B. (1997). Characterization of genotype-specific epitopes of the HN protein of mumps virus. J. Gen. Virol. *78,* 3187–3193.

Orvell, C., Tecle, T., Johansson, B., Saito, H., and Samuelson, A. (2002). Antigenic relationships between six genotypes of the small hydrophobic protein gene of mumps virus. J. Gen. Virol. *83,* 2489–2496.

Overman, J.R. (1958). Viremia in human mumps virus infections. AMA Arch. Intern. Med. *102,* 354–356.

Overman, J.R., Peers, J.H., and Kilham, L. (1953). Pathology of mumps virus meningoencephalitis in mice and hamsters. AMA Arch. Pathol. *55,* 457–465.

Paccaud, M.F., Hazeghi, P., Bourquin, M., Maurer, A.M., Steiner, C.A., Seiler, A.J., Helbling, P., and Zimmermann, H. (1995). [A look back at 2 mumps outbreaks]. Soz. Praventivmed. *40,* 72–79.

Parker, A.A., Staggs, W., Dayan, G.H., Ortega-Sanchez, I.R., Rota, P.A., Lowe, L., Boardman, P., Teclaw, R., Graves, C., and LeBaron, C.W. (2006). Implications of a 2005 measles outbreak in Indiana for sustained elimination of measles in the United States. N. Engl. J. Med. *355,* 447–455.

Parks, G.D., Ward, C.D., and Lamb, R.A. (1992). Molecular cloning of the NP and L genes of simian virus 5: identification of highly conserved domains in paramyxovirus NP and L proteins. Virus Res. *22,* 259–279.

Parks, G.D., Ward, K.R., and Rassa, J.C. (2001). Increased readthrough transcription across the simian virus 5 M-F gene junction leads to growth defects and a global inhibition of viral mRNA synthesis. J. Virol. *75,* 2213–2223.

Patch, J.R., Han, Z., McCarthy, S.E., Yan, L., Wang, L.F., Harty, R.N., and Broder, C.C. (2008). The YPLGVG sequence of the Nipah virus matrix protein is required for budding. Virol. J. *5,* 137–149.

Paterson, R.G., and Lamb, R.A. (1990). RNA editing by G-nucleotide insertion in mumps virus P-gene mRNA transcripts. J. Virol. *64,* 4137–4145.

Patja, A., Makinen-Kiljunen, S., Davidkin, I., Paunio, M., and Peltola, H. (2001). Allergic reactions to measles-mumps-rubella vaccination. Pediatrics *107,* E27.

Patton, J.T., Davis, N.L., and Wertz, G.W. (1984). N protein alone satisfies the requirement for protein synthesis during RNA replication of vesicular stomatitis virus. J. Virol. *49,* 303–309.

Pelosi, J., Meyer, P.A., and Schluter, W.W. (2001). Mumps surveillance: results of improved case investigation and serologic testing of suspected cases, Texas, 1995–1996. J. Public Health Manag. Pract. *7,* 69–74.

Peltola, H. (1993). Mumps vaccination and meningitis. Lancet *341,* 994–995.

Peltola, H., Heinonen, O.P., Valle, M., Paunio, M., Virtanen, M., Karanko, V., and Cantell, K. (1994). The elimination of indigenous measles, mumps, and rubella from Finland by a 12-year, two-dose vaccination programme. N. Engl. J. Med. *331,* 1397–1402.

Peltola, H., Davidkin, I., Paunio, M., Valle, M., Leinikki, P., and Heinonen, O.P. (2000). Mumps and rubella eliminated from Finland. JAMA *284,* 2643–2647.

Peltola, H., Kulkarni, P.S., Kapre, S.V., Paunio, M., Jadhav, S.S., and Dhere, R.M. (2007). Mumps outbreaks in Canada and the United States: time for new thinking on mumps vaccines. Clin. Infect. Dis. *45,* 459–466.

Peltola, H., Jokinen, S., Paunio, M., Hovi, T., and Davidkin, I. (2008). Measles, mumps, and rubella in Finland: 25 years of a nationwide elimination programme. Lancet Infect. Dis. *8,* 796–803.

Philip, R.N., Reinhard, K.R., and Lackman, D.B. (1959). Observations on a mumps epidemic in a virgin population. Am. J. Hyg. *69*, 91–111.

Plotkin, S.A., and Rubin, S.A. (2008). Mumps Vaccine, vol 20 (Saunders Elsevier: Philadelphia).

Poch, O., Blumberg, B.M., Bougueleret, L., and Tordo, N. (1990). Sequence comparison of five polymerases (L proteins) of unsegmented negative-strand RNA viruses: theoretical assignment of functional domains. J. Gen. Virol. *71*, 1153–1162.

Poenisch, M., Wille, S., Staeheli, P., and Schneider, U. (2008). Polymerase read-through at the first transcription termination site contributes to regulation of borna disease virus gene expression. J. Virol. *82*, 9537–9545.

Poggio, G.P., Rodriguez, C., Cisterna, D., Freire, M.C., and Cello, J. (2000). Nested PCR for rapid detection of mumps virus in cerebrospinal fluid from patients with neurological diseases. J. Clin. Microbiol. *38*, 274–278.

Popow-Kraupp, T., Kundi, M., Ambrosch, F., Vanura, H., and Kunz, C. (1986). A controlled trial for evaluating two live attenuated mumps-measles vaccines (Urabe Am. 9-Schwarz and Jeryl Lynn-Moraten) in young children. J. Med. Virol. *18*, 69–79.

Public Health Agency of Canada (2007). Update on Mumps Outbreak in the Maritimes: National Summary, 26 October 2007.

Public Health Agency of Canada (2009). Mumps in Canada, 2007.

Puri, M., Lemon, K., Duprex, W.P., Rima, B.K., and Horvath, C.M. (2009). A point mutation, E95D, in the mumps virus V protein disengages STAT3 targeting from STAT1 targeting. J. Virol. *83*, 6347–6356.

Radecke, F., Spielhofer, P., Schneider, H., Kaelin, K., Huber, M., Dotsch, C., Christiansen, G., and Billeter, M.A. (1995). Rescue of measles viruses from cloned DNA. EMBO J. *14*, 5773–5784.

Rassa, J.C., and Parks, G.D. (1998). Molecular basis for naturally occurring elevated readthrough transcription across the M–F junction of the paramyxovirus SV5. Virology *247*, 274–286.

Ratzmann, K.P., Strese, J., Rjasanowski, I., Witt, S., Kohnert, K.D., Keilacker, H., Richter, K.V., Giebelmann, R., and Schulz, B. (1985). Metabolic, hormonal, and immunological alterations in subjects with antecedent mumps infection. Exp. Clin. Endocrinol. *86*, 323–334.

Raut, S.K., Kulkarni, P.S., Phadke, M.A., Jadhav, S.S., Kapre, S.V., Dhere, R.M., Dhorje, S.P., and Godse, S.R. (2007). Persistence of antibodies induced by measles-mumps-rubella vaccine in children in India. Clin. Vaccine Immunol. *14*, 1370–1371.

Reid, F., Hassan, J., Irwin, F., Waters, A., Hall, W., and Connell, J. (2008). Epidemiologic and diagnostic evaluation of a recent mumps outbreak using oral fluid samples. J. Clin. Virol. *41*, 134–137.

Rima, B.K., Earle, J.A., Baczko, K., ter Meulen, V., Liebert, U.G., Carstens, C., Carabana, J., Caballero, M., Celma, M.L., and Fernandez-Munoz, R. (1997). Sequence divergence of measles virus haemagglutinin during natural evolution and adaptation to cell culture. J. Gen. Virol. *78*, 97–106.

Ritter, B.S. (1958). Mumps meningoencephalitis in children. J. Pediatr. *52*, 424–433.

Roberts, C., Porter-Jones, G., Crocker, J., and Hart, J. (2009). Mumps outbreak on the island of Anglesey, North Wales, December 2008–January 2009. Euro. Surveill. *5*, 19109.

Roberts, W.C., and Fox, S.M., 3rd (1965). Mumps of the heart. Clinical and pathologic features. Circulation *32*, 342–345.

Rosas-Morrieta, N.H., Herrera-Camacho, I., Palma-Ocampo, H., Santos-Lopéz, G., and Reyes-Leyva, J. (2010). Interaction of mumps virus V protein variants with STAT1-STAT2 heterodimer: experimental and theoretical studies. Virol. J. *11*, 263–273.

Rota, J.S., Heath, J.L., Rota, P.A., King, G.E., Celma, M.L., Carabana, J., Fernandez-Munoz, R., Brown, D., Jin, L., and Bellini, W.J. (1996). Molecular epidemiology of measles virus: identification of pathways of transmission and implications for measles elimination. J. Infect. Dis. *173*, 32–37.

Rota, J.S., Turner, J.C., Yost-Daljev, M.K., Freeman, M., Toney, D.M., Meisel, E., Williams, N., Sowers, S.B., Lowe, L., Rota, P.A., *et al.* (2009). Investigation of a mumps outbreak among university students with two measles-mumps-rubella (MMR) vaccinations, Virginia, September–December 2006. J. Med. Virol. *81*, 1819–1825.

Rota, P., Lowe, L., Rota, J., and Bellini, W. (2007). Molecular epidemiology of mumps viruses isolated in the United States, 1960–2006. Paper presented at: American Society for Virology 26th Annual Meeting.

Rozina, E.E., and Hilgenfeldt, M. (1985). Comparative study on the neurovirulence of different vaccine strains of parotitis virus in monkeys. Acta Virol. *29*, 225–230.

Rubin, S.A., Pletnikov, M., and Carbone, K.M. (1998). Comparison of the neurovirulence of a vaccine and a wild-type mumps virus strain in the developing rat brain. J. Virol. *72*, 8037–8042.

Rubin, S.A., Snoy, P.J., Wright, K.E., Brown, E.G., Reeve, P., Beeler, J.A., and Carbone, K.M. (1999). The mumps virus neurovirulence safety test in Rhesus monkeys: a comparison of mumps virus strains. J. Infect. Dis. *180*, 521–525.

Rubin, S.A., Pletnikov, M., Taffs, R., Snoy, P.J., Kobasa, D., Brown, E.G., Wright, K.E., and Carbone, K.M. (2000). Evaluation of a neonatal rat model for prediction of mumps virus neurovirulence in humans. J. Virol. *74*, 5382–5384.

Rubin, S.A., Amexis, G., Pletnikov, M., Li, Z., Vanderzanden, J., Mauldin, J., Sauder, C., Malik, T., Chumakov, K., and Carbone, K.M. (2003). Changes in mumps virus gene sequence associated with variability in neurovirulent phenotype. J. Virol. *77*, *11*,616–11624.

Rubin, S.A., Afzal, M.A., Powell, C.L., Bentley, M.L., Auda, G.R., Taffs, R.E., and Carbone, K.M. (2005). The rat-based neurovirulence safety test for the assessment of mumps virus neurovirulence in humans: an international collaborative study. J. Infect. Dis. *191*, 1123–1128.

Rubin, S., Mauldin, J., Chumakov, K., Vanderzanden, J., Iskow, R., and Carbone, K. (2006). Serological and phylogenetic evidence of monotypic immune responses to different mumps virus strains. Vaccine *24,* 2662–2668.

Rubin, S.A., Qi, L., Audet, S.A., Sullivan, B., Carbone, K.M., Bellini, W.J., Rota, P.A., Sirota, L., and Beeler, J. (2008). Antibody induced by immunization with the Jeryl Lynn mumps vaccine strain effectively neutralizes a heterologous wild-type mumps virus associated with a large outbreak. J. Infect. Dis. *198,* 508–515.

Russell, R.R., and Donald, J.C. (1958). The neurological complications of mumps. Br. Med. J. *2,* 27–30.

Saika, S., Kidokoro, M., Aoki, A., and Ohkawa, T. (2004). Neurovirulence of mumps virus: intraspinal inoculation test in marmosets. Biologicals *32,* 147–152.

Saika, S., Kidokoro, M., Kubonoya, H., Ito, K., Ohkawa, T., Aoki, A., Nagata, N., and Suzuki, K. (2006). Development and biological properties of a new live attenuated mumps vaccine. Comp. Immunol. Microbiol. Infect. Dis. *29,* 89–99.

Saika, S., Kidokoro, M., Ohkawa, T., Aoki, A., and Suzuki, K. (2002). Pathogenicity of mumps virus in the marmoset. J. Med. Virol. *66,* 115–122.

Santos, C.L., Ishida, M.A., Foster, P.G., Sallum, M.A., Benega, M.A., Borges, D.B., Correa, K.O., Constantino, C.R., Afzal, M.A., and Paiva, T.M. (2008). Detection of a new mumps virus genotype during parotitis epidemic of 2006–2007 in the state of Sao Paulo, Brazil. J. Med. Virol. *80,* 323–329.

Santos-Lopez, G., Cruz, C., Pazos, N., Vallejo, V., Reyes-Leyva, J., and Tapia-Ramirez, J. (2006). Two clones obtained from Urabe AM9 mumps virus vaccine differ in their replicative efficiency in neuroblastoma cells. Microbes Infect. *8,* 332–339.

Sartorius, B., Penttinen, P., Nilsson, J., Johansen, K., Jonsson, K., Arneborn, M., Lofdahl, M., and Giesecke, J. (2005). An outbreak of mumps in Sweden, February-April 2004. Euro. Surveill. *10,* 191–193.

Sauder, C.J., Vandenburgh, K.M., Iskow, R.C., Malik, T., Carbone, K.M., and Rubin, S.A. (2006). Changes in mumps virus neurovirulence phenotype associated with quasispecies heterogeneity. Virology *350,* 48–57.

Sauder, C.J., Zhang, C.X., Link, M.A., Duprex, W.P., Carbone, K.M., and Rubin, S.A. (2009). Presence of lysine at aa 335 of the hemagglutinin-neuraminidase protein of mumps virus vaccine strain Urabe AM9 is not a requirement for neurovirulence. Vaccine *27,* 5822–5829.

Savage, E., Ramsay, M., White, J., Beard, S., Lawson, H., Hunjan, R., and Brown, D. (2005). Mumps outbreaks across England and Wales in 2004: observational study. BMJ *330,* 1119–1120.

Schaffzin, J.K., Pollock, L., Schulte, C., Henry, K., Dayan, G., Blog, D., and Smith, P. (2007). Effectiveness of previous mumps vaccination during a summer camp outbreak. Pediatrics *120,* e862–868.

Schlegel, M., Osterwalder, J.J., Galeazzi, R.L., and Vernazza, P.L. (1999). Comparative efficacy of three mumps vaccines during disease outbreak in Eastern Switzerland: cohort study. BMJ *319,* 352.

Schmitt, A.P., and Lamb, R.A. (2004). Escaping from the cell: assembly and budding of negative-strand RNA viruses. In Biology of negative strand RNA viruses: The Power of Reverse Genetics, Kawaoka, Y. ed. (Springer-Verlag: Heidelberg), pp. 145–196.

Schmitt, A.P., Leser, G.P., Morita, E., Sundquist, W.I., and Lamb, R.A. (2005). Evidence for a new viral late-domain core sequence, FPIV, necessary for budding of a paramyxovirus. J. Virol. *79,* 2988–2997.

Schmitt, P.T., Ray, G., Schmitt A.P. (2010). The C-terminal end of PIV5 NP protein is important for virus-like particle production and M-NP protein interaction. J. Virol. *83,* 7261–7272.

Schnell, M.J., Mebatsion, T., and Conzelmann, K.K. (1994). Infectious rabies viruses from cloned cDNA. EMBO J. *13,* 4195–4203.

Schulz, B., Michaelis, D., Hildemann, W., Witt, S., Hemke, B., and Rjasanowski, I. (1987). Islet cell surface antibodies (ICSA) in subjects with a previous mumps infection–a propective study over a 4 year period. Exp. Clin. Endocrin. *90,* 62–70.

Schwarzer, S., Reibel, S., Lang, A.B., Struck, M.M., Finkel, B., Gerike, E., Tischer, A., Gassner, M., Gluck, R., Stuck, B., *et al.* (1998). Safety and characterization of the immune response engendered by two combined measles, mumps and rubella vaccines. Vaccine *16,* 298–304.

Server, A.C., Merz, D.C., Waxham, M.N., and Wolinsky, J.S. (1982). Differentiation of mumps virus strains with monoclonal antibody to the HN glycoprotein. Infect. Immun. *35,* 179–186.

Shah, D., Vidal, S., Link, M.A., Rubin, S.A., and Wright, K.E. (2009). Identification of genetic mutations associated with attenuation and changes in tropism of Urabe mumps virus. J. Med. Virol. *81,* 130–138.

Sidhu, M.S., Menonna, J.P., Cook, S.D., Dowling, P.C., and Udem, S.A. (1993). Canine distemper virus L gene: sequence and comparison with related viruses. Virology *193,* 50–65.

Singh, R., Mostafid, H., and Hindley, R.G. (2006). Measles, mumps and rubella – the urologist's perspective. Int. J. Clin. Pract. *60,* 335–339.

Smallwood, S., Ryan, K.W., and Moyer, S.A. (1994). Deletion analysis defines a carboxyl-proximal region of Sendai virus P protein that binds to the polymerase L protein. Virology *202,* 154–163.

Smorodintsev, A.A., and Klyatchko, N.S. (1954). Specific prophylaxis of mumps. Zh. Mikrobiol. *11,* 6–11.

Smorodintsev, A.A., and Klyatchko, N.S. (1958). Live anti-mumps vaccine. I. Results of tests of the immunogenic properties of live vaccine when administered intradermally to susceptible children. Acta Virol. *2,* 137–144.

Smorodintsev, A.A., Luzyanina, T.Y., and Mikutskaya, B.A. (1965). Data on the Efficiency of Live Mumps Vaccine from Chick Embryo Cell Cultures. Acta Virol. *9,* 240–247.

Somboonthum, P., Yoshii, H., Okamoto, S., Koike, M., Gomi, Y., Uchiyama, Y., Takahashi, M., Yamanishi,

K., and Mori, Y. (2007). Generation of a recombinant Oka varicella vaccine expressing mumps virus hemagglutinin-neuraminidase protein as a polyvalent live vaccine. Vaccine *25*, 8741–8755.

St Geme, J.W., Jr., Peralta, H., and Van Pelt, L.F. (1972). Intrauterine infection of the rhesus monkey with mumps virus: abbreviated viral replication in the immature fetus as an explanation for split immunologic recognition after birth. J. Infect. Dis. *126*, 249–256.

Stillman, E.A., and Whitt, M.A. (1998). The length and sequence composition of vesicular stomatitis virus intergenic regions affect mRNA levels and the site of transcript initiation. J. Virol. *72*, 5565–5572.

Stokes, J., Jr., Weibel, R.E., Buynak, E.B., and Hilleman, M.R. (1967). Live attenuated mumps virus vaccine. II. Early clinical studies. Pediatrics *39*, 363–371.

Sutter, G., and Moss, B. (1995). Novel vaccinia vector derived from the host range restricted and highly attenuated MVA strain of vaccinia virus. Dev. Biol. Stand. *84*, 195–200.

T.E.C., Jr. (1967). Hippocrates describes mumps followed by orchitis. Pediatrics *40*, 420.

Takano, T., Uno, M., Yamano, T., and Shimada, M. (1994). Pathogenesis of cerebellar deformity in experimental Chiari type I malformation caused by mumps virus. Acta Neuropathol. *87*, 168–173.

Takeuchi, K., Hishiyama, M., Yamada, A., and Sugiura, A. (1988). Molecular cloning and sequence analysis of the mumps virus gene encoding the P protein: mumps virus P gene is monocistronic. J. Gen. Virol. *69*, 2043–2049.

Takeuchi, K., Tanabayashi, K., Hishiyama, M., and Yamada, A. (1996). The mumps virus SH protein is a membrane protein and not essential for virus growth. Virology *225*, 156–162.

Takikita, S., Takano, T., Narita, T., Takikita, M., Ohno, M., and Shimada, M. (2001). Neuronal apoptosis mediated by IL-1 beta expression in viral encephalitis caused by a neuroadapted strain of the mumps virus (Kilham Strain) in hamsters. Exp. Neurol. *172*, 47–59.

Takimoto, T., and Portner, A. (2004). Molecular mechanism of paramyxovirus budding. Virus Res. *106*, 133–145.

Takimoto, T., Taylor, G.L., Connaris, H.C., Crennell, S.J., and Portner, A. (2002). Role of the hemagglutinin-neuraminidase protein in the mechanism of paramyxovirus-cell membrane fusion. J. Virol. *76*, 13028–13033.

Tanabayashi, K., Takeuchi, K., Hishiyama, M., Yamada, A., Tsurudome, M., Ito, Y., and Sugiura, A. (1990). Nucleotide sequence of the leader and nucleocapsid protein gene of mumps virus and epitope mapping with the *in vitro* expressed nucleocapsid protein. Virology *177*, 124–130.

Tecle, T., Johansson, B., Jejcic, A., Forsgren, M., and Orvell, C. (1998). Characterization of three co-circulating genotypes of the small hydrophobic protein gene of mumps virus. J. Gen. Virol. *79*, 2929–2937.

Tecle, T., Johansson, B., Yun, Z., and Orvell, C. (2000). Antigenic and genetic characterization of the fusion (F) protein of mumps virus strains. Arch. Virol. *145*, 1199–1210.

Tecle, T., Bottiger, B., Orvell, C., and Johansson, B. (2001). Characterization of two decades of temporal co-circulation of four mumps virus genotypes in Denmark: identification of a new genotype. J. Gen. Virol. *82*, 2675–2680.

Tischer, A., and Gerike, E. (2000). Immune response after primary and re-vaccination with different combined vaccines against measles, mumps, rubella. Vaccine *18*, 1382–1392.

Tsuji, M., Sasaki, K., Nakagawa, M., Mori, T., Nakayama, T., Makino, S., and Okamura, M. (1983). Improved enzyme-linked immunosorbent assay (ELISA) for the detection of mumps virus antibodies. Kitasato Arch. Exp. Med. *56*, 137–147.

Tsurudome, M., Yamada, A., Hishiyama, M., and Ito, Y. (1987). Replication of mumps virus in mouse: transient replication in lung and potential of systemic infection. Arch. Virol. *97*, 167–179.

Uchida, K., Shinohara, M., Shimada, S., Segawa, Y., Doi, R., Gotoh, A., and Hondo, R. (2005). Rapid and sensitive detection of mumps virus RNA directly from clinical samples by real-time PCR. J. Med. Virol. *75*, 470–474.

Ukkonen, P., Granstrom, M.L., and Penttinen, K. (1981). Mumps-specific immunoglobulin M and G antibodies in natural mumps infection as measured by enzyme-linked immunosorbent assay. J. Med. Virol. *8*, 131–142.

Ulane, C.M., Rodriguez, J.J., Parisien, J.P., and Horvath, C.M. (2003). STAT3 ubiquitylation and degradation by mumps virus suppress cytokine and oncogene signaling. J. Virol. *77*, 6385–6393.

Unal, M., Katircioglu, S., Karatay, M.C., Suoglu, Y., Erdamar, B., and Aslan, I. (1998). Sudden total bilateral deafness due to asymptomatic mumps infection. Int. J. Pediatr. Otorhinolaryngol. *45*, 167–169.

Usonis, V., Bakasenas, V., Chitour, K., and Clemens, R. (1998). Comparative study of reactogenicity and immunogenicity of new and established measles, mumps and rubella vaccines in healthy children. Infection *26*, 222–226.

Usonis, V., Bakasenas, V., and Denis, M. (2001). Neutralization activity and persistence of antibodies induced in response to vaccination with a novel mumps strain, RIT 4385. Infection *29*, 159–162.

Utz, J.P., Kasel, J.A., Cramblett, H.G., Szwed, C.F., and Parrott, R.H. (1957). Clinical and laboratory studies of mumps. I. Laboratory diagnosis by tissue-culture technics. N. Engl. J. Med. *257*, 497–502.

Utz, J.P., Szwed, C.F., and Kasel, J.A. (1958). Clinical and laboratory studies of mumps. II. Detection and duration of excretion of virus in urine. Proc. Soc. Exp. Biol. Med. *99*, 259–261.

Utz, J.P., Houk, V.N., and Alling, D.W. (1964). Clinical and laboratory studies of mumps. IV. Viruria and abnormal renal function. N. Engl. J. Med. *11*, 1283–1286.

Vandermeulen, C., Roelants, M., Vermoere, M., Roseeuw, K., Goubau, P., and Hoppenbrouwers, K. (2004). Outbreak of mumps in a vaccinated child population: a question of vaccine failure? Vaccine *22*, 2713–2716.

Vandermeulen, C., Clement, F., Roelants, M., Van Damme, P., Hoppenbrouwers, K., and Leroux-Roels, G. (2009). Evaluation of cellular immunity to mumps in vaccinated individuals with or without circulating antibodies up to 16 years after their last vaccination. J. Infect. Dis. *199,* 1457–1460.

Vandermeulen, C., Verhoye, L., Vaidya, S., Clement, F., Brown, K.E., Hoppenbrouwers, K., and Leroux-Roels, G. (2010). Detection of mumps virus-specific memory B cells by transfer of peripheral blood mononuclear cells into immune-deficient mice. Immunology *131,* 33–39.

Vicens, C.N., Nobrega, F.T., Joseph, J.M., and Meyer, M.B. (1966). Evaluation of tests for the measurement of previous mumps infection and analysis of mumps experience by blood group. Am. J. Epidemiol. *84,* 371–381.

Vulliemoz, D., Cordey, S., Mottet-Osman, G., and Roux, L. (2005). Nature of a paramyxovirus replication promoter influences a nearby transcription signal. J. Gen. Virol. *86,* 171–180.

Waddell, D.J., Wilbur, J.R., and Merigan, T.C. (1968). Interferon production in human mumps infections. Proc. Soc. Exp. Biol. Med. *127,* 320–324.

Warrener, L., and Samuel, D. (2006). Evaluation of a commercial assay for the detection of mumps specific IgM antibodies in oral fluid and serum specimens. J. Clin. Virol. *35,* 130–134.

Watson-Creed, G., Saunders, A., Scott, J., Lowe, L., Pettipas, J., and Hatchette, T.F. (2006). Two successive outbreaks of mumps in Nova Scotia among vaccinated adolescents and young adults. CMAJ *175,* 483–488.

Weibel, R.E., Stokes, J., Jr., Buynak, E.B., Whitman, J.E., Jr., and Hilleman, M.R. (1967). Live attenuated mumps-virus vaccine. 3. Clinical and serologic aspects in a field evaluation. N. Engl. J. Med. *276,* 245–251.

Wellington, K., and Goa, K.L. (2003). Measles, mumps, rubella vaccine (Priorix; GSK-MMR): a review of its use in the prevention of measles, mumps and rubella. Drugs *63,* 2107–2126.

Wilson, R.L., Fuentes, S.M., Wang, P., Taddeo, E.C., Klatt, A., Henderson, A.J., and He, B. (2006). Function of small hydrophobic proteins of paramyxovirus. J. Virol. *80,* 1700–1709.

Wolinsky, J.S., and Stroop, W.G. (1978). Virulence and persistence of three prototype strains of mumps virus in newborn hamsters. Arch. Virol. *57,* 355–359.

Wolinsky, J.S., Klassen, T., and Baringer, J.R. (1976). Persistence of neuroadapted mumps virus in brains of newborn hamsters after intraperitoneal inoculation. J. Infect. Dis. *133,* 260–267.

Wolinsky, J.S., Waxham, M.N., and Server, A.C. (1985). Protective effects of glycoprotein-specific monoclonal antibodies on the course of experimental mumps virus meningoencephalitis. J. Virol. *53,* 727–734.

World Health Organization (2001). Mumps virus vaccines. Wkly. Epidemiol. Rec. *76,* 346–355.

World Health Organization (2005). Global status of mumps immunization and surveillance. Wkly. Epidemiol. Rec. *80,* 418–424.

World Health Organization (2007a). Mumps. Wkly. Epidemiol. Rec. *81,* 51–60.

World Health Organization (2007b). Mumps virus vaccines. Wkly. Epidemiol. Rec. *82,* 51–60.

Yamada, A., Tsurudome, M., Hishiyama, M., and Ito, Y. (1984). Abortive infection of mumps virus in murine cell lines. J. Gen. Virol. *65,* 973–980.

Yan, Y., and Samal, S.K. (2008). Role of intergenic sequences in newcastle disease virus RNA transcription and pathogenesis. J. Virol. *82,* 1323–1331.

Yates, P.J., Afzal, M.A., and Minor, P.D. (1996). Antigenic and genetic variation of the HN protein of mumps virus strains. J. Gen. Virol. *77,* 2491–2497.

Yeo, R.P., Afzal, M.A., Forsey, T., and Rima, B.K. (1993). Identification of a new mumps virus lineage by nucleotide sequence analysis of the SH gene of 10 different strains. Arch. Virol. *128,* 371–377.

Yokosawa, N., Kubota, T., and Fujii, N. (1998). Poor induction of interferon-induced 2′,5′-oligoadenylate synthetase (2–5 AS) in cells persistently infected with mumps virus is caused by decrease of STAT-1 alpha. Arch. Virol. *143,* 1985–1992.

Yoshida, N., Fujino, M., Miyata, A., Nagai, T., Kamada, M., Sakiyama, H., Ihara, T., Kumagai, T., Okafuji, T., and Nakayama, T. (2008). Mumps virus reinfection is not a rare event confirMed. by reverse transcription loop-mediated isothermal amplification. J. Med. Virol. *80,* 517–523.

Young, M.L., Dickstein, B., Weibel, R.E., Stokes, J., Jr., Buynak, E.B., and Hilleman, M.R. (1967). Experiences with Jeryl Lynn strain live attenuated mumps virus vaccine in a pediatric outpatient clinic. Pediatrics *40,* 798–803.

Yuzepchuk, S.A., Rozina, E.E., Kaptsova, T.I., and Kulish, E.A. (1975). Morphological differences in the central nervous system and organs of monkeys inoculated intracerebrally with virulent and attenuated strains of mumps virus. Acta Virol. *19,* 305–310.

Zhou, F., Reef, S., Massoudi, M., Papania, M.J., Yusuf, H.R., Bardenheier, B., Zimmerman, L., and McCauley, M.M. (2004). An economic analysis of the current universal 2-dose measles-mumps-rubella vaccination programme in the United States. J. Infect. Dis. *189* (Suppl. 1), S131–145.

Zimmerman, L., Reef, S., and Wharton, M. (2002). Mumps, vol 7. In manual for the Surveillance of Vaccine-preventable Diseases, Wharton, M., Hugh, H., Reilly, M., eds (Centers for Disease Control and Prevention, Atlanta), pp. 701–712.

The Parainfluenza Virus Simian Virus 5

2

Griffith D. Parks*, Mary J. Manuse and John B. Johnson

Abstract

Many of the fundamental properties of paramyxovirus biology are founded on studies of the prototype negative strand RNA virus Simian virus 5 (SV5). This includes the basic understanding of the structure of the paramyxovirus particle, as well as the structure and function of the viral haemagglutinin–neuraminidase attachment protein and viral fusion protein. The discovery of the unique mechanism by which many paramyxoviruses express the viral P and V proteins by non-templated addition of G residues was first described for SV5. This prototype paramyxovirus also employs unique biological processes to evade host immune responses, including the blocking host cell sensors that recognize virus replication products and the targeted degradation of STAT1, a host cell protein that is integral to interferon signalling. Based on this rich history of fundamental discoveries, future work will be focused on interactions of SV5 with the immune system and development of novel SV5-based vectors.

History and origin of the virus

Simian virus 5 (SV5) has been studied as a prototype negative strand RNA virus for 50 years, dating back to its discovery and description in the 1950s and 1960s (Hull *et al.*, 1956; Choppin, 1964). Studies on SV5 as a prototype virus were facilitated by its rapid growth to very high titres (>10^8 PFU/ml) as well as the strong and clear biological activities of the glycoproteins (e.g. fusion and haemagglutinin-neuraminidase). Many fundamental discoveries in paramyxovirus biology, replication and immunology have been uncovered through the study of SV5. This includes the structure and function of the viral glycoproteins (Scheid *et al.*, 1972; Scheid and Choppin, 1974; Yuan *et al.*, 2005; Yin *et al.*, 2006), the unique mechanism to produce P and V proteins through RNA editing (Thomas *et al.*, 1988), the remarkable mechanisms to block interferon induction and signalling (Didcock *et al.*, 1999a,b; Young *et al.*, 2000; Andrejeva *et al.*, 2004), the structure and function of the paramyxovirus V protein (Poole *et al.*, 2002; Li *et al.*, 2006), identification and function of the SH protein (Hiebert *et al.*, 1985b; He *et al.*, 2001), and the bi-partite structure of the viral promoter (Murphy *et al.*, 1998; Murphy and Parks, 1999). Many of these principles first described for SV5 were subsequently confirmed to be fundamental properties of the paramyxovirus family.

The origin of SV5 is not clear given the available evidence. SV5 can be a common contaminant of tissue culture cell lines (Young *et al.*, 2007; Hsiung, 1972), a finding which complicates the question of whether a primary tissue truly harbours the virus or if it is acquired during *in vitro* cultivation. SV5 was isolated in the mid-1950s from primary cultures of rhesus monkey kidney cells (Hull *et al.*, 1956), and was named accordingly to denote that it was a simian virus. However, it soon became clear that the virus could be recovered from tissue samples from a wide range of animals, including

*Corresponding author

canine, simian, porcine and human (e.g. McCandish *et al.*, 1978; Robbins *et al.*, 1981; Goswami *et al.*, 1984; Tong *et al.*, 2002). Many of these isolates have differences in sequence (Chatziandreou *et al.*, 2004) as well as antigenicity (Randall *et al.*, 1987). This is interpreted to indicate that these isolates are derived from bona fide infections and are probably not lab contaminants. In the lab setting, the WR and W3A strains are most often used for experimental purposes (Hull *et al.*, 1956; Choppin, 1964; Tsurudome *et al.*, 2001). Unless otherwise stated, results discussed here are based on the more widely studied W3A strain.

Natural SV5 variants or very closely related SV5-like viruses have been isolated from human, porcine, and canine tissue samples. For example, SER virus was isolated from a sow showing porcine reproductive and respiratory syndrome (Heinen *et al.*, 1998), and was shown by nucleotide sequencing to be closely related to SV5 (Tong *et al.*, 2002). While SER shares properties with the lab adapted SV5 strains, it has been reported to have an unusual mechanism of entry into cells, which is detailed below (Tong *et al.*, 2002; Seth *et al.*, 2003). As a second example, a well-characterized set of SV5-related viruses has been isolated from canine tissue. The canine parainfluenza virus CPI^+ strain was isolated from the cerebrospinal fluid of a dog with temporary posterior paralysis and was used to experimentally infect a dog to generate CPI^- (Evermann *et al.*, 1980; Baumgartner *et al.*, 1981). These two isolates differ in their ability to counteract type I interferon (Chatziandreou *et al.*, 2004) as described in the section below on SV5 immunity.

Infectious agent

Classification

SV5 is a member of the *Rubulavirus* genus of the Paramyxovirus family (Lamb and Parks, 2007). This classification is based on sequence comparisons among paramyxoviruses, antigenic similarities (Randall and Young, 1988), and the fact that rubulaviruses share an insertion of two G residues in expression of the P mRNA from the P/V gene. SV5 is closely related to other members of the *Rubulavirus* genus including mumps virus (MuV), human parainfluenza virus type 2 (HPIV2) and HPIV4. There is high sequence identity among these viruses, as illustrated by findings that the SV5 NP protein shares 56% and 58% identity with the NP proteins of MuV and HPIV2, respectively (Parks *et al.*, 1992).

The name SV5 has always caused some confusion in the scientific community. The 'simian' name is based on it being originally isolated from monkey kidney cells. However, a wealth of data support the contention that SV5 has been isolated from a wide variety of species and can be a common contaminant of cell lines (Young *et al.*, 2007). Based on this, it has recently been proposed that SV5 be renamed parainfluenza virus 5 or PIV5 (Chatziandreu *et al.*, 2004), as originally proposed in the 1970s (Hsiung, 1972). For the historic purposes of this chapter and the review of published literature, we retain the original name SV5 to help distinguish it from the other parainfluenza viruses addressed by other chapter contributors.

Virion properties

Like other paramyxoviruses, the SV5 virion is composed of a lipid envelope and an internal ribonucleocapsid complex. The virion is pleomorphic in shape and varies in size from 150 to 350 nm (see Fig. 2.1A). Two major integral membrane glycoproteins, the haemagglutinin–neuraminidase (HN) and fusion (F) protein, are found in the viral lipid envelope and project as a halo about 10 nm from the surface of the virion. There are also small amounts of the small hydrophobic (SH) protein associated with virions (He *et al.*, 1998). The nucleocapsid core is ~18 nm in diameter and consists of the single stranded negative sense RNA genome that is tightly encapsidated by multiple copies of the NP protein. Under some conditions, such as that seen in Fig. 2.1B, the nucleocapsid can be visualized within the viral membrane by electron microscopy. As shown in the partially disrupted virion in Fig. 2.1B, the nucleocapsid that has emerged from the particle has a 'herringbone'-like structure that represents turns of the helical nucleocapsid.

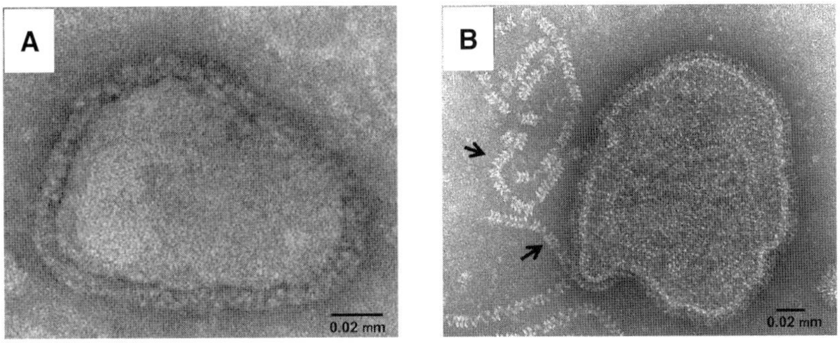

Figure 2.1 Simian virus 5 (SV5) virions. (A) Electron micrograph of intact SV5 particle displaying the characteristic pleomorphic shape and the halo of spike glycoproteins radiating from the virion surface. (B) Electron micrograph of partially disrupted virion displaying the herringbone-like structure of the nucleocapsid that is spilling from the particle and within the intact portion of the particle. Arrow denotes areas where turns of the helical nucleocapsid are evident. 55,000× magnification.

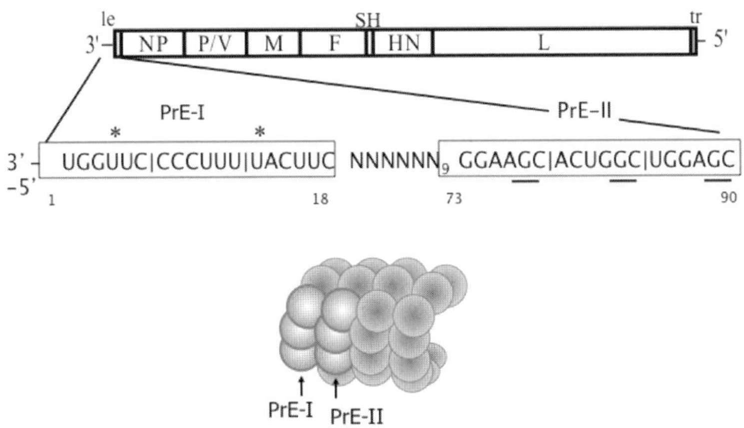

Figure 2.2 Schematic of Simian virus 5 (SV5) genome structure, genomic promoter and helical nucleocapsid. The SV5 genome is depicted at the top of the figure as a rectangle with vertical bars denoting the intergenic regions. Le and tr are the 3′ leader and 5′ trailer regions. The nucleotide sequence of two key elements in the genomic promoter is shown with the position of the internal 5′-CG-3′ motif underlined. The three hexamers that compose promoter element I (PrE-I) and II (PrE-II) are denoted by boxes. Two nucleotides (U5 and A14) that have been shown to enhance transcription (Manuse and Parks, 2009) are highlighted by stars. In the bottom figures, the predicted position of PrE-I and PrE-II along one face of the helical nucleocapsid is shown.

Genome structure

The SV5 genome is a single strand of negative sense RNA of 15,246 bases in length (see references in Lamb and Parks, 2007). As shown in Fig. 2.2, this genome encodes seven tandemly linked genes that are arranged 3′-NP-P/V-M-F-SH-HN-L-5′ and produce eight major proteins as first determined by Paterson *et al.* (1984b). The terminal ends of the genome are non-coding regions that act as bipartite promoters directing viral RNA synthesis. This includes the 55 base leader RNA (le) located at the 3′ end of the genome which directs transcription of viral mRNAs and replication of genomic RNA to produce positive sense antigenomes. The 5′ end of the genome encodes the 31-base trailer (tr), which functions in the context of the antigenomic RNA to direct replication of antigenomic RNA to produce negative-sense progeny genomes. As detailed below, viral genes are separated by highly variable junctions within the genome that can be 1–47 nucleotides in length. These *cis*-acting regions direct polyadenylation of mRNA transcripts as well as capping of nascent transcribed mRNA.

The SV5 genomic and antigenomic promoters have an unusual bi-partite structure first described by Murphy *et al.* (1998). The genomic protomer consists of a 3′ end element (the le) and a second internal element located in the 3′ (upstream) non-coding region of the NP gene. The antigenomic promoter consists of a 3′ end complement to the tr along with a second internal element located in coding region of the L gene. The 3′ promoter element I (PrE-I) and internal PrE-II are two distinct 18 base stretch of bases that are separated by a 54 base sequence-independent spacer region (Murphy *et al.*, 1998). Details of the functions for these elements are described in the RNA synthesis section below.

Infections with many paramyxoviruses can readily generate defective interfering (DI) particles. For SV5, the generation of DI genomes is highly dependent on cell type. A systematic study showed that it was very difficult to generate DI particles in MDBK cells, even after 14 undiluted serial passages in culture (Murphy and Parks, 1997). By contrast, only a few undiluted passages through Vero cells quickly resulted in a spectrum of SV5 DI RNAs. A number of these genomes were cloned and sequenced to reveal necessary cis-acting elements for replication of the viral RNA (Murphy and Parks, 1997). These viral genomes were also used as naturally-occurring templates to systematically test the 'rule of six' requirement for replication (see below).

SV5 proteins

Many of the functions of the SV5 proteins are not unique to this virus and these common functions will only be briefly mentioned for continuity. Excellent reviews with more detailed descriptions of the functions of these proteins are available (e.g. Lamb and Parks, 2007). Here, unique features of the SV5 proteins or their historic value to our general understanding of paramyxoviruses will be highlighted.

Nucleocapsid protein

The SV5 nucleocapsid protein NP is the first gene encoded within the genome and has 502 amino acids in length with a predicted MW of 56,500 (Parks *et al.*, 1992). Nucleotide sequence analysis of SV5 NP indicates that the protein has a core containing a central motif that is highly conserved among paramyxovirus NP sequences (Parks *et al.*, 1992). The C-terminal tail is rich in negatively charged residues such as glutamate and aspartate, but the significance of this for protein function has not been determined.

NP is an unusual RNA-binding protein since it has an overall acidic charge and lacks motifs traditionally found in cellular RNA-binding proteins (Parks *et al.*, 1992). While there is little overall conserved sequence between SV5 NP and paramyxoviruses other than HPIV2 and MuV, the SV5 NP polypeptide has an internal motif of an F-X_4-Y-X_3-φ–S-φ-A-M motif (where X is any residue and φ is an aromatic amino acid) that is conserved among nearly all related paramyxoviruses (Parks *et al.*, 1992). In the case of other paramyxoviruses, this region has been shown to be essential for self assembly of N with RNA and may be involved in N–N or N–RNA interactions.

NP encapsidates the viral genome and antigenome into a highly stable nuclease-resistant structure. Based on correlations with Sendai virus NP, the SV5 NP is thought to bind six nucleotides of the genomic RNA to form a helical structure, but the binding appears to be sequence independent. Some of the biophysical properties of these nucleocapsid structures have been determined (Bhella *et al.*, 2002). It is proposed that the SV5 nucleocapsid is composed of 14 NP subunits per turn, which contrasts with the 13 NP subunits per turn for Sendai virus. Encapsidation of genomic RNA by the NP protects it from degradation, but it also creates a functional promoter (see below), and plays an essential role in assembly of budding virions (Schmitt *et al.*, 2002). In addition to being found as a complex with viral RNA, NP is also found in a soluble form (NP^0) within the cell. SV5 NP^0 has been found associated with the viral P and V proteins and it has been proposed that this binding may aid in keeping NP in a soluble form for use during encapsidation of progeny RNA during replication (Precious *et al.*, 1995).

By contrast, the binding of NP by V protein may serve to inhibit RNA replication (Lin *et al.*, 2005).

Phosphoprotein

The phosphoprotein (P) has 392 amino acids with a predicted MW of 44,000 (Thomas *et al.*, 1988). Similar to other paramyxoviruses, P acts as a non-catalytic subunit of the viral RNA polymerase. Current models support a role for P protein in tethering the L catalytic subunit of the viral polymerase to the genomic template during viral RNA synthesis. However, additional roles have been proposed and are reviewed elsewhere. P is heavily phosphorylated at Ser and Thr residues, mainly within the N terminus. Remarkably, the SV5 P protein has also been demonstrated to play a role in limiting induction of host cell responses during mutant virus infection (Dillon and Parks, 2007; Gainey *et al.*, 2008a). This function is thought to be an indirect consequence of the role that P plays in RNA synthesis: the P protein may function to improve the fidelity of viral RNA synthesis and limit the production of dsRNA (Dillon and Parks, 2007).

Like all non-segmented negative strand RNA viruses, the SV5 P protein is heavily phosphorylated (reviewed in Lamb and Parks, 2007). Recent evidence indicates that phosphorylation at one particular residue (serine 157) is important for controlling viral gene expression (Timani *et al.*, 2008). An SV5 mutant encoding a P protein that could not be phosphorylated in this position exhibited elevated viral RNA synthesis. Similar to the naturally occurring substitutions seen in a chimeric P/V gene mutant (Wansley and Parks, 2002), this point mutant activated host-cell responses not seen with WT SV5 infections. The kinase responsible for this key phosphorylation event is thought to be Polo-like kinase 1, a host cell serine-threonine kinase implicated in regulating the cell cycle (Sun *et al.*, 2009). The relationship between the host cell cycle, viral RNA synthesis, and phosphorylation of the SV5 P protein remains to be determined.

The P protein is expressed by a unique mechanism involving non-templated addition of G residues to mRNA derived from the gene encoding both the P and V proteins. SV5 was the first virus for which this 'RNA Editing' of the P/V gene was discovered (Thomas *et al.*, 1988) and this has now been shown to be a very common feature of many paramyxoviruses. During transcription of the P/V gene by the viral polymerase, mRNAs are produced that encode both the phosphor- and the V proteins. The P and V mRNAs are identical except for an insertion of two non-templated guanosine (G) residues at a precise point in the viral transcripts. The process of RNA editing is thought to involve slippage of the viral polymerase at a precise location within the P/V gene (Thomas *et al.*, 1988), similar in mechanism to that seen at the end of the viral genes where polyadenylation is directed by slippage on a gene end U tract. As shown in Fig. 2.3A, this insertion results in a frame shift in the translational open reading frames downstream of the insertion site. Thus, as shown in Fig. 2.3A, V and P proteins share a 164 residue amino-terminal end, but differ in their C-terminal domains, which are encoded 3′ to the frame shift. The 'hard copy' or accurate transcription of the P/V gene encodes the V protein, while the P protein is encoded by the mRNA with two additional Gs. These unedited and edited mRNAs occur in about a 1:1 ratio (Thomas *et al.*, 1988). Insertions other than two G residues can occur, but they are much less frequently found in mRNA from infected cells. Insertion of either one or four G residues results in mRNA that encodes the W protein (or I protein in some literature). W protein shares the 164 residue N-terminal region but contains only a very short C-terminal domain of a few amino acids. The function of the W protein in the SV5 life cycle is not known, but only very small amounts are usually detected in infected cells.

V protein

The V protein has 222 amino acids with a predicted MW of 24,000 (Thomas *et al.* 1988). V protein plays major roles as an antagonist of host cell antiviral responses, as an inhibitor of viral RNA synthesis and can alter the cell cycle. V protein shares the 164 N-terminal residues with the P protein and this domain is thought to contain an NP-binding domain. In contrast to P protein, V protein contains a unique C-terminal domain containing seven highly conserved cysteine residues. This domain has been shown to bind two zinc

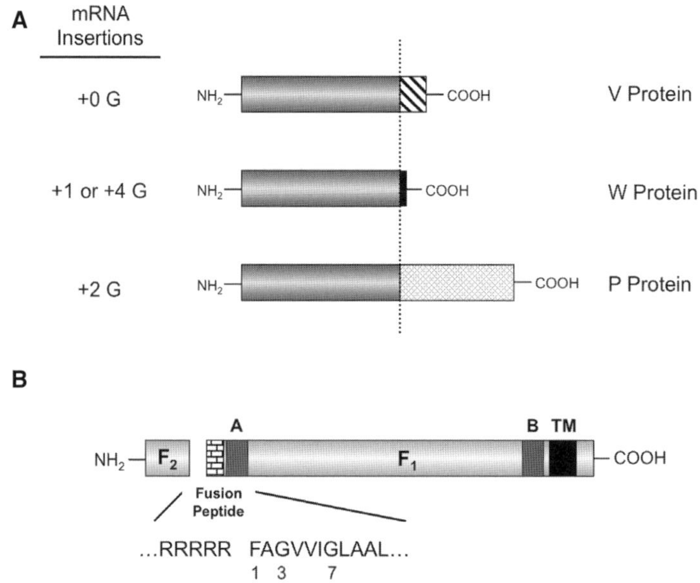

Figure 2.3 Schematic of Simian virus 5 (SV5) P/V gene products and F protein. (A) The V, W and P proteins are shown as boxes with a common N-terminal domain and distinct C-terminal domains that result from addition of 1 or 4 G residues (W/I) or 2 G residues (P). Dotted line indicates the site of divergence for the amino acid sequences. (B) The F protein is shown as a box separated into the N-terminal F2 and C-terminal F1 domains. The stretch of 5 arginine (R) residues at the cleavage site is shown adjacent to the sequence of the fusion peptide. Boxes denote the location of HRs A and B as well as the transmembrane (TM) domain. The glycine (G) residues at positions 3 and 7 which regulate fusion rates are indicated by numbers.

ions (Paterson *et al.*, 1995; Li *et al.*, 2006). SV5 V protein can be found associated with the nucleocapsid of purified virions and is thus a structural component of virions (Paterson *et al.*, 1995). This is very different from most other paramyxoviruses such as Sendai virus and measles virus in which V protein is only a very minor part of the virion structure. Very careful analysis revealed that there were ~350 and ~315 molecules of V associated with purified SV5 particles and nucleocapsids, respectively (Paterson *et al.*, 1995). The function of virion-associated V is unknown, but the location on the nucleocapsid template suggests a role in control of viral RNA synthesis during the initial phases of infection with the incoming particle. This virion-associated V may also play a role in rapid dismantling of the host cell antiviral response, since it has been shown that UV-treated virus can still induce blockage of interferon signalling (Didcock *et al.*, 1999b).

The SV5 V protein has been shown to be responsible for slowing the growth and cell cycle of infected cells (Lin and Lamb, 2000). In SV5 infected cells, there is a delay in progression of the cell cycle through the G1/S boundary and cells remain in S phase for an extended time. This can also be seen in the case of cells expressing V protein, but not P protein. The cellular damage-specific DNA-binding protein (DDB1) is part of a complex, which exhibits high binding affinity for UV-damaged DNA. Over expression of DDB1 partially restored cell cycle progression in V-expressing cells, suggesting that V interaction with DDB1 was responsible at least in part for this alteration of cell growth (Lin and Lamb, 2000). The advantage of delaying the cell cycle for an RNA virus like SV5, which replicates in the cytoplasm is not clear. However, it was speculated that this may be a reflection of constraints on SV5 imposed by a growth cycle of 18–24 h. In this case, cell division would interfere with optimal virus growth (Lin and Lamb, 2000), especially since the dissolution of the Golgi apparatus during mitosis would prevent transport of viral glycoproteins to the cell surface for incorporation into virions.

The SV5 V protein can inhibit viral RNA

synthesis. This was demonstrated using a minigenome system, but the mechanism for this inhibition has not been defined (Lin *et al.*, 2005). Consistent with this property, SV5 mutants with naturally-occurring substitutions in the shared N-terminal domain of P and V have been shown to overexpress viral RNA and protein (Wansley and Parks, 2002). A model has been proposed whereby the binding of V to N^0 competes with the productive P–N^0 interaction that is required to drive encapsidation of nascent RNA during replication of the genome. Alternatively, the V protein can also bind RNA (Lin *et al.*, 1997), which may contribute to inhibition of viral RNA synthesis. Although V binding to RNA *in vitro* did not appear to be specific for viral RNA, this may differ in the context of an infection or when other viral or cellular proteins are available to modulate activity.

A major function of the SV5 V protein is to interact with several host proteins to facilitate abrogation of type I interferon (IFN) induction (Poole *et al.*, 2002; Andrejeva *et al.*, 2004) and IFN signalling (Didcock *et al.*, 1999b; Andrejeva *et al.*, 2002). The role of V protein in counteracting host cell antiviral responses is discussed in a separate section below.

Matrix protein

The matrix (M) protein is the most abundant protein in an SV5 particle. M protein has 377 amino acids with a predicted MW of 42,200. The M protein serves as a major architectural protein for assembly and can be seen as an electron dense layer underlying the lipid bilayer of a virus particle. During assembly, M protein associates peripherally with the inner leaflet of the plasma membrane and is thought to orchestrate assembly of virus particles through coordinating the interactions of host proteins, viral glycoproteins that are integrated into the membrane and cytoplasmic nucleocapsids (Schmidt and Lamb, 2004).

Viral matrix proteins contain a variation on a motif known as the 'late domain', an amino acid sequence or patch, which functions to recruit cellular factors during the budding process. Mutations in late domains can lead to the accumulation of particles at the surface that have failed to pinch off. For the SV5 M protein, no recognizable conventional late domain with homology to other viral matrix proteins has been found. Instead, mutational analyses showed that SV5 M protein has a unique late domain motif with the core sequence 'FPIV' (Schmitt *et al.*, 2005). Viruses harbouring mutations in this motif were shown to be defective in budding, but did not show the accumulation of cell surface buds that had failed to pinch off, which lead Schmitt *et al.* (2005) to surmise that the SV5 'late domain' equivalent may function at a step prior to bud formation. SV5 budding is affected by proteosome inhibitors (Schmitt *et al.*, 2005), similar to viruses that use more conventional late domain motifs. Surprisingly, mutant M proteins with an altered proline in the FPIV motif quickly gave rise to revertants that had regained some budding efficiency. When analysed by sequencing, new proline residues were found substituted in M outside of the FPIV motif. These data indicate that overall structure of SV5 M protein and the location of proline residues may be an important factor in the ability of M protein to coordinate the assembly and budding of SV5 particles.

In addition to coordinating interactions between viral proteins, SV5 M protein also interacts with host cell proteins during assembly and budding. SV5 M binds E3-ubiquitin ligases to initiate budding (Schmitt *et al.*, 2005). Recent data indicate that host cell VPS4A ATPase is required for efficient budding of SV5 particles. Under conditions of expressing a mutant VPS4A, budding of SV5 virus like particles (VLPs) was suppressed and the mutant host protein was associated with particles. Very recently the host protein angiomotin-like 1 (AmotL1) was shown to be an important component of SV5 VLP production (Pei *et al.*, 2010). AmotL1 co-localizes with F-actin and is thought to function in endothelial migration. Pei *et al.* (2010) propose that these interactions between M and host proteins such as AmotL1 are potential targets for antiviral compounds.

Fusion protein

The SV5 fusion protein F has been a prototype for our understanding of membrane fusion proteins since its properties were discovered in the 1970s (Scheid *et al.*, 1972; Scheid and Choppin, 1974). Excellent reviews of the structure, processing and function of the SV5 F protein have been published and provide much more in-depth information (Lamb *et al.*, 2006; Lamb and Parks, 2007). F protein is responsible for promoting two fusion reactions: fusion of the virion envelope with the host cell plasma membrane during the entry process and cell–cell fusion or syncytia formation. The accepted convention is that paramyxoviruses enter cells through fusion at the plasma membrane, which is not dependent on activation by low pH. Recently, the SER porcine isolate of SV5 was reported to require a low-pH step for fusion (Seth *et al.*, 2003), but this was not reproducible in another lab setting (Bissonnette *et al.*, 2006).

The 529 amino acid fusion (F) protein is a type I integral membrane protein of predicted MW of 56,500 (Paterson *et al.*, 1984a). As with most type I membrane proteins, the SV5 F polypeptide contains an N-terminal signal sequence for targeting the endoplasmic reticulum (ER), a large extracellular domain, an unusually long transmembrane domain of ~25 residues, and a short cytoplasmic tail (Paterson *et al.*, 1984a). Alterations to the ectodomain, the transmembrane region or the cytoplasmic tail, can result in mutant F proteins with altered fusion capacity (Bissonnette *et al.*, 2009; Seth *et al.*, 2004)

F protein is initially synthesized as a precursor Fo. In the ER, Fo is glycosylated and forms homotrimers (Russell *et al.*, 1994). F oligomers are transported through the Golgi complex where a host cell furin-like protease cleaves at a stretch of five arginine residues to convert Fo into F1 and F2 (see Fig. 2.3B), which remain held together by disulfide bonds (Paterson *et al.*, 1985). The number of basic residues at an F cleavage site can differ between different paramyxoviruses. For SV5, cleavage occurs at a stretch of five arginine (Arg) residues (Fig. 2.3B), and F is cleaved in almost all tissue culture cell lines. Mutational analyses have shown that reducing the number of Arg residues does not affect transport to the cell surface, but can dramatically change the efficiency of cleavage during transport and the fusion ability at the cell surface (Paterson *et al.*, 1989).

Cleavage into F1 and F2 results in conformational change associated with gaining fusion competence (Dutch *et al.*, 2001). Proteolytic cleavage exposes a new N-terminal hydrophobic peptide on the F1 subunit that is thought to be an amphipathic alpha helix (Fig. 2.3B). Within this fusion peptide, there are a number of residues conserved across the fusion peptides from a range of paramyxoviruses. Horvath and Lamb (1992) systematically analysed the role of a number of conserved residues by mutational analysis. Remarkably, Gly to Ala substitutions in the SV5 fusion peptide at positions 3, 7 and 12 resulted in F proteins that caused massive cell–cell fusion (Horvath and Lamb, 1992). The increase in fusion is thought to reflect not only the final extent of fusion, but also the rate at which membranes fuse (Bagai and Lamb, 1997). These results are consistent with a model that glycine residues normally attenuate the alpha helical nature of the fusion peptide, and that increasing the alpha helical nature by alanine substitutions leads to increased fusion activity. The hyperfusogenic nature of these F mutants has been harnessed for therapeutic applications, as it has been shown that the ability of a cytopathic SV5 mutant to reduce tumour burden in a mouse model system is greatly enhanced when combined with the G3A hyperfusogenic F mutation (Gainey *et al.*, 2008b).

For many paramyxoviruses, the F protein can only carry out fusion reactions when expressed with the virus-specific attachment protein. This has been interpreted in a model for limiting fusion to 'the right time and right place' as this would be dictated by cleavage activation during transport to the surface and by the presence of an attachment protein, which can direct fusion to a target membrane. The F protein from the W3A strain of SV5 is highly unusual among fusion proteins since F protein is fully capable of cell–cell fusion in the absence of the attachment protein HN (Paterson *et al.*, 1985; Horvath and Lamb, 1992). By contrast, the F protein from the WR strain requires HN coexpression for syncytia formation. Exchanges of the three amino acids that differ between these two F proteins showed that a proline residue at position 22 was important for defining the HN-independent fusion activity of the W3A F protein (Ito *et al.*, 1997; Paterson *et al.*, 2000).

The structure of the SV5 F protein indicates that it contains two heptad repeat regions (HRA and HRB), which assemble into a six-helix bundle (reviewed in Dutch *et al.*, 2000; Lamb *et al.*, 2006).

HRA is located near the fusion peptide and HRB is more C-terminally located near the transmembrane domain (Fig. 2.3B). Thus, unlike membrane fusion proteins from other viruses, these HR are separated by very long distances in the SV5 F protein. This is consistent with data indicating large conformational changes that occur when the F protein undergoes the steps in the fusion reaction. Synthetic peptides that mimic these HRs have been used to dissect intermediate steps in F-mediated fusion reactions (Dutch *et al.*, 2000; Russell *et al.*, 2001). Available evidence indicates that the F protein exists in a metastable state that requires energy to drive the structure down a series of changes resulting in a fusion-competent form. The signals that supply this energy to convert the metastable state include cleavage at the stretch of Arg residues flanking HRA and possible HN receptor interactions, but they continue to be of considerable interest and intense study.

Small hydrophobic protein

SV5 was the first negative-strand RNA virus discovered to encode and express a small hydrophobic protein termed SH (Hiebert *et al.*, 1985b). This was subsequently also discovered for both mumps virus and respiratory syncytial virus. The SV5 SH protein derives its name from the amino acid sequence of the polypeptide, which is only 44 residues in size and containing a very large number of hydrophobic residues. SH is predicted to be an integral membrane protein with a short C-terminal ectodomain of five residues, an uncleaved signal/anchor domain of 23 residues, and an N-terminal cytoplasmic domain of 16 residues. The 44 amino acid SH protein is expressed at the cell surface as a type II integral membrane protein (Hiebert *et al.*, 1985a) and is found as a minor component of purified virions (He *et al.*, 1998).

Using reverse genetics to create an SH knockout virus, He and co-workers demonstrated that SH is not essential for virus growth in tissue culture cells (He *et al.*, 1998). In some transformed cell lines (e.g. MDBK), the Δ-SH virus induces apoptosis (He *et al.*, 2001), but interestingly this was not seen in all cell lines. For example, the SV5 Δ-SH virus induced apoptosis in bovine and mouse cell lines, but not in human A549 or Hela cells (He *et al.*, 2001). In a STAT1$^{-/-}$ mouse model for paramyxovirus pathogenesis, the SV5 Δ-SH mutant was attenuated in mouse killing compared with that seen with WT SV5. Thus, the SH protein probably plays an important role in counteracting host immune responses against SV5 infection, although work in mouse cell lines and mice are problematic because of the inability of SV5 to effectively counteract murine antiviral pathways (Didcock *et al.*, 1999a; Capraro *et al.*, 2008).

It has been proposed that the SV5 SH protein blocks induction of apoptosis mediated by TNF-α (Lin *et al.*, 2003; Wilson *et al.*, 2006). This interesting conclusion was based on the finding that in mouse L929 cells, infection with the SV5 Δ-SH mutant resulted in synthesis of TNF-α and induction of apoptosis and that cell killing was reduced by the addition of an antibody against the TNF-α receptor 1 (Lin *et al.*, 2003). However, the conclusion that SH protein blocks TNF-α signalling (Wilson *et al.*, 2006) is at odds with the finding that WT SV5 is not able to block TNF-α signalling when infected human cell lines are treated with exogenous TNF-α (Young *et al.*, 2003b; Arimilli *et al.*, 2007). The exact role of SH in counteracting antiviral responses and in the SV5 life cycle remains unresolved.

Haemagglutinin–neuraminidase protein

The attachment protein for SV5 is the type II integral membrane haemagglutinin–neuraminidase (HN) protein. HN is 565 amino acids in length with a predicted MW of 62,100 (Hiebert *et al.*, 1985a). HN primary structure is composed of a 17 residue cytoplasmic tail, a 19 residue uncleaved signal-anchor domain and a 529 residue ectodomain. SV5 HN is a dimer held together by disulfide bonds (Yuan *et al.*, 2005) and two of these dimers assemble into a non-covalently linked tetramer. HN is thought to form a mushroom like structure with a stalk projecting from the membrane and a large globular head region.

HN has sialic acid-binding activity, which functions during attachment to the surface of host cells to initiate the infectious process. HN also has neuraminidase activity, which carries out enzymatic cleavage of sialic acid most likely to aid in virus release and to prevent self-aggregation of virus particles. X-ray crystallography of SV5 HN has revealed that it harbours the expected sialidase fold containing seven highly conserved active site residues (Yuan *et al.*, 2005).

The SV5 HN protein has been a prototype for the study of the biosynthesis and cell biology of type II integral membrane proteins. The SV5 HN protein was instrumental as a tool for experimental studies during a time when little was known about the biosynthesis, assembly and transport of this class of integral membrane proteins containing uncleaved signal-anchor domains. A type II membrane protein contains an N-terminal uncleaved signal anchor and has a topology such that the N-terminus is cytoplasmic. Using HN and the influenza virus M2 protein, which has an opposite topology, Parks and Lamb carried out detailed mutagenesis studies to define the signals that give rise to the N_{in}–C_{out} orientation of HN and the N_{out}–C_{in} orientation of M2 (Parks and Lamb, 1991, 1993). Positively charged residues in the cytoplasmic tail are important for directing HN into the N_{in}–C_{out} orientation, as this topology can be inverted to the M2-like form of N_{out}–C_{in} through changes from positive to negatively charged amino acids (Parks and Lamb, 1991). Likewise, the sequence, hydrophobicity and length of the HN signal-anchor domain are important for membrane insertion, topology and transport (Parks and Pohlmann, 1995; Parks, 1996). Results from deletion and insertion mutants support a model whereby the helical structure of the HN signal-anchor domain is important for insertion into membranes and for subsequent oligomerization and transport to the cell surface. Residues which map to one face of the predicted transmembrane helix were shown by mutagenesis to be critical determinants of assembly of dimers into a tetramer (Parks and Pohlman, 1995).

To be competent for transport from the ER, integral membrane proteins must oligomerize and fold properly. Both of these basic principles were established using the SV5 HN protein as a model system (Ng *et al.*, 1989). Thus, HN was found to fold through a series of intermediates very shortly after synthesis ($t_{1/2}$ of 25–30 min), but before transport from the ER to the Golgi apparatus (Ng *et al.*, 1989). This folding is aided by a transient, but specific interaction with the cellular chaperone GRP78-BiP, which helps to fold HN into an assembly-competent oligomeric structure. This requirement for GRP78-BiP is in line with previous findings that infection with SV5 upregulates ER stress responses (Peluso *et al.*, 1978), and that the flux of SV5 HN protein transport through the exocytic pathway is sufficient to induce GRP78-BiP transcription (Watowich *et al.*, 1991). With a $t_{1/2}$ of ~90 min, transport of HN from the ER to Golgi is relatively slow compared with many other integral membrane proteins. Alterations to any three of the HN domains (ectodomain, signal-anchor, cytoplasmic tail) can result in assembly and therefore, transport-defective proteins that are either retained in the ER or targeted for degradation.

Folding of the SV5 HN ectodomain is aided by carbohydrates that are added shortly after synthesis in the ER. A detailed analysis of the carbohydrate structure and function for SV5 HN has been carried out (Ng *et al.*, 1990). This study revealed that there are four N-linked oligosaccharides on the HN ectodomain (positions 110, 139, 267 and 504) and that two of the four chains are modified during transport through the Golgi apparatus to a complex form that is operationally defined as resistant to digestion with endoglycosidase H. SV5 HN mutants, which cannot be glycosylated at these positions, are often not transported from the ER, form insoluble aggregates and show an extended interaction with GRP78-BiP (Ng *et al.*, 1990).

Once it is transported to the plasma membrane, SV5 HN is rapidly internalized by clathrin-mediated endocytosis where the polypeptide enters the cellular endocytic pathway (Ng *et al.*, 1989). This internalization of HN is not dependent on antibody and the signal within the HN polypeptide differs significantly from those previously described for targeted endocytosis from the plasma membrane. The HN internalization signals have been mapped to within the stalk region located near the junction of the ectodomain and the transmembrane domain (Leser *et al.*, 1999). While the exact function of this process in the SV5 life cycle is unknown, internalization may contribute to evasion of immune responses against the highly immunogenic HN protein or provide a mechanism to control incorporation of HN into budding virions.

Large protein

The 3′ most distal gene in the SV5 genome encodes the large (L) protein. L protein lives up to its name by being a protein comprising 2255 amino acids and a predicted MW of 255,900 (Parks *et al.*, 1992). By analogy to other paramyxoviruses, the SV5 L is the essential catalytic subunit of the viral RNA dependent RNA polymerase. Small amounts of the L protein exist within virus particles so that primary transcription can be initiated immediately upon infection. By sequence alignment, the SV5 L protein shares six highly conserved domains (I-VI) with L proteins from other negative strand RNA viruses. The putative functions of these domains have been explored by mutational analysis for a number of viruses.

The L protein only functions in conjunction with the P protein, which is thought to tether L to the nucleocapsid template. Using a transfection system, a mutational analysis of the SV5 L protein showed that the N-terminal half of the protein contained regions required for binding to the P protein. Unfortunately, further refinement of these regions by mutational analysis showed that formation of a stable L–P protein complex was very sensitive to even minor sequence changes in the N-terminal half of the L polypeptide (Parks, 1994).

Virus replication cycle – RNA replication, transcription and virion assembly

Viral RNA synthesis

Excellent reviews providing a more in-depth foundation of the fundamental aspects of transcription and genome replication of non-segmented negative strand RNA viruses (Banerjee and Barik, 1992) and of paramyxovirus biology (Lamb and Parks, 2007) have been published previously. Here, we focus on features of these processes for SV5 in particular.

The terminal regions of the SV5 genome and antigenome contain important cis-acting sequences that direct polymerase functions. The 3′ end of the genomic RNA contains a short ~55 base non-coding region called the leader (le). The 5′ end of the genome contains a short 31 base trailer (tr) sequence after the L gene polyA site (Parks *et al.*, 1992). The le region directs the polymerase to initiate transcription to produce viral mRNAs as well as RNA replication to produce a full length complementary copy termed the antigenome. By contrast, the complement to the trailer region (tr′) on the antigenome can only direct RNA replication to produce full-length genomes. For SV5, mutational data and work with minigenome replicons support the proposal that the antigenomic promoter is a stronger promoter for RNA replication than is the genomic promoter (Keller *et al.*, 2001). As discussed below, the le region and tr′ are not sufficient to direct RNA synthesis, but instead require additional discontinuous elements located within the NP and L gene sequences, respectively.

All paramyxovirus RNA synthesis is initiated at or very near the 3′ end of the viral genomic or antigenomic RNA. Both the L and P proteins are required to carry out RNA replication and transcription from the genomic RNA promoter (le region) to yield 5′ capped and 3′ polyadenyated mRNAs. By contrast, the polymerase can carry out replication only from the antigenomic RNA promoter (tr′ region). Viral RNA replication differs from viral transcription in several regards. The most significant difference is that during paramyxovirus RNA replication, synthesis of RNA is tightly coupled with NP encapsidation of the nascent RNA chain into an RNAse-resistant form as it emerges from the viral polymerase (reviewed in Lamb and Parks, 2007). The nascent RNA is encapsidated by a soluble form of NP (NP^0), which is thought to be kept in an assembly competent form by complexing with P protein (Precious *et al.*, 1995; Lamb and Parks, 2007). This encapsidation of the nascent chain confers anti-termination properties to the viral polymerase such that transcription signals in the genome (e.g. at the intergenic junctions) are ignored and the polymerase then synthesizes a full length encapsidated replication product.

RNA genome replication

Recent progress using reverse genetic systems has uncovered several novel aspects of the cis-acting sequences that dictate the efficiency of SV5 RNA replication. These include (1) the rule of six; (2) nucleotide sequence at the promoter; and (3) the spacing of promoter elements on the nucleocapsid template. These requirements reflect the principle that an NP-encapsidated viral RNA is the only template, which functions during paramyxovirus RNA replication.

The 'rule of six'

Calain and Roux (1993) were the first to observe that the overall number of nucleotides that compose the paramyxovirus genome can profoundly affect the efficiency of RNA replication. This genome length requirement, called the 'rule of six', dictates that efficient replication of a viral genome will only occur when the total number of nucleotides in the genome is an even multiple of six (Calain and Roux, 1993). During replication, assembly of the viral genome with NP is thought to initiate at the 5′ end of the nascent RNA chain as it emerges from the polymerase complex while RNA-NP assembly proceeds in a 5′ to 3′ direction. As a result, a genome whose length is an even multiple of six nucleotides will be precisely encapsidated by NP, with no unencapsidated nucleotides protruding from the 3′ end of the nucleocapsid. Consistent with this, the length of the SV5 RNA genome is an even multiple of six at 15,246 bases (Parks *et al.*, 1992).

There is variability in the degree to which replication of a particular paramyxovirus genome adheres to the rule of six requirement. Based on extensive mutational analyses of minigenome RNAs, the rule of six is an apparent strict requisite for Sendai virus (SeV), while there is no advantage to RSV genomes having any of the integer genome lengths tested. Murphy and Parks (1997) carried out a systematic analysis of the rule of six requirement for SV5 RNA replication. Using an 852 base (even multiple of six) naturally-occurring copy-back DI RNA as a starting template, SV5 RNA replication was found to be most efficient for 6N-length genomes, but unlike SeV, this is not an absolute requirement. Thus, the stringency of the rule of six for SV5 is intermediate between that found for SeV and RSV. While the basis for these differences in adherence to the rule of six is not known, it is proposed that this could reflect a function of the viral C proteins, which are expressed in the case of SeV (most stringent), but not from the SV5 genome (less stringent).

Two hypotheses have been put forth to explain the rule of six: congruency and nucleotide 'phasing'. The congruency model dictates that efficient RNA replication requires that the 3′ end nucleotides of the NC template be precisely encapsidated with no 'dangling' free bases (Calain and Roux, 1993). Nucleotide phasing is the concept that critical cis-acting promoter sequences can only be recognized when they are in the correct positions within their encapsidating NP monomers (Kolakofsky *et al.*, 1998). Thus, the phase of a particular base in the promoter can range from position #1 to #6 within a hexamer of NP-bound sequences (see Fig. 2.2 for examples). These two hypotheses have been tested by Vulliemoz and Roux (2001) for SeV copyback DI genomes. Extensions to the 3′ end of SeV DI RNAs did not result in a decrease in RNA replication, regardless of whether these extensions were even multiples of six (6N in length) or not. These results argue against the congruency model. In contrast, altering the predicted phase of nucleotides in the antigenomic promoter reduced RNA replication and this was irrespective of the nature of 3′ end extensions. These data with SeV were interpreted as a need for proper positioning of promoter sequences within an NP-bound hexamer.

Nucleotide sequence

Nucleotide sequence in the promoter is a second factor dictating replication efficiency for SV5. The nucleotide sequence in the le RNA at the 3′ terminus of the genome and in the antitrailer RNA at the 3′ terminus of the antigenome are primary determinants of the efficiency of replication to produce SV5 antigenomes and genomes, respectively (Keller *et al.*, 2001; Keller and Parks, 2003). Hints that additional sequences other than the ultimate 3′ terminal region are also required for RNA replication were evident from the structure of SV5 DI RNAs (Murphy and Parks, 1997). These subgenomic RNAs contain various segments from the standard non-defective viral genome, but for all naturally occurring copyback SV5 DI RNAs examined to date the viral RNA retains 100 or more consecutive bases from the 3′ terminus. The basis for this lower limit on the size of the SV5 promoter became evident from extensive mutational analyses of the replication of SV5 minigenome analogues (Murphy *et al.*, 1998; Murphy and Parks, 1999). Deletion and substitution analysis identified two discontinuous sequence-specific elements within the 3′ terminal 90 bases of the SV5 antigenomic promoter (Fig. 2.2). These elements have been termed promoter element I (PrE-I) and II (PrE-II). Similar cis-acting sequences have been identified for Sendai virus (Tapparell *et al.*, 1998). PrE-I was identified as the 18 nucleotides at the 3′ terminus of the genome and antigenome. This region is highly conserved within members of the rubulaviruses (e.g. SV5 and HPIV-2). PrE-II was first identified by sequence analysis (Murphy *et al.*, 1998) and confirmed by mutational analyses as an essential internal promoter element. PrE-II is located ~100 bases from the 3′ end within the coding region of the L gene for the antigenomic promoter and within the 3′ (upstream) non-coding region of the NP gene in the case of the genomic promoter (see Fig. 2.2 for location and sequence). Thus, the lower limit of ~100 3′ terminal bases for DI RNA replication reflects the need for PrE-II bases 73–90 from the 3′ end.

The essential nature of the internal PrE-II for RNA replication is a shared property of SV5 and SeV. However, features of the PrE-II for these two viruses differ in several regards. First, there is a difference in the position of the critical PrE-II from the 3′ end of the viral RNA. As diagrammed in Fig. 2.2, the SV5 internal PrE-II element is located between 73 and 90 bases from the 3′ end of the viral RNA (Murphy *et al.*, 1998). By contrast, the SeV PrE-II element differs in location by being between 79 and 96 bases from the 3′ end (Tapparell *et al.*, 1998). Secondly, the sequence requirements within PrE-II are remarkably simple, but they differ for SV5 and SeV RNA replication. For SV5, saturation mutagenesis (Murphy and Parks, 1999) indicated that PrE-II requirements consisted of a CG motif located in the first two positions of three sequential hexamers of nucleotides (5′-CGAGGU CGGUCA CGAAGG-3′, genomic promoter template sense, underlined nucleotides, Fig. 2.2). By contrast, SeV PrE-II has simpler requirements, consisting of three copies of a 5′-(NNNNNC)-3′ motif. Finally, SeV and SV5 differ in their sensitivity of RNA replication to mutations in PrE-II. SeV replication shows an absolute requirement for all three of the conserved C residues in CRII (Tapparell *et al.*, 1998) SV5 RNA synthesis appears to be somewhat less sensitive to point mutations in the CG motif (Murphy and Parks, 1999).

A more detailed substitution analysis identified Region III (RIII) as a third sequence-dependent element that was required for optimal SV5 RNA replication. RIII is located 51–66 bases from the 3′ of the antigenome (Keller and Parks, 2003). There is no apparent sequence identity among the RIII elements of members of the *Rubulavirus* genus. However, the HPIV-2 antigenomic promoter, containing a different RIII sequence, is recognized by the SV5 polymerase as efficiently as the wild type SV5 promoter (Keller and Parks, 2003). In addition, our data have indicated that RNA replication is dramatically affected only when the majority of RIII is replaced, while smaller substitutions have a lesser effect (Keller and Parks, 2003). RIII also differs from CRI and CRII by the fact that changes in RIII hexamer phasing do not affect replication. Thus, the RIII element may be a functionally redundant signal distributed across bases 51–66 that is not essential, but rather enhances RNA replication. Taken together, a complete mutational analysis of the SV5 antigenomic promoter shows that optimal RNA replication requires three non-continuous sequence-dependent RNA segments (CRI, CRII and RIII) that are separated by sequence independent spacer regions.

Spacing between the 3′-terminal and internal promoter elements

A third feature of the nucleocapsid important for replication is the position of PrE-I and PrE-II relative to each other. Murphy and co-workers (1998) found that deletion of small segments of RNA located between the 3′ terminal PrE-I and the internal PrE-II in the SV5 antigenome resulted in minigenome analogues that were not competent for replication. The loss of RNA replication for these minigenome templates was not due to removal of an important cis-acting segment, since substitution of non-viral RNA in place of the removed segment restored replication. Likewise this was not due to disruption of the 'rule of six' since insertions or deletions of six bases in the RNA segment separating CRI from CRII also lead to dramatic reductions in RNA replication. It has been proposed that the sensitivity of SV5 RNA replication to changes in the length of the region between PrE-I and PrE-II reflects a requirement for these two RNA segments to align to the same face of the nucleocapsid template (Murphy *et al.*, 1998).

A model for the SV5 promoter

A role for PrE-I and PrE-II elements in SV5 RNA replication has been proposed (Murphy *et al.*, 1998) based on the structure of the SeV NC, which was previously derived from electron micrograph images assuming that SV5 and SeV would have similar overall structures. In its most tightly coiled form, each turn of the left-handed helical SeV NC is thought to consist of 13 NP molecules on average, with each NP binding six nucleotides for a total of 78 nucleotides per turn. Assuming that the SV5 NC has similar properties to that of SeV, we have proposed a hypothetical model for the antigenomic promoter, which predicts that PrE-I and PrE-II are aligned to the same face of the helical NC (Murphy *et al.*, 1998). As shown in Fig. 2.2, the nucleotides in PrE-I and PrE-II are divided into hexamers with each within one NP molecule. While these two elements are distant in linear space, our model indicates that they would align to the same face of the helical nucleocapsid (Fig. 2.2).

Two hypotheses have been proposed for the function of PrE-I and PrE-II in paramyxovirus RNA synthesis (Murphy and Parks, 1999): a nucleation site for assembly and a site for polymerase binding. In the nucleation model, PrE-I and -II serve as a site for initial encapsidation of the nascent RNA by NP as it emerges from the polymerase. Several observations are inconsistent with the nucleation proposal. First, the current model for the differential synthesis of mRNA versus RNA replication from the genomic promoter proposes that NP binding to the nascent chain provides an anti-termination signal to the viral polymerase at the le–NP junction. This site is located ~55 bases from the 3′ end of the genome. Presumably the 'decision' to encapsidate nascent RNA (replication mode) or terminate le' synthesis (mRNA mode) would be made prior to the polymerase synthesizing nascent RNA that contains PrE-II, which is located 73–96 bases from the 3′ end of the genome (within the NP gene). Secondly, it is not clear how changes in the spacing between PrE-I and PrE-II could act in the nascent chain. The alternative hypothesis that PrE-I and PrE-II may act as a site for polymerase binding is supported by work with Sendai virus minigenomes where the polymerase can recognize PrE-I and PrE-II when positioned in an internal site (Vulliemoz and Roux, 2001).

Viral transcription

SV5 employs the same stop–start transcription mechanism that is shared among the non-segmented negative strand RNA viruses (Banerjee and Barik, 1992). As diagrammed in Fig. 2.4 for the SV5 M–F junction, the functions of L-P polymerase complex are directed by cis-acting sequences at the junction. The gene end includes a U tract directing polyadenylation to add more that 200 adenosine residues (Rassa *et al.*, 2000). The polymerase can move across the intergenic region to reinitiate at the downstream gene start site. Alternatively, failure to reinitiate can contribute to the 3′–5′ gradient of transcription. Finally, read-through transcription can occur, whereby the polymerase ignores the gene end signals to produce an M–F dicistronic mRNA. In this case, the polymerase remains bound to the template and can synthesize downstream mRNAs. As shown in Fig. 2.5, the percentage of a given SV5 mRNA that is found as a read-through transcription product ranges from 0 (F–SH) to ~10–15% (NP–P, P–M and SH–HN) of the time as measured by Northern blotting of accumulated mRNAs (Rassa and Parks, 1998). The exception to this occurs with mRNA generated from the M–F junction, which has read-through percentages at about 40% (Rassa and Parks, 1998). This is dependent on both cell type and the time post infection that measurements are made (~50% early, ~20% late). The biological consequences of changes in M-F read-through were evident from a mutant SV5 encoding an altered intergenic region. This M–F

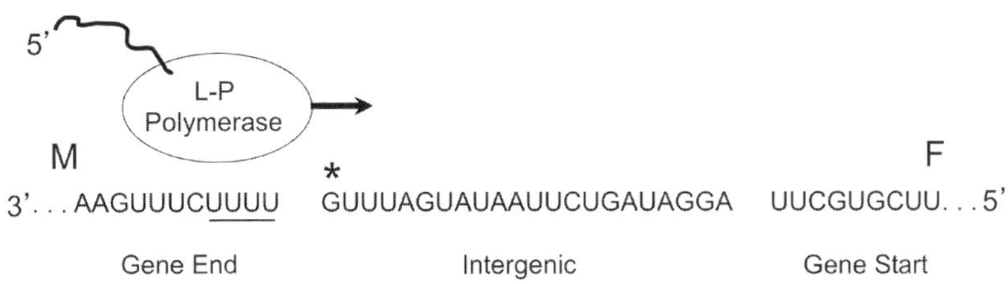

Figure 2.4 Structure of the Simian virus 5 (SV5) M-F gene junction. The viral polymerase is denoted as a circle consisting of the P/L complex with the 5′ end of the nascent mRNA emerging to the left as transcription occurs from left to right across the junction. The four residue U tract for the M gene end is underlined and the critical intergenic G residue that contributes to efficient transcription termination is indicated with a star.

Consensus End	Variable Intergenic	Consensus Start	% Readthrough
...AAANUUCU$_{4-7}$		UCCGGGCUUG....	
NP	A	P	11
P	GCUAAUUGCUAUUUA	M	12
M	GUUUAGUAUAAUUCUGAUAGGA	F	42
F	GCUA	SH	0
SH	A	HN	16
HN	GGUUCUCUUGUUA	L	nd

Figure 2.5 The gene end, intergenic and gene start sequences for each Simian virus 5 (SV5) gene are listed as genomic RNA (3′ to 5′). Consensus sequences are estimates of the most frequent occurrence of a given nucleotide in each position. The percentage read-through (% RT) at each junction is listed from the data in Rassa and Parks (1998).

junction mutant overexpressed L protein and grew faster and to higher titres than WT SV5 (Parks *et al.*, 2001). These data were interpreted to indicate that read-through transcription increases access of the polymerase to 3′ distal genes, which would otherwise be limited owing to efficient stop–start transcription. This mechanism may be important for SV5 to fine tune gene expression for optimal growth.

The SV5 gene junctions are highly diverse, a property shared with the related rubulaviruses human parainfluenza virus type 2, SV41 and Mumps virus (Rassa and Parks, 1998). As shown in Fig. 2.5, this variability includes (1) sequence differences and variable U tract at the gene end segment; (2) differences in length and sequence of the intergenic region; and (3) diverse gene start sequences. The finding of a variable gene end U tract raised the hypothesis that shorter U stretches would direct shorter poly-A tails. This attractive hypothesis was disproven when measurements of the length of the poly-A tail on the M, NP and

SH mRNAs were found to be equivalent, despite having different lengths of the U tract at the M (four Us), NP (seven Us) and SH (six Us) gene ends (Rassa *et al.*, 2000).

An alternative hypothesis for the role of short U tracts in the SV5 genome was that this contributed to read-through transcription, consistent with the finding of high read-through at the M–F junction, which has gene end of only four Us. Extensive mutagenesis studies using a transcription-competent minigenome showed that the gene end U tract worked in conjuction with intergenic sequences to direct polymerase termination and reinitiation at the next gene (Rassa and Parks, 1999). For example, when gene end U tracts were short (e.g. four residues) efficient transcription termination required that the next intergenic residue immediately downstream needed to be a G nucleotide (see starred residue in Fig. 2.4). By contrast, longer U tracts did not require any particular downstream nucleotide for efficient termination and reinitiation. Thus, the signals for termination and reinitiation versus read-through at the SV5 gene junctions consist of specific sequence combinations from the gene end and the intergenic regions (Rassa and Parks, 1999).

The length and sequence of the SV5 intergenic regions are highly variable (see Fig. 2.5). There does not appear to be a requirement for the length of the SV5 intergenic region, since insertions or deletions in this region do not affect polymerase functions (Rassa and Parks, 1999). Interestingly, there is a lower limit to intergenic length that will allow reinitiation at the next downstream gene. This became evident from the analysis of transcription from minigenomes encoding the SH–HN junction, which has a single intergenic A residue (Fig. 2.5). Mutational analyses indicated that efficient reinitiation at the HN site required at least six bases from the end of SH (C residue) to the start of HN (U residue). Shorter lengths did not effect SH termination, but dramatically inhibited reinitiation at SH. These results were interpreted to indicate that the U tract, which directs polyadenylation, could also serve as a spacer region for directing polymerase functions (Rassa *et al.*, 2000). Whether this limit of six nucleotides reflects a constraint from the six bases bound by NP and a variation of the 'rule of six' has not been resolved.

Virion assembly

Assembly of SV5 particles occur by mechanisms that are shared with other negative strand RNA viruses (reviewed in Schmidt and Lamb, 2004). SV5 particles bud from sites where viral proteins have assembled at the host cell plasma membranes. This includes a coordinated assembly of the nucleocapsid with glycoproteins at the surface of the cell, where the M protein serves as an assembly factor. As described above in the section on the M protein, SV5 M encodes a unique 'late domain' motif sequence FPIV, which is essential for the budding process (Schmitt *et al.*, 2005). Many paramyxovirus M proteins have an inherent budding property and can form VLPs in the absence of any other viral component. By contrast, studies with SV5 VLPs have demonstrated that the minimal requirement for efficient budding is expression of M, NP and at least one SV5 glycoprotein (F or HN; Schmitt *et al.*, 2002). Similar studies also confirmed the importance of the viral glycoprotein cytoplasmic tails in budding (Waning *et al.*, 2002). Thus, SV5 has unique requirements for NP and for a redundant contribution of a cytoplasmic tail from one of the glycoproteins.

SV5 was one of the first paramyxoviruses to show the pseudotype assembly paradox. In this phenomenon, budding virus particles appear to selectively exclude the majority of host cell membrane proteins implying a highly specific mechanism for assembly of virus glycoproteins with the budding virion particle. Paradoxically however, glycoproteins from other viruses are often efficiently recruited into budding virions during co-infections, implying a lack of specificity in assembly. This was illustrated in the 1970s by Choppin and Compans (1970) using cells that were co-infected with SV5 and vesicular stomatitis virus (VSV). This paradox remains an open and active area of research, particularly in light of the recent findings from studies with virus-like particles (VLP) and SV5 mutants. Assembly of SV5 particles require NP, which differs significantly from other paramyxoviruses in which the M protein is often sufficient to drive budding of VLPs (Schmitt *et al.*, 2002). The HN and F cytoplasmic tails are required for efficient assembly of VLPs and incorporation of glycoproteins into particles (Waning *et al.*, 2002; Schmitt *et al.*, 2004). This probably reflects interactions of the cytoplasmic tails with the M protein, a proposal that is supported by the finding that cells infected with SV5 recombinants harbouring an F protein cytoplasmic tail mutant show a different localization of the viral M protein compared with WT SV5-infected cells (Waning *et al.*, 2002).

Genetics and reverse genetics

The relationship between paramyxovirus RNA sequences and the polymerase functions that they control is particularly complex, since the cis-acting signals are only recognized by the viral polymerase when they are in the context of the nucleocapsid structure. This feature of the paramyxovirus genome has proven to be a major hurdle in developing genetic systems; whereby, cDNA copies of the viral genome could be manipulated and the consequence of these alterations assayed. However, recently developed reverse genetic systems, which reconstitute RNA replication from cDNA clones or allow the recovery of infectious virus from cDNA has revolutionized the field.

Reverse genetic systems have been developed for SV5. Initially, this involved a minigenome system based on the vaccinia virus-T7 polymerase system (Pattnaik and Wertz, 1990). This system was used for the analysis of cis-acting sequences that directed viral RNA replication and transcription (Murphy and Parks, 1997; Rassa and Parks, 1998). This system was invaluable as a means to demonstrate the biological activity of cDNA-derived 'support' proteins L, P and NP.

Subsequently, the minigenome system was expanded to recover infectious virus from a full length cDNA clone of the SV5 W3A strain (He *et al.*, 1997). The use of vaccinia virus to express T7 polymerase has been replaced by a BHK-derived cell line that constitutively expresses T7 RNA polymerase which has improved the efficiency of recovering SV5 from transfected plasmids (Thompkins *et al.*, 2007). The reverse genetic system has been used to generate a number of SV5 mutants (Young *et al.*, 2001; He *et al.*, 2002; Manuse and Parks, 2010), as well as vectors for expression of foreign proteins (Parks, *et al.*, 2002; Thompkins *et al.*, 2007; Arimilli *et al.*, 2008).

Immune responses and pathogenesis

Innate immune responses to RNA virus infections

RNA virus infections are detected by host cells through a range of RNA-sensing pattern recognition receptors. Fig. 2.6 illustrates pathways that have been analysed for SV5 infections of human cells. TLR3 is localized predominantly to endosomes and recognizes double-stranded RNA (dsRNA; pair of wavy lines, Fig. 2.6) to signal through TIR domain-containing adaptor protein TRIF. This results in activation of the transcription factors IRF-3, AP-1 and NFκB, which can upregulate synthesis of interferon (IFN)-beta or proinflammatory cytokines. Similarly, the two cytoplasmic RNA helicase proteins RIG-I (retinoic acid-inducible gene-1) and MDA-5 (melanoma differentiation-associated gene-5) sense dsRNA in combination with other more specific signals to activate the mitochondria-associated factor MAVS (mitochondrial antiviral signalling). This, in turn, results in induced synthesis of IFN-β and proinflammatory cytokines such as interleukin 6. Activation of these same three host transcription factors may also contribute to apoptosis, but the relationship with each signalling pathway and the influence of different cell types is not clear at this point. Finally, the RNA-activated protein kinase (PKR) can also sense viral dsRNA. Distinct from the other three receptors, activation of PKR occurs through autophosphorylation, which can result in shut-off of host and viral protein synthesis owing to phosphorylation of the translation factor eIF2-α.

Induction of cytokines, interferons and host-cell responses

SV5 has a remarkable ability to establish highly productive long term infections in many tissue culture cell lines with only minimal activation of host cell antiviral responses (Choppin, 1964). Thus, epithelial cells that are infected with SV5 show very low level secretion of type

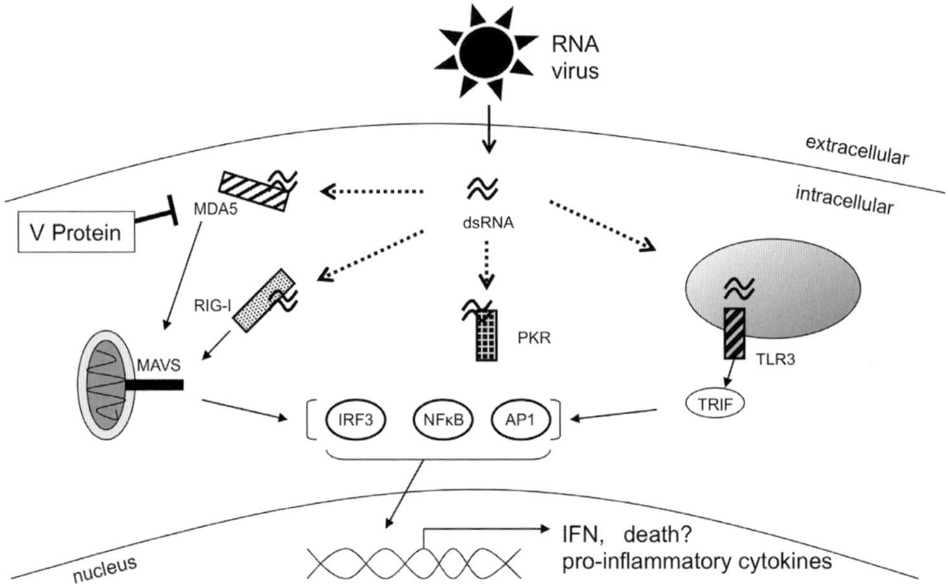

Figure 2.6 RNA-sensing pattern recognition receptors. RNA viruses are sensed by at least four intracellular receptors that recognize various forms of double-stranded RNA, including MDA-5, RIG-I, PKR and TLR3. MDA5 and RIG-I signal interferon (IFN) synthesis through the mitochondrial antiviral signalling molecules MAVS. This results in activation of latent transcription factors IRF-3, NFκB, and AP1. TLR3 signals through the adaptor molecule TRIF to activate similar transcription factors. PKR senses dsRNA to inhibit protein synthesis, although there are controversial data suggesting that PKR can also contribute to cytokine induction. Simian virus 5 (SV5) V protein inhibits MDA-5 activation through direct binding, whereas the other RNA sensing molecules are not directly inhibited by SV5.

I IFN or proinflammatory cytokines, such as IL-6 (Didcock *et al.*, 1999a; He *et al.*, 2002; Poole *et al.*, 2002; Wansley and Parks, 2002). This has also been extended to primary cultures of human epithelial cells (Wansley *et al.*, 2005) as well as monocyte-derived human dendritic cells (Arimilli *et al.*, 2006). Induction of IFN and antiviral genes is not absolute, as evidenced by the finding that SV5 grows better in cell lines engineered to express the V protein (Young *et al.*, 2003a) and by microarray data from SV5 infected IFN-competent cells showing elevated levels of mRNAs for some IFN-stimulated genes (Wansley *et al.*, 2005). Nonetheless, it is remarkable that SV5 can maintain a very high level of virus replication and expression of viral proteins without overt activation of host-cell responses.

A role for the SV5 V protein in blocking cytokine induction has been extensively studied by Randall and colleagues. A major finding was that V protein expressed in transfected cells was able to block induction of the IFN-beta promoter by exogenously-added dsRNA (Poole *et al.*, 2002; Andrejeva *et al.*, 2004). This was consistent with results from mutant SV5 in which the V and P proteins were altered by point mutations (Wansley and Parks, 2002) or by removal of the V protein C-terminal cysteine-rich domain (He *et al.*, 2002). A key mechanistic discovery was the finding that V protein inhibits MDA-5 activation by direct binding through the C-terminal cysteine-rich domain (Andrejeva *et al.*, 2004; Childs *et al.*, 2007, 2009). This binding of MDA-5 inhibits dsRNA-mediated activation of MAVS, NFκB and IRF-3 which would normally lead to cytokine induction.

The V protein cys-rich domain is sufficient to block MDA-5 activation by exogenously added dsRNA. This appeared seemingly at odds with the findings that SV5 variants that encode naturally-occurring substitutions in the N-terminal shared P/V region, which is outside of the cys-rich domain were potent inducer of cytokines (Wansley and Parks, 2002; Dillon *et al.*, 2005). A hypothesis for this discrepancy emerged from two findings. First, V protein blocks MDA-5, but cannot block RIG-I activation (Childs *et al.* 2007). Secondly, cytokine-inducing SV5 mutants with a V containing an intact cys-rich domain activate host-cell responses through intracellular dsRNA and RIG-I (Manuse and Parks, 2009, 2010). Thus, while the cys-rich domain is capable of blocking MDA-5, the V protein mutants with an altered N-terminal domain overexpress viral RNA and activate RIG-I, which cannot be blocked by V protein. These findings raise the question of why SV5 targets inhibition of MDA-5, but not RIG-I, and also why a WT SV5 infection does not activate RIG-I pathways despite a very robust and high level of viral RNA synthesis. One possibility is that the level of viral RNA synthesis is less important for host-cell responses than is the type of RNA produced. In support of this, activation of host-cell responses by SV5 mutants that overexpress viral RNA can be suppressed by expression of a protein which binds dsRNA (Gainey *et al.*, 2008a; Manuse and Parks, 2009). Unexpectedly, expression of the WT P protein can suppress activation of cytokines by an SV5 P/V mutant (Dillon and Parks, 2007). This finding suggests that an important aspect of SV5 being a poor cytokine inducer is the high fidelity of RNA synthesis and that the P protein may provide quality to RNA synthesis by the WT polymerase complex.

The SV5 V protein is also reported to act as a decoy substrate for the TANK-binding kinase 1 (TBK-1)/inhibitor of kappaB kinase epsilon (IKKe) and it was proposed that SV5 V protein could act as a mimic for IRF-3 to limit its activation (Lu *et al.*, 2008). This would predict that SV5 would have the potential to inhibit a large number of innate signalling pathways including TLR3, TLR4, RIG-I, and other activators or IRFs that use TBK1. While intriguing, this would be inconsistent with previous work that shows that SV5 is unable to block signalling through TLR3 and TLR4 (Arimilli *et al.*, 2007; Manuse and Parks, 2010) and that RIG-I also is not blocked in SV5-infected cells (Manuse and Parks, 2009). This may reflect targeting of TBK1 pathways in certain cell types.

SV5 infection does not activate the dsRNA-sensing toll-like receptor 3 (TLR3; Manuse and Parks, 2010). This is not due to an active block in TLR3, since cytokines are induced from SV5 infected human cells by addition of exogenous dsRNA. Interestingly, in SV5 infected cells, activation of TLR3 by added dsRNA leads to higher expressing of RIG-I. Thus, SV5 does not block TLR3 signalling pathways. This dsRNA-induced TLR3-dependent increase in RIG-I in turn sensitizes cells to respond more vigorously to SV5 infection leading to higher levels of cytokine induction through RIG-I pathways (Manuse and Parks, 2010). In a separate study, we have recently found that WT SV5 is a potent activator of IFN-alpha secretion in plasmacytoid dendritic cells through pathways that are independent of virus replication (Manuse *et al.*, 2010). This induction was blocked by inhibitors of endosomal pH, TLR7 signalling and autophagy pathways. Taken together, these data indicate that the level of cytokine induction by SV5 infection of human cells is dictated by at least three parameters: (1) the level and type of viral RNA synthesized (e.g. dsRNA by SV5 mutants), (2) the pathways that are blocked or not targeted by SV5 (e.g. RIG-I active vs. MDA-5 inhibited), and (3) level and type of pattern recognition receptor that is expressed in different cell types (epithelial cell vs. plasmacytoid dendritic cell).

Viral infections can trigger a shutdown of protein synthesis through the activation of the double-stranded RNA (dsRNA)-dependent protein kinase (PKR), a serine/threonine protein kinase that can be induced by IFN. PKR is also constitutively expressed in mammalian tissues in a latent state that can rapidly respond to virus infection. PKR activation involves homodimerization and autophosphorylation, which can lead to phosphorylation of the translation initiation factor eIF2-alpha and an inhibition of translation. It has long been known that SV5 can establish a prolonged and robust replication cycle in epithelial cells without a global shut down in host or viral protein synthesis (Choppin, 1964; Peluso *et al.*, 1977, 1978; Gainey *et al.*, 2008a). This phenotype raises the interesting question of how SV5 can replicate to such high levels (10^8 PFU/ml in 18 h), but can also avoid activation of PKR.

Gainey and coworkers (2008a) examined the status of PKR phosphorylation during infections with WT SV5 or with an engineered mutant that contained six amino acid substitutions in the shared P/V region derived from the CPI⁻ strain of SV5. As predicted from previous work, WT SV5 was a poor inducer of PKR activation and overall translation was virtually unaffected by WT SV5. In fact, it has long been known that SV5 infection actually induces host cell protein synthesis, specifically the synthesis of heat shock proteins (Peluso *et al.*, 1978). In contrast to WT SV5, the P/V mutant that overexpresses viral RNA activated PKR and there was a global inhibition of both viral and cellular translation (Gainey *et al.*, 2008a). Using siRNA knockdown or cell lines engineered to express either the influenza A virus NS1 protein or the reovirus dsRNA-binding protein sigma 3, it was shown that PKR is a key cellular factor involved in the differential shut off of translation in cells infected with WT and the P/V mutant. Remarkably, expression of either the WT P protein or the WT V protein *in trans* corrected the PKR activation and translational defect seen with the P/V mutant. This last result has been misinterpreted by some investigators as evidence that P and V proteins directly inhibit PKR activation. A more likely mechanism proposed by Gainey *et al.*, (2008a) is that the WT P and V proteins act to control viral RNA synthesis which is de-regulated with the P/V mutant, and this indirectly prevents PKR activation. Thus, Gainey *et al.* (2008a) proposed that WT SV5 avoids activation of PKR through P- and V-mediated control of either the level or the type of viral RNA synthesis.

Induction of apoptosis

SV5 infection of most cell types is largely non-cytopathic. Choppin (1964) first described this property of SV5 infections of tissue culture cells 45 years ago. This property is not limited to tissue culture cell lines, but is also seen during long term infections of primary epithelial cells (Wansley *et al.*, 2005). Apoptosis has been reported in SV5-infected epithelial and fibroblast cells only under conditions of mutant viruses that have changes in the P/V gene (He *et al.*, 2002; Wansley and Parks, 2002) or the SH gene (He *et al.*, 2001). In the case of an SV5 V protein mutant, it has been proposed that the V protein blocks apoptosis (Sun *et al.*, 2004). However, this was not directly tested in this study. This is an example of how a phenotype such as apoptosis can be evident when there are alterations in a multifunctional protein such as V protein, but this could be due to an indirect effect such as altering the ability of V to suppress RNA synthesis or to counteract another antiviral pathway (e.g. interferon) that is indirectly connected to apoptosis. Thus, apoptosis can be induced by a virus containing mutations that affect the level, kinetics and type of viral RNA synthesized during an infection.

Support for a role of aberrant viral RNA synthesis in apoptosis comes from an SV5 mutant that harbours two nucleotide substitutions in the genomic promoter PrE-I (see stars in Fig. 2.2), but importantly expresses a functional V protein (Manuse and Parks, 2009). This leader mutant induced a low level of apoptosis at late times post infection which correlated with overexpression of viral mRNAs. Interestingly, IRF-3 nuclear translocation and cytokine secretion induced by the RNA synthesis mutants (leader and P/V mutants) was reduced by expression of a dsRNA-binding protein, but this did not affect apoptosis induction. This result suggests that these two host-cell responses were either through different pathways or that the threshold for activation was lower for apoptosis than cytokine synthesis.

In contrast to the non-cytopathic nature of infections of human epithelial and fibroblast cells, human myeloid dendritic cells (mDC) show massive apoptosis in response to SV5 infection (Arimilli *et al.*, 2006). DC are important antigen presenting cells that link innate and adaptive immunity. Thus, both epithelial cells and mDC are highly permissive to SV5, but only mDC respond to infection by undergoing massive apoptosis. SV5 infected mDC are poor antigen presenting cells (Arimilli *et al.*, 2006), but this is independent of the effect of virus-induced apoptosis. SV5-induced mDC apoptosis correlates with levels of caspase-3. This is evident from studies showing that when SV5 infected mDC are exposed to exogenously added TLR2 or TLR4 agonists, the DC undergo maturation events and levels of caspase-3 and apoptosis are decreased (Arimilli

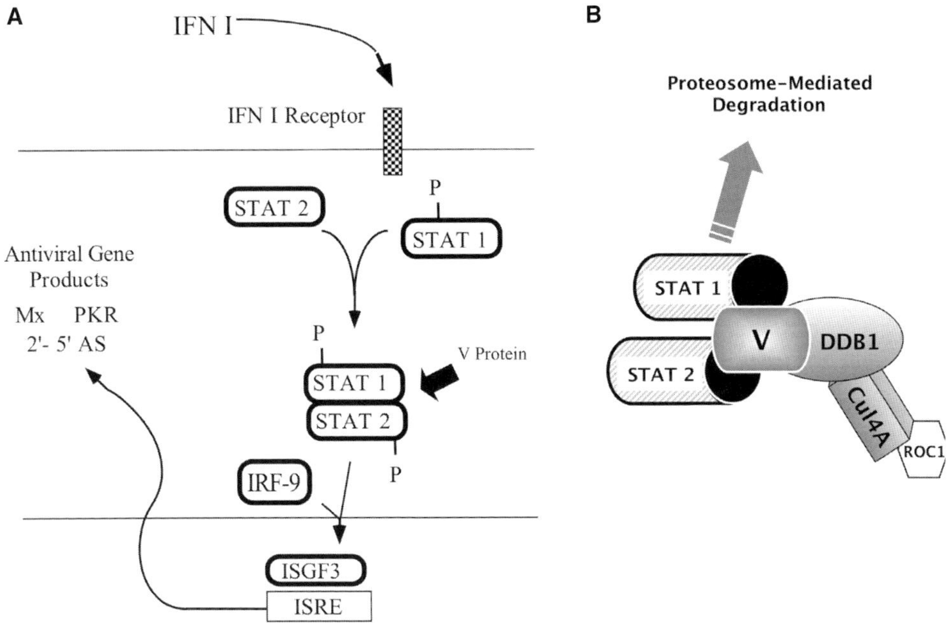

Figure 2.7 Simian virus 5 (SV5) blocks signalling through the IFN receptor. (A) Schematic diagram of interferon (IFN) signalling. IFN binding to the extracellular domain of the receptor results in dimerization and phosphorylation of STAT1 and STAT2, association with IRF9 and translocation of the complex to the nucleus to activate transcription of genes containing an Interferon Stimulated Response Element (ISRE). Examples of IFN-induced gene products are listed on the left. (B) Components of the SV5 V degradation complex. The SV5 V protein targets STAT1 for proteosomal degradation through assembly of a complex containing STAT1, STAT2, DDB1 and Cul4A. The role of ROC1 is not clear. Modified from Horvath (2004) and Ramachandran and Horvath (2009).

et al., 2007). Together, these data indicate that the ability of SV5 to avoid inducing apoptosis is highly dependent on cell type, and that SV5 can block antigen presentation through mechanisms independent of apoptosis.

Inhibition of interferon signalling

Interferons (IFN) are potent antiviral cytokines that can induce a wide range of cellular genes that can inhibit virus replication. As shown in Fig. 2.7A, binding of type I IFN to the extracellular domain of the IFN-I receptor results in heterodimerization of STAT1 (signal transducer and activator of transcription 1) and STAT2, two latent cytoplasmic transcription factors (reviewed in Horvath, 2004). These proteins are phosphorylated by receptor-associated kinases Jak1 and Tyk2. Phosphorylated STAT1 and STAT2 dimers assemble with interferon regulatory factor-9 (IRF-9) to create interferon stimulatory gene factor 3 (ISGF3), which can activate a wide range of cellular genes that are under control of an interferon stimulatory response element (ISRE). Many of these IFN-induced genes have antiviral activity, including Mx, PKR and 2′–5′ OAS (oligoA synthetase). As expected, paramyxoviruses have evolved a range of mechanisms to inhibit IFN signalling at different steps in the pathways (reviewed in Ramachandran and Horvath, 2009).

Randall and colleagues were the first to make the remarkable discovery that SV5 blocks IFN signalling by targeting STAT1 for proteolytic degradation (Didcock *et al.*, 1999b; Young *et al.*, 2000). The SV5 V protein is necessary and sufficient for directing degradation of STAT1, while STAT2 and other factors in the pathway appear to be unaffected. Thus, SV5 infected cells are resistant to signalling in response to both type I and type II IFN, since signalling in response to these two cytokines requires assembly of STAT1 in hetero- and homodimeric forms, respectively.

There has been intense interest in the mechanism by which V protein can redirect cellular machinery to add ubiquitin to STAT1 and target it for degradation. A model for the V-mediated STAT1 degradation complex is shown in Fig. 2.7B and is detailed elsewhere (Andrejeva *et al.*, 2002; Precious *et al.*, 2005b; Ramachandran and Horvath, 2009). The identified cellular components in the SV5 V-dependent degradation complex (VDC; Ulane and Horvath, 2002) include both STAT1 and STAT2, but also the UV-damaged DNA binding protein (DDB1), Cullin 4a and the regulator of cullins 1 (ROC1). DDB1 and Cullin 4a are components of a cellular E3 ubiquitin ligase complex that directs proteosome-mediated degradation of target proteins contained in the complex. DDB1 and Cullin 4a are essential for V-mediated targeting of STAT1. While ROC1 co-purifies with the SV5 VDC, it is thought to be non-essential (Ulane *et al.*, 2005). V-mediated degradation of STAT1 is very efficient, and can be driven by the V protein associated with the incoming virion (Precious *et al.*, 2007). In this regard, it has been proposed that SV5 can persist in cells in cytoplasmic bodies while the antiviral state is disrupted owing to V protein actions (Carlos *et al.*, 2009). A careful analysis of the relative levels of each of the cellular components in the VDC has led to the proposal that SV5 V protein acts catalytically to dismantle the antiviral state of cells (Precious *et al.*, 2007).

Degradation of STAT1 does not require IFN signalling and as such is independent of many components of the IFN signalling pathways, including the IFN receptor, the kinases Jak1 and Tyk2, and IRF-9 (Parisien *et al.*, 2002b). Remarkably, although SV5 infection results in degradation of STAT1 and not STAT2, V-mediated STAT1 degradation requires STAT2 (Parisien *et al.*, 2002b; Precious *et al.*, 2005b). This requirement probably reflects the need to bind to STAT2, which is linked to STAT1 through the DDB1 component of the SV5 VDC. As described below, this requirement is thought to contribute to host range restrictions in mouse cells.

The cysteine-rich C-terminal zinc-binding domain is necessary, but not sufficient for V protein to target STAT1 for degradation (Andrejeva *et al.*, 2002; He *et al.*, 2002). The N-terminal P/V region of V protein also contributes to this function (Andrejeva *et al.*, 2002). This is evident from the naturally-occurring CPI$^-$ strain of SV5, which is defective in inducing STAT1 degradation and blocking type I IFN signalling (Chatziandreou *et al.*, 2002). Mutational analyses have identified amino acid differences in the P/V region between WT SV5 and CPI$^-$ that are responsible for this loss of function in targeting STAT1 for degradation (Chatziandreou *et al.*, 2002). Wansley and Parks (2002) engineered an SV5 mutant (SV5-P/V-CPI$^-$) to encode these same six CPI$^-$ P/V substitutions in the background of the WT genome. These P/V gene substitutions converted WT SV5 into a mutant that failed to target STAT1 for degradation as expected. However, the chimeric virus was also found to be a potent inducer of IFN and proinflammatory cytokines (Wansley and Parks, 2002). The most striking contrast between P/V-CPI$^-$ and WT SV5 was seen in their activation of a cytopathic effect (CPE), since the P/V-CPI$^-$ mutant induced extensive cell death by apoptosis that is not seen with WT SV5 (see above).

While SV5-mediated degradation of STAT1 is very efficient in cells from a wide range of animals (Precious *et al.*, 2005a), this does not apply to infections of murine cells (Young *et al.*, 2001). In SV5-infected mouse cells, STAT1 remains intact and activated IFN signalling pathways are capable of inducing an antiviral state. Murine and human STAT1 are very closely related at the amino acid level, but STAT2 differs significantly between these two species. Parisien *et al.* (2002a) showed that SV5 was capable of blocking IFN signalling and V protein could target STAT1 degradation in mouse cells that expressed human STAT2. Under these conditions of human STAT2 expression, SV5 grew in mouse cells to higher titres. This concept was carried further using a transgenic mouse system. Mice were engineered to constitutively express human STAT2 ubiquitously (Krause *et al.*, 2008). Following intranasal delivery, SV5 replication and viral load were higher in the lungs of hSTAT2-transgenic mice than in control mice. Interestingly, SV5 infected hSTAT2-transgenic mice showed higher levels of proinflammatory cytokines and other host antiviral responses compared with WT mice. This indicates that the ability to block IFN signalling is only one aspect of innate immunity for these viruses. Taken together, these elegant studies indicate that STAT2 and not STAT1 is an intracellular determinant of SV5 host range restriction.

As the counterpart to the host cell protein STAT2, the SV5 V protein can also be a determinant of host range restriction. This is evident from the finding of SV5 variants that were isolated from mouse fibroblast cells persistently infected with WT SV5 (Young *et al.*, 2001). An SV5 variant with an altered P/V gene (among other changes) was found to have a single asparagine-to-aspartate change at residue 100 of the P and V proteins. This change was sufficient to confer on the V protein the ability to target murine STAT1 for degradation and block IFN signalling in transfected murine cells (Young *et al.*, 2001). As expected, a recombinant SV5 harbouring this N100D substitution grew to higher titres in mouse cells in culture. Interestingly, the effect of this N100D mutation on virus growth in whole animals was not evident. Capraro *et al.* (2008) showed that mice infected intranasally with the N100D mutant and WT SV5 viruses showed nearly identical viral loads over time in the lung and nasal tissue. Similarly, antibody responses elicited in mice did not differ between these two viruses despite differing in their ability to disrupt IFN signalling (Capraro *et al.*, 2008). Thus, there are probably additional factors that contribute to restriction of SV5 growth in the respiratory tract of mice. In contrast to mice, analysis of WT and P/V mutant SV5 in ferrets revealed profound differences in growth and antibody responses (Capraro *et al.*, 2008), again supporting the idea of the P/V gene products being determinants of host tropism and restriction.

Complement and SV5

The complement system is an innate immune system comprising both serum and cell-associated components that serves to respond to pathogens, but also to link to adaptive immune responses. It has been previously shown that complement neutralization of mumps virus (MuV) proceeds through the alternative pathway and this was also confirmed for SV5 as well (Hirsch *et al.*, 1986; Johnson *et al.*, 2008). The extent to which the alternative pathway was activated by SV5 and MuV was shown to be inversely related to sialic acid concentrations on either the particle or infected cells owing to the presence of HN neuraminidase activity (McSharry *et al.*, 1981; Hirsch *et al.*, 1986).

Using sera from a range of human donors, it was determined that SV5 was neutralized only in the presence of active complement pathways that were strictly dependent on C3 (Johnson *et al.*, 2008). Only about 10% of donors showed *in vitro* neutralization that was independent of complement. Strikingly, the outcome of complement activation differed between the closely related SV5 and MuV, with SV5 neutralization involving formation of massive aggregates and MuV neutralization including virion lysis. This may reflect differences in the potency of activation, the lipid and protein composition of SV5 versus Mumps virus or the inclusion of host cell inhibitors of complement pathways.

SV5 is sensitive to neutralization by the complement pathways, but also employs mechanisms to counteract or delay complement-mediated neutralization. This was evident by the finding that SV5 incorporates membrane-bound host cell regulators of complement into the virion envelope during budding (Johnson *et al.*, 2009). The host cell regulator CD46 acts as a cofactor to recruit Factor I in the inactivation of C3b. Thus, SV5 particles derived from the appropriate cell lines contains CD46, which is capable of promoting the inactivation of the C3b-dependent pathway and delaying neutralization. A number of other regulators of complement activation can be found associated with purified SV5 particles from tissue culture cells (CD55, CD59, J. Johnson, unpublished). These findings raise the important question of what role these host cell proteins play *in vivo* and their contribution to tropism, neutralization and potential pathogenesis.

Adaptive immune responses

The nature of the adaptive immune response to SV5 during natural infections is unknown since we do not know the natural host for SV5. However, this virus has served as a prototype parainfluenza virus for studies of both humoral and cellular immune responses in a number of animal model systems. For example, early studies showed that SV5 persistent infections in mice were cleared by cellular immune responses and not by neutralizing antibody (Young *et al.*, 1990). SV5 infection of mice results in a burst of virus replication and early antigen detection, which lasts between day 3 and 5 post infection before clearance of infectious virus by day 7 (Randall *et al.*, 1988; Capraro *et al.*, 2008). As described above for the section on V mediated STAT1 degradation, SV5 infection of mice is self limited by a relatively strong IFN response. Consistent with this, replication and the antibody responses elicited by intranasal infection do not differ between WT SV5 and P/V mutants that are defective in blocking IFN responses in human cells (Capraro *et al.*, 2008). In contrast, work in ferrets as an alternative animal model system is more consistent with results from cell culture work as an interferon-sensitive SV5 variant shows attenuated growth in the ferret respiratory tract (Capraro *et al.*, 2008).

The $CD8^+$ T-cell response elicited by intranasal infection with SV5 has been used as a model for study of respiratory tract immune responses to virus infection. Alexander-Miller and co-workers showed that intranasal infection of mice with SV5 elicits an early high avidity $CD8^+$ T-cell response at day 3 post infection in the mediastinal lymph node, which drains the lung environment (Gray *et al.*, 2003). This predominant high avidity response changes to a more typical mixed high and low avidity T-cell profile by day 5. These studies are important for understanding immune responses to viral infections, since high avidity T cells are the most potent cells for clearance of viral infections. Remarkably, the SV5 P protein was recognized as harbouring a $CD8^+$ T-cell epitope that was exclusively high avidity (Gray *et al.*, 2001). A number of factors could dictate why this protein elicits exclusively high avidity $CD8^+$ T-cell responses, such as the high level of P phosphorylation, the efficiency of processing of the P epitopes and low protein turnover owing to the stability of P protein. Support for the last mechanism comes from the finding that the avidity of $CD8^+$ T-cell responses against the SV5 P protein are decreased substantially when proteolytic processing of P is increased by linkage to ubiquitin (Gray *et al.*, 2004). These findings raise the possibility of identifying determinants of antigens that could be used in vaccine preparations to elicit higher and thus more potent T-cell responses.

A remarkable discovery was made on the function of $CD8^+$ T cell in the lung of mice infected with SV5. While the initial population of $CD8^+$ T cells in the lung is fully functional at responding to an immunodominant epitope in the SV5 M protein, T-cell function decreases over time to the point where ~85% of the lung-associated $CD8^+$ T cells are no longer responsive to stimulation by M protein epitopes (Gray *et al.*, 2003). It is thought that this loss of function in cytolytic T cells is a normal mechanism by which potentially damaging $CD8^+$ T-cell responses are reduced to avoid damage to host tissues in the sensitive lung environment. This result has profound implications for immunity to virus infections and future vaccine development.

SV5 infection and immunity has also been studied in ferrets (Capraro *et al.*, 2008). Intranasal infection of ferrets with SV5 leads to high levels of virus growth (~10^5 PFU), which appears to be limited to the trachea. This contrasts sharply with infection of mice where SV5 is recovered from both nasal tissue and lungs. In addition, an SV5 P/V mutant that induces IFN and cannot block IFN signalling grows to lower titres than WT SV5 in ferrets, but this is not seen in the case of mice. Finally, the P/V mutant elicited lower anti-SV5 antibody responses than WT SV5, which differs from the situation in mice where these viruses elicit indistinguishable antibody responses. Taken together, these data are consistent with the proposal that ferrets are a more appropriate animal model system than mice for the study of SV5 respiratory infections and immunity.

Antibody responses to WT and P/V mutant SV5 in the ferret system revealed two interesting properties (Capraro *et al.*, 2008). First, dose–response experiments showed that the P/V mutant elicited higher anti-SV5 antibody titres with increasing dose. In contrast, WT SV5 had the opposite effect of eliciting a decreased response with higher virus doses. This may reflect an increase in the number of types of cells infected with WT SV5 at higher doses coupled with the ability of WT SV5 to block functions in some professional antigen presenting cells, such as dendritic cells (Arimilli *et al.*, 2006). A second interesting property of SV5 immunity emerged from prime-boost experiments. Ferrets given a priming dose of the P/V mutant also responded to secondary boost with the same virus by giving an ~10-fold higher antibody response. This ability to boost animals that were exposed to the P/V mutant in a priming step was not seen with WT SV5. The mechanism behind differential boosting effects of P/V and WT viruses is not known at this time, but could reflect a role for IFN induction by the P/V mutant and not WT SV5 in ferrets (Capraro *et al.*, 2008).

Pathogenesis

SV5 is known to infect a range of animals at least experimentally, but there is only one disease associated with SV5 infection. SV5 has been associated with kennel cough, a common but ill-defined respiratory disease of dogs (Cornwell *et al.*, 1976). The exact relationship and contribution of SV5 and other parainfluenza viruses to disease symptoms in dogs is not clear since these infections are most often also associated with *Bordetella bronchiseptica* infections. There is at least one example of disease associated with canine infections by SV5. A variant of SV5 termed canine parainfluenza virus CPI$^+$ was isolated from a dog with temporary posterior paralysis (Evermann *et al.*, 1980) and when injected into canine cranium caused acute encephalitis (Baumgartner *et al.*, 1981). The mechanisms of SV5-induced disease in these animals are not clear.

In mice, SV5 does not cause any disease symptoms unless the mice are severely immunocompromised, such as the case for STAT1 knockout mice (He *et al.*, 2001). This is consistent with the fact that SV5 is unable to target STAT1 for degradation, in contrast to that seen in cells from most other species of animals (Precious *et al.*, 2005a). The restriction on growth and pathology in mice is not solely due to the inability to target STAT1 as growth in the lungs of mice is identical between WT SV5 and the N100D mutant, which is capable of counteracting IFN signalling (Capraro *et al.*, 2008). Similarly, SV5 grows to very high titres in the respiratory tract of ferrets (Capraro *et al.*, 2008). However, here replication is apparently restricted to the tracheal tissue. Animals do not show overt signs of disease and virus is cleared from animals between seven and eleven days post infection. Examination of ferret respiratory tract tissues after SV5 infection showed a remarkable lack of tissue damage and very low levels of cellular infiltration, inflammation and oedema (Capraro *et al.*, 2008).

SV5 has a remarkable property of establishing highly productive long term persistent infections of human cells in tissue culture with minimal cytopathic effect and limited induction of antiviral responses. This property is thought to reflect two major mechanisms that centre around control of the level and type of viral RNA generated during infection and in the active mechanisms that SV5 employs to counteract host cell antiviral responses. The control of RNA synthesis has been discussed above and this is thought to reflect both the level of mRNAs as well as the generation of dsRNA (Manuse and Parks, 2009, 2010). All of the viral factors that contribute to the fidelity of RNA synthesis for WT SV5 are not known, but so far they include the V protein, P protein, M protein and genomic promoter (Dillon and Parks, 2007; Timani *et al.*, 2008; Manuse and Parks, 2009, 2010; Young *et al.*, 2006).

Syncytia formation is one cytopathic effect that is sometimes evident in SV5 infections, but this is highly dependent on cell type and virus strain. Very early studies showed that in some cell types such as BHK cells, SV5 causes massive cell–cell fusion and the infection is limited because of the response of cell membranes to infection (Compans *et al.*, 1966; Holmes and Choppin, 1966). Different SV5 strains or variants also differ in cytopathic effects owing to syncytia formation. For example, the SV5-related porcine isolate SER causes very little syncytium formation in tissue culture cells (Tong *et al.*, 2002). This is thought to be related to the length or structure of the F protein cytoplasmic tail. Seth and co-workers (2007) passed SER multiple times to isolate variants with enhanced fusogenic properties. Sequence analysis revealed individual point mutations in the F protein and in the HN protein throughout each polypeptide. Interestingly, an M protein mutation was also associated with increased fusogenic properties (Seth *et al.*, 2007), suggesting that M protein interactions with the cytoplasmic tails of F or HN could play a role in modulating fusion.

The ability to counteract IFN responses may be an important aspect of the ability of SV5 to establish persistent infections. Using two variants of SV5 that differ in their sensitivity to IFN, Chatziandreou *et al.* (2002) showed that the IFN-sensitive variant CPI$^-$ established persistent infections of Vero cells but upon exposure to IFN the viral glycoproteins were lost from the cell surface and nucleocapsid was found associated with large inclusion bodies. The authors proposed that *in vivo* SV5 may persist by alternating from IFN-sensitive and -resistant states. Alternating between these states changes the expression of cell surface antigens such as the glycoproteins, which could trigger adaptive immune responses (Chatziandreou *et al.*, 2002; Fearns *et al.*, 1994). In support of this proposal, mouse cells that are persistently infected *in vitro* with SV5 appear to flux between active and quiescent forms, and highly fusogenic variants appear in the population (Young *et al.*, 1997). Thus, the ability to cause cell–cell fusion may be selected during periods of surviving the antiviral effects of interferon. Interferon signalling may be important for shifting between phases of active and repressed replication. This is supported by the finding that when cells infected with an SV5 variant that cannot block IFN signalling are treated with IFN, there is a rapid alteration in virus transcription. Treatment of these infected cells with interferon resulted in an increased steepness of the 3′ to 5′ transcriptional gradient and viral mRNAs contained longer polyA tails (Carlos *et al.*, 2005). This suggested that the processivity of the SV5 polymerase across gene junctions is altered as a result of interferon signalling pathways. *in vivo* evidence in support of this model has not been obtained, but this could reflect inappropriate animal models such as mice for testing persistence of paramyxovirus infections.

Previously, SV5 had been reported to be associated with human diseases, including Paget's disease (Cartwright *et al.*, 1993), and Multiple Sclerosis (Goswami *et al.*, 1987; Vandvik and Norrby, 1989). These proposals were attractive associations, largely owing to SV5's property of readily establishing persistent infections of cells with low levels of pathology (Choppin, 1964). Importantly, all of these claims have been disproven (McClean and Thompson, 1989) and it is currently thought that SV5 is not associated with any human disease. In addition, there are reports of SV5 associated with chronic fatigue syndrome (CFS), based primarily on the finding that cells from CFS patients can have reduced levels of STAT1. Despite this lack of clear association with any human disease, variants of SV5 have been isolated from samples from human patients (Chatziandreou *et al.*, 2004). The divergence of sequence between these isolates indicates that these are not a common lab contaminant, but represent bona fide infections.

Careful studies by Goswami *et al.* (1984) documented that about 20% of human serum samples they had analysed contained neutralizing antibodies against SV5. These authors were careful to control for cross reactivity with human parainfluenza virus type 2 (HPIV2), a common problem in assaying the extent of SV5 neutralization by human serum. Very similar results were obtained recently in a study that indicated only about 10% of healthy donors were positive for neutralizing antibodies (Johnson *et al.*, 2008). In a separate study, it was found that *ex vivo* CD4$^+$ T-cell responses to SV5 antigens were detected in all subjects; however, there was no control for whether these were specific to SV5 or were due to cross-reactivity with HPIV2 antigens (Cohn *et al.*, 1996). Thus, while there is evidence that SV5 can infect humans (Goswami *et al.*, 1984), there is no clear indication of a disease associated with these infections.

Perspectives

SV5 has long served as a prototype paramyxovirus from which a large number of properties and mechanisms common to this family of viruses have been established. As much of the fundamental information on structure, replication and function of paramyxovirus proteins has already been detailed, attention should be focused on at least two future areas. First, the interactions of SV5 with the host immune system at the cellular and the animal level are not well understood. A large number of questions remain as to how this virus can replicate to such high levels and not induce overt disease symptoms. The questions of how SV5 limits antiviral responses can replicate even in IFN treated cells (Wansley *et al.*, 2005) and limits apoptosis remain unanswered.

A second area of future focus included the renewed interest in SV5 as a novel vector for therapeutic purposes and as a platform for delivery of vaccine-associated antigens. The desirable properties of SV5 for these purposes include those shared by other negative strand RNA viruses such as (1) a relatively small RNA genome with no known packaging constraints; (2) the genome does not integrate into host DNA or show evidence of RNA recombination; and (3) respiratory tract infection can elicit strong mucosal immunity.

However, in addition, SV5 has a number of other attractive properties that are desirable in a viral vector, including (1) infections are largely non-cytopathic and are not associated with any disease in humans; (2) SV5 can be grown in Vero cells, an approved cell line for vaccine work and purified to very high titres ($>10^{10}$ PFU/ml); (3) Anti-SV5 seroprevalence is relatively low in humans; and (4) SV5 elicits strong systemic IgG and mucosal IgG and IgA responses even when delivered by a variety of routes. Recently, SV5 has been engineered to express the influenza A virus HA protein for the purposes of demonstrating vaccine potential (Thompkins *et al.*, 2007). Not surprisingly, this HA-expressing vector can elicit protection from lethal influenza virus challenge in mice. Similarly, SV5 has been engineered to express bacterial flagellin (Arimilli *et al.*, 2008), resulting in a novel viral vector that has improved immunostimulatory properties. Future work will build on these findings and will be focused on whether SV5 vectors can be used to elicit protection from more challenging viral infections such as HIV.

SV5 is also being developed as a novel oncolytic virus vector. By contrast to the WT vector, which is non-cytopathic, the P/V mutant containing substitutions in the shared P/V region is a very efficient killer of cells through activation of apoptosis. This P/V mutant-based SV5 vector is capable of reducing tumour burden in a mouse model system of implanted human prostate epithelial cell tumours (Gainey *et al.*, 2008b). The full breadth of which tumours are responsive to SV5 vectors and how we can improve the safety and specificity of infections for tumour versus normal cells will be a focus of future work.

References

Andrejeva, J., Poole, E., Young, D.F., Goodbourn, S., and Randall, R.E. (2002). The p127 subunit (DDB1) of the UV-DNA damage repair binding protein is essential for the targeted degradation of STAT1 by the V protein of the paramyxovirus SV5. J. Virol. *76,* 11379–11386.

Andrejeva, J., Childs, K.S., Young, D.F., Carlos, R.S., Stock, N., Goodbourn, S., and Randall, R.E. (2004). The V proteins of paramyxoviruses bind the IFN-inducible RNA helicase, mda-5, and inhibit its activation of the IFN-beta promoter. Proc. Natl. Acad. Sci. U.S.A. *101,* 17264–17269.

Arimilli S., Alexander-Miller, M.A., and Parks, G.D. (2006). A Simian Virus 5 (SV5) P/V mutant is less cytopathic than wild type SV5 in human dendritic cells and is a more effective activator of dendritic cell maturation and function. J. Virol. *80,* 3416–3427.

Arimilli S., Johnson, J.B., Alexander-Miller, M.A., and Parks, G.D. (2007). TLR-4 and -6 agonists reverse apoptosis and promote maturation of Simian virus 5 infected human dendritic cells through NFκB-dependent pathways. Virology *365,* 144–156.

Arimilli S., Johnson, J.B., Clark, K.M., Alexander-Miller, M.A., Mizel, S.B., and Parks, G.D. (2008). Expression of flagellin enhances paramyxovirus activation of dendritic cells. J. Virol. *82,* 10975–10985.

Bagai, S., and Lamb, R.A. (1997). A glycine to alanine substitution in the paramyxovirus SV5 fusion peptide increases the initial rate of fusion. Virology *238,* 283–290.

Banerjee, A.K., and Barik, S. (1992). Gene expression of vesicular stomatitis virus genome RNA. Virology *188,* 417.

Baumgartner, W., Metzler, A.E., Krakowka, S., and Koestner, A. (1981). *In vitro* identification and characterization of a virus isolated from a dog with neurological disfunction. Infect. Immunity *31,* 1177–1193.

Bhella, D., Ralph, A., Murphy, L.B., and Yeo, R.P. (2002). Significant differences in nucleocapsid morphology within the *Paramyxoviridae*. J. Gen. Virol. *83,* 1831–1839.

Bissonnette, M.L., Connolly, S.A., Young, D.F., Randall, R.E., Paterson, R.G., and Lamb, R.A. (2006). Analysis of the pH requirement for membrane fusion of different isolates of the paramyxovirus parainfluenza virus 5. J. Virol. *80,* 3071–3077.

Bissonnette, M.L., Donald, J.E., DeGrado, W.F., Jardetzky, T.S., and Lamb, R.A. (2009). Functional analysis of the tranmembrane domain in paramyxovirus F protein-mediated membrane fusion. J. Mol. Biol. *386,* 14–36.

Calain, P., and Roux, L. (1993). The rule of six, a basic feature for efficient replication of Sendai virus defective interfering RNA. J. Virol. *67*,4822–4830.

Capraro G.A. Johnson, J.J., Kock N.D., and Parks, G.D. (2008). Growth and antibody responses to respiratory tract infection of ferrets and mice with WT and P/V mutants of the Paramyxovirus Simian virus 5. Virology *376*, 416–428.

Carlos, T.S., Fearns, R., and Randall, R.E. (2005). Interferon-induced alterations in the pattern of parainfluenza virus 5 transcription and protein synthesis and the induction of virus inclusion bodies. J. Virol. *79, 14*,112–14121.

Carlos, T.S., Young, D.F., Schneider, M., Simas, J.P., and Randall, R.E. (2009). Parainfluenza virus 5 genomes are located in viral cytoplasmic bodies whilst the virus dismantles the interferon-induced antiviral state of cells. J. Gen. Virol. *90*, 2147–2156.

Cartwright, E.J., Gordon, M.T., Freemont, A.J., Anderson, D.C., and Sharpe, P.T. (1993). Paramyxoviruses and Paget's disease. J. Med. Virol. *40*, 133–141.

Chatziandreou, N., Young, D., Andrejeva, J., Goodbourn, S., and Randall, R.E. (2002). Differences in interferon sensitivity and biological properties of two related isolates of simian virus 5: a model for virus persistence. Virology *293*, 234–242.

Chatziandreou, N., Stock, N., Young, D., Andrejeva, J., Hagmaier, K., McGeoch, D.J., and Randall, R.E. (2004). Relationships and host range of human, canine, simian and porcine isolate of SV5. J. Gen. Virol. *85*, 3007–3016.

Childs, K.S., Stock, N., Ross, C., Andrejeva, J., Hilton, L., Skinner, M., Randall, R., and Goodbourn S. (2007). Mda-5, but not RIG-I, is a common target for paramyxovirus V proteins. Virology *359*, 190–200.

Childs, K.S., Andrejeva, J., Randall, R.E., and Goodbourn, S. (2009). Mechanism of mda-5 inhibition by paramyxovirus V proteins. J. Virol. *83*, 1465–1473.

Choppin, P.W. (1964). Multiplication of a myxovirus (SV5) with minimal cytopathic effects and without interference. Virology *23*, 224–233.

Choppin, P.W., and Compans, R.W. (1970). Phenotypic mixing of enveloped proteins of the parainfluenza virus SV5 and vesicular stomatitis virus. J. Virol. *5*, 609–616.

Cohn, M.L., Robinson, E.D., Thomas, D., Faerber, M., Carey, S., Sawyer, R., Goswami K.K.A., Johnson, A.H., and Richert, J.R. (1996). T-cell responses to the paramyxovirus SV5: studies in multiple sclerosis and normal populations. Pathobiology *64*, 131–135.

Compans, R.W., Homes, K.V., Dales, S., and Choppin, P.W. (1966). An electron microscopic study of moderate and virulent virus-cell interactions of parainfluenza virus SV5. Virology *30*, 411–426.

Cornwell, H.J., McCandish, I.A., Thompson, H., Laird, H.M., and Wright, N.G. (1976). Isolation of parainfluenza virus SV5 from dogs with respiratory disease. Vet. Rec. *98*, 301–302.

Didcock, L., Young, D.F., Goodbourn, S., and Randall, R.E. (1999a). Sendai virus and SV5 block activation of IFN-responsive genes: importance of virus pathogenesis. J. Virol. *73*, 3125–3133.

Didcock, L., Young, D.F., Goodbourn, S., and Randall, R.E. (1999b). The V protein of SV5 inhibits interferon signaling by targeting STAT1 for proteasome-mediated degradation. J. Virol. *73*, 9928–9933.

Dillon, P.J., and Parks, G.D. (2007). A role for the phosphoprotein P subunit of the paramyxovirus polymerase in limiting induction of host cell antiviral responses. J. Virol. *81*, 11116–11127.

Dillon, P.J., Wansley, E.K., Young, V.A., Alexander-Miller, M.A., and Parks, G.D. (2006). Exchange of P/V genes between two noncytopathic SV5 variants results in a recombinant virus that kill cells through death pathways that are sensitive to caspase inhibitors. J. Gen. Virol. *87*, 3643 – 3648.

Dutch, R.E., Jardetzky, T.S., and Lamb, R.A. (2000). Virus membrane fusion proteins: biological machines that undergo a metamorphosis. Biosci. Rep. *20*, 597–612.

Dutch, R.E., Hagglund, R.N., Nagel, M.A., Paterson, R.G., and Lamb, R.A. (2001). Paramyxovirus fusion protein: a conformational change on cleavage activation. Virology *281*, 138–150.

Evermann, J.F., Lincoln, J.D., and McKiernan, A.J. (1980). Isolation of a paramyxovirus from the cerebrospinal fluid of a dog with posterior paresis. J. Am. Vet. Med. Assoc. *177*, 1132–1134.

Fearns, R., Young, D.F., and Randall, R.E. (1994). Evidence that the paramyxovirus SV5 can establish quiescent infections by remaining inactive in cytoplasmic inclusion bodies. J. Gen. Virol. *75*, 3525–3529.

Gainey M.D., Dillon, P.J., Clark, K.M., Manuse, M.J., and Parks, G.D. (2008a). Paramyxovirus induced shutoff of host and viral protein synthesis: role of the P and V proteins in limiting PKR activation. J. Virol. *82*, 828–839.

Gainey, M.D., Manuse, M.J., and Parks, G.D. (2008b) A hyperfusogenic F protein enhances the oncolytic potency of a paramyxovirus simian virus 5 P/V mutant without compromising sensitivity to type I interferon. J. Virol. *82*, 9369–9380.

Gray, P.M., Parks, G.D., and Alexander-Miller, M.A. (2001). A novel CD8-independent high avidity CTL response directed against an epitope in the phosphoprotein of the paramyxovirus SV5. J. Virol. *75*, *10*,065–10072.

Gray, P.M., Parks, G.D., and Alexander-Miller, M.A. (2003). High avidity CD8 T cells are the initial population elicited following viral infection of the respiratory tract. J. Immunol. *170*, 174 –181.

Gray, P.M., Parks, G.D., and Alexander-Miller, M.A. (2004). Modulation of $CD8^+$ T cell avidity by increasing the turnover of viral antigen during infection. Cell Immunol. *231*, 14–19.

Goswami, K.K.A., Lange, L.S., Mitchell, D.N., Cameron, K.R., and Russell, W.C. (1984). Does SV5 Infect. humans? J. Gen. Virol. *65*, 1295–1303.

Goswami, K.K., Randall, R.E., Lange, L.S., and Russell, W.C. (1987). Antibodies against the paramyxovirus SV5 in the cerebrospinal fluids of some multiple sclerosis patients. Nature *327*, 244–247.

He, B., Paterson, R.G., Ward, C.D., and Lamb, R.A. (1997). Recovery of infectious SV5 from cloned DNA and expression of a foreign gene. Virology *237,* 249–260.

He B., Leser, G.P., Paterson, R.G., and Lamb, R.A. (1998). The paramyxovirus SV5 small hydrophobic (SH) protein is not essential for virus growth in tissue culture cells. Virology *250,* 30–40.

He B., Lin, G.Y., Durbin, J.E., Durbin, R.K., and Lamb, R.A. (2001). The SH integral membrane protein of the paramyxovirus simian virus 5 is required to block apoptosis in MDBK cells. J. Virol. *75,* 4068–4079.

He, B., Paterson, R.G., Stock, N., Durbin, J.E., Durbin, R.K., Goodbourn, S., Randall, R.E., and Lamb, R.A. (2002). Recovery of paramyxovirus simian virus 5 with a V protein lacking the conserved cysteine-rich domain: the multifunctional V protein blocks both interferon-beta induction and interferon signaling. Virology *303,* 15–32.

Heinen, E., Herbst, W., and Schmeer, N. (1998). Isolation of a cytopathogenic virus from a case of porcine reproductive and respiratory syndrome (PRRS) and its characterization as parainfluenza virus type 2. Arch. Virol. *143,* 2233–2239.

Hirsch, R.L., Wolinsky, J.S., and Winkelstein, J.A. (1986). Activation of the alternative complement pathway by mumps infected cells: Relationship to viral neuraminidase activity. Arch. Virol. *87,*181–190.

Hiebert, S.W., Paterson, R.G., and Lamb, R.A. (1985a). Hemagglutinin-neuraminidase protein of the paramyxovirus simian virus 5: nucleotide sequence of the mRNA predicts an N-terminal membrane anchor. J. Virol. *54,* 1–6.

Hiebert, S.W., Paterson, R.G., and Lamb, R.A. (1985b). Identification and predicted sequence of a previously unrecognized small hydrophobic protein, SH, of the paramyxovirus simian virus 5. J. Virol. *55,* 744–751.

Hiebert, S.W., Richardson, C.D., and Lamb, R.A. (1988). Cell surface expression and orientation in membranes of the 44-amino-acid SH protein of simian virus 5. J. Virol. *62,* 2347–2357.

Holmes, K.V., and Choppin, P.W. (1966). On the role of the response of the cell membrane in determining virus virulence. J. Exp. Med. *124,* 501–520.

Horvath, C.M. (2004). Weapons of STAT destruction. Interferon evasion by paramyxovirus V protein. Eur. J. Biochem. *271,* 4621–4628.

Horvath, C.M., and Lamb, R.A. (1992). Studies on the fusion peptide of a paramyxovirus fusion glycoprotein: roles of conserved residues in cell fusion. J. Virol. *66,* 2443–2455.

Hsiung, G.D. (1972). Parainfluenza-5 virus, infection of man and animal. Prog. Med. Virol. *14,* 241–274.

Hull, R., Minner, J.R., and Smith, J.W. (1956). New viral agents recovered from tissue cultures of monkey cells. I. Origin and properties of cytopathogenic agents SVI, SV2, SV4, SV5, SV6, SV11, SVI2 and SVI5. Am. J. Hygiene *63,* 204–205.

Ito, M., Nishio, M., Kawano, M., Kusagawa, S., Komada, H., Ito, Y., and Tsurudome, M. (1997). Role of a single amino acid at the amino terminus of the simian virus 5 F2 subunit in syncytium formation. J. Virol. *71,* 9855–9858.

Johnson, J.B., Capraro, G.A., and Parks, G.D. (2008). Differential mechanisms of complement mediated neutralization of the closely related paramyxoviruses SV5 and Mumps virus. Virology *376,* 112–123.

Johnson, J.B., Grant, K., and Parks, G.D. (2009). SV5 and Mumps virus recruit CD46 to avoid complement mediated neutralization. J. Virol. *83,* 7602–7611.

Keller, M.A., Murphy, S.K., and Parks, G.D. (2001). RNA replication from the SV5 antigenomic promoter requires three sequence-dependent elements separated by spacer regions. J. Virol. *75,* 3993–3998.

Keller M.A., and Parks, G.D. (2003). Positive- and negative-acting signals combine to determine differential RNA replication from the paramyxovirus simian virus 5 genomic and antigenomic promoters. Virology *306,* 347–358.

Kolakofsky, D., Pelet, T., Garcin, D., Hausmann, S., Curran, J., and Roux, L. (1998). Paramyxovirus RNA synthesis and the requirement for hexamer genome length: the rule of six revisited. J. Virol. *72,* 891–899.

Kraus, T.A., Garza, L., and Horvath, C.M. (2008). Enabled interferon signaling evasion in an immune-competent transgenic mouse model of parainfluenza virus 5 infection. Virology *371,* 196–205.

Lamb, R.A., Paterson, R.G., and Jardetzky, T.S. (2006). Paramyxovirus membrane fusion: Lessons from F and HN atomic structures. Virology *344,* 30–37.

Lamb, R.A., and Parks, G.D. (2007). *Paramyxoviridae*: the viruses and their replication. In Fields Virology, 5th ed, Fields, B., Knipe, D., and Howley. P., eds. (Lippincott Williams & Wilkins: Philadelphia), pp. 1449–1496

Leser, G.P., Ector, K.J., Ng, D.T., Shaunessy, M.A., and Lamb, R.A. (1996). The signal for clathrin-mediated endocytosis of the paramyxovirus SV5 HN protein resides at the transmembrane domain-ectodomain boundary region. Virology *262,* 79–92.

Li, T., Chen, X., Garbutt, K.C., Zhou, P., and Zheng, N. (2006). Structure of DDB1 in complex with a paramyxovirus V protein: viral hijack of a propeller cluster in ubiquitin ligase. Cell *124,* 105–117.

Lin, G., and Lamb, R.A. (2000). The paramyxovirus simian virus 5 V protein slows progression of the cell cycle. J. Virol. *74,* 9152–9166.

Lin, G., Paterson, R.G., and Lamb, R.A. (1997). The RNA binding region of the paramyxovirus SV5 V and P proteins. Virology *238,* 460–469.

Lin, Y., Bright, A.C., Rothermel, T.A., and He, B. (2003). Induction of apoptosis by SV5 lacking a small hydrophobic gene. J. Virol. *77,* 3371–3383.

Lin, Y., Horvath, F., Agio, J.A., Wilson, R., and He, B. (2005). The role of simian virus 5 V protein on viral RNA synthesis. Virology *338,* 270–280.

Lu, L.L., Puri, M., Horvath, C.M., and Sen, G.C. (2008). Select paramyxoviral V proteins inhibit IRF3 activation by acting as altenative substrates for inhibitor of kappaB kinase epsilon (IKKe)/TBK1. J. Biol. Chem. *283,* 14269–14276.

McCandish, L.A.P., Thompson, H., Cornwell, H.J.C., and Wright, N.G. (1978). A study of dogs with kennel cough. Vet. Rec. *102,* 298–301.

McLean, B.N., and Thompson, E.J. (1989). Antibodies against the paramyxovirus SV5 are not specific for cerebrospinal fluid from multiple sclerosis patients. J. Neurol. Sci. *92,* 261–266.

McSharry, J.J., Pickering, R.J., and Caliguiri, L.A. (1981). Activation of the alternative complement pathway by enveloped viruses containing limited amounts of sialic acid. Virology *114,* 507–515.

Manuse, M.J., and Parks, G.D. (2009). Role for the paramyxovirus genomic promoter in limiting host cell antiviral responses and cell killing. J. Virol. *83,* 9057–9067.

Manuse, M.J., and Parks, G.D. (2010). TLR3-dependent upregulation of RIG-I leads to enhanced cytokine production from cells infected with the parainfluenza virus SV5. Virology *397,* 231–241.

Manuse, M.J., Briggs C.M., and Parks, G.D. (2010). Replication-independent activation of human plasmacytoid dendritic cells by the paramyxovirus SV5 requires TLR7 and autophagy pathways. Virology *405,* 383–389.

Murphy, S.K., and Parks, G.D. (1997). Genome nucleotide lengths that are divisible by six are not essential but enhance replication of defective interfering RNAs of the paramyxovirus simian virus 5. Virology *232,* 145–157.

Murphy, S.K., and Parks, G.D. (1999). RNA replication for the paramyxovirus simian virus 5 requires an internal repeated (CGNNNN) sequence motif. J. Virol. *73,* 805–809.

Murphy, S.K., Ito,Y., and Parks, G.D. (1998). A functional antigenomic promoter for the paramyxovirus simian virus 5 requires proper spacing between an essential internal segment and the 3′ terminus. J. Virol. *72,* 10–19.

Ng, D.T., Randall, R.E., and Lamb, R.A. (1989). Intracellular maturation and transport of the SV5 type II glycoprotein hemagglutinin-neuraminidase: specific and transient association with GFP78 BiP in the ER and extensive internalization from the cell surface. J. Cell Biol. *109,* 3273–3289.

Ng, D.T., Hiebert, S.W., and Lamb, R.A. (1990). Different roles of individual N-linked oligosaccharide chains in folding, assembly, and transport of the simian virus 5 hemagglutinin neuraminidase. Mol. Cell Biol. *10,* 1989–2001.

Parisien, J.P., Lau, J.F., and Horvath, C.M. (2002a). STAT2 acts as a host range determinant for species-specific paramyxovirus interferon antagonism and SV5 replication. J. Virol. *76,* 6435–6441.

Parisien, J.P., Lau, J.F., Rodriguez, J.J., Ulane, C.M., and Horvath, C.M. (2002b). Selective STAT protein degradation induced by paramyxoviruses requires both STAT1 and STAT2 but is independent of alpha/beta interferon signal transduction. J. Virol. *76,* 4190–4198.

Parks, G.D. (1994). Mapping of a region of the paramyxovirus L protein required for the formation of a stable complex with the viral phosphoprotein P. J. Virol. *68,* 4862–4872.

Parks, G.D. (1996). Differential effects of changes in the length of a signal/anchor domain on membrane insertion, subunit assembly and intracellular transport of a type II integral membrane protein. J. Biol. Chem. *271,* 7187–7195.

Parks, G.D., and Alexander-Miller, M.A. (2002). High avidity CTL to a foreign antigen are efficiently activated following immunization with a recombinant SV5. J. Gen. Virol. *83,* 1167–1172.

Parks, G.D., and Lamb, R.A. (1991). Topology of eukaryotic type II membrane proteins: importance of N-terminal positively charged residues flanking the hydrophobic domain. Cell *64,* 777–787.

Parks, G.D., and Lamb, R.A. (1993). Role of NH2-terminal positively charged residues in establishing membrane protein topology. J. Biol. Chem. *268,* 19101–19109.

Parks, G.D., and Pohlmann, S. (1995). Structural requirements in the membrane-spanning domain of the paramyxovirus HN protein for the formation of a stable tetramer. Virology *213,* 263–270.

Parks, G.D., Ward, C.D., and Lamb, R.A. (1992). Molecular cloning of NP and L genes of simian virus 5: Identification of conserved domains in paramyxovirus NP and L. Virus Res. *22,* 259–279.

Parks, G.D., Ward, K.R., and Rassa, J.C. (2001). Increased readthrough transcription across the simian virus 5 M-F gene junction leads to growth defects and a global inhibition of viral mRNA synthesis. J. Virol. *75,* 2213–2223.

Parks, G.D., Young, V.A., Koumenis, C., Wansley, E.K., J.L. Layer, and Cooke, K.M. (2002). Controlled cell killing by a recombinant nonsegmented negative strand RNA virus. Virology *293,* 192–203.

Paterson, R.G., Harris, T.J., and Lamb, R.A. (1984a). Fusion protein of the paramyxovirus simian virus 5: nucleotide sequence of mRNA predicts a highly hydrophobic protein. Proc. Natl. Acad. Sci. U.S.A. *81,* 6706–6710.

Paterson, R.G., Harris, T.J., and Lamb, R.A. (1984b). Analysis and gene assignment of mRNAs of a paramyxovirus, simian virus 5. Virology *138,* 310–323.

Paterson, R.G., Hiebert, S.W., and Lamb, R.A. (1985). Expression at the cell surface of biologically active fusion and hemagglutinin/neuraminidase proteins of the paramyxovirus simian virus 5 from cloned cDNA. Proc. Natl. Acad. Sci. U.S.A. *82,* 7520–7524.

Paterson, R.G., Shaunessy, M.A., and Lamb, R.A. (1989). Analysis of the relationship between cleavability of a paramyxovirus fusion protein and length of the connecting peptide. J. Virol. *63,* 1293–1301.

Paterson, R.G., Leser, G.P., Shaunessy, M.A., and Lamb, R.A. (1995). The paramyxovirus SV5 V protein binds two atoms of zinc and is a structural component of virions. Virology *208,* 121–131.

Paterson, R.G., Russell, C.J., and Lamb, R.A. (2000). Fusion protein of the paramyxovirus SV5: destabilizing and stabilizing mutants of fusion activation. Virology *270,* 17–30.

Pattnaik A.K., and Wertz G.W. (1990). Replication and amplification of defective interfering particle RNAs of vesicular stomatitis virus in cells expressing viral proteins from vectors containing cloned cDNAs. J. Virol. *64,* 2948–57.

Pei, Z., Bai, Y., and Schmitt, A.P. (2010). PIV5 M protein interaction with host protein angiomotin like-1. Virology *397,* 155–166.

Peluso, R.W., Lamb, R.A., and Choppin, P.W. (1977). Polypeptide synthesis in SV5-infected cells. J. Virol. *23,* 177–187.

Peluso, R.W., Lamb, R.A., and Choppin, P.W. (1978). Infection with paramyxoviruses stimulates synthesis of cellular polypeptides that are also stimulated in cells transforMed. by Rous sarcoma virus or deprived of glucose. Proc. Natl. Acad. Sci. U.S.A. *75,* 6120–6124.

Poole, E., He, B., Lamb, R.A., Randall, R.E., and Goodbourn, S. (2002). The V proteins of simian virus 5 and other paramyxoviruses inhibit induction of interferon-beta. Virology *303,* 33–46.

Precious, B., Young, D.F., Bermingham, A., Fearns, R., Ryan, M., and Randall, R.E. (1995). Inducible expression of the P, V, and NP genes of the paramyxovirus SV5 in cell lines and examination of NP-P and NP-V interactions. J. Virol. *69,* 8001–8010.

Precious, B., Young, D.F., Andrejeva, L., Goodbourn, S., and Randall, R.E. (2005a). *In vitro* and *in vivo* specificity of ubiquitination and degradation of STAT1 and STAT2 by the V proteins of the paramyxoviruses SV5 and human parainfluenza virus type 2. J. Gen Virol. *86,* 151–158.

Precious, B., Childs, K., Fitzpatrick-Swallow, V., Goodbourn, S., and Randall, R.E. (2005b). Simian Virus 5 V protein acts as an adaptor, linking DDB1 to STAT2, to facilitate the ubiquitination of STAT1. J. Virol. *79,* 13434–13441.

Precious, B.L., Carlos, T.S., V., Goodbourn, S., and Randall, R.E. (2007). Catalytic turnover of STAT1 allows PIV5 to dismantle the interferon-induced anti-viral state of cells. Virology *368,* 114–121.

Ramachandran, A., and Horvath, C.M. (2009). Paramyxovirus disruption of interferon signal transduction: STATus report. J. Interferon Cytokine Res. *29,* 531–537.

Randall, R.E., and Young, D.F. (1988). Comparison between parainfluenza virus type 2 and simian virus 5: monoclonal antibodies reveal major antigenic differences. J. Gen. Virol. *69,* 2051–2060.

Randall, R.E., Young, D.F., Goswami, K.K.A., and Russell, W.C. (1987). Isolation and characterization of monoclonal antibodies to SV5 and their use in revealing antigenic differences between human, canine and simian isolates. J. Gen. Virol. *68,* 2769–2780.

Randall, R.E., Young, D.F., and Southern, J.A. (1988). Immunization with solid matrix-antibody antigen complexes containing surface or internal virus structural proteins protects mice from infection with the paramyxovirus, simian virus 5. J. Gen. Virol. *69,* 2517–2526.

Rassa, J.C., and Parks, G.D. (1998). Molecular basis for naturally occurring elevated readthrough transcription across the M–F junction of the paramyxovirus SV5. Virology *247,* 274–286.

Rassa, J.C., and Parks, G.D. (1999). Highly diverse intergenic regions of the paramyxovirus simian virus 5 cooperate with the gene end U tract in viral transcription termination and can influence reinitiation at a downstream gene. J. Virol. *73,* 3904–3912.

Rassa, J.C., Wilson, G.M., Brewer, G.A., and Parks, G.D. (2000). Spacing constraints on reinitiation of paramyxovirus transcription: the gene end U tract acts as a spacer to separate gene end from gene start sites. Virology *274,* 438–449.

Robbins, S.J., Wrzos, J. Kline, A.L. Tenser, R.B., and Rapp, F. (1981). Rescue of a cytopathic paramyxovirus from peripheral blood leukocytes in subacute sclerosing panencephalitis. Journal of Infectious Diseases *143,* 396–402.

Russell, R., Paterson, R.G., and Lamb, R.A. (1994). Studies with cross-linking reagents on the oligomeric form of the paramyxovirus fusion protein. Virology *199,* 160–168.

Russell, C.J., Jardtzky, T.S., and Lamb, R.A. (2001). Membrane fusion machines of paramyxoviruses: capture of intermediates of fusion. EMBO J. *20,* 4024–4034.

Scheid, A., and Choppin, P.W. (1974). Heamagglutinating and neuraminidase protein of a paramyxovirus: interaction with neuraminic acid in affinity chromatography. Virology *62,*125–133.

Scheid, A., Caliguiri, L.A., Compans, R.W., and Choppin, P.W. (1972). Isolation of paramyxovirus glycoproteins. Association of both hemagglutinating and neuraminidase activites with the larger SV5 glycoprotein. Virology *50,* 640–652.

Schmitt, A.P., and Lamb, R.A. (2004). Escaping the cell: assembly and budding of negative-strand RNA viruses. Curr. Top. Microbiol. Immunol. *283,* 145–196.

Schmitt, A.P., Leser, G.P., Waning, D.L., and Lamb, R.A. (2002). Requirements for budding of paramyxovirus SV5-like particles. J. Virol. *76,* 3952–3964.

Schmitt, A.P., Leser, G.P., Morita, E., Sundquist, W.I., and Lamb, R.A. (2005). Evidence for a new viral late-domain core sequence, FPIV, necessary for budding of a paramyxovirus. J. Virol. *79,* 2988–2997.

Seth, S., Vincent, A., and Compans, R.W. (2003). Activation of fusion by the SER virus F protein: a low pH-dependent paramyxovirus entry process. J. Virol. *77,* 6520–6527.

Seth, S., Goodman, A.L., and Compans, R.W. (2004). Mutations in multiple domains activate paramyxovirus F protein–induced fusion. J. Virol. *78,* 8513–8523.

Seth, S., Skountzou, I., Gernert, K.M., and Compans, R.W. (2007). Fusogenic variants of a noncytopathic paramyxovirus. J. Virol. *81,* 4286–4297.

Sun M., Rothermel, T.A., Shuman, L., Aligo, J.A., Xu, S., Lin, Y., Lamb, R.A., and He, B. (2004). Conserved cysteine-rich domain of paramyxovirus simian virus 5 V protein plays an important role in blocking apoptosis. J. Virol. *78,* 5068–5078.

Sun, D., Luthra, P., Li, Z., and He, B. (2009). PLK1 down-regulates parainfluenza virus 5 gene expression. Plos Pathogen *5,* e1000525.

Tapparel, C., Maurice, D., and Roux, L. (1998). The activity of Sendai virus genomic and antigenomic promoters requires a second element past the leader template regions: a motif (GNNNNN)3 is essential for replication. J. Virol. *72,* 3117–3128.

Thomas, S.M., Lamb, R.A., and Paterson, R.G. (1988). Two mRNAs that differ by two nontemplated nucleotides encode the amino co-terminal proteins P and V of paramyxovirus SV5. Cell *54,* 891–902.

Timani, K.A., Sun, D., Sun, M., Keim, C., Lin, Y., Schmitt, P.T., Schmitt, A.P., and He, B. (2008). A single amino acid residue change in the P protein of parainfluenza virus 5 elevates viral gene expression. J. Virol. *82,* 9123–9133.

Tompkins S.M., Lin Y., Leser G.P., Kramer K.A., Haas D.L., Howerth E.W., Xu J., Kennett M.J., Durbin, R.K., Durbin, J.E., *et al.* (2007). Recombinant parainfluenza virus 5 (PIV5) expressing the influenza A virus hemagglutinin provides immunity in mice to influenza A virus challenge. Virology *362,*139–50.

Tong, S., Li, M., Vincent, A., Compans, R.W., Fritsch, E., Beier, R., Klenk, C., Ohuchi, M., and Klenk, H.D. (2002). Regulation of fusion activity by the cytoplasmic domain of a paramyxovirus F protein. Virology *301,* 322–333.

Tsurudome M., Ito M., Nishio M., Kawano M., Komada H., and Ito Y. (2001). Hemagglutinin–neuraminidase-independent fusion activity of simian virus 5 fusion (F) protein: difference in conformation between fusogenic and nonfusogenic F proteins on the cell surface. J. Virol. *75,* 8999–9009.

Ulane, C.M., and Horvath, C.M. (2002). Paramyxoviruses SV5 and HPIV2 assemble STAT protein ubiquitin ligase complexes from cellular components. Virology *304,* 160–166.

Ulane, C.M., Kentsis, A., Cruz, C.D., Parisien, J.P. Schneider, K.L., and Horvath, C.M. (2005). Composition and assembly of STAT-targeting ubiquitin ligase complexes: paramyxovirus V protein carboxyl terminus is an oligomerization domain. J. Virol. *79,* 10180–10189.

Vandvik, B., and Norrby, E. (1989). Paramyxovirus SV5 and multiple sclerosis. Nature *338,* 769–771.

Vulliemoz, D., and Roux, L. (2001). Rule of six: How doe the Sendai virus RNA polymerase keep count? J. Virol., *75,* 4506.

Waning, D.L., Schmitt, A.P., Leser, G.P., and Lamb, R.A. (2002). Roles for the cytoplasmic tails of the F and HN proteins in budding of the paramyxovirus SV5. J. Virol. *76,* 9284–9297.

Wansley, E.K., and Parks, G.D. (2002). Naturally occurring substitutions in the P/V gene convert the noncytopathic paramyxovirus SV5 into virus that induces type I interferon synthesis and cell death. J. Virol. *76,* 10109–10121.

Wansley, E.K., Dillon, P.J., Gainey, M.D., Tam, J., Cramer, S.D., and Parks, G.D. (2005). Growth sensitivity of a recombinant simian virus 5 P/V mutant to type I interferon differs between tumor cell lines and normal primary cells. Virology *335,* 131–144.

Watowich, S.S., Morimoto, R.I., and Lamb, R.A. (1991). Flux of the paramyxovirus hemagglutinin-neuraminidase glycoprotein through the endoplasmic reticulum activates transcription of the GRP78-BiP gene. J. Virol. *65,* 3590–3597.

Wilson, R.L., Fuentes, S.M., Wang, P., Taddeo, E.C., Klatt, A., Henderson, A.J., and He, B. (2006). Function of small hydrophobic proteins of paramyxovirus. J. Virol. *80,* 1700–1709.

Yin, H.-S., Wen, X., Paterson, R.G., Lamb, R.A., and Jardetzsky, T.S. (2006). Structure of the parainfluenza virus 5 F protein in its metastable, prefusion conformation. Nature *439,* 38–44.

Young, V.A., and Parks, G.D. (2003). Simian virus 5 is a poor inducer of chemokine secretion from human lung epithelial cells: identification of viral mutants that activate interleukin-8 secretion by distinct mechanisms. J. Virol. *77,* 7124–30.

Young, D.F., Randall, R.E., Hoyle, J.A., and Souberbielle, B.E. (1990). Clearance of a persistent paramyxovirus infection is mediated by cellular immune responses but not by serum-neutralizing antibody. J. Virol. *64,* 5403–5411.

Young, D.F., Didcock, L., and Randall R.E. (1997). Isolation of highly fusogenic variants of simian virus 5 from persistently infected cells that produce and respond to interferon. J. Virol. *71,* 9333–9342.

Young, D.F., Didcock, L., Goodbourn, S., and Randall R.E. (2000). Paramyxoviruses use distinct virus-specific mechanisms to circumvent the interferon response. Virology *269,* 383–390.

Young, D.F., Chatziandreou, N., He, B., Goodbourn, S., Lamb, R.A., and Randall, R.E. (2001). Single amino acid substitution in the V protein of SV5 differentiates its ability to block interferon signaling in human and murine cells. J. Virol. *75,* 3363–3370.

Young, D.F., Andrejeva, L., Livingston, A., Goodbourn, S., Lamb, R.A., Collins, P.L., Elliott, R.M., and Randall, R.E. (2003). Virus replication in engineered human cells that do not respond to interferons. J. Virol. *77,* 2174–2181.

Young, D.F., Carlos, T.S., Hagmaier, K., Fan, L., and Randall, R.E. (2007). AGS and other issue culture cells can unknowingly be persistently infected with PIV5, a virus that blocks interferon signaling by degrading STAT1. Virology *365,* 238–240.

Young, V.A., Dillon, P.J., and Parks, G.D. (2006). Variants of the paramyxovirus Simian virus 5 with accelerated or delayed viral gene expression activate proinflammatory cytokine synthesis. Virology *350,* 90–102.

Yuan, P., Thompson, T., Wurzburg, B.A., Paterson, R.G., Lamb, R.A., and Jardetzsky, T.S. (2005). Structural studies of the parainfluenza virus 5 hemagglutinin-neuraminidase tetramer in complex with its receptor, sialyllactose. Structure *13,* 1–13.

Newcastle Disease and Related Avian Paramyxoviruses

3

Siba K. Samal

Abstract

Newcastle disease (ND) is one of the most important diseases of poultry worldwide. Virulent ND virus (NDV) infections may have flock mortality rates up to 100%. The economic impact is not only due to loss of birds, but also due to trade restrictions and embargoes placed on areas and countries where the outbreaks have occurred. In many developing countries ND is endemic, and the disease has the greatest impact on villages where the livelihood of people depend on poultry farming. ND probably has the most impact on the world's economy than any other animal disease. NDV belongs to avian paramyxovirus serotype-1 (APMV-1) and is the most characterized member among the nine APMV serotypes. In recent years, NDV has drawn a lot of research interest not only because it is an important pathogen of poultry, but it is also an oncolytic agent and is a potential vaccine vector for human and animal uses. Currently, very little is known about the molecular biology and pathogenicity of APMV-2 to -9, but our knowledge on these serotypes is rapidly expanding with the availability of complete genome sequences of all APMV serotypes.

History

Newcastle disease was first recorded on the island of Java in Indonesia by Kraneveld in 1926. A disease with similar characteristics was reported the same year on a farm near the seaport town of Newcastle-upon-Tyne in England by Doyle (1927). He showed that the disease was caused by a filterable virus and was different from fowl plague (now known as highly pathogenic avian influenza). He named the disease 'Newcastle disease' (ND) temporarily, because he wanted to avoid a descriptive name that might lead to confusion with other poultry diseases (Doyle, 1935). He also showed that the experimental disease could be subacute and that a common feature was nervous involvement with lameness, twitching movement and paralysis of one or both legs.

ND was also recognized in Korea in 1926 (Kanno *et al.*, 1929); in Ranikhet, India, in 1927 (Edwards, 1928); in Ceylon (now Sri Lanka) in 1927 (Crawford, 1930); in Manila, Philippines, in 1927 (Rodier, 1928); in Fusan, Japan, in 1929 (Ochi and Hashimoto, 1929); and in Melbourne, Australia, in 1930 (Albiston and Gorrie, 1942). In a short time, the disease had spread all over the world and was recognized as a new poultry disease. In the majority of cases, a local name was given to identify the disease, but the name 'Newcastle disease' remained most popular and is still used.

The exact origin of ND and the virus is not known. There are evidences that ND might have occurred much sooner than 1926. According to MacPherson (1956), the mysterious disease resulting in the death of all domestic fowl in the Western Isles of Scotland in 1898 was ND. It is possible that ND was present in the chicken population before 1926, but was recognized when large-scale commercial poultry farming developed in the first half of the twentieth century.

In the U.S., a new respiratory and nervous systems disease in poultry was first reported in California by Stover (1940), which was later recognized as ND (Beach, 1944). It was interesting that the disease was much milder and had a mortality rate of about 15% compared with 95–100% mortality rates reported from other parts of the world. The first recorded appearance of ND in the east coast occurred in New Jersey in 1945. But retroactive identification of viruses showed that ND was present in the east coast as early as 1938 (Beaudette and Hudson, 1956).

Shope suggested that strains of ND virus (NDV), which caused the two major clinically known forms of ND in chickens (Asiatic and American) originally, existed in different species of wild birds (Shope, 1964). From those wild species there were probably separate 'escapes' into domestic chickens, the virulent form in most parts of the world, and the mild form in the USA.

Geographic distribution

NDV is distributed throughout the world, but an exact assessment of the true geographic distribution of ND (caused by virulent strains of NDV) is difficult because official reporting does not take into account strains of low pathogenicity and the widespread use of live attenuated NDV vaccines, which also mask the presence of virulent strains. Furthermore, many village and backyard outbreaks of ND are not always reported. Currently, ND is present on at least six of the seven continents of the world and is either endemic or cause of frequent outbreaks throughout Africa, Asia, the Middle East, Central America and parts of South America. The countries of Oceania are relatively free from ND, probably as a result of effective quarantine and geographic isolation.

In the USA, virulent NDV strains are not endemic and the disease caused by virulent NDV is referred to as exotic ND (END). The first case of virulent NDV in the USA was reported in California in 1950. Since then, virulent NDV has been introduced into the U.S. on several occasions, but each outbreak was subsequently eradicated. The last END outbreak was reported in California in 2002, and approximately 3.5 million birds were depopulated in an effort to eradicate the disease. Sequencing of M and F genes showed the California END virus to be most closely related to viruses isolated from recent outbreaks in Mexico and Central America (Pedersen *et al.*, 2004). Historically, virulent NDV has been introduced into the USA through the importation of exotic avian species (Utterback and Schwartz, 1973; Senne *et al.*, 1983; Bruning-Fann *et al.*, 1992; Panigrahy *et al.*, 1993) and water birds (Wobeser *et al.*, 1993; Banerjee *et al.*, 1994). In Western Europe, sporadic epizootics of ND regularly occur in many countries despite widespread vaccination programmes. From 1932 to 1998, Australia has been free of virulent NDV. However, six outbreaks caused by virulent NDV occurred between 1998 and 2002 in localized areas of New South Wales. In each case, the NDV is thought to have evolved from certain low-virulence Australian strains. There is no evidence that Australian outbreaks have been caused by viruses introduced from overseas (Westbury, 2001).

Natural and experimental host range

NDV infections have been reported in at least 241 species of birds representing 27 of the 50 orders of birds (Kaleta and Baldauf, 1988). It is probable that all species of birds are susceptible to NDV infection. However, the disease may vary greatly depending upon the virus strain and host species. Infection in most species do not result in disease. Although avian species are the natural host for NDV, many non-avian species have also been shown to be naturally or experimentally susceptible to infection with NDV. Both avirulent and virulent strains of NDV can infect and cause clinical signs of disease in humans (Chang, 1981). The virus usually causes only mild, transient conjunctivitis or flu-like symptoms. However, natural human infection with NDV is extremely rare. ND is now recognized as an occupational hazard to those people associated with poultry farming, production of vaccines and laboratory personnel. There is no report of human-to-human transmission. NDV has also been shown to experimentally infect calves, swine, sheep, mice, guinea pigs, rabbits, ferrets, hamsters and non-human primates such as monkeys (Reagan *et al.*, 1947; DiNapoli *et al.*, 2007a; Subbiah *et al.*, 2008b).

Infectious agent

Virus classification

NDV is a member of the genus *Avulavirus* in the subfamily *Paramyxovirinae* of the family *Paramyxoviridae* (Lamb and Parks, 2007). All members of the genus *Avulavirus* have haemagglutinin and neuraminidase activities. They share greater sequence relatedness among them than with members of other genera. Currently, all avian paramyxoviruses (APMVs) that have been isolated from avian species are placed in the genus *Avulavirus* except for avian metapneumovirus, which is classified in the genus *Metapneumovirus*. The viruses in the genus *Avulavirus* are classified into nine serotypes (APMV-1 to -9) based on haemagglutination inhibition (HI) and neuraminidase inhibition (NI) assays (Alexander, 1988). NDV is the type species for the genus *Avulavirus*. All strains of NDV comprise APMV-1. Since NDV is an important poultry pathogen, APMV-1 is the most extensively characterized serotype in the genus *Avulavirus*.

Virion morphology

Virions are 100 nm or more in diameter, pleomorphic, but mostly spherical in shape (Fig. 3.1). Occasionally, filamentous particles of variable length are present. The virion is enveloped with a lipid membrane derived from the host cell plasma membrane. The envelope contains two transmembrane glycoproteins – the haemagglutinin–neuraminidase (HN) and fusion (F) proteins. These proteins are present as homooligomers and form spike-like projections of 8 nm length on the outer surface of the envelope. Beneath the envelope lies a non-glycosylated membrane protein called matrix (M) protein. Inside the virion particle is the nucleocapsid, which has the classical herringbone morphology with a diameter of 18 nm. The viral nucleocapsid consists of a single species of viral RNA and replication complex proteins – the nucleocapsid (N), phosphoprotein (P) and large polymerase protein (L).

Virion properties

The virion has a molecular weight of about 500×10^6 and a buoyant density of 1.18–1.20 g/ml in sucrose. NDV is relatively stable in nature. The virus can remain infective for weeks at low temperatures and for several hours over a wide range of pH. In association with organic matter, it can survive for days in litter, water, carcass, eggs and feathers. The virus can remain infectious on feather down for 123 days at 20–30°C, 255 days at 11–36°C, and 538 days at 3–6°C (Olesiuk, 1951). NDV is very sensitive to heat, lipid solvents, ionic and non-ionic detergents,

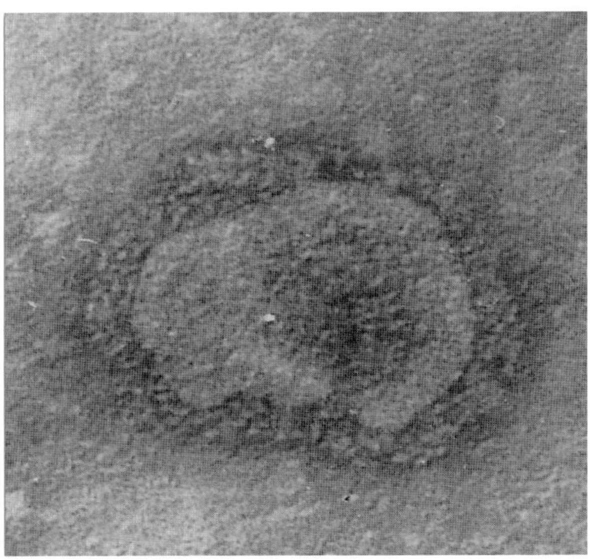

Figure 3.1 Negative contrast electron micrograph of intact Newcastle disease virus particle.

formaldehyde and oxidizing agents. It has been shown that the thermal stability varies among NDV strains. Some strains are more resistant to heat than others (Pierce and Haywood, 1973; Kim *et al.*, 1978). It was found that the infectivity of NDV strains in buffer was reduced by 99.9% between 10 min and 20 min at 56°C (Pierce and Haywood, 1973). However, the virus survival during heat stress is dependent on the medium involved and thus an important factor in determining the temperature and time needed to inactivate NDV (Alexander and Manvell, 2004). The knowledge of heat resistance of the circulating virulent NDV strain is important because it will indicate how long that virus is likely to remain infectious in meat, and other contaminated materials of infected poultry.

Viral genome structure, transcription, and RNA replication

The genome of NDV is a non-segmented negative-sense single-stranded RNA. The complete genome sequence for many strains of NDV are available on the web at www.ncbi.nlm.nih.gov/. Three different categories of genome size exist among NDV strains. Initially, 15,186 nucleotides (nt) was reported for genomes of several NDV strains, most of which were isolated before 1960 (Krishnamurthy and Samal, 1998; de Leeuw and Peeters, 1999; Romer-Oberdorfer *et al.*, 1999). A second genome size category, 15,192 nt, was originally based on the genome sequence of an isolate from geese in China (Huang *et al.*, 2004a). The increase in length was due to presence of a 6 nt insert in the 5′ (downstream) non-coding region of the N gene after genome position 1647. Later, two other isolates, one from goose (Zou *et al.*, 2005) and a pigeon-type strain (Ujvari, 2006), were also found to have genomes of 15,192 nt. Sequence analysis showed that the 6 nt insert is present in many other recent NDV strains at the same genome position, but the sequence of the insert varies among NDV strains. A third genome size category, containing 15,198 nt, was first recognized in an avirulent isolate from ducklings in Germany in 1999 (Czegledi *et al.*, 2006). The increase in length was due to presence of a 12 nt insert in the open reading frame (ORF) of P and V proteins after genome position 2381. The addition of 12 nt did not change the reading frames of P and V proteins, but increased their lengths by four amino acids. The sequence of the 12 nt insert also varies among NDV strains. The 15,192 and 15,198 nt genome length viruses emerged after 1960. Sequence analysis indicates that 15,192 nt genome length viruses are more prevalent than 15,198 nt genome length viruses. The evolutionary significance of these inserts is not known. Recently, the effects of these inserts on replication and pathogenicity of NDV was evaluated by adding them to two NDV strains that have natural genome of 15,186 nt but represent two different pathotypes. The results showed that the 6 nt insert slightly increases the replication and virulence of NDV, whereas the 12 nt insert slightly attenuates NDV (Paldurai *et al.*, 2010a). All three genome sizes of NDV are consistent with the 'rule of six' that is a characteristic of the subfamily *Paramyxovirinae* (Calain and Roux, 1993). According to this rule, the genome length must be an even multiple of six for efficient RNA replication. The 'rule of six' is thought to arise from the need to fully encapsidate the entire length of the genome with a chain of N protein monomers that span exactly 6 nt. A genome whose length is an even multiple of six will be precisely encapsidated by N proteins, with no unencapsidated nt protruding from the 3′ end of the nucleocapsid. However, the 'rule of six' does not apply to the members of the subfamily *Pneumovirinae* (Samal and Collins, 1996).

A phylogenetic tree using complete genome sequences of 105 NDV strains, representing different geographic and chronological origins, is shown in Fig. 3.2. The genomic RNA contains a 55 nt 3′ extragenic region known as the leader and a 114 nt 5′ extragenic region known as the trailer. The leader and trailer are the control regions essential not only for transcription and replication, but also for encapsidation of newly synthesized RNAs into virus particles. The sequences at the 3′ and 5′ termini exhibit a high level of complementarity, suggestive of conserved elements in the 3′ promoters of the genome and antigenome. They flank the six genes (3′-N-P/V-M-F-HN-L-5′) that encode the nucleocapsid protein (N), phosphoprotein (P), V protein (V), matrix protein (M), fusion protein (F), haemagglutinin–neuraminidase

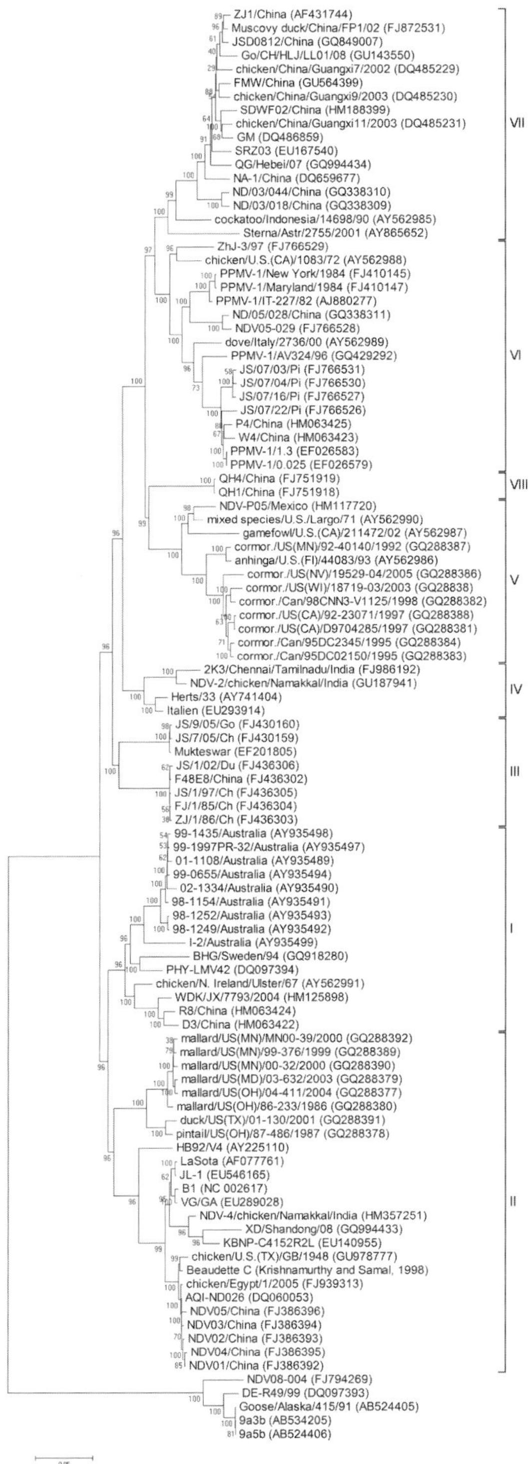

Figure 3.2 Phylogenetic tree illustrating relationship among 105 Newcastle disease virus (NDV) strains based on complete genome sequences (some have genome sequence without leader and trailer sequence). Shown are class I and class II (I-VIII genotypes) of NDV following the neighbour-joining method using MEGA4 software. Parameters include: pairwise deletion, 1000 replicates for Bootstrap analysis and Kimura 2-parameter substitution model.

protein (HN), and the large polymerase protein (L). On an average 91% of the NDV genome carry coding sequences.

The beginning and end of each gene contain conserved transcriptional control sequences known as the gene-start (3′-UGCCCAUCU/CU-5′) and gene-end (3′-AAUU/CC/UU_{5-6}-5′), respectively. Between the gene boundaries are non-coding intergenic sequences (IGSs), which vary in length from 1 to 47 nt. Each of the first three IGSs, the N-P, P-M, and M-F gene junctions, has only 1 nt, while the other two IGSs, the F-HN and HN-L gene junctions, are 31 nt and 47 nt, respectively. The lengths of IGSs are generally conserved among NDV strains, with the exception that the IGS in the N-P gene junction is 2 nt long in some strains; for example, D26, LaSota and Texas GB. It has been shown that the IGS length modulates the transcription of the downstream gene (Kim and Samal, 2010). The same study suggested that the length of IGS at each junction is conserved evolutionarily because these lengths probably produce the optimal amount of transcripts necessary for efficient NDV replication. Any increase or decrease in the length of IGS affected the replication of the viruses and severely attenuated the pathogenicity of NDV in chickens (Yan and Samal, 2008). The sequence of IGS vary among NDV strains, indicating that the sequence of IGS is probably not important for modulation of NDV transcription.

The transcription process of NDV is similar to that of other non-segmented negative-sense RNA viruses (Lamb and Parks, 2007). Transcription begins at a single promoter that is present in the leader region. First a leader RNA of 55 nt is synthesized, which is followed by sequential transcription of the genes in the 3′–5′ order to yield individual mRNAs by a stop–start mechanism guided by the conserved gene-start and gene-end signals. Gene-start initiates synthesis of the mRNA and gene-end directs polyadenylation and termination of the mRNA, after which the polymerase remains attached to the template and passes over the IGS. Therefore, the IGSs are not copied into mRNAs. The leader RNA is neither capped nor polyadenylated and is not functional as mRNA. The mRNAs are capped, methylated and polyadenylated by the viral polymerase. The poly(A) tail is synthesized by stuttering on the stretch of U residues in the GE signal. Reinitiation of transcription at the gene-start site is not perfect, which leads to a gradient of mRNA abundance that decreases according to the distance from the 3′ end of the genome. Replication occurs when the polymerase complex ignores the transcription stop signals at the 3′ end of each gene and a full-length positive-sense antigenome is synthesized. Like the full-length genomic RNA, these antigenomic replicative intermediates are totally encapsidated by viral nucleocapsid proteins. In turn, the antigenomic RNA functions as template for synthesis of full-length progeny genomes. How the switch from transcription to replication occurs is not clearly known, but it has been proposed that when sufficient amount of N protein accumulates in the cytoplasm, it binds to the P protein to form a soluble complex, which is used for replication of the progeny genome RNA. The viral transcription and replication occurs within a nucleocapsid that consists of N, P, and L proteins. It is thought that the promoter present in the antigenome is stronger than the promoter present in the genome, resulting in synthesis of more genome sense RNAs from replication.

Each gene of the NDV genome encodes a single major protein with the exception of the P gene, which encodes three proteins by a process of mRNA editing. The mRNA editing, which involves insertion of non-templated G residue(s), occurs by a polymerase stuttering mechanism (Lamb and Parks, 2007). Like other members of the subfamily *Paramyxovirinae*, the P gene mRNA of NDV is edited by insertion of one or two G residues into a stretch of G residues within a conserved editing site 5′AAAAAGGG (mRNA sense) at position 484 in the P gene (Steward *et al.*, 1993). As a result, three P gene derived mRNA species are produced which encode for the P, V, and W ORFs with unedited, +1 and +2 frameshifts, respectively. The V and W proteins are colinear to the P protein at the N-terminal half, but differ in their carboxy-terminal parts. Analysis of mRNAs produced from the P gene showed that 68% were P-encoding mRNA, 29% were V-encoding mRNA, and 2% were W-encoding mRNA (Mebatsion *et al.*, 2001).

Each NDV mRNA contains a major ORF with short 5′ and 3′ untranslated regions (UTRs). The sequence and length of the UTRs vary among the different mRNA species. It has been shown that the UTRs modulate the level of transcription and translation of NDV genes (Yan *et al.*, 2009).

Viral proteins

All NDV strains encode seven proteins: N, P, V, M, F, HN, and L. All seven proteins are found in virus particles. The V protein is not essential for virus replication *in vitro* and serves as an accessory protein. Three proteins (N, P, and L), together with the RNA genome form the nucleocapsid, which is the minimum infectious unit necessary for transcription and replication of the viral genome. Three proteins (M, F, and HN) are associated with the envelope. The M protein forms the inner layer of the envelope and is not an integral membrane protein (Fig. 3.3). The F and HN proteins are the only two integral membrane proteins and are the virus neutralizing antigens. The F and HN proteins also are the major protective antigens. Comparison of deduced amino acid sequences of N, P, V, M, F, HN and L proteins among 50 NDV strains, representing different chronological and geographic origins, from GenBank showed 93.9%, 83.9%, 81.8%, 92.1%, 91.3%, 90.4% and 94.5% amino acid identity, respectively.

The nucleocapsid protein

The N protein of NDV is 489 amino acids long and has a molecular weight of 55 kDa. It is the most abundant protein in infected cells and in virus particles. N is an RNA binding protein that coats the full-length genomic (–) sense and antigenomic (+) sense RNAs to protect them from nuclease digestion and form biologically active templates. Recombinant N protein of NDV expressed in *Escherichia coli* (Kho *et al.*, 2003) or the baculovirus expression system (Errington and Emmerson, 1997) is able to self-assemble to form herringbone-like structures, which morphologically resemble the authentic nucleocapsid structures. The N-terminal 375 amino acids of N protein is required for proper folding to form a herringbone-like structure (Kho *et al.*, 2003). The N protein forms nucleocapsids only when

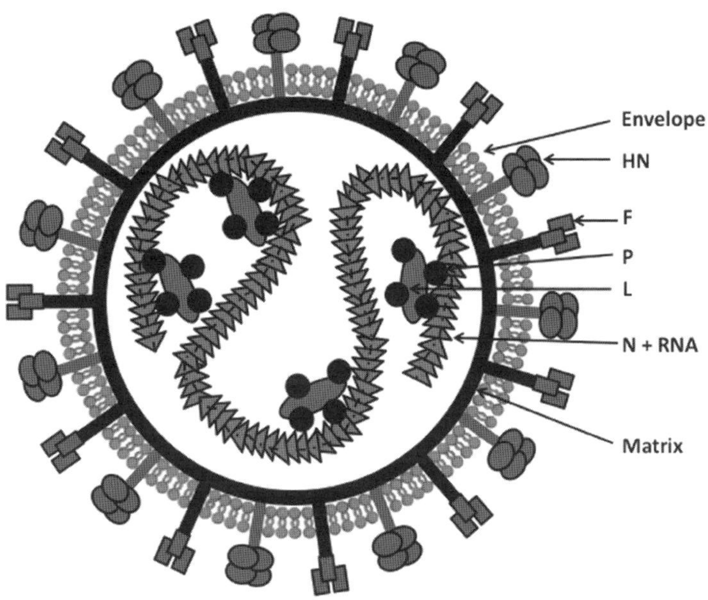

Figure 3.3 Schematic diagram of a Newcastle disease virus particle. The lipid bilayer derived from host cell membrane is shown as envelope. The viral matrix protein is shown as a black concentric circle beneath the envelope. The haemagglutinin–neuraminidase (HN) and the fusion (F) glycoproteins are shown imbedded inside the viral envelope. The inside of virion is made up of negative-strand RNA encapsidated with nucleocapsid protein (N) and associated with the phosphoprotein (P) and the large polymerase protein (L).

it is available in a soluble complex with the P protein. The association of the N and P proteins prevents non-specific binding of N protein to cellular RNA and allowing the N protein to bind selectively to the genome leader and antigenome trailer sequences. It has been shown that the first 25 amino acids of N-terminus of NDV N protein is involved in forming soluble complex with the P protein (Kho *et al.*, 2004).

The phosphoprotein

The P protein of NDV is 395 amino acids long and is heavily phosphorylated at specific serine and threonine residues. The P protein functions as a homooligomer. The calculated molecular weight of P protein is ~42 kDa, but by SDS-PAGE analysis, the P protein forms multiple bands with molecular weight ranging from 50 to 55 kDa. These observations are the result of multiple phosphorylated forms of the P protein. But, how phosphorylation is involved in determining the functions of the P protein still remains unknown. The P protein is a non-catalytic subunit of the viral RNA polymerase. It interacts with the N and the L protein and serves as a bridge between the L protein and the N-RNA template. The P protein also acts as a chaperone to prevent uncontrolled encapsidation of non-viral RNA by the N protein. The complex of P with unassembled N monomer is believed to regulate the switch from transcription to replication. The C-terminal 45 amino acids (residues 247–291) of the NDV P protein have been shown to be involved in P–P and P–N interactions (Jahanshiri *et al.*, 2005).

The V protein

The V protein of NDV is 239 amino acids long and has a molecular weight of 36 kDa. The P and V proteins are amino coterminal, but vary at their carboxy-terminal ends in length and amino acid composition (Fig. 3.4). The NDV V protein, similar to other paramyxovirus V proteins, has a cysteine-rich carboxy-terminal domain, which binds to two molecules of Zinc (Steward *et al.*, 1993). NDV V protein can be found in very small quantity in association with the nucleocapsid of the purified virions and thus is a minor structural component of the virion. Recombinant NDV lacking V protein has impaired growth in cell culture and in embryonated chicken eggs and are highly attenuated in chickens (Mebatsion *et al.*, 2001; Huang *et al.*, 2003). These mutant viruses also exhibited increased sensitivity to exogenous interferon (IFN) (Huang *et al.*, 2003; Elankumaran *et al.*, 2006;). The V protein of NDV has been shown to inhibit IFN response in two ways. Huang *et al.* (2003) found that the V protein of NDV inhibits IFN signalling by targeting STAT1 for degradation and this function lies in the C-terminal domain. Another study showed that the V protein of NDV interacts with MDA5 via the C-terminal domain and this interaction correlates with inhibition of IFNβ induction (Park *et al.*, 2003). The IFN antagonistic activities of the V proteins from two strains of NDV were shown to correlate with their known virulence properties, indicating that the V protein plays a role in NDV virulence which is probably because of its ability to antagonize IFN (Alamares *et al.*, 2010). These

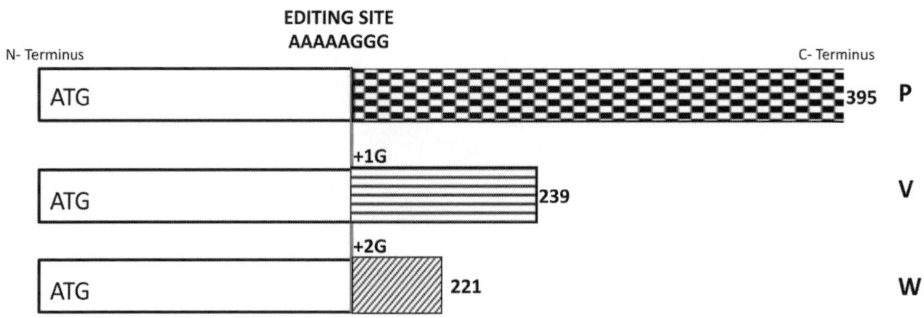

Figure 3.4 Schematic representation of the phosphoprotein (P) gene of Newcastle disease virus. The unedited version of P mRNA encodes P protein (chequered rectangle). The RNA editing site (AAAAAGGG), at which insertion of one or two non-template G nucleotides to mRNA encodes V protein (parallel striped rectangle) and W protein (oblique striped rectangle), respectively.

results suggest that the V is an accessory protein that is involved in antagonizing the host antiviral response to infection.

Insertion of two G residues at the P gene editing site produces W mRNA, but insertion of two G residues into the mRNA is relatively rare. Although W mRNA has been detected in NDV infected cells, a W protein has never been found. Therefore, the role of W protein in NDV replication cycle has not been established.

The large protein

The L protein is the largest structural protein of NDV, comprising 2204 amino acids with a molecular weight of 250 kDa. The L protein is the least abundant protein in infected cells or in virions. The L protein of NDV also contains the six highly conserved domains (I–VI) that have been identified in the L proteins of other non-segmented negative-sense RNA viruses and believed to carry out distinct functions related to RNA synthesis, post-transcriptional modifications and stability of the newly synthesized RNA (Poch *et al.*, 1990). The L protein possesses all the enzymatic activities necessary for synthesis of viral mRNAs and genomic RNA replication, including nucleotide polymerization, mRNA capping, methylation, and polyadenylation of mRNAs. The L protein only functions in association with the P protein. It does not utilize naked RNA genome as a template, but recognizes it only when the genomic RNA is tightly bound to the N protein. It is thought that the P protein tethers L protein to the nucleocapsid. The L protein has been associated with the virulence of NDV (Rout and Samal, 2008).

The matrix protein

The M protein of NDV is 364 amino acids long and has a molecular weight of 40 kDa (Seal *et al.*, 2000). The M protein is somewhat hydrophobic and contains many basic amino acid residues. Sequence analysis of the M protein revealed that it does not contain hydrophobic regions long enough to span a lipid bilayer, indicating that it is not an intrinsic membrane protein (Li *et al.*, 1980). It is located on the inner face of the virion envelope. The M protein plays an important role in the assembly of the virion by interacting with the nucleocapsid, the lipid bilayer and the cytoplasmic region of the HN protein. The domains of M protein involved in all these interactions have not yet been identified. The M protein may also play a role in transport of viral components to the plasma membrane (Takimoto and Portner, 2004). The M protein is essential for forming virus-like particle (Pantua *et al.*, 2006).

The haemagglutinin–neuraminidase protein

The HN protein of NDV is a multifunctional protein. It is responsible for attachment of the virion to sialic acid-containing cell surface receptors. In addition, HN has neuraminidase activity (NA) that cleaves sialic acid from sugar side chains, thereby, releasing progeny virions from the surface of infected cells. The NA also cleaves sialic acid from surface of virions to prevent self-association (Lamb and Parks, 2007). The HN protein is also required to interact with the F protein for fusion promotion activity (Morrison *et al.*, 1991).

The HN of NDV is a type II integral membrane protein that contains an uncleaved signal/anchor sequence located near the N-terminus (Fig. 3.5). The HN exists on the surface of virions and in virus infected cells as a homotetramer composed of two disulfide linked dimers (McGinnes *et al.*, 1993). The HN ectodomain is a mushroom like structure consisting of a long stalk that supports a terminal globular head. The attachment, NA activities and all known antibody binding sites reside in the globular head (Mirza *et al.*, 1993). The predicted molecular weight of the HN protein is 74 kDa. Comparison of the nucleotide sequences of HN genes of 50 NDV isolates demonstrated HN proteins of at least seven different lengths according to differences in the position of the stop codon of the HN ORF, yielding HN proteins of 571, 572, 577, 578, 580, 585 or 616 amino acids. The majority of NDV strains have HN protein of 571 amino acids. Only in the case of the HN protein with 616 amino acids (HN_0) post translational cleavage is required for activation of the HN protein, but not for the proteins of other lengths (Nagai *et al.*, 1976). It has been shown that this variation in protein lengths has no impact on NDV pathogenicity (Romer-Oberdorfer *et al.*, 2003). It is possible that the shorter HN proteins of NDV strains that lack the carboxy

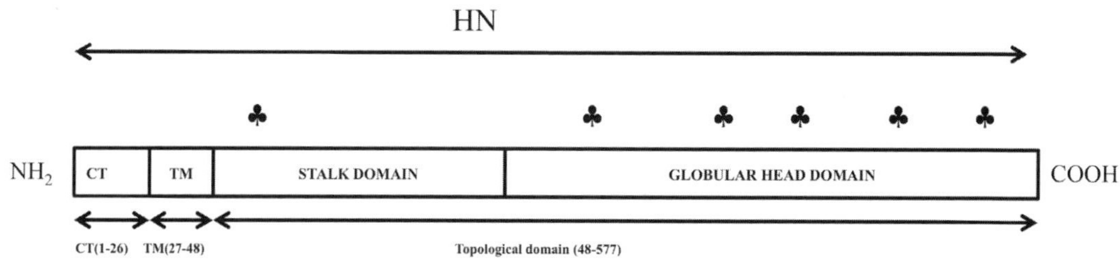

Figure 3.5 Schematic diagram representing different domains of the Newcastle disease virus attachment protein haemagglutinin–neuraminidase (HN): The position of the cytoplasmic tail (CT) domain, the transmembrane (TM) domain, and the stalk and globular head domains are indicated in the parenthesis. Tree represents the sites used for addition of N-linked carbohydrate.

terminus extension arose evolutionarily from a longer ancestral HN protein by introduction of translational stop codons.

The crystal structure of the HN protein of NDV showed that the C terminal globular head (residues 124–570) is composed of four identical monomers arranged in four-fold symmetry, with each of the monomers consisting of a six-blade β-propeller motif with a NA active site at its centre. The crystal structure of the globular head of the HN dimer originally suggested that receptor recognition and NA activities are mediated by a single binding site that adopts two different conformations (Crennell *et al.*, 2000; Connaris *et al.*, 2002). However, a second sialic acid binding site was later discovered at the membrane-distal end of the dimer interface in the globular head (Zaitsev *et al.*, 2004). Studies using receptor-binding inhibition experiments also confirmed the presence of a second receptor binding site without NA activity (Porotto *et al.*, 2004). The second site enhances, but is not required for fusion (Bousse *et al.*, 2004). In addition to its receptor binding activity, HN contributes to fusion in a virus-specific manner. It has been shown for many paramyxoviruses, including NDV, that the F protein can only carry out fusion in conjunction with the virus-specific attachment protein. It is thought that the attachment protein directs the F protein to initiate fusion only at the target membrane (Sergel *et al.*, 1993; Takimoto *et al.*, 2002; Lamb and Parks, 2007). This is consistent with an interaction between HN and F that is mediated by specific domains on the two proteins (Hu *et al.*, 1992). The stalk region of NDV HN protein has been implicated in both fusion promotion and its specificity for the F protein (Deng *et al.*, 1995). The stalk region of NDV HN protein contains two heptad repeats. These heptad repeats interact with the heptad repeats of F protein for fusion promotion activity (Stone-Hulslander and Morrison, 1999). However, the detailed molecular mechanism of how these two surface glycoproteins, HN and F, work together in viral membrane penetration is not well understood.

The HN glycoprotein of NDV contains six potential N-linked glycosylation sites at residues 119, 341, 433, 481, 508, and 538 (Fig. 3.5). These sites are conserved among NDV strains, indicating that they might be important for the structure and function of HN protein. It is known that four of these glycosylation sites at residues 119, 341, 433, and 481 are used (McGinnes and Morrison, 1995). It was shown that viable virus could be recovered from cDNA clones lacking carbohydrate at these four glycosylation sites (Panda *et al.*, 2004a). In all mutant viruses loss of carbohydrate had no effect on receptor recognition of HN protein. It was found that removal of carbohydrate at residues 119, 341, and 433 did not affect cell surface expression, neuraminidase activity and cell fusion, but the mutant viruses were attenuated in chickens. However, removal of carbohydrate at residue 481 affected cell surface expression, neuraminidase activity, cell fusion, virus replication and the mutant virus was highly attenuated in chickens, suggesting that carbohydrate at this site might be involved in folding of the HN protein and thus essential for virus replication and pathogenicity (Panda *et al.*, 2004a).

The HN protein of NDV has 14 cysteine residues. Twelve of 14 cysteine residues are conserved and are involved in intramolecular disulfide bond formation (McGinnes *et al.*, 1987). It was found that the second cysteine (amino acid 123) is responsible for the covalent linkage of the HN protein dimer. Mutation in different cysteine residues can block folding at different stages of maturation of the HN protein, indicating that they are important for the structural integrity of the HN protein (McGinnes and Morrison, 1994). It has also been indicated that mutating certain key amino acids in the globular head region of the HN protein can attenuate the virulence of NDV (Khattar *et al.*, 2009). The cytoplasmic tail contains 26 highly conserved amino acids of the cytoplasmic tail and has been shown to interact with the M protein (Garcia-Sastre *et al.*, 1989). Mutational analysis showed that only the first two amino acids of the glycoprotein could be deleted without affecting the viability of NDV. Any further deletion or substitution of amino acid affected the viability of the virus, indicating that these residues are critical for its incorporation into virions (Kim *et al.*, 2009). The HN protein has been shown to determine the tropism and virulence of NDV (Huang *et al.*, 2004c; de Leeuw *et al.*, 2005).

The fusion protein

The F glycoprotein mediates entry of the virus into the host cell by fusion of the viral envelope to the plasma membrane. During the fusion process the F protein undergoes extensive structural alterations (Lamb and Parks, 2007). However, the HN protein is required to facilitate fusion by promoting conformational changes in the F protein. Like the F proteins of other paramyxoviruses, the NDV F protein does not require the acid pH of endosomes to activate its fusion function. Hence, the newly synthesized F protein expressed on the plasma membrane of infected cells can mediate fusion at neutral pH with neighbouring cells to form syncytia, or multinucleated giant cells. This cytopathic effect can lead to tissue necrosis *in vivo* and may also be a mechanism of virus spread.

The NDV F protein is a homotrimer and is a typical type I integral membrane protein with an amino terminus containing a cleavable signal sequence, a hydrophobic transmembrane domain (TM) located near the carboxy terminus and a 31 amino acid cytoplasmic tail (CT) (Russell *et al.*, 1994; Morrison, 2003; Fig. 3.6). Another topological form of NDV F protein has also been found in which approximately 200 amino acids of the amino terminus as well as the CT domain of the protein were translocated across membranes (McGinnes *et al.*, 2003). It has been suggested that this second form of the F protein may be involved in cell–cell fusion (Pantua *et al.*, 2005). The F protein is synthesized as a 553 amino acid inactive precursor, F_0, which to be biologically active is cleaved in the trans-Golgi membranes to F_1 and F_2 subunits derived from the carboxyl terminal and amino terminal domains, respectively. The F_1 and F_2 subunits remain covalently linked to each other by disulfide bonds (Morrison, 2003). During intracellular transport, the F protein

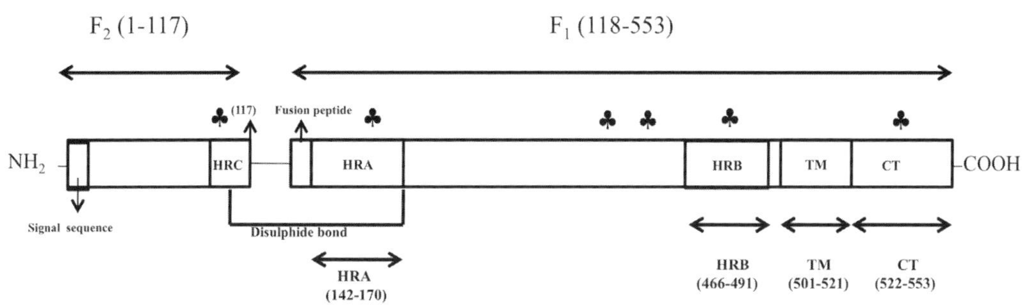

Figure 3.6 Schematic diagram showing important domains and features of the Newcastle disease virus fusion (F) protein: The position of the signal sequence, the transmembrane domain (TM), the cleavage site, the hydrophobic fusion peptide, the heptad repeats (HRA, HRB and HRC) and cytoplasmic tail (CT) are indicated. The sites used for addition of N-linked carbohydrate (tree) are also indicated with reference to the putative protein. Position 117 indicates the amino acid position for cleavage-activation. The amino acids sequences for major domains are indicated in the parenthesis.

undergoes post translational modifications, such as glycosylation and fatty acid acylation (Morrison, 2003). The amino terminus of F_1 subunit created by cleavage is a hydrophobic region called the 'fusion peptide' that is thought to be inserted into the plasma membrane to initiate fusion. In case of NDV, the fusion peptide is 25 amino acids in length. Cleavage of the F_0 protein is required for initiation of infection and is a well-characterized determinant of NDV pathogenicity (Nagai, 1995; Peeters *et al.*, 1999; Panda *et al.*, 2004b; de Leeuw *et al.*, 2005; Wakamatsu *et al.*, 2006; see section 'Determinants of NDV virulence').

The F glycoprotein of NDV contains six potential N-linked glycosylation sites at residues 85, 191, 366, 447, 471 and 541 (Fig. 3.6). All potential N-glycosylation sites are conserved among different NDV strains, suggesting that N-glycosylation at these sites might be important for the structural and functional integrity of the F protein. It is known that all six potential sites for the addition of N-linked carbohydrate are used (McGinnes *et al.*, 2001; Samal *et al.*, 2010). Mutational analysis showed that a live virus could not be rescued when the putative glycosylation site at residue 541 was removed, indicating that presence of carbohydrate at this site is essential for virus viability. Removal of carbohydrate at each of the remaining five glycosylation sites did not affect the cleavage of F protein or cell surface expression, but affected fusion activity and pathogenicity in different ways. Loss of carbohydrate at residues 85, 119, 366, and 471 did not have any effect on syncytia formation but the mutant viruses were attenuated. Loss of carbohydrate at residue 447 increased syncytia formation and the mutant virus was more virulent (Samal *et al.*, 2010).

Like other paramyxoviruses, NDV F protein also has several important domains which have been suggested to participate in fusion process. The TM domain is important for the conformation of the ectodomain, interaction with the HN protein, and fusion activity (Gravel *et al.*, 2011) Sequences adjacent to the fusion peptide and immediately upstream to the TM domain contain heptad repeats designated as heptad repeat A (HRA) and heptad repeat B (HRB), respectively. Mutational analysis in these heptad repeats abolished fusion activity suggesting that they play an important role in fusion process (Sergel *et al.*, 2001). Furthermore, HRB domain consists of a series of highly conserved leucine or isoleucine residues similar to leucine zipper motif found in DNA-binding proteins. Heptad repeat motifs mediate protein–protein interactions by enabling the formation of coiled coils. For fusion activation, the F trimer is thought to refold, inserting fusion peptides into target membranes and forming a very stable, six stranded coiled-coil domain composed of HRA and HRB complex. NDV F protein also has a second leucine zipper-like motif (HRD) located between amino acids 265–296, which consists of four heptads. It has been reported that single amino acid changes in this motif affect either folding of the molecule or the fusion activity (Reitter *et al.*, 1995). There is also the presence of HRC domain which extends from amino acid 81 to 102 in the F2 domain. However, its role in fusion process is yet to be determined.

Crystal structure of the NDV F protein revealed that the F protein trimer consists of a globular head held distal from the viral membrane by stalk and neck regions (Chen *et al.*, 2001). High resolution structures of the NDV F protein, a prototype in the family *Paramyxoviridae,* in its pre-fusion and post-fusion conformations now provide a structural basis for the analysis of the membrane fusion mechanism of the paramyxoviruses.

Overview of NDV replication

NDV attaches to respiratory epithelial cells by complexing of viral HN protein to sialic acid containing cell surface receptors such as gangliosides and N-glycoproteins. *In vivo*, NDV attaches first to cilia to infect respiratory epithelium. This is followed by activation of the F protein which leads to fusion of the viral and the host cell membranes. NDV can enter the cell via two main pathways: (i) pH-independent fusion of the viral envelope with the plasma membrane of host cell as discussed before, and (ii) receptor-mediated endocytosis. It has been shown that NDV can also infect cells through a caveolae-dependent endocytic pathway as an alternative route (Cantin *et al.*, 2007). In the cytoplasm of the host cell, the viral genome is transcribed into mRNAs and translated to viral proteins. The proteins N, P, and L are required for nucleocapsid assembly.

The positive-sense nucleocapsid 'antigenome' is then used as template for synthesis of progeny genomic RNAs. The assembly and release of infectious NDV particles has been shown to depend on membrane lipid rafts, where HN, F, and N proteins accumulate (Laliberte *et al.*, 2006). The nucleocapsid is formed by interactions of the P protein with N-RNA template and L protein. It has been shown that M protein is necessary and sufficient for NDV budding (Pantua *et al.*, 2006). It is thought that M–N interaction is responsible for incorporation of nucleocapsids into virus particles. The HN protein is incorporated into the envelope by interaction with the M protein, whereas; the F protein is incorporated into the envelope by interaction with the HN and/or N protein.

Genetics

Mutation rates for genomes of RNA viruses are in the range of 10^{-3}–10^{-5} substitutions per nucleotide (Moya *et al.*, 2000). This means each progeny genomic molecule of NDV will contain an average of 0.15 to several mutations. This rate of mutation suggests that viral progeny of a single NDV infected cell will be genetically heterogeneous or quasispecies. However, NDV appears to evolve very slowly in nature. Sequence analysis of NDV strains isolated over a 50 year period identified progressive accumulation of mutations, but this was not accompanied by progressive antigenic change (Sakaguchi *et al.*, 1989; Toyoda *et al.*, 1989). The slow accumulation of mutations indicates that NDV has already reached a high fitness value and rapid accumulation of mutations has no additional growth advantage for this virus in nature.

There is evidence of recombination in non-segmented negative-strand RNA viruses, but it occurs at very low frequency (Chare *et al.*, 2003). Although there have been many attempts to demonstrate recombination in negative-strand RNA viruses *in vitro*, only a single report of clear recombination is available (Spann *et al.*, 2003). It is thought that rapid association of the N protein with the genomic and antigenomic RNAs prevents interactions between RNA molecules, which might result in recombination.

Genomic recombination involving NDV has become a subject of recent debate (Afonso, 2008; Collins *et al.*, 2008; Han *et al.*, 2008b). Many attempts have been made using temperature sensitive mutants to identify NDV recombinants *in vitro*, but all have been unsuccessful (Granoff, 1959a,b; Dahlberg and Simon, 1969a,b). Phylogenetic analysis of F and HN genes of NDV strains isolated over a period of 50 years have shown existence of distinct lineages (Sakaguchi *et al.*, 1989; Toyoda *et al.*, 1989). It was concluded that the different lineages evolved by accumulation of point mutations and no gene exchange by recombination had occurred. A separate analysis of M gene among NDV strains also showed no evidence of recombination (Seal *et al.*, 2000).

Recently, there have been reports showing indirect evidence of natural recombination in NDV involving vaccine and circulating viruses (Han *et al.*, 2008a; Qin *et al.*, 2008a; Chong *et al.*, 2010). These studies were performed by phylogenetic and recombination analyses of a single gene or full NDV genome. These analyses occasionally identified a virus with sequence discontinuity suggestive of a recombination breakpoint. The putative parents for the recombinant were identified from the sequence database. Using this method recombination was found to occur in all coding genes of NDV (Zhang *et al.*, 2010). Although the possibility of recombination between vaccine and circulating viruses exist, the evidence based on phylogenetic analysis should be considered speculative and it will require *in vitro* and *in vivo* recombination experiments with known parents to unequivocally demonstrate recombination in NDV.

Reverse genetics

Reverse genetics is a method by which infectious viruses can be generated entirely from cloned DNA and provides a method for introducing desired changes into the viral genome. The ability to recover infectious negative-strand RNA viruses, including NDV, from cDNA clones has not only allowed investigation of the structure and function of viral genes and their proteins, but also to engineer live vaccines and vaccine vectors. To initiate an infectious cycle of negative-strand

RNA viruses, intracellular reconstitution of an ribonucleoprotein complex entirely from cloned cDNAs is required. Most of the systems for recovery of NDV are based on co-transfection of cell with a plasmid expressing the full-length antigenomic RNA with three other support plasmids encoding viral N, P and L proteins all under the control of bacteriophage T7 RNA polymerase promoter. The positive-sense antigenomic RNA instead of negative-sense genomic RNA was used to avoid hybridization with the positive-sense mRNAs transcribed from the support plasmids. The T7 RNA polymerase was provided by either a recombinant vaccinia virus expressing the T7 gene or a cell line constitutively expressing the T7 polymerase.

This system has been used for recovery of different pathotypes of NDV strains: (1) lentogenic strains LaSota (Peeters *et al.*, 1999; Romer-Oberdorfer *et al.*, 1999; Huang *et al.*, 2001;) and B1 (Nakaya *et al.*, 2001); (2) mesogenic strains Beaudette C (Krishnamurthy *et al.*, 2000) and Anhinga (Estevez *et al.*, 2007); and (3) velogenic strains Herts 33 (de Leeuw *et al.*, 2005), ZJ1 (Liu *et al.*, 2007) and Texas GB (Paldurai *et al.*, 2010b). A pigeon paramyxovirus-1 (strain AV324) has also been rescued (Dortmans *et al.*, 2009). Reverse genetics has greatly benefited our understanding of NDV pathogenicity by introducing specific mutations into the genome of the virus and then analysing the phenotype of the rescued virus. For example, the role of the F protein cleavage site in NDV pathogenicity was determined (Peeters *et al.*, 1999; Panda *et al.*, 2004b; de Leeuw *et al.*, 2005). The F and HN proteins were identified as major determinants of NDV virulence (Peeters *et al.*, 1999; Huang *et al.*, 2004c). The V protein was identified as an IFN antagonist (Huang *et al.*, 2003; Park *et al.*, 2003). The role of the internal proteins (N, P and L) in NDV virulence was determined by exchanging the N, P, and L genes individually between avirulent and virulent NDV strains (Rout and Samal, 2008; Dortmans *et al.*, 2010b). These results demonstrated that the virulence of NDV is a multigenic trait. In addition, the reverse genetics system has made it possible to engineer improved NDV vaccines (Peeters *et al.*, 2001). Recombinant NDVs have been used to evaluate the contributions of the avian influenza virus HA, NA, and M2 surface proteins to the induction of neutralizing antibodies and protective immunity (Nayak *et al.*, 2010).

NDV as a vaccine vector

NDV has a number of characteristics that make it an attractive candidate as a vaccine vector for both human and animal uses. A wide range of NDV strains exist that can be used as vectors. NDV replicates efficiently in embryonated chicken eggs, cell culture and in the respiratory tract of avian and non-avian species. NDV has a simple genome that encodes only seven proteins. NDV does not integrate into host cell genome. Recombination involving NDV is extremely rare. NDV has a modular genome that facilitates genetic manipulation. A number of strategies exist for attenuating NDV by reverse genetics. The level of expression of a foreign gene can be modified by altering its position in the gene order. NDV infects via the intranasal route and therefore induces both local and systemic immune responses. Local immunity is particularly important since most pathogens enter through the respiratory tract.

To construct an NDV vaccine vector, by reverse genetics the ORF encoding the foreign antigen of interest is engineered so that it is flanked by NDV-specific gene-start and gene-end sequences. The transcription cassette is then inserted into the NDV genomic or antigenomic cDNA at a gene junction and this becomes an additional gene that is transcribed into a separate mRNA (Krishnamurthy *et al.*, 2000; Huang *et al.*, 2001; Bukreyev *et al.*, 2006). Although the upper limit of the foreign gene that can be accommodated in the NDV genome is not known, a foreign gene of 3.8 kb in length has been inserted into the NDV genome without severely affecting virus replication efficiency (DiNapoli *et al.*, 2007a). Based on studies, the transcription cassette can be inserted between any two genes, before N gene or behind L gene, but the level of expression of the foreign gene is very low when the transcription cassette is inserted between N and P genes (Zhao and Peeters, 2003; Carnero *et al.*, 2009). The level of expression of the foreign gene is higher when it is inserted closer to the 3′end of the genome (Huang *et al.*, 2001; Carnero *et al.*, 2009).

NDV is an ideal vaccine vector for poultry. Live attenuated NDV vaccines are widely used all over the world. Therefore, a live attenuated NDV carrying the protective antigen of another avian pathogen can be used as a bivalent vaccine. Such a vaccine will be economical for the poultry industry. NDV replicates efficiently in the respiratory tract of chickens and induces strong local and systemic immune responses. Therefore, it will also induce a strong immune response to the foreign antigen. The ability of NDV to infect a wide variety of non-avian species makes it also a promising vaccine vector for other animal species. NDV can be a vaccine delivery vector for foreign animal diseases or diseases for which currently vaccines are not available.

Several special features make NDV a promising vaccine vector for use in humans. NDV is not a natural human pathogen and is highly attenuated in humans owing to natural host range restriction. Clinical trials for cancer treatment have also shown that NDV is safe in humans with very few side effects. Another advantage is that most humans do not have pre-existing immunity to NDV.

NDV recombinants expressing influenza virus HA or Simian immunodeficiency virus (SIV) Gag protein were tested in mice and were found to be immunogenic (Nakaya *et al.*, 2001, 2004). The mouse model was also used to evaluate NDV recombinants expressing the F glycoprotein of human respiratory syncytial virus, which provided only partial reduction of challenge virus replication (Martinez-Sobrido *et al.*, 2006). Vaccination of mice with NDV expressing the HA of highly pathogenic avian influenza virus (HPAIV) conferred a high level of protection against challenge (Ge *et al.*, 2007). As mouse model may not provide a reliable indication of the level of attenuation and immunogenicity that might be observed in humans, recombinant NDV expressing the HN gene of human parainfluenza virus type 3 (HPIV3) was tested in monkeys and was found to be highly immunogenic (Bukreyev *et al.*, 2005). NDV-vectored vaccine expressing the HA protein of H5N1 HPAIV was found to elicit high levels of serum and mucosal antibodies against HPAIV in monkeys (DiNapoli *et al.*, 2007b) and were protected against HPAIV challenge (DiNapoli *et al.*, 2010). In another study, recombinant NDV expressing the spike (S) glycoprotein of SARS-CoV induced a high level of SARS-CoV neutralizing antibodies and restricted the replication of the challenge virus (DiNapoli *et al.*, 2007a). These studies indicate that NDV has good potential for use as a vaccine vector in humans.

The potential of NDV as a vaccine vector for veterinary pathogens has also been explored. A recombinant NDV expressing the VP2 protein of infectious bursal disease virus (IBDV) provided dual protection against NDV and IBDV in chickens (Huang *et al.*, 2004b). NDV-vectored vaccines expressing the HA protein of H5N1 HPAIV (Park *et al.*, 2006; Ge *et al.*, 2007; Romer-Oberdorfer *et al.*, 2008; Nayak *et al.*, 2009;) and H7N7 HPAIV (Park *et al.*, 2006; Schroer *et al.*, 2009) have been shown to provide dual protection against NDV and HPAIV in chickens. Recombinant NDV expressing the gD glycoprotein of bovine herpesvirus-1 (BHV-1) conferred partial protection against virulent BHV-1 challenge in calves (Khattar *et al.*, 2010). Recently, a NDV-vectored vaccine expressing the Gn and Gc glycoproteins of Rift valley fever virus (RVFV) protected mice against RVFV challenge and elicited neutralizing antibody response in lambs (Kortekaas *et al.*, 2010a). It was further shown that a NDV vectored vaccine only expressing RVFV Gn glycoprotein elicited neutralizing antibodies against RVFV in calves (Kortekaas *et al.*, 2010b). These results indicate that NDV is a promising vaccine vector for veterinary pathogens.

NDV as an oncolytic agent

NDV selectively replicates in tumour cells and induces death while sparing normal cells (Reichard *et al.*, 1992). Owing to this property, NDV has been evaluated as a potential anti-cancer agent in humans. The anti-cancer effect of NDV was first reported by Cassel and Garret in 1965 . Since then many preclinical and clinical trials using NDV for cancer therapy have been performed (Schirrmacher *et al.*, 2001). Some of the clinical trials reported positive results and some did not. NDV has been labelled as a complementary and alternative medicine by the National Cancer Institute in the USA. Several approaches have been used during clinical investigations, such as direct injection of live virus to patients through various routes of administration. The development of anti-NDV antibodies with subsequent re-growth of tumours has limited the applicability of this approach. Viral oncolysates, preparations containing plasma membrane fractions from NDV-infected cancer cells, were tested as anti-cancer vaccines. Oncolysate-based vaccines are injected under or into the skin. A new approach has also been used, which involves the use of intact cancer cells infected with NDV as whole-cell vaccines. The tumour cells used in the vaccine are changed in laboratory so that they can not multiply in the patients. Whole-cell vaccines stimulate the immune system better than oncolysates. They are given only by injection under the skin.

NDV has a number of desirable properties that make it an attractive oncolytic agent. NDV is a fast growing virus with progeny virions detectable as early as 3 h post-infection and is highly fusogenic in tumour cells. NDV has the ability to replicate up to 10,000-fold faster in human cancer cells than in most normal human cells. NDV is known to enter cells by binding to sialic acid residues, which are present on most human cancer cells, making it suitable for use in a broad range of cancer cell types. There is no pre-existing immunity to NDV in the general human population. NDV is safe and highly tolerated in cancer patients. From all clinical studies, it is known that NDV when applied to humans, usually induces only mild fever for a day or conjunctivitis.

NDV strains suitable for anti-cancer therapy have been classified as either lytic or non-lytic for human cells. Both lytic and non-lytic strains have been investigated as potential anti-cancer agents. One major difference between lytic and non-lytic strains is that lytic strains are able to produce infectious progeny in human cancer cells and have the ability to spread in tumour tissues, whereas non-lytic strains do not produce infectious progeny and cannot spread in tumour tissues. The reason being the progeny of lytic strains contain active F protein, whereas the progeny of non-lytic strains contain inactive F protein. Lentogenic strains behave as non-lytic virus, whereas mesogenic and velogenic strains behave as lytic virus. Lytic strains of NDV selectively replicate in and rapidly kill human tumour cells. Non-lytic strains of NDV also kill infected cells, but more slowly with death apparently because of the result of abnormal host cell metabolism. The NDV strains that have been evaluated most widely for treatment of human cancer are PV701, 73-T, MTH-68 and Ulster (Cassel and Murray, 1992; Ahlert *et al.*, 1997). Strains PV701, 73-T, and MTH-68 are lytic, while strain Ulster is non-lytic.

The inherent oncolytic properties of NDV are believed to derive from defective interferon (IFN) signalling pathways in tumour cells and its immune-stimulatory capacity in human cells (Stojdl *et al.*, 2000). NDV is a potent inducer of type I IFN by viral protein and RNA (Fournier *et al.*, 2003). Thus, in tumour cells with a defective antiviral response, NDV replicates efficiently, but in normal cells with an effective antiviral response, the replication of NDV is compromised. Oncolytic NDV strains are cytotoxic to human tumour cell lines of ectodermal, endodermal and mesodermal origin (Fabian *et al.*, 2007). They exert oncolysis by both intrinsic and extrinsic caspase-dependent pathways of cell death (Elankumaran *et al.*, 2006).

Recently, reverse genetics technology has been used to enhance the oncolytic property of NDV (Vigil *et al.*, 2007). It was shown that the oncolytic activity of recombinant NDV was enhanced either by containing a highly fusogenic F protein (Bian *et al.*, 2005; Elankumaran *et al.*, 2010), expressing immunostimulatory molecules, such as GMCSF, IFN-γ, IL-2 and TNFα (Vigil *et al.*, 2007; Zamarin *et al.*, 2009) or IgG monoclonal antibody (Puhler *et al.*, 2008). It was also shown that recombinant NDV expressing green fluorescent protein could be used to detect peritoneal gastric cancer cells and provide important prognostic information (Song *et al.*, 2010). These results collectively suggest that reverse genetic approaches can improve the safety and oncolytic property of NDV for effective cancer treatment in humans.

Antigenic and genetic variations of NDV isolates

NDV is monotypic in nature, i.e. only a single serotype of the virus has been described. However, both antigenic and genetic diversity are recognized for NDV isolates. Monoclonal antibodies have been used to demonstrate antigenic variation among NDV strains and isolates (Russell and Alexander, 1983; Abenes *et al.*, 1986; Alexander *et al.*, 1987; Erdei *et al.*, 1987; Lana *et al.*, 1988). On the basis of monoclonal antibody reactivity, NDV strains and isolates were placed into 10 antigenic groups (Alexander *et al.*, 1997b). Viruses within a monoclonal antibody group shared biological and epizootiological properties (Russell and Alexander, 1983). Monoclonal antibodies have also been used to distinguish between specific strains (Russell and Alexander, 1983), and between vaccine and virulent viruses (Srinivasappa *et al.*, 1986). However, monoclonal antibody binding tests have failed to differentiate viruses that are antigenically similar, but not genetically identical.

Availability of automated DNA sequencing techniques and NDV sequence information in computer database have allowed classification of NDV isolates based on sequence relatedness. Currently, two different classification systems are used worldwide to group NDV isolates. Both systems use sequence of the F gene because the amino acid sequence at the F protein cleavage site is a major determinant of NDV pathogenicity. One system is based on a 375-nucleotide sequence of the F gene, which includes the F protein cleavage site (Aldous *et al.*, 2003). This system has classified NDV isolates into six lineages and 16 sublineages (Snoeck *et al.*, 2009). The other system, using genome length and sequence of F and L genes, classified NDV into two major subdivisions, Class I and Class II (Ballagi-Pordany *et al.*, 1996; Czegledi *et al.*, 2006; Kim *et al.*, 2007b). The Class I viruses are generally avirulent to chickens and have been recovered primarily from waterfowl and are frequently isolated from U.S. live bird market samples (Alexander *et al.*, 1992; Kim *et al.*, 2007a,b). Class I viruses have a genome consisting of 15,198 nt (Czegledi *et al.*, 2006). The Class II viruses are commonly isolated from poultry and from pet and wild birds (Seal *et al.*, 2005; Kim *et al.*, 2007b). Class I viruses are further divided into nine genotypes and Class II viruses into 11 genotypes (Czegledi *et al.*, 2006; Kim *et al.*, 2007b; Maminiaina *et al.*, 2010). Class II viruses are most characterized because the vast majority of sequence data currently available are for Class II viruses. Class II genotypes (I–IV and IX) emerged 'early' (1930–1960) and have a genome consisting of 15,186 nt. Genotypes (V–VIII, X and XI) that emerged 'late' (after 1960) have a genome consisting of 15,192 nt (Czegledi *et al.*, 2006; Maminiaina *et al.*, 2010). Genotype I consists of viruses that are of low virulence and some are used as vaccines. Genotype I also includes the virulent NDV responsible for 1998–2000 Australian outbreaks. Genotype II includes viruses of low virulence that are used as vaccines worldwide, such as LaSota and B1. It also includes the neurotropic virulent NDV strain Texas GB. Genotype III viruses were mostly isolated in Japan before 1960, but have also been isolated in Taiwan in 1969 and in Zimbabwe in 1990 (Yu *et al.*, 2001). Genotype IV viruses were predominantly isolated in Europe before 1970 and recently isolated in India (Czegledi *et al.*, 2006; Kapgate *et al.*, 2010a,b). Genotypes V-VIII were first recognized in the mid-1960s (Czegledi *et al.*, 2006) and contain only virulent viruses (Miller *et al.*, 2010b). Genotype V viruses were recognized in South and Central America in 1970 and have caused outbreaks in Europe (Ballagi-Pordany *et al.*, 1996) and in North America (Wise *et al.*, 2004a). These viruses are still circulating in Mexico (Perozo *et al.*, 2008). Genotype VI viruses circulated in Asia between 1960 and 1985 (Mase *et al.*, 2002). Genotype VII and VIII were reported in the 1990s from several countries (Aldous *et al.*, 2003; Abolnik *et al.*, 2004; Bogoyavlenskiy *et al.*, 2009; Snoeck *et al.*, 2009). Genotype IX includes the first virulent outbreak virus from China in 1948 and occasional isolates from China (Wang *et al.*, 2006). Genotype X viruses have been isolated exclusively from Taiwan in 1969 and 1981 (Tsai *et al.*, 2004). Recently, class II viruses belonging to a new genotype XI were isolated from apparently healthy chickens in Madagascar (Maminiaina *et al.*, 2010). All genotypes are still in circulation. The genotyping of NDV has proven to be a useful method for investigating epidemiological relationship among the NDV isolates present in various parts of the world.

Pathogenicity of NDV isolates

NDV isolates cause a broad range of clinical disease in chickens, varying from asymptomatic to 100% mortality. Historically, NDV isolates have been classified into three major clinicopathological groups (pathotypes) based on their disease signs and lesions produced in chickens: low virulent (lentogenic), moderately virulent (mesogenic), and highly virulent (velogenic). Lentogenic viruses produce a mild respiratory disease or subclinical infection. Mesogenic viruses produce more severe respiratory diseases with mortality only in very young birds, while velogenic viruses produce severe disease with high mortality. Velogenic viruses are further divided into viscerotropic, which cause mortality with haemorrhagic lesions in the intestines of dead birds and neurotropic, when neurological diseases predominate without haemorrhagic lesions in the intestine (Hanson and Brandly, 1955). Three internationally accepted pathogenicity tests are devised to determine the pathotype of NDV isolates. These tests are mean death time (MDT) in embryonated chicken eggs, intracerebral pathogenicity index (ICPI) in day-old chicks and the intravenous pathogenicity index (IVPI) in 6-week-old chickens (Alexander, 2009). Of the three tests, ICPI test is required by the World Organization for Animal Health (formerly known as OIE) for the *in vivo* determination of NDV virulence.

Determinants of NDV virulence

The virulence of NDV strains vary widely in chickens. The viral determinants responsible for the wide range of virulence among NDV strains are not fully known. To understand the molecular basis of virulence of NDV, the focus has been on the role of the F protein. In paramyxoviruses, the F protein is expressed as an inactive precursor (F_0) that is cleaved into two biologically active disulfide bonded F1-F2 subunits by host cell proteases (Lamb and Parks, 2007a). Activation of the viral infectivity is dependent upon cleavage of the F protein. The amino acid sequence at the F protein cleavage site determines the substrate specificity for different types of cellular proteases (Kawahara *et al.*, 1992). For this reason, the F protein cleavage site has been postulated as the primary determinant of NDV virulence (Nagai *et al.*, 1976; Toyoda *et al.*, 1987). The consensus sequence of the F protein cleavage site of velogenic and mesogenic strains is ^{112}R/K-R-Q-R/K-R↓F^{117}. The consensus sequence of the F protein cleavage site of lentogenic strains is ^{112}G/E-K/R-Q-G/E-R↓L^{117}. The F protein cleavage site of velogenic and mesogenic strains contains polybasic amino acids that is the preferred recognition site for furin (R-X-K/R-R↓F), which is an intracellular protease that is present in most cells. This provides for efficient cleavage of F protein in a wide range of tissues, making it possible for virulent strains to spread systemically, resulting in fatal infection. In contrast, lentogenic NDV strains have one or two basic residues at the –1 and –4 positions relative to the cleavage site. These cleavage sequences are insensitive to intracellular proteases and depend on secretory proteases (or, in cell culture, added trypsin) for cleavage. This limits the replication of avirulent strains to the respiratory and enteric tracts where the secretory protease is found. In addition, the F1 subunits of virulent NDV strains begin with a phenylalanine residue, whereas most avirulent NDV strains have a leucine residue at this position. The presence of phenylalanine at the beginning of F1 subunit has been suggested as a possible contributor to the neurological effects (Kattenbelt *et al.*, 2006). However, there are cases of NDV found in China that had lentogenic motif G-R-Q-G-R↓L at their F protein cleavage site, but were highly virulent in chickens (Tan *et al.*, 2008), and also there are cases in Africa where NDV with virulent motif R-R-Q-K-R↓F at their F protein cleavage site were isolated from chickens apparently not vaccinated and in good health (Servan de Almeida *et al.*, 2009). Furthermore, the mesogenic and velogenic NDV strains have the same amino acid sequence at their F protein cleavage site but they differ in virulence, indicating that additional viral factors contribute to the virulence of the NDV.

The role of F protein cleavage site in virulence has been evaluated using reverse genetics techniques. The F protein cleavage site of the lentogenic strain LaSota (G-R-Q-G-R↓L) was changed to the F protein cleavage site of virulent NDV strains (R-R-Q-R-R↓F) and (R-R-Q-K-R↓F). Pathogenicity tests in chickens showed that the amino acid change at the F protein cleavage site resulted in a dramatic increase in virulence of the mutated viruses (Peeters *et al.*, 1999; Romer-Oberdorfer *et al.*, 2003; Panda *et al.*, 2004b).

However, the modification of the cleavage site did not completely convert the lentogenic strain to a highly virulent velogenic strain, suggesting the involvement of additional virulence factors. The individual amino acids at the F protein cleavage site have also been examined for their requirement to virulence. It was found that F at position 117, R at 116, K or R at 115 and R at 113 are required for virulence of NDV (de Leeuw *et al.*, 2003). Interestingly, Glutamine (Q) is present at position 114 of most avirulent and virulent NDV strains, except some recently reported isolates from Africa (Servan de Almeida *et al.*, 2009) and Madagascar (Maminiaina *et al.*, 2010). These isolates contain R-R-R-K-R↓F or R-R-R-R-R↓F at their F protein cleavage site but were isolated from unvaccinated apparently healthy chickens. Recently, the role of Q at position 114 was examined by changing the F protein cleavage site of a mesogenic strain Beaudette C (R-R-Q-K-R↓F) to those of the African and Madagascar strains (R-R-R-K-R↓F) or (R-R-R-R-R↓F). It was found that change of Q to R at position 114 decreased virus replication and pathogenicity of the mesogenic strain Beaudette C, suggesting that the presence of Q at this position is necessary for efficient cleavage by cellular proteases (Samal *et al.*, unpublished results).

The HN protein, which possesses both the receptor recognition and neuraminidase activities of the virus, has also been evaluated for its role in virulence by using reverse genetics approach (Huang *et al.*, 2004c; de Leeuw *et al.*, 2005; Kim *et al.*, 2009). It was shown that the HN protein determines tropism and contributes to virulence (Huang *et al.*, 2004c; de Leeuw *et al.*, 2005). However, in another study, replacement of the HN gene of a mesogenic NDV strain by the HN gene of a neurotropic or viscerotropic NDV strain failed to enhance the pathogenicity of the chimeric virus (Estevez *et al.*, 2007). Furthermore, replacement of both the F and HN genes of a mesogenic NDV strain by F and HN genes of velogenic NDV strains also did not increase the virulence of the chimeric viruses indicating that the F and HN genes and their homotypic interactions are not responsible for the wide range of virulence observed among mesogenic or velogenic strains (Estevez *et al.*, 2007). These results suggest that additional factors other than F and HN contribute to the virulence of NDV.

The individual contribution of internal proteins N, P, and L to the virulence of NDV has also been examined using reverse genetics methods (Rout and Samal, 2008; Dortmans *et al.*, 2010b). Results of a study, in which individual genes were exchanged between an avirulent NDV strain and a virulent NDV strain belonging to the same phylogenetic lineage and genotype, showed that the N and P proteins play a minimal role in NDV virulence, but the L protein is associated with virulence (Rout and Samal, 2008). However, in another study, in which the genes were exchanged between an avirulent NDV strain and a virulent NDV strain belonging to different phylogenetic lineage and genotype, showed that the whole viral replication complex (N, P, and L) contributes to the virulence of NDV (Dortmans *et al.*, 2010b). This study also showed that the M protein played a minor role in NDV virulence. The V protein of NDV has also been suggested as a determinant of virulence (Alamares *et al.*, 2010). Taken together, these observations suggest that virulence of NDV is a complex trait determined by multiple genetic factors. It is conceivable that the virulence of NDV is due to both F and HN proteins, but the degree of virulence is determined by the viral replication complex.

Emergence of virulent NDV

It is not fully understood how virulent NDV is introduced into a geographic area. However, three possibilities have been proposed to explain the emergence of virulent NDV: (1) virulent NDV resides unnoticed in vaccinated poultry population until development of an outbreak; (2) virulent viruses are transmitted to chickens from another species in which it shows less severe disease; and (3) virulent viruses arise from low virulent viruses by mutation (Hanson, 1972).

The first explanation may be possible in some parts of the world where ND is endemic. There is now increased evidence that current NDV vaccines may not be very effective under certain circumstances (Kapczynski and King, 2005; Miller *et al.*, 2007; Hu *et al.*, 2009). In these areas, current vaccines do not stop viral shedding, allowing unnoticed circulation of the virulent virus in the vaccinated bird population.

Several evidences exist that support the second explanation. It was found that during 1970–73 ND outbreaks in the US, virulent viruses were introduced by the movement of infected captive caged birds of mostly psittacine species (Walker *et al.*, 1973). Virulent NDV have frequently been isolated from dead, sick or asymptomatic psittacines and other species of pet birds (Senne *et al.*, 1983; Panigrahy *et al.*, 1993). Therefore, wild populations of tropical birds, particularly, psittacines, are thought to be natural reservoirs of virulent NDV. However, virulent NDV has never been isolated from wild birds belonging to these species. It is also possible that wild birds captured for the pet bird trade were not previously infected, but got infected during transportation, marketing or at holding facilities (Kaleta and Baldauf, 1988). Importation of infected pet birds remain the major source of introduction of virulent NDV into the U.S. Parrots and other psittacine birds are of special concern because they can carry virulent NDV without showing any clinical signs. Wild birds have been suspected as another source of introduction of virulent NDV into poultry populations. Pigeons and cormorants represent two known natural reservoirs of virulent NDV among wild birds (Alexander, 1988; Johnston and Key, 1992; Kuiken *et al.*, 1998). Pigeons and cormorants have been suspected of being the source of outbreaks in poultry (Alexander *et al.*, 1984a, 1985; Banerjee *et al.*, 1994; Heckert *et al.*, 1996a).

The APMV-1 isolated from pigeons are variants of NDV and are known as pigeon paramyxovirus-1 (PPMV-1). These viruses cause neurological disease in pigeons and doves (*Columbiformes*), have specific monoclonal antibody binding patterns, and often do not haemagglutinate chicken red blood cells well (Alexander, 1988; Alexander *et al.*, 1984b). All PPMV-1 isolates have a multibasic F protein cleavage site motif and are moderately virulent to chickens by ICPI test (Collins *et al.*, 1994). In many countries, including the U.S., the virus is believed to be endemic in pigeon populations (Alexander, 1988; Kim *et al.*, 2008a). PPMV-1 strains have been shown to infect poultry and several ND outbreaks in chickens have been attributed to PPMV-1 (Pearson *et al.*, 1987; Alexander *et al.*, 1998). The virulence of PPMV-1 strains has been reported to be variable in chickens (Alexander and Parsons, 1984; Kommers *et al.*, 2001). However, increased pathogenicity in chickens has been identified when PPMV-1 is passaged in chickens or embryonated eggs, indicating that PPMV-1 is a threat to poultry (Kommers *et al.*, 2001; Alexander and Parsons, 1984; Fuller *et al.*, 2007). Sequence analysis of chicken passaged PPMV-1 showed that the increase in virulence was not associated with changes in the F protein (Kommers *et al.*, 2003; Dortmans *et al.*, 2009), but by mutations in the L and P proteins (Dortmans *et al.*, 2010a).

Virulent NDV can arise by mutation from viruses of low virulence, which is a concern for poultry producers worldwide. Wild waterfowl, especially ducks and geese, appear to be natural reservoirs for avirulent NDV. It has been shown that the virulent NDVs responsible for outbreaks in Ireland in 1990 were antigenically and genetically very similar to avirulent viruses isolated from wild waterfowl, suggesting that virulent viruses originated from wild waterfowl (Collins *et al.*, 1993, 1998). It has also been experimentally demonstrated that the lentogenic NDV isolates present in wild waterfowl have the potential to become velogenic in chickens as a result of amino acid substitutions at the F protein cleavage site (Shengqing *et al.*, 2002). Another evidence where a virulent virus emerged from an avirulent virus by mutation came from 1998–2000 ND outbreaks in Australia (Gould *et al.*, 2001; Westbury, 2001). Nucleotide sequence analysis of the HN and F genes demonstrated that the virulent viruses isolated from the outbreak were closely related to a recognized Australian lentogenic strain and only two nucleotide changes at the F protein cleavage site were necessary to change the lentogenic virus to a velogenic virus. However, nucleotides present at the F protein cleavage site of lentogenic strains are highly stable and exceptional circumstances may be necessary to allow mutations to change a low virulent cleavage site to a high virulent cleavage site (Miller *et al.*, 2009).

Transmission and spread

NDV is a highly infectious virus. When introduced into an avian population, NDV is rapidly transmitted among susceptible birds by either inhalation or ingestion (Alexander, 1988). The virus is present at high concentrations in secretions and excretions from actively infected birds and also in all tissues of dead birds. Inhalation of infectious virus occurs via aerosol and is probably the major mode of transmission. Transmission also occurs by ingestion of faeces and contaminated food and water. This is the most likely method of transmission of avirulent enteric NDV. Once introduced into poultry, backyard birds, or other captive bird populations, the virus spreads rapidly between premises by the movement of apparently healthy but infected birds, movement of people and contaminated equipment, movement of contaminated poultry products, contaminated food and water, and by airborne spread from one premise to another.

Clinical signs

Clinical signs in birds infected with NDV vary widely and are dependent on factors such as: the virulence of the virus, host species, age of host, route of exposure, immune status and environmental conditions. In susceptible hosts, such as chickens, the NDV strains show a broad range of virulence. The disease is more severe in younger birds. The incubation period for ND after natural exposure varies from 2–15 days (average 5–6 days). The generalized signs of ND include depression, loss of appetite, abnormal thirst, severe dehydration, emaciation and fever. Infection with velogenic strains may result in sudden high mortality with few clinical signs. The velogenic viscerotropic Newcastle disease (VVND) pathotype viruses affect the viscera or internal organs. Although the most noticeable signs are in the digestive tract; the heart, liver, and kidneys are also affected. VVND virus usually produces marked congestion in the trachea, ulcer and erosions on the soft palate and upper oesophagus, and inflammation and haemorrhages in the proventriculus (Alexander, 1998). The VVND virus may cause severe respiratory signs, oedema of the face (especially the eyelids), greenish diarrhoea, muscular tremors, torticollis, paralysis of legs and wings, and opisthotonos. Mortality can reach up to 100% in susceptible chickens. In neurotropic velogenic forms of ND, the neurological signs are more prominent. Usually, other signs of VVND are absent. The morbidity may be 100%, but the mortality is generally lower, up to 50% in adult birds and 90% in young chickens (Alexander, 1988).

Mesogenic strains of NDV usually cause respiratory disease in adult chickens. Neurological signs are rarely seen in adult chickens. These strains sometimes kill young chickens, but rarely kill older birds. Lentogenic viruses usually do not cause disease in young and adult chickens. But in rare cases some lentogenic strains, such as LaSota, can cause serious respiratory disease in young birds (Alexander, 2003).

The clinical signs of ND vary widely in other avian species. Turkeys are highly susceptible to NDV, but the generalized signs are less severe with predominating respiratory and nervous signs (Box *et al.*, 1970; Alexander *et al.*, 1997a). Ducks and geese show subclinical signs when infected with mesogenic or velogenic strains of NDV (Higgins, 1971). The signs of ND vary widely in wild birds. Psittacine species are highly susceptible to ND, but the signs are always confined to the nervous system. Pigeons and cormorants commonly show central nervous system signs, which include paralysis of legs and wings (Barton *et al.*, 1992; Kuiken *et al.*, 1998). Pheasants of all ages are susceptible to ND. The clinical disease in pheasants include nervous and respiratory signs (Aldous and Alexander, 2008).

Pathological lesions

VVND viruses can produce a wide range of gross lesions. Although no pathognomonic lesions are found in any form of the disease, presence of haemorrhagic lesions in the intestine of infected chickens has been used to distinguish viscerotropic viruses from neurotropic viruses (Hanson *et al.*, 1973). These haemorrhagic lesions are often prominent in the mucosa of proventriculus and caecal tonsils and other lymphoid patches in the intestines. Other lesions include enlarged and mottled spleen, haemorrhages in the caudal part of the pharynx and proximal trachea, and pulmonary oedema. In case of velogenic neurotropic strains, even though neurological signs are present, all tissues including brain, may be grossly normal. NDV produces a marked effect on lymphocytes, lymphoid organs and vascular tissues. Microscopically, necrosis of the lymphatic cells at the caecal tonsil and spleen are seen with viscerotropic strains. With neurotropic strains, microscopic lesions are present most regularly in the brain stem and cerebellum (Alexander, 1988).

Diagnosis

A definitive diagnosis of ND requires isolation and characterization of the virus. However, in an endemic area, clinical signs and gross lesions may be highly suggestive of the disease. For virus isolation, tracheal and cloacal swabs are good sources of virus from living birds. Spleen, lung, trachea, brain and intestines (especially caecal tonsils) are usually collected for virus isolation from dead or moribund birds. Although NDV grows well in primary chicken cells and many established cell lines, embryonated chicken eggs (ECEs) are universally used for virus isolation. ECE is an extremely sensitive and convenient system for propagation of all NDV isolates. Although all NDV isolates replicate well in chicken embryo kidney cells, lentogenic isolates require supplementation of exogenous protease in the medium for replication in avian fibroblast and established cell lines. In most instances, acetylated typsin (1 μg/ml) or 10% fresh allantoic fluid is used as the source of exogenous protease. Nine-day-old ECEs are injected with 0.1 ml of the sample into the allantoic cavity. The eggs are incubated at 37°C and examined twice daily. Dead eggs as well as eggs after 5–7 days of incubation are chilled at 4°C and the allantoic fluid is harvested. The presence of virus in the allantoic fluid is determined by the HA test (Alexander, 2009).

Virus characterization

The HI test is commonly used to identify NDV. However, the pathogenicity of the isolate must be determined to associate the virus with the disease. In addition, panels of monoclonal antibodies can be used to characterize and group isolates (Alexander *et al.*, 1987; Lana *et al.*, 1988). Although monoclonal antibodies are very helpful in diagnosis of NDV, their ability to detect a broad range of NDV isolates is often limited. Other serological tests, such as ELISA (Snyder *et al.*, 1984; Adair *et al.*, 1989), virus neutralization assay in chicken embryos (Beard, 1980) and plaque neutralization test (Beard, 1984) are available, but they have limited value in assessing the pathogenicity of the isolate.

Three *in vivo* tests have been used for the assessment of pathogenicity of NDV isolates: (1) MDT in 9-day-old eggs, (2) ICPI in 1-day-old chicks, and (3) IVPI in 6-week-old chickens (Alexander, 2009). The MDT is the mean time in hours for the minimum lethal dose to kill chicken embryos. The MDT has been used to classify NDV isolates as velogenic (taking less than 60 h to kill), mesogenic (taking between 60 and 90 h to kill), and lentogenic (taking more than 90 h to kill; Alexander, 2009). ICPI is a weighted score of clinical signs after intracerebral injection of the virus into each of 10-day-old chicks. The velogenic viruses give indices that approach the maximum score of 2.0, lentogenic viruses give values close to 0.0. IVPI is a weighted score of clinical signs after intravenous injection of the virus into 10 6-week-old chickens. Lentogenic and some mesogenic strains have IVPI values of 0.0, the indices for velogenic viruses approach the maximum score of 3.0. Although these conventional pathogenicity tests have been invaluable in determining the virulence of chicken isolates, sometimes these tests do not reflect the true pathogenicity of viruses isolated from other species. For example, many pigeon isolates (PPMV-1) have a MDT value typical of lentogenic strains and an ICPI value of mesogenic strains even though the isolate caused severe disease and mortality in pigeon (Alexander, 1998). Furthermore, increased pathogenicity in chickens has been identified when PPMV-1 is passaged in chickens or embryonated eggs (Alexander and Parsons, 1984; King, 1996; Kommers *et al.*, 2003).

The conventional pathogenicity tests are time consuming and expensive; hence, several rapid pathotyping methods based on reverse transcription (RT)-PCR have been developed (Miller *et al.*, 2010b). Of these assays the real-time RT-PCR (RRT-PCR) offers the highest sensitivity. Several RRT-PCR assays have been developed in the last decade around the world to detect viruses circulating in their locations (Miller *et al.*, 2010b). In the US, two different RRT-PCR assays have been extensively used (Wise *et al.*, 2004b). An M-gene based assay was developed for the detection of NDV isolates (Kim *et al.*, 2007a,b). However, owing to sequence variation of the M gene, some NDV isolates are not detected by this assay. Therefore, a L-gene based RRT-PCR assay in conjunction with a new M-gene assay is used to detect a broad range of isolates (Kim *et al.*, 2008b; Miller *et al.*, 2010b). The F-gene based assay was developed specifically to detect virulent NDV from field swabs during the 2002 ND outbreak in the US (Wise *et al.*, 2004b). Although the assay is able to identify a wide range of isolates from different parts of the world, it fails to identify some PPMV-1 isolates (Kim *et al.*, 2008a). These results suggest that because NDV isolates show considerable variation in the nucleotide sequence around the F protein cleavage site, the designed F gene probe may not recognize all NDV isolates.

Currently, the World Organization for Animal Health (OIE) uses two parameters to differentiate between virulent isolate (notifiable disease), and avirulent isolate (not notifiable). An isolate with an ICPI value equal to or greater than 0.7 and/or having a multiple basic amino acid motif at the F protein cleavage site is considered as virulent and its presence should be reported to the OIE (Alexander and Senne, 2008). However, there are a number of reports questioning the reliability of the OIE virulence criteria by showing that the ICPI values are not always consistent with the amino acid sequence at the F protein cleavage site. For example, some PPMV-1 isolates have low pathogenicity index, but a virulent F protein cleavage site (Collins *et al.*, 1994; Meulemans *et al.*, 2002). There are also NDV isolates found in nature that have the lentogenic motif in their F protein cleavage site, but ICPI and IVPI values of virulent strain (Tan *et al.*, 2008). Recent studies have shown that the F protein cleavage site is not the sole determinant of virulence; other viral genes such as the HN and L also play important roles in determining the pathogenicity of NDV (Huang *et al.*, 2004c; Rout and Samal, 2008; Dortmans *et al.*, 2010b).

Immune response to NDV

The immune response to NDV varies depending on the infecting strain and the host species. NDV infections caused by lentogenic or mesogenic strains are usually resolved by host innate and adaptive immunity. But in the case of velogenic strains, the virus might simply outrace host defence and quickly overwhelm the host leading to death. NDV infection results in widespread changes in cellular gene expression and up-regulation of expression of a variety of factors, including cytokines, chemokines and immune molecules at cell surface (Munir *et al.*, 2005). Some of these factors have direct antiviral properties, while others activate natural killer (NK) cells, granulocytes, monocytes, macrophages, and T lymphocytes to produce direct antiviral activities and initiate an adaptive immune response.

The host innate immune response is the immediate reaction to NDV infection, which is aimed at retarding the virus growth and aid the host in developing specific protection from the adaptive immune responses. Virulent NDV is a strong inducer of innate immune response in chickens. NDV has been shown *in vitro* to induce nitric oxide (NO) in chicken heterophils (Sick *et al.*, 2000) and in chicken peripheral blood mononuclear cells (Ahmed *et al.*, 2007), interferon alpha (IFNα) and beta (IFNβ) mRNAs in chicken macrophages (Sick *et al.*, 1998), and interferon gamma (IFNγ) mRNA in peripheral chicken blood mononuclear cells (Ahmed *et al.*, 2007). The release of NDV dsRNA into cytoplasm (Levy and Marie, 2004), expression of HN on the cell surface (Zeng *et al.*, 2002a), or only binding of NDV HN to the cell surface receptor can induce type I IFN (Zeng *et al.*, 2002b). NDV strains differ widely in their capacity to induce IFN. Therefore, the IFN inducing or anti-IFN capacity of a NDV isolate could play an important role in determining its pathogenicity. However, innate immune response alone is not sufficient to protect chickens from ND.

The cell mediated immunity (CMI) is the initial immune response that is detected as early as 2–3 days after infection with live vaccine strains (Ghumman and Bankowski, 1976; Timms and Alexander, 1977). The role of CMI response to NDV in non-vaccinated chickens is not clear. Depletion of B cells by cyclophosphamide resulted in undetectable levels of IgG, IgM, and IgA antibodies, but NDV was still cleared (Lam and Hao, 1987). Furthermore, bursectomized chickens vaccinated with a live virus vaccine had little humoral response, but were fully protected from a virulent NDV challenge (Marino and Hanson, 1987). These results suggest that CMI plays an important role in virus clearance, but alone it may not be sufficient for protection against virulent NDV infection (Reynolds and Maraqa, 2000).

The humoral immune response plays a major role in virus clearance and in protection against re-infection. Passive transfer of immunoglobulin from vaccinated chickens to unvaccinated chickens induces protection against virulent NDV challenge indicating that antibodies are capable of providing protection. Serum IgM antibodies are detected as early as 4 days post infection. The IgG antibodies are detected between 6 and 10 days with peak responses at about 3–4 weeks. The antibodies may last up to one year in the blood of infected birds, but their titre may not be sufficient to provide protection against virulent NDV infection. The protective immune response to NDV is usually assessed by HI test because the HI antibody titre parallels well with the virus neutralization antibody titre. The neutralizing antibody responses to NDV are directed against both the surface glycoproteins F and HN. Antibodies against either F or HN protein provide protection against ND (Heckert *et al.*, 1996b). Although monoclonal antibodies to F protein neutralizes NDV better than monoclonal antibodies to HN protein (Russell, 1988), it is still not clear which surface glycoprotein is the major protective antigen. Passive immunity is conferred by transfer of maternal antibodies to chicks via egg yolk (Heller *et al.*, 1977). These maternal antibodies provide protection in the first 2 weeks of life, but during vaccination can neutralize the vaccine virus rendering the vaccine ineffective (Awang *et al.*, 1992). Levels of maternal antibody in day-old chicks directly correlate to the antibody titre of the parent.

The mucosal immune responses are characterized by the production of secretory IgA that plays an important role in providing protection to NDV both by restricting primary replication at the site of entry and also preventing shedding of the virus (Holmes, 1979). Vaccination with live virus induces IgA response at mucosal surfaces of respiratory and gastrointestinal tracts, and in the Harderian gland of eye, whereas parenteral administration of vaccine rarely induces local IgA response (Parry and Aitken, 1977; Powell *et al.*, 1979). Secretory IgA is particularly important in protecting the upper respiratory tract, which is accessed inefficiently by serum IgG. In summary, although it is not known which arm of the immune response nor which viral envelope glycoprotein is more important in providing protection against ND, it is known that a robust immune response is required to completely cease virus replication and provide long-term protection, which can only be induced by both the envelope glycoproteins and participation of local, CMI and humoral immune responses.

Vaccination

Vaccination against ND is performed using either live or inactivated vaccines. Both mesogenic and lentogenic viruses are used as live vaccines. Since mesogenic vaccines are considered virulent NDV under the current OIE classification, they are used only in countries where ND is endemic. Normally mesogenic vaccines, such as Roakin, Mukteswar, and Komarov are used for secondary vaccination because of their low virulence. Current live lentogenic vaccines were derived from field viruses that were shown to have low pathogenicity for poultry, but produced an adequate protective immune response. Several lentogenic strains have been used as vaccines, but B1 and LaSota strains have proven highly efficient worldwide. B1 has very mild vaccinal reaction and is used widely for initial vaccination of intensive poultry. In general, LaSota vaccines give better protection than B1 vaccines, but produces moderate vaccinal reactions. All conventional live vaccines have the disadvantage of needing to be kept at low temperature because of their thermolability. This is particularly a problem in village flocks in tropical countries. Several thermostable vaccines have been developed, but the most used thermostable vaccine has been a variant of the avirulent Australian V4 strain (Spradbrow and Sabine, 1995). The most common method of mass application of live vaccines is via drinking water. Mass application of live vaccines by sprays and aerosols is also very popular. Vaccination of individual bird with live vaccines is carried out by intranasal instillation, eye drop, and beak-dipping method.

Inactivated vaccines are produced by treating infective allantoic fluid with an inactivating agent, such as formalin or betapropiolactone. An adjuvant, such as mineral oil is added to increase the immunogenicity of the inactivated virus. However, most adjuvants cause tissue destruction and irritation leading to condemnation of meat; hence, they are not suitable for broilers. Inactivated vaccines are administered either intramuscularly or subcutaneously. Because of the cost of the vaccine and the cost of individual vaccination, inactivated vaccines are used mainly for revaccinating layers, breeders and some turkeys. Inactivated vaccines are extensively used in village poultry. Inactivated vaccines licensed for use in pigeons are widely used, particularly in racing pigeons. Several recombinant vaccines have been developed, which provide various levels of protection against NDV in experimental conditions. Pox virus (Boursnell *et al.*, 1990a,b; King, 1999), vaccinia virus (Meulemans *et al.*, 1988), pigeon pox (Letellier *et al.*, 1991), herpesvirus of turkeys (Morgan *et al.*, 1993; Sakaguchi *et al.*, 1998), Marek's disease virus, and retrovirus (Morrison *et al.*, 1990; Sonoda *et al.*, 2000) have been used as vectors to express F and/or HN proteins of NDV. The F and HN proteins expressed individually using a baculovirus expression system were also found protective (Meulemans *et al.*, 1988; Kamiya *et al.*, 1994). Naked DNA plasmids expressing the F glycoprotein were also protective (Loke *et al.*, 2005). These recombinant vaccines are not used in the poultry industry because of high cost and unsuitability for mass application. Recently, a recombinant ND vaccine 'Innovax-ND' was approved by the USDA for commercial use. Innovax-ND has the NDV F gene inserted in the herpesvirus of turkey (HVT) vector. However, the efficacy of this vaccine in field condition is currently not known.

The current NDV vaccines have been used for more than 50 years with proven track record of safety and efficacy. However, they do not completely prevent infection or virus shedding, nor do they possess genetic markers to allow differentiation between infected and vaccinated birds, which are critical elements of a vaccine used for control and eradication purposes. They have also reduced efficacy when used in chickens with high level of maternal antibodies. Furthermore, there are reports from different countries suggesting that there has been a major antigenic drift in the types of NDV strains that have been identified as circulating in poultry (Yu *et al.*, 2001; Tsai *et al.*, 2004; Kapczynski and King, 2005; Miller *et al.*, 2007; Qin *et al.*, 2008b). These new strains are not completely protected by the current vaccine strains LaSota and B1, which were isolated more than 60 years ago (Hitchner and Johnson, 1948; Goldhaft, 1980). This concern has been supported by a study showing that vaccines from virus phylogenitically closer to the potential outbreak virus provide better control by reducing virus shedding from infected birds (Miller *et al.*, 2007). Recently, it was also shown that an attenuated genotype VII NDV vaccine generated by reverse genetics effectively protected the vaccinated birds against homologous virulent genotype VII virus challenge and significantly reduced virus shedding from the vaccinated birds when compared with LaSota vaccinated animals, reinforcing the importance of matching vaccines (Hu *et al.*, 2009). Therefore, there is a great need to improve the current NDV vaccines.

Avian paramyxovirus serotypes 2–9

APMVs other than NDV (APMV-1) have been isolated worldwide from many species of domestic, pet and wild birds. Many of these viruses were isolated as a result of influenza virus surveillance programmes. These viruses have been separated into eight serotypes (APMV-2 to -9) by HI and NI assays (Alexander, 2003). However, antigenic cross relationships by HI and virus neutralization tests exist between APMV serotypes. APMV-1 cross reacts strongly by HI and virus neutralization tests with APMV-9 followed by APMV-3. Prior infection of chickens with either APMV-3 or APMV-9 provide some degree of protection from challenge by virulent NDV. APMV-2 shows weak cross reaction by HI and virus neutralization tests with APMV-3 and -8, APMV-5 also shows weak cross reaction by virus neutralization test with APMV-6 (Nayak and Samal unpublished results). Recently, APMVs were isolated from Rockhopper Penguins in the Falkland Islands. Serological and genome characterizations suggested that these viruses probably represent a new serotype (APMV-10; Miller *et al.*, 2010a). NDV (APMV-1) is the best characterized member among APMVs. Very little is known about APMV-2 to -9, but our knowledge on these viruses is rapidly expanding owing to availability of full genome sequence of at least one member of each serotype (Chang *et al.*, 2001; Kumar *et al.*, 2008; Nayak *et al.*, 2008; Subbiah *et al.*, 2008a; Paldurai *et al.*, 2009; Samuel *et al.*, 2009, 2010; Xiao *et al.*, 2009). Comparison of the genome maps of all nine APMV serotypes

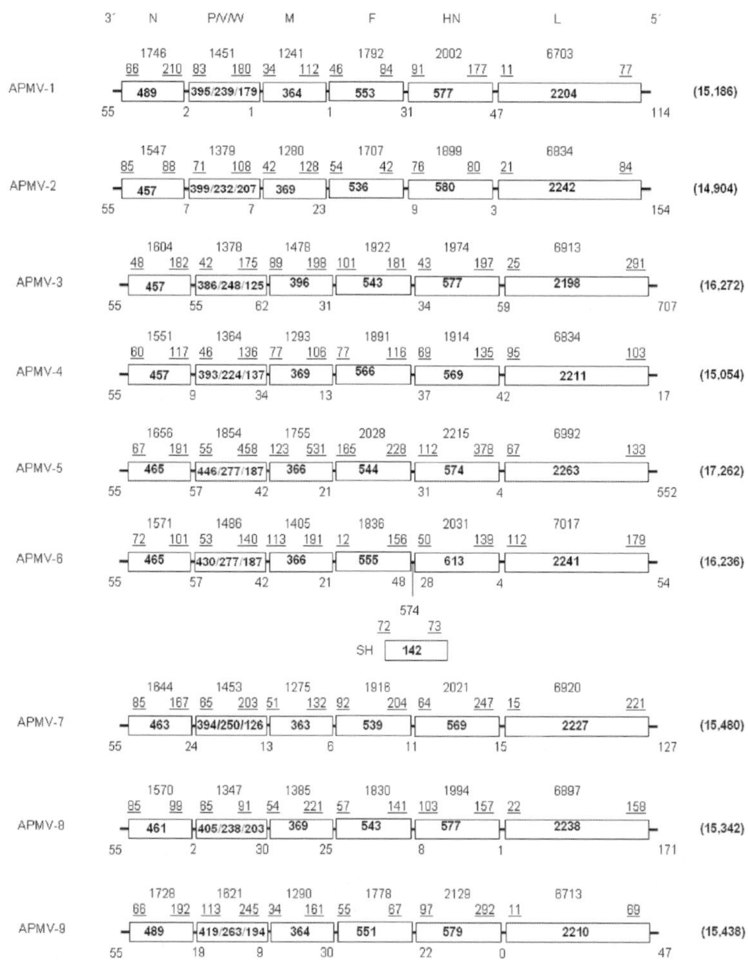

Figure 3.7 Genome maps of all the serotypes of genus *Avulavirus.* The genes size, the intergenic lengths, the putative amino acid lengths, and the complete genome sequence of representative members are drawn to the scale. All the members contains typically the six genes in the order of 3′-N-P/V/W-M-F-HN-L-5′ with the exception of avian paramyxovirus type-6 (APMV-6), which is shown to contain an additional SH gene in their genome.

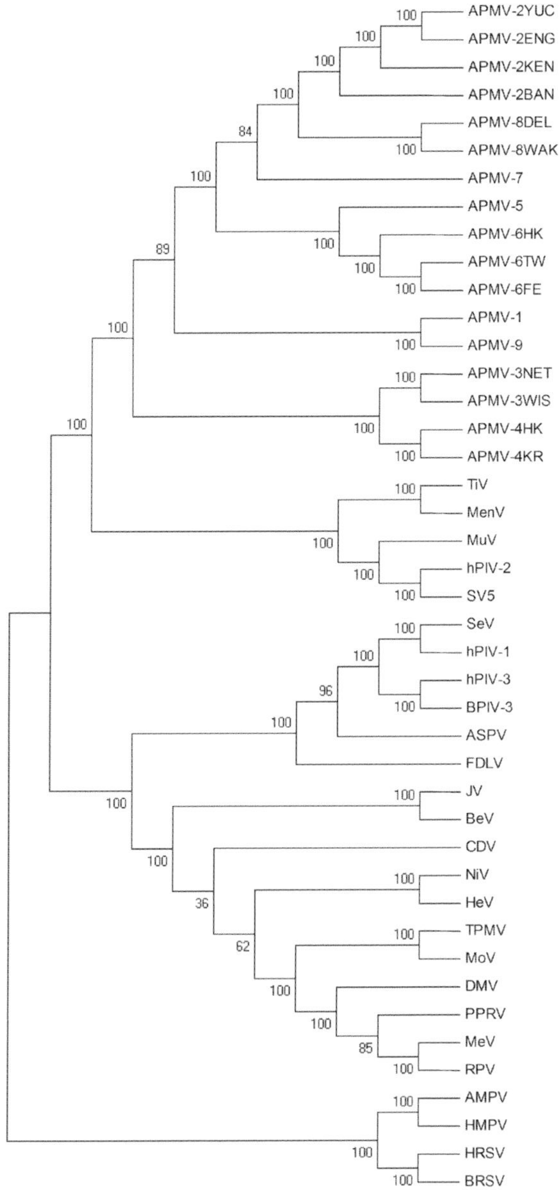

Figure 3.8 Phylogenetic tree of representative members of family *Paramyxoviridae*. Accession numbers for the complete genome sequence used in the tree are as follows. *Avulaviruses*: APMV-1, $AF_0$77761; APMV-2 YUC, EU338414; APMV-2 BAN, HM159995; APMV-2 KEN, HM159994; APMV-2 ENG, HM159993; APMV-3 NET, EU403085; APMV-3 WIS, EU782025; APMV-4 KR, EU877976; APMV-4HK, FJ177514; APMV-5, GU206351; APMV-6TWN, NC 00,3043; APMV-6 HK, EU622637; APMV-6 FE, EF569970; APMV-6 ITY, GQ406232; APMV-7, FJ231524; APMV-8 DEL, FJ215863; APMV-8 WAK, FJ215864; APMV-9, EU910942. *rubulaviruses*: hPIV-2 NC_003443; PIV5 (also known as SV-5) NC_006430; MuV NC_002200; simian virus 41 (SV41) NC_006428. respiroviruses: hPIV-1, NC_003461; hPIV-3, NC_001796; SeV, NC_001552, BPIV-3, NC_002161. Henipaviruses: NiV, NC_002728; HeV, NC_001906. Morbilliviruses: CDV, NC_001921; MeV, AF266288; phocine distemper virus (PDV), NC_006383; rinderpest virus (RPV), NC_006296; peste des petits ruminants virus (PPRV), NC_006383; dolphin *Morbillivirus* (DMV), NC_005283; other paramyxovirus: Atlantic salmon paramyxovirus (ASPV), EF646380; Beilong virus (BeV), NC_007803; Fer-de-Lance virus (FDLV), NC_005084; J virus (JV), NC_007454; Menangle virus (MenV), NC_007620; Mossman (MoV), NC_005339; Tupaia paramyxovirus (TpV), NC_002199; *Pneumoviruses*: HRSV, NC001781; BRSV, NC001989. *MetaPneumoviruses*: AMPV, NC007652; HMPV, NC004148.

Table 3.1 Per cent nucleotide sequence identities between the complete genomes of prototype strains of the nine avian paramyxovirus (APMV) serotypes

	APMV-2	APMV-3	APMV-4	APMV-5	APMV-6	APMV-7	APMV-8	APMV-9
APMV-1	46	44	43	44	45	45	46	60
APMV-2		43	44	48	50	51	60	45
APMV-3			46	42	44	43	44	43
APMV-4				42	44	43	43	43
APMV-5					55	48	48	44
APMV-6						49	49	45
APMV-7							51	45
APMV-8								46

Table 3.2 Per cent amino acid sequence identities between the proteins of avian paramyxovirus serotype-1 (APMV-1) strain LaSota and those of prototype strains of the other APMV serotypes

	Amino acid sequence identity (%)					
APMV	N	P	M	F	HN	L
APMV-2	41	23	31	41	35	38
APMV-3	38	23	28	32	35	33
APMV-4	38	24	29	33	35	33
APMV-5	38	20	33	41	35	38
APMV-6	40	21	31	39	30	39
APMV-7	41	20	33	39	36	38
APMV-8	40	25	28	41	34	38
APMV-9	67	46	57	56	61	60

Table 3.3 Comparison of the putative cleavage sites of the F_0 proteins of prototypes strains of the 10 avian paramyxovirus (APMV) serotypes, and requirements for exogenous protease. APMV-1, -3 and -5 produce syncytia in cell culture while the other serotypes produce single-cell infection

APMVs	Cleavage site[1]	Exogenous protease[2]	Syncytium	Plaque formation
APMV-1/Newcastle disease virus	GGRQGR↓L (avirulent)	+	+	+
	GRRQKR↓F (virulent)	–	+	+
APMV-2	DKPASR↓F (Yuc)	–	–	–
	TLPSAR↓F (Ban)	–	–	–
APMV-3	PRPSGR↓L (Wis)	+	+	+
	ARPRGR↓L (Net)	+	+	+
APMV-4	VDIQPR↓F	–	–	–
APMV-5	GKRKKR↓F	–	+	+
APMV-6	PAPEPR↓L (HK)	–/+	–	–
	SIREPR↓L (IT)	–/+	–	–
APMV-7	TLPSSR↓F	–	–	–
APMV-8	TYPQTR↓L	–/+	–	–
APMV-9	RIREGR↓I	+	–	–
APMV-10?	DKPSQR↓I	+	?	?

[1]Parenthesis represent different strains of APMV serotypes.

[2]Requirement for exogenous protease in cells (–/+ depends on the cell line used).

is shown in Fig. 3.7, and the phylogenetic analysis of complete genomes of APMV strains is shown in Fig. 3.8. The nucleotide sequence identities, amino acid sequence identities and the putative cleavage sites of F protein of all nine APMV serotypes are presented in Tables 3.1, 3.2 and 3.3, respectively.

The disease potential of APMV-2 to -9 is not well known because many of these viruses were isolated from birds dying in quarantine, hunter killed, trapped wild birds, apparently healthy poultry or exotic birds. In these cases it is difficult to associate the virus to a specific disease. Only the disease potential of APMV-5 is clearly known because it causes a severe disease in budgerigars (Nerome *et al.*, 1978). APMV-2, -3, -6 and -7 have been associated with disease in domestic poultry (Alexander and Senne, 2008). Specifically, APMV-2 and -3 can cause mild respiratory disease and drop in egg production in turkeys and chickens (Tumova *et al.*, 1979; Kumar *et al.*, 2010a; Subbiah *et al.*, 2010b). There are no reports of isolation of APMV-5, -8 and -9 from poultry. But recent seroprevalence of commercial poultry farms in USA indicated the possible prevalence of all APMV serotypes excluding APMV-5 in chickens (Warke *et al.*, 2008a). The replication and pathogenicity of APMV-2 to -9 in non-avian species has also been studied (Khattar *et al.*, 2011; Samuel *et al.*, 2011). Experimental infection of hamsters by intranasal route with prototype strain of each APMV serotype showed that each of the viruses replicated in hamsters and produced mild or in apparent clinical signs (Samuel *et al.*, 2011). Intranasal inoculation of BALB/c mice with prototype strain of each APMV serotype showed that all APMV serotypes except APMV-5 are capable of replicating in mice with minimal disease and pathology (Khattar *et al.*, 2011).

Avian paramyxovirus type 2

APMV-2 was first isolated in 1956 in Yucaipa, California, from a diseased chicken that was also infected with infectious laryngotracheitis virus (Bankowski *et al.*, 1960). Since then, many APMV-2 viruses have been isolated worldwide from chickens, turkeys and wild birds (Alexander and Senne, 2008). However, isolation of APMV-2 from domestic poultry has been rare. APMV-2 viruses are frequently isolated from passerine and psittacine species suggesting that these birds may represent the reservoir species for APMV-2 (Bankowski *et al.*, 1968; Bradshaw and Jensen, 1979; Senne *et al.*, 1983; Warke *et al.*, 2008b). Serological surveys of poultry have indicated that this virus is widespread. APMV-2 infection was shown to affect hatchability and poult yield in turkeys (Bankowski *et al.*, 1981). Experimental infection of 4-week-old chickens and turkeys by two different strains of APMV-2 did not induce overt clinical disease, although all birds were seroconverted. Large amounts of viral antigens were present in epithelial linings of respiratory and alimentary tracts. Intracerebral inoculation of 1-day-old chicks showed that APMV-2 strains neither caused disease nor replicated detectably in the brain tissues. Chickens are comparatively more susceptible than turkeys to APMV-2 infections (Subbiah *et al.*, 2010a). APMV-2 grows to high titre (2^{10}–2^{12} HA units) in embryonated chicken eggs. The virus also grows in many mammalian and avian cell lines. The virus titres are the same with or without 10% allantoic fluid supplementation in the media. The general CPE in all cell types involves rounding of single cells without any syncytia formation (Subbiah *et al.*, 2008a).

The complete genome sequences of APMV-2 strains Yucaipa, Bangor, England, and Kenya have been determined (Subbiah *et al.*, 2008a, 2010a). The genome lengths of Yucaipa, Bangor, England, and Kenya are 14,904, 15,024, 14,904 and 14,916 nt, respectively. Each genome consists of six non-overlapping genes in the order of 3′-N-P/V/W-M-F-HN-L-5′, with a 55-nt leader at the 3′ end. The length of the trailer at the 5′ end of strain Bangor was 173 nt, compared with 154 nt for strains England, Kenya, and Yucaipa. In general, comparison of sequences of APMV-2 strains England and Kenya with strain Yucaipa showed 94.5 and 88% nt and 96 and 92% aggregate amino acid identity, respectively. In contrast, strain Bangor has a much lower per cent nt identity (70.4, 69.4, and 70.8%) and amino acid identity (75.3, 76.2, and 76.3%) with strains Yucaipa, England, and Kenya, respectively. Furthermore, strain Bangor has a single basic amino acid residue (T-L-P-S-A-R↓F) at the fusion protein cleavage site compared with the two basic amino acids (D-K-P-A-S-R↓F) found in those of the other three strains. Reciprocal cross-HI and cross-neutralization assays using post-infection chicken sera indicated that strain Bangor is antigenically related to the other APMV-2 strains, but with a 4- to 8-fold difference in homologous versus heterologous HI titre. These differences in antigenic relatedness suggest that these four APMV-2 strains represent a single serotype with two antigenic subgroups, and this is strongly supported by the dimorphism in nt and amino acid sequence identity (Subbiah *et al.*, 2010a).

In a recent study, comparisons of nucleotide sequences of HN genes of 22 APMV-2 isolates showed five different lengths of the HN proteins; each of these involves extensions of the C-terminus of the ORF owing to change in position of the stop codon. On the basis of phylogenetic analysis of HN gene sequences, five genetic groups could be established (Sahar *et al.*, 2010).

Avian paramyxovirus type 3

APMV-3 were first isolated from turkeys in Ontario in 1967 and Wisconsin in 1968 (Tumova *et al.*, 1979). Since then, many APMV-3 strains have been isolated from turkeys in different parts of the world including England (Macpherson *et al.*, 1983), France (Andral and Toquin, 1984) and Germany (Zeydanli *et al.*, 1988). The APMV-3 strain parakeet/Netherlands/449/75, isolated from parakeets in the Netherlands, is the prototype for the entire serotype (Alexander and Chettle, 1978). APMV-3 was also isolated from non-domesticated species such as *Psittaciformes* and *Passeriformes* (Alexander, 1980). APMV-3 has been associated with encephalitis and high mortality in caged birds, and with respiratory disease in turkeys (Tumova *et al.*, 1979). The virus causes acute pancreatitis and CNS symptoms in psittacine and *Passeriformes* birds (Beck *et al.*, 2003). Experimentally, APMV-3 can infect chickens at an early age with evidence of stunting growth that may be more marked in broiler chicken breeds (Alexander and Collins, 1982), but natural infection of chickens with APMV-3 has never been reported.

In a recent study, APMV-3 strains, Netherlands and Wisconsin, were shown to cause mild respiratory disease in 1-day-old chickens and turkeys. Both these strains replicated well in the brain tissues when inoculated intracerebrally in 1-day-old chicks. There were no overt signs of illness in 2-week-old chickens and turkeys by either strain, although all the birds seroconverted after infection. The viruses were isolated predominantly from brain, lungs, spleen, trachea, pancreas and kidneys. Large amounts of viral antigens were present in both epithelial and sub-epithelial lining of respiratory and alimentary tracts (Kumar *et al.*, 2010a). In terms of pathogenicity for domestic poultry, APMV-3 is probably second in importance only to NDV.

APMV-3 grows to a titre of 2^8–2^{10} HA units in 9-day-old embryonated chicken eggs. The virus also grows in many mammalian and avian cell lines, but requires supplementation of 10% allantoic fluid in the medium for replication. Viral CPE involves rounding, detachment, and syncytia formation. APMV-3 produces distinct plaques under methylcellulose overlay in DF1 cells in the presence of 10% allantoic fluid (Kumar *et al.*, 2008).

The complete genome sequences of APMV-3 strains Netherlands and Wisconsin have been determined (Kumar *et al.*, 2008, 2010a). The genome lengths of Netherlands and Wisconsin are 16,272 and 16,182 nt, respectively. Each genome consists of six non-overlapping genes in the order of 3′-N-P/V/W-M-F-HN-L-5′, with a 55 nt leader at the 3′ end. The length of the trailer at the 5′ end of strain Netherlands is 707 nt, compared with 681 nt for strain Wisconsin. These trailer lengths are longest in the family *Paramyxoviridae*. The intergenic regions are 31–63 nt in length. Comparison of nt and amino acid sequences of strain Netherlands with those of strain Wisconsin revealed 67% and 78% identity, respectively. Furthermore, strain Wisconsin has a single basic amino acid residue (P-R-P-S-G-R↓L) at the fusion protein cleavage site compared with the polybasic amino acids (A-R-P-R-G-R↓L) at the fusion protein cleavage site of strain Netherlands. Reciprocal cross-HI and cross-neutralization assays using post-infection sera from chickens indicated that strains Wisconsin and Netherlands are highly related antigenically, with only a 2- to 4-fold difference in antibody reactivity between the homologous and heterologous strains. Taken together, these results indicate that the two APMV-3 strains represent a single serotype with two subgroups that differ substantially based on nt and amino acid sequences, but with only a modest antigenic difference (Kumar *et al.*, 2010b).

Avian paramyxovirus type 4

APMV-4 viruses were isolated from apparently healthy wild ducks in the Mississippi flyway (Webster *et al.*, 1976) and from apparently healthy domestic ducks, chickens and geese in Hong Kong (Shortridge and Alexander, 1978). Since then, APMV-4 viruses have been isolated predominantly from feral birds of the order *Anseriformes* (Stanislawek *et al.*, 2002), and from commercial ducks and geese (Shortridge and Alexander, 1978; Turek *et al.*, 1984). Experimental infection of one-day-old and three-week-old chickens with APMV-4 strains failed to show any signs of disease (Shortridge and Alexander, 1978; Warke *et al.*, 2008b).

The strain duck/Hong Kong/D3/75 is the prototype strain for APMV-4. APMV-4 grows efficiently in embryonated chicken eggs and in many cell lines without the addition of exogenous proteases (Nayak *et al.*, 2008). The addition of exogenous protease did not increase the virus titre. The CPE observed was rounding, detachment of cells, but no syncytia formation.

The complete genome sequences for APMV-4 prototype strain duck/Hong Kong/D3/75 and strain KR/YJ/06, isolated from a mallard duck in Korea, have been determined (Jeon *et al.*, 2008; Nayak *et al.*, 2008). The genome length of both APMV-4 strains is 15,054 nt, with a 55 nt leader region at the 3′ end and 17 nt trailer region at the 5′ end. The genomes contain six non-overlapping genes in the order 3′-N-P/V/W-M-F-HN-L-5′. The cleavage site of the F protein (D-I-Q-P-R↓F) does not conform to the preferred cleavage site of the ubiquitous intracellular protease furin. The intergenic regions are 9–42 nt in length. Comparison of nt and the aggregate amino acid sequences of strain Hong Kong with those of strain Korea revealed 91%, and 96% identity, respectively.

Avian paramyxovirus type 5

APMV-5 was first isolated from an epizootic outbreak involving budgerigars (*Melopsittacus undulatus*) at Kunitachi, Tokyo, Japan, in 1974 (Mustaffa-Babjee *et al.*, 1974). This virus was serologically and antigenically distinct from the other previously known APMV serotypes and serves as the prototype virus for this serotype. Natural and experimental APMV-5 infections in budgerigars are characterized by depression, dyspnoea, diarrhoea, torticollis and very high mortality (95–100%). A second outbreak occurred among budgerigars in Brisbane in Queensland, Australia. Experimental infection of APMV-5 isolated from this outbreak caused no signs of disease in young and adult chickens and pigeons, but caused acute fatal enteritis among immature budgerigars (Mustaffa-Babjee *et al.*, 1974). The third known outbreak of APMV-5 disease occurred in budgerigars from UK in 1993 (Gough *et al.*, 1993). The clinical signs consisted mainly of vomiting and diarrhoea followed by death. Of all the APMVs examined to date, only virulent APMV-1 and APMV-5 have been associated with 100% mortality in chickens and budgerigars, respectively (Nerome *et al.*, 1978). Budgerigars are considered the natural host of APMV-5.

APMV-5 has not been well characterized, but appears to differ from all the other APMV serotypes in two major attributes: (1) the inability to grow in allantoic cavity of embryonated chicken eggs and (2) the failure to cause haemagglutination with chicken RBC. APMV-5 grows in many cell lines, but grows most efficiently in Vero, DF1, and BHK-21 cells. Viral CPE involved rounding, detachment and syncytia formation. The virus produces distinct plaques in Vero cells under methylcellulose overlay. Supplementation with 10% allantoic fluid, acetyl typsin (1–5 µg/ml) or α-chymotrypsin (1–5 µg/ml) did not enhance the growth of the virus indicating the lack of a requirement for external protease for efficient cleavage of the F protein. APMV-5 has NA activity, which can be used to detect virus presence in cell culture or tissue samples. APMV-5 can be serotyped by virus neutralization assay using either $TCID_{50}$ or plaque reduction method in Vero cells.

The complete genome sequence of APMV-5 prototype strain APMV-5/budgerigar/Kunitachi/74 has been determined (Samuel *et al.*, 2010). The genome is 17,262 nt long, which is the longest among members of the genus *Avulavirus* and encodes six non-overlapping genes in the order of 3′-N-P/V/W-M-F-HN-L-5′ with intergenic regions of 4–57 nt. The genome contains a 55 nt leader sequence at the 3′ end and a 552 nt trailer sequence at the 5′ end. The P gene contains a conserved RNA editing site and is predicted to encode P, V, and W proteins. The cleavage site of the F protein (G-K-R-K-K-R↓F) conforms to the cleavage site motif of the ubiquitous cellular protease furin. Consistent with this, exogenous protease was not required for virus replication *in vitro*. However, the intracerebral pathogenicity index of APMV-5 strain Kunitachi in 1-day-old chicks was found to be zero indicating that the virus is avirulent for chickens despite the presence of a polybasic F protein cleavage site. Oculonasal inoculation of 2-week-old SPF chickens with APMV-5 caused no clinical signs but showed virus replication in trachea.

Avian paramyxovirus type 6

APMV-6 strain duck/Hong Kong/18/199/77 was first isolated from an apparently healthy domestic duck in Hong Kong (Shortridge *et al.*, 1980) in 1977 and is considered the prototype for the entire serotype. Since then, APMV-6 viruses are frequently isolated from ducks, geese and turkeys around the world (Stanislawek *et al.*, 2002; Warke *et al.*, 2008b). The virus appears to be apathogenic to chickens, but may cause mild respiratory disease and decrease in eggs production in turkeys (Alexander, 2003).

APMV-6 grows efficiently in embryonated chicken eggs and in many established cell lines (Xiao *et al.*, 2010). It was found that supplementation of trypsin (1 μg/ml) resulted in a marginal (2-fold) increase in the yield of viruses in some cell lines for some isolates indicating that replication is substantially trypsin independent. The CPE involves rounding and detachment of cells without any syncytia formation. The MDT of four APMV-6 strains in chicken eggs were more than 168 h, indicating that they are avirulent for chickens.

The complete genome sequences of prototype strain duck/Hong Kong/18/199/77 (HK), strain duck/Taiwan/Y1/99 (TW), strain goose/Far East/4440/2003 (FE) and strain duck/Italy/4524-2/2007 (IT4524–2) have been determined (Chang *et al.*, 2001; Xiao *et al.*, 2010). The genome length of strains HK, TW and FE is 16,236 nt, whereas that of strain IT4524-2 is 16,230 nt. The genome length difference in strain IT4524-2 is due to a 6 nt deletion in the downstream untranslated region of the F gene. Each genome consists of seven genes in the order of 3′-N-P/V/W-M-F-SH-HN-L-5′, which differs from other APMV serotypes in containing an additional gene encoding the small hydrophobic (SH) protein. The biological function of SH protein is not known. Sequence comparisons revealed that strain IT4524-2 shares an unexpectedly low level of genome nt sequence identity (70%) and aggregate amino acid sequence identity (79%) with other three strains, which in contrast are more closely related to each other with nt sequence 94–98% nt identity and 90–100% aggregate amino acid identity. Sequence analysis of the F-SH-HN genome region of two other recent Italian isolates showed that they fall in the HK/FE/TW group. The F protein cleavage site of strain IT4524-2, R-E-P-R↓L, has two basic amino acids (arginine, R) compared with the monobasic F protein cleavage site of P-E-P-R↓L in other strains. Reciprocal cross-HI assays using post-infection chicken sera indicated that strain IT4524-2 is antigenically related to the other APMV-6 strains, but with 4- to 8-fold lower HI titres for the test sera between strain IT4524-2 and the other APMV-6 strains. Taken together, these results indicated that the APMV-6 strains represent a single serotype with two subgroups that differ substantially based on nt and amino acid sequences and can be distinguished by HI assay (Xiao *et al.*, 2010).

Avian paramyxovirus type 7

APMV-7 was first isolated from a hunter killed dove in 1975 in Tennessee, USA (Alexander *et al.*, 1981). Subsequently, APMV-7 was isolated from a natural outbreak of respiratory rales and mortality (0.9%) in commercial turkey breeder flocks in Ohio in 1997, but it has not been associated with severe disease in avian species. The experimental infection of this APMV-7 isolate in turkeys caused respiratory disease that could be complicated by secondary infection (Saif *et al.*, 1997). APMV-7 has been isolated from ostriches (Woolcock *et al.*,

1996). APMV-7 viruses are prevalent in pigeons and turkeys (Alexander and Senne, 2008).

APMV-7 strain dove/Tennessee/4/75 is considered the prototype strain for the entire serotype. This virus grows efficiently in embryonated chicken eggs. However, the virus did not cause death of chicken embryo until 144 h, indicating that it is avirulent in chickens. APMV-7 replicated in BHK-21, HEp-2, Vero and MDCK cells, but failed to grow in several other cell lines. The supplementation of 10% allantoic fluid, as an exogenous source of protease, did not enhance the growth of the virus. The CPE observed was rounding, detachment of cells and no syncytia formation (Xiao *et al.*, 2009).

The complete genome sequence of APMV-7 strain dove/Tennessee/4/75 has been determined (Xiao *et al.*, 2009). The genome size is 15,480 nt long and contains six non-overlapping genes in the order of 3′-N-P/V/W-M-F-HN-L-5′. The 3′-leader and 5′-trailer sequences of the genome are 55 and 127 nt, respectively. The IGS are between 11 and 70 nt. The P gene contains a conserved RNA editing site presumed to be involved in the production of V and W proteins. The viral F protein has a single basic amino acid at the putative cleavage site (T-L-P-S-S-R↓F); however, the virus did not require exogenous protease for *in vitro* replication.

Avian paramyxovirus type 8

APMV-8 strain goose/Delaware/1053/76 was first isolated in 1976 from a hunter killed feral Canadian goose (*Branta canadensis*) in the Atlantic flyway in the US (Cloud and Rosenberger, 1980). APMV-8 strain pintail/Wakuya/20/78 was isolated from an apparently healthy feral pintail duck (*Anas acuta*) in Wakuya, Japan, in 1978 (Yamane *et al.*, 1982). Strain Delaware is considered to be the prototype strain of APMV-8. Very little is known about the molecular characteristics and pathogenicity of APMV-8.

Complete genome sequences were determined for APMV-8 strains goose/Delaware/1053/76 (prototype strain) and pintail/Wakuya/20/78 (Paldurai *et al.*, 2009). The genome of each strain is 15,342 nt long. The genome consists of six genes in the order of 3′-N-P/V/W-M-F-HN-L-5′. The genes are flanked on either side by conserved transcription start and stop signals, and have intergenic regions ranging from 1 to 30 nt. The genome contains a 55 nt leader region at the 3′-end and a 171 nt trailer region at the 5′-end. Comparison of sequences of strains Delaware and Wakuya showed nucleotide identity of 96.8% at the genome level and amino acid identities of 96–99% for the predicted proteins. Both strains grew in embryonated chicken eggs to titres of 2^7–2^8 HA units. Both strains contain only a single basic residue at the cleavage activation site of the F protein (T-Y-P-Q-T-R↓L) and their efficiency of replication *in vitro* depended on and was augmented by the presence of exogenous protease in most cell lines (Paldurai *et al.*, 2009). APMV-8 grew in Vero, DF-1, QT-35 and primary chicken embryo fibroblast cells in the presence of acetyl trypsin (1 μg/ml). The primary chicken kidney and 293T cells support the growth of APMV-8 strains without exogenous protease. The CPE observed were rounding and detachment of cells. There was no evidence of syncytium and plaque formation for the APMV-8 strains Delaware and Wakuya (Paldurai *et al.*, 2009). Phylogenitically and antigenically APMV-8 is more closely related to APMV-2 than to other APMV serotypes.

Avian paramyxovirus type 9

APMV-9 strain domestic duck/New York/22/78 was first isolated from a domestic duck during routine surveillance in New York in 1978 (Sandhu and Hinshaw, 1981). Since then, many APMV-9 strains have been isolated in Italy from feral ducks indicating world-wide distribution (Capua *et al.*, 2004). Strain domestic duck/New York/22/78 serves as the prototype strain for the serotype.

The complete genome sequence was determined for APMV-9 strain domestic duck/New York/22/78 (Samuel *et al.*, 2009). The genome is 15,438 nt long and encodes six non-overlapping genes in the order of 3′-N-P/V/W-M-F-HN-L-5′ with intergenic regions of 0–30 nt. The genome contains a 55 nt leader sequence at the 3′ end and a 47 nt trailer sequence at the 5′ end. The cleavage site of the F protein is R-I-R-E-G-R↓I, which does not conform to the conventional cleavage site of the ubiquitous cellular protease furin. The virus required exogenous protease for *in vitro* replication and grew only in a few established cell lines, such as Vero, DF1 and BHK21. It was necessary to include 10% allantoic fluid for replication of the virus in cell culture, which could not be met by chymotrypsin or by trypsin. The virus grew efficiently in primary CEK cells but not in CEF cells. The CPE observed was rounding, detachment of cells and no syncytia formation. The virus failed to produce plaques under methylcellulose overlay. Phylogenetic and antigenic analysis showed that APMV-9 is more closely related to APMV-1 than to other APMVs. The MDT in embryonated chicken eggs was found to be more than 120 h indicating APMV-9 to be avirulent for chickens (Samuel *et al.*, 2009).

A comparison of F and HN sequences among four Italian APMV-9 isolates and the prototype strain revealed significant sequence variation suggesting the existence of different lineages among APMV-9 isolates (Dundon *et al.*, 2010).

Future perspectives

ND remains as a major problem for the poultry industry around the world. Disease free countries are vulnerable to accidental or intentional ND outbreaks. In developing countries ND continues to affect the socio-economic life of the people. Many NDV strains with varying levels of virulence circulate through wild birds and poultry populations. The ecology of NDV is poorly understood and there are no predictive tools to determine when and where new virulent strains will emerge. Depopulation still remains the primary method to eradicate ND. Priority should be given to conduct research on (1) improving diagnostic tests to rapidly identify new disease outbreaks and predict susceptibility levels in commercial flocks; (2) epidemiological research to better understand virus transmission in breeds that comprise backyard flocks, identify NDV reservoirs in nature, create an epidemiological database of circulating NDV strains, and improve the current surveillance programme; and (3) develop safe and effective vaccines specifically designed for control and eradication. It is likely that answers to these challenges will come from both basic and applied research.

The use of reverse genetics technology can be expected to lead to the production of safer and more effective vaccines, as they can be readily modified as required. Reverse genetics also offer the opportunity to engineer NDV as a vaccine vector for human and veterinary uses. Availability of complete genome sequences of APMV-2 to -9 will not only allow us to understand the molecular biology and pathogenesis of these viruses, but also engineer them as vaccine vectors for human and animal pathogens.

References

Abenes, G., Kida, H., and Yanagawa, R. (1986). Biological activity of monoclonal antibodies to hemagglutinin-neuraminidase (HN) protein of Newcastle disease virus. Jpn. J. Vet. Sci. *48*, 353–362.

Abolnik, C., Horner, R.F., Bisschop, S.P., Parker, M.E., Romito, M., and Viljoen, G.J. (2004). A phylogenetic study of South African Newcastle disease virus strains isolated between 1990 and 2002 suggests epidemiological origins in the Far East. Arch. Virol. *149*, 603–619.

Adair, B.M., McNulty, M.S., Todd, D., Connor, T.J., and Burns, K. (1989). Quantitative estimation of Newcastle disease virus antibody levels in chickens and turkeys by ELISA. Avian Pathol. *18*, 175–192.

Afonso, C.L. (2008). Not so fast on recombination analysis of Newcastle disease virus. J. Virol. *82*, 9303.

Ahlert, T., Sauerbrei, W., Bastert, G., Ruhland, S., Bartik, B., Simiantonaki, N., Schumacher, J., Hacker, B., Schumacher, M., and Schirrmacher, V. (1997). Tumor-cell number and viability as quality and efficacy parameters of autologous virus-modified cancer vaccines in patients with breast or ovarian cancer. J. Clin. Oncol *15*, 1354–1366.

Ahmed, K.A., Saxena, V.K., Ara, A., Singh, K.B., Sundaresan, N.R., Saxena, M., and Rasool, T.J. (2007). Immune response to Newcastle disease virus in chicken lines divergently selected for cutaneous hypersensitivity. Int J. Immunogenet. *34*, 445–455.

Alamares, J.G., Elankumaran, S., Samal, S.K., and Iorio, R.M. (2010). The interferon antagonistic activities of the V proteins from two strains of Newcastle disease virus correlate with their known virulence properties. Virus Res. *147*, 153–157.

Albiston, H., and Gorrie, C. (1942). Newcastle disease on Victoria. Aust. Vet. J. *18*, 75–79.

Aldous, E.W., and Alexander, D.J. (2008). Newcastle disease in pheasants (*Phasianus colchicus*): a review. Vet. J. *175*, 181–185.

Aldous, E.W., Mynn, J.K., Banks, J., and Alexander, D.J. (2003). A molecular epidemiological study of avian paramyxovirus type 1 (Newcastle disease virus) isolates by phylogenetic analysis of a partial nucleotide sequence of the fusion protein gene. Avian Pathol. *32*, 239–256.

Alexander, D. (2003). *Paramyxoviridae*, 11th edn (Iowa State University Press: Iowa).

Alexander, D., and Senne, D. (2008). Newcastle disease and other avian paramyxovirus and *Pneumovirus* infection. In Diseases of Poultry, 12th edn, Saif, Y.M., ed. (Blackwell Publishing Ltd, Oxford), pp. 75–115.

Alexander, D., Manvell, R., Banks, J., Collins, M., Parsons, G., Cox, B., Frost, K., Speidel, C., Ashman, S., and Aldous, E. (1997a). Esperimental assesment of the pathogenicity of the Newcastle disease viruses outbreaks in Great Britain in 1997 for chickens and turkeys and the protection afforded by vaccinaation. Avian Pathol. *28,* 501–512.

Alexander, D.J. (1980). Avian paramyxoviruses. Vet. Bull. *50,* 737–752.

Alexander, D.J. (1988). Newcastle disease virus- An avian paramyxovirus. (Boston, MA, Kulwer Acedemic Press).

Alexander, D.J. (1998). Newcastle disease and other avian paramyxoviruses, 4th edn (University of Pennsylvania, Kennett Square, PA, American Association of Avian Pathologists).

Alexander, D.J. (2009). Newcastle, disease: OIE Terrestrial Manual 2009. In In: Manual of Diagnostic Tests and Vaccines for Terrestrial Animals.

Alexander, D.J., and Chettle, N.J. (1978). Relationship of parakeet/Netherlands/449/75 virus to other avian paramyxoviruses. Res. Vet. Sci. *25,* 105–106.

Alexander, D.J., and Collins, M.S. (1982). Pathogenicity of PMV-3/parakeet/Netherlands/449/75 for chickens. Avian Pathol. *11,* 179–185.

Alexander, D.J., and Parsons, G. (1984). Avian paramyxovirus type 1 infections of racing pigeons: 2 pathogenicity experiments in pigeons and chickens. Vet. Rec. *114,* 466–469.

Alexander, D.J., Hinshaw, V.S., and Collins, M.S. (1981). Characterization of viruses from doves representing a new serotype of avian paramyxoviruses. Arch. Virol. *68,* 265–269.

Alexander, D.J., Parsons, G., and Marshall, R. (1984a). Infection of fowls with Newcastle disease virus by food contaminated with pigeon faeces. Vet. Rec. *115,* 601–602.

Alexander, D.J., Russell, P.H., and Collins, M.S. (1984b). Paramyxovirus type 1 infections of racing pigeons: 1 characterisation of isolated viruses. Vet. Rec. *114,* 444–446.

Alexander, D.J., Wilson, G.W., Russell, P.H., Lister, S.A., and Parsons, G. (1985). Newcastle disease outbreaks in fowl in Great Britain during 1984. Vet. Rec. *117,* 429–434.

Alexander, D.J., Manvell, R.J., Kemp, P.A., Parsons, G., Collins, M.S., Brockman, S., Russell, P.H., and Lister, S.A. (1987). Use of monoclonal antibodies in the characterisation of avian paramyxovirus type 1 (Newcastle disease virus) isolates submitted to an international reference laboratory. Avian Pathol. *16,* 553–565.

Alexander, D.J., Campbell, G., Manvell, R.J., Collins, M.S., Parsons, G., and McNulty, M.S. (1992). Characterisation of an antigenically unusual virus responsible for two outbreaks of Newcastle disease in the Republic of Ireland in 1990. Vet. Rec. *130,* 65–68.

Alexander, D.J., Manvell, R.J., Lowings, J.P., Frost, K.M., Collins, M.S., Russell, P.H., and Smith, J.E. (1997b). Antigenic diversity and similarities detected in avian paramyxovirus type 1 (Newcastle disease virus) isolates using monoclonal antibodies. Avian Pathol. *26,* 399–418.

Alexander, D.J., Morris, H.T., Pollitt, W.J., Sharpe, C.E., Eckford, R.L., Sainsbury, R.M., Mansley, L.M., Gough, R.E., and Parsons, G. (1998). Newcastle disease outbreaks in domestic fowl and turkeys in Great Britain during 1997. Vet. Rec. *143,* 209–212.

Alexander, D.J., and Manvell, R.J. (2004). Heat inactivation of Newcastle disease virus (strain Herts 33/56) in artificially infected chicken meat homogenate. Avian Pathol. *33,* 222–225.

Andral, B., and Toquin, D. (1984). Infectious a myxovirus: chutes de ponte chez les dindes reproducterices I infections par les paramyxovirus aviaires de type III. Recl. Med. Vet. 43–48.

Awang, I., Wan, W., and Abdurazak, J. (1992). Detection of maternal antibodies against Newcastle disease virus in chicks using an indirect immunoperoxidate test. J. Vet. Malaysia *4,* 19–23.

Ballagi-Pordany, A., Wehmann, E., Herczeg, J., Belak, S., and Lomniczi, B. (1996). Identification and grouping of Newcastle disease virus strains by restriction site analysis of a region from the F gene. Arch. Virol. *141,* 243–261.

Banerjee, M., Reed, W.M., Fitzgerald, S.D., and Panigrahy, B. (1994). Neurotropic velogenic Newcastle disease in cormorants in Michigan: pathology and virus characterization. Avian Dis. *38,* 873–878.

Bankowski, R.A., Corstvet, R.E., and Clark, G.T. (1960). Isolation of an unidentified agent from the respiratory tract of chickens. Science *132,* 292–293.

Bankowski, R.A., Conrad, R.D., and Reynolde, B. (1968). Avian influenza A and paramyxo viruses complicating respiratory disease diagnosis in poultry. Avian Dis. *12,* 259–278.

Bankowski, R.A., Almquist, J., and Dombrucki, J. (1981). Effect of paramyxovirus yucaipa on fertility, hatchability, and poult yield of turkeys. Avian Dis. *25,* 517–520.

Barton, J.T., Bickford, A.A., Cooper, G.L., Charlton, B.R., and Cardona, C.J. (1992). Avian paramyxovirus type 1 infections in racing pigeons in California. I. Clinical signs, pathology, and serology. Avian Dis. *36,* 463–468.

Beach, J.R. (1944). The Neutralization *in vitro* of Avian Pneumoencephalitis Virus by Newcastle Disease Immune Serum. Science *100,* 361–362.

Beard, C. (1980). Isolation and Identification of Avian Pathogens. (American Association of Avian Pathologists: Kennett Square, PA).

Beard, C. (1984). Newcastle Disease (Iowa State University Press: Ames, IA).

Beaudette, F., and Hudson, C. (1956). Evidence of Newcastle disease in the eastern U.S. as early as 1938. Cornell Vet. *46,* 227–244.

Beck, I., Gerlach, H., Burkhardt, E., and Kaleta, E.F. (2003). Investigation of several selected adjuvants regarding their efficacy and side effects for the production of a vaccine for parakeets to prevent a disease caused by a paramyxovirus type 3. Vaccine *21,* 1006–1022.

Bian, H., Fournier, P., Peeters, B., and Schirrmacher, V. (2005). Tumor-targeted gene transfer *in vivo* via recombinant Newcastle disease virus modified by a bispecific fusion protein. Int. J. Oncol. *27*, 377–384.

Bogoyavlenskiy, A., Berezin, V., Prilipov, A., Usachev, E., Lyapina, O., Korotetskiy, I., Zaitceva, I., Asanova, S., Kydyrmanov, A., Daulbaeva, K., *et al.* (2009). Newcastle disease outbreaks in Kazakhstan and Kyrgyzstan during 1998, 2000, 2001, 2003, 2004, and 2005 were caused by viruses of the genotypes VIIb and VIId. Virus Genes *39*, 94–101.

Boursnell, M.E., Green, P.F., Campbell, J.I., Deuter, A., Peters, R.W., Tomley, F.M., Samson, A.C., Chambers, P., Emmerson, P.T., and Binns, M.M. (1990a). Insertion of the fusion gene from Newcastle disease virus into a non-essential region in the terminal repeats of fowlpox virus and demonstration of protective immunity induced by the recombinant. J. Gen. Virol. *71*, 621–628.

Boursnell, M.E., Green, P.F., Samson, A.C., Campbell, J.I., Deuter, A., Peters, R.W., Millar, N.S., Emmerson, P.T., and Binns, M.M. (1990b). A recombinant fowlpox virus expressing the hemagglutinin-neuraminidase gene of Newcastle disease virus (NDV) protects chickens against challenge by NDV. Virology *178*, 297–300.

Bousse, T.L., Taylor, G., Krishnamurthy, S., Portner, A., Samal, S.K., and Takimoto, T. (2004). Biological significance of the second receptor binding site of Newcastle disease virus hemagglutinin-neuraminidase protein. J. Virol. *78*, *13*,351–13355.

Box, P., Halliwell, B., and Halliwell, P. (1970). Newcastle disease in turkeys. Vet. Rec. *86*, 524–527.

Bradshaw, G.L., and Jensen, M.M. (1979). The epidemiology of Yucaipa virus in relationship to th eacute respiratory disease syndrome in turkeys. Avian Dis. *23*, 539–542.

Bruning-Fann, C., Kaneene, J., and Heamon, J. (1992). Investigation of an outbreak of velogenic viscerotropic Newcastle disease in pet birds in Michigan, Indiana, Illinois, and Texas. J. Am. Vet. Med. Assoc. *201*, 1709–1714.

Bukreyev, A., Huang, Z., Yang, L., Elankumaran, S., St Claire, M., Murphy, B.R., Samal, S.K., and Collins, P.L. (2005). Recombinant newcastle disease virus expressing a foreign viral antigen is attenuated and highly immunogenic in primates. J. Virol. *79*, 13275–13284.

Bukreyev, A., Skiadopoulos, M.H., Murphy, B.R., and Collins, P.L. (2006). Nonsegmented negative-strand viruses as vaccine vectors. J. Virol. *80*, 10293–10306.

Calain, P., and Roux, L. (1993). The rule of six, a basic feature for efficient replication of Sendai virus defective interfering RNA. J. Virol. *67*, 4822–4830.

Cantin, C., Holguera, J., Ferreira, L., Villar, E., and Munoz-Barroso, I. (2007). Newcastle disease virus may enter cells by caveolae-mediated endocytosis. J. Gen. Virol. *88*, 559–569.

Capua, I., De Nardi, R., Beato, M.S., Terregino, C., Scremin, M., and Guberti, V. (2004). Isolation of an avian paramyxovirus type 9 from migratory waterfowl in Italy. Vet. Rec. *155*, 156.

Carnero, E., Li, W., Borderia, A.V., Moltedo, B., Moran, T., and Garcia-Sastre, A. (2009). Optimization of human immunodeficiency virus gag expression by newcastle disease virus vectors for the induction of potent immune responses. J. Virol. *83*, 584–597.

Cassel, W.A., and Garrett, R.E. (1965). Newcastle Disease Virus as an Antineoplastic Agent. Cancer *18*, 863–868.

Cassel, W.A., and Murray, D.R. (1992). A 10-year follow-up on stage II malignant melanoma patients treated postsurgically with Newcastle disease virus oncolysate. Med. Oncol. Tumor Pharmacother. *9*, 169–171.

Chang, P.W. (1981). Section B: Viral Zoonosis, vol II (CRC Press, Boca Raton), pp. 261–274.

Chang, P.C., Hsieh, M.L., Shien, J.H., Graham, D.A., Lee, M.S., and Shieh, H.K. (2001). Complete nucleotide sequence of avian paramyxovirus type 6 isolated from ducks. J. Gen. Virol. *82*, 2157–2168.

Chare, E.R., Gould, E.A., and Holmes, E.C. (2003). Phylogenetic analysis reveals a low rate of homologous recombination in negative-sense RNA viruses. J. Gen. Virol. *84*, 2691–2703.

Chen, L., Gorman, J.J., McKimm-Breschkin, J., Lawrence, L.J., Tulloch, P.A., Smith, B.J., Colman, P.M., and Lawrence, M.C. (2001). The structure of the fusion glycoprotein of Newcastle disease virus suggests a novel paradigm for the molecular mechanism of membrane fusion. Structure *9*, 255–266.

Chong, Y.L., Padhi, A., Hudson, P.J., and Poss, M. (2010). The effect of vaccination on the evolution and population dynamics of avian paramyxovirus-1. PLoS Pathog. *6*, e1000872.

Cloud, S., and Rosenberger, J. (1980). Characterization of nine avian paramyxoviruses. Avian Dis. *24*, 139–152.

Collins, M.S., Bashiruddin, J.B., and Alexander, D.J. (1993). Deduced amino acid sequences at the fusion protein cleavage site of Newcastle disease viruses showing variation in antigenicity and pathogenicity. Arch. Virol. *128*, 363–370.

Collins, M.S., Strong, I., and Alexander, D.J. (1994). Evaluation of the molecular basis of pathogenicity of the variant Newcastle disease viruses terMed. 'pigeon PMV-1 viruses'. Arch. Virol. *134*, 403–411.

Collins, M.S., Franklin, S., Strong, I., Meulemans, G., and Alexander, D.J. (1998). Antigenic and phylogenetic studies on a variant Newcastle disease virus using anti-fusion protein monoclonal antibodies and partial sequencing of the fusion protein gene. Avian Pathol. *27*, 90–96.

Collins, P.L., Bukreyev, A., and Murphy, B.R. (2008). What are the risks–hypothetical and observed–of recombination involving live vaccines and vaccine vectors based on nonsegmented negative-strain RNA viruses? J. Virol. *82*, 9805–9806.

Connaris, H., Takimoto, T., Russell, R., Crennell, S., Moustafa, I., Portner, A., and Taylor, G. (2002). Probing the sialic acid binding site of the hemagglutinin-neuraminidase of Newcastle disease virus: identification of key amino acids involved in cell binding, catalysis, and fusion. J. Virol. *76,* 1816–1824.

Crawford, M. (1930). Rannikhet Ann. Rept. Govt. Vet. Surg. Colombo. Ceylon.

Crennell, S., Takimoto, T., Portner, A., and Taylor, G. (2000). Crystal structure of the multifunctional paramyxovirus hemagglutinin-neuraminidase. Nat. Struct. Biol. *7,* 1068–1074.

Czegledi, A., Ujvari, D., Somogyi, E., Wehmann, E., Werner, O., and Lomniczi, B. (2006). Third genome size category of avian paramyxovirus serotype 1 (Newcastle disease virus) and evolutionary implications. Virus Res. *120,* 36–48.

Dahlberg, J.E., and Simon, E.H. (1969a). Physical and genetic studies of Newcastle disease virus: evidence for multiploid particles. Virology *38,* 666–678.

Dahlberg, J.E., and Simon, E.H. (1969b). Recombination in Newcastle disease virus (NDV): the problem of complementing heterozygotes. Virology *38,* 490–493.

Deng, R., Wang, Z., Mirza, A.M., and Iorio, R.M. (1995). Localization of a domain on the paramyxovirus attachment protein required for the promotion of cellular fusion by its homologous fusion protein spike. Virology *209,* 457–469.

DiNapoli, J.M., Kotelkin, A., Yang, L., Elankumaran, S., Murphy, B.R., Samal, S.K., Collins, P.L., and Bukreyev, A. (2007a). Newcastle disease virus, a host range-restricted virus, as a vaccine vector for intranasal immunization against emerging pathogens. Proc. Natl. Acad. Sci. U.S.A. *104,* 9788–9793.

DiNapoli, J.M., Yang, L., Suguitan, A., Jr., Elankumaran, S., Dorward, D.W., Murphy, B.R., Samal, S.K., Collins, P.L., and Bukreyev, A. (2007b). Immunization of primates with a Newcastle disease virus-vectored vaccine via the respiratory tract induces a high titer of serum neutralizing antibodies against highly pathogenic avian influenza virus. J. Virol. *81, 11,*560–11568.

DiNapoli, J.M., Nayak, B., Yang, L., Finneyfrock, B.W., Cook, A., Andersen, H., Torres-Velez, F., Murphy, B.R., Samal, S.K., Collins, P.L., *et al.* (2010). Newcastle disease virus-vectored vaccines expressing the hemagglutinin or neuraminidase protein of H5N1 highly pathogenic avian influenza virus protect against virus challenge in monkeys. J. Virol. *84,* 1489–1503.

Dortmans, J.C., Koch, G., Rottier, P.J., and Peeters, B.P. (2009). Virulence of pigeon paramyxovirus type 1 does not always correlate with the cleavability of its fusion protein. J. Gen. Virol. *90,* 2746–2750.

Dortmans, J.C., Rottier, P.J., Koch, G., and Peeters, B.P. (2010). The viral replication complex is associated with virulence of Newcastle disease virus (NDV). J. Virol. *84,* 10113–10120.

Dortmans, J.C., Rottier, P.J., Koch, G., and Peeters, B.P. (2011). Passaging of a Newcastle disease virus pigeon variant in chickens results in selection of viruses with mutations in the polymerase complex enhancing virus replication and virulence. J. Gen Virol. *92,* 336–345.

Doyle, T. (1927). A hitherto unrecorded disease of fowls due to a filter-passing virus. J. Comp. Pathol. Ther. *40,* 144–169.

Doyle, T. (1935). Newcastle disease of fowls. J. Comp. Pathol. Ther. *48,* 1–20.

Dundon, W.G., Heidari, A., De Nardi, R., Terregino, C., Capua, I., and Cattoli, G. (2010). Genetic variation of Italian avian paramyxoviruses serotype 9. Virus Genes *41,* 43–46.

Edwards, J. (1928). A new fowl disease. Mukteswar Ann. Rep. Inp. Inst. Vet. Res. *31,* 14–15.

Elankumaran, S., Rockemann, D., and Samal, S.K. (2006). Newcastle disease virus exerts oncolysis by both intrinsic and extrinsic caspase-dependent pathways of cell death. J. Virol. *80,* 7522–7534.

Elankumaran, S., Chavan, V., Qiao, D., Shobana, R., Moorkanat, G., Biswas, M., and Samal, S.K. (2010). Type I interferon-sensitive recombinant newcastle disease virus for oncolytic virotherapy. J. Virol. *84,* 3835–3844.

Erdei, J., Bachir, K., Kaleta, E.F., Shortridge, K.F., and Lomniczi, B. (1987). Newcastle disease vaccine (La Sota) strain specific monoclonal antibody. Brief report. Arch. Virol. *96,* 265–269.

Errington, W., and Emmerson, P.T. (1997). Assembly of recombinant Newcastle disease virus nucleocapsid protein into nucleocapsid-like structures is inhibited by the phosphoprotein. J. Gen. Virol. *78,* 2335–2339.

Estevez, C., King, D., Seal, B., and Yu, Q. (2007). Evaluation of Newcastle disease virus chimeras expressing the hemagglutinin-neuraminidase protein of velogenic strains in the context of a mesogenic recombinant virus backbone. Virus Res. *129,* 182–190.

Fabian, Z., Csatary, C.M., Szeberenyi, J., and Csatary, L.K. (2007). p53-independent endoplasmic reticulum stress-mediated cytotoxicity of a Newcastle disease virus strain in tumor cell lines. J. Virol. *81,* 2817–2830.

Fournier, P., Zeng, J., and Schirrmacher, V. (2003). Two ways to induce innate immune responses in human PBMCs: paracrine stimulation of IFN-alpha responses by viral protein or dsRNA. Int. J. Oncol. *23,* 673–680.

Fuller, C.M., Collins, M.S., Easton, A.J., and Alexander, D.J. (2007). Partial characterisation of five cloned viruses differing in pathogenicity, obtained from a single isolate of pigeon paramyxovirus type 1 (PPMV-1) following passage in fowls' eggs. Arch. Virol. *152,* 1575–1582.

Garcia-Sastre, A., Cabezas, J.A., and Villar, E. (1989). Proteins of Newcastle disease virus envelope: interaction between the outer hemagglutinin-neuraminidase glycoprotein and the inner non-glycosylated matrix protein. Biochim. Biophys. Acta *999,* 171–175.

Ge, J., Deng, G., Wen, Z., Tian, G., Wang, Y., Shi, J., Wang, X., Li, Y., Hu, S., Jiang, Y., *et al.* (2007). Newcastle disease virus-based live attenuated vaccine completely protects chickens and mice from lethal challenge of homologous and heterologous H5N1 avian influenza viruses. J. Virol. *81,* 150–158.

Ghumman, J.S., and Bankowski, R.A. (1976). *In vitro* DNA synthesis in lymphocytes from turkeys vaccinated with LaSota, TC, and inactivated Newcastle disease vaccines. Avian Dis. *20,* 18–31.

Goldhaft, T.M. (1980). Historical note on the origin of the LaSota strain of Newcastle disease virus. Avian Dis. *24,* 297–301.

Gough, R.E., Manvell, R.J., Drury, S.E., Naylor, P.F., Spackman, D., and Cooke, S.W. (1993). Deaths in budgerigars associated with a paramyxovirus-like agent. Vet. Rec. *133,* 123.

Gould, A.R., Kattenbelt, J.A., Selleck, P., Hansson, E., Della-Porta, A., and Westbury, H.A. (2001). Virulent Newcastle disease in Australia: molecular epidemiological analysis of viruses isolated prior to and during the outbreaks of 1998–2000. Virus Res. *77,* 51–60.

Granoff, A. (1959a). Studies on mixed infection with Newcastle disease virus. I. Isolation of Newcastle disease virus mutants and tests for genetic recombination between them. Virology *9,* 636–648.

Granoff, A. (1959b). Studies on mixed infection with Newcastle disease virus. II. The occurrence of Newcastle disease virus heterozygotes and study of phenotypic mixing involving serotype and thermal stability. Virology *9,* 649–670.

Gravel, K.A., McGinnes, L.W., Reitters, J., and Morrison, T.G. (2011). The transmembrane domain sequence affects the structure and function of the Newcastle disease virus fusion protein. J. Virol. doi: 1128/JVI.02308–10.

Han, G.Z., He, C.Q., Ding, N.Z., and Ma, L.Y. (2008a). Identification of a natural multirecombinant of Newcastle disease virus. Virology *371,* 54–60.

Han, G.Z., Liu, X.P., and Li, S.S. (2008b). Caution about Newcastle disease virus-based live attenuated vaccine. J. Virol. *82,* 6782.

Hanson, R. (1972). World wide spread of viscerotropic Newcastle disease. Proceedings of the 76th meeting of the US animal health assoication, Florida, 5–10 November (US Animal Health Association), pp. 276–279.

Hanson, R.P., and Brandly, C.A. (1955). Identification of vaccine strains of Newcastle disease virus. Science *122,* 156–157.

Hanson, R.P., Spalatin, J., and Jacobson, G.S. (1973). The viscerotropic pathotype of Newcastle disease virus. Avian Dis. *17,* 354–361.

Heckert, R.A., Collins, M.S., Manvell, R.J., Strong, I., Pearson, J.E., and Alexander, D.J. (1996a). Comparison of Newcastle disease viruses isolated from cormorants in Canada and the USA in 1975, 1990 and 1992. Can. J. Vet. Res. *60,* 50–54.

Heckert, R.A., Riva, J., Cook, S., McMillen, J., and Schwartz, R.D. (1996b). Onset of protective immunity in chicks after vaccination with a recombinant herpesvirus of turkeys vaccine expressing Newcastle disease virus fusion and hemagglutinin-neuraminidase antigens. Avian Dis. *40,* 770–777.

Heller, E.D., Nathan, D.B., and Perek, M. (1977). The transfer of Newcastle serum antibody from the laying hen to the egg and the chick. Res. Vet. Sci. *22,* 376–379.

Higgins, D.A. (1971). Nine disease outbreaks associated with myxoviruses among ducks in Hong Kong. Trop. Anim. Health Prod. *3,* 232–240.

Hitchner, S.B., and Johnson, E.P. (1948). A virus of low virulence for immunizing fowls against Newcastle disease; avian pneumoencephalitis. Vet. Med. *43,* 525–530.

Holmes, H.C. (1979). Resistance of the respiratory tract of the chicken to Newcastle disease virus infection following vaccination: the effect of passively acquired antibody on its development. J. Comp. Pathol. *89,* 11–19.

Hu, S., Ma, H., Wu, Y., Liu, W., Wang, X., Liu, Y., and Liu, X. (2009). A vaccine candidate of attenuated genotype VII Newcastle disease virus generated by reverse genetics. Vaccine *27,* 904–910.

Hu, X.L., Ray, R., and Compans, R.W. (1992). Functional interactions between the fusion protein and hemagglutinin-neuraminidase of human parainfluenza viruses. J. Virol. *66,* 1528–1534.

Huang, Z., Krishnamurthy, S., Panda, A., and Samal, S.K. (2001). High-level expression of a foreign gene from the most 3′-proximal locus of a recombinant Newcastle disease virus. J. Gen. Virol. *82,* 1729–1736.

Huang, Z., Krishnamurthy, S., Panda, A., and Samal, S.K. (2003). Newcastle disease virus V protein is associated with viral pathogenesis and functions as an alpha interferon antagonist. J. Virol. *77,* 8676–8685.

Huang, Y., Wan, H.Q., Liu, H.Q., Wu, Y.T., and Liu, X.F. (2004a). Genomic sequence of an isolate of Newcastle disease virus isolated from an outbreak in geese: a novel six nucleotide insertion in the non-coding region of the nucleoprotein gene. Brief Report. Arch. Virol. *149,* 1445–1457.

Huang, Z., Elankumaran, S., Yunus, A.S., and Samal, S.K. (2004b). A recombinant Newcastle disease virus (NDV) expressing VP2 protein of infectious bursal disease virus (IBDV) protects against NDV and IBDV. J. Virol. *78,* 10054–10063.

Huang, Z., Panda, A., Elankumaran, S., Govindarajan, D., Rockemann, D.D., and Samal, S.K. (2004c). The hemagglutinin-neuraminidase protein of Newcastle disease virus determines tropism and virulence. J. Virol. *78,* 4176–4184.

Jahanshiri, F., Eshaghi, M., and Yusoff, K. (2005). Identification of phosphoprotein:phosphoprotein and phosphoprotein:nucleocapsid protein interaction domains of the Newcastle disease virus. Arch. Virol. *150,* 611–618.

Jeon, W.J., Lee, E.K., Kwon, J.H., and Choi, K.S. (2008). Full-length genome sequence of avain paramyxovirus type 4 isolated from a mallard duck. Virus Genes *37,* 342–350.

Johnston, K.M., and Key, D.W. (1992). Paramyxovirus-1 in feral pigeons (Columba livia) in Ontario. Can. Vet. J. *33,* 796–800.

Kaleta, E., and Baldauf, C. (1988). Newcastle Disease in Free-living and Pet Birds (Kluwer Acedemic Press: Boston, MA).

Kamiya, N., Niikura, M., Ono, M., Kai, C., Matsuura, Y., and Mikami, T. (1994). Protective effect of individual glycoproteins of Newcastle disease virus expressed in insect cells: the fusion protein derived from an avirulent strain had lower protective efficacy. Virus Res. *32*, 373–379.

Kanno, T., Ochi, Y., and Hashimoto, K. (1929). N Geflugelseuche in Korea. Dtsch. Tierarzhl. Wshr. *37*, 515–517.

Kapczynski, D.R., and King, D.J. (2005). Protection of chickens against overt clinical disease and determination of viral shedding following vaccination with commercially available Newcastle disease virus vaccines upon challenge with highly virulent virus from the California 2002 exotic Newcastle disease outbreak. Vaccine *23*, 3424–3433.

Kapgate, S., Kumanan, K., Vijayarani, K., and Tirumurugaan, K. (2010a). Complete genome sequence of a chicken isolate of Newcastle disease virus from India. Available at: http://wwwncbi..nlom.nih.gov/GU187941.1.

Kapgate, S., Kumanan, K., Vijayarani, K., and Tirumurugaan, K. (2010b). Whole genome sequence of a chicken isolate of Newcastle disease virus from Tamil Nadu, India. Available at http://www.ncbi.nlom.nih.gov/HM357251.1.

Kattenbelt, J.A., Stevens, M.P., and Gould, A.R. (2006). Sequence variation in the Newcastle disease virus genome. Virus Res. *116*, 168–184.

Kawahara, N., Yang, X.Z., Sakaguchi, T., Kiyotani, K., Nagai, Y., and Yoshida, T. (1992). Distribution and substrate specificity of intracellular proteolytic processing enzyme(s) for paramyxovirus fusion glycoproteins. J. Gen. Virol. *73*, 583–590.

Khattar, S.K., Yan, Y., Panda, A., Collins, P.L., and Samal, S.K. (2009). A Y526Q mutation in the Newcastle disease virus HN protein reduces its functional activities and attenuates virus replication and pathogenicity. J. Virol. *83*, 7779–7782.

Khattar, S.K., Collins, P.L., and Samal, S.K. (2010). Immunization of cattle with recombinant Newcastle disease virus expressing bovine herpesvirus-1 (BHV-1) glycoprotein D induces mucosal and serum antibody responses and provides partial protection against BHV-1. Vaccine *28*, 3159–3170.

Khattar, S.K., Kumar, S., Xiao, S., Collins, P.L., and Samal, S.K. (2011). Experimental infection of mice with avian paramyxovirus serotypes 1 to 9. PLoS One *6*, e16776.

Kho, C.L., Tan, W.S., Tey, B.T., and Yusoff, K. (2003). Newcastle disease virus nucleocapsid protein: self-assembly and length-determination domains. J. Gen. Virol. *84*, 2163–2168.

Kho, C.L., Tan, W.S., Tey, B.T., and Yusoff, K. (2004). Regions on nucleocapsid protein of Newcastle disease virus that interact with its phosphoprotein. Arch. Virol. *149*, 997–1005.

Kim, S.H., and Samal, S.K. (2010). Role of untranslated regions in regulation of gene expression, replication, and pathogenicity of Newcastle disease virus expressing green fluorescent protein. J. Virol. *84*, 2629–2634.

Kim, S.J., Spradbrow, P.B., and MacKenzie, M. (1978). The isolation of lentogenic strains of Newcastle disease virus in Australia. Aust. Vet. J. *54*, 183–187.

Kim, L.M., King, D.J., Curry, P.E., Suarez, D.L., Swayne, D.E., Stallknecht, D.E., Slemons, R.D., Pedersen, J.C., Senne, D.A., Winker, K., *et al.* (2007a). Phylogenetic diversity among low-virulence newcastle disease viruses from waterfowl and shorebirds and comparison of genotype distributions to those of poultry-origin isolates. J. Virol. *81*, 12641–12653.

Kim, L.M., King, D.J., Suarez, D.L., Wong, C.W., and Afonso, C.L. (2007b). Characterization of class I Newcastle disease virus isolates from Hong Kong live bird markets and detection using real-time reverse transcription-PCR. J. Clin. Microbiol. *45*, 1310–1314.

Kim, L.M., King, D.J., Guzman, H., Tesh, R.B., Travassos da Rosa, A.P., Bueno, R., Jr., Dennett, J.A., and Afonso, C.L. (2008a). Biological and phylogenetic characterization of pigeon paramyxovirus serotype 1 circulating in wild North American pigeons and doves. J. Clin. Microbiol. *46*, 3303–3310.

Kim, L.M., Suarez, D.L., and Afonso, C.L. (2008b). Detection of a broad range of class I and II Newcastle disease viruses using a multiplex real-time reverse transcription polymerase chain reaction assay. J. Vet. Diagn. Invest. *20*, 414–425.

Kim, S.H., Yan, Y., and Samal, S.K. (2009). Role of the cytoplasmic tail amino acid sequences of Newcastle disease virus hemagglutinin-neuraminidase protein in virion incorporation, cell fusion, and pathogenicity. J. Virol. *83*, *10*, 250–10255.

King, D.J. (1996). Avian paramyxovirus type 1 from pigeons: isolate characterization and pathogenicity after chicken or embryo passage of selected isolates. Avian Dis. *40*, 707–714.

King, D.J. (1999). A comparison of the onset of protection induced by Newcastle disease virus strain B1 and a fowl poxvirus recombinant Newcastle disease vaccine to a viscerotropic velogenic Newcastle disease virus challenge. Avian Dis. *43*, 745–755.

Kommers, G.D., King, D.J., Seal, B.S., and Brown, C.C. (2001). Virulence of pigeon-origin Newcastle disease virus isolates for domestic chickens. Avian Dis. *45*, 906–921.

Kommers, G.D., King, D.J., Seal, B.S., and Brown, C.C. (2003). Pathogenesis of chicken-passaged Newcastle disease viruses isolated from chickens and wild and exotic birds. Avian Dis. *47*, 319–329.

Kortekaas, J., de Boer, S.M., Kant, J., Vloet, R.P., Antonis, A.F., and Moormann, R.J. (2010a). Rift Valley fever virus immunity provided by a paramyxovirus vaccine vector. Vaccine *28*, 4394–4401.

Kortekaas, J., Dekker, A., de Boer, S.M., Weerdmeester, K., Vloet, R.P., de Wit, A.A., Peeters, B.P., and Moormann, R.J. (2010b). Intramuscular inoculation of calves with an experimental Newcastle disease virus-based vector vaccine elicits neutralizing antibodies against Rift Valley fever virus. Vaccine *28*, 2271–2276.

Kraneveld, F. (1926). Over een in Ned-Indie heerschende Ziete under het pluimves. Ned. Indisch Bl. Diergeneesk *38*, 448–450.

Krishnamurthy, S., and Samal, S.K. (1998). Nucleotide sequences of the trailer, nucleocapsid protein gene and intergenic regions of Newcastle disease virus strain

Beaudette C and completion of the entire genome sequence. J. Gen. Virol. *79,* 2419–2424.

Krishnamurthy, S., Huang, Z., and Samal, S.K. (2000). Recovery of a virulent strain of newcastle disease virus from cloned cDNA: expression of a foreign gene results in growth retardation and attenuation. Virology *278,* 168–182.

Kuiken, T., Leighton, F.A., Wobeser, G., Danesik, K.L., Riva, J., and Heckert, R.A. (1998). An epidemic of Newcastle disease in double-crested cormorants from Saskatchewan. J. Wildl. Dis. *34,* 457–471.

Kumar, S., Nayak, B., Collins, P.L., and Samal, S.K. (2008). Complete genome sequence of avian paramyxovirus type 3 reveals an unusually long trailer region. Virus Res. *137,* 189–197.

Kumar, S., Militino Dias, F., Nayak, B., Collins, P.L., and Samal, S.K. (2010a). Experimental avian paramyxovirus serotype-3 infection in chickens and turkeys. Vet. Res. *41,* 72.

Kumar, S., Nayak, B., Samuel, A.S., Xiao, S., Collins, P.L., and Samal, S.K. (2010b). Complete genome sequence of avian paramyxovirus-3 strain Wisconsin: evidence for the existence of subgroups within the serotype. Virus Res. *149,* 78–85.

Laliberte, J.P., McGinnes, L.W., Peeples, M.E., and Morrison, T.G. (2006). Integrity of membrane lipid rafts is necessary for the ordered assembly and release of infectious Newcastle disease virus particles. J. Virol. *80,* 10652–10662.

Lam, K.M., and Hao, Q. (1987). Vaccination of cyclophosphamide-treated chickens against Newcastle disease virus infection. Vet. Microbiol. *15,* 41–48.

Lamb, R., and Parks, G. (2007). In *Paramyxoviridae*: the viruses and their replication. In Field's Virology, 5th edn, Knipe, D.M., Howley, P.M., Griffin, D.E., Lamb, R.A., Martin, M.A., Roizman, B., and Straus, S.E., eds. (Lippincott Williams & Wilkins, Philadelphia), pp. 1444–1496.

Lana, D.P., Snyder, D.B., King, D.J., and Marquardt, W.W. (1988). Characterization of a battery of monoclonal antibodies for differentiation of Newcastle disease virus and pigeon paramyxovirus-1 strains. Avian Dis. *32,* 273–281.

Letellier, C., Burny, A., and Meulemans, G. (1991). Construction of a pigeonpox virus recombinant: expression of the Newcastle disease virus (NDV) fusion glycoprotein and protection of chickens against NDV challenge. Arch. Virol. *118,* 43–56.

de Leeuw, O., and Peeters, B. (1999). Complete nucleotide sequence of Newcastle disease virus: evidence for the existence of a new genus within the subfamily *Paramyxovirinae*. J. Gen. Virol. *80,* 131–136.

de Leeuw, O.S., Hartog, L., Koch, G., and Peeters, B.P. (2003). Effect of fusion protein cleavage site mutations on virulence of Newcastle disease virus: non-virulent cleavage site mutants revert to virulence after one passage in chicken brain. J. Gen. Virol. *84,* 475–484.

de Leeuw, O.S., Koch, G., Hartog, L., Ravenshorst, N., and Peeters, B.P. (2005). Virulence of Newcastle disease virus is determined by the cleavage site of the fusion protein and by both the stem region and globular head of the haemagglutinin–neuraminidase protein. J. Gen. Virol. *86,* 1759–1769.

Levy, D.E., and Marie, I.J. (2004). RIGging an antiviral defense–it's in the CARDs. Nat. Immunol. *5,* 699–701.

Li, J.K., Miyakawa, T., and Fox, C.F. (1980). Protein organization in Newcastle disease virus as revealed by perturbant treatment. J. Virol. *34,* 268–271.

Liu, Y.L., Hu, S.L., Zhang, Y.M., Sun, S.J., Romer-Oberdorfer, A., Veits, J., Wu, Y.T., Wan, H.Q., and Liu, X.F. (2007). Generation of a velogenic Newcastle disease virus from cDNA and expression of the green fluorescent protein. Arch. Virol. *152,* 1241–1249.

Loke, C.F., Omar, A.R., Raha, A.R., and Yusoff, K. (2005). Improved protection from velogenic Newcastle disease virus challenge following multiple immunizations with plasmid DNA encoding for F and HN genes. Vet. Immunol. Immunopathol. *106,* 259–267.

McGinnes, L.W., and Morrison, T.G. (1994). The role of the individual cysteine residues in the formation of the mature, antigenic HN protein of Newcastle disease virus. Virology *200,* 470–483.

McGinnes, L.W., and Morrison, T.G. (1995). The role of individual oligosaccharide chains in the activities of the HN glycoprotein of Newcastle disease virus. Virology *212,* 398–410.

McGinnes, L.W., Wilde, A., and Morrison, T.G. (1987). Nucleotide sequence of the gene encoding the Newcastle disease virus hemagglutinin-neuraminidase protein and comparisons of paramyxovirus hemagglutinin-neuraminidase protein sequences. Virus Res. *7,* 187–202.

McGinnes, L., Sergel, T., and Morrison, T. (1993). Mutations in the transmembrane domain of the HN protein of Newcastle disease virus affect the structure and activity of the protein. Virology *196,* 101–110.

McGinnes, L., Sergel, T., Reitter, J., and Morrison, T. (2001). Carbohydrate modifications of the NDV fusion protein heptad repeat domains influence maturation and fusion activity. Virology *283,* 332–342.

McGinnes, L.W., Reitter, J.N., Gravel, K., and Morrison, T.G. (2003). Evidence for mixed membrane topology of the newcastle disease virus fusion protein. J. Virol. *77,* 1951–1963.

Macpherson, L.W. (1956). Some Observations On The Epizootiology Of NewCastle Disease. Can J. Comp Med. Vet. Sci. *20,* 155–168.

Macpherson, I., Watt, R.G., and Alexander, D.J. (1983). Isolation of avian paramyxovirus other than Newcastle disease virus from commercial poultry in Great Britain. Vet. Rec. *112,* 479–480.

Maminiaina, O.F., Gil, P., Briand, F.X., Albina, E., Keita, D., Andriamanivo, H.R., Chevalier, V., Lancelot, R., Martinez, D., Rakotondravao, R., *et al.* (2010). Newcastle disease virus in Madagascar: identification of an original genotype possibly deriving from a died out ancestor of genotype IV. PLoS One *5,* e13987.

Marino, O.C., and Hanson, R.P. (1987). Cellular and humoral response of in ovo-bursectomized chickens to experimental challenge with velogenic Newcastle disease virus. Avian Dis. *31,* 293–301.

Martinez-Sobrido, L., Gitiban, N., Fernandez-Sesma, A., Cros, J., Mertz, S.E., Jewell, N.A., Hammond, S., Flano, E., Durbin, R.K., Garcia-Sastre, A., *et al.* (2006). Protection against respiratory syncytial virus by a recombinant Newcastle disease virus vector. J. Virol. *80,* 1130–1139.

Mase, M., Imai, K., Sanada, Y., Sanada, N., Yuasa, N., Imada, T., Tsukamoto, K., and Yamaguchi, S. (2002). Phylogenetic analysis of Newcastle disease virus genotypes isolated in Japan. J. Clin. Microbiol. *40,* 3826–3830.

Mebatsion, T., Verstegen, S., De Vaan, L.T., Romer-Oberdorfer, A., and Schrier, C.C. (2001). A recombinant newcastle disease virus with low-level V protein expression is immunogenic and lacks pathogenicity for chicken embryos. J. Virol. *75,* 420–428.

Meulemans, G., Letellier, C., Gonze, M., Carlier, M.C., and Burny, A. (1988). Newcastle disease virus F glycoprotein expressed from a recombinant vaccinia virus vector protects chickens against live-virus challenge. Avian Pathol. *17,* 821–827.

Meulemans, G., van den Berg, T.P., Decaesstecker, M., and Boschmans, M. (2002). Evolution of pigeon Newcastle disease virus strains. Avian Pathol. *31,* 515–519.

Miller, P.J., King, D.J., Afonso, C.L., and Suarez, D.L. (2007). Antigenic differences among Newcastle disease virus strains of different genotypes used in vaccine formulation affect viral shedding after a virulent challenge. Vaccine *25,* 7238–7246.

Miller, P.J., Kim, L.M., Ip, H.S., and Afonso, C.L. (2009). Evolutionary dynamics of Newcastle disease virus. Virology *391,* 64–72.

Miller, P.J., Afonso, C.L., Spackman, E., Scott, M.A., Pedersen, J.C., Senne, D.A., Brown, J.D., Fuller, C.M., Uhart, M.M., Karesh, W.B., *et al.* (2010a). Evidence for a new avian paramyxovirus serotype-10 detected in rockhopper penguins from the Falkland Islands. J. Virol. *84,* 11496–11504.

Miller, P.J., Decanini, E.L., and Afonso, C.L. (2010b). Newcastle disease: evolution of genotypes and the related diagnostic challenges. Infect. Genet. Evol. *10,* 26–35.

Mirza, A.M., Sheehan, J.P., Hardy, L.W., Glickman, R.L., and Iorio, R.M. (1993). Structure and function of a membrane anchor-less form of the hemagglutinin-neuraminidase glycoprotein of Newcastle disease virus. J. Biol. Chem. *268,* 21425–21431.

Morgan, R.W., Gelb, J., Jr., Pope, C.R., and Sondermeijer, P.J. (1993). Efficacy in chickens of a herpesvirus of turkeys recombinant vaccine containing the fusion gene of Newcastle disease virus: onset of protection and effect of maternal antibodies. Avian Dis. *37,* 1032–1040.

Morrison, T.G. (2003). Structure and function of a paramyxovirus fusion protein. Biochim. Biophys. Acta *1614,* 73–84.

Morrison, T., Hinshaw, V.S., Sheerar, M., Cooley, A.J., Brown, D., McQuain, C., and McGinnes, L. (1990). Retroviral expressed hemagglutinin-neuraminidase protein protects chickens from Newcastle disease virus induced disease. Microb. Pathog. *9,* 387–396.

Morrison, T., McQuain, C., and McGinnes, L. (1991). Complementation between avirulent Newcastle disease virus and a fusion protein gene expressed from a retrovirus vector: requirements for membrane fusion. J. Virol. *65,* 813–822.

Moya, A., Elena, S.F., Bracho, A., Miralles, R., and Barrio, E. (2000). The evolution of RNA viruses: A population genetics view. Proc. Natl. Acad. Sci. U.S.A. *97,* 6967–6973.

Munir, S., Sharma, J.M., and Kapur, V. (2005). Transcriptional response of avian cells to infection with Newcastle disease virus. Virus Res. *107,* 103–108.

Mustaffa-Babjee, A., Spradbrow, P.B., and Samuel, J.L. (1974). A pathogenic paramyxovirus from a budgerigar (*Melopsittacus undulatus*). Avian Dis. *18,* 226–230.

Nagai, Y. (1995). Virus activation by host proteinases. A pivotal role in the spread of infection, tissue tropism and pathogenicity. Microbiol. Immunol. *39,* 1–9.

Nagai, Y., Klenk, H.D., and Rott, R. (1976). Proteolytic cleavage of the viral glycoproteins and its significance for the virulence of Newcastle disease virus. Virology *72,* 494–508.

Nakaya, T., Cros, J., Park, M.S., Nakaya, Y., Zheng, H., Sagrera, A., Villar, E., Garcia-Sastre, A., and Palese, P. (2001). Recombinant Newcastle disease virus as a vaccine vector. J. Virol. *75,* 11868–11873.

Nakaya, Y., Nakaya, T., Park, M.S., Cros, J., Imanishi, J., Palese, P., and Garcia-Sastre, A. (2004). Induction of cellular immune responses to simian immunodeficiency virus gag by two recombinant negative-strand RNA virus vectors. J. Virol. *78,* 9366–9375.

Nayak, B., Kumar, S., Collins, P.L., and Samal, S.K. (2008). Molecular characterization and complete genome sequence of avian paramyxovirus type 4 prototype strain duck/Hong Kong/D3/75. Virol. J. *5,* 124.

Nayak, B., Rout, S.N., Kumar, S., Khalil, M.S., Fouda, M.M., Ahmed, L.E., Earhart, K.C., Perez, D.R., Collins, P.L., and Samal, S.K. (2009). Immunization of chickens with Newcastle disease virus expressing H5 hemagglutinin protects against highly pathogenic H5N1 avian influenza viruses. PLoS One *4,* e6509.

Nayak, B., Kumar, S., DiNapoli, J.M., Paldurai, A., Perez, D.R., Collins, P.L., and Samal, S.K. (2010). Contributions of the avian influenza virus HA, NA, and M2 surface proteins to the induction of neutralizing antibodies and protective immunity. J. Virol. *84,* 2408–2420.

Nerome, K., Nakayama, M., Ishida, M., and Fukumi, H. (1978). Isolation of a new avian paramyxovirus from budgerigar (Melopsittacus undulatus). J. Gen. Virol. *38,* 293–301.

Ochi, Y., and Hashimoto, K. (1929). Uber eine neue Geflugelseuche in Korea. 6th Rept, Govt. Inst. Vet. Res. 16.

Olesiuk, O.M. (1951). Influence of environmental factors on viability of Newcastle disease virus. Am. J. Vet. Res. *12,* 152–155.

Paldurai, A., Subbiah, M., Kumar, S., Collins, P.L., and Samal, S.K. (2009). Complete genome sequences of avian paramyxovirus type 8 strains goose/ Delaware/1053/76 and pintail/Wakuya/20/78. Virus Res. *142,* 144–153.

Paldurai, A., Kumar, S., Kim, S.H., Nayak, B., and Samal, S.K. (2010a). Effects of naturally occurring six- and twelve-nucleotide inserts on Newcastle disease virus replication and pathogenesis. J. Virol. The 29th Annual American Society for Virology Meeting, Montana, 17–21 July 2010, (Virology), p. 176.

Paldurai, A., Kumar, S., Nayak, B., and Samal, S.K. (2010b). Complete genome sequence of highly virulent neurotropic Newcastle disease virus strain Texas GB. Virus Genes *41,* 67–72.

Panda, A., Elankumaran, S., Krishnamurthy, S., Huang, Z., and Samal, S.K. (2004a). Loss of N-linked glycosylation from the hemagglutinin-neuraminidase protein alters virulence of Newcastle disease virus. J. Virol. *78,* 4965–4975.

Panda, A., Huang, Z., Elankumaran, S., Rockemann, D.D., and Samal, S.K. (2004b). Role of fusion protein cleavage site in the virulence of Newcastle disease virus. Microb. Pathog. *36,* 1–10.

Panigrahy, B., Senne, D.A., Pearson, J.E., Mixson, M.A., and Cassidy, D.R. (1993). Occurrence of velogenic viscerotropic Newcastle disease in pet and exotic birds in 1991. Avian Dis. *37,* 254–258.

Pantua, H., McGinnes, L.W., Leszyk, J., and Morrison, T.G. (2005). Characterization of an alternate form of Newcastle disease virus fusion protein. J. Virol. *79,* 11660–11670.

Pantua, H.D., McGinnes, L.W., Peeples, M.E., and Morrison, T.G. (2006). Requirements for the assembly and release of Newcastle disease virus-like particles. J. Virol. *80,* 11062–11073.

Park, M.S., Shaw, M.L., Munoz-Jordan, J., Cros, J.F., Nakaya, T., Bouvier, N., Palese, P., Garcia-Sastre, A., and Basler, C.F. (2003). Newcastle disease virus (NDV)-based assay demonstrates interferon-antagonist activity for the NDV V protein and the Nipah virus V, W, and C proteins. J. Virol. *77,* 1501–1511.

Park, M.S., Steel, J., Garcia-Sastre, A., Swayne, D., and Palese, P. (2006). Engineered viral vaccine constructs with dual specificity: avian influenza and Newcastle disease. Proc. Natl. Acad. Sci. U.S.A. *103,* 8203–8208.

Parry, S., and Aitken, I. (1977). Local immunity in the respiratory tract of the chicken. II The secretory immune response to Newcastle disease virus and the role of IgA. Vet. Microbiol. *2,* 143–165.

Pearson, J.E., Senne, D.A., Alexander, D.J., Taylor, W.D., Peterson, L.A., and Russell, P.H. (1987). Characterization of Newcastle disease virus (avian paramyxovirus-1) isolated from pigeons. Avian Dis. *31,* 105–111.

Pedersen, J.C., Senne, D.A., Woolcock, P.R., Kinde, H., King, D.J., Wise, M.G., Panigrahy, B., and Seal, B.S. (2004). Phylogenetic relationships among virulent Newcastle disease virus isolates from the 2002–2003 outbreak in California and other recent outbreaks in North America. J. Clin. Microbiol. *42,* 2329–2334.

Peeters, B.P., de Leeuw, O.S., Koch, G., and Gielkens, A.L. (1999). Rescue of Newcastle disease virus from cloned cDNA: evidence that cleavability of the fusion protein is a major determinant for virulence. J. Virol. *73,* 5001–5009.

Peeters, B.P., de Leeuw, O.S., Verstegen, I., Koch, G., and Gielkens, A.L. (2001). Generation of a recombinant chimeric Newcastle disease virus vaccine that allows serological differentiation between vaccinated and infected animals. Vaccine *19,* 1616–1627.

Perozo, F., Merino, R., Afonso, C.L., Villegas, P., and Calderon, N. (2008). Biological and phylogenetic characterization of virulent Newcastle disease virus circulating in Mexico. Avian Dis. *52,* 472–479.

Pierce, J.S., and Haywood, A.M. (1973). Thermal inactivation of Newcastle disease virus. I. Coupled inactivation rates of hemagglutinating and neuraminidase activities. J. Virol. *11,* 168–176.

Poch, O., Blumberg, B.M., Bougueleret, L., and Tordo, N. (1990). Sequence comparison of five polymerases (L proteins) of unsegmented negative-strand RNA viruses: theoretical assignment of functional domains. J. Gen. Virol. *71,* 1153–1162.

Porotto, M., Murrell, M., Greengard, O., Lawrence, M.C., McKimm-Breschkin, J.L., and Moscona, A. (2004). Inhibition of parainfluenza virus type 3 and Newcastle disease virus hemagglutinin-neuraminidase receptor binding: effect of receptor avidity and steric hindrance at the inhibitor binding sites. J. Virol. *78,* 13911–13919.

Powell, J., Aitken, I., and survashe, B. (1979). The response of the Herderian gland of the fowl to antigen given by the ocular route. II antibody production. Avian Pathol. *8,* 363–373.

Puhler, F., Willuda, J., Puhlmann, J., Mumberg, D., Romer-Oberdorfer, A., and Beier, R. (2008). Generation of a recombinant oncolytic Newcastle disease virus and expression of a full IgG antibody from two transgenes. Gene Ther. *15,* 371–383.

Qin, Z., Sun, L., Ma, B., Cui, Z., Zhu, Y., Kitamura, Y., and Liu, W. (2008a). F gene recombination between genotype II and VII Newcastle disease virus. Virus Res. *131,* 299–303.

Qin, Z.M., Tan, L.T., Xu, H.Y., Ma, B.C., Wang, Y.L., Yuan, X.Y., and Liu, W.J. (2008b). Pathotypical characterization and molecular epidemiology of Newcastle disease virus isolates from different hosts in China from 1996 to 2005. J. Clin. Microbiol. *46,* 601–611.

Reagan, R.L., Lillie, M.G., and *et al.* (1947). Transmission of the virus of Newcastle disease to the Syrian hamster. Am. J. Vet. Res. *8,* 136–138.

Reichard, K.W., Lorence, R.M., Cascino, C.J., Peeples, M.E., Walter, R.J., Fernando, M.B., Reyes, H.M., and Greager, J.A. (1992). Newcastle disease virus selectively kills human tumor cells. J. Surg. Res. *52,* 448–453.

Reitter, J.N., Sergel, T., and Morrison, T.G. (1995). Mutational analysis of the leucine zipper motif in the Newcastle disease virus fusion protein. J. Virol. *69,* 5995–6004.

Reynolds, D.L., and Maraqa, A.D. (2000). Protective immunity against Newcastle disease: the role of cell-mediated immunity. Avian Dis. *44,* 145–154.

Rodier, E. (1928). Philippines fowl disease. Proc. Soc. Exptl. Biol. Med. *25,* 781–783.

Romer-Oberdorfer, A., Mundt, E., Mebatsion, T., Buchholz, U.J., and Mettenleiter, T.C. (1999). Generation of recombinant lentogenic Newcastle disease virus from cDNA. J. Gen. Virol. *80,* 2987–2995.

Romer-Oberdorfer, A., Werner, O., Veits, J., Mebatsion, T., and Mettenleiter, T.C. (2003). Contribution of the length of the HN protein and the sequence of the F protein cleavage site to Newcastle disease virus pathogenicity. J. Gen. Virol. *84,* 3121–3129.

Romer-Oberdorfer, A., Veits, J., Helferich, D., and Mettenleiter, T.C. (2008). Level of protection of chickens against highly pathogenic H5 avian influenza virus with Newcastle disease virus based live attenuated vector vaccine depends on homology of H5 sequence between vaccine and challenge virus. Vaccine *26,* 2307–2313.

Rout, S.N., and Samal, S.K. (2008). The large polymerase protein is associated with the virulence of Newcastle disease virus. J. Virol. *82,* 7828–7836.

Russell, P. (1988). Monoclonal Antibodies in Research, Diagnosis and Epizootiology of Newcastle Disease. (Kluwer Acedemic Press: Boston, MA).

Russell, P.H., and Alexander, D.J. (1983). Antigenic variation of Newcastle disease virus strains detected by monoclonal antibodies. Arch. Virol. *75,* 243–253.

Russell, R., Paterson, R.G., and Lamb, R.A. (1994). Studies with cross-linking reagents on the oligomeric form of the paramyxovirus fusion protein. Virology *199,* 160–168.

Sahar, M., Alexander, D.J., Merek, J., Ruth, J., Amanda, H., Chad, M., and Brown, I. (2010). Phylogenitic analysis of the nucleotide sequences of the HN gene of 22 avian paramyxovirus type-2 viruses reveals marked heterogeneity. Avian Pathol. *39,* 453–458.

Saif, Y.M., Mohan, R., Ward, L., Senne, D.A., Panigrahy, B., and Dearth, R.N. (1997). Natural and experimental infection of turkeys with avian paramyxovirus-7. Avian Dis. *41,* 326–329.

Sakaguchi, T., Toyoda, T., Gotoh, B., Inocencio, N.M., Kuma, K., Miyata, T., and Nagai, Y. (1989). Newcastle disease virus evolution. I. Multiple lineages defined by sequence variability of the hemagglutinin-neuraminidase gene. Virology *169,* 260–272.

Sakaguchi, M., Nakamura, H., Sonoda, K., Okamura, H., Yokogawa, K., Matsuo, K., and Hira, K. (1998). Protection of chickens with or without maternal antibodies against both Marek's and Newcastle diseases by one-time vaccination with recombinant vaccine of Marek's disease virus type 1. Vaccine *16,* 472–479.

Samal, S.K., and Collins, P.L. (1996). RNA replication by a respiratory syncytial virus RNA analog does not obey the rule of six and retains a nonviral trinucleotide extension at the leader end. J. Virol. *70,* 5075–5082.

Samal, S., Khattar, S., Kumar, S., and Samal, S. (2010). Role of N linked glycosylation of Newcastle Disease virus fusion protein in virulence and viral pathogenicity. The 29th Annual American Society for Virology Meeting (Montana) 17–21 July 2010, (Abstract), p. 111.

Samuel, A.S., Kumar, S., Madhuri, S., Collins, P.L., and Samal, S.K. (2009). Complete sequence of the genome of avian paramyxovirus type 9 and comparison with other paramyxoviruses. Virus Res. *142,* 10–18.

Samuel, A.S., Paldurai, A., Kumar, S., Collins, P.L., and Samal, S.K. (2010). Complete genome sequence of avian paramyxovirus (APMV) serotype 5 completes the analysis of nine APMV serotypes and reveals the longest APMV genome. PLoS One *5,* e9269.

Samuel, A.S., Subbiah, M., Shive, H., Collins, P.L., and Samal, S. (2011). Experimental infection of hamsters with avian paramyxovirus serotypes 1 to 9. Vet. Res. *42,* 38.

Sandhu, T., and Hinshaw, V. (1981). Influenza A virus infection of domestic ducks. In Proceedings of the First International Symposium on Avian Influenza, April 1981, (US Animal Health Association, Richmond, VA), pp. 93–99.

Schirrmacher, V., Griesbach, A., and Ahlert, T. (2001). Antitumor effects of Newcastle Disease Virus *in vivo*: local versus systemic effects. Int. J. Oncol. *18,* 945–952.

Schroer, D., Veits, J., Grund, C., Dauber, M., Keil, G., Granzow, H., Mettenleiter, T.C., and Romer-Oberdorfer, A. (2009). Vaccination with Newcastle disease virus vectored vaccine protects chickens against highly pathogenic H7 avian influenza virus. Avian Dis. *53,* 190–197.

Seal, B.S., King, D.J., and Meinersmann, R.J. (2000). Molecular evolution of the Newcastle disease virus matrix protein gene and phylogenetic relationships among the *Paramyxoviridae*. Virus Res. *66,* 1–11.

Seal, B.S., Wise, M.G., Pedersen, J.C., Senne, D.A., Alvarez, R., Scott, M.S., King, D.J., Yu, Q., and Kapczynski, D.R. (2005). Genomic sequences of low-virulence avian paramyxovirus-1 (Newcastle disease virus) isolates obtained from live-bird markets in North America not related to commonly utilized commercial vaccine strains. Vet. Microbiol. *106,* 7–16.

Senne, D.A., Pearson, J.E., Miller, L.D., and Gustafson, G.A. (1983). Virus isolations from pet birds submitted for importation into the United States. Avian Dis. *27,* 731–744.

Sergel, T., McGinnes, L.W., Peeples, M.E., and Morrison, T.G. (1993). The attachment function of the Newcastle disease virus hemagglutinin-neuraminidase protein can be separated from fusion promotion by mutation. Virology *193,* 717–726.

Sergel, T.A., McGinnes, L.W., and Morrison, T.G. (2001). Mutations in the fusion peptide and adjacent heptad repeat inhibit folding or activity of the Newcastle disease virus fusion protein. J. Virol. *75,* 7934–7943.

Servan de Almeida, R., Maminiaina, O.F., Gil, P., Hammoumi, S., Molia, S., Chevalier, V., Koko, M., Andriamanivo, H.R., Traore, A., Samake, K., *et al.* (2009). Africa, a reservoir of new virulent strains of Newcastle disease virus? Vaccine *27,* 3127–3129.

Shengqing, Y., Kishida, N., Ito, H., Kida, H., Otsuki, K., Kawaoka, Y., and Ito, T. (2002). Generation of velogenic Newcastle disease viruses from a nonpathogenic waterfowl isolate by passaging in chickens. Virology *301,* 206–211.

Shope, R. (1964). The Birth of a New Disease (The University of Wisconsin, in press).

Shortridge, K.F., and Alexander, D.J. (1978). Incidence and preliminary characterisation of a hitherto unreported, serologically distinct, avian paramyxovirus isolated in Hong Kong. Res. Vet. Sci. *25,* 128–130.

Shortridge, K.F., Alexander, D.J., and Collins, M.S. (1980). Isolation and properties of viruses from poultry in Hong Kong which represent a new (sixth) distinct group of avian paramyxoviruses. J. Gen. Virol. *49,* 255–262.

Sick, C., Schultz, U., Munster, U., Meier, J., Kaspers, B., and Staeheli, P. (1998). Promoter structures and differential responses to viral and nonviral inducers of chicken type I interferon genes. J. Biol. Chem. *273,* 9749–9754.

Sick, C., Schneider, K., Staeheli, P., and Weining, K.C. (2000). Novel chicken CXC and CC chemokines. Cytokine *12,* 181–186.

Snoeck, C.J., Ducatez, M.F., Owoade, A.A., Faleke, O.O., Alkali, B.R., Tahita, M.C., Tarnagda, Z., Ouedraogo, J.B., Maikano, I., Mbah, P.O., *et al.* (2009). Newcastle disease virus in West Africa: new virulent strains identified in non-commercial farms. Arch. Virol. *154,* 47–54.

Snyder, D.B., Marquardt, W.W., Mallinson, E.T., Savage, P.K., and Allen, D.C. (1984). Rapid serological profiling by enzyme-linked immunosorbent assay. III. Simultaneous measurements of antibody titers to infectious bronchitis, infectious bursal disease, and Newcastle disease viruses in a single serum dilution. Avian Dis. *28,* 12–24.

Song, K.Y., Wong, J., Gonzalez, L., Sheng, G., Zamarin, D., and Fong, Y. (2010). Antitumor efficacy of viral therapy using genetically engineered Newcastle disease virus [NDV(F3aa)-GFP] for peritoneally disseminated gastric cancer. J. Mol. Med. *88,* 589–596.

Sonoda, K., Sakaguchi, M., Okamura, H., Yokogawa, K., Tokunaga, E., Tokiyoshi, S., Kawaguchi, Y., and Hirai, K. (2000). Development of an effective polyvalent vaccine against both Marek's and Newcastle diseases based on recombinant Marek's disease virus type 1 in commercial chickens with maternal antibodies. J. Virol. *74,* 3217–3226.

Spann, K.M., Collins, P.L., and Teng, M.N. (2003). Genetic recombination during coinfection of two mutants of human respiratory syncytial virus. J. Virol. *77, 11,*201–11211.

Spradbrow, P.B., and Sabine, M. (1995). Australian studies on Newcastle disease virus. The French heritage. Vet. Microbiol. *46,* 11–*12,* 15–19.

Srinivasappa, G.B., Snyder, D.B., Marquardt, W.W., and King, D.J. (1986). Isolation of a monoclonal antibody with specificity for commonly employed vaccine strains of Newcastle disease virus. Avian Dis. *30,* 562–567.

Stanislawek, W.L., Wilks, C.R., Meers, J., Horner, G.W., Alexander, D.J., Manvell, R.J., Kattenbelt, J.A., and Gould, A.R. (2002). Avian paramyxoviruses and influenza viruses isolated from mallard ducks (Anas platyrhynchos) in New Zealand. Arch. Virol. *147,* 1287–1302.

Steward, M., Vipond, I.B., Millar, N.S., and Emmerson, P.T. (1993). RNA editing in Newcastle disease virus. J. Gen. Virol. *74,* 2539–2547.

Stojdl, D.F., Lichty, B., Knowles, S., Marius, R., Atkins, H., Sonenberg, N., and Bell, J.C. (2000). Exploiting tumor-specific defects in the interferon pathway with a previously unknown oncolytic virus. Nat. Med. *6,* 821–825.

Stone-Hulslander, J., and Morrison, T.G. (1999). Mutational analysis of heptad repeats in the membrane-proximal region of Newcastle disease virus HN protein. J. Virol. *73,* 3630–3637.

Stover, D. (1940). Report of the poultry pathological laboratory, petaluma, Calif. Calif. Dept. Agr. Bull. *20,* 367–368.

Subbiah, M., Xiao, S., Collins, P.L., and Samal, S.K. (2008a). Complete sequence of the genome of avian paramyxovirus type 2 (strain Yucaipa) and comparison with other paramyxoviruses. Virus Res. *137,* 40–48.

Subbiah, M., Yan, Y., Rockemann, D., and Samal, S.K. (2008b). Experimental infection of calves with Newcastle disease virus induces systemic and mucosal antibody responses. Arch. Virol. *153,* 1197–1200.

Subbiah, M., Nayak, S., Collins, P.L., and Samal, S.K. (2010a). Complete genome sequences of avian paramyxovirus serotype 2 (APMV-2) strains Bangor, England and Kenya: evidence for the existence of subgroups within serotype 2. Virus Res. *152,* 85–95.

Subbiah, M., Sa, X., Sunil K. Khattar, Dias, F.M., Collins, P.L., and Samal, S.K. (2010b). Pathogenesis of two strains of avian Paramyxovirus serotype 2, Yucaipa and Bangor, in chickens and turkeys. Avian Dis. *54,* 1050–1057.

Takimoto, T., and Portner, A. (2004). Molecular mechanism of paramyxovirus budding. Virus Res. *106,* 133–145.

Takimoto, T., Taylor, G.L., Connaris, H.C., Crennell, S.J., and Portner, A. (2002). Role of the hemagglutinin-neuraminidase protein in the mechanism of paramyxovirus-cell membrane fusion. J. Virol. *76,* 13028–13033.

Tan, L.T., Xu, H.Y., Wang, Y.L., Qin, Z.M., Sun, L., Liu, W.J., and Cui, Z.Z. (2008). Molecular characterization of three new virulent Newcastle disease virus variants isolated in China. J. Clin. Microbiol. *46,* 750–753.

Timms, L., and Alexander, D.J. (1977). Cell-mediated immune response of chickens to Newcastle disease vaccines. Avian Pathol. *6,* 51–59.

Toyoda, T., Sakaguchi, T., Imai, K., Inocencio, N.M., Gotoh, B., Hamaguchi, M., and Nagai, Y. (1987). Structural comparison of the cleavage-activation site of the fusion glycoprotein between virulent and avirulent strains of Newcastle disease virus. Virology *158,* 242–247.

Toyoda, T., Sakaguchi, T., Hirota, H., Gotoh, B., Kuma, K., Miyata, T., and Nagai, Y. (1989). Newcastle disease virus evolution. II. Lack of gene recombination in generating virulent and avirulent strains. Virology *169,* 273–282.

Tsai, H.J., Chang, K.H., Tseng, C.H., Frost, K.M., Manvell, R.J., and Alexander, D.J. (2004). Antigenic and genotypical characterization of Newcastle disease viruses isolated in Taiwan between 1969 and 1996. Vet. Microbiol. *104,* 19–30.

Tumova, B., Robinson, J.H., and Easterday, B.C. (1979). A hitherto unreported paramyxovirus of turkeys. Res. Vet. Sci. *27,* 135–140.

Turek, R., Gresikova, M., and Tumova, B. (1984). Isolation of influenza A virus and paramyxoviruses from sentinel domestic ducks. Acta Virol. *28,* 156–158.

Ujvari, D. (2006). Complete nucleotide sequence of IT-227/82, an avian paramyxovirus type-1 strain of pigeons (Columba livia). Virus Genes *32,* 49–57.

Utterback, W.W., and Schwartz, J.H. (1973). Epizootiology of velogenic viscerotropic Newcastle disease in southern California, 1971–1973. J. Am. Vet. Med. Assoc. *163,* 1080–1088.

Vigil, A., Park, M.S., Martinez, O., Chua, M.A., Xiao, S., Cros, J.F., Martinez-Sobrido, L., Woo, S.L., and Garcia-Sastre, A. (2007). Use of reverse genetics to enhance the oncolytic properties of Newcastle disease virus. Cancer Res. *67,* 8285–8292.

Wakamatsu, N., King, D.J., Seal, B.S., Peeters, B.P., and Brown, C.C. (2006). The effect on pathogenesis of Newcastle disease virus LaSota strain from a mutation of the fusion cleavage site to a virulent sequence. Avian Dis. *50,* 483–488.

Walker, J.W., Heron, B.R., and Mixson, M.A. (1973). Exotic Newcastle disease eradication programme in the United States. Avian Dis. *17,* 486–503.

Wang, Z., Liu, H., Xu, J., Bao, J., Zheng, D., Sun, C., Wei, R., Song, C., and Chen, J. (2006). Genotyping of Newcastle disease viruses isolated from 2002 to 2004 in China. Ann. N. Y. Acad. Sci. *1081,* 228–239.

Warke, A., Appleby, L., and Mundt, E. (2008a). Prevalence of antibodies to different avian paramyxoviruses in commercial poultry in the United States. Avian Dis. *52,* 694–697.

Warke, A., Stallknecht, D., Williams, S.M., Pritchard, N., and Mundt, E. (2008b). Comparative study on the pathogenicity and immunogenicity of wild bird isolates of avian paramyxovirus *2, 4,* and 6 in chickens. Avian Pathol. *37,* 429–434.

Webster, R.G., Morita, M., Pridgen, C., and Tumova, B. (1976). Ortho- and paramyxoviruses from migrating feral ducks: characterization of a new group of influenza A viruses. J. Gen. Virol. *32,* 217–225.

Westbury, H. (2001). Newcastle disease virus: an evolving pathogen? Avian Pathol. *30,* 5–11.

Wise, M.G., Sellers, H.S., Alvarez, R., and Seal, B.S. (2004a). RNA-dependent RNA polymerase gene analysis of worldwide Newcastle disease virus isolates representing different virulence types and their phylogenetic relationship with other members of the *Paramyxoviridae.* Virus Res. *104,* 71–80.

Wise, M.G., Suarez, D.L., Seal, B.S., Pedersen, J.C., Senne, D.A., King, D.J., Kapczynski, D.R., and Spackman, E. (2004b). Development of a real-time reverse-transcription PCR for detection of newcastle disease virus RNA in clinical samples. J. Clin. Microbiol. *42,* 329–338.

Wobeser, G., Leighton, F.A., Norman, R., Myers, D.J., Onderka, D., Pybus, M.J., Neufeld, J.L., Fox, G.A., and Alexander, D.J. (1993). Newcastle disease in wild water birds in western Canada, 1990. Can. Vet. J. *34,* 353–359.

Woolcock, P.R., Moore, J.D., McFarland, M.D., and Panigrahy, B. (1996). Isolation of paramyxovirus serotype 7 from ostriches (*Struthio camelus*). Avian Dis. *40,* 945–949.

Xiao, S., Paldurai, A., Nayak, B., Subbiah, M., Collins, P.L., and Samal, S.K. (2009). Complete genome sequence of avian paramyxovirus type 7 (strain Tennessee) and comparison with other paramyxoviruses. Virus Res. *145,* 80–91.

Xiao, S., Subbiah, M., Kumar, S., De Nardi, R., Terregino, C., Collins, P.L., and Samal, S.K. (2010). Complete genome sequences of avian paramyxovirus serotype 6 prototype strain Hong Kong and a recent novel strain from Italy: evidence for the existence of subgroups within the serotype. Virus Res. *150,* 61–72.

Yamane, N., Arikawa, J., Odagiri, T., and Ishida, N. (1982). Characterization of avian paramyxoviruses isolated from feral ducks in northern Japan: the presence of three distinct viruses in nature. Microbiol. Immunol. *26,* 557–568.

Yan, Y., and Samal, S.K. (2008). Role of intergenic sequences in newcastle disease virus RNA transcription and pathogenesis. J. Virol. *82,* 1323–1331.

Yan, Y., Rout, S.N., Kim, S.H., and Samal, S.K. (2009). Role of untranslated regions of the hemagglutinin-neuraminidase gene in replication and pathogenicity of newcastle disease virus. J. Virol. *83,* 5943–5946.

Yu, L., Wang, Z., Jiang, Y., Chang, L., and Kwang, J. (2001). Characterization of newly emerging Newcastle disease virus isolates from the People's Republic of China and Taiwan. J. Clin. Microbiol. *39,* 3512–3519.

Zaitsev, V., von Itzstein, M., Groves, D., Kiefel, M., Takimoto, T., Portner, A., and Taylor, G. (2004). Second sialic acid binding site in Newcastle disease virus hemagglutinin-neuraminidase: implications for fusion. J. Virol. *78,* 3733–3741.

Zamarin, D., Martinez-Sobrido, L., Kelly, K., Mansour, M., Sheng, G., Vigil, A., Garcia-Sastre, A., Palese, P., and Fong, Y. (2009). Enhancement of oncolytic properties of recombinant newcastle disease virus through antagonism of cellular innate immune responses. Mol. Ther. *17,* 697–706.

Zeng, J., Fournier, P., and Schirrmacher, V. (2002a). Induction of interferon-alpha and tumor necrosis factor-related apoptosis-inducing ligand in human blood mononuclear cells by hemagglutinin-neuraminidase but not F protein of Newcastle disease virus. Virology *297,* 19–30.

Zeng, J., Fournier, P., and Schirrmacher, V. (2002b). Stimulation of human natural interferon-alpha response via paramyxovirus hemagglutinin lectin–cell interaction. J. Mol. Med. *80,* 443–451.

Zeydanli, M., Redmann, T., Kaleta, E., and Alexander, D.J. (1988). Paramyxoviruses (PMV) isolated from turkeys with respiratory diseases. Proc. 37th West Poultry Disease Conference, Davis, California, 29 February–2 March 1988 (University of California), pp. 46–50.

Zhang, R., Wang, X., Su, J., Zhao, J., and Zhang, G. (2010). Isolation and analysis of two naturally-occurring multirecombination Newcastle disease viruses in China. Virus Res. *151*, 45–53.

Zhao, H., and Peeters, B.P. (2003). Recombinant Newcastle disease virus as a viral vector: effect of genomic location of foreign gene on gene expression and virus replication. J. Gen. Virol. *84*, 781–788.

Zou, J., Shan, S., Yao, N., and Gong, Z. (2005). Complete genome sequence and biological characterizations of a novel goose paramyxovirus-SF_02 isolated in China. Virus Genes *30*, 13–21.

Sendai Virus: Evolution from Mouse Pathogen to a State-of-the-art Tool in Virus Research and Biotechnology

4

Yoshiyuki Nagai*, Akira Takakura, Takashi Irie, Yoshikazu Yonemitsu and Bin Gotoh

Abstract

Sendai virus (SeV) is not just an old mouse pathogen. SeV has been an irreplaceable tool in basic research to understand paramyxovirus replication and pathogenesis. SeV has further made an entrance into the scene of technological innovation in delivering foreign genes of interest to target cells and tissues. In this context, there are two different approaches. One is to make good use of the fusogenic capacity of inactivated SeV virions and the other is to generate a non-genotoxic (non-integrating) cytoplasmic RNA vector by reverse genetics. Here, we review the long history of SeV research focusing on its contribution to basic virology and technological development. We also highlight some old discoveries that have mostly vanished from recent textbooks but provided crucial momentum to the development of the current concepts and technologies.

Early history

Discovery

Sendai virus (SeV) was first isolated in 1953 at Tohoku University Hospital in Sendai, Japan, during an outbreak of fatal pneumonia of newborn children (Kuroya *et al.*, 1953). The materials used for virus isolation were the homogenates of autopsy lung tissues taken from the newborn victims of the disease. The initial step of isolation was intranasal inoculation of the materials into mice. The inoculated mice developed influenza-like lesions in the lungs. The homogenates of those mouse lung tissues were then transferred to the amniotic cavities of chicken embryonated eggs and the amniotic fluid harvested several days later displayed a high level of agglutination of chicken erythrocytes. Since the laboratory tests available for identification of known influenza and other haemagglutinating viruses were all negative, the isolate was concluded to be a new virus (reviewed in Ishida and Homma, 1978). The paper reporting this episode was entitled 'Newborn virus pneumonitis (type Sendai). II. The isolation of a new virus possessing hemagglutinating activity' (Kuroya *et al.*, 1953).

One year later an opposite interpretation was put forth by Fukumi and coworkers. They became aware that passaging in mouse lungs of influenza virus isolates from various regions of Japan quite often generated a virus antigenically totally different from the original influenza virus (Fukumi *et al.*, 1954). This new agent was later identified as SeV. It was even said that dropping physiological saline alone into the noses of apparently healthy mice could sometimes activate the persistently infecting SeV, resulting in induction of severe pneumonia. These observations suggested a nationwide enzootic (subclinical) spread of SeV among mice during the 1950s (reviewed in Ishida and Homma, 1978). The virus appeared to be readily activated by various exogenous stimuli to the airway of mouse.

On the other hand, several groups reported that isolation of SeV was also possible by inoculating

*Corresponding author

the human materials directly into chick embryos or cells in culture, not via intranasal inoculation into mice (reviewed in Ishida and Homma, 1978), still leaving open the possibility that the virus would be indeed the causative agent for newborn pneumonia at least in the 1950s. However, this was also questioned in view of the ease with which SeV could contaminate a laboratory engaged in experimental studies with the agent (Chanock *et al.*, 1963; reviewed in Ishida and Homma, 1978). Infection of humans with SeV might be proven by detecting anti SeV antibodies or seroconversion. However, this has been unfortunately quite difficult because of serological cross reactivity of SeV with human parainfluenza virus 1, a causative agent of infant pneumonia, that was isolated in 1956 to 1958 (Chanock *et al.*, 1958). Moreover, there have been no subsequent reports of SeV isolation from humans into eggs or cell cultures over the past five decades until now. Under these circumstances, SeV isolation from humans now appears to be regarded as a result of laboratory pick-up. Then, what was the actual cause for that newborn pneumonia? This question remains unanswered but may be addressed, for example, by a metagenomic approach, if frozen or paraffin-embedded lung tissues from those victims are still available.

Seroepidemiological survey by haemagglutination inhibition (HI) and other serological tests suggested the spread of SeV among pigs in the 1950s throughout Japan (reviewed in Ishida and Homma, 1978). The infection appeared to cause influenza-like illness for pigs. Moreover, the offspring of infected pregnant pigs were sometimes still-born. There were also inapparent infections since healthy pigs were often seropositive. However, no seropositive pigs have been found since the early 1960s and thus SeV, which once circulated in pigs, appears to have become extinct in the pig population in Japan. Overall, the prologue of the tale of SeV was full of wonder.

Nomenclature and taxonomy

In a situation where the researchers' views were controversial concerning whether or not SeV would be involved in a human disease, the Japanese Society for Virology decided to name the virus haemagglutinating virus of Japan (HVJ) at its third annual meeting held in Kyoto in April 1955. It was just a crude word-for-word translation. Moreover, HVJ sounded quite clumsy as an acronym of a virus, so it did not become on everyone's lips. Instead, the name 'Sendai virus' was already in frequent use in papers published in the 1950s. However, the name HVJ is still favoured by certain Japanese researchers. The International Committee on Taxonomy of Viruses has now formally adopted the name 'Sendai virus' with the synonym, murine parainfluenza virus 1 (MPIV1). SeV belongs to the genus *Respirovirus* of the subfamily *Paramyxovirinae* together with human parainfluenza 1 (HPIV1) and 3 (HPIV3), and bovine parainfluenza virus 3 (BPIV3).

Host range, pathogenicity and surveillance

SeV has a broad host range in experimental animals. These include mouse, rat, guinea pig, rabbit, ferret and marmoset. The pig may be also susceptible as noted above. Experimentally, SeV was found to replicate to a level similar to the human counterpart, HPIV1 in the upper and lower respiratory tracts of African green monkeys and chimpanzees (Skiadopoulos *et al.*, 2002). While in enzootic infections, SeV usually produces a subclinical or asymptomatic outcome, epizootic infections as well as experimental infections cause acute pneumonia in mouse and rat (reviewed in Homma and Tashiro, 1994). The symptoms are more severe for mouse than for rat. The remaining animals listed above are less susceptible and infections are usually inapparent. However, great caution is needed because inapparent infections are less easy to notice than apparent infections and because the virus readily jumps the species barrier and spreads from these animals to mouse and rat.

The disease course has been well documented in experimental infections of mice by Homma and Tashiro (1994) and by others (Kiyotani *et al.*, 1990; Kato *et al.*, 1997a; see below). It begins with symptoms characteristic of the upper respiratory illness including moderate fever, ruffled furs and nasal discharge, which appear 2 or 3 days after exposure. Subsequently, within several days, infection advances to the lower respiratory tracts causing bronchopneumonia symptomatically characterized by dyspnoea, cough and anorexia.

In the second week, infected animals either die of bronchopneumonia or begin to recover from infection (Homma and Tashiro, 1994). Susceptibility to SeV differs among mouse strains. A comparative study using six inbred strains revealed a pattern of continuous variation with relatively resistant (e.g. BALB/c), intermediate, and susceptible (e.g. 129Sv) strains (Faisca *et al.*, 2005). Also to be noted is that SeV usually used for pathogenicity studies has been serially passaged in eggs, which results in profound attenuation. For instance, a field isolate became attenuated by as much as 400-fold on mouse LD_{50} (50% lethal dose) basis after 50 passages in eggs (Kiyotani *et al.*, 2001; see below).

SeV is highly contagious and transmission rapidly occurs via airway tracts through close contact with or an airborne route from an infected host, and has been a devastating threat to mice and rats in animal experiment facilities. The Central Institute for Experimental Animals in Kawasaki, Japan has conducted extensive national surveys of SeV and other microbes for the mouse and rat colonies of many animal research facilities belonging to universities, pharmaceutical industries and animal breeders since 1980 (Table 4.1). Contamination of mouse and rat colonies with SeV was found in as high as 25.0% and 31.8%, respectively, of the total colonies tested in 1980. Although considerably decreased, the contamination rates were still high in 1990. The rates further decreased to 1.3% and 0.7%, respectively, in 2000 and even lower in 2002 to 2007. The rate became zero for both mouse and rat colonies in 2009. This success of SeV control was brought about by introduction of barrier systems into the facilities, which had previously maintained animals under open and conventional conditions and by careful monitoring of animals to be brought in and maintained (Kagiyama *et al.*, 1987). Similar serological monitoring studies also revealed decreasing trends of SeV contamination rates during the period 1988–1997 in French facilities (Zenner and Regnault, 2000), during 2004–2007 in Taiwan (Liang *et al.*, 2009) and during 2000–2003 in Western Europe (Schoondermark-van de Ven *et al.*, 2006). However, even though it once became negative in mouse colonies in 2004, contamination re-emerged in 2006 and 2007 (Table 4.1). Thus, although negative in both mouse and rat colonies in 2009, we should not be relieved. In this regard, a survey of rodent-borne pathogens carried out for a total of 162 wild-caught Norway rats showed that 4.1% of them were SeV antibody positive (Easterbrook *et al.*, 2008). These results indicate that SeV, indeed, has not become extinct. We must continue to maintain our caution and vigilance.

Ultrastructure

SeV is an enveloped virus with a negative-strand RNA genome. The virus particle (virion) consists of three major structural components; the inner ribonucleoprotein complex (RNP), the outer envelope and the matrix (M) protein layer between the internal and external structures. The

Table 4.1 Changes in the prevalence of serum anti-SeV antibodies detected by HI test in mouse and rat colonies in Japan over the past three decades

Year	Mouse	Rat
1980	9/36 (25.0%)[1]	14/44 (31.8%)
1990	19/209 (9.1%)	15/253 (5.9%)
2000	10/785 (1.3%)	2/299 (0.7%)
2002	0/1446 (0.0%)	2/412 (0.5%)
2004	0/2074 (0.0%)	1/278 (0.4%)
2006	3/1719 (0.2%)	1/384 (0.3%)
2007	3/2392 (0.1%)	1/568 (0.2%)
2009	0/1786 (0.0%)	0/403 (0.0%)

[1]Number of antibody-positive colonies/total number of colonies tested with per cent positivity in the parentheses.

SeV, Sendai virus; HI, haemagglutination inhibition.

ultrastructure has been revealed using mainly egg grown viruses. It is spherical, with a diameter of approximately 200 nm. With freshly prepared, intact virions, which were harvested from eggs relatively early (~24 h) after infection and immediately processed for negative staining, uranyl acetate hardly penetrated through the envelope, and no RNP was visible (Shimizu *et al.*, 1976; also see Fig. 4.1A). This type of fresh preparation was fully infectious and fusogenic but lacked haemolytic activity (Homma *et al.*, 1976). The internal RNP structure can be seen only with virions with envelopes partially damaged owing to late harvesting (at ~72 h), freezing and thawing or storage in a refrigerator (at 4°C) for a considerable period (1 week) (Fig. 4.1B). Here, morphological changes are remarkable and the surface glycoproteins have become more clearly visible for unknown reasons (Fig. 4.1B, cf. Fig. 4.1A). Haemolysis was exhibited by such partially damaged virions (Homma *et al.*, 1976). It is believed that unlimited influx of extracellular ions into erythrocytes through a damaged envelope integrated into the cell membrane causes haemolysis (rupture of erythrocytes).

The RNP is best studied after isolation from detergent-solubilized virions. The RNP consists of the RNA genome that is encapsidated with the nucleocapsid (N) proteins. It has a left-handed helical structure 1.0 μm in length and 18 nm in diameter, with a pitch of 5.5 nm, and contains 13 N proteins per turn (Hosaka, 1968; Egelman *et al.*, 1989; marked by na in Fig. 4.1C). Because one N protein is thought to bind to every six RNA bases (Egelman *et al.*, 1989; Calain and Roux, 1993) and because the SeV genome consists of 15,384 nucleotides (nt), a single RNP is estimated to contain 2564 units of N protein. The paramyxovirus L (large) and P (phospho) proteins constitute the RNA polymerase complex together (Hamaguchi *et al.*, 1983; Horikami *et al.*, 1992) and must be bound to the RNP. Using immunogold labelling the P proteins were found to be dispersed uniformly along the SeV RNPs isolated from the virions, while they were present

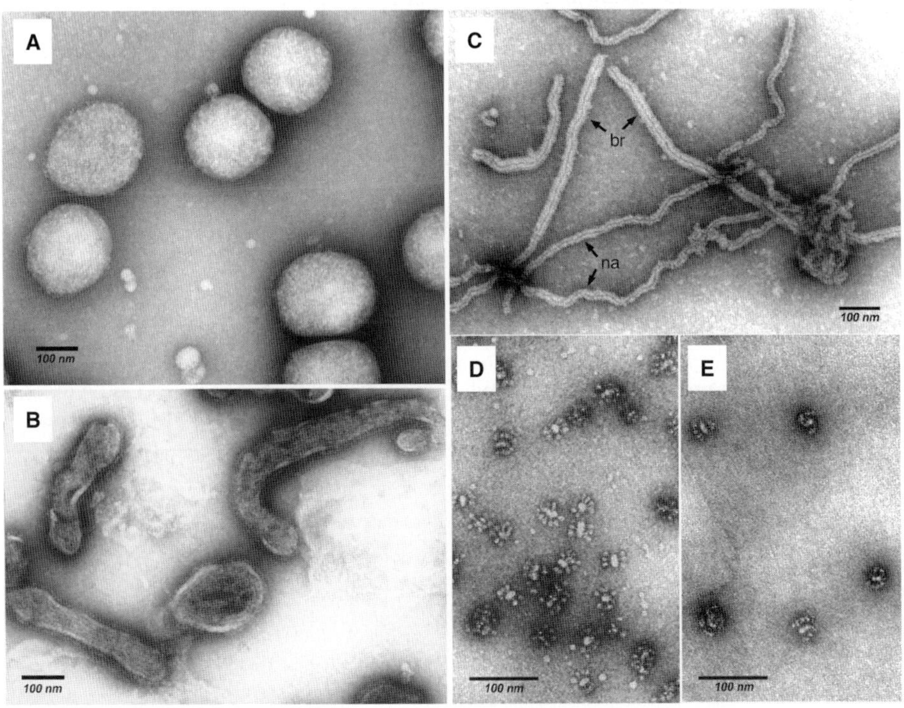

Figure 4.1 Sendai virus (SeV) virions freshly prepared (A), SeV virions after storage at 4°C for 7 days (B), SeV nucleocapsids isolated from the virions (C), and petal-like structures formed by the F (D) and HN (E) glycoproteins after their purification by electrofocusing of the proteins solubilized by alkali-Tween 20 followed by dialysis to remove the detergent. All samples were stained with 1% uranyl acetate. Courtesy of K. Shimizu.

in discrete clusters at varying locations of the RNPs isolated from the cytosol (Portner and Murti, 1986). The L proteins were colocalized with those P protein clusters, suggesting that the P and L are co-mobile along the intracellular RNP to act as the RNA polymerase (Portner *et al.*, 1988). The isolated RNPs are sometimes broader (40 nm) in diameter (see br in Fig. 4.1C). This appeared to be due to additional binding of the M proteins. The broader RNP was abundant in the cytosol, suggesting the role of M protein in the recognition between the RNP and the envelope proteins in the process of virus assembly (Portner and Murti, 1986).

The outermost structure, the envelope, contains two types of glycoproteins protruding in the form of spikes, anchored within the lipid bilayer derived from the plasma membrane of host cells. One of them has haemagglutination (HA) and neuraminidase (NA) activities and is named HN. The discovery that a single glycoprotein possesses both HA and NA activities, which was first made for PIV5 in the genus *Rubulavirus* (Scheid *et al.*, 1972) and SeV (Tozawa *et al.*, 1973) and then generalized among paramyxoviruses, was somewhat unexpected in view of the just preceding discovery in the late 1960s that they are encoded by separate glycoproteins in the case of influenza A viruses. The HN protein recognizes terminal sialic acids in the sugar chains as the receptor on the target cell surface. It also contributes to the release of progeny virus from the cell surface. The other spike glycoprotein is called F protein because it induces fusion of the viral envelope with the host cell plasma membrane, thereby introducing the viral RNP into the cytoplasm. The F protein is synthesized as an inactive precursor molecule F_0, which is cleaved by host proteases to F_1 and F_2 to be biologically active (Homma and Ohuchi, 1973; Scheid and Choppin, 1974; see below). The HN protein exists as a tetramer, and the F protein as a trimer (reviewed in Lamb and Parks, 2007). The HN and F glycoproteins were individually isolated from detergent-solubilized virions by isoelectric focusing. When the detergent was removed by gel filtration, they assembled through hydrophobic interactions between the transmembrane regions, forming petal-like structures (Fig. 4.1D and E).

The matrix (M) protein located between the envelope and RNP is thought to serve as an anchor to stabilize the spike glycoprotein molecules floating in the lipid bilayer by binding to their cytoplasmic tails; it also binds to RNP, crosslinking the internal and external structures (Shimizu and Ishida, 1975; Yoshida *et al.*, 1976). This M protein is indispensable for structural proteins to assemble into virus particles as well as for viral budding from the plasma membrane (see below).

Genome nucleotide sequencing and establishment of reverse genetics

Nucleotide sequencing

The nucleotide sequencing of the entire paramyxovirus genome was first achieved for SeV in the mid 1980s (Shioda *et al.*, 1983, 1986; Hidaka *et al.*, 1984). This achievement proved to be a milestone in genome-wide explorations of paramyxovirus gene expression and genome replication mechanisms. The genome consists of 15,384 nt. The genes coding for the above described six structural proteins lie in the order of 3′-*N-P-M-F-HN-L*-5′. The lengths of the encoded proteins in the numbers of amino acids (aa) are 524 (N), 568 (P), 348 (M), 565 (F), 575 (HN) and 2228 (L). Each gene begins with the 10-nt-long transcription start signal (3′-UCCCNNUUNC-5′) and ends with the 11-nt-long transcription stop signal (3′-UNAUUCUUUUU-5′). The stop signal of the upstream gene is connected to the start signal of the just downstream gene with an intergenic three nucleotides (3′-GNN-5′). The genome further contains a 3′-extracistronic region, the leader sequence (55 nt) and a 5′-extracistronic region, the trailer sequence (57 nt).

SeV further encodes the V protein (384 aa) and the C proteins of four different lengths (215–175 aa) in addition to the six major gene products. Although widespread among the members of the subfamily *Paramyxovirinae*, the V and C proteins are not as ubiquitous as the six major structural proteins, and therefore have been regarded as nonessential 'accessory' gene products. Nucleotide sequencing revealed that the *P* gene encodes both of them.

Reverse genetics

Plasmid-based expression of the individual genes has enhanced our understanding of paramyxovirus protein functions. However, mere summation of the functions of individual genes does not guarantee an accurate overall picture of the virus, a viable organism in which various gene functions are intertwined with one another in a complex and systematic manner. Forward genetics fell short of providing methods of reproducing (and thus confirming) phenotypes identified as corresponding to certain genotypes. Furthermore, there were a number of unsolved issues in paramyxovirus biology that had been unapproachable by conventional virology, including gene expression studies. Typical of such problems are the accessory genes encoding the V and C proteins because the virological definition of an accessory gene is that the gene can be deleted (knocked out) without destroying infectivity at the cellular level and because the actual roles of the accessory gene can be defined by comparing the phenotypes of the knockout virus with those of the parental virus at the cellular level and in the susceptible host organism.

Under these circumstances, post-genome research looked forward to the establishment of reverse genetics, which would allow gene manipulation at will and assessment of its outcome not only in terms of cellular-level proliferation but also in the context of viral pathogenesis at the level of the entire organism. For SeV, reverse genetics was established independently by two research groups (Garcin *et al.*, 1995; Kato *et al.*, 1996) sharing the same plasmid encoding the full-length SeV genome developed by the latter group. The protocol essentially followed that established for the reverse genetics of rabies virus in the *Rhabdoviridae* (Schnell *et al.*, 1994).

Briefly, four plasmids encoding the entire genome, the N protein, P protein and L protein were transfected into cells and driven by bacteriophage T7 polymerase expressed from vaccinia virus vector vTF7–3 (Fig. 4.2A). Hepatitis delta virus ribozyme sequence was placed immediately downstream of the viral genome sequence to generate the correct 3′ end of the full-length viral RNA. The efficiency of virus recovery was one infectious unit from 10^4 to 10^5 transfected cells in our system (Kato *et al.*, 1996), whereas the efficiencies were 50- to 100-fold lower in others established those days for a variety of non-segmented negative strand RNA viruses including rabies virus

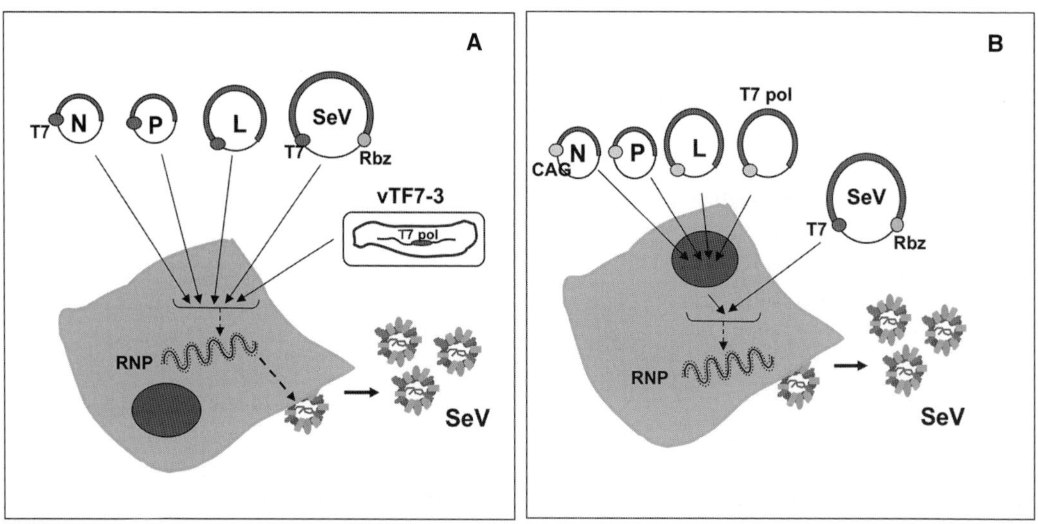

Figure 4.2 Sendai virus (SeV) recovery from cDNA. (A) Method with T7 RNA polymerase (T7 pol) expressing vaccinia virus (vTF7–3). (B) Method without vTF7–3. T7, T7 promoter; Rbz, ribozyme sequence derived from hepatitis delta virus; CAG, CAG promoter. Adapted from Nagai *et al.* (2007).

(Schnell *et al.*, 1994) and another SeV system (Garcin *et al.*, 1995; reviewed in Nagai, 1999). Both genomic and antigenomic RNAs must be encapsidated with the N proteins throughout the viral lifecycle (see below). It was thought that the recovery of non-segmented negative strand RNA viruses from the (–)DNA plasmid directing genome (–)RNA would generally be difficult as the encapsidation of the intracellularly expressed (–)RNA would be prevented by prior hybridization with the abundantly expressed mRNAs from the three supporting plasmids. However, our system allowed SeV rescue even from (–)DNA, albeit at a 50-times lower frequency, compared with that from (+)DNA (Kato et al., 1996). Moreover, transfection with in vitro-synthesized RNA again, either (–) or (+) sense, was able to initiate virus rescue (Kato et al., 1996). This does not appear to meet the concept that, in the natural life cycle, encapsidation/replication occurs through concomitant, continuous association of the de novo-synthesized N proteins with the nascent, growing RNA chains in the direction from the 5′ to 3′ end (see below; reviewed in Nagai, 1999).

The remarkably high virus rescue rates in our SeV system probably result from fine-tuning of the system (Palese *et al.*, 1996). Particularly important could be constructing plasmids giving rise to full-length viral RNA with the 5′-end identical to that of the wild-type SeV RNA without extra G residues templated by the non-viral T7 promoter sequence so that the expressed RNA meets the rule of six (Kato *et al.*, 1996; reviewed in Nagai, 1999; see below). Indeed, three such non-viral Gs attached to the 5′ end of antigenomic RNA appeared to break the rule of six and be deleterious for SeV recovery, as no appreciable titre was recovered until they were deleted to create the correct 3′ end of genomic RNA during serial passages of the recovered virus in eggs (Garcin *et al.*, 1995). Later, however, Leyrer *et al.* (1998) also established a rapid and efficient SeV recovery system from cDNA and indicated that if T7 promoter activity is strong enough, the resulting large amounts of transcripts can outweigh the inhibitory effect of the extra three Gs for replication.

Psoralen compound covalently binds to double-stranded DNA under UV light. UV irradiation of vTF7–3 in the presence of psoralen under certain conditions caused reduction of the vaccinia virus-associated cytopathogenicty, while T7 RNA polymerase expression remained little changed. As a result, recovery efficiency increased by 100- to 1000-fold (1 infectious unit from 10^2–10^3 transfected cells) (Li *et al.*, 2000). We further established a vaccinia virus-free system. Here, all supporting plasmids to express the N, P and L proteins, respectively, and an additional plasmid to express T7 polymerase are driven by the cellular RNA polymerase II under the control of a CAG promoter (Fig. 4.2B). The plasmid to generate SeV genome RNA was left unchanged, being under the control of T7 promoter. The resulting all-plasmid-driven method demonstrated virus recovery efficiency similar to or even higher than that using UV irradiated vTF7-3 (reviewed in Nagai *et al.*, 2007).

SeV reverse genetics now started to play a prominent role in settling outstanding questions in paramyxovirus life cycle and pathogenicity (reviewed in Nagai, 1999; Nagai and Kato, 1999, 2004; Sakaguchi *et al.*, 2008).

Steps in the SeV life cycle and the involved proteins and *cis*-acting genome elements

Prominent features of SeV life cycle are illustrated below according to our knowledge obtained by conventional virology and reverse genetics.

Initiation of infection

Attachment to the target cell surface

SeV attachment protein is the HN glycoprotein (575 aa). It is a type II integral membrane protein, spanning the membrane once and possessing an N-terminal cytoplasmic tail, a single transmembrane domain, a membrane-proximal stalk domain, and a large C-terminal globular head domain. It exists on the virion surface as a tetramer consisting of a pair of disulfide-bonded dimmers (reviewed in Lamb and Parks, 2007). While the stalk domain is important for tetramer formation, the head domain contains the sites for receptor binding, HA activity and NA activity.

As noted below, the NA activity is thought to play an essential role in removing the terminal sialic acids of the sugar chains of both the HN and F glycoproteins and thereby preventing aggregation of the mature virions and facilitating their release from the cell surface. The HN protein of paramyxoviruses is required for the fusion process of virus entry or makes fusion more efficient.

It has been an issue of debate for years whether paramyxovirus HN protein possess combined or separate sites for HA (receptor binding) and NA activities. Identification of the second sialic acid binding site, in addition to the primary NA catalytic site, on the crystal structure of the HN of Newcastle disease virus (NDV) in the genus *Avulavirus* (Zaitsev *et al.*, 2004) was a clue to resolving this issue. A compound named BCX-2798 inhibited NA activity but not HA activity of NDV (Bousse *et al.*, 2004). This compound also blocked NA activity, but not HA activity of SeV, while it inhibited both the HA and NA activities of HPIV1, suggesting that the SeV HN has a second receptor binding site, while HPIV1 HN does not (Bousse and Takimoto, 2006). This lack of a second site in HPIV1 HN could be due to the presence of Asn at residue 524 instead of the Asp in the case of SeV HN. Later, however, it was found that a secondary site is present in HPIV1, but is masked by a nearby *N*-glycan at residue 173 (Alymova *et al.*, 2008).

Sialo-oligosaccharides of both glycoproteins and glycolipids serve as the host cell receptors of SeV (Suzuki *et al.*, 1983, 1985). The specificity of SeV towards receptor recognition was examined in a bovine cell line. The cells were first treated with *Vibrio cholerae* neuraminidase to remove cell surface sialic acids and to render them insusceptible to the viral infection and then modified with sialyltransferases to contain sialyloligosaccharides of defined sequences and examined for restoration of susceptibility to SeV (Markwel and Paulson, 1980). Sialylation, by incubation of the cells with CMP-sialic acid and β-galactoside α2,3-sialyltransferase, fully restored susceptibility to infection while no restoration occurred, when sialylated by β-galactoside α2,6-sialyltransferase. This suggested preferential use of sialic acid bound to galactose by α2,3 linkage (SAα2,3Gal). However, SeV is usually passaged in the amnion or allantois of chick embryo. SAα2,3Gal was found to be much more abundant than SAα2,6Gal on the amniotic and allantoic cells (Sriwilaijaroen *et al.*, 2009). Avian influenza viruses show specificity to SAα2,3Gal, while human influenza viruses to SAα2,6Gal. Passage of human viruses in the amnion or allantois resulted in a shift of receptor specificity from SAα2,6 to SAα2,3 (Ito *et al.*, 1997). Thus, SeV receptor specificity has to be revisited by using fresh virus isolates from mouse or rat.

In a special case, SeV attachment, followed by entry, was demonstrated to occur by means of the asialoglycoprotein receptor using an SeV temperature-sensitive (ts) mutant in the HN protein (Markwell *et al.*, 1985). This mutant (HN^-) grown at non-permissive temperature (38°C) lacked HA and NA activities and was unable to infect or agglutinate conventional host cells, but did attach to and infect Hep G2 cells that are known to express asialoglycoprotein receptor (ASGP-R). The ASGP-R is a special type of mammalian lectin that recognizes galactose- or *N*-acetyl galactosamine-terminal proteins. That the ASGP-R permitted the HN^- virus (possessing only the F protein) infection was verified by the observations that the infection was abolished by the competitors and antibodies specific to ASGP-R. The authors speculated that the infection would be mediated through the interaction of a terminal galactose moiety of the F protein (Yoshima *et al.*, 1981; see below) with ASGP-R. However, it is unlikely that galactoses become exposed as the terminal moiety for this particular ts mutant that lacks NA activity and hence is unable to remove the terminal sialic acids of the complex-type N-linked-glycans of F protein. SeV entry into cells takes place through fusion of the envelope with the plasma membrane, and this fusion needs the help of the HN protein (see below). How the entry is executed by the HN-deficient mutant remains to be elucidated.

Entry into the cytoplasm

SeV life cycle has no nuclear phase, being accomplished in the cytoplasm. Two different pathways of entry of the viral genomic material into the cytoplasm are known for envelope viruses; one is direct fusion of the viral envelope with the plasma membrane and the other is the clathrin-mediated endocytosis followed by fusion between the viral envelope and the endosomal membrane. The latter route is dependent on a low pH, while the former occurs at neutral pH. Thus, the entry via the endocytosis pathway is inhibited by lysosomotropic agents such as chloroquine and NH_4Cl that raise the endosomal pH, while direct fusion with the plasma membrane is not. SeV and NDV entry was not inhibited by those agents (pH independent) under the conditions where infections by influenza A virus and vesicular stomatitis virus (VSV), which are known to use the endocytosis route, were inhibited (Nagai *et al.*, 1983a). Morphological studies revealed fusion of the SeV envelope with the plasma membrane of the endodermal allantoic cells of chick embryo (Morgan and Howe, 1968) and human erythrocyte membranes (Howe and Morgan, 1969). SeV and paramyxovirus entry, thus, has been believed to occur through direct fusion at the plasma membrane.

However, those data obtained by classical methods are too indirect to draw any conclusions. Indeed, more recent studies on respiratory syncytial virus (RSV), a member of the subfamily *Pneumovirinae*, showed that the infection was inhibited by (1) small interfering (si) RNAs targeting genes associated with clathrin-mediated endocytosis, (2) the expression of well-characterized dominant-negative mutants of genes in the endocytosis pathway and (3) the clathrin endocytosis inhibitor chlorpromazine (Kolokoltsov *et al.*, 2007). Human immunodeficiency virus (HIV) entry is also pH independent and has been thought to occur via direct fusion at the plasma membrane. However, recent live cell imaging that enabled tracking single HIV particles demonstrated that the endocytosis route is the major pathway (Miyauchi *et al.*, 2009). SeV and paramyxovirus entry has to be re-examined using more refined experimental procedures.

SeV fusion with the cellular membrane is mediated by the F glycoprotein. The fusion event needs the help of the HN glycoprotein. The F is a type I transmembrane glycoprotein, which possesses the N terminal signal peptide and the C-terminal membrane anchoring domain, leaving a short cytoplasmic tail. It is synthesized as an inactive precursor F_0 (565 aa). The biologically active F is generated through posttranslational cleavage between the Arg116 and Phe117 of the F_0 into two disulfide-linked subunits, the N terminal (extracellular) F_2 and C-terminal (transmembrane) F_1 subunits (Homma and Ohuchi, 1973; Scheid and Choppin, 1974). Cleavage releases a hydrophobic fusion peptide (26 aa) at the N terminus of F_1 (Gething *et al.*, 1978). The 4-3 (heptad) repeat patterns of hydrophobic repeats, HRA and HRB, were identified just after the fusion peptide and just before the transmembrane domain, respectively. These structural features are conserved in the primary sequences of paramyxovirus F protein (reviewed in Lamb and Parks, 2007).

The Cys-70, the only cysteine residue in the SeV F_2 and Cys-199, which is the most upstream cysteine in the F_1 are involved in the interchain bond and identification of pairing of most other cysteines (Cys-338, -347,–362, -394, -399, -399, -401 and -424) suggested their contribution to the formation of a bunched structure in the middle of F_1 subunit (Iwata *et al.*, 1994). These structural features also appear to be shared with other paramyxovirus F proteins because of prominent conservation of the cysteine residues.

The paramyxovirus F glycoproteins exist as a homotrimer and undergo dynamic conformational changes from the pre-fusion state to the post-fusion conformation (reviewed in Lamb and Parks, 2007). The F protein in the pre-fusion state consists of a large globular head domain attached to a three-helix coiled-coil formed by HRB followed by the transmembrane (TM) domain. The HRB stalks then melt (are separated), and the HRA within the head of the pre-fusion state forms a coiled-coil, extends and inserts the fusion peptide into the target cell membrane (pre-hairpin intermediate conformation). Finally, the HRBs are positioned over the coiled-coil HRA domain in an antiparallel fashion, resulting in the formation of the final post-fusion six-helix bundle (hairpin). This final step allows juxtaposing the fusion-peptide containing target cell membrane and the TM-domain containing viral membrane so that they fuse, ultimately introducing the viral genomic material into the cytoplasm. It is believed that a virus species specific interaction between the HN and F proteins is required for or facilitates the fusion process. Upon binding the receptor, the HN protein may undergo a conformational change that in turn could trigger the conformational change of the F protein (from the pre-fusion state to pre-hairpin intermediate conformation) (reviewed in Lamb and Parks, 2007).

Transcription

General features

After entering the cytoplasm, the SeV and other non-segmented negative-strand (–)RNA virus (NNV) genomes do not become naked but remain as the encapsidated (–)RNP to be transcribed. This transcription is mediated by the virion-associated, virus-specific RNA polymerase comprising the L and P proteins (Hamaguchi *et al.*, 1983; Horikami *et al.*, 1992). The SeV L protein is believed to possess all the enzymatic activities required for RNA synthesis, including nucleotide polymerization as well as 5′ end capping and methylation of the mRNAs (Ogino *et al.*, 2005).

Our knowledge on the transcription of paramyxoviruses is largely adopted from the studies on the transcription of VSV in the *Rhabdoviridae*, a prototypic NNV (reviewed in Whelan *et al.*, 2004). Like VSV, there are *cis*-acting elements to terminate (stop) the upstream gene transcription and to reinitiate (start) the downstream next gene transcription for paramyxoviruses. The entry at the very 3′ end of genome of the RNA polymerase (L–P complex) and polar, sequential transcription of each gene by the stop and restart mechanism towards the 5′ end (the VSV paradigm of transcription) has been adopted to SeV and other NNVs.

According to this VSV paradigm, it is thought that the infecting SeV virion-associated RNA polymerase first transcribes the 3′-terminal leader RNA (55 nt) and then sequentially the *N, P, M, F, HN* and *L* genes by recognizing the stop and (re) start signals (the primary transcription). Polyadenylation of each mRNA is thought to occur through polymerase stuttering on the U repeats in the stop motif (the last five U residues in the 11-nt-long stop signal). The same mechanism could operate in the secondary transcription templated by the newly replicated full-length negative strand genome RNA to amplify the gene products in infected cells. Both the stop and restart signals are highly conserved within a given NNV. However, this polar transcription does not generate each mRNA in equimolar quantities in infected cells perhaps because the reinitiation efficiency of transcription at each gene boundary is high, but not perfect. Thus, it has been believed that there is linear attenuation of transcription towards the 5′ end, where the leader transcript (Le^+) is the most abundant, while the *L* gene product the least abundant (reviewed in Whelan *et al.*, 2004).

For VSV transcription, a two-entry site model has also been proposed. In this model, the RNA polymerase is thought to initiate synthesis of the Le^+ and the full length antigenome at the very 3′ end (position 1); however, for mRNA synthesis, the polymerase initiates by entry directly at the first gene-start site (Whelan and Wartz, 2002; reviewed in Whelan *et al.*, 2004). Alternatively, the polymerase was thought to enter the genome at the very 3′ end, but for transcription scans through the leader region to initiate mRNA synthesis at the first gene start site. Qanungo *et al.* (2004) isolated two different kinds of RNA polymerase complex from infected cells. One complex was the transcriptase that contained a host protein EF1 and started RNA synthesis exclusively at the first *N* gene start. The other was the replicase that contained soluble N but not EF1 and started RNA synthesis exclusively at the genome 3′ end. Based on these findings it was proposed that the presence of EF1 would prevent RNA synthesis initiation at the 3′ end (position 1) and permit the polymerase access to the N start site by scanning, whereas the presence of soluble N helps the polymerase initiate at the 3′ end (Curran and Kolakofsky, 2008). Curran and Kolakofsky (2008) further presented evidence suggesting that SeV soluble N is not only required for replication but also stimulates transcription. These issues are now a focus of active discussion (Banerjee, 2008; Whelan, 2008).

Non-linear attenuation

For SeV, the *P, M,* and *HN* gene-starts are 3′-UCCCACUUUC-5′, whereas the *N* start is 3′-UCCCAgUUUC-5′, the *F* start is 3′-UCCCuaUUUC-5′, and the *L* start is 3′-UCCCCACUUaC-5′. The *F* gene start was found to direct a lower level (about one-fifth) of transcription, compared with the other three whose initiation capacities were almost equivalent (Kato *et al.*, 1999). Thus, the expression of the *F* gene and consequently the downstream *HN* and *L* genes was remarkably down-regulated in the natural life cycle of the parental SeV in cells, whereas the expression levels of the first three genes (N, P and M) were nearly equivalent. The recombinant (r) SeV in which the authentic *F* gene start of wild type was converted to that shared with the *P*, *M* and *HN* genes expressed significantly more F proteins as well as the two downstream gene products, HN and L, replicated with faster kinetics in cultured cells and eggs and was more virulent for mice, exhibiting a 20-fold lower LD_{50} (Kato *et al.*, 1999). Therefore, although probably polar, SeV transcription appears to undergo not linear, but rather biphasic attenuation with a remarkable down modulation between the *M* and *F* genes. Identical sequence differences in the start signals are seen in all SeV isolates sequenced to date, suggesting that the variations are locus specific (Kato *et al.*, 1999). The gene-start signal mediated moderation of F expression thus appears to be universal for SeV, suggesting its selective advantage for the virus in nature. The non-conserved positions within the conserved *cis*-acting elements appear to be highly relevant to the biology of the virus in the natural settings.

Dispensability of the accessory *V* and *C* genes for the cellular-level proliferation

Expression strategies of the *V* and *C* genes

The paramyxovirus V and C proteins were first identified for SeV (Lamb and Choppin, 1977a). They were originally considered to be non-structural proteins expressed in infected cells, but not present in the virions. Later, however, the presence of C protein in a small, but appreciable amount both in the SeV virion and RNP was demonstrated (Portner *et al.*, 1986; Yamada *et al.*, 1990).

The SeV V protein is produced by a transcriptional frame shift (co-transcriptional RNA editing). The P mRNA is a faithful copy of the *P* gene, accounting for about three-quarters of the total transcripts in cells and giving rise to the P protein (568 aa), the smaller subunit of the RNA polymerase complex. The remaining transcripts (one-fourth) are the V mRNA, which carry a single guanine residue (+1G; at nucleotide position 1052) inserted by polymerase stuttering at a specific editing site (3′-UUUUUCCC-5′) located midway down the *P* gene; this insertion shifts the

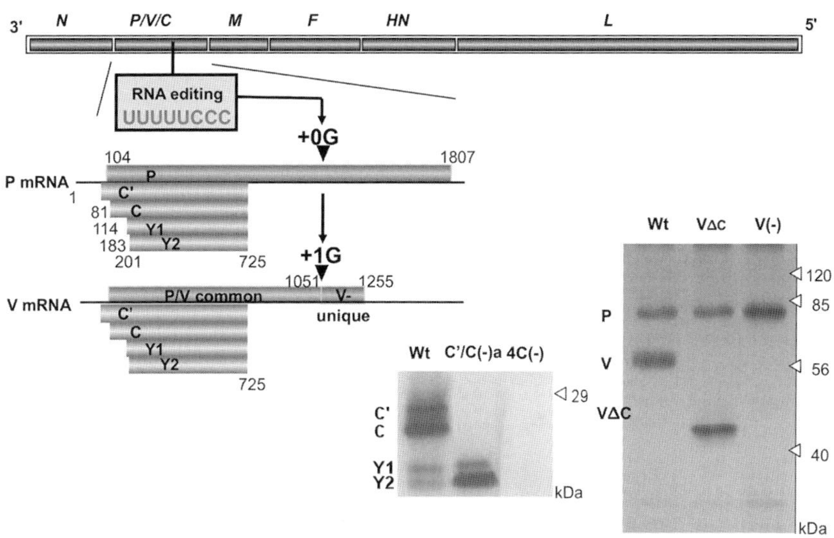

Figure 4.3 Expression of V and C proteins from Sendai virus (SeV) *P* gene and the Western blotting of the cells infected with the respective knockout viruses. Wt, parental wild-type SeV. For details see the text. Adapted from Kato *et al.* (1997a,b), Kurotani *et al.* (1998) and Nagai *et al.* (2007).

reading frame to fuse the upstream P/V common open reading frame (ORF) encoding 316 residues to the downstream −1 ORF of V-unique (Vu) 68 residues (Vidal *et al.*, 1990ab; Fig. 4.3). The V protein, therefore, consists of the P/V common region (the N-terminal approximately four-fifths before the frame shift) and the Vu region (the C-terminal approximately one-fifth after the frame shift). Two Gs insertion can also occur to yield the W protein that terminates three codons downstream. This transcript accounted for less than 5% of the total (Kato *et al.*, 1997b) and is not addressed here.

All five genera of the subfamily *Paramyxovirinae* (*Respirovirus, Morbillivirus, Henipavirus, Rubulavirus and Avulavirus*) express the V protein from the respective *P* genes with the exception of HPIV1 (Matsuoka *et al.*, 1991) and HPIV3 (Galinsky *et al.*, 1992) in the *Respirovirus*. The Vu region of paramyxoviruses exhibits a perfect preservation of seven cysteine residues that form a zinc finger-like motif and indeed was demonstrated to bind Zn^{2+} for at least some members including SeV (Fukuhara *et al.*, 2002).

An ORF, which overlaps the N-terminal P (P/V common) ORF in the +1 frame gives rise to four proteins: C′ (215 aa), C (204 aa), Y1(181 aa), and Y2 (175 aa), collectively referred to as the C proteins that initiate at a non-AUG codon, ACG/81 and AUGs/114, 183, 201, respectively and terminate at 725 (Curran and Kolakofsky, 1988a, 1989; Gupta and Patwardhan, 1988; Fig. 4.3). The C′ and C proteins are translated by a leaky scanning mechanism (Curran and Kolakofsky, 1988a), while the Y1 and Y2 are translated through a scanning independent ribosomal shunting (Curran and Kolakofsky, 1989: Latorre *et al.*, 1998b). The Y1 and Y2 were suggested to occur not only by *de novo* translation initiation, but also by proteolytic processing of the C′ (de Breyne *et al.*, 2004). Among the four C proteins, C is the major species expressed in infected cells at a molar ratio several-fold higher than those of the other three (Curran and Kolakofsky, 1989; Kurotani *et al.*, 1998). Three genera, *Respirovirus, Morbillivirus* and *Henipavirus,* express the C protein. Its homology is remarkably high within each genus, but quite poor between the different genera; in common, however, they are relatively small (~200 residues) and highly basic with the pIs around 10 (reviewed in Nagai, 1999).

One additional protein called X is synthesized from the P frame that is encoded more than 1500 nt downstream of the 5′ end of P mRNA and expressed by scanning independent ribosomal initiation (Curran and Kolakofsky, 1987, 1988b).

Evolutionary aspects in generating the two overlapping ORFs in the *P* gene

Because of the lack of proofreading and repair mechanisms, RNA virus genomes are error-prone. It is thus even surprising that SeV *P* gene of only 1893 nt opens overlapping frames to express at least three different protein species, P, V and C.

Non-synonymous nucleotide substitutions (Ka) are caused by various selection pressures, while synonymous substitutions (Ks) are largely neutral, free from selection pressures, and accumulate as the rounds of replication increase. Ks occur at a nearly constant rate in the variable and conserved regions for paramyxovirus monocistronic genes (Sakaguchi *et al.*, 1989). The ratio of Ka to Ks (Ka/Ks) represents how an ORF has been under strong or weak selection pressure. To address how SeV evolved overlapping ORFs to encode the V and C proteins despite rapid nucleotide changes characteristic of RNA genomes, Ks and Ka were determined between the two SeV strains representative of each of two distinct lineages. They were strain Hamamatsu isolated in 1976 from an ill mouse in an epizootic infection of a mouse colony and not adapted to growth in eggs and the strain Z isolated more than 50 years ago and egg-adapted, and used as the backbone of the reverse genetics described above. Table 4.2 summarizes Ks, Ka and Ka/Ks values in the SeV ORFs based upon the published data (Fujii *et al.*, 2001) and the unpublished supplementary data kindly provided by T. Sakaguchi at Hiroshima University, Hiroshima, Japan.

As expected, Ks values were within a relatively narrow range (0.351–0.450) for the five (N, M, F, HN and L) monocistronic ORFs, indicating that neutral changes have accumulated at a nearly constant rate over these different ORFs. Note that the Ks of the two non-overlapping P unique frames, P1 and P2, were within the same range

Table 4.2 Ks, Ka and Ka/Ks values in the SeV ORFs[1]

Frame	Region (aa residues)	Ks	Ka	Ka/Ks
P	P (1–568)	0.285	0.066	0.232
	PC (5–208)	0.097	0.085	0.876
	PC1 (5–75)	0.232	0.049	0.211
	PC2 (76–208)	0.046	0.105	2.283
	P1 (209–317)	0.417	0.137	0.329
	PVu (318–370)	0.168	0.000	0.000
	P2 (371–568)	0.439	0.029	0.066
C	C (1–204)	0.288	0.029	0.101
	C1 (1–72)	0.133	0.080	0.602
	C2 (73–204)	0.388	0.003	0.008
Vu[1]	Vu (318–370)	0.000	0.040	–
N	N (1–524)	0.412	0.024	0.058
M	M (1–348)	0.351	0.014	0.040
F	F (1–565)	0.408	0.028	0.069
HN	HN (1–575)	0.450	0.031	0.069
L	L (1–2228)	0.376	0.011	0.029

[1]The Vu region of Hamamatsu strain was 53 residues long, 15 residues shorter than that (68 residues) of strain Z.

SeV, Sendai virus; ORF, open reading frame.

(0.417 and 0.439, respectively). The Ks values of the entire P ORF (0.285) and C ORF (0.288) were obviously lower than those of the five monocistronic ORFs and the P1 and P2. Mainly owing to this difference and in part owing to a high Ka of P ORF, the Ka/Ks ratio was significantly higher in the P ORF (0.232) and C ORF (0.101) than those in the other five ORFs ranging from 0.029 to 0.069. Regarding the non-overlapping regions, P1 and P2, P2 displayed Ks, Ka and Ka/Ks profiles similar to those of the five monocistronic ORFs, while P1 had a higher Ka/Ks ratio (0.329) owing to increased Ka value (0.137) (Table 4.2). P1 thus appeared to be somewhat variable, while P2 (Ka/Ks, 0.066) was less variable. This could be related to the fact that most of P protein functions have been mapped to the C-terminal half as noted below.

The counterpart in P ORF of the entire C ORF was designated PC. The C ORF was divided into two parts, its N-terminal one-third, C1 and its C-terminal two-thirds, C2, and their counterparts in P ORF were designated PC1 and PC2, respectively. The Ks value of C2 (0.388) was within a normal range, while its Ka value was extremely low (0.003), giving rise to an extremely low Ka/Ks (0.008) (Table 4.2). On the other hand, the trend was reversed in the PC2, showing an unusually low Ks (0.046), considerably high Ka (0.105) and extremely high Ka/Ks (2.283) (Table 4.2). These results indicated that the C-terminal two-thirds (C2) had been subject to little change and that the accumulated synonymous changes in the C2 had been dictated largely as non-synonymous, but yet tolerable changes by the PC2 frame. In other words, PC2 was flexible enough to buffer the nucleotide changes in the C2; thereby, stably maintaining the two overlapping frames C2 and PC2. Although not as remarkable as this case, the two other overlapping frames C1 and PC1 also were found to have evolved in a similar way; however, C1 (Ka/Ks, 0.602) appeared to have acted as a buffer to allow relative conservation of the PC1 (Ka/Ks, 0.211) (Table 4.2). Because of the short length, a statistically reliable interpretation was difficult for the Vu and its P ORF counterpart, PVu. Although Vu appeared to be a little more variable than PVu on the Ka basis (Table 4.2), the seven cysteine residues were perfectly conserved in the Vu of the two strains. The structural constraint of

PVu (Ka; zero) could be due to its involvement, in part, in the oligomerization domain of the P protein (see below). Taken together, the results indicated that the two overlapping frames in the SeV *P* gene might have evolved by positioning a conserved ORF in parallel with a variable ORF.

Deletion of the *V* or *C* genes without destroying infectivity

Without affecting the P ORF, changes in two nucleotides (underlined) were introduced into the editing site (U<u>C</u>UUUC<u>U</u>C) or a stop codon introduced just downstream of the editing site. Mutants with the total V protein deleted [V(–)] and only the C-terminal V-unique region deleted (VΔC) were respectively recovered without any difficulty just after inoculation of transfected cells into eggs (Kato *et al.*, 1997ab; Expt. 1 in Table 4.3 and Fig. 4.3). These mutants replicated in cells as efficiently as the parental SeV and their ratios of infectivity (PFU/ml) to the physical particle numbers [substituted by the haemagglutination units (HAU)] were as high as that of the wild type recovered from the cDNA (Table 4.3), establishing that the V protein is non-essential and that both of the V mutants are fully infectious. The role of V protein, if any, in the life cycle therefore remains to be elucidated. However, the V protein appeared to encode a luxury function required for *in vivo* pathogenesis (Kato *et al.*, 1997a,b). The *in vivo* role of the V protein is discussed below.

Through a difficult process including at least three passages in eggs, a viable clone, 4C(–), which expressed none of the four C proteins (C′, C, Y1, and Y2), was recovered (Expt. 2 in Table 4.3 and Fig. 4.3) by disrupting the respective ORFs, indicating that SeV C proteins also fall into the category of non-essential accessory proteins (Kurotani *et al.*, 1998). The difficulty of recovery was the result of a reduction of 4C(–) virus infectivity by about five logs (Table 4.3), and the success of recovery probably was attributable to the high efficiency of our SeV rescue system. At least four successive egg passages were required to obtain detectable HAU for 4C (–) (Table 4.3). After the 4th passage, the PFU to HAU ratio was nearly 100-fold lower than that of the wild type (Table 4.3), suggesting that the stock contained considerable amounts of non-infectious physical particles. The ratio appeared to become a little higher after an additional passage in eggs.

Three clones, C′/C(–)a, C′/C(–)b and C′/C(–)c, that expressed Y1 and Y2, but not C′ and C were isolated with relative ease, compared with 4C(–) (Expt. 2 in Table 4.3). Their infectiveness was, however, two logs lower than that of

Table 4.3 rSeV recovery following egg-passages of transfected LLCMK2 cells[1]

Virus	Inoculation size (no. of cells)/egg	Passage no. in eggs	HAU	Infectivity (pfu/ml)	Infectivity/HAU
Expt. 1					
Wide type	10^5	1st	384	6.5×10^8	1.7×10^6
V(–)	10^5	1st	384	7.5×10^8	2.0×10^6
VΔC	10^5	1st	180	4.0×10^8	2.2×10^6
Expt. 2					
Wild type	10^5	3rd	1024	1.7×10^{10}	1.7×10^7
C′/C(–)a	10^6	3rd	16	2.3×10^8	1.4×10^6
C′/C(–)b	10^6	3rd	16	1.7×10^8	1.1×10^6
C′/C(–)c	10^6	3rd	16	2.1×10^8	1.3×10^6
4C(–)	10^7	3rd	<2	4.2×10^4	-
	10^7	4th	4	9.8×10^5	2.5×10^5
	10^7	5th	8	7.0×10^6	9.0×10^5

[1]Adapted from Kato *et al.* (1997a,b) and Kurotani *et al.* (1998).

rSeV, recombinant Sendai virus; HAU, haemagluttination unit.

the wild type and their PFU to HAU ratios were nearly one-tenth of the wild type (Table 4.3). The elimination of C′ and C expression was done by introducing stop codons just downstream of their initiation codons that were left unchanged. The C′ and C expression was indeed successfully eliminated, but the levels of Y1 and Y2 expression were remarkably increased (Kurotani *et al.*, 1998; representatively shown with C′/C(−)a in Fig. 4.3). This result could provide a remarkable case in which translation at an internal AUG codon in mammalian cells can be facilitated by the introduction of a stop codon within the upstream ORF (in this case the C′ and C ORFs) (Hughes *et al.*, 1984; Liu *et al.*, 1984; Kozak 1984; Kurotani *et al.*, 1998). Facilitation was more significant for Y2 than for Y1, in agreement with the notion that the efficiency of reinitiation steadily improves as the distance from the upstream ORF increases (Kozak, 1987). All the phenotypes displayed by the three C′/C(−) clones in cell culture and in mice have been so far essentially identical, and clone C′/C (−)a has been solely used as the C′/C(−) virus in all subsequent studies.

Such severely debilitated viruses as 4C(−) and C′/C(−) are obviously under strong pressures and are prone to accumulate unintended mutations. So, sufficient caution is needed for preparation of the virus stocks and interpretation of the data obtained using those stocks.

Genome replication

Replication promoters

After accumulation of viral proteins translated from the primary transcripts, the 15,384-nt-long genome (−) RNA is replicated to a full-length exact copy, the antigenomic (+) RNA, which is in turn replicated to the genomic (−) RNA. This newly synthesized (−) RNA is used as the genome RNA to be incorporated into the progeny or as the template for secondary transcription as well as for replication to amplify the synthesis of viral genomes and gene products. The (−) and (+) RNA templates are tightly encapsidated with the N proteins during the rounds of replication. The same L–P complex as used for transcription is believed to act as the replicase, but the former is finely tuned by the transcription regulatory signals, while the latter must ignore these signals and read through the intergenic sequences. The mechanism of switching from transcription has been a central issue in the NNV biology and a focus of active discussion (Banerjee, 2008; Curran and Kolakofsky, 2008; Whelan, 2008).

One of the key factors that are involved in this switching is the concomitant, continuous association of *de novo* synthesized soluble N proteins with the nascent, growing chain of RNA in the direction from the 5′ to the 3′ end (Gubbay *et al.*, 2001). The P protein is essential for maintaining the N protein in a soluble form and chaperones its association with the nascent RNA chain (Curran *et al.*, 1994). However, as described above, soluble N also appears to be required for transcription, although the amount required for this is lower than that required for replication and the transcription probably requires host factors such as EF1, while replication perhaps does not.

The genomic and antigenomic RNAs are tightly encapsidated with the N proteins. The RNAs have to be locally and transiently exposed in order to become the template for the RNA polymerase (L-P complex) during replication as well as transcription. However, the RNAs still must remain close to the N proteins because the polymerase stays attached to the N proteins. How such subtle conformational changes are made is not known for SeV. For the rhabdovirus rabies virus it was suggested that a local dissociation of the RNA from one or several N protomers could provide sufficient surface for binding of the polymerase complex (Albertini *et al.*, 2006).

Each N protein of SeV recognizes six nucleotides (hexamer) (Egelman *et al.*, 1989) and 2564 monomers of N protein are required to finish the assembled 15,384-nt-long SeV genome or antigenome. Mutagenesis studies on SeV model minigenomes revealed that efficient replication occurs only when the total number of nucleotides of a genomic or antigenomic RNA is an even multiple of six (rule of six) (Calain and Roux, 1993; Kolakofsky *et al.*, 1998). There are two discontinuous replication promoters PrE-1 and PrE-II (Fig. 4.4A, reviewed in Lamb and Parks 2007). The 3′ terminal 12 residues (2 hexamers) in PrE-1 are identical except for the fifth position between the genome and antigenome (Fig. 4.4A). PrE-II

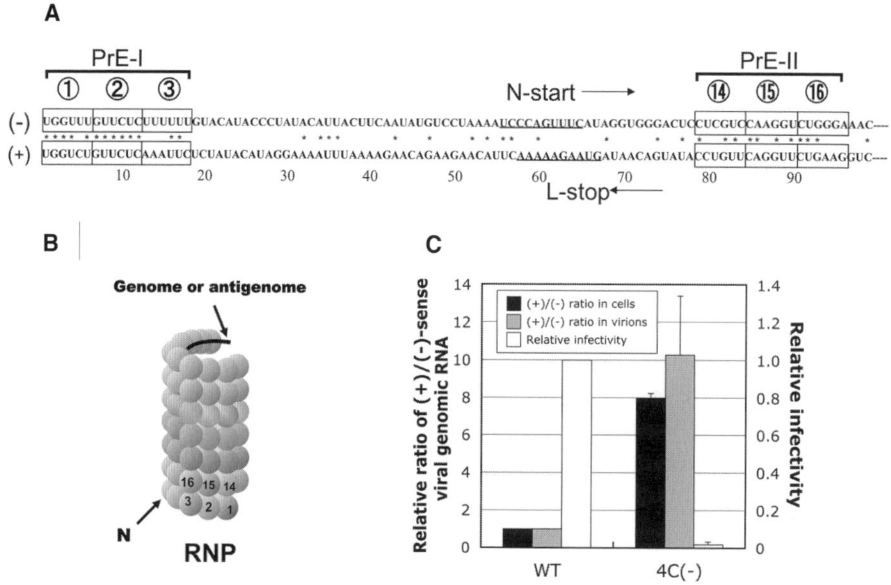

Figure 4.4 (A) The 3′-terminal nucleotide sequences of Sendai virus (SeV) genome (–) and antigenome (+) (Shioda *et al.*, 1983, 1986), two separate promoters PrE-I and PrE-II (Tapparel *et al.*, 1998) and the hexamers (boxed) constituting the promoters. Asterisks indicate nucleotide identity between the genome and antigenome. N gene start motif in the genome and L gene stop motif in the antigenome are underlined. (B) Positioning of the hexamers constituting the two separate promoters on the helical RNP. Adapted from Nagai *et al.* (2007). (C) Relative ratios of antigenome to genome and relative infectivities. Adapted from Irie *et al.* (2008a).

consists of (3′-CNNNNN-5′)x3 positioning at the 14th, 15th and 16th hexamers in both the genome and antigenome (Tapparel *et al.*, 1998; Fig. 4.4A). PrE-1 and PrE-II are, thus, separated by 60 nucleotides (10 hexamers). Changing the length of this spacer impaired replication, probably because the N bound hexamers 1st to 3rd and 14th to 16th have to be juxtaposed on the same surface of coiled RNP with a turn of 13 Ns for the RNA polymerase to recognize the template and initiate replication of the genome and antigenome (reviewed in Lamb and Parks, 2007; Fig. 4.4B).

How can SeV be an NNV? Possible involvement of the accessory C proteins

How an NNV can be an NNV predominantly possessing (–) RNA genome is a central question in virology. Is there a mechanism to selectively incorporate only the (–) RNP into the virions during the process of assembly and budding? No, there is no such selectivity, at least for SeV, because genomic as well as antigenomic RNA containing SeV virions were shown to be produced in proportion to the intracellular ratio of genomic and antigenomic RNA, although antigenome containing virions were not infectious (Kolakofsky *et al.*, 1974; Kolakofsky and Bruschi, 1975; Mottet and Roux, 1989).

Then, there should be a mechanism leading to the predominance of (–) sense genomic RNA late in the life cycle. It has been postulated that the 3′-end promoter in the antigenome, which directs the synthesis of genome RNA, would be stronger than the 3′-end promoter in the genomic RNA to direct the synthesis of antigenome and mRNAs, raising the hypothesis that the relative strength of genomic and antigenomic promoters defines the relative ratio of these full length RNAs in infected cells (reviewed in Lamb and Parks, 2007).

The idea of different promoter strength appears to have come from the following experimental results obtained using internal-deletion and copy-back DI (defective interfering) particles. The internal-deletion DI particle genomes contain the authentic leader and trailer sequences present in the full-length (–) RNA, while the leader sequence in the copy-back DI RNAs is the exact copy of the 5′ terminal trailer sequence of the genome. Both of the DI types contained the PrE-II as well as PrE-I. Mixed infections of eggs with SeV stocks containing internal-deletion DI and copy-back DI genomes followed by serial egg passage out-competed internally deleted DI particle genomes, resulting in predominance of copy-back DI genomes and extinction of internal-deletion DI genomes (Re and Kingsbury, 1986; reviewed in Re, 1991). This out-competition was not due to size differences between the internal-deletion DI and copy-back DI genomes nor to whether the transcription was inert (copy-back DI) or active (internal-deletion DI). Late in the natural life cycle of SeV, eight to 10 times as many genomes as antigenomes are produced. More recently, an rSeV was generated whose trailer sequence was replaced with the leader sequence. In the cells infected with this rSeV, the ratio of genome to antigenome was now about 0.8, indicating roughly equal levels of genome and antigenome (Le Mercier *et al.*, 2002). Taken together, these results appeared to provide strong support for the idea that the differences in the promoter strength is a critical determinant for (–) genome predominance.

However, the interpretation of the results cannot be that simple because the analyses were carried out under the conditions where the C proteins were expressed from either the helper virus to replicate the DI genomes or the rSeV itself and because the C proteins were found to have a capacity to down-modulate SeV mRNA synthesis (Curran *et al.*, 1991, 1992) and RNA replication in a promoter selective fashion (Cadd *et al.*, 1996; Horikami *et al.*, 1997; Tapparel *et al.*, 1997). The promoter specific inhibition was suggested by the observation that the inhibition by the *trans*-supplied C proteins was stronger on an internally deleted DI genome than on a copy-back DI genome. However, the copy-back DI RNA replication was also inhibited under the conditions where the ratio of C to copy-back template was increased (Tapparel *et al.*, 1997). The promoter selectivity by the C protein, thus, does not appear to be all-or-none.

The question of how SeV can be an NNV was revisited using rSeVs with the expression of the C proteins variously silenced. It was first found that 4C(–) SeV expressing none of the four C proteins directed elevated levels of both antigenome RNA and mRNA synthesis in infected CV1 cells, compared with the parental SeV (Hasan *et al.*, 2000). In addition, the PFU to HAU ratio of 4C(–) harvested at 60 h.p.i, was as low as about 10^4, while that of the wild type was above 10^6. Large differences in the PFU to HAU ratio were also seen between egg-grown 4C(–) and wild type (Table 4.3). Then, a remarkable difference in the intracellular proportion of antigenome to genome was found between the 4C(–) infected and wild-type infected LLCMK2 cells (Irie *et al.*, 2008a). The actual ratio of antigenome to genome in cells was approximately 1:8 at 48 h after infection with the wild type. This ratio was set to 1.0, and the relative ratio was determined for 4C(–) infected cells. It was about 8-fold higher than that in the wild-type infected cells and this proportion reflected that in the virions (Fig. 4.4C). The infectivities of wild-type and 4C(–) virions relative to the respective N protein amounts were determined, and wild-type infectivity was set to 1.0. The relative infectivity of the virions released from 4C(–) infected cells was less than one-tenth of that of the wild type (Irie *et al.*, 2008a; Fig. 4.4C).

The C′/C(–) virus expressing the Y1 and Y2, but not C′ and C exhibited essentially the same phenotype as that of 4C(–) (Irie *et al.*, 2008a). This appears to be contradictory to the previous results indicating that the RNA synthesis inhibiting capacity of the SeV C protein was mapped to its C-terminal half (Grogan and Moyer, 2001) and that the smallest C protein, the Y2, still retained RNA synthesis inhibiting capacity in cells (Kato *et al.*, 2001). One plausible explanation for this discrepancy could be that the levels of Y1 and Y2 expression from C′/C(–) virus were extremely low in this particular experiment (Irie *et al.*, 2008a), in contrast to enhanced Y1 and Y2 expression in other studies (Kurotani *et al.*, 1998; also see Fig. 4.3) and thus might not be quantitatively sufficient to exhibit that promoter specific inhibition of RNA replication.

SeV induces IFNs and the SeV C proteins counteract the IFNα/β responses as noted below. However, it was unlikely that IFN-induced signalling had affected the genome (wild-type) or antigenome [4C(–) and C′/C(–)] predominance in LLCMK2 cells because essentially the same results were obtained in Vero cells, which are known to be totally unable to produce IFNs (Irie *et al.*, 2008a).

Taken together, these results suggested that SeV promoter activity measured in the presence of L-P RNA polymerase and N protein, but in the absence of the RNA synthesis inhibiting C proteins, i.e. the intrinsic promoter activity, is rather stronger for the genomic promoter than for the antigenome promoter (or at least nearly equivalent between the two). Moreover, the C proteins could reverse this relation by inhibiting genomic promoter-dependent replication. The levels of C proteins synthesized in cells are lower early in life cycle than in a later phase. It is thus probable that the synthesis of antigenome as well as mRNAs could be dominant early in the life cycle, while the synthesis of genome could become dominant later.

It is surprising that just the accessory gene products, which although widespread, are not ubiquitous in paramyxoviruses, play such an important role in leading to predominance of genomic RNA, thereby rendering SeV an NNV. However, it is premature to draw a conclusion solely from the results obtained using C knockout rSeVs because the C knockout viruses, particularly the 4C(–) virus, are severely impaired in replication in cells and eggs and could be prone to accumulate unintended mutations during the preparation of virus stocks. Various supplementary approaches will be necessary to reach a firm conclusion. These may include the development of the way enabling nearly complete knockdown of the C proteins in the context of wild-type replication as well as sufficient *trans*-supply of the C proteins during replication of the 4C(–) virus. These approaches will also help answer the question of whether or not the smaller Y1 and Y2 proteins still function in the predominance of genome RNA.

Functional and interacting domains in the P, N and L proteins

The SeV P protein (568 aa) is divided into two segments, the N-terminal PNT (residues 1–319) and the C-terminal PCT (residues 320–568) (Blanchard *et al.*, 2004). The PNT is unstructured and contains the site (residues 33–41) to bind newly synthesized, soluble N protein, chaperoning the unassembled N protein during the nascent chain assembly step of genome replication (Curran *et al.*, 1995). All other known activities have been mapped to the PCT. These include the homotetramer formation domain (PMD; residues 320–433) (Tarbouriech *et al.*, 2000b), the L binding domain (residues 412–445) (Curran *et al.*, 1994; Smallwood *et al.*, 1994), which is in part within the PMD and a predicted triple α-helical bundle of the C terminus (residues 479–568) that is suggested to bind the N protein of the N:RNA template (Ryan *et al.*, 1991; Curran, 1998). This module of N:RNA binding domain is expressed as the X protein in the natural life cycle through scanning independent initiation of the P mRNA (Curran and Kolakofsky, 1988b). When expressed alone in *Escherichia coli*, the X protein is extremely soluble and monomeric (Tarbourriech *et al.*, 2000a). NMR revealed that the X protein consisted of N- and C-terminal subdomains linked by an 11-residue-long linker and that the latter domain contained three α-helices (α1, α2 and α3) (Blanchard *et al.*, 2004). The surface created by the α2 and α3 of X binds a region in the tail of the N protein (see below), which also has a helical propensity, with a 1:1 stoichiometry (Houben *et al.*, 2007). The C-terminal ~40% (residues 344–568) is competent for mRNA synthesis *in vitro*, but incapable of replication (Curran, 1996).

The N protein (524 aa) is also divided into two segments, the N-terminal Ncore (residues 1–399) and the C-terminal Ntail (residues 400–524) (Buchholz *et al.*, 1993). An N-terminal domain (residues 114–129) is required for RNP to function as a template in viral replication (Myers and Moyer, 1997). The Ncore is involved in N–N and N–RNA interactions as well as RNA replication capacity (Buchholz *et al.*, 1993). A central region of Ncore (residues 258–357) contains the 10-residue long motif (residues 262–271) that is highly conserved among paramyxoviruses and the N-terminal 255 residues are essential for self assembly of N with RNA (Myers *et al.*, 1997). The Ntail is also required for template function (Curran *et al.*, 1993). As noted above, it contains the site to bind the P (in the L-P complex) (residues 462–471) (Çevik *et al.*, 2004), which is followed by the two predicted α helices (residues 477–493 and 513–520, respectively) (Houben *et al.*, 2007). Treatment of SeV RNP with trypsin removed a C-terminal portion of the Ntail. This yielded a more rigid and yet RNase-insensitive structure (Heggeness *et al.*, 1981), suggesting that the Ntail confers flexibility of the coiling of the native RNP and its functional diversity.

The SeV L protein (2228 aa) is also oligomerized and exhibits all the catalytic activities including nucleotide polymerization, 5′ end capping and methylation and polyadenylation of mRNAs (reviewed in Lamb and Parks, 2007). These functions are executed with association of the P protein. Polyadenylation occurs through a mechanism of polymerase (L–P complex) stuttering at the 5Us stretch in the end of transcription stop signals. Six domains (I-VI) are conserved among the L proteins of NNV (Poch *et al.*, 1990). For SeV they are domains I (residues 225–416), II (503–607), III (653–876), IV (927–1128), V (1129–1378) and VI (1770–1847) (Feller *et al.*, 2000).

A series of extensive mutagenesis including point mutations, deletions, insertions and swapping with the corresponding measles virus residues within each of the six domains indicated that the activities of SeV L protein are not simply compartmentalized in a particular domain, but rather that each domain contributes to multiple steps in viral RNA synthesis (Chandrika *et al.*, 1995; Cortese *et al.*, 2000; Feller *et al.*, 2000; Smallwood *et al.*, 2002b). Thus, pairs of mutated inactive L proteins could sometimes restore viral RNA synthesis. This type of complementation occurred with mutants in five of the six domains, the only exception being domain III (Smallwood *et al.*, 2002a). However, a catalytic residue for mRNA cap formation was pinpointed at the Lys1782 in domain VI and the rSeV with the K1782A change completely abolished cap methylation (Murphy and Grdzelishvili, 2009). It is interesting to note that the transcription and replication were uncoupled for considerably many mutants.

The RNA synthesis inhibition and the inhibition of complex formation with the P protein are largely independent since many of the L mutants incapable of RNA synthesis were capable of P binding. The P binding site was later mapped to residues 1 to 350 (Holmes and Moyer, 2002). The N-terminal 178 residues preceding the domain I was required for homooligomerization. Two regions (residues 2–19 and 20–178, respectively) appeared to independently mediate homooligomerization since deletion of either region still supported oligomerization, although abolished RNA synthesis (Çevik *et al.*, 2007). The interaction of L protein and C protein could be relevant to the promoter-dependent RNA synthesis inhibition by the latter. The region in the L protein involved in this interaction was mapped to the N-terminal 895 residues containing domains I to III (Horikami *et al.*, 1997), and that in the C protein to the C-terminal half (Grogen and Moyer, 2001). Thus, the smallest C protein, Y2, retained RNA synthesis inhibition capacity in cells (Kato *et al.*, 2001).

Posttranslational modifications

Phosphorylation

Because of heavy phosphorylation, the paramyxovirus phosphoprotein (P) is so named. The primary phosphorylation site is the serine residue at 249 in the SeV P protein. This phosphorylation is thought to be carried out by protein kinase C ζ (Huntley *et al.*, 1997). The SeV L protein may also be involved as it was found to phosphorylate the P protein as well as the N protein *in vitro* (Einberger *et al.*, 1990). Phosphorylation-deficient viruses were generated by reverse genetics that had the mutation, S249A, S249D or P250A (because P250A also abrogated phosphorylation at S249, the enzyme involved could be proline-directed). These mutants were comparable to the wild type with respect to replication in cell cultures and pathogenicity for mice as well as oligomerization (Hu *et al.*, 1999). Since S249 is localized within the P/V common region, V protein phosphorylation at S249 was also abrogated in these mutants. Thus, primary phosphorylation at residue 249 of the P and V proteins appeared to have no essential role in SeV transcription and replication. However, the situation is complex because alternative serine residues have been phosphorylated in these mutants. These alternative sites were further knocked out so that the net phosphorylation was reduced to about 10% of the wild type. These mutants were found to be fully competent at least in supporting minigenome replication (Hu and Gupta, 2000). Thus, the biological significance of the SeV P protein phosphorylation remains to be addressed.

At least a portion of intracellular M proteins was found to be phosphorylated, while those in the virions were not (Lamb and Choppin, 1977b). This suggested the possibility that the phosphorylated M was selectively eliminated from progeny virions during the process of assembly or that M was dephosphorylated during or after the process. After identifying the serine residue at residue 70 as the phosphorylation site, an rSeV S70A was generated, which no longer underwent M protein phosphorylation. This mutant exhibited no particular phenotype in cell cultures (growth kinetics and virus yield) and in mice (killing ability). Thus, it appeared that phosphorylation of SeV M protein is just an epiphenomenon irrelevant to either the life cycle or pathogenesis (Sakaguchi *et al.*, 1997).

The N-terminal regions of C′ and C extending from the Y1 have a serine residue phosphorylated (Hendricks *et al.*, 1993). In infected cells, they are present in phosphorylated and unphosphorylated forms. The SeV N and HN proteins are also phosphorylated in cells (Einberger *et al.*, 1990; Takimoto *et al.*, 1998). The biological significance for these phosphorylation events remains to be elucidated.

Glycosylation

The carbohydrate structures of egg-grown SeV HN and F glycoproteins were analysed. Both glycoproteins contained fucose, mannose, galactose and glucosamine, but not galactosamine, indicating that their sugar chains are exclusively of asparagine-linked (*N*-linked) type (Yoshima *et al.*, 1981). These authors further showed that two main classes of oligosaccharides, complex-type and high mannose-type, are present in both of the glycoproteins and that the terminal moiety of the former is exclusively a galactose, but not a sialic acid since the pre-existing terminal sialic acid has been removed by the viral NA activity.

N-linked oligosaccharide chains are known to affect many aspects of protein structure and function, including proper folding, intracellular transport, conformation maintenance, stability, protection from proteolysis and inherent biological activities of the proteins. The SeV HN protein contains four potential *N*-glycosylation sites, N77, N448, N499 and N511. Site directed mutagenesis and plasmid-based expression revealed that oligosaccharide chains are attached to three (N77, N499 and N511) of them, but not to N448 and that elimination of the sites N499 and N511 by replacement of the asparagine residue with glutamine residue (N499Q and N511Q) resulted in the most detrimental outcomes, significantly decreasing cell surface expression, haemadsorption, and neuraminidase activity (Segawa *et al.*, 2003). Moreover, they were unable to assist syncytia formation by the F glycoprotein coexpressed.

By a similar approach, all the three potential *N*-glycosylation sites, N104, N245 and N449, of the SeV F protein were shown to be glycosylated. Using an anti F monoclonal antibody specific for native conformation, cell fusion assay with HN coexpression and glycosylation-defective mutants, deglycosylation at the site N245 was found to critically impair efficient intracellular transport of the F protein, its cell surface expression and cell fusing capacity, whereas the mutations of the other two sites were not (Segawa *et al.*, 2000). The *N*-linked oligosaccharide chains of SeV F protein affected the F protein interaction with the endoplasmic molecular chaperons in a position dependent manner and the altered interactions appeared to affect the folding and intracellular transport of the F protein (Tamura *et al.*, 2002).

These results suggested that the N-linked-glycans of both the HN and F proteins of SeV indeed play important roles in their quality control in the endoplasmic reticulum, intracellular transport, and cell surface expression in biologically active forms and that these roles are played in a position-dependent manner.

Cleavage-activation

SeV is routinely propagated in the allantoic cavity or amniotic cavity of embryonated chicken eggs. The egg-grown SeV is highly infectious to a wide variety of cells of mammalian and avian origin in culture as well as to eggs, and possesses cell fusing and erythrocyte-lysing (haemolytic) capacity. However, when propagated in cultured cells, the virus does not display any infectivity, haemolytic activity or fusion activity. These capacities were restored just after a single passage of the inactive tissue-culture-grown SeV in eggs (Ishida and Homma, 1961; Matsumoto and Maeno, 1962). This phenomenon, called 'host-controlled variation or modification', remained an enigma of SeV for a long time and attracted many researchers in the field. A clue to solving this enigma came from the discovery that tissue-culture-grown SeV came to life again, becoming fully infectious, haemolytic and fusogenic, by treatment with a low concentration of trypsin (Homma, 1971, 1972; reviewed in Ishida and Homma, 1978). It was established that tissue-culture-grown SeV possesses the biologically inactive precursor F_0 protein. The F_0 is cleaved at a specific site (between Arg116 and Phe117) by trypsin into the biologically active F protein consisting of the F_1 and F_2 subunits in cells in culture, while cleavage occurs naturally in eggs (Homma and Ohuchi, 1973; Scheid and Choppin, 1974). Therefore, SeV cannot undergo multicycle infection nor produce plaques in cell culture unless trypsin is added to the culture media, and, as noted below, its tissue tropism *in vivo* depends upon whether or not an activating protease is available in the tissue. This discovery of cleavage-activation of SeV was a clue to establishing a general concept shared with most enveloped viruses in different virus families that the envelope protein responsible for initiation of infection through fusion of the viral envelope with the target cell membrane is synthesized as a biologically inactive precursor and converted to an active form by proteolytic cleavage (reviewed in Nagai, 1993, 1995).

As there is a single arginine residue at the cleavage site of SeV F_0, it is not cleavable by a ubiquitous intracellular, *trans*-Golgi localized endoprotease such as furin that preferentially recognizes and cleaves a multibasic motif (reviewed in Nagai, 1993). Accordingly, SeV F_0 is cleaved by an extracellular protease. As a naturally occurring SeV-activating endoprotease, the blood clotting factor Xa was first isolated from the allantoic and amniotic fluids of chicken eggs (Gotoh *et al.*, 1990; Suzuki *et al.*, 1991). It further was able to cleave and activate the F_0 of another paramyxovirus, NDV (avirulent strains) and the haemagglutinin (HA) of influenza virus strain H1N1 A/PR/8/34, which share a single arginine cleavage-motif. A protease with similar substrate specificity to that of FXa, tryptase Clara, was then suggested to be a responsible protease for SeV activation on the epithelial surface of the respiratory tracts of rat (Kido *et al.*, 1992; Tashiro *et al.*, 1992).

More recently, human type II transmembrane serine proteases with known airway localization, TMPRSS2 and HAT, were molecular-cloned and their expression in a canine kidney cell line, MDCK cells enabled cleavage-activation of influenza HA of different subtypes, which has a single-arginine-cleavage motif and facilitated multicycle replication of these viruses in cells in the absence of exogenous trypsin (Böttcher *et al.*, 2006). Similarly, TMPRSS2 as well as TMPRSS4 might be the responsible proteases for cleavage-activation of the 1918 pandemic influenza virus HA (Chaipan *et al.*, 2009). The possibility remains that these newly identified proteases are also able to activate SeV.

Assembly, budding and release

The M protein as the central player in assembly and budding

All viral structural components of a paramyxovirus meet at the plasma membrane and are organized into a mature virion, which buds from the plasma membrane. For SeV and many paramyxoviruses the M protein has been thought to be the central player in this assembly and budding process. Straightforward evidence for this has been obtained by SeV reverse genetics and gene silencing. It was found that virus particle formation was abrogated in cells infected with rSeV/Δ*M*, with the *M* gene deleted (Inoue *et al.*, 2003a; see below) or by suppression of the M protein expression by an siRNA during the viral life cycle (Mottet-Osman *et al.*, 2007).

The SeV M protein contains five cysteine residues at positions 83, 106, 158, 251 and 295, all of which are not involved in disulfide bond formation but are present in a free form throughout the life cycle. Three recombinant viruses (C83S, C106S and C295S) were successfully generated, and the virus particles collected from culture media of infected cells were examined for their size distributions (Sakaguchi *et al.*, 2002). The parental virus particles displayed a typical normal distribution with the diameter of 200 ± 48 nm, whereas the patterns of the three mutants significantly deviated from the normal distribution. They yielded either predominantly smaller virus particles (C83S and C106S), or larger and size-heterogeneous particles (C295S). Moreover, C106S yielded a significant amount of empty particles lacking the RNPs. These results further substantiate the critical importance of M protein for SeV morphogenesis.

How the M protein plays this role in virus assembly and budding has been a central question. Earlier studies by *in vitro* reconstitution of isolated SeV glycoproteins, M protein and RNP demonstrated that association of the glycoproteins with the RNP occurred only if M protein was present (Shimizu and Ishida, 1975; Yoshida *et al.*, 1976), suggesting that bridging the viral envelope and the core is one of the roles played by the M protein. The SeV M protein was demonstrated to interact with the cytoplasmic tails of F and HN proteins in the absence of the other viral components including RNP (Ali and Nayak, 2000). These interactions were virus specific because no interaction occurred when influenza virus glycoproteins were used in place of the homologous SeV glycoproteins. Using freshly prepared and thus structurally firm virions (like those shown in Fig. 4.1A) and chemical cross-linkers, Markwell and Fox (1980) demonstrated that the M proteins lie in close proximity (within 1.1 nm) to the RNP. Thus virus-specific physical HN/F-M-RNP interactions appear to have been well established.

In the process of assembly during the life cycle, the glycoproteins of SeV and NDV appear to be initially dispersed over the entire cell surface and then condensed into a patch of viral membrane with morphologically visible spikes in the regions of membrane where the M proteins are present (Nagai *et al.*, 1976b, 1983b). The RNPs align beneath those modified membrane areas. Also, the membrane sites to which RNPs adhere are modified by virus specific particles arranged in a crystalline orthogonal array (Bächi, 1980), which strikingly resemble the cylindrical structure formed by the M proteins (Hewitt and Nermut, 1977; Heggeness *et al.*, 1982). Cell fractionation studies following NDV infection suggested that M protein became associated with the patch of viral membrane and incorporated into a virion immediately after synthesis, while the HN and F took a longer time to reach the plasma membrane (Nagai *et al.*, 1976b).

However, it is unknown whether the association of M protein promotes condensation of the glycoproteins into a patch or if patch formation is a precondition for attachment of M protein. The former possibility appeared to be supported by the studies with an SeV mutant, which was temperature sensitive in M protein synthesis (Yoshida *et al.*, 1979). The glycoproteins exhibited much higher mobility in the plane of membrane, as detected by glycoprotein capping by the specific antibody in the absence of M protein synthesis at a non-permissive temperature than when M protein synthesis was allowed to occur at a permissive temperature. Using NDV-infected subcellular fractions, it was shown that the higher the density of M protein relative to the glycoproteins, the lower the lateral mobility of glycoproteins in the plane of membrane (Nagai *et al.*, 1975, 1976b).

On the other hand, like other enveloped viruses, SeV proteins (F, HN, M, N and P) became, at least in part, associated with a mild detergent resistant fraction, the lipid raft, which is thought to be a platform of virus assembly (Gosselin-Grenet *et al.*, 2006). Thus, to know whether or not association of M protein with detergent resistant raft fraction requires the prior presence there of glycoproteins would help answer the question of whether M proteins promote patch formation or the patch formation is a precondition of M protein attachment. Unfortunately, the experimental results appear to be conflicting; in one study, it was found that when expressed alone by transfection, the HN, F or M protein partially became detergent resistant (Sanderson *et al.*, 1995), while in another similar transfection experiment, M protein was found associated with detergent resistant fraction only when coexpressed with the HN and F proteins (Ali and Nayak, 2000). A more straightforward approach is to use the *M* gene deleted rSeV, SeV/Δ*M*. It can be decided whether or not patch formation occurs in SeV/Δ*M* infected cells and if not, whether or not *trans*-supplied M proteins promote patch formation.

The C protein as a potential supporting player

The SeV accessory C proteins are essentially nonstructural proteins because they are detectable in very few copies in the virions. It is thus surprising that the C proteins probably contribute greatly to the structural assembly process. This was first suggested by the experiments using four C(–) virus which expressed none of the four C proteins and when propagated in eggs, replicated to the titres several logs lower than that of the wild type (Table 4.3) (Kurotani *et al.*, 1998). In CV1 cell culture, the peak titre of 4C(–) was about two logs lower than that of the wild type, despite efficient synthesis of mRNAs, proteins and genomic RNA, and the virions produced were highly heterogeneous in size and shape (Hasan *et al.*, 2000). The defective assembly was also observed in Vero cell line that is defective in IFN-α/β production (Hasan *et al.*, 2000), indicating that the virus assembly function of the C protein is separate from its IFN antagonism (see below).

The VLP (virus-like particle) production in cells transfected with the plasmids encoding the N, M, HN and F proteins or with the M protein encoding plasmid alone was enhanced by an additional expression of the C protein (Sugahara *et al.*, 2004; Irie *et al.*, 2008b). The longer C proteins (C′ and C) contain N-terminal membrane-targeting sequences, while the shorter Y1 and Y2 lack these sequences (Marq *et al.*, 2007) and the former two, but not the latter two were capable of facilitating VLP production (Sakaguchi *et al.*, 2005; Irie *et al.*, 2008b).

However, such a supporting role of C protein in SeV budding was not confirmed by others (Gosselin-Grenet *et al.*, 2007). The virus used in this experiment was a 4C knockout virus into which a separate transcription unit encoding the C protein fused with GFP was introduced. Even when C/GFP expression was silenced by siRNA, no apparent inhibition of the production of the rSeV was found. However, siRNA-induced depletion might not be sufficient for inhibiting virus production, as about 30% of C/GFP protein level, compared with that of the control cells, remained expressed.

Are Alix and Vps4 involved in budding?

Evidence is accumulating that suggests that the M protein and Gag protein of enveloped viruses utilize the cellular endosomal sorting complexes required for transport (ESCRTs), which are involved in multi vesicular body formation and cytokinesis, for the final (late) step of virus assembly (release of the budding particles from cellular membrane by the process of fission). Three major viral L (late)-domain motifs (PPxY, PT/SAP and YPxL) have been identified as the responsible sequences for that interaction with the ESCRTs (reviewed in Demirov and Freed, 2004; Bieniasz, 2006; Calistri *et al.*, 2009; Harrison *et al.*, 2010). A potential L-domain 49-YLDL-52 was identified for SeV M protein and its interaction with Alix/AIP1 (at the N-terminal residues 1–211) appeared to be important for SeV M protein VLP production (Irie *et al.*, 2007). The Alix is associated with the two (ESCRT-I and -III) of the three ESCRTs and was demonstrated to be the interacting partner of the lentivirus YPxL (reviewed in Demirov and Freed, 2004). Although SeV YLDL could not be replaced by YPxL, SeV M protein budding was inhibited by the overexpression of some deletion mutants of Alix and depletion of endogenous Alix by siRNA (Irie *et al.*, 2007). The requirement of YLDL motif and its association with Alix was further substantiated by characterization of rSeVs with point mutations in the motif, which exhibited poor replication and the emergence of a revertant, which restored replication capacity (Irie *et al.*, 2010a).

The SeV C protein was also found to physically interact with the C-terminal region (residues 212–357) of Alix (Sakaguchi *et al.*, 2005; Irie *et al.*, 2007). The mutant C proteins that were unable to bind Alix (Sakaguchi *et al.*, 2005) or that failed to bind the plasma membrane (Irie *et al.*, 2008b) could not accelerate the release of VLP from M-, N-, HN- and F-expressing cells or from only M protein-expressing cells. Overexpression of Alix enhanced SeV virion release from infected cells (Sakaguchi *et al.*, 2005). Thus, SeV M and C proteins appeared to synergistically facilitate M protein budding through the interaction with different Alix domains.

In the final step of protein sorting, AAA-type ATPase, Vps4 interacts with ESCRT-III to catalyse disassembly of the ESCRT machinery to recycle its components (Babst *et al.*, 1998). Dominant negative (DN) Vps4 mutants interfered with the budding of many enveloped viruses including PIV5 and mumps virus in the genus *Rubulavirus* (Schmitt *et al.*, 2005; Li *et al.*, 2009; reviewed in Bieniasz, 2006 and Harrison *et al.*, 2010). Sakaguchi *et al.* (2005) reported that transfection of a DN Vps4 yielded significantly lower amounts of infectious SeV in cells transfected with infectious SeV nucleocapsid. DN Vps protein expression was found to affect the release of SeV M protein VLP that occurred in the presence of the C protein expression, but not M-alone VLP production (Irie *et al.*, 2008b).

Overall, the above results suggested that SeV budding and release from cells involve recruitment of Alix and ESCRT machinery by the viral M and C proteins. However, in contrast to those 'positive' results, Gosselin-Grenet *et al.* (2007) presented 'negative' data that in the SeV natural life cycle neither the depletion of Alix expression nor the DN Vps4 expression affected virus production. In this study on DN Vps4 expression, the experimental system appeared to be well validated in a parallel setting of PIV5 whose replication depends upon Vps4 and is inhibited by a DN Vps4 (Schmitt *et al.*, 2005). Thus, currently available concepts are sharply conflicting regarding not only the contribution of the C protein, but also the involvement of ESCRT machinery (reviewed in Harrison *et al.*, 2010).

There are many other important aspects to be addressed regarding SeV assembly, including the involvement of cytoskeletal framework and budding from the apical surface of polarized cells (Takimoto and Portner, 2004; Harrison *et al.*, 2010).

Release from cells

After fission of the bud, the virus is liberated from cells and initiates the next round of the life cycle in a new cell. The NA of the HN glycoprotein removes the terminal sialic acid residues from the HN glycoprotein itself and the F glycoprotein, thereby, preventing particle aggregation and facilitating virus release from cells. The NA has acidic pH optima and perhaps acts primarily in the Golgi apparatus. However, the fact that the erythrocytes once agglutinated by the exogenous SeV readily become liberated in haemagglutination tests at neutral pH on ice suggests that the SeV NA would further act exogenously, for instance, to remove cell surface sialic acids to prevent reattachment of progeny virus.

Persistent infection

A number of paramyxoviruses can establish long-term persistent infections in cultured cells (reviewed in Rima and Martin, 1976; Youngner and Preble, 1980). Paramyxovirus persistent infections are also established *in vivo* and the resulting host responses and pathological outcomes are complex (reviewed in Randall and Russell, 1991). A typical example of human disease caused by a persistently infecting paramyxovirus is subacute sclerosing panencephalitis (SSPE), a disease of the human central nervous system, which is triggered by persistently infecting measles virus many years after the onset of acute measles. Despite its very rare occurrence, SSPE has been intensively studied to reveal the molecular bases underlying the pathogenesis (reviewed in Billeter and Cattaneo, 1991).

Although its clinical impact is not known, SeV has been intensively used to understand the process leading to the establishment of a stable persistent infection at the cellular level. Roux and Holland (1979, 1980) showed that SeV persistent infection could be established by co-infection with a standard, non-defective (ND) virus and DI particles. DI RNAs persisted in the cells and maintained their interfering capacity with the replication of ND virus. On the other hand, temperature-sensitive (ts) ND mutants were frequently isolated from cells persistently infected with SeV (Nagata *et al.*, 1972). The ts mutants grew well at permissive temperature (33°C), but did not grow at non-permissive temperature (38°C), although the majority of cells expressed viral antigens at the non-permissive temperature. Such SeV ts mutants were capable of establishing a persistent infection (Kimura *et al.*, 1975). Thus, two factors, the presence of DI particles and evolution of ts mutants, are thought to be critical for establishing a stable SeV persistent infection.

SeV causes lytic infection for cultured cells. Routinely, a small fraction of cells that survived this lethal infection have to be carefully cultivated and expanded to initiate persistent infection. They are then subjected to numerous rounds of passage until they become stably maintained (steady state) and completely resistant to the original SeV. During serial passages, however, these cells encounter occasional crisis (massive cell death). Yoshida *et al.* (1982) characterized SeV genomes, which persisted in a stable culture established after several hundreds passages and found that both full-length and smaller-sized DI genomes coexisted. In addition, a ts mutant isolated from such stable lines was found to be able to immediately establish a stable persistent infection without detectable DI genomes. Thereafter, no DI genome evolved and the cells no longer encountered sudden crisis.

Because of widespread occurrence of DI particles among RNA viruses, the stock SeV initially used could contain a trace level of DI particles. This might be sufficient for protecting a small cell population from lytic infection and thus initiating a persistent infection. DI particles require standard wild-type virus to reproduce themselves. Thus, the persistent infection initiated by the aid of DI particles could be maintained by a subtle balance of intracellular levels of lytic wild-type and non-lytic DI particles. Once such a balance shifts towards the predominance of wild type, the system probably undergoes crisis. Persistent infection then enters the steady state no longer showing crisis. Just concurrently with, but never antecedently to, this stable phase, ts mutants began to arise. They were always little cytopathic, multiplied without wild-type and importantly, interfered strongly with the replication of wild type (Yoshida *et al.*, 1982; Kiyotani *et al.*, 1990). Once evolved, the ts mutants could replace the wild type and ultimately expel it from the system. Thus a two-step theory has emerged that SeV persistent infection is initiated by DI particles followed by stabilization by endogenously generated non-cytopathic ts mutants with homologous interfering capacity (reviewed in Nagai and Yoshida, 1984).

The nucleotide sequence of a persistent–infection derived ts mutant (named cl. 151) with homologous interfering capacity and non-cytopathic nature was compared with that of the parental SeV strain Nagoya. Numerous non-synonymous and synonymous mutations were identified throughout the cl. 151 genome (Nishimura *et al.*, 2007). The two nucleotide changes, A to G at position 34 and G to A at position 47, were found in the leader sequence. These two changes were introduced into the parental virus. The resulting recombinant virus displayed increased homologous-interfering capacity (note that interference occurs to some extent between standard laboratory strains of SeV) (Shimazu *et al.*, 2008). The L1618V mutation in the L protein of the ts mutant also appeared to be relevant to the establishment of persistent infection but did not confer temperature sensitivity (Nishio *et al.*, 2004). On the other hand, several amino acid changes in the M and HN proteins of cl. 151 (Kondo *et al.*, 1993) were found to contribute to its temperature sensitivity as well as poor cytopathogenicity (Inoue *et al.*, 2003b).

Pathogenesis: general features

Disease course in an experimental infection

An experimental infection with SeV to study the disease course is usually performed with an egg-adapted virus strain and a susceptible mouse strain. A typical disease course was depicted using strain Z and ICR mouse (Kato *et al.*, 1997a) (see also Fig. 4.5). The tissues targeted by SeV were confined to the epithelial surfaces of the upper respiratory tracts, trachea, bronchi and bronchioles. Subepithelial tissues were not involved. Infiltration into the alveoli of leukocytes including neutrophils took place already on day 1, rapidly progressed thereafter, and was seen in the entire lung by day 5 (Fig. 4.5A). Detachment of epithelial cells became distinct on day 3 and was seen even at the trachea on day 5 (Fig. 4.5A). Macroscopically, the lungs became hyperemic and there was lung consolidation that made the lung look like the liver. Consolidation is usually scored from 1 to 4 according to the spaces that it

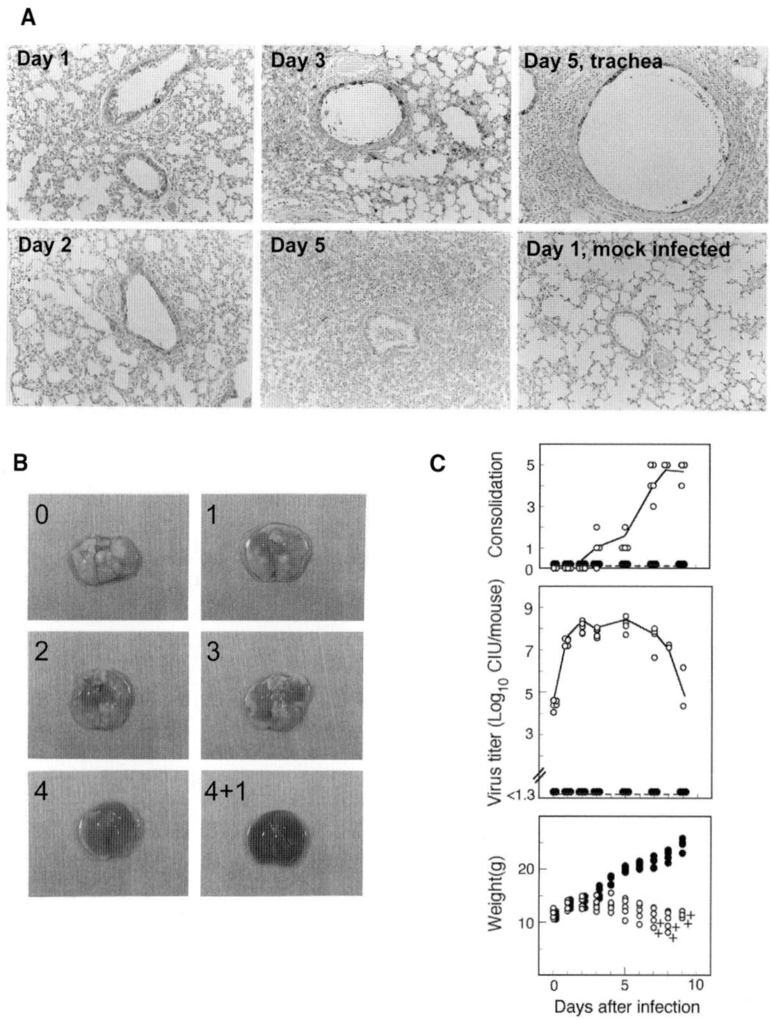

Figure 4.5 Progress of Sendai virus (SeV) infection in mice. 3-week-old ICR mice were intranasally inoculated with 10^7 CIU (cell infectious units) of SeV/mouse. (A) Immunohistochemical staining with mouse anti-SeV antibody and peroxidase conjugated anti-mouse IgG followed by haematoxylin-eosin staining. (B) Macroscopic view of the entire lung with consolidation scores. (C) Consolidation scores, virus growth and body weights of mice during the course of infection. Open circle, mouse infected; filled circle, uninfected; +, the mouse that died. For details see the text. Adapted from Kato *et al.* (1997a). A colour version of this figure is located in the plate section at the back of the book.

occupies over the entire lung surface (Fig. 4.5B). One point is added (4+1) when mice die (score 5), where, however, post-mortem changes are added. Consolidation began to appear on day 3 and advanced thereafter (Fig. 4.5C). Viral load in the lung rapidly increased within the initial day of infection and was maintained at a high level throughout the experiment until the mice died (Fig. 4.5C). Body weight loss or little gain was a good marker for disease progression (Fig. 4.5C). In this experiment, all of the mice died. Resolution, if it occurs, begins on about the tenth day and proceeds rapidly, but complete resolution takes more than a month (Homma and Tashiro, 1994).

Effect of egg passages on pathogenicity

It has to be taken into consideration that numerous rounds of egg-passage result in significant attenuation of SeV for the original host, the mouse.

SeV strain Hamamatsu was isolated in 1976 from an epizootic infection in a mouse colony by inoculating the lung homogenates into the allantois of 10-day-old eggs. During various rounds of egg-passage of this isolate up to the 50th passage, the infectivities released into the allantoic fluids and the viral LD_{50} values were determined (Kiyotani *et al.*, 2001). In its initial propagation in eggs, the titre in the allantoic fluid reached ca 10^9 infectious units (IU)/ml, and after several passages 10^{10} IU/ml followed by comparable titres throughout the experiment until the 50th passage. On the other hand, LD_{50} increased incrementally from the lowest, 7.9×10 IU/mouse to the highest, 3.2×10^4 IU/mouse at the 50th passage, indicating attenuation by about 400-fold through 50 passages. The degrees of attenuation closely paralleled the decreases in the viral loads in the mouse lungs.

The viruses of different egg-passage history (i.e. different virulence) and the parental virus were compared for growth in the primary mouse pulmonary epithelial cells (MPE) and the primary chick embryo cells (CE). The growth capacity in MPE was decreased in parallel with the attenuation, while that in CE was comparable regardless of the different attenuation levels. The analyses of the synthesis of viral macromolecules in MPE and CE cells infected with those viruses with remarkably divergent virulence suggested that poor genome replication in mouse cells could be the primary cause for attenuation through egg-passage (Kiyotani *et al.*, 2001).

The parental and variously attenuated viruses were compared for their entire nucleotide sequences and point mutations were found to accumulate in the leader sequence and the HN and L proteins as attenuation advanced (Fujii *et al.*, 2002a). Furthermore, when the egg-passaged, attenuated virus was back-adapted to the growth in MPE, virulence was restored remarkably and the nucleotide sequencing of this revertant indicated that some of those mutations found in the leader, *HN* and *L* genes of the attenuated virus reverted to those found in the original virus. The two mutations (U to A at positions 20 and 24) identified in the leader sequence of the attenuated virus were found to indeed contribute, at least in part, to the attenuation, because introduction of these two mutations into the original virus resulted in remarkable attenuation (Fujii *et al.*, 2002b).

Protease dependent virus tropism

There are at least two categories of paramyxovirus F_0 activation by host proteases and they decide whether the infection is pantropic or limited to special tissues (reviewed in Nagai, 1993, 1995). This paradigm of protease dependent virus tropism emerged from comparisons between NDV virulent strains, which are pantropic and cause a systemic lethal infection for chicken and chick embryo, and avirulent strains, which target limited types of tissue and cause a non- or slightly lethal localized infection (Nagai *et al.*, 1976a). The receptor of NDV is the terminal sialic acid residues of gangliosides and glycoproteins that are ubiquitous throughout the susceptible hosts. The F_0 of virulent strains has multiple basic residues at the cleavage site and is activated by a ubiquitous intracellular protease such as *trans*-Golgi-localized furin, while that of avirulent strains has a single arginine cleavage motif and is resistant to the intracellular proteases and activated by extracellular, highly tissue specific protease such as the FXa in eggs (Toyoda *et al.*, 1987, 1989; Glickman *et al.*, 1988; Gotoh *et al.*, 1990, 1992; Ohnishi *et al.*, 1995; reviewed in Nagai 1993, 1995). The critical requirement of multibasic cleavage motif for high cleavability of the F_0 protein and high virulence of NDV was confirmed by reverse genetics to swap the monobasic cleavage site of an avirulent strain with a multibasic cleavage site characteristic of virulent strains (Peeters *et al.*, 1999). Thus, NDV paradigm of viral pathogenicity has been well established. The paradigm has been extended to other paramyxoviruses and other virus families (reviewed in Nagai, 1993, 1995). Most remarkably, essentially the same relationship was found between the cleavability of the haemagglutinin and pathogenicity of avian influenza viruses (reviewed in Nagai, 1993, 1995; Klenk and Garten, 1994). More recently, the third category of activation was noted for the *Henipavirus* (Hendra and Nipah viruses). They cause a systemic infection, but have only a single lysine residue at the F_0 cleavage site. Interestingly, this lysine residue is not essential for cleavage. It was found that inhibition of cathepsin L blocks cleavage (Pager *et al.*, 2005), suggesting the involvement of the endocytosis pathway in *Henipavirus* cleavage-activation (Diederich *et al.*, 2005; Meulendyke *et al.*, 2005).

SeV receptor is also widespread terminal sialic acids of glycolipids and glycoproteins as noted above, but the virus displays limited tissue tropism. In embryonated chicken eggs, SeV activating FXa appeared to be synthesized by the allantoic and amniotic cells and released into the allantoic and amniotic fluids, respectively. The tissues targeted by SeV in egg were limited to those in direct contact with these FXa containing fluids (Ogasawara *et al.*, 1992), embodying for the first time the concept of a protease dependent tropism using that simple organism, the chick embryo. Tryptase Clara was found to be the responsible protease for SeV activation and spreading in the respiratory tracts of rat (Kido *et al.*, 1992; Tashiro *et al.*, 1992). SeV does not appear to spread to other organs possibly because they express no such activating enzyme. Infection with protease-activation mutants whose F_0 is cleavable by chymotrypsin, but no longer by trypsin could not spread in the lung nor cause pneumonia in mice, confirming the requirement of a trypsin-like protease for SeV tropism and pathogenicity (Itoh *et al.*, 1987).

On the other hand, a laboratory-selected SeV mutant named F1-R displayed multistep replication in tissue culture cells without trypsin and pantropic phenotype *in vivo* probably because its F_0 was rendered cleavable by a ubiquitous protease. The increased cleavability appeared to be due to a single point mutation, Ser115 to Pro, adjacent to the Arg116 cleavage site (Okada *et al.*, 1998). In addition, F1-R buds bipolarly from both apical and basolateral surfaces of epithelial cells. This phenotype could also contribute to its pantropic nature and was thought to be attributable to Asp128 to Gly and Ile 210 to Thr changes in the M protein (Okada *et al.*, 1998). However, the postulated phenotype–genotype relationships for the F1-R F and M proteins could not be confirmed by rSeVs with the respective mutations (Hou *et al.*, 2005).

Pathogenesis: Immune evasion by the accessory proteins

Paramyxovirus technology innovation, especially the generation of V- and C-knockout viruses by reverse genetics and the concomitant big-bang of our understanding on innate immunity of the host have opened a new research field, the viral evasion of various facets of innate immunity by the accessory proteins.

Discovery of paramyxovirus interferon antagonism

A new field was opened up in 1999 by the earliest discovery of the ability of SeV and PIV5 to antagonize interferon (IFN) actions (Didcock *et al.*, 1999a,b; Garcin *et al.*, 1999; Gotoh *et al.*, 1999; Yokoo *et al.*, 1999). Infection with these viruses renders cells unresponsive to IFNs (IFN-α/β and IFN-γ), allowing highly IFN sensitive viruses such as VSV to replicate in the presence of IFNs. Analysis of this property eventually uncovered that the accessory C protein of SeV and V protein of PIV5 block the IFN signalling pathway required for induction of an antiviral state.

Host cells sense virus invasion by recognizing microbe-specific pathogen-associated molecular patterns (PAMPs), such as double-stranded (ds) RNA generated in the course of an infection, and this recognition activates the signalling pathway leading to IFN-β production. Autocrine or paracrine IFNs (IFN-α/β and IFN-γ) bind to the cell surface receptors and its signal is transmitted to the nucleus through IFN signalling (JAK/STAT) pathway, resulting in activation of IFN-stimulated genes (ISGs). The products of ISGs include antiviral proteins, such as dsRNA dependent protein kinase (PKR), which participate in establishment of the antiviral state. Since signal transduction in both the IFN-β production pathway and the IFN signalling pathway is crucial for accomplishing IFN actions, they are targeted by various viruses (reviewed in Randall and Goodbourn, 2008). For SeV, it has been found primarily using a series of C and V knockout viruses that the former pathway is targeted by both of the accessory proteins, C and V and the latter only by the C protein. A recent study has further uncovered another function of V protein essential for SeV pathogenicity in mice, apart from the IFN antagonism (Kiyotani *et al.*, 2007; see below). Fig. 4.6 summarizes the plausible sites of intervention by the SeV C and V proteins in the RLR-dependent signalling pathway to

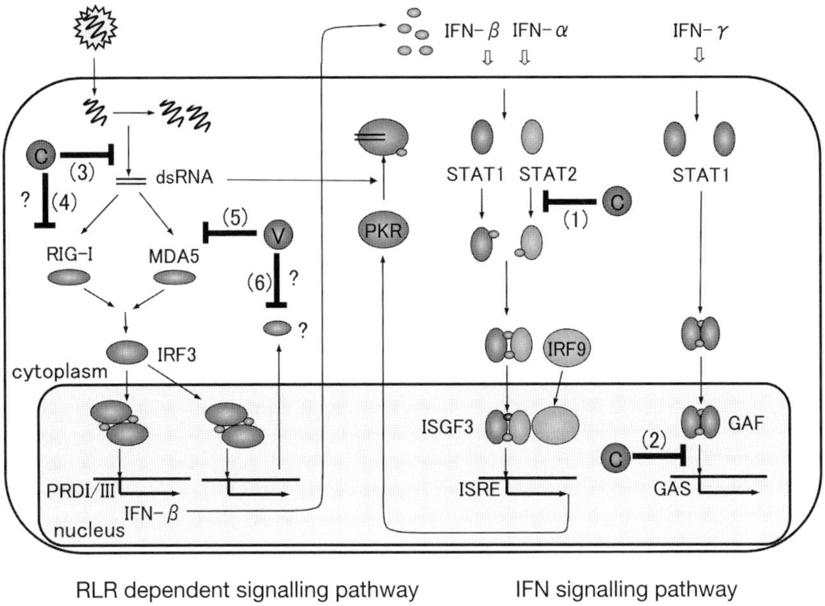

Figure 4.6 The actions of Sendai virus (SeV) accessory C and V proteins in the interferon (IFN) signalling pathway and RLR-dependent signalling pathway. The steps of intervention by the C and V proteins are shown together with the related signalling molecules on the signalling pathways. (1) inhibition of tyrosine phosphorylation of STAT2, (2) prevention of interaction between GAF and GAS, (3) suppression of generation of dsRNA during viral genome replication, (4) blockade of RIG-I dependent signalling pathway, (5) inhibition of interaction of MDA5 with dsRNA and (6) inhibition of function of unknown host factor induced by IRF3. For details see the text.

produce IFN and the IFN signalling pathway to induce an antiviral state.

Mechanisms for blocking IFN signalling pathway

SeV proteins responsible for preventing IFN signalling

In SeV infected cells, neither IFN-α/β nor IFN-γ can activate transcription of various ISGs and a reporter gene under the control of the IFN responsive promoter (Didcock *et al.*, 1999a; Garcin *et al.*, 1999; reviewed in Gotoh *et al.*, 2001, 2002), indicating SeV's ability to prevent IFN signalling. The V- and C-knockout rSeVs exhibited highly attenuated phenotypes in mice (Kato *et al.*, 1997ab; Delenda *et al.*, 1997, 1998; Kurotani *et al.*, 1998; Latorre *et al.*, 1998a). This raised the possibility that these accessory proteins would be responsible for evading host immunity. Using V-knockout and C-knockout rSeVs, there were studies (Garcin *et al.*, 1999; Gotoh *et al.*, 1999), which sought to identify the viral factors responsible for the blockade of IFN signalling. Of the three different rSeVs, V(–), C/C′(–) and 4C(–), only 4C(–) completely lost the ability to prevent IFN signalling and permitted the induction of the full antiviral state (Gotoh *et al.*, 1999). Plasmid-based expression of the individual proteins indicated that not only the longer C and shorter Y1 and Y2, but also the C-terminal half of the C protein (residues 99–204) was able to inhibit IFN-mediated induction of ISGs and subsequent establishment of the antiviral state, whereas the expression of V protein did not at all (Kato *et al.*, 2001, 2002, 2004; Gotoh *et al.*, 2003ab). It was concluded that the C-terminal half of the SeV C protein encodes the IFN antagonism. The same was true for inhibition of viral RNA synthesis (Kato *et al.*, 2001).

An SeV field isolate, Ohita strain, exhibits higher virulence in mice compared with egg-adapted laboratory strains. Itoh *et al.* (1997) isolated an avirulent clone (MVC11) by adaptation of this strain for growth in LLCMK2 cells.

MVC11 contained only two amino acid substitutions relative to the parental strain. They are F170S in the C protein, and E2050A in the L protein. Introduction of F170S mutation into the parental SeV genome significantly reduced virus virulence for mice, demonstrating that the C protein is a determinant of virulence (Garcin *et al.*, 1997). It was, however, difficult to determine whether the IFN antagonistic ability of C protein functions as a determinant of virulence, since C protein is a multifunctional protein, which not only antagonizes IFN actions, but also regulates viral RNA synthesis (Curran *et al.*, 1991, 1992; Cadd *et al.*, 1996; Tapparel *et al.*, 1997), inhibits virus-induced apoptosis (Koyama *et al.*, 2003), and facilitates virus budding (Kurotani *et al.*, 1998; Hasan *et al.*, 2000; Sakaguchi *et al.*, 2005; Irie *et al.*, 2007, 2008b). Indeed, F170S mutation resulted in a dual abrogation of the functions to prevent IFN signalling and to regulate viral RNA synthesis (Tapparel *et al.*, 1997; Garcin *et al.*, 1999).

By introducing three amino acid substitutions into the C protein, an rSeV(Cm*) was generated. Cm* was found to have lost the ability to prevent IFN-α/β signalling, but still retain the other functions including RNA synthesis inhibition and facilitation of virus budding (Kato *et al.*, 2007). Compared with the wild-type parental SeV, Cm* replicated poorly in IFN-responsive, 2fTGH parental human fibrosarcoma cells, whereas it replicated well in IFN-non-responsive, STAT1 deficient U3A cells. When intranasally inoculated, Cm* could propagate in the lungs of $STAT1^{-/-}$ mice and cause pneumonia, but was cleared from those of $STAT1^{+/+}$ mice without any appreciable propagation and pathogenicity. Thus, the critical requirement of virus-coded IFN antagonism for *in vivo* replication and pathogenesis was proved for the first time with the SeV C protein.

Inhibition of IFN-α/β-stimulated tyrosine phosphorylation of STAT2 by the SeV C protein

Binding of autocrine or paracrine IFN-α/β to the type I IFN receptor on cell surface triggers the assembly of receptor subunits, IFNAR1 and IFNAR2, causing the cross-activation of receptor-associated kinases, TYK2 and JAK1. These activated kinases sequentially phosphorylate STAT2 and then STAT1 on their specific tyrosine residues. The tyrosine-phosphorylated (pY) STAT2 and pY-STAT1 form a heterodimer, translocate into the nucleus and combine with IFN regulatory factor (IRF) 9 to activate ISGs by binding to IFN stimulated response elements (ISREs) (reviewed in Stark *et al.*, 1998) (Fig. 4.6).

The *Rubulavirus* V proteins target STAT1 or STAT2 for proteasome mediated degradation and thereby block IFN signalling (Randall and Goodbourn, 2008). In contrast, no degradation of signalling molecules including STAT1 and STAT2 was observed for not only SeV infected HeLa cells, but also the established HeLa cell lines constitutively expressing the C, Y1, or Y2 (Komatsu *et al.*, 2000; Young *et al.*, 2000; Kato *et al.*, 2001). Instead, phosphorylation of STAT1 and STAT2 was markedly affected (Komatsu *et al.*, 2000, 2002; Gotoh *et al.*, 2003b). When cells were continuously stimulated with IFN-α in the presence of SeV C protein, tyrosine phosphorylation of STAT1 was initiated first with slower kinetics, but eventually reached a high level (Komatsu *et al.*, 2000, 2002), while tyrosine phosphorylation of STAT2 remained almost completely suppressed (Gotoh *et al.*, 2003b). Analysis of established HeLa cell lines, each of which constitutively expresses a truncated C or a mutated C with substitutions of one to three amino acids, revealed a strict correlation between the ability to block IFN-α/β signalling and the ability to inhibit STAT2 phosphorylation (Gotoh *et al.*, 2003b; Kato *et al.*, 2004), demonstrating that inhibition of tyrosine phosphorylation of STAT2 was a crucial step for SeV-mediated blockade of IFN-α/β signalling (Fig. 4.6).

Although the C proteins (C, Y1, and Y) bind to STAT1, their direct interactions with STAT2 were undetectable (Takeuchi *et al.*, 2001; Garcin *et al.*, 2002; Gotoh *et al.*, 2003b). The domains involved in the C–STAT1 interaction were mapped to the C-terminal half (residues 85–204) of the 204-residue-long C (Gotoh *et al.*, 2003b), and the N-terminal region (residues 1–136) of the 750-residue-long STAT1 (Gotoh *et al.*, 2003a). A study using several C deletion mutants and a C mutant with substitution of F170S (C^{F170S}) showed a correlation between the ability to block

IFN-α/β signalling and the STAT1 binding capacity. Since STAT1 and STAT2 are pre-associated with the cytoplasmic tail of IFNAR2 prior to IFN-α stimulation, C protein was assumed to affect STAT2 phosphorylation events by employing STAT1 molecules as a scaffold. However, the subsequent study revealed that several C mutants prevented IFN-α/β signalling, although exhibited decreased STAT1-binding capacity, raising the possibility that C protein targets an unknown molecule other than STAT1 for blockade of IFN-α/β signalling (Kato *et al.*, 2004).

Destabilization of STAT1 by the C protein in certain cell types

In contrast to the above notions, a series of studies by Garcin and coworkers proposed that the ability to prevent establishment of the antiviral state was the characteristics of only longer C proteins (C and C′), which could induce STAT1 instability in certain kinds of cells such as mouse embryo fibroblasts (MEFs) and human fibrosarcoma 2fTGH cells. This view emerged from the early finding that rSeV C/C′(–), which does not express longer C proteins but expresses shorter Y1 and Y2 proteins, and rSeV-CΔ10–15, which carries a deletion of residues 10 to 15 of the C protein, were unable to prevent the establishment of antiviral state induced by IFN-α/β (Garcin *et al.*, 1999, 2000, 2001). They also found that the shorter C proteins (Y1 and Y2) as well as the longer C proteins (C′ and C) prevented IFN-α/β signalling and therefore, put forward a hypothesis that IFN-α induces the antiviral state by an unknown parallel pathway, which may be targeted by only longer C proteins, independent of the well-established IFN-α/β signalling (JAK-STAT) pathway (Garcin *et al.*, 2001). The longer C proteins induced STAT1 instability accompanied by an increase in the level of mono-ubiquitinated STAT1. Treatment with a proteasome inhibitor MG132 prevented longer C proteins from dismantling the antiviral state (Garcin *et al.*, 2002). From these results obtained mainly using MEFs, it was concluded that longer C proteins act to dismantle the IFN-induced antiviral state by inducing STAT1 instability. The C-mediated STAT1 instability was also seen in 2fTGH cells and this activity was mapped to the first 23 amino acids of C (Garcin *et al.*, 2004). The C^{1-23} peptide fused to GFP (C^{1-23}/GFP) could induce not only STAT1 instability but also degradation of C^{1-23}/GFP protein itself. It appeared probable that the C^{1-23} peptide acted as an independent viral element that induces STAT1 instability.

At the same time, however, the results reported by the same group appear to be conflicting; it was shown, on one hand, that C^{F170S} could induce STAT1 instability as efficiently as the wild-type C, but failed to prevent IFN signalling in 2fTGH cells (Garcin *et al.*, 2003, 2004). On the other, the same C^{F170S} was reported to be a representative null mutant that has lost both the ability to induce STAT1 instability and the ability to prevent IFN signalling in MEFs and other cell lines (Garcin *et al.*, 2000, 2001). C-induced STAT1 instability has not been confirmed, either, in 2fTGH cells infected with wild-type SeV or several rSeVs expressing mutated C proteins (Kato *et al.*, 2007) nor in cells infected with other cognate respiroviruses, HPIV1 (Bousse *et al.*, 2006; Chambers and Takimoto, 2009), HPIV3 (Malur *et al.*, 2005) or BPIV3 (unpublished). To settle those conflicts it is hoped that evidence will be obtained that will support the presence of a parallel pathway independent of the JAK-STAT pathway in MEFs and 2fTGH cells.

Expression of the C, but not the shorter Y1 and Y2 proteins induced tyrosine-phosphorylation of STAT1 without IFN treatment (IFN-independent phosphorylation), and further IFN treatment greatly elevated the pY-STAT1 level as described above (Komatsu *et al.*, 2002; Saito *et al.*, 2002; Garcin *et al.*, 2003; Kato *et al.*, 2004). The SeV C protein is localized at the plasma membrane through its membrane targeting N-terminal 15 residues (aa 1–15) and this was shown to be essential for IFN-independent tyrosine phosphorylation of STAT1 (Marq *et al.*, 2007). However, the significance of this IFN-independent tyrosine phosphorylation of STAT1 remains to be elucidated.

C protein-mediated blockade of IFN-γ signalling pathway

The SeV C protein prevents not only IFN-α/β signalling, but also IFN-γ signalling (Garcin *et al.*, 1999; Gotoh *et al.*, 1999; Young *et al.*, 2000; Kato *et al.*, 2001; Takeuchi *et al.*, 2001). Since STAT2 does not participate in IFN-γ signalling, the mechanism by which C protein prevents IFN-γ signalling must be different from that for IFN-α/β signalling. IFN-γ signalling initiates with binding of IFN-γ to the specific receptor, which causes activation of the receptor-associated kinases, JAK1 and JAK2 (reviewed in Stark *et al.*, 1998). The activated JAKs phosphorylate STAT1 on a specific tyrosine residue (Schindler *et al.*, 1992; Shuai *et al.*, 1992, 1993), resulting in the formation of a homodimer of pY-STAT1, termed γ-activated factor (GAF). The GAF then translocates into the nucleus, and binds to γ-activated sequence (GAS) elements in the promoter of ISGs (Fig. 4.6). In addition to tyrosine phosphorylation, phosphorylation of STAT1 on a specific serine residue is essential for the maximal activation of transcription by GAF (Wen *et al.*, 1995).

IFN-γ mediated tyrosine and serine phosphorylation of STAT1 and translocation of pY-STAT1 into the nucleus were not prevented by the expression of C protein (Takeuchi *et al.*, 2001; Gotoh *et al.*, 2003a). It was found that STAT1 and pY-STAT1 in the extracts from cells expressing C protein was present in a form of high molecular weight complexes (HMWCs), which are too large to be analysed by usual electrophoretic mobility shift assay (EMSA) (Takeuchi *et al.*, 2001). This HMWC formation induced by the C protein made it difficult to assess whether or not the pY-STAT1 normally forms a functional homodimer that could bind a GAS element. An N-terminal deletion mutant C (residues 85–204) retained both abilities to bind to STAT1 and to prevent IFN-γ signalling, but lost the ability to mobilize STAT1 or pY-STAT1 into HMWCs. EMSA with this deletion mutant revealed that the mutant prevented pY-STAT1 from binding to a GAS element (Fig. 4.6) (Gotoh *et al.*, 2003a).

Subsequent studies, however, led to an unexpected finding; C^{F170S} retained the ability to inhibit the IFN-γ stimulated response without binding to STAT1, although it is inferior to the wild-type C in the inhibitory activity (Gotoh *et al.*, 2004). Like the wild-type C, expression of C^{F170S} inhibited neither IFN-γ mediated phosphorylation of STAT1 nor its translocation into the nucleus. Furthermore, the GAF–GAS complex was detected by EMSA. These results raise the possibility that some other IFN-mediated modification of STAT1 or a factor cooperating with STAT1 is the second target of the SeV C protein.

Suppression of IFN-β production by the C protein

SeV is one of the most potent inducers of IFN-β. However, this ability of SeV is not attributed solely to infectious SeV, but rather largely to the coexisting DI genomes, especially copy-back DI genomes, in SeV stocks (Johnston, 1981; Poole *et al.*, 2002; Strahle *et al.*, 2006; Yount *et al.*, 2008). Although copy-back DI genomes, if present as naked RNA, probably have a strong potential to form dsRNA that triggers induction of IFN-β production, immunofluorescent experiments with a specific monoclonal antibody recognizing dsRNA of more than 40 bp in length gave no visible signal in cells infected with not only the non-defective SeV (Weber *et al.*, 2006), but also a mixture of SeV and DIH4, one of the well-characterized copy-back DI genomes (Takeuchi *et al.*, 2008). DI genomes appear to be sensed by at least MDA5 (melanoma differentiation-associated gene 5), since MDA5-deficient dendritic cells respond inefficiently to SeV virus stocks containing DI particles (Yount *et al.*, 2008). However, it remains to be elucidated how copy-back DI genomes induce IFN-β production.

When IFN-inducing activities of various rSeV stocks containing little DI genomes were compared, 4C(–) and V(–) exhibited strong IFN inducing activity compared with the parental SeV that induces little IFN-β. This indicates important roles of C and V proteins in regulating IFN-β production (Komatsu *et al.*, 2004; Strahle *et al.*, 2006). Since C/C′(–) induces IFN-β as poorly as the parental SeV does, the shorter C proteins (Y1 and Y2) seem to be able to substitute for the longer C proteins in this regulation (Komatsu *et al.*, 2004). The IFN-β promoter consists of four overlapping positive regulatory domains (PRDs), designated PRDI through PRDIV (reviewed in Goodbourn *et al.*, 2000). PRDI and PRDIII are bound and activated by IRF3 and IRF7 (Lin *et al.*, 1998; Sato *et al.*, 1998ab; Schafer *et al.*, 1998; Wathelet *et al.*, 1998; Weaver *et al.*, 1998; Yoneyama *et al.*, 1998). PRDII is activated by NF-κB (Lenardo *et al.*, 1989; Visvanathan and Goodbourn, 1989), while PRDIV by ATF-2/c-Jun (Du and Maniatis, 1992; Du *et al.*, 1993). The 4C(–) infection activated a luciferase reporter gene under the control of the PRDI/III, slightly activated the PRDII driven reporter and did not activate the PRDIV driven reporter at all (Komatsu *et al.*, 2004). A dominant negative mutant of IRF3 (IRF3ΔN), which lacks most of its DNA binding domain, significantly suppressed 4C(–)-induced activation of the IFN-β promoter. These results indicate that the C protein mainly targets the IRF3-dependent signalling pathway and/or the production of PAMPs required for activation of its pathway (Komatsu *et al.*, 2004). Indeed, the latter potential to decrease PAMPs has been suggested by the discovery of a novel function of C protein that limits generation of dsRNA (Takeuchi *et al.*, 2008) (Fig. 4.6).

PAMPs are sensed by pattern-recognition receptors (PRRs), which include toll-like receptors (TLRs) and retinoic acid-inducible gene I (RIG-I)-like receptors (RLRs). This recognition initiates signal transduction leading to IRF3 activation. Major viral PAMPs are viral nucleic acids. TLR3 recognizes viral nucleic acids localized to the endosomes, whereas RLRs recognize those present in the cytoplasm. The RLR family includes MDA5 and RIG-I, which are cytoplasmic DExH/D-box helicases with tandem caspase recruitment domain (CARD)-like regions at the N terminus (reviewed in Fujita, 2009; Rehwinkel, 2010; Rehwinkel and Reis e Sousa, 2010; Rehwinkel *et al.*, 2010; Yoneyama and Fujita, 2010; Fig. 4.6). Both MDA5 and RIG-I can be activated by the synthetic dsRNA poly(I:C), but the length of dsRNA affects their recognition. Longer dsRNAs are recognized by MDA5, while shorter ones by RIG-I. Recent studies uncovered that MDA5 is triggered by only a high-molecular-weight RNA composed of both single- and double-stranded portions rather than perfect dsRNA (Pichlmair *et al.*, 2009), whereas RIG-I recognizes 5′-triphosphated RNA with the base-paired structure represented by the panhandle region of NNV RNA genomes (Schlee *et al.*, 2009; Schmidt *et al.*, 2009). Upon detection of viral PAMPs, the CARD-like structure of the RIG-I or MDA5 interacts with the CARD-like domain of IPS-1/Cardiff/MAVS/VISA, which is present in the mitochondrial membrane. This CARD–CARD interaction is thought to lead to the recruitment and activation of TBK1, IKKi and other IKK kinases that activate NF-κB and IRF3; thereby, activating the IFN-β promoter. On the other hand, TLR3-mediated recognition of viral nucleic acids induces activation of TBK1 and IKKi through the adaptor molecule, Toll-interleukin-1 receptor domain-containing adaptor-inducing beta interferon (TRIF).

Independent expression of the SeV C protein suppressed IFN-β production induced by treatment with poly(I:C) or by overexpression of any signalling molecule upstream of IRF3 (MDA5, RIG-I, TBK1, IKKi or TRIF). However, assays with luciferase reporter under the control of the IFN-β promoter did not provide evidence for specific ability of the C protein to prevent the overexpressed signalling molecules from inducing activation of the IFN-β promoter in 293T cells (Komatsu *et al.*, 2007), suggesting no specific ability of the C protein to block the RLR dependent signalling pathway at least in this cell line. Instead, the C protein was found to exert moderate inhibitory effect on the gene expression independently of promoter species. This promoter-independent inhibition of transcription may in part affect IFN-β production. Measles virus C protein also lacks the specific ability to block the RLR dependent signalling pathway and acts by blocking the production of PAMPs (Nakatsu *et al.*, 2008). Thus SeV C protein seems to be similar to measles virus C protein in the mechanism that suppresses IFN-β production.

There are several conflicting data whether the SeV C protein has the ability to prevent the RLR dependent signalling pathway (Fig. 4.6). Phosphorylation and dimer formation of IRF3 in response to poly(I:C) treatment or NDV infection was suppressed in a HeLa cell clone that expressed Y1 (Komatsu *et al.*, 2004). In addition, it was reported that independent expression of C protein prevented 5′-triphosphate RNA or overexpressed RIG-I from activating the IFN-β promoter in MEFs (Strahle *et al.*, 2007).

Keeping PKR inactive, an additional role of the C protein

In an attempt to isolate a mouse mutant cell line resistant to NDV infection Aoki and Kawakita (1996) established the cell line Had-2, which constitutively releases IFN-α and IFN-β. Had-2 was resistant to NDV as expected, whereas SeV was found to replicate well in Had-2. This suggested that SeV has a distinctive ability to antagonize antiviral proteins, such as dsRNA dependent PKR. PKR is activated upon binding dsRNA, which is generated during virus replication and transcription, by autophosphorylation on specific threonine residues (Garcia *et al.*, 2006). Activated PKR phosphorylates the alpha subunit of eukaryotic initiation factor 2 (eIF2α) on a specific serine residue, resulting in inhibition of translation of mRNAs. PKR expression levels are generally low in many cell types.

The intracellular protein synthesis level was decreased by 4C(–) infection, whereas it was maintained in the parental SeV infection of U118 human glioblastoma cells (which have a global defect in type I IFN-related genes), irrespective of whether cells were pre-treated with IFN-α or not (Takeuchi *et al.*, 2008). Phosphorylation of PKR and its substrate, eIF2α, was not observed following SeV infection, whereas 4C(–) infection induced a significant level of phosphorylation of both PKR and eIF2α. Nevertheless, independent expression of C did not inhibit poly(I:C)-stimulated or NDV-induced phosphorylation of PKR and eIF2α. Immunofluorescent experiments with a specific monoclonal antibody against dsRNA revealed significant dsRNA signal in 4C(–) infected cells, but not in the parental SeV infected cells (Takeuchi *et al.*, 2008). Constitutive expression of C in U118 cells abolished this phenotype characteristic of 4C(–). Thus, the C protein limits generation of dsRNA in SeV infected cells; thereby, keeping PKR inactive and maintaining intracellular protein synthesis (Takeuchi *et al.*, 2008) (Fig. 4.6). A correlation between the dsRNA generation and the IFN-β induction has been shown by a recent study using mutant rSeVs, in which one to three amino acid substitutions are introduced into the C protein (Irie *et al.*, 2010b). It is probable that this ability to limit dsRNA generation contributes to a decrease in the total amount of viral PAMPs recognized by RIG-I and MDA5 as described above.

It remains to be elucidated how the C protein limits generation of dsRNA in SeV infected cells, where viral genome/antigenome synthesis and encapsidation are tightly coupled (Gubbay *et al.*, 2001). The leader, trailer RNAs, and viral mRNAs can be naked RNA of viral origin, but appear to have little potential to generate dsRNA. This may explain why little dsRNA signal is seen in SeV infected cells. Assuming that tight coupling of genome/antigenome RNA synthesis and encapsidation would be loosened, the potential to generate dsRNA would increase. This speculation leads to one working hypothesis that C protein plays a key role in the precise coupling of genome/antigenome synthesis and encapsidation. This hypothesis may be applicable to measles virus C protein since a C-knockout measles virus exhibits similar phenotypes in IFN-β production, phosphorylation of PKR and eIF2α (Nakatsu *et al.*, 2006, 2008). Indeed, it has been recently reported that dsRNA is detected by the dsRNA specific monoclonal antibody in cells infected with C-knockout, but not wild-type measles virus (Ikegame *et al.*, 2010).

Anti-apoptotic function of the C protein

SeV infection triggers cellular apoptosis (Tropea *et al.*, 1995; Garcin *et al.*, 1998; Bitzer *et al.*, 1999). It takes 2 to 5 days for completion depending on the virus strains and cell lines. The cellular apoptosis is thought to be an important host defence mechanism, leading to the reduction of the total virus burden of the host organism. It was found that apoptosis induced by SeV required activation of IRF3 (Heylbroeck *et al.*, 2000). In the absence of IRF3 activation, the cells become persistently infected and continuously produce infectious virions (Peters *et al.*, 2008). The IRF3 dependent apoptosis is also caused by transfected dsRNA and requires TRAF2 and TRAF6 in addition to the members of RIG-I pathway (RIG-I, IPS1, TRAF3 and TBK1) leading to IRF3 activation (Chattopadhyay *et al.*, 2010). The pro-apoptotic action of IRF3 is independent of its transcriptional activity and is triggered by its direct interaction with the pro-apoptotic protein Bax, their co-translocation to the mitochondria and the resulting activation of the mitochondrial apoptotic pathway. The IRF3-dependent apoptosis includes activation of caspase 8 and caspase 9, both of which converge at activation of caspase 3, the executioner caspase (Bitzer *et al.*, 1999, 2002; Heylbroeck *et al.*, 2000; Peters *et al.*, 2008; Chattopadhyay *et al.*, 2010). However, conflicting data exist as to which caspase plays an essential role in the IRF3-dependent apoptosis.

In contrast to VSV-triggered apoptosis that is almost completed within one day, SeV-induced apoptosis is relatively slow, raising the possibility that SeV encodes an anti-apoptotic factor (Koyama *et al.*, 2001). Indeed, with all the four Cs knocked out, 4C(–) virus was found to cause much faster and more massive apoptosis than the parental SeV, indicating that C protein acts as an anti-apoptotic factor during the course of SeV infection (Koyama *et al.*, 2003). However, C protein had no specific ability to prevent the apoptotic signalling pathway common to various apoptotic stimuli, since pre-infection with SeV could not inhibit apoptosis triggered by VSV infection or sorbitol treatment (high-osmolality stress) (Koyama *et al.*, 2001). 4C(–) induces phosphorylation and dimer formation of IRF3 earlier in infection than does the parental SeV (Komatsu *et al.*, 2004). It is probable that the early activation of IRF3 causes faster apoptosis seen in 4C(–) infected cells. Accordingly, the mechanism underlying the anti-apoptotic function of the C protein may be linked to that for suppression of IFN-β production.

SeV-triggered apoptosis may also be controlled by trailer RNAs, which are short transcripts generated during abortive genome synthesis (Garcin *et al.*, 1998; Iseni *et al.*, 2002; Wiegand *et al.*, 2005). A recombinant SeV (rSeVGP42) was created, in which the first 42 nt of leader sequence of the genome was replaced with the sequence corresponding to the trailer, so as to express trailer-like RNAs from the leader region. In contrast to the parental SeV (Z strain), rSeVGP42 did not induce apoptosis and further prevented SeV from inducing apoptosis in a mixed infection. It was found that the trailer RNA binds to a cellular T-cell-activated intracellular antigen-related (TIAR) protein via the AU rich sequence 5′UUUUAAAUUUU3′ (Iseni *et al.*, 2002). Introduction of this AU-rich sequence alone into the leader RNA could confer a non-cytopathic phenotype on SeV. Overexpression of TIAR and SeV infection act synergistically in inducing cellular apoptosis. From these results, Iseni *et al.* (2002) proposed that the trailer RNA modulated SeV-induced apoptosis by sequestering TIAR. A similar GP42-like recombinant SeV, whose genome backbone is F strain that induces more rapid apoptosis than SeV Z strain, also exhibited reduced apoptosis compared with the parental F strain, but its degree was cell-type dependent (Wiegand *et al.*, 2005). It is, therefore, probable that the trailer RNA, when expressed from the leader region, exerts an anti-apoptotic effect independently of SeV strains. However, SeV-AGP1–55, in which the entire trailer region was replaced with the leader sequence, grew similarly to rSeVGP42 and was not more apoptotic than the parental SeV (Iseni *et al.*, 2002), still leaving open the question concerning the actual contribution of the trailer RNA to apoptosis regulation in the natural life cycle of SeV.

Unique but yet enigmatic role of the SeV V protein in pathogenesis

Growth of SeV V(–) and VΔC in the mouse lung

Both V(–) and VΔC mutants recovered by reverse genetics exhibited no particular phenotype in cells in culture except for the increased cytopathogenicity characteristic of V(–) but not VΔC (Kato *et al.*, 1997a,b). So far no apoptosis markers were associated with this increased cytopathogenicity of V(–). When inoculated into mice under the same conditions as in Fig. 4.5, the virus growth and disease course differed dramatically from those seen with the parental virus that are shown in Fig. 4.5. Both of the mutants replicated as efficiently as the parental SeV up to 1 day after infection, but were then rapidly cleared, produced only moderate consolidation scores and killed none of the mice, in contrast to the parental virus, which, as shown in Fig. 4.5, maintained a high viral load and high consolidation scores up to 7 to 9 days and killed all the mice (Kato *et al.*, 1997ab). Similar attenuation was observed for several mutants with the zinc-finger-like motif in the Vu region disrupted and no capacity to bind zinc (Huang *et al.*, 2000; Fukuhara *et al.*, 2002).

V(–) and VΔC clearance through IRF3 activation but not via known pathways including IFNs

Unlike the V proteins of other genera, SeV V protein lacks the ability to prevent IFN signalling (Kato *et al.*, 2004). SeV is essentially a poor inducer of IFN-β. The IFN-β production was elevated by several-fold by the V(–) mutant *in vitro* and *in vivo*, compared with the parental SeV (Komatsu *et al.*, 2004; Kiyotani *et al.*, 2007). The V protein was found to mainly target the IRF3-dependent IFN production pathway (Komatsu *et al.*, 2004). The V proteins from more than 10 different paramyxoviruses commonly interact with MDA5 and inhibit its function (Poole *et al.*, 2002; Andrejeva *et al.*, 2004; Childs *et al.*, 2007, 2009; Goodbourn and Randall, 2009). This inhibition is dependent on the highly conserved Vu region. The V protein binds to the helicase domain of MDA5 and prevents its activation by inhibiting dsRNA binding and consequent homooligomer formation (Childs *et al.*, 2009). In contrast to inhibition of MDA5 dependent activation of the IFN-β promoter, the V proteins are not able to bind to RIG-I (Childs *et al.*, 2007; Komatsu *et al.*, 2004; Strahle *et al.*, 2007) and exhibit little (Komatsu *et al.*, 2007) or only moderate inhibition of RIG-I dependent activation of the IFN-β promoter (lowered to about 60%; Childs *et al.*, 2007; Strahle *et al.*, 2007), suggesting that the V protein only exerts a marginal effect on the RIG-I dependent signalling pathway. However, the possibility remains that the V protein, which influences the control of viral transcription and replication, downregulates generation of PAMPs that are sensed by RIG-I, since V(–)-induced production of IFN-β is severely impaired in RIG-$I^{-/-}$ MEFs (Kato *et al.*, 2006).

To decide whether or not the ability to prevent MDA5 dependent signalling pathway leading to production of IFN-β is responsible for the rapid clearance of V(–) and VΔC from the mouse lung, their replication was examined in various kinds of mice defective in IFN systems, i.e. IFN-α/β-receptor, STAT1, and IRF3 knockout mice (Kiyotani *et al.*, 2007). Of these kinds, only the IRF3 knockout mice allowed V(–) and VΔC to replicate as efficiently as the parental SeV at least until 5 days with disease manifestation typical of SeV. Rapid clearance of V(–) and VΔC was clearly observed in IFN-α/β receptor or STAT1 knockout mice. Exactly the same rapid clearance occurred in IFN-γ knockout mice, the beige mice, as well as nude mice, indicating the involvement of neither IFN-γ, natural killer (NK) cells nor adaptive immune response in the rapid clearance of V(–) and VΔC.

These results demonstrate that SeV V protein counteracts innate virus clearance through IRF3 activation, but not via IFN system. Activated IRF3 up-regulates some chemokine genes such as RANTES, and other ISGs such as ISG15. Further, IRF3/Box pathway is involved in SeV-induced apoptosis (Heylbroeck *et al.*, 2000; Peters *et al.*, 2008; Chadpadhyay, 2010). It would be intriguing to learn whether one of those host factors is responsible for rapid clearance of V(–) and VΔC. It may be even possible that there is an IRF3 dependent, yet hitherto unknown, antiviral mechanism (Fig. 4.6; reviewed in Sakaguchi *et al.*, 2008).

Taken together, the results presented in this section suggested that the SeV V and C proteins interact with very many different facets of innate immunity. However, what has been verified for their significance in the context of *in vivo* pathogenesis so far is limited at this moment to the antagonism to IFN signalling and to an IRF3 dependent, yet unknown antiviral mechanism, that are encoded by the C and V proteins, respectively.

SeV technologies

Generation of hybrid cells and macromolecule delivery systems based upon SeV fusogenic capacity

Use of inactivated SeV whole virions to produce hybrid cells

SeV represents the first virus that was demonstrated to be able to induce cell fusion and generate multinucleated giant cells (Okada *et al.*, 1957; Okada, 1958). The SeV cell fusion was found to be induced by UV-inactivated virions that could not replicate and thus was non-cytopathic (Okada and Tadokoro, 1962). This discovery, coincided with the establishment of methods for *in vitro* culture of somatic cells of various origins or species, broke the new ground to use UV inactivated SeV as a fusogenic reagent for somatic cell genetics (reviewed in Okada, 1993). Remarkable progress in this field included the success of formation of heterokaryons derived from cells of two different species, such as mouse and human (Harris and Watkins, 1965; Okada and Murayama, 1965) and the establishment of a system for selection of hybrid cells using tk- and hprt- mutants (Littlefield, 1966). Also interestingly, following fusion with human HeLa cells, even dormant nuclei of chicken erythrocytes were reactivated, initiating synthesis of chicken proteins (Harris, 1965; Harris *et al.*, 1969). Thus two nuclei of different species were shown to be able to communicate with each other. Another good example of signalling cross-talk was activation of tumour virus genomes integrated into transformed non-producer cells by their fusion with uninfected permissive cells (Gerver, 1966)

Application of SeV fusion technology was further extended (reviewed in Okada, 1993). For example, selective disappearance of human chromosomes in mouse–human hybrid cells led to the field of gene mapping of human chromosomes (Weiss and Ephrussi, 1966; Weiss and Green, 1967; reviewed in Ruddle and Creagan, 1975) and complementary grouping became possible for a hereditary disease, xenoderma pigmentosum (deWeerd-Kastelein *et al.*, 1972; Tanaka *et al.*, 1975). Köhler and Milstein (1975) succeeded in generating cell lines, which secreted antibodies of predefined specificity [in this case, anti sheep red blood cells (SRBC)] by fusing a mouse myeloma cell line with mouse spleen cells from an SRBC immunized donor, opening the gate to monoclonal antibody technology, one of the biggest innovations in modern biotechnology.

Use of inactivated SeV for molecular delivery from erythrocyte ghosts into target cells

Efficient delivery of a foreign substance of interest into a target cell was developed by fusing the target cells with erythrocyte ghosts into which the substance had been sealed (Furusawa *et al.*, 1974). Using this novel approach, the transfer of the introduced nucleoproteins from the cytoplasm to the nucleus could be analysed, opening the research area of nuclear import (Yamaizumi *et al.*, 1978b). Diphtheria toxin fragment A shuts off cellular protein synthesis machinery by inactivating peptide elongation factor 2 (EF-2) and has been a useful tool in cell biology. Yamaizumi *et al.* (1978a) demonstrated that only one molecule of fragment A introduced into a living cell from an erythrocyte ghost was sufficient to kill the cell.

From envelope reconstitution, via HVJ (SeV) liposome, to HVJ-envelope (E) vector

A glycoprotein-enriched fraction was obtained after solubilization of SeV virions with a mild detergent. When the detergent was removed by gel filtration, reassembled virus particles were obtained with morphologically distinct spike glycoproteins. These reassembled envelope particles exhibited HA, cell-fusion and haemolytic activities (Shimizu *et al.*, 1972). When the detergent was removed after adding a foreign substance to the solution, the substance was expected to be incorporated into the reassembled envelope particles. Feasibility of using this type of reconstituted envelope as a novel delivery system was verified by using diphtheria toxin fragment A as the foreign substance because successful introduction of fragment A into a target cell could be readily judged by its death (Uchida *et al.*, 1977). However, this reconstitution technology involves a complex process including solubilization of virions, separation of envelope components from the viral cores by ultracentrifugation and dialysis to remove the detergent. Moreover, the nature and size of the substances to be incorporated are limited.

UV-inactivated SeV was found to fuse with a liposome, even though the latter did not express the SeV receptor. Thus, the contents within the liposome could be introduced into target cells through the usual SeV infection pathway (Nakanishi *et al.*, 1985; Kato *et al.*, 1991). This technology was called HVJ (SeV) liposome, and is far easier to employ, compared with envelope reconstitution. Additional merits were that SeV liposome was able to seal substances of diverse nature, such as nucleic acids, proteins, glycans and lipids and that their size limitation is not as critical as in the case of the reconstituted envelope. Conventional liposome technology is widely used for macromolecule delivery into cells. It probably involves endocytosis of the liposome followed by the content delivery to the lysosome, raising the fear of degradation or digestion of the content before it would work in the target cells. In contrast, the HVJ liposome, as well as reconstituted SeV envelope, makes use of the SeV infection mechanism that is believed to involve envelope fusion with the plasma membrane; thus, delivering the substance directly into the cytosol without its non-specific digestion in the lysosome. However, efficiency of fusion of SeV with liposomes not expressing SeV receptor was perhaps low. In addition, perhaps viral glycoproteins spread laterally from the SeV virion to the entire liposome after fusion, being diluted and thus not as dense as on the intact virion. Therefore, the efficiency of substance delivery (i.e. *infection efficiency* of SeV liposome) was not satisfactorily high.

An invention was made in order to avoid such dilution of SeV glycoproteins. UV-inactivated SeV particles were mixed with a plasmid DNA in the presence of a low concentration of detergent. The mixture was then centrifuged to sediment the virus particles. Through this process exchange of the viral RNP and the plasmid DNA occurred, generating SeV particles containing the DNA. It was able to trap plasmid DNAs up to 14 kb into the virus particles. The efficiency of DNA trapping into SeV particles was estimated to be 15 to 20% of the DNA contained in the initial mixture under optimal conditions and the enveloped DNA can be delivered into cells through the infection pathway of SeV (Kaneda *et al.*, 2002; reviewed in Kaneda *et al.*, 2005; Kaneda, 2008). This novel delivery system was named HVJ-E (envelope) vector and has been able to envelope various macromolecules including DNA, protein and siRNA. This HVJ-E vector that mediated gene delivery into target cells appeared to be approximately 10-fold higher in efficiency than that by a standard cationic liposome.

The HVJ-E vector was developed using egg-grown viruses as the starting material. However, use of egg-derived virus might encounter limitations in application in clinical settings, although virus titres are quite high when grown in eggs. Thus, the technology to produce large amounts of SeV in human cells with animal product-free medium was developed (Kaneda *et al.*, 2005). A pilot plant for commercial production of a human cell derived, clinical-grade HVJ-E vector has been established at GenomIdea, Inc. in Ibaragi, Osaka, Japan (Kaneda *et al.*, 2005).

Application of HVJ-E vector

A wide range of HVJ-E vector application has been described (reviewed in Kaneda, 2008), although to our knowledge, the application appears to be limited at the preclinical (cellular and animal) level. Here, just some remarkable cases are illustrated.

Gene delivery to various brain regions of rats was explored after intrathecal injection into cerebrospinal fluid (CSF) of the HVJ-E vector carrying *LacZ* or luciferase gene. The gene expression was seen in the brain stem, cerebellum, cerebral cortex, medulla and inner ear tissue including spinal ganglion cells (SGCs) (Shimamura *et al.*, 2003). Kanamycin is known to cause loss of SGCs and hair cells owing to apoptosis, damaging hearing function. The therapeutic potential of hepatocyte growth factor (HGF) for hearing loss was examined following intrathecal administration of HVJ-E vector containing the human *HGF* gene (HVJ-E/HGF) (Oshima *et al.*, 2004). It was found that the administration of HVJ-E/HGF prior to kanamycin insult significantly prevented the loss of SGCs and hair cells within the cochlea, compared with the empty HVJ-E vector. Furthermore, when HVJ-E/HGF was given 2 weeks after kanamycin insult, the number of SGCs returned to nearly half of the normal level. The *HGF* gene delivery into cerebrospinal fluid by HVJ-E vector appeared to have the potential for both prevention and treatment of hearing loss. By essentially the same approach (administration into CSF via the cisterna magna), the therapeutic potential of HVJ-F/HGF on brain injury due to infarction was suggested (Kaneda, 2008).

The application of HVJ-E vector has included therapeutic gene silencing *in vivo* by siRNAs and decoy oligonucleotides, protein delivery enabling therapy and bioimaging *in vivo*, and high throughput functional screening of candidate therapeutic genes from a cDNA library (reviewed in Kaneda, 2008). The empty HVJ-E vector alone injected into murine colon carcinoma tumours of CT26 cells growing in syngenetic BALB/c mice eradicated 60 to 80% of the tumour and inhibited the growth of the remainder, suggesting that HVJ-E vector itself induces anti-tumour immunity of T-cell mediated and non-T cell mediated facets through multiple pathways (Kaneda, 2008). It stimulated the maturation of human and murine dendritic cells. It also caused systemic activation of NK cells and enhanced their cytotoxicity against tumour cells (Kaneda, 2008). Whatever the anti-cancer drug used, its delivery by HVJ-E vector would produce a significant synergistic outcome due to a cancer specific molecular therapy combined with non-specific immunotherapy.

It is worthy to note the demonstration of the great potential of HVJ-E vector as a tuberculosis vaccine vector. The DNAs encoding the heat shock protein 65 (HSP65) of *Mycobacterium tuberculosis* (TB) and interleukin 12 (IL-12) were delivered via HVJ-E vector as well as HVJ liposome, and their protective efficacy was evaluated in a mouse model on the basis of TB load, animal survival, induction of CD8 positive CTL activity and the improvement of histopathological tuberculosis lesions. Compared with the conventional BCG vaccine, the prophylactic efficacy of HVJ liposome was 100-fold higher and that of HVJ-E vector was as much as 10,000-fold higher (Okada and Kita, 2010). The protective efficacy was also demonstrated in guinea pigs and cynomolgus monkeys. Furthermore, the HVJ-E vector version conferred on mice a remarkable therapeutic efficacy against multidrug resistant (MDR) TB and extremely drug-resistant (XDR) TB.

Further engineering of HVJ-E vector has been attempted. It includes conjugation of biocompatible polymers, such as cationized gelatin, which resulted in an enhanced level of gene expression, primarily within tumour deposits (Mima *et al.*, 2006) and a vector construction with a chimera of SeV F protein and a special cell surface recognizing molecule to target the respective special cell type *in vivo* (Kawachi *et al.*, 2007). These efforts would expand the utility of HVJ-E vector.

Development of live SeV vector

'Send in a sendai vector' as the beginning

An 18-nt-long sequence containing a unique *NotI* site was inserted 41 nucleotides downstream of PrE-II and just before the initiation codon for the N ORF of SeV cDNA backbone to insert a *NotI* tagged, PCR-amplified foreign gene of interest such that the authentic *N* gene start signal was now used for initiation of the inserted-gene transcription. The sequence to be inserted contained in the 5′ terminus (in negative sense) the stop signal to finish the inserted gene transcription and the start signal to initiate the next N gene transcription, which are connected with the intergenic three nucleotides. The pilot studies to insert and express luciferase or the gp120 of HIV-1 showed that not only the inserted genes were stably maintained but also their expression was quite robust (Hasan *et al.*, 1997; Yu *et al.*, 1997). For instance, the level of gp120 in the culture medium reached as high as 6 mg/l, probably the highest from *in vitro* mammalian cell cultures using other expression vectors (Yu *et al.*, 1997). It was possible to accommodate a gene up to 4.5 kb into this additive type vector in which a transcription unit is added to the entire SeV genome of ca 15.4 kb (Sakai *et al.*, 1999).

The studies on the reporter *LacZ* gene delivery into and expression in the airway epithelial cells of various animal origins and human origin *in vivo* or *ex vivo* demonstrated that the expression levels were several thousand-fold higher than those obtained by conventionally employed adenovirus vector or cationic liposomes (Yonemitsu *et al.*, 2000). These results appeared to gain wide attention; an article (p. 11 in *News and Views of Nature Genetics*, September 2000) stated 'Send in a sendai vector'. Surprisingly high performance of the additive-type SeV was found in the vessel wall (Masaki *et al.*, 2001), skeletal muscle (Masaki *et al.*, 2002) and joint synovium (Yamashita *et al.*, 2002). Independently reported were efficient gene transfer to human airway epithelium by rSeV (Pinkenburg *et al.*, 2004) and systemic delivery of therapeutic genes following intravenous administration of rSeV (Bitzer *et al.*, 2003b).

Gene therapy using live viral vectors is a concept that has been expected to successfully treat intractable diseases over many decades; however, the technologies are still in their infancy. The retroviruses including lentiviruses represent one of the most extensively studied viral vectors. These vectors achieve transgene expression through the insertion of proviral DNA into the host genome. Thus, the major concern in using retrovirus vectors has been the potential of protooncogene activation. Indeed, such concern became reality during clinical follow-up of the successful regimen for X-linked SCID therapy as indicated by the rather frequent occurrence of T-cell leukaemia through activation of LMO-2 protooncogene (Hacein-Bey-Abina *et al.*, 2003). On the other hand, SeV life cycle has no nuclear phase and is completed solely in the cytoplasm, theoretically excluding any risk due to protooncogene activation and other possible harmful outcomes on the chromosome (genotoxicity). In addition, SeV is believed to undergo no recombinational event with the homologous or related viruses, infects, cell cycle independently, both dividing and non-dividing cells, exhibits quite a broad host range, and so far has not been linked to any human diseases. In view of these considerations together with the above described stable and robust transgene expression and target range breadth prompted us and others to develop SeV into a novel class of expression vector that would be used safely and effectively in clinical settings (reviewed in Nagai, 1999; Nagai and Kato, 1999; Bitzer *et al.*, 2003a; Nagai *et al.*, 2007).

Generation of advanced versions

Although SeV is thought to have little or no implication in human diseases, the additive type SeV vectors with a full-length viral genome may be inappropriate for pharmaceutical use under certain situations because they can be propagated within the targeted patient and further transmitted to other persons. The use of SeV vectors in various clinical settings would therefore necessitate construction of non-propagating, non-transmissible versions through deletion of one or more viral genes other than the *N, P,* and *L* genes essential to maintain transgene expression.

Initially, an F-expressing packaging cell line with a Cre/*lox P*-inducible expression system was established for *trans* supply of F proteins in recovering SeV with the *F* gene deleted (SeV/Δ*F*). GFP-encoding gene-inserted SeV/Δ*F* vector was readily recovered. The transgene expression was remarkably high in efficiency in a wide variety of cells including primary hepatocytes, primary lung epithelial cells, primary smooth muscle cells, CD34-positive blood stem cells, primary cerebral cortex neurons, primary dorsal root ganglion neurons, and embryonic stem (ES) cells of human or animal origin (Li *et al.*, 2000). Moreover, *in vivo* GFP expression was also remarkable in the ependymal cells and pyramidal neurons, when the vector was stereotaxically introduced into the intraventricular and hippocampal regions, respectively, of rats, as well as into the airway epithelial cells, myocytes, vascular endothelial cells, nasal mucosa epithelial cells, and subretinal neurons of rats or mice (Li *et al.*, 2000). Also to be noted is that the titres of Δ*F* vectors generated in cell cultures were as high as those of the additive-type vectors, reaching 10^8 PFU/ml in the culture media (Li *et al.*, 2000). Subsequently, other single-gene deletion vectors, Δ*M* and Δ*HN*, double gene-deletion vectors, Δ*F*Δ*M* and Δ*F*Δ*HN*, and a triple gene-deletion vector, Δ*M*Δ*F*Δ*HN* were successfully generated (Inoue *et al.*, 2003a, 2004; Yoshizaki *et al.*, 2006). To minimize cytopathogenicity, an additional version was constructed by introducing ts mutations found in the M, HN, P, and L proteins of cl. 151 and other known ts mutants into the SeV/Δ*F* (Inoue *et al.*, 2003b; unpublished). The cl. 151 was derived from a steady-state persistent infection, temperature sensitive and was marginally cytopathic, as noted above. This version is designated here tsSeVΔ*F*. These vectors were developed at DNAVEC corporation in Tukuba, Ibaragi, Japan, which is the patentee of SeV vector technology. An additive type tsSeV was independently established also based upon cl. 151 by Nishimura *et al.* (2007). It is now possible to choose the most appropriate one from various additive and defective versions, as the situation and purpose demand (reviewed in Nagai *et al.*, 2007). The additive type vector is simply termed rSeV hereafter.

Application to preclinical and clinical studies

Use as vaccine vectors

Simian immunodeficiency virus (SIV) infection of non-human primates (macaques) provides virtually the sole available model for research and development of AIDS vaccines. The strain SHIV89.6P has most often been used as a challenge virus. It is a chimera SIV in which the envelope and some accessory genes are replaced with the HIV-1 counterparts, while the major internal proteins, such as Gag and Pol are derived from SIV and uses CXCR4 as the co-receptor. The reason for the use of this strain as a challenge virus was simply that experimental outcomes are obtained in a much shorter time frame because CD4-positive T cells in simian peripheral blood drastically decline within about two weeks following injection. In the natural realm, however, HIV-1 and SIV use co-receptor CCR5 and infected hosts develop AIDS following a long period (over years) of chronic infection. Moreover, SHIV89.6P tends to be readily controlled by the simian immune system for unknown reasons despite its highly acute and highly pathogenic nature. SHIV89.6P may be compared with 'a sheep in wolf's clothing' (Feinberg and Moore, 2002). Thus, the use of SHIV89.6P as a challenge virus is not justified for efficacy tests in SIV models and the use of CCR5-tropic, chronic SIV strains, such as SIVmac239 have been strongly recommended, which reflect more closely the natural course of human AIDS (Feinberg and Moore, 2002).

An rSeV expressing the SIVmac239 Gag protein, which was intranasally administered, was virtually the first SIV vaccine vector with a satisfactory protection against the homologous SIVmac239 challenge in macaques (Kano *et al.*, 2000; Matano *et al.*, 2004). However, the protection was not perfect in that a portion of vaccinees probably escaped protection because it was based on cytotoxic T cells (CTL) and hence restricted to particular classes of MHC haplotypes (Matano *et al.*, 2004). Feasibility of using a combination of rSeVs expressing the Gag, Pol and Nef of HIV-1 is now being studied on a collaborative basis between the International Aids Vaccine Initiative (IAVI), NY, USA and the DNAVEC corp.

The researchers at St. Jude Children's Hospital, Memphis, TN, USA have been engaged in using SeV vector for protection against acute respiratory viral infections of children. They found that rSeV expressing the G protein of respiratory syncytial virus (RSV) intranasally inoculated into cotton rats elicited RSV-specific antibody and protected the animals from RSV challenge (Takimoto *et al.*, 2004). The research team at Tierärztlich Hochschule, Hannover, Germany demonstrated that rSeV expressing RSV F protein-induced T cell immunity against RSV that was protective in the absence of antibodies (Voges *et al.*, 2007). The former group further found that a mixed rSeV-RSV-F, rSeV-HPIV3-HN, and rSeV-HPIV2-HN vaccine conferred protection against RSV, HPIV1, HPIV2 and HPIV3 challenges in the cotton rat model (Jones *et al.*, 2009). Thus, a mixture of the three different rSeVs was expected to prevent pulmonary infections with four different viruses. Here, protection from HPIV1 was obviously conferred by its immunological close relatedness to SeV. These rSeV vaccines are reported to be under testing in the clinic, and thus far to have been well tolerated. The researchers are also working to formulate an SeV-based cocktail vaccine designed to prevent several HPIVs and RSV infections (Hurwitz, 2008).

Protective immunity against *Trypanosoma cruzi* (*T. cruzi*) that causes Chagas' disease could be induced by plasmid DNA vaccination encoding the amastigote surface protein-1 (ASP-1), and this protective efficacy was found to be enhanced by fusing ASP-1 with ubiquitin sequences, which facilitated the Th1-dependent ubiquitin-fusion degradation pathway (Chou *et al.*, 2010). Immunization of ASP-1 without ubiquitin sequences via nasal epithelium using SeV/Δ*F* evoked CD8 T cell-dependent, strong protective immunity against *T. cruzi* without requirement of ubiquitin-fusion construct (Duan *et al.*, 2009).

Taken together, these findings indicate that SeV-mediated antigen expression in the nasal epithelium would be useful to induce both humoral and cellular protective responses against diverse pathogens.

Use for cancer therapy

No virus budding occurs in SeVΔM–infected cells. Therefore, the viral HN and F_0 glycoproteins accumulate at the cell surface. When the F_0 on the cell surface is activated by trypsin, these cells efficiently fuse with the adjacent cells, leading to the formation of large syncytia of several tens or more cells in the entire culture. At the same time, the RNPs are transferred from cell to cell and are further multiplied to incur cell injury, leading to lysis of the syncytia. The ability of such massive syncytia formation and cell lysis characteristic of SeV/ΔM appeared to be well suited for the treatment of solid cancers that form a mass of malignant cells.

For the SeV/ΔM to target tumours, the tryptic cleavage site of the F_0 protein was converted to a site cleavable by matrix metalloproteinases-2/9 (MMP-2/9) that are known to be often overproduced by and secreted from malignant cells. The resulting SeV/F(MMP)ΔM was able to spread and induce massive cell fusion in certain cancer cells cultured *in vitro* without trypsin, while it no longer spread nor induced fusion in normal cells in the presence or absence of trypsin. When injected into a mass of fibrosarcoma HT1080 transplanted into nude mice, this mutant manifested marked tumour-growth suppressing activity (Kinoh *et al.*, 2004).

However, the efficacy of tumour killing by this mutant was relatively limited to special tumour types, because of the requirement of a relatively high level expression of MMPs for efficient virus spread and cell killing and because the types of MMP-overexpressing tumour cells are limited in number (~20%). To overcome these problems, attempts were made that included the deletion of 14 amino acids from the cytoplasmic tail of the F_0 protein, hoping for more efficient cell fusion, and retargeting to urokinase-type plasminogen activator (uPA) but no longer to MMPs. This new construct [SeV/Fct14(uPA2)ΔM] exhibited several times greater tumour cell killing ability, compared with the MMP-targeted version (Kinoh *et al.*, 2009). Furthermore, incorporation of IFN-β gene drastically enhanced its anti-tumour effect on an orthotropic rat model of glioblastoma without any damage to normal brain cells (Hasegawa *et al.*, 2010). Owing to such cancer-specific killing activity without damage to normal tissues, SeV/Fct14(uPA2)ΔM was nicknamed 'BioKnife' by the researchers.

It was recently found that rSeVs activate DCs and improve DC-based cancer immunotherapy (Shibata *et al.*, 2006; Yoneyama *et al.*, 2007). rSeV-modified DCs exhibited highly effective prevention of lung metastasis in a rat model of prostate cancer (Komaru *et al.*, 2009); the new concept 'immunostimulatory virotherapy' was raised (Yonemitsu *et al.*, 2008). Thus, the synergy between specific 'BioKnife' and non-specific 'immunostimulation' is expected for SeV/Fct14(uPA2)ΔM.

Use for induction of human pluripotent stem cells

The discovery that human somatic cells are reprogrammed to become induced pluripotent stem cells (iPSCs) by intracellular delivering and expression of the four special reprogramming factors in combination has opened the gate to a new and exciting era in regenerative medicine (Takahashi and Yamanaka, 2006). However, the introduction of those reprogramming factors into cells has most often been done by using a retrovirus (including lentivirus) vector, which is integrated into host chromosomes. This would be one of the factors increasing the risk of malignant transformation. The risk is thought to come from two possible pathways, the activation of protooncogenes by the integrated retrovirus and the constitutive expression of the reprogramming factors, some of which (e.g. c-Myc) are potentially oncogenic.

One of the plausible means to avoid this risk is to use non-integrating SeV and related viral vectors that have no nuclear phase in their life cycle and so far have not been shown to have link to any human diseases. Further advantages of SeV vector are its extraordinary high capacity in foreign gene expression and broad range of host cells. Its disadvantage is, however, considerable cytotoxic activity. Thus, a defective SeV vector with minimal cytotoxicity, tsSeVΔF, was used to generate human iPSCs. Because iPSCs are established after numerous rounds of cell proliferation and perhaps because of their temperature sensitivity, the vectors expressing the reprogramming factors were expected to be diluted out and eventually eliminated from the system. Indeed, the results demonstrated an extremely high efficiency of generation of integration-free iPSCs and complete elimination of the entire vector virus sequences encoding the reprogramming factors, probably minimizing the risk of malignant transformation (Fusaki *et al.*, 2009).

To date, human skin fibroblasts have most often been used as the source to generate patient-specific iPSCs. The cells are obtained by biopsy, which is invasive and has various potential risks including infections. Moreover, a small number of cells obtained by biopsy have to be expanded before use, which takes a long time, perhaps at least a month, is expensive, and may cause accumulation of unintended mutations. In contrast, taking blood samples is far less invasive, and the blood cells can be immediately processed, without expansion, for transduction. In a recent issue of *Cell Stem Cell*, three groups reported the generation of human iPSCs from terminally differentiated peripheral T cells using the same original four reprogramming factors but different viral vectors. The vector infection could be initiated within several days after blood sampling and picking up iPSCs colonies after ca 25 days in all the three cases. When lentivirus vectors were used, a relatively large volume of blood samples (30–350 ml) was required to generate the iPSCs (Loh *et al.*, 2010; Staerk *et al.*, 2010). On the other hand, just one ml of blood was sufficient when the tsSeVΔF was used (Seki *et al.*, 2010). More recently, this latter research team has succeeded in producing the same outcome from as little as 0.1 ml of blood (K. Fukuda, personal communication). These results indicate an extremely high capacity of SeV vector for generating integration-free iPSCs (reviewed in Yamanaka, 2010).

Phase I/IIa trial using SeV/ΔF-hFGF2 for the peripheral arterial disease (PAD)

Peripheral arterial disease (PAD) is mainly caused by atherosclerosis and becomes quite common as one gets older, being found in about 12–20% of the population over 65 years old. Diagnosis is critical, as people with PAD are at 4- to 5-fold higher risk of heart attack or stroke. The symptomatic stage of PAD includes intermittent claudication (IC) and critical limb ischaemia (CLI). IC is the most prominent symptom of PAD and occurs during exercise and walking, indicating that IC would be 'stress-induced' ischaemia. About one-third to one-half of patients with PAD manifest this symptom. Well-established and evidence-based treatments for IC are exercise rehabilitation and administration of a few limited kinds of drugs, including cilostazol (Norgren *et al.*, 2007). CLI represents a more advanced stage and devastating manifestation of PAD. The diagnosis is established when patients present with ischemic rest pain, ulcerations, or gangrene of the leg associated with evidence of reduced arterial blood flow to the foot. Within the first year of illness, 30% suffer a major leg amputation, 25% die, and 20% endure with unsolved pain or gangrene (Feiring *et al.*, 2010).

In cases of relatively serious IC (i.e. a walking distance of usually less than 200 m) and CLI, re-vascularization by bypass surgery or vascular intervention should be encountered. However, the indication of re-vascularization is relatively limited owing to multiple or longer obstruction as well as severe calcification of the arteries. No other option has been available in these cases; therefore, physicians and scientists have looked forward to developing an alternative approach, namely 'angiogenic therapy' in the last decade.

The effects of various relevant recombinant proteins and plasmids intramuscularly administered into upstream regions of diseased arteries were examined for their capacity to induce neovascularization for PAD patients, but the outcomes so far have not been very beneficial. Since SeV-based vector was found to have dramatically high gene transfer and expression capacity in the muscle *in vivo*, expression of angiogenic growth factors from rSeVs has been studied. A preclinical study using a mouse model with severe critical limb ischaemia followed by autoamputation of the leg demonstrated that SeV/ΔF carrying the fibroblast growth factor (FGF)-2 gene prevented loss of the leg, suggesting a notable therapeutic potential of FGF-2 for CLI (Masaki *et al.*, 2002). Neither injection of recombinant FGF-2 proteins nor their plasmid-based expression could rescue the ischemic limb in this model. Taken together, the results strongly suggested the feasibility of using SeV/Δ*F* expressing the human (h) FGF2 as a novel therapeutic means for PAD.

The study has now moved on to phase I/IIa clinical trial at Kyushu University Hospital in Fukuoka, Japan. As of the end of April 2010, a total of 12 patients were recruited. They received 5×10^7 to 5×10^9 infectious units/60 kg of SeV/Δ*F*-hFGF through intramuscular injection to multiple points in the upstream region of the leg arteries of occlusion. Eleven of the 12 patients have completed all the scheduled visits during the study period of six months. The remaining one was lost owing to acute exacerbation of pre-existing interstitial pneumonia two years after gene transfer.

This therapy so far has not given rise to any drug-/therapy-related severe adverse event. Seven of the eight patients, who were able to adhere to the treadmill test, showed improved walking ability by over 50% at the end of observation. Peak walking distance exceeded 200 m in four patients. The temperature increase of the sole and arterial flow wave in the 1st toe were apparent for the treated leg, but not for the opposite limb in some cases. A case showing such improvement is illustrated in

A

Case 201	Pre	1Mo	3Mo	5Mo
Peak walking distance (m)	74	138 (186%)	235 (318%)	>300m (complete)
Claudication onset distance (m)	43	118 (274%)	208 (483%)	>300m (complete)

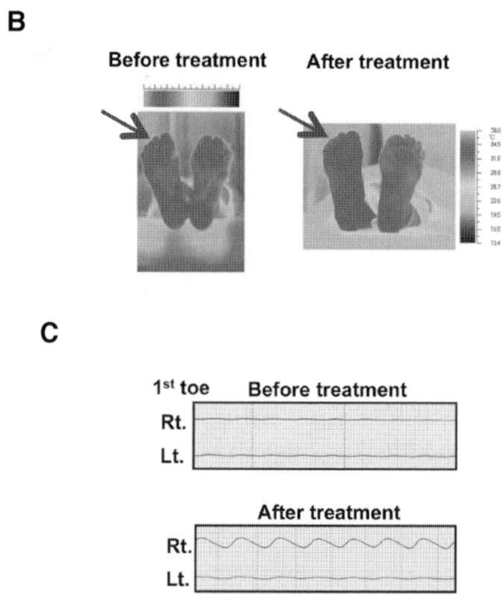

Figure 4.7 Improvement of efficacy parameters in a patient (Case No. 201) treated with Sendai virus (SeV) SeV/Δ*F*-hFGF2. This patient was a 65 year-old woman who had been suffering from rest pain of the right limb. Her left limb was also diseased, but did not show either rest pain or gangrene. (A) Time course of peak walking distance and claudication onset distance examined by treadmill test immediately before (pre) and 1, 3 and 5 months (Mo) after treatment. (B) Thermography assessing foot (sole) temperature. Both of the limbs before treatment showed similar temperature (left). In contrast, the gene transferred right limb, but not the untreated opposite limb, showed apparent increase of temperature by 3°C. Arrows indicate the right limbs before and after treatment. (C) Detection of flow wave on the first toe. At the time immediately before treatment, blood flow at the first toe was not detected in either the right (Rt) or left (Lt) limbs. Two months after gene transfer it became evident at the first toe of the treated Rt limb, but not the untreated Lt limb. A colour version of this figure is located in the plate section at the back of the book.

Fig. 4.7. These data suggested the potential utility of SeV/Δ*F*-hFGF for treating PAD. The efficacy will be evaluated in the upcoming double-blinded and placebo-controlled trial.

Concluding remarks

Looking back, SeV has played a great role in advancing basic virology and developing novel technologies beyond its primitive status just as a mouse pathogen. Many of the prominent theories and technologies established in the field find their origin in rather old crucial discoveries. At the same time, they sometimes emerged in coincidence with the great leaps in other fields. For instance, the early discovery of cell fusion by SeV coincided with the establishment of cell culture technology; thereby, breaking the new ground of somatic cell genetics and cell engineering. This was followed by the development of HVJ-E vector for therapeutic gene delivery. Similarly, the establishment of reverse genetics coincided with the big-bang of our understanding of innate immunity. In this context, it is important to note that the outcomes of SeV reverse genetics can be assessed with relative ease using laboratory mice with a wide variety of genetic backgrounds. SeV reverse genetics swiftly advanced to the development of a novel class of non-integrating vector. Diverse applications are now in the pipeline and poised for testing in clinical trials. A safe and efficient transduction procedure is being eagerly awaited to generate patient-specific iPSCs. SeV vector appears to be able to fulfil this requirement.

However, the research outcomes are not only fragmentary and at the phenomenological level, but also sometimes in serious conflict as illustrated in many sections. More penetrating research is needed to gain deeper understanding at the molecular level as well as to resolve these inconsistencies. Furthermore, it is certainly a long journey hereafter with twists and turns in which the HVJ-E vector and live SeV vector will move on to diverse clinical trials and ultimately demonstrate their therapeutic or prophylactic utility for specific diseases of various kinds. Therefore, in a very true sense, the next exciting chapter in the tale of SeV is really just beginning.

Acknowledgements

We thank K. Shimizu for providing electron micrographs in Fig. 4.1, K. Kiyotani for preparing Fig. 4.5, and T. Sakaguchi for providing unpublished data. We also thank A. Kato, T. Sakaguchi, Y. Yoshida, K. Kiyotani, Y. Suzuki, T. Nakajima and H. Taira for critical reading of the manuscript and/or kindly providing valuable information and suggestions, and M. Sumi for preparing the manuscript. Finally we thank Paul Langman Ph.D. for his careful reading of the manuscript and kind assistance with English usage.

References

Albertini, A.A.V., Wernimont, A.K., Muziol, T., Ravelli, R.B.G., Clapier, C.R., Schoehn, G., Weissenhorn, W., and Ruigrok, R.W.H. (2006). Crystal structure of the rabies virus nucleoprotein–RNA complex. Science *313,* 360–363.

Ali, A., and Nayak, D.P. (2000). Assembly of Sendai virus: M protein interacts with F and HN proteins and with the cytoplasmic tail and transmembrane domain of F protein. Virology *276,* 289–303.

Alymova, I.V., Taylor, G., Mishin, V.P., Watanabe, M., Murti, K.G., Boyd, K., Chand, P., Babu, Y.S., and Portner, A. (2008). Loss of the *N*-linked glycan at residue 173 of human parainfluenza virus type 1 hemagglutinin-neuraminidase exposes a second receptor-binding site. J. Virol. *82,* 8400–8410.

Andrejeva, J., Childs, K.S., Young, D.F., Carlos, T.S., Stock, N., Goodbourn, S., and Randall, R.E. (2004). The V proteins of paramyxoviruses bind the IFN-inducible RNA helicase, mda-5, and inhibit its activation of the IFN-beta promoter. Proc. Natl. Acad. Sci. U.S.A. *101,* 17264–17269.

Aoki, K., and Kawakita, M. (1996). Differential sensitivity of two related viruses, Newcastle disease virus and Sendai virus, to interferon in mouse Had-2 cells: selective inhibition of translation of NDV mRNA. Arch. Virol. *141,* 1847–1862.

Babst, M., Wendland, B., Estepa, E.J., and Emr, S.D. (1998). The Vps4p AAA ATPase regulates membrane association of a Vps protein complex required for normal endosome function. EMBO J. *17,* 2982–2993.

Bächi, T. (1980). Intramembrane structural differentiation in Sendai virus maturation. Virology *106,* 41–49.

Banerjee, A.K. (2008). Response to 'Non-segmented negative-strand RNA virus RNA synthesis *in vivo*'. Virology *371,* 231–233.

Bieniasz, P.D. (2006). Late budding domains and host proteins in enveloped virus release. Virology *344,* 55–63.

Billeter, M.A., and Cattaneo, R. (1991). Molecular biology of defective measles viruses persisting in the human central nervous system. In The Paramyxoviruses, Kingsbury, D.W., ed. (Plenum Press: New York), pp. 323–345.

Bitzer, M., Prinz, F., Bauer, M., Spiegel, M., Neubert, W.J., Gregor, M., Schulze-Osthoff, K., and Lauer, U. (1999). Sendai virus infection induces apoptosis through activation of caspase-8 (FLICE) and caspase-3 (CPP32). J. Virol. *73*, 702–708.

Bitzer, M., Armeanu, S., Prinz, F., Ungerechts, G., Wybranietz, W., Spiegel, M., Bernlohr, C., Cecconi, F., Gregor, M., Neubert, W.J., *et al.* (2002). Caspase-8 and Apaf-1-independent caspase-9 activation in Sendai virus-infected cells. J. Biol. Chem. *277, 29*,817–29824.

Bitzer, M., Armeanu, S., Lauer, U.M., and Neubert, W.J. (2003a). Sendai virus vectors as an emerging negative-strand RNA viral vector system. J. Gene Med. *5*, 543–553.

Bitzer, M., Ungerechts, G., Bossow, S., Graepler, F., Sedlmeier, R., Ameanu, S., Bernloehr, C., Spiegel, M., Gross, C.D., Gregor, M., *et al.* (2003b). Negative-strand RNA viral vectors: intravenous application of Sendai virus vectors for the systemic delivery of therapeutic genes. Mol. Ther. *7*, 210–217.

Blanchard, L., Tarbouriech, N., Blackledge, M., Timmins, P., Burmeister, W.P., Ruigrok, R.W.H., and Marion, D. (2004). Structure and dynamics of the nucleocapsid-binding domain of the Sendai virus phosphoprotein in solution. Virology *319*, 201–211.

Böttcher, E., Freuer, c., Steinmetzer, T., Klenk, H.D., and Garten, W. (2006). MDCK cells that express proteases TMPRSS2 and HAT provide a cell system to propagate influenza viruses in the absence of trypsin and to study cleavage of HA and its inhibition. Vaccine *27*, 6324–6329.

Bousse, T., and Takimoto T. (2006). Mutation at residue 523 creates a second receptor binding site on human parainfluenza virus type 1 hemagglutinin-neuraminidase protein. J. Virol. *80*, 9009–9016.

Bousse, T.L., Taylor, G., Krishnamurthy, S., Portner, A., Samal, S.K., and Takimoto, T. (2004). Biological significance of the second receptor biding site of Newcastle disease virus hemagglutinin-neuraminidase protein. J. Virol. *78*, 13,351–13355.

de Breyne, S., Monney, R.S., and Curran, J. (2004). Proteolytic processing and translation initiation: two independent mechanisms for the expression of the Sendai virus Y proteins. J. Biol. Chem. *279, 16*,571–16580.

Buchholz, C.J., Spehner, D., Drillien, R., Neubert, W.J., and Homann, H.E. (1993). The conserved N-terminal region of Sendai virus nucleocapsid protein NP is required for nucleocapsid assembly. J. Virol. *67*, 5803–5812.

Cadd, T., Garcin, D., Tapparel, C., Itoh, M., Homma, M., Roux, L. Curran, J., and Kolakofsky, D. (1996). The Sendai paramyxovirus accessory C proteins inhibit viral genome amplification in a promoter-specific fashion. J. Virol. *70*, 5067–5074.

Calain, P., and Roux, L. (1993). The rule of six, a basic feature for efficient replication of Sendai virus defective interfering RNA. J. Virol. *67*, 4822–4830.

Calistri, A., Salata, C., Parolin, C., and Palù, G. (2009). Role of multivesicular bodies and their components in the egress of enveloped RNA viruses. Rev. Med. Virol. *19*, 21–45.

Çevik, B., Kaesberg, J., Smallwood, S., Feller J.A., and Moyer, S.A. (2004). Mapping the phosphoprotein binding site on Sendai virus NP protein assembled into nucleocapsids. Virology *325*, 216–224.

Çevik, B., Smallwood, S., and Moyer, S.A. (2007). Two N-terminal regions of the Sendai virus L RNA polymerase protein participate in oligomerization. Virology *363*, 189–197.

Chaipan, C., Kabasa, D., Bertram, S. Glowacka, I., Steffen, I., Tsegaye, T.S., Takeda, M., Bugge, T.H., Kim, S., Park, Y., Marzi, A., and Pöhlmann, S. (2009). Proteolytic activation of the 1918 influenza virus hemagglutinin. J. Virol. *83*, 3200–3211.

Chambers, R., and Takimoto, T. (2009). Host specificity of the anti-interferon and anti-apoptosis activities of parainfluenza virus P/C gene products. J. Gen. Virol. *90*, 1906–1915.

Chandrika, R., Horikami, S.M., Smallwood, S., and Moyer, S.A. (1995). Mutations in conserved domain I of the Sendai virus L polymerase protein uncouple transcription and replication. Virology *213*, 352–363.

Chanock, R.M., Parrott, R.H., Cook, K., Andrews, B.E., Bell, A., Reichelderfer, T., Kapikian, A.Z., Mastrota, F.M., and Huebner, R.J. (1958). Newly recognized myxoviruses from children with respiratory disease. New Engl. J. Med. *258*, 207–213.

Chanock, R.M., Parrott, R.H., Johnson, K.M., Kapikian, A.Z., and Bell, J.A. (1963). Myxoviruses: parainfluenza. Am. Rev. Respir. Dis. *88* (*Suppl.*), 152–166.

Chattopadhyay, S., Marques, J.T., Yamashita, M., Peters, K.L., Smith, K., Desai, A., Williams, B.R., Sen, G.C. (2010). Viral apoptosis is induced by IRF-3-mediated activation of Bax. EMBO J. *29*, 1762–1773.

Childs, K., Stock, N., Ross, C., Andrejeva, J., Hilton, L., Skinner, M., Randall, R., and Goodbourn, S. (2007). mda-5, but not RIG-I, is a common target for paramyxovirus V proteins. Virology *359*, 190–200.

Childs, K.S., Andrejeva, J., Randall, R.E., and Goodbourn, S. (2009). Mechanism of mda-5 Inhibition by paramyxovirus V proteins. J. Virol. *83*, 1465–1473.

Chou, B., Hiromatsu, K., Hisaeda, H., Duan, X., Imai, T., Murata, S., Tanaka, K., and Himeno, K. (2010). Genetic immunization based on the ubiquitin-fusion degradation pathway against Trypanosoma cruzi. Biochem. Biophys. Res. Commun. *392*, 277–282.

Cortese, C.K., Feller, J.A., and Moyer, S.A. (2000). Mutations in domain V and the Sendai virus L polymerase protein uncouple transcription and replication and differentially affect replication *in vitro* and *in vivo*. Virology *277*, 387–396.

Curran, J. (1996). Reexamination of the Sendai virus P protein domains required for RNA synthesis: a possible supplemental role for the P protein. Virology *221*, 130–140.

Curran, J. (1998). A role for the Sendai virus P protein trimer in RNA synthesis. J. Virol. *72*, 4274–4280.

Curran, J., and Kolakofsky, D. (1987). Identification of an additional Sendai virus non-structural protein encoded by the P/C mRNA. J. Gen. Virol. *68*, 2515–2519.

Curran, J., and Kolakofsky, D. (1988a). Ribosomal initiation from an ACG codon in the Sendai virus P/C mRNA. EMBO J. *7*, 245–251.

Curran, J., and Kolakofsky, D. (1988b). Scanning independent ribosomal initiation of the Sendai virus X protein. EMBO J. *7,* 2869–2874.

Curran, J., and Kolakofsky, D. (1989). Scanning independent ribosomal initiation of the Sendai virus Y proteins *in vitro* and *in vivo*. EMBO J. *8,* 521–526.

Curran, J., and Kolakofsky, D. (2008). Nonsegmented negative-strand RNA virus RNA synthesis *in vivo*. Virology *371,* 227–230.

Curran, J., Boeck, R., and Kolakofsky, D. (1991). The Sendai virus P gene expresses both an essential protein and an inhibitor of RNA synthesis by shuffling modules via mRNA editing. EMBO J. *10,* 3079–3085.

Curran, J., Marq, J.-B., and Kolakofsky, D. (1992). The Sendai virus nonstructural C proteins specifically inhibit viral mRNA synthesis. Virology *189,* 647–656.

Curran, J., Homann, H., Buchholz, C., Rochat, S., Neubert, W., and Kolakofsky, D. (1993). The hypervariable C-terminal tail of the Sendai paramyxovirus nucleocapsid protein is required for template function but not for RNA encapsidation. J. Virol. *67,* 4358–4364.

Curran, J., Pelet, T., and Kolakofsky, D. (1994). An acidic activation-like domain of the Sendai virus P protein is required for RNA synthesis and encapsidation. Virology *202,* 875–884.

Curran, J., Marq, J.-B., and Kolakofsky, D. (1995). An N-terminal domain of the Sendai paramyxovirus P protein acts as a chaperone for the NP protein during the nascent chain assembly step of genome replication. J. Virol. *69,* 849–855.

Delenda, C., Hausmann, S., Garcin, D., and Kolakofsky, D. (1997). Normal cellular replication of Sendai virus without the trans-frame, nonstructural V protein. Virology *228,* 55–62.

Delenda, C., Taylor, G., Hausmann, S., Garcin, D., and Kolakofsky, D. (1998). Sendai viruses with altered P, V, and W protein expression. Virology *242,* 327–337.

Demirov, D.G., and Freed, E.O. (2004). Retrovirus budding. Virus Res. *106,* 87–102.

Didcock, L., Young, D.F., Goodbourn, S., and Randall, R.E. (1999a). Sendai virus and simian virus 5 block activation of interferon-responsive genes: importance for virus pathogenesis. J. Virol. 73, 3125–3133.

Didcock, L., Young, D.F., Goodbourn, S., and Randall, R.E. (1999b). The V protein of simian virus 5 inhibits interferon signalling by targeting STAT1 for proteasome-mediated degradation. J. Virol. *73,* 9928–9933.

Diederich, S., Moll, M., Klenk, H.D., and Maisner, A. (2005). The Nipah virus fusion protein is cleaved within the endosomal compartment. J. Biol. Chem. *280,* 29899–29903.

Du, W., and Maniatis, T. (1992). An ATF/CREB binding site is required for virus induction of the human interferon beta gene [corrected]. Proc. Natl. Acad. Sci. USA *89,* 2150–2154.

Du, W., Thanos, D., and Maniatis, T. (1993). Mechanisms of transcriptional synergism between distinct virus-inducible enhancer elements. Cell *74,* 887–898.

Duan, X., Yonemitsu, Y., Chou, B., Yoshida, K., Tanaka, S., Hasegawa, M., Tetsutani, K., Ishida, H., Himeno, K., and Hisaeda, H. (2009). Efficient protective immunity against Trypanosoma cruzi infection after nasal vaccination with recombinant Sendai virus vector expressing amastigote surface protein-2. Vaccine *27,* 6154–6159.

Easterbrook, J.D., Kaplan, J.B., Glass, G.E., Watson, J., and Klein, S.L. (2008). A survey of rodent-borne pathogens carried by wild-caught Norway rats: a potential threat to laboratory rodent colonies. Lab. Anim. *42,* 92–98.

Egelman, E.H., Wu, S.-S., Amrein, M., Portner, A., and Murti, G. (1989). The Sendai virus nucleocapsid exists in at least four different helical states. J. Virol. *63,* 2233–2243.

Einberger, H., Mertz, B., Hofschneider, P.H., and Neubert, W.J. (1990). Purification, renaturation, and reconstituted protein kinase activity of the Sendai virus large (L) protein: L protein phosphorylates the NP and P protein *in vitro*. J. Virol. *64,* 4274–4280.

Faisca, P., Anh, D.B., and Desmecht, D.J. (2005). Sendai virus-induced alterations in lung structure/function correlate with viral loads and reveal a wide resistance/susceptibility spectrum among mouse strains. Am. J. Physiol. Lung Cell Mol. Physiol. *289,* L777–787.

Feinberg, M.B., and Moore, J.P. (2002). AIDS vaccine models: challenging challenge viruses. Nat. Med. *8,* 207–210.

Feiring, A.J., Krahn, M., Nelson, L., Wesolowski, A., Eastwood, D., and Szabo, A. (2010). Preventing leg amputations in critical limb ischemia with below-the-knee drug-eluting stents: the PaRADISE (PReventing Amputations using Drug eluting StEnts) trial. J. Am. Coll. Cardiol. *55,* 1580–1589.

Feller, J.A., Smallwood, S., Horikami, S.M., and Moyer, S.A. (2000). Mutations in conserved domains IV and VI of the large (L) subunit of the Sendai virus RNA polymerase give a spectrum of defective RNA synthesis phenotypes. Virology *269,* 426–439.

Fujii, Y., Kiyotani, K., Yashida, T., and Sakaguchi, T. (2001). Conserved and non-conserved regions in the Sendai virus genome: evolution of a gene possessing overlapping reading frames. Virus Genes *22,* 47–52.

Fujii, Y., Sakaguchi, T., Kiyotani, K., Huang, C., Fukuhara, N., Egi, Y., and Yoshida, T. (2002a). Involvement of the leader sequence in Sendai virus pathogenesis revealed by recovery of a pathogenic field isolate from cDNA. J. Virol. *76,* 8540–8547.

Fujii, Y., Sakaguchi, T., Kiyotani, K., Huang, C., Fukuhara, N., and Yoshida, T. (2002b). Identification of mutations associated with attenuation of virulence of a field Sendai virus isolate by egg passage. Virus Genes *25,* 189–193.

Fujita, T. (2009). A nonself RNA pattern: tri-*p* to panhandle. Immunity *31,* 4–5.

Fukuhara, N., Huang, C., Kiyotani, K., Yoshida, T., and Sakaguchi, T. (2002). Mutational analysis of the Sendai virus V protein: importance of the conserved residues for Zn binding, virus pathogenesis and efficient RNA editing. Virology *299,* 172–178.

Fukumi, H., Nishikawa, F., and Kitayama, T. (1954). A pneumotropic virus from mice causing haemagglutination. Jap. J. Med. Sci. Biol. *7*, 345–363.

Furusawa, M., Nishimura, T., Yamaizumi, M., and Okada, Y. (1974). Injection of foreign substances into single cells by cell fusion. Nature *249*, 449–450.

Fusaki, N., Ban, H., Nishiyama, A., Saeki, K., and Hasegawa, M. (2009). Efficient induction of transgene-free human pluripotent stem cells using a vector based on Sendai virus, an RNA virus that does not integrate into the host genome. Proc. Jpn. Acad., Ser.. B *85*, 348–362.

Galinski, M.S., Troy, R.M., and Banerjee, A.K. (1992). RNA editing in the phosphoprotein gene of the human parainfluenza virus type 3. Virology *186*, 543–550.

Garcia, M.A., Gil, J., Ventoso, I., Guerra, S., Domingo, E., Rivas, C., and Esteban, M. (2006). Impact of protein kinase PKR in cell biology: from antiviral to antiproliferative action. Microbiol. Mol. Biol. Rev. *70*, 1032–1060.

Garcin, D., Pelet, T., Calain, P., Roux, L., Curran, J., and Kolakofsky, D. (1995). A highly recombinogenic system for the recovery of infectious Sendai paramyxovirus from cDNA: generation of a novel copy-back nondefective interfering virus. EMBO J. *14*, 6087–6094.

Garcin, D., Itoh, M., and Kolakofsky, D. (1997). A point mutation in the Sendai virus accessory C proteins attenuates virulence for mice, but not virus growth in cell culture. Virology *238*, 424–431.

Garcin, D., Taylor, G., Tanebayashi, K., Compans, R., and Kolakofsky, D. (1998). The short Sendai virus leader region controls induction of programmed cell death. Virology *243*, 340–353.

Garcin, D., Latorre, P., and Kolakofsky, D. (1999). Sendai virus C proteins counteract the interferon-mediated induction of an antiviral state. J. Virol. *73*, 6559–6565.

Garcin, D., Curran, J., and Kolakofsky, D. (2000). Sendai virus C proteins must interact directly with cellular components to interfere with interferon action. J. Virol. *74*, 8823–8830.

Garcin, D., Curran, J., Itoh, M., and Kolakofsky, D. (2001). Longer and shorter forms of Sendai virus C proteins play different roles in modulating the cellular antiviral response. J. Virol. *75*, 6800–6807.

Garcin, D., Marq, J.B., Strahle, L., le Mercier, P., and Kolakofsky, D. (2002). All four Sendai virus C proteins bind Stat1, but only the larger forms also induce its mono-ubiquitination and degradation. Virology *295*, 256–265.

Garcin, D., Marq, J.B., Goodbourn, S., and Kolakofsky, D. (2003). The amino-terminal extensions of the longer Sendai virus C proteins modulate pY701-Stat1 and bulk Stat1 levels independently of interferon signaling. J. Virol. *77*, 2321–2329.

Garcin, D., Marq, J.B., Iseni, F., Martin, S., and Kolakofsky, D. (2004). A short peptide at the amino terminus of the Sendai virus C protein acts as an independent element that induces STAT1 instability. J. Virol. *78*, 8799–8811.

Gerber, P. (1966). Studies on the transfer of subviral infectivity from SV40-induced hamster tumor cells to indicator cells. Virology *28*, 501–509.

Gething, M.J., White, J.M., and Waterfield, M.D., (1978). Purification of the fusion protein of Sendai virus: analysis of the NH_2-terminal sequence generated during precursor activation. Proc. Natl. Acad. Sci. U.S.A. *75*, 2737–2740.

Glickman, R.L., Syddall, R.J., Iorio, R.M., Sheehan, J.P., and Bratt, M.A. (1988). Quantitative basic residue requirements in the cleavage activation site of the fusion glycoprotein as a determinant of virulence of Newcastle disease virus. J. Virol. *62*, 354–356.

Goodbourn, S., Didcock, L., and Randall, R.E. (2000). Interferons: cell signalling, immune modulation, antiviral response and virus countermeasures. J. Gen. Virol. *81*, 2341–2364.

Goodbourn, S., and Randall, R.E. (2009). The regulation of type I interferon production by paramyxoviruses. J. Interferon Cytokine Res. *29*, 539–547.

Gosselin-Grenet, A.S., Marq, J.B., Abrami, L., Garcin, D., and Roux, L. (2007). Sendai virus budding in the course of an infection does not require Alix and VPS4A host factors. Virology *365*, 101–112.

Gosselin-Grenet, A.S., Mottet-Osman, G., and Roux, L. (2006). From assembly to virus particle budding: pertinence of the detergent resistant membranes. Virology *344*, 296–303.

Gotoh, B., Ogasawara, T., Toyoda, T., Inocencio, N.M., Hamaguchi, M., and Nagai, Y. (1990). An endoprotease homologous to the blood clotting factor X as a determinant of viral tropism in chick embryo. EMBO J. *9*, 4189–4195.

Gotoh, B., Ohnishi, Y., Inocencio, N.M., Esaki, E., Nakayama, K., Barr, P.J., Thomas, G., and Nagai, Y. (1992). Mammalian subtilisin-related proteinases in cleavage-activation of the paramyxovirus fusion glycoprotein. Superiority of furin/PACE to PC2 or PC1/PC3. J. Virol. *66*, 6391–6397.

Gotoh, B., Takeuchi, K., Komatsu, T., Yokoo, J., Kimura, Y., Kato, A., Kurotani, A., and Nagai, Y. (1999). Knockout of the Sendai virus C genes eliminates the viral ability to prevent the interferon-α/β mediated responses. FEBS Lett. *459*, 205–210.

Gotoh, B., Komatsu, T., Takeuchi, K., and Yokoo, J. (2001). Paramyxovirus accessory proteins as interferon antagonists. Microbiol. Immunol. *45*, 787–800.

Gotoh, B., Komatsu, T., Takeuchi, K., and Yokoo, J. (2002). Paramyxovirus strategies for evading the interferon response. Rev. Med. Virol. *12*, 337–357.

Gotoh, B., Komatsu, T., Takeuchi, K., and Yokoo, J. (2003a). The C-terminal half-fragment of the Sendai virus C protein prevents the gamma-activated factor from binding to a gamma-activated sequence site. Virology *316*, 29–40.

Gotoh, B., Takeuchi, K., Komatsu, T., and Yokoo, J. (2003b). The STAT2 activation process is a crucial target of Sendai virus C protein for the blockade of alpha interferon signaling. J. Virol. *77*, 3360–3370.

Gotoh, B., Takeuchi, K., and Komatsu, T. (2004). Inhibition of the gamma interferon response by a

Sendai virus C protein mutant with no STAT1-binding ability. FEBS Lett. *567,* 291–296.

Grogan, C.C., and Moyer, S.A. (2001). Sendai virus wild-type and mutant C proteins show a direct correlation between L polymerase binding and inhibition of viral RNA synthesis. Virology *288,* 96–108.

Gubbay, O., Curran, J., and Kolakofsky, D. (2001). Sendai virus genome synthesis and assembly are coupled: a possible mechanism to promote viral RNA polymerase processivity. J. Gen. Virol. *82,* 2895–2903.

Gupta, K.C., and Patwardhan, S. (1988). ACG, the initiator codon for Sendai virus C protein. J. Biol. Chem. *263,* 8553–8556.

Hacein-Bey-Abina, S., Von Kalle, C., Schmidt, M., McCormack, M.P., Wulffraat, N., Leboulch, P., Lim, A., Osborne, C.S., Pawliuk, R., Morillon, E., *et al.* (2003). LMO2-associated clonal T cell proliferation in two patients after gene therapy for SCID-X1. Science *302,* 415–419.

Hamaguchi, M., Yoshida, T., Nishikawa, K., Naruse, H., and Nagai, Y. (1983). Transcriptive complex of Newcastle disease virus. I. Both L and P proteins are required to constitute an active complex. Virology *128,* 105–117.

Harris, H. (1965). Behaviour of differentiated nuclei in heterokaryons of animal cells from different species. Nature *206,* 583–588.

Harris, H., and Watkins, J.F. (1965). Hybrid cells derived from mouse and man: artificial heterokaryons of mammalian cells from different species. Nature *205,* 640–646.

Harris, H., Sidebottom, E., Grace, D.M., and Bramwell, M.E. (1969). The expression of genetic information: a study with hybrid animal cells. J. Cell Sci. *4,* 499–525.

Harrison, M.S., Sakaguchi, T., and Schmitt, A.P. (2010). Paramyxovirus assembly and budding: building particles that transmit infections. Int. J. Biochem. Cell Biol. *42,* 1416–1429.

Hasan, M.K., Kato, A., Shioda, T., Sakai, Y., Yu, D., and Nagai, Y. (1997). Creation of an infectious recombinant Sendai virus expressing the firefly luciferase gene from the 3′ proximal first locus. J. Gen. Virol. *78,* 2813–2820.

Hasan, M.K., Kato, A., Muranaka, M., Yamaguchi, R., Sakai, Y., Hatano, I., Tashiro, M., and Nagai, Y. (2000). Versatility of the accessory C proteins of Sendai virus. Contribution to virus assembly as an additional role. J. Virol. *74,* 5619–5628.

Hasegawa, Y., Kinoh, H., Iwadate, Y., Ueda, Y., Harada, Y., Saito, S., Furuya, A., Saegusa, K., Morodomi, Y., Hasegawa, M., *et al.* (2010). Urokinase-targeted cell–cell fusion by oncolytic Sendai virus eradicates orthotropic model of glioblastomas by a pronounced synergy with interferon-β. Mol. Ther. *18,* 1778–1786.

Heggeness, M.H., Scheid, A., and Choppin, P.W. (1981). The relationship of conformational changes in the Sendai virus nucleocapsid to proteolytic cleavage of the NP polypeptide. Virology *114,* 555–562.

Heggeness, M.H., Smith, P.R., and Choppin, P.W. (1982). *In vitro* assembly of the nonglycosylated membrane protein (M) of Sendai virus. Proc. Natl. Acad. Sci. USA *79,* 6232–6236.

Hendricks, D.D., Ono, E., Seyer, J.M., and Gupta, K.C. (1993). Phosphorylation of the Sendai virus C proteins. Virology *197,* 471–474.

Heylbroeck, C., Balachandran, S., Servant, M.J., DeLuca, C., Barber, G.N., Lin, R., and Hiscott, J. (2000). The IRF-3 transcription factor mediates Sendai virus-induced apoptosis. J. Virol. *74,* 3781–3792.

Hewitt, J.A., and Nermut, M.V. (1977). A morphological study of the M-protein of Sendai virus. J. Gen. Virol. *34,* 127–136.

Hidaka, Y., Kanda, T., Iwasaki, K., Nomoto, A., Shioda, T., and Shibuta, H. (1984). Nucleotide sequence of Sendai virus genome region covering the entire M gene and the 3′ proximal 1013 nucleotides of the F gene. Nucl. Acids Res. *12,* 7965–7973.

Holmes, D.E., and Moyer, S.A. (2002). The phosphoprotein (P) binding site resides in the N terminus of the L polymerase subunit of Sendai virus. J. Virol. *76,* 3078–3083.

Homma, M. (1971). Trypsin action on the growth of Sendai virus in tissue culture cells. I. Restoration of the infectivity for L cells by direct action of trypsin on L cell-borne Sendai virus. J. Virol. *8,* 619–629.

Homma, M. (1972). Trypsin action on the growth of Sendai virus in tissue culture cells. II. Restoration of the hemolytic activity of L cell-borne Sendai virus by trypsin. J. Virol. *9,* 829–835.

Homma, M., and Ohuchi, M. (1973). Trypsin action on the growth of Sendai virus in tissue culture cells. III. Structural difference of Sendai viruses grown in eggs and tissue culture cells. J. Virol. *12,* 1457–1465.

Homma, M., and Tashiro, M. (1994). Sendai virus. In Encyclopedia of Virology, vol. 3, Webster, R., and Granoff, A., eds. (Academic Press: San Diego), pp. 1299–1304.

Homma, M., Shimizu, K., Shimizu, Y.K., and Ishida, N. (1976). On the study of Sendai virus hemolysis. I. Complete Sendai virus lacking in hemolytic activity. Virology *71,* 41–47.

Horikami, S.M., Curran, J., Kolakofsky, D., and Moyer, S.A. (1992). Comlpexes of Sendai virus NP-P and P-L proteins are required for defective interfering particle genome replication *in vitro*. J. Virol. *66,* 4901–4908.

Horikami, S.M., Hector, R.E., Smallwood, S., and Moyer, S.A. (1997). The Sendai virus C protein binds the L polymerase protein to inhibit viral RNA synthesis. Virology *235,* 261–270.

Hosaka, Y. (1968). Isolation and structure of the nucleocapsid of HVJ. Virology *35,* 445–457.

Hou, X., Suquilanda, E., Zeledon, A., Kacsinta, A., Moore, A. Seto, J., and McQueen, N. (2005). Mutations in Sendai virus variant F1-R that correlate with plaque formation in the absence of trypsin. Med. Microbiol. Immunol. *194,* 129–136.

Houben, K., Marion, D., Tarbouriech, N., Ruigrok, R.W.H., and Blanchard, L. (2007). Interaction of the C-terminal domains of Sendai virus N and P proteins: comparison of polymerase–nucleocapsid interactions within the paramyxovirus family. J. Virol. *81,* 6807–6816.

Howe, C., and Morgan, C. (1969). Interactions between Sendai virus and human erythrocytes. J. Virol. *3,* 70–81.

Hu, C., Kato, A., Bowman, M.C., Kiyotani, K., Yoshida, T., Moyer, S.A., Nagai, Y., and Gupta, K.C. (1999). Role of primary constitutive phosphorylation of Sendai virus P and V proteins in viral replication and pathogenesis. Virology *263,* 195–208.

Hu, C.-J., and Gupta, K.C. (2000). Functional significance of alternate phosphorylation in Sendai virus P protein. Virology *268,* 517–532.

Huang, C., Kiyotani, K., Fujii, Y., Fukuhara, N., Kato, A., Nagai, Y., Yoshida, T., and Sakaguchi, T. (2000). Involvement of the zinc-binding capacity of Sendai virus V protein in viral pathogenesis. J. Virol. *74,* 7834–7841.

Hughes, S., Mellstrom, K., Kosik, E., Tamanoi, F., and Brugge, J. (1984). Mutation of a termination codon affects *src* initiation. Mol. Cell. Biol. *4,* 1738–1746.

Huntley, C.C., De, B.P., and Banerjee, A.K. (1997). Phosphorylation of Sendai virus phosphoprotein by cellular protein kinase C ζ. J. Biol. Chem. *272, 16,*578–16584.

Hurwitz, J.L. (2008). Development of recombinant Sendai virus vaccines for prevention of human parainfluenza and respiratory syncytial virus infections. Pediatr. Infect. Dis. J. *27,* S126-S128.

Ikegame, S., Takeda, M., Ohno, S., Nakatsu, Y., Nakanishi, Y., and Yanagi, Y. (2010). Both RIG-I and MDA5 RNA helicases contribute to the induction of alpha/beta interferon in measles virus-infected human cells. J. Virol. *84,* 372–379.

Inoue, M., Tokusumi, Y., Ban, H., Kanaya, T., Shirakura, M., Tokusumi, T., Hirata, T., Nagai, Y., Iida, A., and Hasegawa, M. (2003a). A new type of Sendai virus vector deficient in the matrix gene has lost virus particle formation and gained extensive cell-to-cell spreading. J. Virol. *77,* 6419–6429.

Inoue, M., Tokusumi, Y., Ban, H., Kanaya, T., Tokusumi, T., Nagai, Y., Iida, A., and Hasegawa, M. (2003b). Nontransmissible virus-like particle formation by F-deficient Sendai virus is temperature sensitive and reduced by mutations in M and HN proteins. J. Virol. *77,* 3238–3246.

Inoue, M., Tokusumi, Y., Ban, H., Shirakura, M., Kanaya, T., Yoshizaki, M., Hironaka, T., Nagai, Y., Iida, A., and Hasegawa, M. (2004). Recombinant Sendai virus vectors deleted in both the matrix and the fusion genes: efficient gene transfer with preferable properties. J. Gene Med. *10,* 1069–1081.

Irie, T., Shimazu, Y., Yoshida, T., and Sakaguchi, T. (2007). The YLDL sequence within Sendai virus M protein is critical for budding of virus-like particles and interacts with Alix/AIP1 independently of C protein. J. Virol. *81,* 2263–2273.

Irie, T., Nagata, N., Yoshida, T., and Sakaguchi, T. (2008a). Paramyxovirus Sendai virus C proteins are essential for maintenance of negative-sense RNA genome in virus particles. Virology *374,* 495–505.

Irie, T., Nagata, N., Yoshida, T., and Sakaguchi, T. (2008b). Recruitment of Alix/AIP1 to the plasma membrane by Sendai virus C protein facilitates budding of virus-like particles. Virology *371,* 108–120.

Irie, T., Inoue, M., and Sakaguchi, T. (2010a). Significance of the YLDL motif in the M protein and Alix/AIP1 for Sendai virus budding in the context of virus infection. Virology *405,* 334–341.

Irie, T., Nagata, N., Igarashi, T., Okamoto, I., and Sakaguchi, T. (2010b). Conserved charged amino acids within Sendai virus C protein play multiple roles in the evasion of innate immune responses. PLoS One 5, e10719

Iseni, F., Garcin, D., Nishio, M., Kedersha, N., Anderson, P., and Kolakofsky, D. (2002). Sendai virus trailer RNA binds TIAR, a cellular protein involved in virus-induced apoptosis. EMBO J. *21,* 5141–5150.

Ishida, N., and Homma, M. (1961). Host-controlled variation observed with Sendai virus grown in mouse fibroblast (L) cells. Virology *14,* 486–488.

Ishida, N., and Homma, M. (1978). Sendai virus. Adv. Virus Res. *23,* 349–383.

Ito, T., Suzuki, Y., Takada, A., Kawamoto, A., Otsuki, K., Masuda, H., Yamada, M., Suzuki, T., Kida, H., and Kawaoka, Y. (1997). Differences in sialic acid-galactose linkages in the chicken egg amnion and allantois influence human influenza virus receptor specificity and variant selection. J. Virol. *71,* 3357–3362.

Itoh, M., Shibuta, H., and Homma, M. (1987). Single amino acid substitution of Sendai virus at the cleavage site of the fusion protein confers trypsin resistance. J. Gen. Virol. *68,* 2939–2944.

Itoh, M., Isegawa, Y., Hotta, H., and Homma, M. (1997). Isolation of an avirulent mutant of Sendai virus with two amino acid mutations from a highly virulent field strain through adaptation to LLC-MK2 cells. J. Gen. Virol. *78,* 3207–3215.

Iwata, S., Schmidt. A.C., Titani, K., Suzuki, M., Kido, H., Gotoh, B., Hamaguchi, M., and Nagai, Y. (1994). Assignment of disulfide bridges in the fusion glycoprotein of Sendai virus. J. Virol. *68,* 3200–3206.

Johnston, M.D. (1981). The characteristics required for a Sendai virus preparation to induce high levels of interferon in human lymphoblastoid cells. J. Gen. Virol. *56,* 175–184.

Jones, B., Zhan, X., Mishin, V., Slobod, K.S., Surman, S., Russell, C.J., Portner, A., and Hurwitz, J.L. (2009). Human PIV-2 recombinant Sendai virus (rSeV) elicits durable immunity and combines with two additional rSeVs to protect against hPIV-1, hPIV-2, hPIV-3, and RSV. Vaccine *27,* 1848–1857.

Kagiyama, N., Itoh, T., Takakura, A., Yoshimura, M., and Nomura, T. (1987). Microbiological monitoring in inbred mouse foundation stocks in Japan. Exp. Anim. *36,* 135–142.

Kaneda, Y. (2008). Applications of hemagglutinating virus of Japan in therapeutic delivery systems. Expert Opin. Drug Deliv. *5,* 221–233.

Kaneda, Y., Nakajima, T., Nishikawa, T., Yamamoto, S., Ikegami, H., Suzuki, N., Nakamura, H., Morishita, R., and Kotani, H. (2002). Hemagglutinating virus of Japan (HVJ) envelope vector as a versatile gene delivery system. Mol. Ther. *6,* 219–226.

Kaneda, Y., Yamamoto, S., and Nakajima, T. (2005). Development of HVJ envelope vector and its application to gene therapy. In Non-Viral Vectors for Gene Therapy, Huang, L., Huang, M.-C., Wagner, E., eds. (Elsevier Academic Press, Amsterdam), pp. 307–332.

Kano, M., Matano, T., Nakamura, H., Takeda, A., Kato, A., Ariyoshi, K., Mori, K., Sata, T., and Nagai, Y. (2000). Elicitation of protective immunity against simian immunodeficiency virus by a recombinant Sendai virus expressing the Gag protein. AIDS *14,* 1281–1282.

Kato, A., Sakai, Y., Shioda, T., Kondo, T., Nakanishi, M., and Nagai, Y. (1996). Initiation of Sendai virus multiplication from transfected viral cDNA or RNA with negative or positive sense. Genes Cells *1,* 569–579.

Kato, A., Kiyotani, K., Sakai, Y., Yoshida, T., and Nagai, Y. (1997a). The paramyxovirus, Sendai virus, V protein encodes a luxury function required for viral pathogenesis. EMBO J. *16,* 578–587.

Kato, A., Kiyotani, K., Sakai, Y., Yoshida, T., Shioda, T., and Nagai, Y. (1997b). Importance of the cysteine-rich carboxyl-terminal half of V protein for Sendai virus pathogenesis. J. Virol. *71,* 7266–7272.

Kato, A., Kiyotani, K., Hasan, M.K., Shioda, T., Sakai, Y., Yoshida, T., and Nagai, Y. (1999). Sendai virus gene start signals are not equivalent in reinitiation capacity: moderation at the F gene. J. Virol. *73,* 9273–9246.

Kato, A., Ohnishi, Y., Kohase, M., Saito, S., Tashiro, M., and Nagai, Y. (2001). Y2, the smallest of the Sendai virus C proteins, is fully capable of both counteracting the antiviral action of interferons and inhibiting viral RNA synthesis. J. Virol. *75,* 3802–3810.

Kato, A., Ohnishi, Y., Hishiyama, M., Kohase, M., Saito, S., Tashiro, M., and Nagai, Y. (2002). The amino-terminal half of Sendai virus C protein is not responsible for either counteracting the antiviral action of interferons or down-regulating viral RNA synthesis. J. Virol. *76,* 7114–7124.

Kato, A., Cortese-Grogan, C., Moyer, S.A., Sugahara, F., Sakaguchi, T., Kubota, T., Otsuki, N., Kohase, M., Tashiro, M., and Nagai, Y. (2004). Characterization of the amino acid residues of Sendai virus C protein that are critically involved in its interferon antagonism and RNA synthesis down-regulation. J. Virol. *78,* 7443–7454.

Kato, A., Kiyotani, K., Kubota, T., Yoshida, T., Tashiro, M., and Nagai, Y. (2007). Importance of the anti-interferon capacity of Sendai virus C protein for pathogenicity in mice. J. Virol. *81,* 3264–3271.

Kato, H., Takeuchi, O., Sato, S., Yoneyama, M., Yamamoto, M., Matsui, K., Uematsu, S., Jung, A., Kawai, T., Ishii, K.J., *et al.* (2006). Differential role of MDA5 and RIG-I in the recognition of RNA viruses. Nature *441,* 101–105.

Kato, K., Nakanishi, M., Kaneda, Y., Uchida, T., and Okada, Y. (1991). Expression of hepatitis B virus surface antigen in adult rat liver; co-introduction of DNA and nuclear protein by a simplified liposome method. J. Biol. Chem. *266,* 3361–3364.

Kawachi, M., Tamai, K., Saga, K., Yamazaki, T., Fujita, H., Shimbo, T., Kikuchi, Y., Nimura, K., Nishifuji, K., Amagai, M., *et al.* (2007). Development of tissue-targeting hemagglutinating virus of Japan envelope vector for successful delivery of therapeutic gene to mouse skin. Hum. Gene Ther. *18,* 881–894.

Kido, H., Yokogoshi, Y., Sakai, K., Tashiro, M., Kishino, Y., Fukutomi, A., and Katunuma, N. (1992). Isolation and characterization of a novel trypsin-like protease found in rat bronchiolar epithelial Clara cells: a possible activator of the viral fusion glycoprotein. J. Biol. Chem. *267,* 13573–13579.

Kimura, Y., Ito, Y., Shimokata, K., Nishiyama, Y., Nagata, I., and Kitoh, J. (1975). Temperature-sensitive virus derived from BHK cells persistently infected with HVJ (Sendai virus). J. Virol. *15,* 55–63.

Kinoh, H., Inoue, M., Washizawa, K., Yamamoto, T., Fujikawa, S., Tokusumi, Y., Iida, A., Nagai, Y., and Hasegawa, M. (2004). Generation of a recombinant Sendai virus that is selectively activated and lyses human tumor cells expressing matrix metalloproteinases. Gene Ther. *11,* 1137–1145.

Kinoh, H., Inoue, M., Komaru, A., Ueda, Y., Hasegawa, M., and Yonemitsu, Y. (2009). Generation of optimized and urokinase-targeted oncolytic Sendai virus vectors applicable for various human malignancies. Gene Ther. *16,* 392–403.

Kiyotani, K., Takao, S., Sakaguchi, T., and Yoshida, T. (1990). Immediate protection of mice from lethal wild-type Sendai virus (HVJ) infections by a temperature-sensitive mutant, HVJpi, possessing homologous interfering capacity. Virology *177,* 65–74.

Kiyotani, K., Sakaguchi, T., Fujii, Y., and Yoshida, T (2001). Attenuation of a field Sendai virus isolate through egg-passages is associated with an impediment of virus genome replication in mouse respiratory cells. Arch. Virol. *146,* 893–908.

Kiyotani, K., Sakaguchi, T., Kato, A., Nagai, Y., and Yoshida, T. (2007). Paramyxovirus Sendai virus V protein counteracts innate virus clearance through IRF-3 activation, but not via interferon, in mice. Virology *359,* 82–91.

Klenk, H.-D., and Garten, W. (1994). Host cell proteases controlling virus pathogenicity. Trends Microbiol. *2,* 39–43.

Köhler, G., and Milstein, C. (1975). Continuous cultures of fused cells secreting antibody of predefined specificity. Nature *256,* 495–497.

Kolakofsky, D., Boy de la Tour, E., and Bruschi, A. (1974). Self-annealing of Sendai virus RNA. J. Virol. *14,* 33–39.

Kolakofsky, D., and Bruschi, A. (1975). Antigenomes in Sendai virions and Sendai virus-infected cells. Virology *66,* 185–191.

Kolakofsky, D., Pelet, T., Garcin, D., Hausmann, S., Curran, J., and Roux, L. (1998). Paramyxovirus RNA synthesis and the requirement for hexamer genome length: the rule of six revisited. J. Virol. *72,* 891–899.

Kolokoltsov, A.A., Deniger, D., Fleming, E.H., Roberts, N.J., Jr., Karpilow, J.M., and Davey, R.A. (2007). Small interfering RNA profiling reveals key role of clathrin-mediated endocytosis and early endosome formation for infection by respiratory syncytial virus. J. Virol. *81,* 7786–7800.

Komaru, A., Ueda, Y., Furuya, A., Tanaka, S., Yoshida, K., Kato, T., Kinoh, H., Harada, Y., Suzuki, H., Inoue, M., *et al.* (2009). Sustained and NK/CD4$^+$ T cell-dependent efficient prevention of lung metastasis induced by dendritic cells harboring recombinant Sendai virus. J. Immunol. *183,* 4211–4219.

Komatsu, T., Takeuchi, K., Yokoo, J., Tanaka, Y., and Gotoh, B. (2000). Sendai virus blocks alpha interferon signaling to signal transducers and activators of transcription. J. Virol. *74,* 2477–2480.

Komatsu, T., Takeuchi, K., Yokoo, J., and Gotoh, B. (2002). Sendai virus C protein impairs both phosphorylation and dephosphorylation processes of Stat1. FEBS Lett. *511,* 139–144.

Komatsu, T., Takeuchi, K., Yokoo, J., and Gotoh, B. (2004). C and V proteins of Sendai virus target signaling pathways leading to IRF-3 activation for the negative regulation of interferon-beta production. Virology *325,* 137–148.

Komatsu, T., Takeuchi, K., and Gotoh, B. (2007). Bovine parainfluenza virus type 3 accessory proteins that suppress beta interferon production. Microbes Infect. *9,* 954–962.

Kondo, T., Yoshida, T., Miura, N., and Nakanishi, M. (1993). Temperature sensitive phenotype of a mutant Sendai virus strain is caused by its insufficient accumulation of the M protein. J. Biol. Chem. *268,* 21924–21930.

Koyama, A.H., Ogawa, M., Kato, A., Nagai, Y., and Adachi, A. (2001). Lack of apoptosis in Sendai virus-infected Hep-2 cells without participation of viral antiapoptosis gene. Microbes Infect. *3,* 1115–1121.

Koyama, A.H., Irie, H., Kato, A., Nagai, Y., and Adachi, A. (2003). Virus multiplication and induction of apoptosis by Sendai virus: role of the C proteins. Microbes Infect. *5,* 373–378.

Kozak, M. (1984). Selection of initiation sites by eucaryotic ribosomes: effect of inserting AUG triplets upstream from the coding sequence for preproinsulin. Nucl. Acids Res. *12,* 3873–3893.

Kozak, M. (1987). Effects of intercistronic length on the efficiency of reinitiation by eucaryotic ribosomes. Mol. Cell. Biol. *7,* 3438–3445.

Kurotani, A., Kiyotani, K., Kato, A., Shioda, T., Sakai, Y., Mizumoto, K., Yoshida, T., and Nagai, Y. (1998). The Sendai virus, C proteins are categorically nonessential gene products but silencing their expression severely impairs viral replication and pathogenesis. Genes Cell *3,* 111–124.

Kuroya, M., Ishida, N., and Shiratori, T. (1953). Newborn virus pneumonitis (Type Sendai) II. Report: The isolation of a new virus possessing hemagglutinin activity. Yokohama Med. Bull. *4,* 217–233.

Lamb, R.A., and Choppin, P.W. (1977a). The synthesis of Sendai virus polypeptides in infected cells. II. Intracellular distribution of polypeptides. Virology *81,* 371–381.

Lamb, R.A., and Choppin, P.W. (1977b). The synthesis of Sendai virus polypeptides in infected cells. III. Phosphorylation of polypeptides. Virology *81,* 382–397.

Lamb, R.A., and Parks, G.D. (2007). *Paramyxoviridae*: the viruses and their replication. In Fields Virology, vol. 1, 5th ed., Knipe, D.M., and Howley, P.M., eds. (Lippincott, Williams and Wilkins: Philadelphia), pp. 1449–1496.

Latorre, P., Cadd, T., Itoh, M., Curran, J., and Kolakofsky, D. (1998a). The various Sendai virus C proteins are not functionally equivalent and exert both positive and negative effects on viral RNA accumulation during the course of infection. J. Virol. *72,* 5984–5993.

Latorre, P., Kolakofsky, D., and Curran, J. (1998b). Sendai virus Y proteins are initiated by a ribosomal shunt. Mol. Cell. Biol. *18,* 5021–5031.

Le Mercier, P., Garcin, D., Hausmann, S., and Kolakofsky, D. (2002). Ambisense Sendai viruses are inherently unstable but are useful to study viral RNA synthesis. J. Virol. *76,* 5492–5502.

Lenardo, M.J., Fan, C.M., Maniatis, T., and Baltimore, D. (1989). The involvement of NF-kappa B in beta-interferon gene regulation reveals its role as widely inducible mediator of signal transduction. Cell *57,* 287–294.

Leyrer, S., Neubert, W.J., and Sedlmeier, R. (1998). Rapid and efficient recovery of Sendai virus from cDNA: factors influencing recombinant virus rescue. J. Virol. Meth. *75,* 47–58.

Li, H.-O., Ueda, Y., Asakawa, M., Kuma, H., Hirata, T., Lee, Y.-S., Fukumura, M., Iida, A., Kato, T., Nagai, Y., *et al.* (2000). A cytoplasmic RNA vector derived from non-transmissible Sendai virus with efficient gene transfer and expression. J. Virol. *74,* 6564–6569.

Li, M., Schmitt, P.T., Li, Z., McCrory, T.S., He, B., and Schmitt, A.P. (2009). Mumps virus matrix, fusion, and nucleocapsid proteins cooperate for efficient production of virus-like particles. J. Virol. *83,* 7261–7272.

Liang, C.T., Shih, A., Chang, Y.H., Liu, C.W., Lee, Y.T., Hsieh, W.C., Huang, Y.L., Huang, W.T., Kuang C.H., Lee, K.H., *et al.* (2009). Microbial contaminations of laboratory mice and rats in Taiwan from 2004 to 2007. J. Am. Assoc. Lab. Anim. Sic. *48,* 381–386.

Lin, R., Heylbroeck, C., Pitha, P.M., and Hiscott, J. (1998). Virus-dependent phosphorylation of the IRF-3 transcription factor regulates nuclear translocation, transactivation potential, and proteasome-mediated degradation. Mol. Cell. Biol. *18,* 2986–2996.

Littlefield, J.W. (1966). The use of drug-resistant markers to study the hybridization of mouse fibroblasts. Exp. Cell Res. *41,* 190–196.

Liu, C.-C., Simonsen, C.C., and Levinson, A.D. (1984). Initiation of translation at internal AUG codons in mammalian cells. Nature *309,* 82–85.

Loh, Y.-H., Hartung, O., Guo, C., Sahalie, J.M., Manos, P.D., Urbach, A., Heffner, G.C., Grskovic, M., Vigneault, F., Lensch, M.W., *et al.* (2010). Reprogramming of T cells from human peripheral blood. Cell Stem Cell. *7,* 15–19.

Malur, A.G., Chattopadhyay, S., Maitra, R.K., and Banerjee, A.K. (2005). Inhibition of STAT 1 phosphorylation by human parainfluenza virus type 3 C protein. J. Virol. *79,* 7877–7882.

Markwell, M.A., and Fox, C.F. (1980). Protein–protein interactions within paramyxoviruses identified by native disulfide bonding or reversible chemical cross-linking. J. Virol. *33,* 152–166.

Markwell, M.A.K., and Paulson, J.C. (1980). Sendai virus utilizes specific sialyloligosaccharides as host cell receptor determinants. Proc. Natl. Acad. Sci. U.S.A. *77,* 5693–5697.

Markwell, M.A., Portner, A., and Schwartz, A.L. (1985). An alternative route of infection for viruses: entry by means of the asialoglycoprotein receptor of a Sendai virus mutant lacking its attachment protein. Proc. Natl. Acad. Sci. U.S.A. *82,* 978–982.

Marq, J.-B., Brini, A., Kolakofsky, D., and Garcin, D. (2007). Targeting of the Sendai virus C protein to the plasma membrane via a peptide-only membrane anchor. J. Virol. *81,* 3187–3197.

Masaki, I., Yonemitsu, Y., Komori, K., Ueno, H., Nakashima, Y., Nakagawa, K., Fukumura, M., Kato, A., Hasan, M.K., Nagai, Y., *et al.* (2001). Recombinant Sendai virus-mediated gene transfer to vasculature: a new class of efficient gene transfer vector to the vascular system. FASEB J. *15,* 1294–1296.

Masaki, I., Yonemitsu, Y., Yamashita, A., Sata, S., Tanii, M., Komori, K., Nakagawa, K., Hou, X., Nagai, Y., Hasegawa, M., *et al.* (2002). Angiogenic gene therapy for experimental critical limb ischemia: acceleration of limb loss by overexpression of vascular endothelial growth factor 165 but not of fibroblast growth factor-2. Circ. Res. *90,* 966–973.

Matano, T., Kobayashi, M., Igarashi, H., Takeda, A., Nakamura, H., Kano, M., Sugimoto, C., Mori, K., Iida, A., Hirata, T., *et al.* (2004). Cytotoxic T lymphocyte-based control of simian immunodeficiency virus replication in a preclinical AIDS vaccine trial. J. Exp. Med. *199,* 1709–1718.

Matsumoto, T., and Maeno, K. (1962). A host-induced modification of hemagglutinating virus of Japan (HVJ, Sendai virus) in its hemolytic and cytopathic activity. Virology *17,* 563–570.

Matsuoka, Y., Curran, J., Pelet, T., Kolakofsky, D., Ray, R., and Compans, R.W. (1991). The P gene of human parainfluenza virus type 1 encodes P and C proteins but not a cysteine-rich V protein. J. Virol. *65,* 3406–3410.

Meulendyke, K.A., Wurth, M.A., McCann, R.O., and Dutch, R.E. (2005). Endocytosis plays a critical role in proteolytic processing of the Hendra virus fusion protein. J. Virol. *79,* 12643–12649.

Mima, H., Yamamoto, S., Ito, M., Tomoshige, R., Tabata, Y., Tamai, K., and Kaneda, Y. (2006). Targeted chemotherapy against intraperitoneally disseminated colon carcinoma using a cationized gelatin-conjugated HVJ envelope vector. Mol. Cancer Ther. *5,* 1021–1028.

Miyauchi, K., Kim, Y., Latinovic, O., Morozov, V., and Melikyan, G.B. (2009). HIV enters cells via endocytosis and dynamin–dependent fusion with endosomes. Cell *137,* 433–444.

Morgan, C., and Howe, C. (1968). Structure and development of viruses as observed in the electron microscope. IX. Entry of parainfluenza I (Sendai) virus. J. Virol. *2,* 1122–1132.

Mottet, G., and Roux, L. (1989). Budding efficiency of Sendai virus nucleocapsids: influence of size and ends of the RNA. Virus Res. *14,* 175–187.

Mottet-Osman, G., Iseni, F., Pelet, T., Wiznerowics, M., Garcin, D., and Roux, L. (2007). Suppression of the Sendai virus M protein through a novel short interfering RNA approach inhibits viral particle production but does not affect viral RNA analysis. J. Virol. *81,* 2861–2868.

Murphy, A.M., and Grdzelishvili, V.Z. (2009). Identification of Sendai virus L protein amino acid residues affecting viral mRNA cap methylation. J. Virol. *83,* 1669–1681.

Myers, T.M., and Moyer, S.A. (1997). An amino-terminal domain of the Sendai virus nucleocapsid protein is required for template function. J. Virol. *71,* 918–924.

Myers, T.M., Pieters A., and Moyer, S.A. (1997). A highly conserved region of the Sendai virus nucleocapsid protein contributes to the NP-NP binding domain. Virology *229,* 322–335.

Nagai, Y. (1993). Protease-dependent virus tropism and pathogenicity. Trends Microbiol. *1,* 81–87.

Nagai, Y. (1995). Virus activation by host proteinases. A pivotal role in the spread of infection, tissue tropism and pathogenicity. Microbiol. Immunol. *39,* 1–9.

Nagai, Y. (1999). Paramyxovirus replication and pathogenesis. Reverse genetics transforms understanding. Rev. Med. Virol. *9,* 83–99.

Nagai, Y., and Kato, A. (1999). Paramyxovirus reverse genetics is coming of age. Microbiol. Immunol. *43,* 613–624.

Nagai, Y., and Kato, A. (2004). Accessory genes of the *Paramyxoviridae*, a large family of nonsegmented negative-strand and RNA viruses, as a focus of active investigation by reverse genetics. Curr. Top. Microbiol. Immunol. *283,* 197–248.

Nagai, Y., and Yoshida, T. (1984). Viral pathogenesis: mechanism of acute and persistent infections with paramyxoviruses. Nagoya J. Med. Sci. *46,* 1–17.

Nagai, Y., Yoshida, T., Yoshii, S., Maeno, K., and Matsumono, T. (1975). Modification of normal cell surface by smooth membrane preparations from BHK-21 cells infected with Newcastle disease virus. Med. Microbiol. Immunol. *161,* 175–188.

Nagai, Y., Klenk, H.-D., and Rott, R. (1976a). Proteolytic cleavage of the viral glycoproteins and its significance for the virulence of Newcastle disease virus. Virology *72,* 494–508.

Nagai, Y., Ogura, H., and Klenk, H.-D. (1976b). Studies on the assembly of the envelope of Newcastle disease virus. Virology *69,* 523–538.

Nagai, Y., Hamaguchi, M., Toyoda, T., and Yoshida, T. (1983a). The uncoating of paramyxoviruses may not require a low pH mediated step. Virology *130,* 263–268.

Nagai, Y., Yoshida, T., Hamaguchi, M., Nagura, H., Hasegawa, H., Yoshimura, S., and Watanabe,

K. (1983b). Subcellular location of the major protein antigens of paramyxoviruses revealed by immunoperoxidase cytochemistry. Microbiol. Immunol. *27,* 531–545.

Nagai, Y., Inoue, M., Iida, A., Zhu, Y.-F., Hasegawa, M., Kato, A., and Matano, T. (2007). Sendai virus engineering: from reverse genetics to vector development. In Virus Expression Vectors, Hefferon, K.L., ed. (Transworld ReseArch. Network: Kerala, India), pp. 123–146.

Nagata, I., Kimura, Y., Ito, Y., and Tanaka, T. (1972). Temperature-sensitive phenomenon of viral maturation observed in BHK cells persistently infected with HVJ. Virology *49,* 453–461.

Nakanishi, M., Uchida, T., Sugawa, H., Ishiura, M., and Okada, Y. (1985). Efficient introduction of contents of liposomes into cells using HVJ (Sendai virus). Exp. Cell Res. *159,* 399–409.

Nakatsu, Y., Takeda, M., Ohno, S., Koga, R., and Yanagi, Y. (2006). Translational inhibition and increased interferon induction in cells infected with C protein-deficient measles virus. J. Virol. *80,* 11861–11867.

Nakatsu, Y., Takeda, M., Ohno, S., Shirogane, Y., Iwasaki, M., and Yanagi, Y. (2008). Measles virus circumvents the host interferon response by different actions of the C and V proteins. J. Virol. *82,* 8296–8306.

Nishimura, K., Segawa, H., Goto, T., Morishita, M., Masago, A., Takahashi, H., Ohmiya, Y., Sakaguchi, T., Asada, M., Imamura, T., *et al.* (2007). Persistent and stable gene expression by a cytoplasmic RNA replicon based on a noncytopathic variant Sendai virus. J. Biol. Chem. *282,* 27383–27391.

Nishio, M., Nagata, A., Tsurudome, M., Ito, M., Kawano, M., Komada, H., and Ito, Y. (2004). Recombinant Sendai viruses with L1618V mutation in their L polymerase protein establish persistent infection, but not temperature sensitivity. Virology *329,* 289–301.

Norgren, L., Hiatt, W.R., Dormandy, J.A., Harris, K.A., Fowkes, F.G., and TASC II Working Group. (2007). J. Vasc. Surg. 45, (*Suppl.*) S5–67.

Ogasawara, T., Gotoh, B., Suzuki, H., Asaka, J., Shimokata, K., Rott, R., and Nagai, Y. (1992). Expression of factor X and its significance for the determination of paramyxovirus tropism in the chick embryo. EMBO J. *11,* 467–472.

Ogino, T., Kobayashi, M., Iwama, M., and Mizumoto, K. (2005). Sendai virus RNA-dependent RNA polymerase L protein catalyzes cap methylation of virus-specific mRNA. J. Biol. Chem. *280,* 4429–4435.

Ohnishi, Y., Shioda, T., Nakayama, K., Iwata, S., Gotoh, B., Hamaguchi, M., and Nagai, Y. (1995). A furin-defective cell line is able to process correctly the gp160 of human immunodeficiency virus type 1. J. Virol. *68,* 4075–4079.

Okada, H., Seto, J.T., McQueen, N.L., Klenk H.D., Rott, R., and Tashiro, M. (1998). Determinants of pantropism of the F1-R mutant of Sendai virus: specific mutations involved are in the F and M genes. Arch. Virol. *143,* 2343–2352.

Okada, M., and Kita, Y. (2010). Tuberculosis vaccine development. The development of novel (preclinical) DNA vaccine. Human Vaccines *6,* 1–12.

Okada, Y. (1958). The fusion of Ehrlich's tumor cells caused by HVJ virus *in vitro*. Biken J. *1,* 103–110.

Okada, Y. (1993). Sendai virus-induced cell fusion. Methods Enzymol. *221,* 18–41.

Okada, Y., and Murayama, F. (1965). Multinucleated giant cell formation by fusion between cells of different strains. Biken J. *8,* 7–21.

Okada, Y., and Tadokoro, J. (1962). Analysis of giant polynuclear cell formation caused by HVJ virus from Ehrlich's ascites tumor cells. II. Quantitative analysis of giant polynuclear cell formation. Exp. Cell Res. *26,* 108–118.

Okada, Y., Suzuki, T., and Hosaka, Y. (1957). Interaction between influenza virus and Ehrlich's tumor cells. III. Fusion phenomenon of Ehrlich's tumor cell by the action of HVJ Z strain. Med. J. Osaka Univ. *7,* 709–717.

Oshima, K., Shimamura, M., Mizuno, S., Tamai, K., Doi, K., Morishita, R., Nakamura, T., Kubo, T., and Kaneda, Y. (2004). Intrathecal injection of HVJ-E containing HGF gene to cerebrospinal fluid can prevent and ameliorate hearing impairment in rats. FASEB J. *18,* 212–214.

Pager, C.T., and Dutch, R.E. (2005). Cathepsin L is involved in proteolytic processing of the Hendra virus fusion protein. J. Virol. *79, 12,*714–12720.

Palese, P., Zheng, H., Engelhardt, O.G., Pleschka, S., and García-Sastre, A. (1996). Negative-strand RNA viruses: genetic engineering and applications. Proc. Natl. Acad. Sci. U.S.A. *93,* 11354–11358.

Peeters, B.P.H., de Leeuw, O.S., Koch, G., and Gielkens, A.L.J. (1999). Rescue of Newcastle disease virus from cloned cDNA: Evidence that cleavability of the fusion protein is a major determinant of virulence. J. Virol. *73,* 5001–5009.

Peters, K., Chattopadhyay, S., and Sen, G.C. (2008). IRF-3 activation by Sendai virus infection is required for cellular apoptosis and avoidance of persistence. J. Virol. *82,* 3500–3508.

Pichlmair, A., Schulz, O., Tan, C.P., Rehwinkel, J., Kato, H., Takeuchi, O., Akira, S., Way, M., Schiavo, G., and Reis e Sousa, C. (2009). Activation of MDA5 requires higher-order RNA structures generated during virus infection. J. Virol. *83,* 10761–10769.

Pinkenburg, O., Vogelmeier, C., Bossow, S., Neubert, W.J., Lutz, R.B., Ungerechts, G., Lauer, U.M., Bitzer, M., and Bals, R. (2004). Recombinant Sendai virus for efficient gene transfer to human airway epithelium. Exp. Lung Res. *30,* 83–96.

Poch O., Blumberg, B.M., Bougueleret, L., and Tordo, N. (1990). Sequence comparison of five polymerases (L proteins) of unsegmented negative-strand RNA viruses: theoretical assignment of functional domains. J. Gen. Virol. *71,* 1153–1162.

Poole, E., He, B., Lamb, R.A., Randall, R.E., and Goodbourn, S. (2002). The V proteins of simian virus 5 and other paramyxoviruses inhibit induction of interferon-beta. Virology *303,* 33–46.

Portner, A., and Murti, K.G. (1986). Localization of P, NP, and M proteins on Sendai virus nucleocapsid using immunogold labeling. Virology *150,* 469–478.

Portner, A., Gupta, K.C., Seyer, J.M., Beachey, E.H., and Kingsbury, D.W. (1986). Localization and characterization of Sendai virus nonstructural C and C′ proteins by antibodies against synthetic peptides. Virus Res. *6,* 109–121.

Portner, A., Murti, K.G., Morgan, E.M., and Kingsbury, D.W. (1988). Antibodies against Sendai virus L protein: distribution of the protein in nucleocapsids revealed by immunoelectron microscopy. Virology *163,* 236–239.

Qanungo, K.R., Shaji, D., Mathur, M., and Banerjee, A.K. (2004). Two RNA polymerase complexes from vesicular stomatitis virus-infected cells that carry out transcription and replication of genome RNA. Proc. Natl. Acad. Sci. U.S.A. *101,* 5952–5957.

Randall, R.E., and Goodbourn, S. (2008). Interferons and viruses: an interplay between induction, signalling, antiviral responses and virus countermeasures. J. Gen. Virol. *89,* 1–47.

Randall, R.E., and Russell, W.C. (1991). Paramyxovirus persistence: consequences for host and virus. In The Paramyxoviruses, Kingsbury, D.W., ed. (Plenum Press: New York), pp. 299–321.

Re, G.G. (1991). Deletion mutants of paramyxoviruses. In The Paramyxoviruses, Kingsbury, D.W., ed. (Plenum Press: New York), pp. 275–298.

Re, G.G., and Kingsbury, D.W. (1986). Nucleotide sequences that affect replicative and transcriptional efficiencies of Sendai virus deletion mutants. J. Virol. *58,* 578–582.

Rehwinkel, J. (2010). Exposing viruses: RNA patterns sensed by RIG-I-like receptors. J. Clin. Immunol. *30,* 491–495.

Rehwinkel, J., and Reis e Sousa, C. (2010). RIGorous detection: exposing virus through RNA sensing. Science *327,* 284–286.

Rehwinkel, J., Tan, C.P., Goubau, D., Schulz, O., Pichlmair, A., Bier, K., Robb, N., Vreede, F., Barclay, W., Fodor, E., *et al.* (2010). RIG-I detects viral genomic RNA during negative-strand RNA virus infection. Cell *140,* 397–408.

Rima, B.K., and Martin, S.J. (1976). Persistent infection of tissue culture cells by RNA viruses. Med. Microbiol. Immunol. *162,* 89–118.

Roux, L., and Holland, J.J. (1979). Role of defective interfering particles of Sendai virus in persistent infections. Virology *93,* 91–103.

Roux, L., and Holland, J.J. (1980). Viral genome synthesis in BHK-21 cells persistently infected with Sendai virus. Virology *100,* 53–64.

Ruddle, F.H., and Creagan, R.P. (1975). Parasexual approaches to the genetics of man. Annu. Rev. Genet. *9,* 407–486.

Ryan, K.W., Morgan, E.M., and Portner, A. (1991). Two noncontiguous regions of Sendai virus P protein combine to form a single nucleocapsid binding domain. Virology *180,* 126–134.

Saito, S., Ogino, T., Miyajima, N., Kato, A., and Kohase, M. (2002). Dephosphorylation failure of tyrosine-phosphorylated STAT1 in IFN-stimulated Sendai virus C protein-expressing cells. Virology *293,* 205–209.

Sakaguchi, T., Toyoda, T., Gotoh, B., Inocencio, N.M., Kuma, K., Miyata, T., and Nagai, Y. (1989). Newcastle disease virus evolution. I. Multiple lineages defined by sequence variability of the hemagglutinin-neuraminidase gene. Virology *169,* 260–272.

Sakaguchi, T., Kiyotani, K., Kato, A., Asakawa, M., Fujii, Y., Nagai, Y., and Yoshida, T. (1997). Phosphorylation of the Sendai virus M protein is not essential for virus replication either *in vitro* or *in vivo*. Virology *235,* 360–366.

Sakaguchi, T., Uchiyama, T., Huang, C., Fukuhara, N., Kiyotani, K., Nagai, Y., and Yoshida, T. (2002). Alteration of Sendai virus morphogenesis and nucleocapsid incorporation due to mutation of cysteine residues of the matrix protein. J. Virol. *76,* 1682–1690.

Sakaguchi, T., Kato, A., Sugahara, F., Shimazu, Y., Inoue, M., Kiyotani, K., Nagai, Y., and Yoshida, T. (2005). AIP1/Alix is a binding partner of Sendai virus C protein and facilitates virus budding. J. Virol. *79,* 8933–8941.

Sakaguchi, T., Kato, A., Kiyotani, K., Yoshida, T., and Nagai, Y. (2008). Studies on the paramyxovirus accessory genes by reverse genetics in the Sendai virus-mouse system. Proc. Jpn. Acad. Ser. B. *84,* 439–451.

Sakai, Y., Kiyotani, K., Fukumura, M., Asakawa, M., Kato, A., Shioda, T., Yoshida, T., Tanaka, A., Hasegawa, M., and Nagai, Y. (1999). Accommodation of foreign genes into Sendai virus genome. Sizes of inserted genes and viral replication. FEBS Lett. *456,* 221–226.

Sanderson, C.M., Avalos, R., Kundu, A., and Nayak, D.P. (1995). Interaction of Sendai virus F, HN, and M proteins with host cytoskeletal and lipid components in Sendai virus-infected BHK cells. Virology *209,* 701–707.

Sato, M., Hata, N., Asagiri, M., Nakaya, T., Taniguchi, T., and Tanaka, N. (1998a). Positive feedback regulation of type I IFN genes by the IFN-inducible transcription factor IRF-7. FEBS Lett. *441,* 106–110.

Sato, M., Tanaka, N., Hata, N., Oda, E., and Taniguchi, T. (1998b). Involvement of the IRF family transcription factor IRF-3 in virus-induced activation of the IFN-beta gene. FEBS Lett. *425,* 112–116.

Schafer, S.L., Lin, R., Moore, P.A., Hiscott, J., and Pitha, P.M. (1998). Regulation of type I interferon gene expression by interferon regulatory factor-3. J. Biol. Chem. *273,* 2714–2720.

Scheid, A., and Choppin, P.W. (1974). Identification of biological activities of paramyxovirus glycoproteins. Activation of cell fusion, hemolysis, and infectivity of proteolytic cleavage of an inactive precursor protein of Sendai virus. Virology *57,* 475–490.

Scheid, A., Caliguili, L.A., Compans, R.W., and Choppin, P.W. (1972). Isolation of paramyxovirus glycoproteins. Association of both hemagglutinating and neuraminidase activities with the larger SV5 glycoprotein. Virology *50,* 640–652.

Schindler, C., Shuai, K., Prezioso, V.R., and Darnell, J.E., Jr. (1992). Interferon-dependent tyrosine phosphorylation of a latent cytoplasmic transcription factor. Science *257,* 809–813.

Schlee, M., Roth, A., Hornung, V., Hagmann, C.A., Wimmenauer, V., Barchet, W., Coch, C., Janke, M., Mihailovic, A., Wardle, G., *et al.* (2009). Recognition of 5′ triphosphate by RIG-I helicase requires short blunt double-stranded RNA as contained in panhandle of negative-strand virus. Immunity *31,* 25–34.

Schmidt, A., Schwerd, T., Hamm, W., Hellmuth, J.C., Cui, S., Wenzel, M., Hoffmann, F.S., Michallet, M.C., Besch, R., Hopfner, K.P., *et al.* (2009). 5′-triphosphate RNA requires base-paired structures to activate antiviral signaling via RIG-I. Proc. Natl. Acad. Sci. U.S.A. *106,* 12067–12072.

Schmitt, A.P., Leser, G.P., Morita, E., Sundquist, W.I., and Lamb, R.A. (2005). Evidence for a new viral late domain core sequence, FPIV, necessary for budding of a paramyxovirus. J. Virol. *79,* 2988–2997.

Schnell, M.J., Mebatsion, T., and Conzelmann, K.-K. (1994). Infectious rabies virus from cloned cDNA. EMBO J. *13,* 4195–4203.

Schoondermark-van de Ven, E.M., Philipse-Bergmann, I.M., and van der Logt, J.T. (2006). Prevalence of naturally occurring viral infections, Mycoplasma pulmonis and Clostridium piliforme in laboratory rodents in Western Europe screened from 2000 to 2003. Lab. Anim. *40,* 137–143.

Segawa, H., Yamashita, T., Kawakita, M., and Taira, H. (2000). Functional analysis of the individual oligosaccharide chains of Sendai virus fusion protein. J. Biochem. *128,* 65–72.

Segawa, H., Inakawa, A., Yamashita, T., and Taira, H. (2003). Functional analysis of individual oligosaccharide chains of Sendai virus hemagglutinin-neuraminidase protein. Biosci. Biotechnol. Biochem. *67,* 592–598.

Seki, T., Yuasa, S., Oda, M., Egashira, T., Yae, K., Kusumoto, D., Nakata, H., Tohyama, S., Hashimoto, H., Kodaira, M., *et al.* (2010). Generation of induced pluripotent stem cells from human terminally differentiated circulating T cells. Cell Stem Cell. *7,* 11–14.

Shibata, S., Okano, S., Yonemitsu, Y., Onimaru, M., Sata, S., Nagata-Takeshita, H., Inoue, M., Zhu, T., Hasegawa, M., Moroi, Y., *et al.* (2006). Induction of efficient antitumor immunity using dendritic cells activated by Sendai virus and its modulation of exogenous interferon-beta gene. J. Immunol. *177,* 3564–3576.

Shimamura, M., Morishita, R., Endoh, M., Oshima, K., Aoki, M., Waguri, S., Uchiyama, Y., and Kaneda, Y. (2003). HVJ-envelope vector for gene transfer into central nervous system. Biochem. Biophys. Res. Commun. *300,* 464–471.

Shimazu, Y., Takao, S.I., Irie, T., Kiyotani, K., Yoshida, T., and Sakaguchi, T. (2008). Contribution of the leader sequence to homologous viral interference among Sendai virus strains. Virology *372,* 64–71

Shimizu, K., and Ishida, N. (1975). The smallest protein of Sendai virus: its candidate function of binding nucleocapsid to envelope. Virology *67,* 427–437.

Shimizu, K., Hosaka, Y., and Shimizu, Y.K. (1972). Solubilization of envelopes of HVJ (Sendai virus) with alkali-emasol treatment and reassembly of envelope particles with removal of the detergent. J. Virol. *9,* 842–850.

Shimizu, Y.K., Shimizu, K., Ishida, N., and Homma, M. (1976). On the study of Sendai virus hemolysis. II. Morphological study of envelop fusion and hemolysis. Virology *71,* 48–60.

Shioda, T., Hidaka, Y., Kanda, T., Shibuta, H., Nomoto, A., and Iwasaki, K. (1983). Sequence of 3687 nt from the 3′ end of Sendai virus genome RNA and the predicted amino acid sequences of viral NP, P and C proteins. Nucl. Acids Res. *11,* 7317–7330.

Shioda, T., Iwasaki, K., and Shibuta, H. (1986). Determination of the complete nucleotide sequence of the Sendai virus genome RNA and the predicted amino acid sequence of the F, HN, and L proteins. Nucl. Acids Res. *14,* 1545–1563.

Shuai, K., Schindler, C., Prezioso, V.R., and Darnell, J.E., Jr. (1992). Activation of transcription by IFN-gamma: tyrosine phosphorylation of a 91-kDa DNA binding protein. Science *258,* 1808–1812.

Shuai, K., Stark, G.R., Kerr, I.M., and Darnell, J.E., Jr. (1993). A single phosphotyrosine residue of Stat1 required for gene activation by interferon-gamma. Science *261,* 1744–1746.

Skiadopoulos, M.H., Surman, S.R., Riggs, J.M., Elkins, W.R., St. Claire, M., Nishio, M., Garcin, D., Kolakofsky, D., Collins, P.L., and Murphy, B.R. (2002). Sendai virus, a murine parainfluenza virus type *1,* replicates to a level similar to human PIV1 in the upper and lower respiratory tract of African green monkeys and chimpanzees. Virology *297,* 153–160.

Smallwood, S., Ryan, K.W., and Moyer, S.A. (1994). Deletion analysis defines a carboxyl-proximal region of Sendai virus P protein that binds to the polymerase L protein. Virology *202,* 154–163.

Smallwood, S., Çevik, B., and Moyer, S.A. (2002a). Intragenic complementation and oligomerization of the L subunit of the Sendai virus RNA polymerase. Virology *304,* 235–245.

Smallwood, S., Hövel, T., Neubert, W.J., and Moyer S.A. (2002b). Different substitutions at conserved amino acids in domains II and III in the Sendai L RNA polymerase protein inactivate viral RNA synthesis. Virology *304,* 135–145.

Sriwilaijaroen, N., Kondo, S., Yagi, H., Wilairat, P., Hiramatsu, H., Ito, M., Ito, Y., Kato, K., and Suzuki, Y. (2009). Analysis of *N*-glycans in embryonated chicken egg choriollantoic and amniotic cells responsible for binding and adaptation of human and avian influenza viruses. Glycoconj. J. *26,* 433–443.

Staerk, J., Dawlaty, M.M., Gao, Q., Maetzel, D., Hanna, J., Sommer, C.A., Mostoslavasky, G., and Jaenisch, R. (2010). Reprogramming of human peripheral blood cells to induced pluripotent stem cells. Cell Stem Cell. *7,* 20–24.

Stark, G.R., Kerr, I.M., Williams, B.R., Silverman, R.H., and Schreiber, R.D. (1998). How cells respond to interferons. Annu. Rev. Biochem. *67,* 227–264.

Strahle, L., Garcin, D., and Kolakofsky, D. (2006). Sendai virus defective-interfering genomes and the activation of interferon-beta. Virology *351,* 101–111.

Strahle, L., Marq, J.B., Brini, A., Hausmann, S., Kolakofsky, D., and Garcin, D. (2007). Activation of the beta interferon promoter by unnatural Sendai virus infection requires RIG-I and is inhibited by viral C proteins. J. Virol. *81,* 12227–12237.

Sugahara, F., Uchiyama, T., Watanabe, H., Shimazu, Y., Kuwayama, M., Fujii, Y., Kiyotani, K., Adachi, A., Kohno, N., Yoshida, T., *et al.* (2004). Paramyxovirus Sendai virus-like particle formation by expression of multiple viral proteins and acceleration of its release by C protein. Virology *325,* 1–10.

Suzuki, H., Harada, A., Hayashi, Y., Wada, K., Asaka, J., Gotoh, B., Ogasaweara, T., and Nagai, Y. (1991). Primary structure of the virus activating protease from chick embryo; its identity with the blood clotting factor Xa. FEBS Lett. *283,* 281–285.

Suzuki, Y., Suzuki, T., and Matsumoto, M. (1983). Isolation and characterization of receptor sialoglycoprotein for hemagglutinating virus of Japan (Sendai virus) from bovine erythrocyte membrane. J. Biochem. *93,* 1621–1633.

Suzuki, Y., Suzuki, T., Matsunaga, M., and Matsumoto, M. (1985). Gangliosides as paramyxovirus receptor. Structural requirement of sialo-oligosaccharides in receptors for hemagglutinating virus of Japan (Sendai virus) and Newcastle disease virus. J. Biochem. *97,* 1189–1199.

Takahashi, K., and Yamanaka, S. (2006). Induction of pluripotent stem cells from mouse embryonic and adult fibroblast cultures by defined factors. Cell *126,* 663–676.

Takeuchi, K., Komatsu, T., Yokoo, J., Kato, A., Shioda, T., Nagai, Y., and Gotoh, B. (2001). Sendai virus C protein physically associates with Stat1. Genes Cells *6,* 545–557.

Takeuchi, K., Komatsu, T., Kitagawa, Y., Sada, K., and Gotoh, B. (2008). Sendai virus C protein plays a role in restricting PKR activation by limiting the generation of intracellular double-stranded RNA. J. Virol. *82,* 10102–10110.

Takimoto, T., and Portner, A. (2004). Molecular mechanism of paramyxovirus budding. Virus Res. *106,* 133–145.

Takimoto, T., Bousse, T., Coronel, E.C., Scroggs, R.A., and Portner, A. (1998). Cytoplasmic domain of Sendai virus HN protein contains a specific sequence required for its incorporation into virions. J. Virol. *72,* 9747–9754.

Takimoto, T., Hurwitz, J.L., Coleclough, C., Prouser, C., Krishnamurthy, S., Zhan, X., Boyd, K., Scroggs, R.A., Brown, B., Nagai, Y., *et al.* (2004). Recombinant Sendai virus expressing the G glycoprotein of respiratory syncytial virus (RSV) elicits immune protection against RSV. J. Virol. *78,* 6043–6047.

Tamura, T., Yamashita, T., Segawa, H., and Taira, H. (2002). *N*-linked oligosaccharide chains of Sendai virus fusion protein determine the interaction with endoplasmic reticulum molecular chaperones. FEBS Lett. *513,* 153–158.

Tanaka, K., Sekiguchi, M., and Okada, Y. (1975). Restoration of ultraviolet induced unscheduled DNA synthesis of xeroderma pigmentosum cells by the concomitant treatment with T4 endonuclease V and HVJ (Sendai virus). Proc. Natl. Acad. Sci. U.S.A. *72,* 4071–4075.

Tapparel, C., Hausmann, S., Pelet, T., Curran, J., Kolakofsky, D., and Roux, L. (1997). Inhibition of Sendai virus genome replication due to promoter-increased selectivity: a possible role for the accessory C proteins. J. Virol. *71,* 9588–9599.

Tapparel, C., Maurice, D., and Roux, L. (1998). The activity of Sendai virus genomic and antigenomic promoters requires a second element past the leader template regions: a motif, $(GNNNNN)_3$, is essential for replication. J. Virol. *72,* 3117–3128.

Tarbouriech, N., Curran, J., Ebel, C., Ruigrok, R.W.H., and Burmeister, W.P. (2000a). On the domain structure and the polymerization state of the Sendai virus P protein. Virology *266,* 99–109.

Tarbouriech, N., Curran, J., Ruigrok, R.W.H., and Burmeister, W.P. (2000b). Tetrameric coiled coil domain of Sendai virus phosphoprotein. Nat. Struct. Biol. *7,* 777–781.

Tashiro, M. Yokogoshi, Y., Tobita, K., Seto, J.T., Rott, R., and Kido, H. (1992). Tryptase Clara, an activating protease for Sendai virus in rat lungs, is involved in pneumopathogenicity. J. Virol. *66,* 7211–7216.

Toyoda, T., Sakaguchi, T., Imai, K., Inocencio, N.M., Gotoh, B., Hamaguchi, M., and Nagai, Y. (1987). Structural comparison of the cleavage-activation site of the fusion glycoprotein between virulent and avirulent strains of Newcastle disease virus. Virology *158,* 242–247.

Toyoda, T., Sakaguchi, T., Hirota, H., Gotoh, B., Kuma, K., Miyata, T., and Nagai, Y. (1989). Newcastle disease virus evolution. II. Lack of gene recombination in generating virulent and avirulent strains. Virology *169,* 273–282.

Tozawa, H., Watanabe, M., and Ishida, N. (1973). Structural components of Sendai virus. Serological and physicochemical characterization of hemagglutinin subunit associated with neuraminidase activity. Virology *55,* 242–253.

Tropea, F., Troiano, L., Monti, D., Lovato, E., Malorni, W., Rainaldi, G., Mattana, P., Viscomi, G., Ingletti, M.C., Portolani, M., *et al.* (1995). Sendai virus and herpesvirus type 1 induce apoptosis in human peripheral blood mononuclear cells. Exp. Cell Res. *218,* 63–70.

Uchida, T., Yamaizumi, M., and Okada, Y. (1977). Reassembled HVJ (Sendai virus) envelopes containing non-toxic mutant proteins of diphtheria toxin show toxicity to mouse L cell. Nature *266,* 839–840.

Vidal, S., Curran, J., and Kolakofsky, D. (1990a). Editing of the Sendai virus P/C mRNA by G insertion occurs during mRNA synthesis via a virus-encoded activity. J. Virol. *64,* 239–246.

Vidal, S., Curran, J., and Kolakofsky, D. (1990b). A stuttering model for paramyxovirus P mRNA editing. EMBO J. *9,* 2017–2022.

Visvanathan, K.V., and Goodbourn, S. (1989). Double-stranded RNA activates binding of NF-κB to an inducible element in the human β-interferon promoter. EMBO J. *8,* 1129–1138.

Voges, B., Vallbracht, S., Zimmer, G., Bossow, S., Neubert, W.J., Richter, K., Hobeika, E., Herrler, G., and Ehl, S. (2007). Recombinant Sendai virus induces T cell immunity against respiratory syncytial virus that is protective in the absence of antibodies. Cell. Immunol. *247,* 85–94.

Wathelet, M.G., Lin, C.H., Parekh, B.S., Ronco, L.V., Howley, P.M., and Maniatis, T. (1998). Virus infection induces the assembly of coordinately activated transcription factors on the IFN-beta enhancer *in vivo*. Mol. Cell *1,* 507–518.

Weaver, B.K., Kumar, K.P., and Reich, N.C. (1998). Interferon regulatory factor 3 and CREB-binding protein/p300 are subunits of double-stranded RNA-activated transcription factor DRAF1. Mol. Cell. Biol. *18,* 1359–1368.

Weber, F., Wagner, V., Rasmussen, S.B., Hartmann, R., and Paludan, S.R. (2006). Double-stranded RNA is produced by positive-strand RNA viruses and DNA viruses but not in detectable amounts by negative-strand RNA viruses. J. Virol. *80,* 5059–5064.

de Weerd-Kastelein, E.A., Keijzer, W., and Bootsma, D. (1972). Genetic heterogeneity of xeroderma pigmentosum demonstrated by somatic cell hybridization. Nat. New Biol. *238,* 80–83.

Weiss, M.C., and Ephrussi, B. (1966). Studies of interspecific (rat x mouse) somatic hybrids. I. isolation, growth and evolution of the karyotype. Genetics *54,* 1095–1109.

Weiss, M.C., and Green, H. (1967). Human–mouse hybrid cell lines containing partial complements of human chromosomes and functioning human genes. Proc. Natl. Acad. Sci. U.S.A. *58,* 1104–1111.

Wen, Z., Zhong, Z., and Darnell, J.E., Jr. (1995). Maximal activation of transcription by Stat1 and Stat3 requires both tyrosine and serine phosphorylation. Cell *82,* 241–250.

Whelan, S.P.J. (2008). Response to 'Non-segmented negative-strand RNA virus RNA synthesis *in vivo*'. Virology *371,* 234–327.

Whelan, S.P.J., and Wertz, G.W. (2002). Transcription and replication initiate at separate site on the vesicular stomatitis virus genome. Proc Natl. Acad. Sci. U.S.A. *99,* 9178–9183.

Whelan, S.P.J., Barr, J.N., and Wertz, G.W. (2004). Transcription and replication of nonsegmented negative-strand RNA viruses. Curr. Top. Microbiol. Immunol. *283,* 61–119.

Wiegand, M., Bossow, S., and Neubert, W.J. (2005). Sendai virus trailer RNA simultaneously blocks two apoptosis-inducing mechanisms in a cell type-dependent manner. J. Gen. Virol. *86,* 2305–2314.

Yamada, H., Hayata, S., Omata-Yamada, T., Taira, H., Mizumoto, K., and Iwasaki, K. (1990). Association of the Sendai virus C protein with nucleocapsids. Arch. Virol. *113,* 245–253.

Yamaizumi, M., Mekada, E., Uchida, T., and Okada, Y. (1978a). One molecule of diphtheria toxin fragment A introduced into a cell can kill the cell. Cell *15,* 245–250.

Yamaizumi, M., Uchida, T., Okada, Y., Furusawa, M., and Mitsui, Y. (1978b). Rapid transfer of non-histone chromosomal proteins to the nucleus of living cells after their introduction by cell fusion with erythrocyteghosts. Nature *273,* 782–784.

Yamanaka, S. (2010). Patient-specific pluripotent stem cells become even more accessible. Cell Stem Cell. *7,* 1–2.

Yamashita, A., Yonemitsu, Y., Okano, S., Nakagawa, K., Nakashima, Y., Irisa, T., Iwamoto, Y., Nagai, Y., Hasegawa, M., and Sueishi, K. (2002). Fibroblast growth factor-2 determines severity of joint disease in adjuvant-induced arthritis in rats. J. Immunol. *168,* 450–457.

Yokoo, J., Gotoh, B., Komatsu, T., Takeuchi, K., and Miyadai, T. (1999). Replication-incompetent Sendai virus can suppress the antiviral action of type I interferon. Arch. Virol. *144,* 1043–1055.

Yonemitsu, Y., Kitson, C., Ferrari, S., Farley, R., Griesenbach, U., Judd, D., Steel, R., Scheid, P., Zhu, J., Jeffery, P.K., *et al.* (2000). Efficient gene transfer to airway epithelium using recombinant Sendai virus. Nat. Biotechnol. *18,* 970–973.

Yonemitsu, Y., Ueda, Y., Kinoh, H., and Hasegawa, M. (2008). Immunostimulatory virotherapy using recombinant Sendai virus as a new cancer therapeutic regimen. Front. Biosci. *13,* 1892–1898.

Yoneyama, M., and Fujita, T. (2010). Recognition of viral nucleic acids in innate immunity. Rev. Med. Virol. *20,* 4–22.

Yoneyama, M., Suhara, W., Fukuhara, Y., Fukuda, M., Nishida, E., and Fujita, T. (1998). Direct triggering of the type I interferon system by virus infection: activation of a transcription factor complex containing IRF-3 and CBP/p300. EMBO J. *17,* 1087–1095.

Yoneyama, Y., Ueda, Y., Akutsu, Y., Matsunaga, A., Shimada, H., Kato, T., Kubota-Akizawa, M., Okano, S., Shibata, S., Sueishi, K., *et al.* (2007). Development of immunostimulatory virotherapy using non-transmissible Sendai virus-activated dendritic cells. Biochem. Biophys. Res. Commun. *355,* 129–135.

Yoshida, T., Nagai, Y., Yoshii, S., Maeno, K., Matsumoto, T., and Hoshino, M. (1976). Membrane (M) protein of HVJ (Sendai virus): its role in virus assembly. Virology *71,* 143–161.

Yoshida, T., Nagai, Y., Maeno, K., Iinuma, M., Hamaguchi, M., and Matsumoto, T. (1979). Studies on the role of M protein in virus assembly using a TS mutant of HVJ (Sendai virus). Virology *92,* 139–154.

Yoshida, T., Hamaguchi, M., Naruse, H., and Nagai, Y. (1982). Persistent infection by a temperature-sensitive mutant isolated from a Sendai virus (HVJ) carrier culture: its initiation and maintenance without aid of defective interfering particles. Virology *120,* 329–339.

Yoshima, H., Nakanishi, M., Okada, Y., and Kobata, A. (1981). Carbohydrate structures of HVJ (Sendai virus) glycoproteins. J. Biol. Chem. *256,* 5355–5361.

Yoshizaki, M., Hironaka, T., Iwasaki, H., Ban, H., Tokusumi, Y., Iida, A., Nagai, Y., Hasegawa, M., and Inoue, M. (2006). Naked Sendai virus vector lacking all of the envelope-related genes: reduced cytopathogenicity and immunogenicity. J. Gene Med. *8,* 1151–1159.

Young, D.F., Didcock, L., Goodbourn, S., and Randall, R.E. (2000). *Paramyxoviridae* use distinct virus-specific mechanisms to circumvent the interferon response. Virology *269,* 383–390.

Youngner, J.S., and Preble, O.T. (1980). Viral persistence: Evolution of viral populations. In Comprehensive Virology, vol. 16, Frankel-Conrat, H., and Wagner, R.R., eds. (Plenum Press: New York/London), pp. 73–135.

Yount, J.S., Gitlin, L., Moran, T.M., and Lopez, C.B. (2008). MDA5 participates in the detection of paramyxovirus infection and is essential for the early activation of dendritic cells in response to Sendai virus defective interfering particles. J. Immunol. *180,* 4910–4918.

Yu, D., Shioda, T., Kato, A., Hasan, M.K., Sakai, Y., and Nagai, Y. (1997). Sendai virus-based expression of HIV-1 gp120: reinforcement by the V(–) version. Genes Cells *2,* 457–466.

Zaitsev, V., von Itzstein, M., Groves, D., Kiefel, M., Takimoto, T., Portner, A., and Taylor, G. (2004). Second sialic acid binding site in Newcastle disease virus hemagglutinin-neuraminidase: implications for fusion. J. Virol. *78,* 3733–3741.

Zenner, L., and Regnault, J.P. (2000). Ten-year long monitoring of laboratory mouse and rat colonies in French facilities: a retrospective study. Lab. Anim. *34,* 76–83.

Parainfluenza Viruses

5

Santanu Chattopadhyay, Frank Esper and Amiya K. Banerjee*

Abstract

Parainfluenza viruses (PIVs) include several members of the *Paramyxoviridae* family of the *Mononegavirales* order ranging from *Respirovirus, Avulavirus* and *Rubulavirus* genera. These group of viruses consist of several species that causes upper and lower respiratory illness in humans, especially among children and in a variety of animals. These medium-sized, pleomorphic, enveloped viruses attach to the host cells' membranes and get internalized at a neutral pH with the help of haemagglutinin-nuraminidase (HN) and fusion (F) – the two surface glycoproteins. Inside the cytoplasm, the ~15 kb of non-segmented negative-sense RNA (NNS) genomes that are encapsidated by the viral nucleocapsid protein (N) are transcribed and replicated by the associated viral polymerase heterocomplex – the large protein (L) and the phosphoprotein (P). The newly replicated RNA genomes are concomitantly encapsidated by the N protein and with the help of the matrix proteins (M), assemble and bud out as mature virions containing the viral glycoproteins and the polymerase along with the host cell membrane envelope. The P gene of PIVs characteristically encodes several accessory proteins, viz C, V, W, D, which are expressed by mRNA editing or alternative translation mechanisms. Despite significant ongoing efforts, no suitable vaccine or antiviral compounds are available to date against these clinically and economically important viruses.

History

The first parainfluenza virus (PIV) was discovered from the outbreak of a severe respiratory, alimentary and neuronal disease of poultry that occurred in Java, Indonesia, in 1926 (Kraneveld, 1926) and in Newcastle-upon-Tyne, England, in 1927 (Doyle, 1927). This virus was named as Newcastle disease virus (NDV) by Doyle. Since then, eight closely related avian viruses, with distinct serotypes, have been discovered (Alexander, 1988). These viruses are collectively termed as avian paramyxoviruses (APMVs) 1–9, NDV being the serotype 1 (Alexander, 1988). Many years later, in March 1952, an outbreak of pneomonitis among newborns occurred in Sendai, Japan. The autopsy specimens from the infected infants were collected and a virus possessing haemagglutinin activity was isolated by perinasal instillation of mice (Kuroya and Ishida, 1953). This virus, which could be differentiated from influenza viruses, mumps virus (MuV), NDV and pneumonia virus of mice (PVM) was named as Sendai virus (SeV) and later identified as a murine PIV. Subsequently, three human PIVs (HPIVs) were recovered from children with lower respiratory tract illnesses (Chanock, 1956, 1958). In 1959, the term 'Parainfluenza' was first used to refer to HPIV1–3 and SeV owing to their structural and functional similarities and dissimilarities with influenza virus (Andrewes *et al.*, 1959). Apart from causing respiratory illnesses among the infected individuals, the PIV virions, like influenza virus, possess lipid

*Corresponding author

envelope and haemagglutinin–neuraminidase activities (Andrewes *et al.*, 1959). These viruses grow poorly in embryonic eggs and share few antigenic sites with influenza virus (Henrickson, 2003). Soon after another human virus that possesses these properties were identified and named as HPIV4 (Johnson *et al.*, 1960). Bovine parainfluenza virus 3 (BPIV3) was isolated in 1959 from cattle with respiratory diseases (shipping fever) and identified as closely related to HPIV3 by serological and biochemical properties (Abinanti and Huebner, 1959; Reisinger *et al.*, 1959; Abinanti *et al.*, 1961; Marshall, 1964). Parainfluenza virus 5 (PIV5), a canine virus responsible for *kennel cough*, was originally isolated from rhesus and cynomolgus monkey kidney cell cultures and hence named as Simian virus 5 (SV5; Hull *et al.*, 1956; Hsiung, 1972). Although SV5 was initially thought to be the animal counterpart of HPIV2, another animal virus, Simian virus 41 (SV41), was also isolated from primary monkey kidney cell culture and shown to be more closely related to HPIV2 (Miller *et al.*, 1964; Tsurudome *et al.*, 1990, 1991; Karron and Collins, 2007). The timeline of the identifications and isolations of the PIVs are shown in Fig. 5.1. This group of viruses, apart from being important human and animal pathogens, served as models for studying the biology of NNS viruses.

Infectious agent

Classification

As shown in Table 5.1, the 17 PIV members are scattered in three genera (*Avulavirus, Respirovirus* and *Rubulavirus*) of the *Paramyxoviridae* family (subfamily *Paramyxovirinae*) of the *Mononegavirales* order. The genus *Avulavirus* contains nine members – APMV1 (or NDV) and APMV2–9; the genus *Respirovirus* contains four members – SeV, HPIV1, HPIV3 and BPIV3; and the genus *Rubulavirus* contains four members HPIV2, HPIV4, PIV5 (or SV5) and SV41. As discussed in more detail below, these 17 members of PIV infect primarily the lung epithelium of different species and share similar virion morphology, biochemical properties and genome structure, but also bear some distinct features. Phylogenetic analyses show that the nine serogroups of APMV cluster together, but are distinguishable from other genera (Xiao *et al.*, 2009). APMV7 is

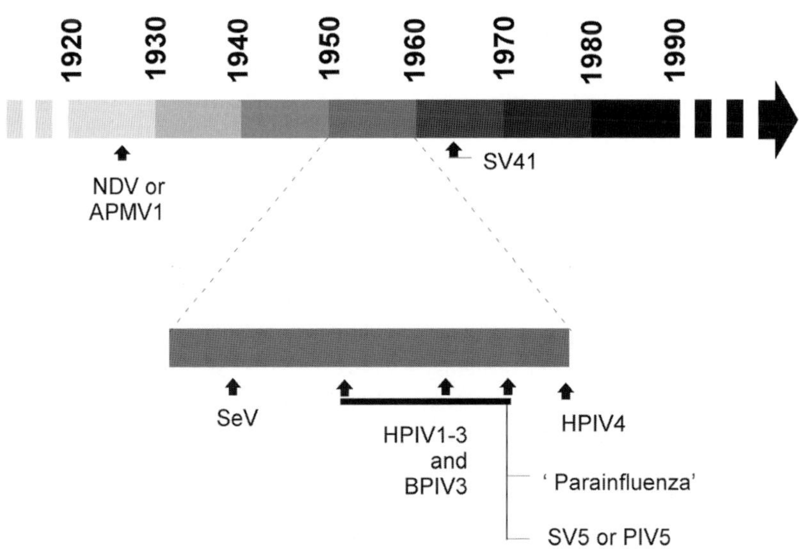

Figure 5.1 Timeline of the discoveries of Parainfluenza viruses (PIVs). Newcastle disease virus (NDV) is the first PIV discovered in 1926 and 1927although several reports suggest that similar infection among birds prior to that date. From 1953 to 1960, most of the PIVs (Sendai virus [SeV], HPIV1–4, BPIV3 and SV5 or PIV5) have been discovered. SV41 was discovered in 1964.

Table 5.1 Classification of parainfluenza viruses (PIV) – modified from the International Committee of Taxonomy of Viruses (ICTV), 2009

Order: *Mononegavirales*
Family: *Paramyxoviridae*
Subfamily: *Paramyxovirinae*
Genus: *Avulavirus* (type species: Newcastle disease virus)
Species: Avian paramyxovirus 1 (APMV1) or Newcastle disease virus (NDV)
Species: Avian paramyxovirus 2 (APMV2)
Species: Avian paramyxovirus 3 (APMV3)
Species: Avian paramyxovirus 4 (APMV4)
Species: Avian paramyxovirus 5 (APMV5)
Species: Avian paramyxovirus 6 (APMV6)
Species: Avian paramyxovirus 7 (APMV7)
Species: Avian paramyxovirus 8 (APMV8)
Species: Avian paramyxovirus 9 (APMV9)
Genus: *Respirovirus* (type species: Sendai virus)
Species: Human parainfluenza virus 1 (HPIV1)
Species: Human parainfluenza virus 3 (HPIV3)
Species: Bovine parainfluenza virus 3 (BPIV3)
Species: Sendai virus (SeV)
Genus: *Rubulavirus* (type species: Mumps virus[1])
Species: Human parainfluenza virus 2(HPIV2)
Species: Human parainfluenza virus 4 (HPIV4; antigenic subgroups A and B)
Species: Parainfluenza virus 5 (PIV5) or Simian virus 5 (SV5)
Species: Simian virus 41(SV41)

[1]Mumps virus (MuV), type species of the *Rubulavirus* genus does not belong to the PIV.

more closely related to APMV2, -6 and -8 than to APMV1, -3, -4 and -9 (Xiao *et al.*, 2009). Recent genome analysis of APMV5 suggest that the virus is more closely related to the AMPV6 than other members of the genus (Samuel *et al.*, 2010). Two of the four HPIVs belong to the Respirovirus genus. HPIV1 is closely related to the murine PIV, SeV while BPIV3 is considered as the bovine counterpart of HPIV3. Recent sequence analyses of the M genes and the entire genome RNAs suggest that the BPIV3 may exist in two distinct genotypes, viz. BPIV3a and BPIV3b (Horwood *et al.*, 2008). The two HPIVs belonging to the *Rubulavirus* genus are related to PIV5 (canine) and SV41 (simian). Nucleocapsid (N) gene sequence analyses revealed that HPIV2 is more closely related to SV41 (70.5% amino acid identity) than that of SV5 (57%), while SV5 and SV41 share 63.3% homology (Tsurudome *et al.*, 1991). The relatedness among these three PIVs belonging to the *Rubulavirus* genus were further consolidated by phylogenetic analysis using L gene sequences (Ogawa *et al.*, 1992). Interestingly, HPIV4, the other HPIV that belongs to the *Rubulavirus* genus, is antigenically heterogeneous and classified as HPIV4A and HPIV4B (Canchola *et al.*, 1964; Komada *et al.*, 1989). Subsequent phylogenetic analysis using fusion (F) gene sequence showed that apart from having strong homology between the two serogroups and among other members of the *Rubulavirus* genus HPIV4A and HPIV4B are more closely related to NDV (*Avulavirus*) than HPIV1 and HPIV3 (*Respirovirus*; Komada *et al.*, 1995).

Propagation

PIVs can be propagated in embryonated chicken eggs, although some altered physicochemical and biological properties of the haemagglutinin–neuraminidase (HN) protein and infectivity for SeV were noted (Bukrinskaya *et al.*, 1968; Brugh *et al.*, 1980; Popa *et al.*, 1985; Itoh *et al.*, 1992). Primary rhesus monkey kidney cells were historically used for propagation of the HPIV1–4 in spite of its high cost and low supply (Frank *et al.*, 1979; Meguro *et al.*, 1979). Moreover, these cell lines, are often contaminated with PIV5. Later, continuous line of monkey kidney cell culture (LLC-MK2) in the presence of trypsin provided an alternative to primary monkey kidney cells for the isolation of PIVs (and also influenza virus) from patients (Orstavik, 1981). In order to select a suitable human cell line for the propagation of PIV, a series of cells of human origin were tested and NCI-H292, human lung mucoepidermoid carcinoma cells, proved to be effective (Castells *et al.*, 1990). RPMI-1640 with 1.5 μg/ml trypsin or Eagle's MEM supplemented with 1.5 μg/ml trypsin proved to be highly satisfactory for viral propagation in culture (Castells *et al.*, 1990). Interestingly, trypsin is required for the growth of HPIV1 and HPIV4, but not for HPIV2 and 3. Virus infection in cultured cells is generally monitored by cytopathic effect, hemadsorption, haemagglutination (Castells *et al.*, 1990). Another human malignant melanoma cell line, HMV-II, was also used for the propagation of HPIV with success; however, in most laboratories, HPIV3 is now propagated in A549 cells (human lung carcinoma) or CV-1 cells (African green monkey kidney) with good virus yield (Moriuchi *et al.*, 1990).

Virion morphology and ultrastructure

The PIV virions in electron micrograph appear pleomorphic and mostly spherical with diameter ranging from 150–600 nm (Tsai and Thomson, 1975; Karron and Collins, 2007). Although mostly spherical, several interesting exceptions were also reported, e.g. tailed forms of egg-adapted purified NDV were initially noticed (Bang, 1948; Reagan *et al.*, 1951). Later, thin sections of cells infected with several PIV belonging to *Rubulavirus*, but not to *Respirovirus* have shown to release filamentous particles (Compans *et al.*, 1966; Howe *et al.*, 1967; Bachi and Howe, 1973; Tsai and Thomson, 1975; Yao and Compans, 2000). The filamentous forms of HPIV2 is inhibited by cytochalasin D or jasplakinolide treatment suggesting host cell components, particularly actin microfilaments, play a critical role in viral morphology (Yao and Compans, 2000). A recent study showed that in a batch of purified PIV5 virion 70.9% of the virions have spherical; while 29.1% have elongated morphologies as revealed by negatively stained preparation in conventional TEM. Interestingly, the elongated virion morphology became predominant (52.05%) after heat treatment (Terrier *et al.*, 2009). However, when the ice embedded, hydrated PIV5 virions were observed by cryo-electron microscopy (cryo-EM), no such elongated or tailed virions were observed, suggesting that the PIV5 virions are predominantly spherical and those deformed structures may have arisen from the drying process involved in conventional TEM (Terrier *et al.*, 2009). Similarly, cryo-EM tomography of SeV virions purified from embryonated eggs revealed that the virions are mostly spherical (Loney *et al.*, 2009). This latter study also revealed some unexpected features, viz. blisters in the viral envelope, vesicles inside the virions and uncharacterized, cylindrical or rectangular objects (90 nm in length and 30 nm in diameter) within the virions (Loney *et al.*, 2009). Although these cylindrical structures resembled M proteins stored under low salt conditions, no immunological staining was performed to confirm if the structures were really composed of M protein (Hewitt and Nermut, 1977; Loney *et al.*, 2009).

Important ultra structures within the PIV virions as revealed by EM and biochemical studies are represented in a schematic diagram (Fig. 5.2). All PIV virions contain a viral envelope of 6–12 nm thickness, derived from host cell membrane at the time of budding (Karron and Collins, 2007; Loney *et al.*, 2009). The HN and F glycoproteins, which are embedded on the outer surface of the membrane, have spike-like projections of ~8–12 nm from the surface of the membrane (Lamb and Parks, 2007). The PIV virions, owing to the HN proteins possess haemagglutination as well as neuraminidase activity (Karron and Collins, 2007). Both of these glycoproteins are distributed uniformly (~170 glycoprotein spikes/10,000 nm^2 of viral surface; or 2000 spikes for a 200-nm-diameter virion) on the membrane and are important for attachment and internalization of the virus to

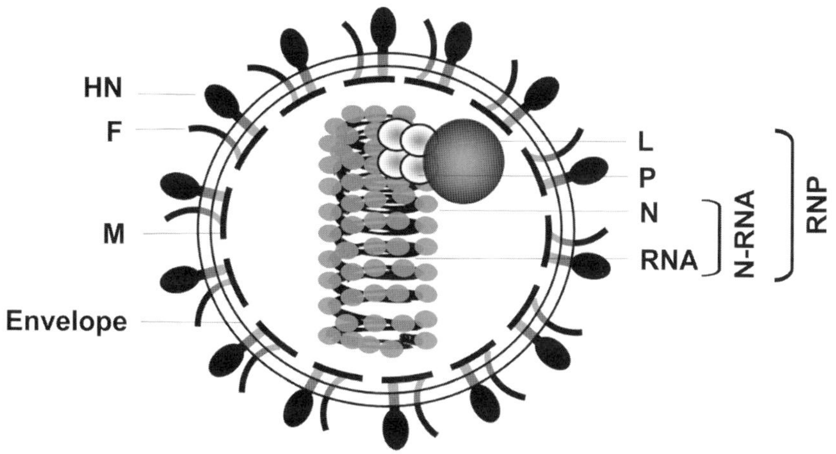

Figure 5.2 Schematic representation of a Parainfluenza virus (PIV) virion. Each virion contains an envelope derived from the host membrane. The HN and F glycoproteins are embedded in the envelope. The PIV5 and the APMV6 also contain the SH glycoprotein. Beneath the envelope the M protein forms electron dense patches. The ribonucleoprotein (RNP) is composed of the N–RNA complex (~15 kb genome RNA encapsidated by the N protein), the P protein (co-factor) and the L protein (polymerase).

the host cells (Terrier *et al.*, 2009). The envelope of PIV5 and APMV6 also contains a small hydrophobic (SH) protein whose role is not precisely known (Karron and Collins, 2007). The M protein molecules, which lie just beneath the envelope, interact with each other and are the most abundant protein in the virion (Karron and Collins, 2007; Terrier *et al.*, 2009). Recent cryo-EM analysis suggests that the M proteins instead of forming a continuous layer, rather forms electron-opaque patches (Terrier *et al.*, 2009). The M protein molecules interact with the glycoproteins as well as the viral core nucleocapsid/ribonucleoprotein (RNP) and are important in viral assembly, budding and morphogenesis (Stricker *et al.*, 1994; Sanderson *et al.*, 1995; Takimoto and Portner, 2004). The RNP (~1.1 μm in length) is composed of ~15 Kb NNS genome RNA, encapsidated by the N protein (a full length viral genome RNA and ~2500 molecules of N protein forms the N–RNA complex); the L protein (~40 molecules), the polymerase; and the P protein (~300 molecules), which interacts with both L and N proteins, mediates the contact between N–RNA complex and the L protein (Finch and Gibbs, 1970; Egelman *et al.*, 1989; Karron and Collins, 2007). The nucleocapsid, which takes up 9–24% (average 14%) of the virion interior, is helical but there is no consistent order to its packaging and appears as a tangled knot (Loney *et al.*, 2009). Although most virions classically contain one nucleocapsid, multiple nucleocapsids (up to six genomes) having variable lengths within a virion are also not uncommon (Karron and Collins, 2007; Loney *et al.*, 2009). On the other hand, virions containing extremely small nucleocapsids were also observed, which does not possibly carry a full length genome (Loney *et al.*, 2009). These smaller sized nucleocapsids may carry defective interfering (DI) genomes (partially deleted), which interfere with replication of the full length genome (see later) (Loney *et al.*, 2009). The presence of several accessory proteins that are encoded by the P gene in the virion has not been firmly established.

Genome organization

Genome of the PIVs belonging to all three genera (*Avulavirus*, *Respirovirus* and *Rubulavirus*) are now sequenced that provide us an opportunity to carry out comparative sequence analyses (Kawano *et al.*, 1991, 1993; Ogawa *et al.*, 1992, 1998; He *et al.*, 1997; Itoh *et al.*, 1997; Durbin *et al.*, 1999; Bailly *et al.*, 2000; Newman *et al.*, 2002). The complete nucleotide sequence of PIV genomes can be obtained from database (www.ncbi.nlm.nih.gov/).

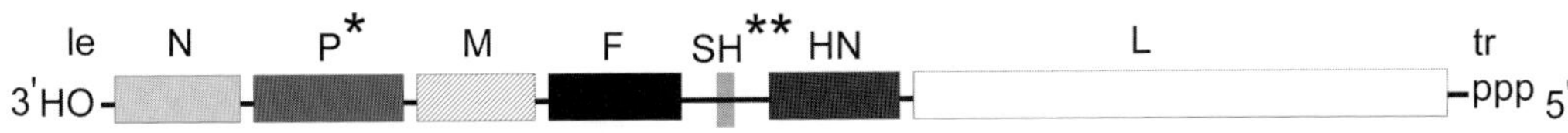

Figure 5.3 Schematic representation (not to the scale) of Parainfluenza virus (PIV) genome. PIV carry a non-segmented negative-sense genome of ~15 Kb. For most PIV, the gene order from the 3′ to 5′ end is N-P-M-F-HN-L. *The P gene also encodes C/V/D/W/I proteins. **The PIV5 or SV5 and the APMV6 also contain the SH gene in between the F and HN genes. The termini of the genome carry complementary sequences. Each gene starts with the gene start sequence and ends with the gene end sequence. Each gene end also carry the signal for polyadenylation. An intergenic sequence is present between the genes.

Fig. 5.3 and Table 5.2 summarize some important features of PIV genomes. The average GC content is 40.7% and the average length of the PIV genome is 15,518.4 nt. Interestingly, the number of nucleotides present in the PIV genomes are always a multiple of 6. This phenomenon, termed as the 'rule of six', is a hallmark for all PIVs as well as for all viruses belonging to subfamily *Paramyxovirinae,* but does not follow for viruses that belong to subfamily *Pneumovirinae* (Samal and Collins, 1996) and other families (Bornaviridae, Filoviridae and *Rhabdoviridae*) of the *Mononegavirales* (Kolakofsky *et al.*, 1998; Vulliemoz and Roux, 2001, 2002). Recombinant PIV genomes having odd multiples of six, generated by reverse genetics approach reverts back to even multiple of six (Kolakofsky *et al.*, 1998; Vulliemoz and Roux, 2001, 2002; Karron and Collins, 2007). On an average 91.6% of the PIV genomes carry coding sequences. The PIV genomes typically consist of six genes in the order of 3′-N-P-M-F-HN-L-5′. However, PIV5 and APMV6 genomes also carry the SH (small hydrophobic) gene between F and HN genes (Fig. 5.3). The genes are separated by conserved intergenic sequences and for PIVs it is GAA. Each gene begins with a conserved 'gene start' and ends with conserved 'gene end' sequences. Every gene in the genome, except the P gene, is transcribed to synthesize a single mRNA, which is translated into a single polypeptide. The P gene, as will be discussed later in more detail, is capable of encoding multiple proteins, which are expressed by unique means including different mRNA editing and alternative translation mechanisms. The viral mRNAs are 5′-capped and methylated as well as 3′-polyadenylated. In contrast, the genome RNAs are not capped or polyadenylated and contains 5′-free tri-phosphate (ppp) and 3′- hydroxyl (OH) group (Fig. 5.3). PIV genomes also carry variable stretches of 3′ and 5′ non-coding sequences. Interestingly, 3′ and the 5′ ends of the genome contain inverted repeats, but possible circularization of the genome RNA has not been demonstrated so far. The transcription process is initiated at the 3′ end of the genome RNA, first with the synthesis of a 40–70 nt long leader RNA (neither capped nor polyadenylated), followed by the sequential synthesis of the mRNAs in the order 3′-N-P-M-F-HN-L-5′.

Transcription and replication

Following fusion and entry into the cell, the PIV ribonucleoprotein (RNP), composed of the N-protein encapsidated (–) genome RNA (N-RNA) and the associated RdRp hetero-complex, the L and P proteins, is released in the cytoplasm. In the cytoplasm, the L protein (polymerase), which is active only in the presence of the P protein (co-factor), carries out viral transcription and replication using the N–RNA complex as template. It is believed that the PIVs as well as all members of the NNS virus follow a common transcription and replication strategy (Banerjee *et al.*, 1977; Barr *et al.*, 2002; Kolakofsky *et al.*, 2004; Whelan *et al.*, 2004; Cowton *et al.*, 2006; Karron and Collins, 2007; Lamb and Parks, 2007). Most of the biochemical data regarding the proposed model for the transcription and replication of NNS virus were obtained by using VSV and SeV as model viruses (Abraham *et al.*, 1975; Both *et al.*, 1975; Abraham and Banerjee, 1976; Ball and White, 1976; Colonno and Banerjee, 1976; Wertz, 1978; Emerson, 1982; Dowling *et al.*, 1983; Patton *et al.*, 1984; Baker and Moyer, 1988; Hercyk *et al.*, 1988; Vidal and Kolakofsky, 1989;Curran and Kolakofsky, 1991; Horikami

Table 5.2 Important features of the genomes of parainfluenza viruses belonging to *Avulavirus, Respirovirus*, and *Rubulavirus* genera

Genus	Species	Length (nt):	GC Content (%):	% Coding:	Genes	Protein coding	Ref. seq.	GenBank	Reference
Avulavirus[1]	NDV	15,186	46	90	6	6	NC_002617	AF309418	Sellers *et al.* (unpublished)
Avulavirus[1]	APMV6	16,236	46	89	7	7	NC_003043	AY029299	Chang *et al.* (unpublished)
Respirovirus	SeV	15,384	46	94	6	10	NC_001552	AB005795	Itoh *et al.* (1997)
Respirovirus	HPIV1	15,600	37	92	6	10	NC_003461	AF457102	Newman *et al.* (2002)
Respirovirus	HPIV3	15,462	34	93	6	8/9 (?)	NC_001796	AB012132	Durbin *et al.* (1999), Ohsawa *et al.* (1998)
Respirovirus	BPIV3	15,456	35	93	6	6	NC_002161	AF178654	Bailly *et al.* (2000)
Rubulavirus	HPIV2	15,646	38	90	6	7	NC_003443	X57559	Kawano *et al.* (1991)
Rubulavirus	PIV5	15,246	42	92	7	8	NC_006430	AF$_{0}$52755	He *et al.* (1997)
Rubulavirus	SV41	15,450	42	91	6	7	NC_006428	X64275	Kawano *et al.* (1993), Ogawa *et al.* (1992)

[1]For detailed analyses of *Avulavirus* (AMPV1–9) genomes (see Xiao *et al.*, 2009 and Samuel *et al.*, 2010)

et al., 1992; Whelan and Wertz, 1999; Barr and Wertz, 2001; Le Mercier *et al.*, 2003; Plattet *et al.*, 2007). Based on many such experiments carried out *in vitro* and *in vivo*, it is proposed that the RdRp heterocomplex (L and P proteins) gains access to the genome RNA at the 3′ end and initiates transcription to synthesize the leader RNA and the mRNAs by a stop–start mechanism (Fig. 5.4; Emerson, 1982). The RdRp is also responsible for capping, cap-methylation and addition of polyA tail at the 3′ end of the nascent mRNA (Banerjee *et al.*, 1974; Rhodes *et al.*, 1974; Moyer *et al.*, 1975; Grdzelishvili *et al.*, 2005, 2006; Li *et al.*, 2005, 2006, 2008; Ogino *et al.*, 2005, 2010; Ogino and Banerjee, 2007). At some point when sufficient amount of N protein accumulates in the cytoplasm, it complexes with the P protein to form a soluble (N^0–P) complex, which is used for replication of the progeny genome RNA (Fig. 5.4). How exactly the switch from transcription to replication occurs is not clear at present, but it may involve participation of several host factors. Recently, in the VSV system a second transcription complex (L–N–P) has been shown to be involved in the replication process (Qanungo *et al.*, 2004). It is believed that the strength of the promoter present in the anti-genome is stronger than that of the promoter present in the genome. As a result, more (−) sense genome RNA is generated from replication, which then becomes encapsidated, packaged and released as virions. The readers are directed to several recent articles and reviews that deal with the complex mechanism of transcription and replication of NNS viruses (Fearns *et al.*, 1997; Whelan and Wertz, 2002; Qanungo *et al.*, 2004; Wiegand *et al.*, 2007; Banerjee, 2008; Curran and Kolakofsky, 2008; Whelan, 2008).

Several host proteins have been shown to be involved in transcription of PIV genome RNA. For example, the cellular actin, particularly filamentous actin, is required for HPIV3 transcription and replication (Moyer *et al.*, 1986; De and Banerjee, 1999; De *et al.*, 1991, 1993; Gupta *et al.*, 1998). For SeV, another host cytoskeleton protein, tubulin interacts with the M protein and stimulates viral mRNA synthesis (Moyer *et al.*, 1986; Ogino *et al.*, 2003). Two cellular glycolytic enzymes, phosphoglycerate kinase and enolase, are also required for efficient transcription of SeV genome (Ogino *et al.*, 1999, 2001). Similarly, cellular β-catenin associates with the RNP of HPIV3 and activates viral transcription (Bose and Banerjee, 2004). On the other hand, specific phosphorylated forms of glyceraldehyde-3-phosphate dehydrogenase (GAPDH) associate with HPIV3 and inhibit viral transcription (Choudhary *et al.*, 2000). The GAPDH and LA protein also interact with the cis-acting RNAs of HPIV3, but their precise role in the virus' life cycle remains unclear (De *et al.*, 1996).

Viral polypeptides

The nucleocapsid protein (N)

The (N) protein is the most abundant viral polypeptide in the infected cell since the N gene is first from the 3′ end of the genome to be transcribed. The PIV N protein (MW ~50 kDa) is highly acidic in nature and despite lacking any known RNA binding motif, can interact with cellular RNA non-specifically when expressed alone (Lamb and Parks, 2007). However, when co-expressed with the P protein, the N-terminal ~40 amino acids of the P protein and N-terminal ~400 amino acid of the N protein interact with each other to form a soluble N^0–P complex (Lamb and Parks, 2007). In NNS viruses, this complex formation is necessary for preventing the non-specific interaction between the N protein and cellular RNA followed by aggregation (Masters and Banerjee, 1988a,b; Chen *et al.*, 2007). The N^0–P complex is also required for the specific encapsidation of the viral genomic and antigenomic RNAs by the N protein during replication. Recently, for VSV, a tripartite complex (L–N–P) along with the (N^0–P) complex has been shown to be involved in the replication process (Qanungo *et al.*, 2004). Similar complex in PIV-infected cells has not been reported so far. Biochemical and genetic analyses revealed that each PIV N molecule binds six nucleotides, first shown in SeV (Kolakofsky *et al.*, 1998; Vulliemoz and Roux, 2001, 2002). The SeV N protein possesses two major domains: the N-terminal N_{core} constitutes ~75% of the protein and the N_{tail} constitutes the C-terminal 25% of the protein (Karron and Collins, 2007). The central region of the N_{core}, containing a conserved

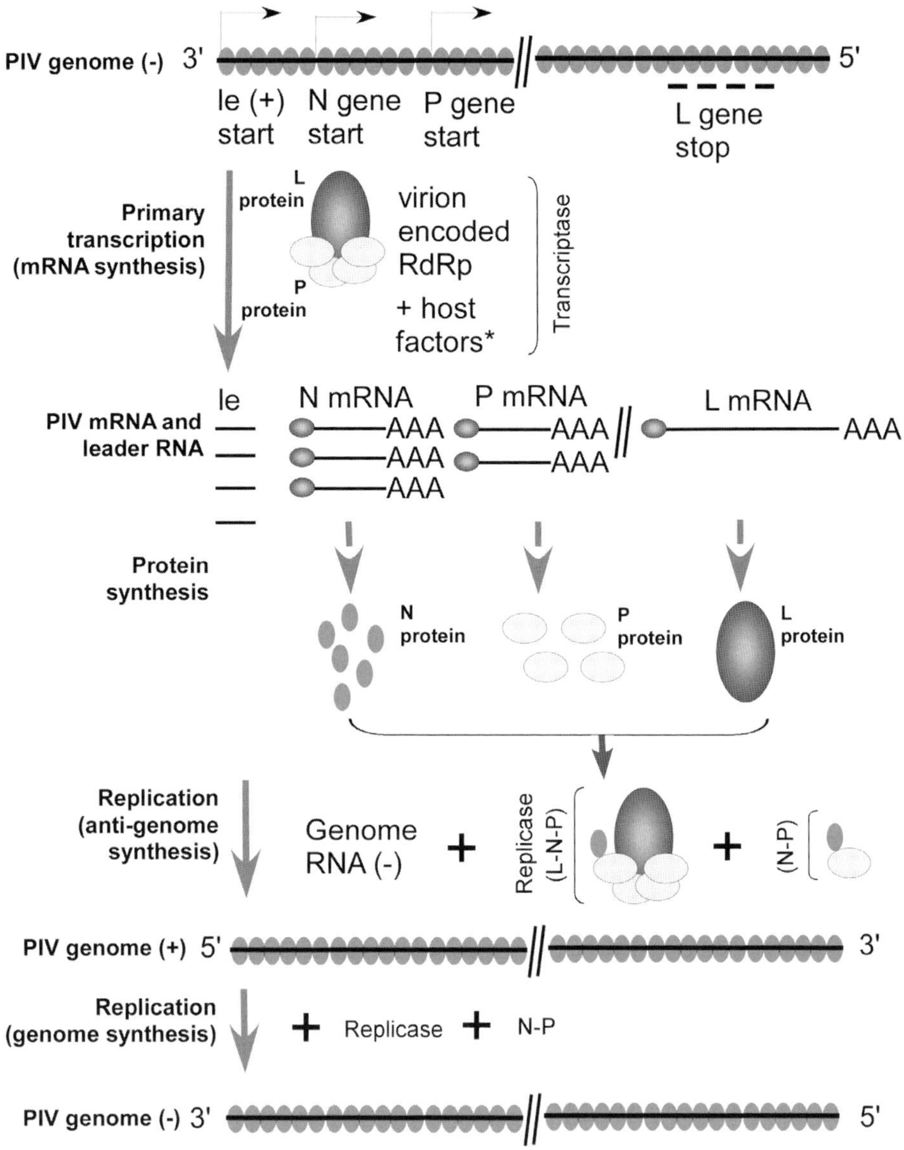

Figure 5.4 Schematic representation of the Parainfluenza virus (PIV) replication strategy. The NNS genome is encapsidated by the N protein to form the N-RNA complex. The virion, encoded RdRp (L protein that remained associated with the P protein) carry out primary transcription that leads to the synthesis of leader (le) and mRNAs. When there is sufficient amount of N protein present, the polymerase switches from transcriptase to replicase mode, and copy the full length (+) strand genome RNA, which in turn is copied to synthesize the (–) strand genome RNA. *The VSV transcriptase contains host factors like transcription elongation factor-1α and heat shock protein 60.

F-X4-Y-X3-ϕ-S-ϕ-A-M motif is important for encapsidation (Myers and Moyer, 1997). Interestingly, the C-terminal 25% of the N protein – the Ntail, which can be digested by limited proteolysis, is not necessary for protecting the genome RNA from nuclease action (Mountcastle *et al.*, 1974; Heggeness *et al.*, 1980). However, the N_{tail} plays a key role in the interaction between the N-RNA template and the RdRp heterocomplex (Buchholz *et al.*, 1994). The N_{tail}, which is intrinsically unstructured, becomes folded when it interacts with the C-terminus of P protein (Lamb and

Parks, 2007). The P protein, which also interacts with the L protein, mediates contact between the N-RNA and the L protein via this N_{tail}–P interaction. The dynamic make–break type of interaction between the N_{tail} and the P protein (or the L–P complex) is necessary for the movement of the polymerase along the N-RNA template (Lamb and Parks, 2007). It is still unclear, nevertheless, how the polymerase gets access to the genome RNA when the latter is tightly encapsidated by the N protein. It is believed that the N molecules open up transiently so that the polymerase gets access to the RNA template.

The phosphoprotein (P)

The P gene of PIVs (and Paramyxoviruses), unlike other NNS viruses, is unique since it is capable of expressing multiple proteins by a variety of mechanisms. The P protein (~400–600 amino acids) is expressed from the full length P mRNA using the first AUG initiation codon for P and terminates at the UAA codon located at 3′end. It is the most abundant among the P gene products and is an obligatory component of the viral transcription and replication machinery. The P protein, which forms oligomer and becomes heavily phosphorylated at specific serine and threonine residues, lacks enzymatic activity, but interacts to the N and the L proteins in order to perform several functions in viral life cycle (Karron and Collins, 2007; Lamb and Parks, 2007). There is a lack of sequence homology among P proteins of PIV, but contain some common modules that carry out specific functions. The extreme N-terminal ~40 amino acids are required for binding to the free N proteins (N^0; Curran *et al.*, 1995; De *et al.*, 2000; Horikami *et al.*, 1992; Zhao and Banerjee, 1995). This N^0–P complex formation is necessary for the prevention of non-specific interaction of N protein to the cellular RNA and also for the encapsidation of the newly synthesized genome RNA (Horikami *et al.*, 1992). The next module (~40–300 amino acids) is highly acidic and undergoes post translational modification, since it becomes heavily phosphorylated by cellular kinases. Several cellular kinases are shown to phosphorylate at specific residues of the P protein of several PIVs. For example, the P proteins of HPIV3 and SeV are phosphorylated by the cellular protein kinase C (PKC) (De *et al.*, 1995; Huntley *et al.*, 1995, 1997). Interestingly, unlike VSV (rhabdovirus), where the casein kinase mediated phosphorylation of the P protein is essential for transcription, the precise role of phosphorylation at specific residues of the PIV P proteins remained elusive (Chattopadhyay and Banerjee, 1987; Barik and Banerjee, 1992a,b; Das *et al.*, 1995; Jackson *et al.*, 1995; Chen *et al.*, 1997; Pattnaik *et al.*, 1997; Hwang *et al.*, 1999; Das and Pattnaik, 2004). For example, the PKC mediated phosphorylation has been shown to regulate the replication of HPIV3, while for SeV, removal of most of the phosphorylation sites does not seem to have a significant effect in viral RNA synthesis (De *et al.*, 1995; Hu *et al.*, 1999; Hu and Gupta, 2000). Recently, Akt1 and polo-like kinase 1 (PLK1) have been shown to phosphorylate specific residues of PIV5 P and regulate viral replication (Sun *et al.*, 2008, 2009). Another significant difference between the rhabdovirus and PIV (and paramyxovirus) P proteins is that phosphorylated forms at specific sites of the rhabdovirus P proteins are observed, while such distinguishable multiple phosphorylated forms of P protein have not been identified for PIV (and paramyxovirus; Clinton *et al.*, 1979; Barik and Banerjee, 1992b; Chen *et al.*, 1997; Gupta *et al.*, 2000). Complete identification of cellular kinases and detail analyses of the role of phosphorylated form(s) of different PIV P proteins are necessary to understand the exact function of the P protein in viral life cycle. Interestingly, recent analyses suggested that the module (~41–300 amino acids) is intrinsically unstructured, but possibly required for the optimum function of the P protein (Karlin *et al.*, 2003; Gerard *et al.*, 2009). But, how the phosphorylation is involved in determining the magnitude of disorder in this domain or the entire protein still remains unknown.

The C-terminal half of the PIV P protein (and possibly all NNS virus P protein), on the other hand, possesses several distinct structural domains that can fold independently. The P-P oligomerization domain of SeV has been expressed in bacteria and the crystal structure has been solved (Tarbouriech *et al.*, 2000b). In SeV, the P protein forms a tetramer and this oligomerization domain is essential for transcription (Tarbouriech *et al.*, 2000a,b). Although no other crystal structure has been solved for other PIV P proteins distinct oligomerization domains have been identified and alterations of the domain have been shown to render the P protein inactive in transcription (Choudhary *et al.*, 2002). In order to be catalytically active, the L protein needs to interact with the P protein. The L–P interaction is necessary to prevent the non-specific L–L self-interaction followed by aggregation (Chattopadhyay and Banerjee, 2009). The domain responsible for L–P interaction on P protein remains proximal to the P-P oligomerization domain, while the extreme C-terminal domain of the P protein is required for the interaction of the P protein to the N-RNA template (Curran *et al.*, 1993; Zhao and Banerjee, 1995; Tuckis *et al.*, 2002). The C-terminus of the P protein interacts to the C-terminal tail of the N-protein (Ntail) and helps the polymerase to access the N-RNA template. Apart from its role in RNA synthesis, recent data suggest that the PIV P protein, together with other accessory proteins encoded by P gene (C and V proteins), may play an important role in counteracting the host antiviral mechanisms (Dillon and Parks, 2007).

The C protein

Viruses belonging to Respirovirus but not Avulavirus and Rubulavirus express C protein. It is expressed from the P mRNA using a +1 reading frame with respect to the P ORF. In HPIV3, only one C protein (199 amino acids) is produced from the P mRNA, while for HPIV1 and SeV, four C proteins, C′, C, Y1 and Y2 are translated (215–175 amino acids) by two different mechanisms. The C′ and C proteins are expressed by using leaky ribosomal scanning mechanism, while the Y1 and Y2 are expressed by using the ribosomal shunt mechanism (Latorre *et al.*, 1998). C′, C, Y1 and Y2 proteins have common C-terminal end, but varies at length owing to the differences at the N-termini. The C proteins, although present in very low amount in the virions and virus infected cells, are involved in regulating the viral transcription and replication as well as in counteracting host innate immunity (Kurotani *et al.*, 1998). Apart from these two primary functions, the C protein has been shown to be involved in viral assembly, budding and suppressing virus-induced apoptosis (Hasan *et al.*, 2000; Koyama, 2003; Sakaguchi *et al.*, 2005).

Although expression of C protein is not essential for viral RNA synthesis, the recombinant virus that does not express C protein is highly attenuated. Interestingly, the C protein can inhibit viral RNA synthesis although the mechanism is not precisely understood. The C protein has been shown to interact to the N-terminus of the L protein (a site different from the P interacting domain on L protein) and suggestion was made that the L–C interaction results in inhibition of the RNA synthesis (Horikami *et al.*, 1997; Grogan and Moyer, 2001; Smallwood and Moyer, 2004). On the other hand, several reports suggest that the C protein can inhibit viral RNA synthesis in a promoter dependent manner (Cadd *et al.*, 1996; Tapparel *et al.*, 1997). Prediction analysis showed that the HPIV3 C protein may contain a coiled-coil domain, which might be essential for the C protein mediated inhibition of the viral transcription (Malur *et al.*, 2004). Interestingly, recent results suggest that the removal of N-terminal 25 amino acids of the C protein of HPIV3 renders the truncated protein an extremely potent inhibitor of viral transcription (Mao *et al.*, 2009). This truncated C protein has been suggested to be a candidate for antiviral agent against clinically important respiratory viruses (Mao *et al.*, 2009). Another important function of the C protein is the inhibition of anti-viral action of host cell induced by interferon (IFN).

The V protein

A unique V mRNA is generated by the insertion of a non-templated G residue at a specific site on the P mRNA during transcription of the genome RNA (Vidal *et al.*, 1990; Jacques *et al.*, 1994; Hausmann *et al.*, 1996, 1999). Upon translation of the extra G containing transcripts, a V protein is synthesized (25–30 kDa) that has its N-terminus common to the P protein while the C-terminus differs completely. Similar to the C protein, the V proteins also are present in lesser amount in virion, but in infected cells carry out critical functions like regulation of viral RNA synthesis and inhibition of host anti-viral response.

The C-terminal regions of V proteins carry a conserved Cys and His rich domain, which is responsible for interacting with zinc (Kato *et al.*, 1997b; Huang *et al.*, 2000; Fukuhara *et al.*, 2002). This region of the V protein interacts with the cellular damage-specific DNA-binding protein 1 (DDB1) and also self-assembles to form high molecular weight particles (Lin *et al.*, 1998; Andrejeva *et al.*, 2002; Lamb and Parks, 2007). These interactions lead to higher order structure of the V protein and the overall structure of the V protein varies significantly with the P protein (Lamb and Parks, 2007). However, owing to its common N-terminal region with the P protein, the V protein of several PIV have been shown to interact to the N protein to form V–N^0 complex (Horikami *et al.*, 1996). Possibly owing to its ability to interact with the N and L proteins and RNA the V protein is capable of inhibiting the viral RNA synthesis (Lin *et al.*, 1997; Lamb and Parks, 2007; Nishio *et al.*, 2008). Consistent with this phenomenon, mutation in the V protein gives rise to elevated viral RNA synthesis (Delenda *et al.*, 1997, 1998; Kato *et al.*, 1997b; Durbin *et al.*, 1999; Huang *et al.*, 2000; He *et al.*, 2002; Nishio *et al.*, 2008). Virus with mutation at V ORF grows well in tissue culture cells but remains highly attenuated *in vivo* (Delenda *et al.*, 1997; Kato *et al.*, 1997a). This latter phenomenon might be due to the fact that the V protein is involved in antagonizing the host antiviral response (see below).

The D, W and I proteins

For BPIV3 and HPIV3 the D transcript is generated owing to the insertion of two G residues in the P mRNA (Lamb and Parks, 2007). The N terminus (241 amino acids) of the D protein is same as the P protein while the C terminus (131 amino acids) is unique (Wells and Malur, 2008). A recent report suggests that the D protein of HPIV3 may localize to the nucleus as well as cytoplasm (Wells and Malur, 2008). However, the function of this protein is not known. The W protein is also generated from insertion of two G residues in the P mRNA, but at a different site (Lamb and Parks, 2007). The N-terminus of W protein, since it is same as the P protein, can interact with the N^0 protein and inhibits viral RNA synthesis (Horikami *et al.*, 1996). In *Rubulavirus* the I protein is generated due the insertion of either one or four G residues to the P mRNA during RNA editing, but the function of this protein remains unknown (Thomas *et al.*, 1988).

The matrix protein (M)

The M proteins (~38–41 kDa) of the PIV are conserved, basic, hydrophobic and non-glycosylated proteins (Karron and Collins, 2007; Lamb and Parks, 2007). It is the most abundant protein packaged in the virion and is peripherally associated with the membrane. The molecules of M protein lie beneath the envelope and interact with each other (Karron and Collins, 2007; Terrier *et al.*, 2009). Earlier studies with purified protein indicate that in low salt condition the M protein molecules self-interact to form tube or sheet like structure beneath the membrane (Bachi, 1980; Heggeness *et al.*, 1982). However, recent cryo-EM analysis suggests that the M proteins forms electron-opaque patches rather than a continuous sheet (Terrier *et al.*, 2009). They interact with the glycoproteins as well as the viral core nucleocapsid/ribonucleoprotein (RNP) and are important in viral assembly, budding and morphogenesis (Stricker *et al.*, 1994; Sanderson *et al.*, 1995; Takimoto and Portner, 2004). SeV M protein is phosphorylated, but mutant virus lacking M protein phosphorylation does not show any altered phenotype compared with the wild type (Sakaguchi *et al.*, 1997).

The fusion protein (F)

The PIV F is a type I integral membrane protein that contains a proteolytic cleavage site at the N-terminus and span the membrane once. The F protein is first translated as a biologically inactive precursor form, F_0 (540–580 amino acids), which upon cleavage by the host cell protease produces F1 and F2 polypeptides that are linked by the disulfide bond (Lamb and Parks, 2007). For *Avulavirus, Respirovirus* and *Rubulavirus*, the F1 and F2 are glycosylated (Lamb and Parks, 2007). Crystallographic and biophysical data revealed that the PIV F protein forms a trimeric, coiled-coil structure and can be divided as head, neck and stalk (Russell *et al.*, 1994; Chen *et al.*, 2001; Yin *et al.*, 2005, 2006; Swanson *et al.*, 2010). The PIV F protein mediates the fusion of viral envelope to the host cell membrane and that allows the viral RNP to be internalized and subsequently replication within the cytoplasm ensures (Lamb and Parks, 2007). During the fusion process the F protein undergoes extensive structural alterations (Dutch *et al.*, 2001; Lamb and Jardetzky, 2007; Bissonnette *et al.*, 2009). Moreover, it is thought that for most PIV, the HN glycoprotein facilitates fusion by promoting conformational changes in the fusion (F) protein at neutral pH (Porotto *et al.*, 2003). Late in infection newly synthesized F proteins become associated with the cell membrane and mediate cell–cell fusion to form syncytia (Lamb and Parks, 2007). This cytopathic effect is possibly necessary for the transmission of PIV (Lamb and Parks, 2007).

The haemagglutinin–neuraminidase (HN) glycoprotein

The PIV HN glycoproteins (565–582 amino acids), which contains dual haemagglutinin and neuraminidase activities are type II glycoproteins containing uncleaved signal or anchor sequence. The crystal structures of HN proteins of NDV and HPIV3 have been solved (Bousse *et al.*, 2004; Lawrence *et al.*, 2004). The HN proteins covalently associate to form a homotetrameric protein containing N-terminal cytoplasmic tail, N-terminal transmembrane domain, membrane proximal stalk domain and C-terminal globular head domain (Hiebert *et al.*, 1985a). The globular head domain contains major antigenic sites and also all the biological activities. This region is involved in attachment to the host cell by interacting to the sialic acid and also responsible for haemagglutinin activity. Moreover, the HN protein can physically interact with the F protein and may help the F protein in the fusion process. It is believed that the HN tetramer associates with the F protein trimer laterally in the absence of ligand but upon interaction with the ligand, the HN as well as F proteins undergoes structural alteration, which is necessary for fusion (Moscona, 2005; Lamb and Parks, 2007). Both haemagglutinin and neuraminidase activities are dependent on halide ion concentration and pH (Merz *et al.*, 1981). The neuraminidase activity of the HN glycoprotein plays a crucial role in the release of the virion (Moscona, 2005).

The small hydrophobic protein (SH)

In PIV5 and APMV6, a gene located between HN and F genes, expresses a type II integral membrane protein, the SH protein (44 amino acids for PIV5 and 142 amino acids for APMV6; Hiebert *et al.*, 1985b, 1988; Karron and Collins, 2007). Recombinant PIV5 that does not express SH protein grows well in tissue culture, but is attenuated *in vivo* (He *et al.*, 1998). This recombinant PIV5 elicits strong apoptosis through TNFα pathway, suggesting that this protein is required to block the host cell apoptosis (He *et al.*, 2001; Lin *et al.*, 2003).

The large protein (L)

The gene encoding PIV L protein (~2200 amino acids) is located close to the 3′-end of genome RNA, thus, expressed least during infection; ~50 copies of the protein are packaged in the virion (Karron and Collins, 2007; Lamb and Parks, 2007). The PIV L proteins, like all NNS virus L proteins, act as RdRp and also possess enzymatic activities required for mRNA capping, methylation and polyadenylation (Banerjee, 1987; Banerjee *et al.*, 1991; Karron and Collins, 2007; Lamb and Parks, 2007). The NNS virus L proteins possess six distinct regions containing conserved amino acid sequences that are believed to carry out distinct functions related to RNA synthesis, post-transcriptional modifications and stability to the newly synthesized RNA (Fig. 5.5; Poch *et al.*, 1990; Sidhu *et al.*, 1993; Muller *et al.*, 1994). Homology region II for example, because of its highly charged residues is believed to interact with RNA (Karron and Collins, 2007; Lamb and Parks, 2007). The homology region III acts as polymerase module and mutations at the highly conserved GDN residues in this region lead to total loss of RNA synthesis (Sleat and Banerjee, 1993; Malur *et al.*, 2002b). The GXGXG motif present in the homology region VI is necessary for cap methylation (Grdzelishvili *et al.*, 2005, 2006; Li *et al.*, 2005, 2006; Ogino *et al.*, 2005; Murphy and Grdzelishvili, 2009; Rahmeh *et al.*, 2009). Within homology region V, a specific HR motif is required for the capping of nascent viral RNA as shown for VSV and Chandipura viruses (Ogino and Banerjee, 2007, 2010; Li *et al.*, 2008; Ogino *et al.*, 2010). The exact functions of the other conserved

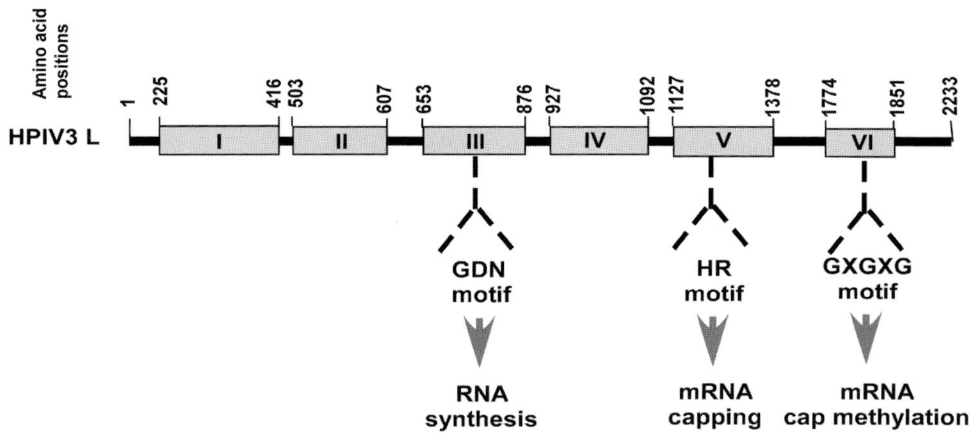

Figure 5.5 Schematic representation (not to the scale) of full length HPIV3 L protein (2,233 amino acids). The boxes indicate the conserved regions I-VI, which were identified by aligning HPIV3 and SeV L proteins and by comparison with other L proteins (Poch *et al.*, 1990). The GDN motif in the conserved region III is important for RNA synthesis; the HR motif in the conserved region V is important for viral mRNA capping; the GXGXG motif present in the conserved region VI is important for cap methylation.

modules remain unknown, but several mutational analyses within blocks I, II and IV have been carried out for SeV, which gave rise to a spectrum of variable polymerase activity ranging from partial to complete inactivation of transcription and/or replication (Chandrika *et al.*, 1995; Smallwood *et al.*, 1999, 2002b; Feller *et al.*, 2000). The L protein of PIV (and also paramyxoviruses) strongly interacts with its cognate cofactor, the P protein, via the N-terminal half (Parks, 1994; Malur *et al.*, 2002a; Cevik *et al.*, 2004; Chattopadhyay and Banerjee, 2009). In contrast, rhabdovirus L protein interacts with its cognate P protein through its C-terminus (Canter and Perrault, 1996; Chenik *et al.*, 1998). Another important difference between the two genera is that for the PIV (and paramyxoviruses) the L and P proteins need to be co-expressed in the same cell in order to be functionally active, while for rhabdovirus, the two proteins can be expressed separately and can still be active when reconstituted *in vitro* (Horikami *et al.*, 1992; Smallwood *et al.*, 1994). Interestingly, several reports show that the L protein of PIVs (and paramyxoviruses) seem to oligomerize in order to be functionally active (Smallwood *et al.*, 2002a; Cevik *et al.*, 2003, 2004, 2007; Smallwood and Moyer, 2004). Through C-terminal deletions of the HPIV3 L protein, it was initially postulated that the L-L oligomerization domain is located at the N-terminus of the protein (Smallwood *et al.*, 2002a; Cevik *et al.*, 2003, 2004, 2007; Smallwood and Moyer, 2004). Recently, for HPIV3 it was shown that the N-terminally deleted L proteins self-interact and no specific oligomerization domain could be identified, suggesting that the L protein by itself self-aggregates non-specifically, which is prevented in the presence of P protein, indicating that the P protein might act as chaperone for the L protein (Chattopadhyay and Banerjee, 2009).

Genetics and reverse genetics

With the advent of cloning and sequencing techniques the functions of PIV encoded proteins and the PIV cis-acting elements were studied extensively by both forward and reverse genetics. The genomes of PIV from all genera have been sequenced (Kawano *et al.*, 1991, 1993; Ogawa *et al.*, 1992; He *et al.*, 1997; Itoh *et al.*, 1997; Ohsawa *et al.*, 1998; Durbin *et al.*, 1999; Bailly *et al.*, 2000; Newman *et al.*, 2002). Nucleotide sequence analysis of PIV encoded genes have been employed in studying strain variability (Canchola *et al.*, 1964; Sanchez *et al.*, 1986; Ohgimoto *et al.*, 1990; Yuasa *et al.*, 1990; Tsurudome *et al.*, 1991; Ogawa *et al.*, 1992; Horwood *et al.*, 2008; Xiao *et al.*, 2009; Samuel *et al.*, 2010). Sequence analyses also revealed that the PIV genome, in spite of high rate of misincorporation of nucleotide by polymerases, appear to evolve slowly over time (Sanchez *et al.*, 1986; Sakaguchi *et al.*, 1989; Ohgimoto *et al.*, 1990; Yuasa *et al.*, 1990; Karron and Collins, 2007). However, it is not uncommon to identify PIV strain by nucleotide sequencing that is serologically distinguishable from other strains of the same species (Swierkosz *et al.*, 1995). Forward genetics, by expression of the PIV proteins using corresponding cDNA clones, have been used to assess the function of the viral proteins (Galinski *et al.*, 1986a,b, 1988). Also, mutational studies showed the importance of specific region of the PIV proteins in relation to specific functions (Chandrika *et al.*, 1995; Cevik *et al.*, 2003, 2004, 2007; Chattopadhyay and Banerjee, 2009; De *et al.*, 2000). Several of the experiments studying the functions of the cis-acting elements were carried out by using defective interfering (DI) particles, in which the genomes lack most of the viral genes, but carry the cis-acting elements (Carlsen *et al.*, 1985; Hsu *et al.*, 1985; Baker and Moyer, 1988; Horikami *et al.*, 1992; Horikami *et al.*, 1996; Myers *et al.*, 1999).

The major breakthrough in NNS virus genetics occurred when infectious full length cDNA clone of the genome of rabies followed by VSV was constructed in the presence of helper plasmids encoding N, P and L proteins which produced infectious virus in cultured cells (Fig. 5.6A; Schnell *et al.*, 1994; Lawson *et al.*, 1995; Whelan *et al.*, 1995). Since then full length infectious cDNA clones have been constructed for most of the PIVs (Garcin *et al.*, 1995; Durbin *et al.*, 1997; Hasan *et al.*, 1997; Hoffman and Banerjee, 1997; Leyrer *et al.*, 1998; Krishnamurthy *et al.*, 2000; Kawano *et al.*, 2001; He *et al.*, 2002; Mao *et al.*, 2008; Roth *et al.*, 2010) and subsequently a vast majority of NNS viruses (Conzelmann, 1998, 2004; Nagai, 1999; Neumann *et al.*, 2002; Neumann and Kawaoka, 2004). This approach has proved to be invaluable in subsequent development of chimeric viruses for vaccine development (Bailly *et al.*, 2000; Haller *et al.*, 2000; Skiadopoulos *et al.*, 2001, 2002, 2003; Newman *et al.*, 2004; Tang *et al.*, 2004, 2008; Zhan *et al.*, 2008; Bukreyev *et al.*, 2009).

Using similar technique, minireplicon systems were developed that provided a large body of experimental findings to determine the functions of the cis-regulatory elements with relative ease by manipulating such elements by mutation (Park *et al.*, 1991; De and Banerjee, 1993; Dimock and Collins, 1993; Hoffman and Banerjee, 2000a,b; Malur *et al.*, 2002a; Hoffman *et al.*, 2006). In this approach, practically all of the genes are deleted from genome cDNA leaving only the cis-elements flanked by reporter genes, like luciferase (LUC), chloramphenicol acetyltransferase (CAT) or green fluorescent protein (GFP). This minigenome cDNA is then transfected in cells along with support plasmids expressing N, P and L proteins (Fig. 5.6B). After the first round of transcription, the newly synthesized minireplicon is encapsidated by the N protein (in the presence of P protein) and transcribed by the L protein (in the presence of P protein) to form mRNA of the reporter gene. The activities of the newly synthesized reporter proteins were measured in order to study the function of the viral proteins (primarily N, P and L; also several accessory proteins like C and V) and the functions of the cis-acting elements determined (Malur *et al.*, 2002b, 2004; Mao *et al.*, 2009). Biochemical studies using minireplicons studies have also been informative in viral protein–protein interactions and encapsidation of the RNA by the N protein (Malur *et al.*, 2002a; Chen *et al.*, 2007; Chattopadhyay and Banerjee, 2009).

Pathogenesis

Host range

HPIV1–4 are a common cause of rhinitis, pharyngitis, laryngitis, tracheobronchitis, bronchiolitis and pneumonia in humans especially in children and those with chronic lung disease or altered immunity. They can infect non-human primates and rodents (Karron and Collins, 2007). SeV, although was initially discovered from human patient, is a mouse (or rodent) virus and owing to its close relationship with HPIV1, is also known as murine PIV1 (Kuroya and Ishida, 1953). Cattle are the natural host for BPIV3 and the infection can cause respiratory diseases but may also infect a wide range of animals (Henrickson, 2003). PIV5 is primarily a canine virus and is responsible for *kennel cough* (Hsiung, 1972; Hull *et al.*, 1956).

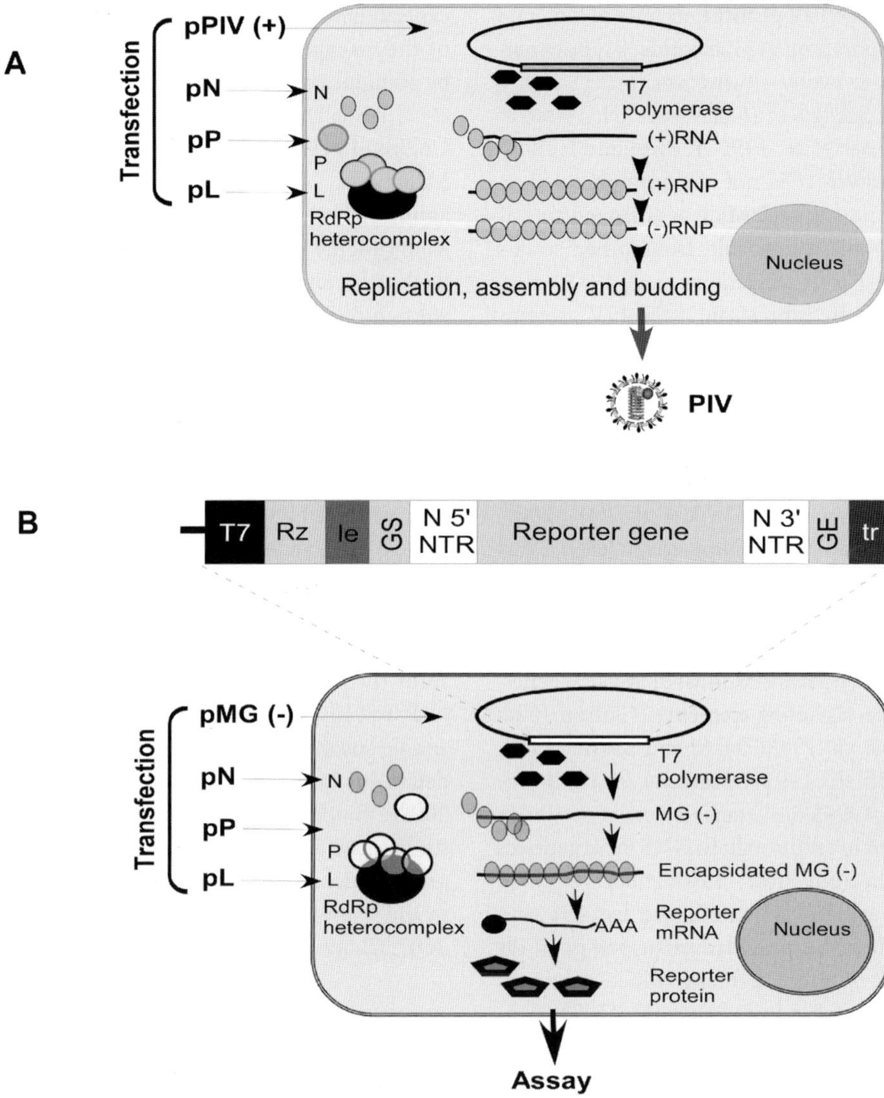

Figure 5.6 Schematic representation of the methods used in reverse genetics in Parainfluenza virus (PIV) and non-segmented negative-sense RNA viruses (NNS viruses) viruses (A) Schematic representation of the generation of infectious PIV from full length infectious cDNA clone. The plasmid encoding the full length infectious cDNA is transfected with the helper plasmids, viz. N, P and L. Prior to the transfection the cells can be superinfected with the recombinant vaccinia virus for the expression of the T7 polymerase. Alternatively, cells that constitutively express T7 polymerase can be used. The (+) RNA, synthesized by the T7 polymerase will be encapsidated by the recombinant N protein and associated to the recombinant polymerase heterocomplex, L-P, to form (+) ribonucleoprotein (RNP) complex. The (+) RNP will subsequently be copied to form (–) RNP. The (–) RNP will be further used for transcription and replication and the newly synthesized (–) RNP will be packaged into PIV virion for budding out. (B) The schematic representation showing the strategy used in minigenome or minireplicon (MG) assay. The plasmid encoding the MG(–) is transfected with the plasmids encoding the L, P and N proteins. In the cytoplasm the viral proteins and the minireplicon will be synthesized with the help of the T7 polymerase (see above) using the T7 promoter present in each of the plasmids. The minireplicons usually contain a reporter gene, flanked by the cis acting elements like, the leader (le), trailer (tr), N gene 5′ end and L gene 3′ end NTRs, gene start (GS, gene end (GE) and hepatitis delta virus ribozyme (Rz). At the second step, the viral polymerase uses the cis-acting elements to synthesize the reporter mRNAs and the reporter mRNAs or the reporter proteins are used for assays.

Several studies indicate that both SV5 and SV41, although initially isolated as contaminants from monkey kidney cell cultures, are closely related to HPIV2 and four and have the potential to infect humans (Miller *et al.*, 1964; Goswami *et al.*, 1984a,b; Tsurudome *et al.*, 1990, 1991; Karron and Collins, 2007). APMV1–9 primarily infect birds but are also capable of causing conjunctivitis in bird handlers and laboratory workers (Alexander, 1988; Henrickson, 2003).

Tissue tropism, entry and spread

Respiratory epithelium is the major site of binding and infection of PIV, although infections in the nervous tissues have also been reported (Mori *et al.*, 1995; Henrickson, 2003). The tissue tropism is mostly directed by the HN protein (Henrickson, 2003; Huang *et al.*, 2004). The entry and release of HPIV3 in A549 cell appears to occur through the epical surface than the basolateral surface and in cotton rats, viral infection leads to localization of antigens mostly in the apical surface (Bose *et al.*, 2001; Porotto *et al.*, 2001). The structural analyses revealed that the HN protein has N-terminal cytoplasmic domain, transmembrane domain and C-terminal globular head domain and forms tetramers (Russell *et al.*, 1994; Crennell *et al.*, 2000; Takimoto *et al.*, 2000; Moscona, 2005). The head domain, which is responsible for the haemagglutinin and neuraminidase activities, interacts with the host cell sialic acid. This interaction possibly leads to structural alteration of the HN protein, which is necessary for the fusion promoting activity of the HN protein (Moscona, 2005). For most PIVs, co-expression of homotypic HN and F proteins are required for fusion since HN can promote the structural changes in the F protein required for the fusion (Russell *et al.*, 2001; Porotto *et al.*, 2003). After fusion, the viral RNP is exposed to the cytoplasm, transcribed and replicated using the virus encoded and the newly synthesized RdRp. The newly synthesized F proteins interact with the membrane and results in the formation of giant cell or syncytia (Karron and Collins, 2007). The HPIV3 interacts with heparan sulfate (HS) and cell surface HS appears to play an important role in efficient viral entry (Bose and Banerjee, 2002). The cell surface nucleolin seems also to be required as cofactor for efficient cellular entry of HPIV3 (Bose *et al.*, 2004). It is believed that the neuraminidase activity of the HN protein present in the budding virus particle is necessary for cleaving the attached receptors, which is necessary for release of the virus (Moscona, 2005). The newly released PIV infects new cells and spreads by using similar replication cycle. There is very little evidence for spread of virus beyond the ciliated cell types (Zhang *et al.*, 2005). Antibodies against the F glycoprotein can prevent the spread of PIV (Merz *et al.*, 1980).

Animal models

Hamster, guinea pigs, cotton rats and ferrets are permissive to infection by HPIV1 and 2, but the infection is mostly asymptomatic. Experimental infection with HPIV3, leads to bronchiolitis in cotton rat and induces hyperreactivity in the trachea of guinea pigs, similar to the human disease, (Porter *et al.*, 1991). This model was proven to be useful particularly in elucidating the role of HN protein in HPIV3 entry (Prince *et al.*, 2001). HPIV3 can cause symptomatic infection in chimpanzees and African green monkeys, while the infection of these animals by other HPIV remains asymptomatic (Karron and Collins, 2007). HPIV3 induces hyperreactivity in the trachea of guinea pigs (Henrickson, 2003). Mice are poorly permissive for HPIV1–3 (Karron and Collins, 2007), whereas mice are routinely used for SeV and African green monkey and rhesus monkey serve as experimental models for BPIV3.

Immunity

The innate immune response

Substantial progress has been made in recent years in understanding the innate immune system including recognition of pathogen and the signalling pathways that mediate the anti-viral responses. Interestingly, several PIV, particularly SeV, PIV5 and NDV have served as excellent models in elucidating these cellular events (Bose and Banerjee, 2003). The recognition of PIV infection is primarily made by pattern-recognition receptors (PRR) viz. retinoic acid-inducible gene I (RIG-I)-like receptors (RLR; it includes RIG-I,

MDA5) and toll-like receptors (TLR; Schroder and Bowie, 2007; Takeuchi and Akira, 2009). The RLR and TLR are involved in the production of type I IFN and cytokines (Takeuchi and Akira, 2009). The IFN-inducible genes like protein kinase R (PKR) and 2′,5′-oligoadenylate synthase are also involved in the elimination of viral components from infected cells by inducing apoptosis as well as protection of uninfected cells (Takeuchi and Akira, 2009).

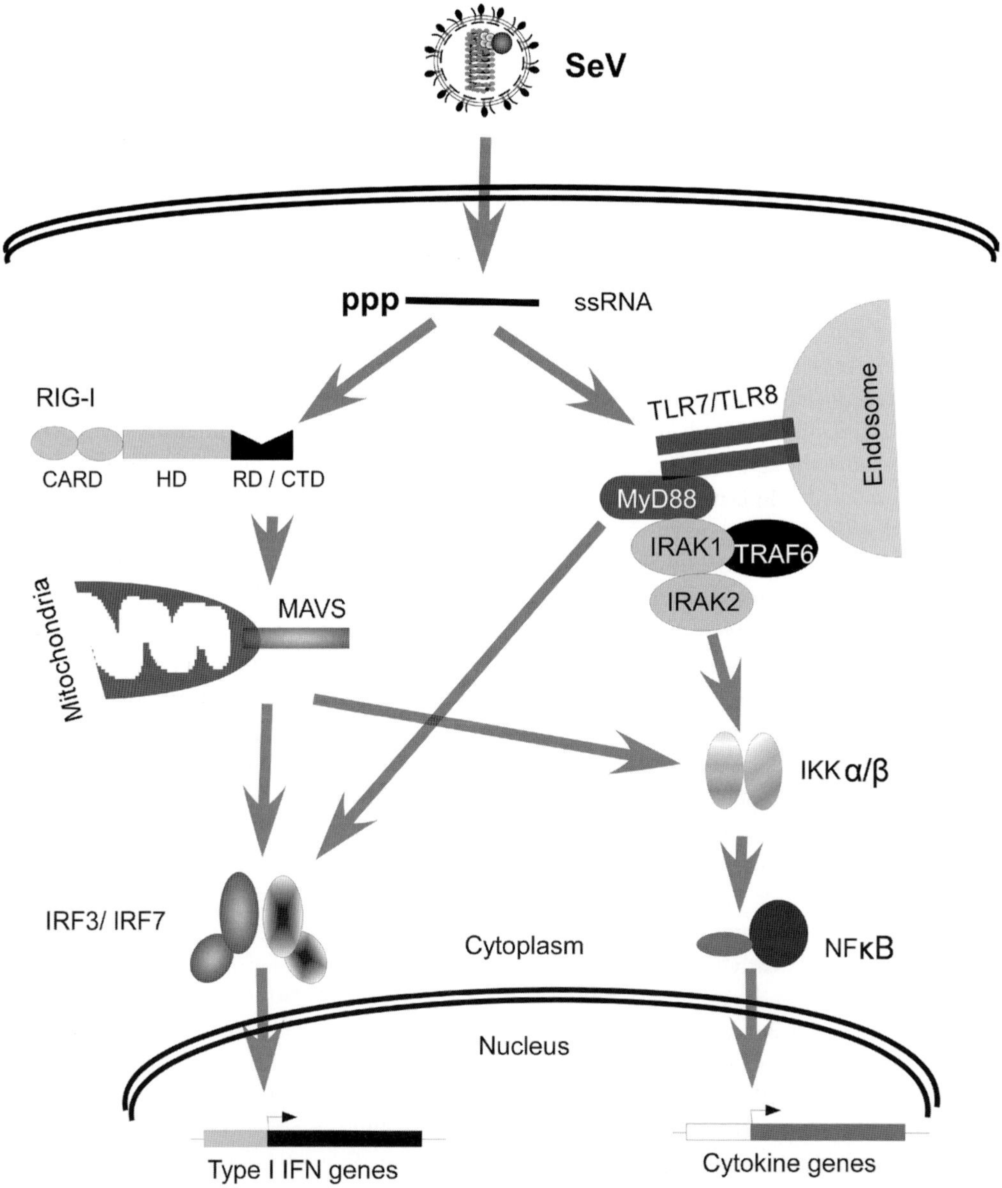

Figure 5.7 Schematic representation of the parainfluenza virus (PIV) ssRNA recognition by RIG-I and TLR7/TLR8. The 5′ end triphosphate (ppp) containing viral RNA can be recognized by the cleft like structure present in the C-terminal domain of RIG-I. This event activates the mitochondrial adaptor protein, MAVS. The MAVS in turn activates IRF3/IRF7, leading to the induction of type I IFN genes. The viral ssRNA can also activate TLR7/TLR8, present in the endosome and gives rise to the induction of type I IFN. Both RIG-I and TLR7/TLR8 can activate NF-κB leading to the induction of cytokine genes.

Recognition of PIV infection by RLR

SeV and possibly other PIV infection can lead to the production of type-I IFN and induction of cytokine expression. These events involve activation of both RLR and TLR, depending on the cell type (Melchjorsen *et al.*, 2005; Fig. 5.7). In non-immune cells, RLR are activated by SeV and this activation process requires transcription/replication competent virion since UV inactivated SeV fails to activate RLR (Melchjorsen *et al.*, 2005). Initially, this activation process was thought to be similar to the activation of RIG-I by synthetic and viral double stranded RNA (dsRNA) (Melchjorsen *et al.*, 2005). Interestingly, unlike the positive strand RNA viruses, the NNS viruses do not produce detectable amount of dsRNA (Weber *et al.*, 2006). Later it was found that the uncapped, 5′- triphosphate (5′-ppp) ended RNA, produced in cells by NNS virus can interact with the C-terminal regulatory domain of RIG-I leading to the subsequent activation (Hornung *et al.*, 2006; Pichlmair *et al.*, 2006; Cui *et al.*, 2008). In fact, for measles virus (MeV), a NNS virus belonging to the Morbillivirus genus, the 5′-ppp end containing leader RNA, generated during transcription, has been shown to activate RIG-I (Plumet *et al.*, 2007). In contrast, for SeV, the 5′-ppp ended genome RNA, rather than any transcription intermediate, appears to be responsible for the activation of RIG-I (Rehwinkel *et al.*, 2010). Cells lacking RIG-I fail to produce type I IFN in response to SeV and several RNA viruses including VSV and influenza virus as well as RIG-I$^{-/-}$ mice are highly susceptible to VSV infection (Kato *et al.*, 2005, 2006; Melchjorsen *et al.*, 2005; Takeuchi and Akira, 2009). Recently, it was found that RIG-I can act as PRR in recognizing HPIV3 infection (Sabbah and Bose, 2009). Although a growing body of literature suggests mechanism of RIG-I activation by PIV ssRNA, the precise mechanism of IFN induction through MDA-5 in response to viral RNA remains unclear. Upon activation, RIG-I and MDA-5 interact with mitochondrial antiviral signalling adaptor (MAVS), which triggers the IRF-3 and IRF-7, which in turn induce IFN-α and IFN-β expression and activate NF-κB in a TBK1-dependent manner (Fig. 5.7; Nakhaei *et al.*, 2009; Yoneyama and Fujita, 2009; Rehwinkel *et al.*, 2010).

Recognition of PIV infection by TLR

While recognition of SeV infection in non-immune cells depend mostly on RLR, in specialized innate immune cells recognition is mediated by TLRs (Melchjorsen *et al.*, 2005; Fig. 5.7). The TLRs constitute a class of membrane-bound PRR present either in the cell-surface or in the endosome (Melchjorsen *et al.*, 2005). Although it was initially thought that the ligands for TLRs are mostly of bacterial origin, like lipopolysaccharide, later it was found that a subclass of TLRs can recognize viral components, particularly viral RNA (Schroder and Bowie, 2007). Myeloid cells can detect SeV infection by utilizing TLR7 and TLR8 (Melchjorsen *et al.*, 2005). TLR7 and TLR8 activates myeloid differentiation factor 88 (MyD88), which in turn activates IL-1R-associated kinase (IRAK) and IRAKs activate TNF-receptor-associated factor 6 (TRAF6), which finally leads to the activation of NFκB through the activation of IKKα/β (Takeuchi and Akira, 2009). Activation of the TLRs, apart from the activation of NFκB followed by cytokine gene expression, also leads to the induction of type I IFN gene expression through IRF3/IRF7 activation (Fig. 5.7). Recently, it has been shown that the L protein of PIV5 can interact with Akt1 and upregulate the phosphorylated state of Akt1, which in turn can activate NF-κB (Luthra *et al.*, 2008). PKR can also activate NF-κB following viral infection, which plays a critical role in innate as well as adaptive immune responses (Bose and Banerjee, 2003). The NF-κB also regulates an interferon-independent innate antiviral response against HPIV3 (Bose *et al.*, 2003).

Upon recognition of PIV infection, the activated cells produce IFN, which in turn leads to the development of anti-viral state primarily by JAK-STAT pathway (Fig. 5.8; Shuai *et al.*, 1993; Novick *et al.*, 1994; Colamonici *et al.*, 1994; Yan *et al.*, 1996; Gotoh *et al.*, 2002). The IFN-inducible protein, MxA, in addition to other IFN-inducible proteins restricts replications of several RNA viruses (Bose and Banerjee, 2003; Zhao *et al.*, 1996).

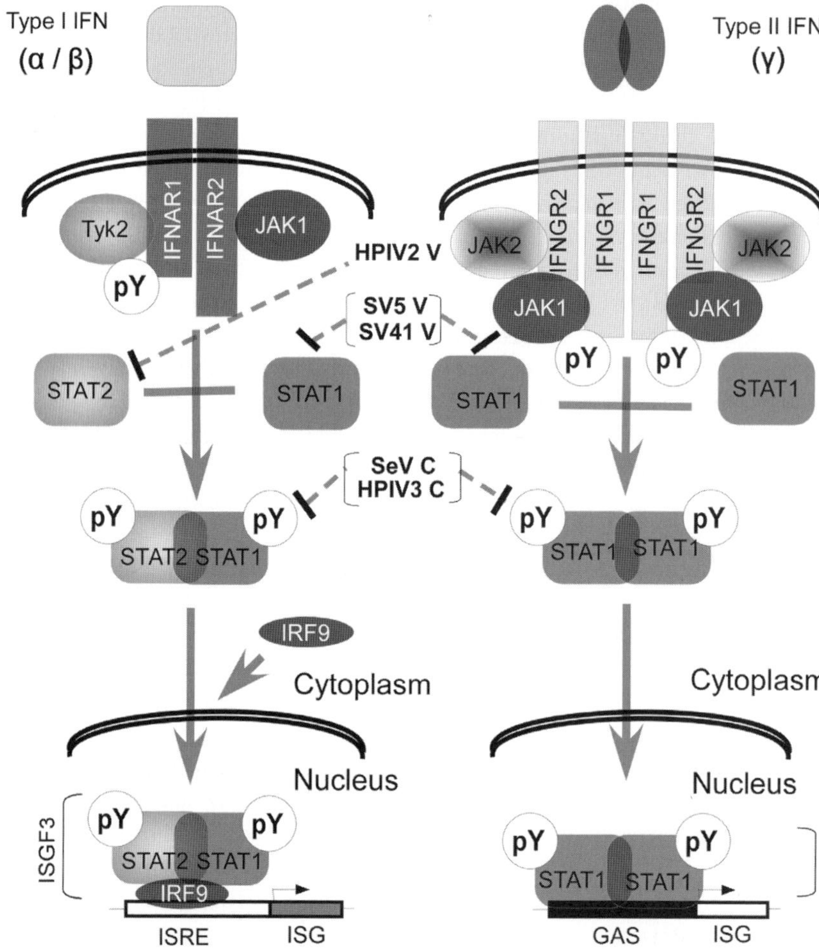

Figure 5.8 Schematic representation showing the components of the JAK-STAT pathway and its interference with Parainfluenza virus (PIV) accessory proteins. In response to the interferon (IFN) α or IFNβ the cytosolic STAT1 and STAT2 becomes phosphorylated and heterodimerized and forms complex with the IFN regulatory factor (IRF) 9. This complex, termed as ISGF3 complex migrates to the nucleus and binds to the interferon-stimulated response element (ISRE). On the other hand, the STAT1 gets phosphorylated and homodimerized in response to the IFNγ to form GAF complex, which binds to the GAS DNA element present in the nucleus. Several PIV accessory proteins interfere with these pathways. The C proteins of SeV and HPIV3 can downregulate the phosphorylation of STAT1. The SeV V protein is responsible for the degradation of STAT1 in a proteasome-dependent pathway, while the V protein of HPIV2 can degrade the STAT2.

Antagonism of host defence mechanisms by PIV

PIVs use several strategies to counteract the host innate immune response (Gotoh *et al.*, 2002; Bose and Banerjee, 2003). Early studies showed that pre-infection with SeV or HPIV3 enhances the growth of IFN sensitive virus (Hermodsson, 1963; Valle and Cantell, 1965). These results indicate that the infection with PIVs render cells unresponsive to IFN (Hermodsson, 1963; Valle and Cantell, 1965). Similar anti-IFN effect was also observed for other PIV, like HPIV3, PIV5, SV41, HPIV2 (Gotoh *et al.*, 2002). Interestingly, SeV infection prior to IFN treatment allows VSV growth even in the presence of IFN, but this effect is not apparent when SeV infection was carried out after IFN treatment (Didcock *et al.*, 1999a; Komatsu *et al.*, 2000). Later it was found that the accessory proteins expressed from P gene of PIV mediate the anti-IFN action (Gotoh *et al.*, 2002). For example, the C protein of SeV can suppress the IFNα mediated induction of ISG products. The 4C(−) SeV fail to rescue VSV growth in the presence of IFN (Gotoh *et al.*, 1999). However, the C′(−) SeV retains the anti-IFN activity since this action is mediated through Y2, the shortest among the four SeV C proteins (Gotoh *et al.*, 2002). In SeV infected cells, both STAT1 and STAT2 becomes underphosphorylated in response to IFNα or IFNγ and this suppressive effect on phosphorylation is attributed to the C protein (Gotoh *et al.*, 2002). Furthermore, the C protein of SeV can physically associate with the phosphorylated as well as unphosphorylated STAT1 and inhibits transcriptional activation of ISG (Fig. 5.8; Gotoh *et al.*, 2002). Interestingly, the SeV C protein appears also to interact with the STAT1 leading to monoubiquitination and degradation of STAT1 and only the larger forms of SeV C proteins can mediate this function (Garcin *et al.*, 2002). The SeV C protein also interferes with the RIG-I mediated induction of Type I IFN and restrict PKR activation by limiting the generation of intracellular double-stranded RNA (Strahle *et al.*, 2007; Takeuchi *et al.*, 2008). Similarly, HPIV3 C protein, like SeV C protein, inhibits the phosphorylation of STAT1 in response to IFNα and IFNγ (Fig. 5.8; Malur *et al.*, 2005). HPIV3 infection also regulates MHC class I and II expression (Gao *et al.*, 1999, 2001, 2002).

For PIV5, a *Rubulavirus*, which does not express C protein, the V protein can interact to the cellular DDB1 and STAT1 resulting in proteasome mediated degradation of STAT1 and inhibition IFN signalling (Fig. 5.8; Didcock *et al.*, 1999b; Goodbourn *et al.*, 2000; Andrejeva *et al.*, 2002; Randall and Goodbourn, 2008; Goodbourn and Randall, 2009). As expected, this degradation of STAT1 is blocked in the presence of proteasome inhibitor, MG132 (Gotoh *et al.*, 2002). Furthermore, the V protein of SV5, with its C-terminal Cys-reach domain can interact with MDA-5 and inhibit the MDA-5 mediated induction of Type I IFN (Andrejeva *et al.*, 2004). The similar repression of MDA-5 activity was also found when the V protein of HPIV2 or SeV was overexpressed (Andrejeva *et al.*, 2004). The V protein of BPIV3 also interacts to the MDA-5 and suppresses IFN-β production enhanced by MDA-5, but not by RIG-I (Komatsu *et al.*, 2007). The V protein of SV41 is capable of degrading the STAT1 by similar mechanism (Gotoh *et al.*, 2002). Interestingly, the V protein of HPIV2, blocks IFN signalling by degrading the STAT2 instead of STAT1 and hence fails to inhibit IFNγ signalling (Fig. 5.8; Young *et al.*, 2000; Gotoh *et al.*, 2002). Several recent reports suggest that both the P and V proteins of SV5 contribute to counteract the host innate immune response (Dillon and Parks, 2007; Gainey *et al.*, 2008).

Clinical feature

Respiratory tract infections are a leading cause of morbidity and mortality worldwide and Parainfluenza are some of the most common causes of respiratory tract illness in children and adults with most human infections occurring from one of the four major serotypes of HPIV. While most infections are limited to the upper respiratory tract, a substantial percentage cause life threatening lower respiratory tract disease in vulnerable populations including those with chronic pulmonary disease and immunocompromised patients. Infections due to these viruses lead to over 250,000 emergency room visits costing close to 200 million each year in estimated health care costs (Henrickson *et al.*, 1994). Rarely other PIV including avian and simian PIV have been implicated in human disease.

HPIV1–3 are well described causes of acute respiratory tract infections. Recent reports note between 7% and 17% of all hospitalizations for respiratory illness in children <5 years and up to 3% of adults are due to one of these three subtypes (Reed *et al.*, 1997; Marx *et al.*, 1999; Weinberg *et al.*, 2009). The most distinctive clinical presentation for HPIV1 (and less often HPIV2) is croup whereby patients present with stridor, cough, fevers and respiratory distress. HPIV3 is most often associated with viral bronchiolitis and pneumonia often within the first six months of life (Weinberg *et al.*, 2009). Only a few studies have investigated HPIV4 and most report mild upper respiratory illness with severe illness found only in patients with underlying disease (Chiu *et al.*, 2006; Lau *et al.*, 2009).

Although most PIV infections are mild, serious complications can occur, particularly in immunocompromised patients. Immunocompromised children and adults can progress from mild upper respiratory illness symptoms to severe pneumonia, with prolonged viral shedding and dissemination. Severe pneumonia from PIV is seen in haematopoietic stem cell transplant recipients and is associated with high mortality rates and graft rejection (Nichols *et al.*, 2001a,b).

Rarely disease has been reported from non-human subtypes of PIV. Several reports of human respiratory disease and death have been reported from APMV1 or NDV (Goebel *et al.*, 2007). NDV may cause conjunctivitis among bird handlers (Alexander, 1988; Henrickson, 2003). SeV has been reported as a cause of an outbreak of pneumonitis in newborn children. However, serological studies show that human infection with SeV seldom occurs. SV5-related paramyxoviruses have been isolated from a variety of tissues associated with human diseases, however no causal relationship has ever been established (Goswami *et al.*, 1984b).

Epidemiology

Different serotypes of parainfluenza have distinct epidemiological patterns which vary depending on geographical location (Karron and Collins, 2007). HPIV1 causes biennial outbreaks during odd number years in North America whereas HPIV2 may be annual or biennial (Marx *et al.*, 1997). HPIV1 and HPIV2 both occur predominantly in the fall. In contrast, HPIV3 activity peaks in the spring and early summer with sporadic activity throughout the remainder of the year. Few clinical or epidemiological data are available regarding HPIV4 but several studies find two distinct genotypes circulate in the fall to early winter months (Lau *et al.*, 2009).

Most parainfluenza viruses are spread through respiratory secretions through close contact with infected individuals and enter through mucous membranes of eyes, nose and mouth. Parainfluenza virus particles have also been demonstrated in large respiratory droplets allowing for inhalational spread. While parainfluenza particles can be identified on contaminated surfaces, these viruses do not persist for long in the environment (Karron and Collins, 2007).

Parainfluenza viruses are predominantly identified in children between the ages of 2 and 6 years. Serological studies have demonstrated that 50% of infants have experienced HPIV-3 by the time they reach one year of age, while HPIV-1 and -2 most often affect preschool-aged children (Parrott *et al.*, 1962). Serological evidence of infection from non-human paramyxoviruses has been reported (Fleury *et al.*, 1984). However, in these studies, the number of people with detectable antibodies is small and there is little evidence of symptomatic disease.

Diagnosis

The identification of human parainfluenza types 1, 2, and 3 are standard on most viral screening panels in use today. Rapid antigen detection by direct immunofluorescence is available and widely used in viral screening, however the sensitivity is less than that offered by genomic amplification (Landry and Ferguson, 2000). Viral culture from the nasopharynx is also employed in many hospitals and the growth of these viruses has been established in numerous cell lines. The use of molecular identification for respiratory viruses has become more routine health care settings owing to the increased sensitivity of detection.

Identification of HPIV4 and other non-human paramyxoviruses is not often employed in the clinical setting. While HPIV4 can grow on LLC-MK2 cell lines, these cell lines are not routinely used in clinical diagnosis (Komada *et al.*, 2000). PCR analysis is available at various research laboratories.

Prevention and control

Most parainfluenza infections are self limiting and require only supportive care until disease resolution. Croup caused by HPIV1 and HPIV2 is often managed with the use of nebulized adrenaline and glucocorticoids (Brown, 2002). For patients with pneumonia, supplemental oxygen, intravenous fluids and occasionally mechanical ventilation is required.

The use of antiviral agents for parainfluenza disease is not routine and reserved for severe cases predominantly in the immunocompromised populations. Ribavirin has activity *in vitro* against RNA viruses and reduces PIV titres in animal models (Canonico, 1985; Gilbert and Knight, 1986). Both oral and inhaled Ribavirin has been used in severely immunocompromised patients with HPIV3 pneumonia with some success (Lewis *et al.*, 1996; Shima *et al.*, 2008). However, these reports are mostly anecdotal and controversial (Nichols *et al.*, 2001a). Large blinded studies on Ribavirin use have not been performed. Recent focus on the use of inhaled siRNA has produced promising results *in vitro* and in animal studies (Barik, 2004; Bitko *et al.*, 2005).

Development of an effective vaccine against human parainfluenza virus (specifically HPIV3) has been ongoing since the early 1980s often in conjunction with the development of a vaccine against respiratory syncytial virus. Initial work with live attenuated bovine parainfluenza 3 did not demonstrate adequate immunogenicity (Greenberg *et al.*, 2005). More recent investigations have used cold-adapted parainfluenza virus and bPIV expressing hPIV3 HN and F genes (Sato and Wright, 2008). Work on these vaccines show promise and are ongoing.

Perspective

The viruses that are collectively called as PIVs (*Paramyxoviridae* family of the *Mononegavirales* order ranging from *Respirovirus, Avulavirus* and *Rubulavirus*) are important human (HPIV1–4), bird (APMV1–9) and animal (SeV, PIV5 or SV5 and SV41) pathogens and are a serious concern for healthcare (primarily HPIV1–3) and poultry (APMV1–9). The HPIV1–3 as a group are responsible for ~7–17% of all hospitalizations related to respiratory illness in children below five years, which is second only to RSV. They spread through respiratory secretions by close contact with infected individuals and 50% of infants experience HPIV3 infection by the age of 1 year and reinfection among children are very common. Unfortunately, no suitable antiviral or vaccine against these viruses is available to date despite considerable attempts. Since the discovery of the first PIV ~80 years ago, significant amount of work were carried out on of these viruses, which also helped in understanding the biology of other NNS RNA viruses. Moreover, the PIVs, like other NNS viruses, were extensively used to identify the mechanisms of host innate immune response pathways against viral infections. Further studies on the biology of PIVs and on the host immune response against PIVs infection are necessary for the development of vaccines against these viruses.

References

Abraham, G., and Banerjee, A.K. (1976). Sequential transcription of the genes of vesicular stomatitis virus. Proc. Natl. Acad. Sci. U.S.A. *73,* 1504–1508.

Abraham, G., Rhodes, D.P., and Banerjee, A.K. (1975). Novel initiation of RNA synthesis *in vitro* by vesicular stomatitis virus. Nature *255,* 37–40.

Alexander, D. (1988). Newcastle disease virus-An avian paramyxovirus. In Newcastle Disease Alexander, D., ed. (Kluwer Academic: Boston, MA), pp. 11–22.

Andrejeva, J., Poole, E., Young, D.F., Goodbourn, S., and Randall, R.E. (2002). The p127 subunit (DDB1) of the UV-DNA damage repair binding protein is essential for the targeted degradation of STAT1 by the V protein of the paramyxovirus simian virus 5. J. Virol. *76,* 11379–11386.

Andrejeva, J., Childs, K.S., Young, D.F., Carlos, T.S., Stock, N., Goodbourn, S., and Randall, R.E. (2004). The V proteins of paramyxoviruses bind the IFN-inducible RNA helicase, mda-5, and inhibit its activation of the IFN-beta promoter. Proc. Natl. Acad. Sci. U.S.A. *101,* 17264–17269.

Andrewes, C.H., Bang, F.B., Chanock, R.M., and Zhdanov, V.M. (1959). Para-influenza viruses 1, 2, and 3: suggested names for recently described myxoviruses. Virology *8,* 129–130.

Bachi, T. (1980). Intramembrane structural differentiation in Sendai virus maturation. Virology *106,* 41–49.

Bachi, T., and Howe, C. (1973). Morphogenesis and ultrastructure of respiratory syncytial virus. J. Virol. *12,* 1173–1180.

Bailly, J.E., McAuliffe, J.M., Skiadopoulos, M.H., Collins, P.L., and Murphy, B.R. (2000). Sequence determination and molecular analysis of two strains of bovine parainfluenza virus type 3 that are attenuated for primates. Virus Genes *20,* 173–182.

Baker, S.C., and Moyer, S.A. (1988). Encapsidation of Sendai virus genome RNAs by purified NP protein during *in vitro* replication. J. Virol. *62,* 834–838.

Ball, L.A., and White, C.N. (1976). Order of transcription of genes of vesicular stomatitis virus. Proc. Natl. Acad. Sci. U.S.A. *73,* 442–446.

Banerjee, A.D., Abraham, G., and Colonno, R.J. (1977). Vesicular stomatitis virus: mode of transcription. J. Gen. Virol. *34,* 1–8.

Banerjee, A.K. (1987). Transcription and replication of rhabdoviruses. Microbiol. Rev. *51,* 66–87.

Banerjee, A.K. (2008). Response to 'Non-segmented negative-strand RNA virus RNA synthesis *in vivo*'. Virology *371,* 231–233.

Banerjee, A.K., Moyer, S.A., and Rhodes, D.P. (1974). Studies on the *in vitro* adenylation of RNA by vesicular stomatitis virus. Virology *61,* 547–558.

Banerjee, A.K., Barik, S., and De, B.P. (1991). Gene expression of nonsegmented negative strand RNA viruses. Pharmacol. Ther. *51,* 47–70.

Bang, F.B. (1948). Studies on Newcastle disease virus; characters of the virus itself with particular reference to electron microscopy. J. Exp. Med. *88,* 251–266.

Barik, S. (2004). Control of nonsegmented negative-strand RNA virus replication by siRNA. Virus Res. *102,* 27–35.

Barik, S., and Banerjee, A.K. (1992a). Phosphorylation by cellular casein kinase II is essential for transcriptional activity of vesicular stomatitis virus phosphoprotein P. Proc. Natl. Acad. Sci. U.S.A. *89,* 6570–6574.

Barik, S., and Banerjee, A.K. (1992b). Sequential phosphorylation of the phosphoprotein of vesicular stomatitis virus by cellular and viral protein kinases is essential for transcription activation. J. Virol. *66,* 1109–1118.

Barr, J.N., and Wertz, G.W. (2001). Polymerase slippage at vesicular stomatitis virus gene junctions to generate poly(A) is regulated by the upstream 3′-AUAC-5′ tetranucleotide: implications for the mechanism of transcription termination. J. Virol. *75,* 6901–6913.

Barr, J.N., Whelan, S.P., and Wertz, G.W. (2002). Transcriptional control of the RNA-dependent RNA polymerase of vesicular stomatitis virus. Biochim. Biophys. Acta *1577,* 337–353.

Bissonnette, M.L., Donald, J.E., DeGrado, W.F., Jardetzky, T.S., and Lamb, R.A. (2009). Functional analysis of the transmembrane domain in paramyxovirus F protein-mediated membrane fusion. J. Mol. Biol. *386,* 14–36.

Bitko, V., Musiyenko, A., Shulyayeva, O., and Barik, S. (2005). Inhibition of respiratory viruses by nasally administered siRNA. Nat. Med. *11,* 50–55.

Bose, S., and Banerjee, A.K. (2002). Role of heparan sulfate in human parainfluenza virus type 3 infection. Virology *298,* 73–83.

Bose, S., and Banerjee, A.K. (2003). Innate immune response against nonsegmented negative strand RNA viruses. J. Interferon Cytokine Res. *23,* 401–412.

Bose, S., and Banerjee, A.K. (2004). Beta-catenin associates with human parainfluenza virus type 3 ribonucleoprotein complex and activates transcription of viral genome RNA *in vitro*. Gene Expr. *11,* 241–249.

Bose, S., Malur, A., and Banerjee, A.K. (2001). Polarity of human parainfluenza virus type 3 infection in polarized human lung epithelial A549 cells: role of microfilament and microtubule. J. Virol. *75,* 1984–1989.

Bose, S., Kar, N., Maitra, R., DiDonato, J.A., and Banerjee, A.K. (2003). Temporal activation of NF-kappaB regulates an interferon-independent innate antiviral response against cytoplasmic RNA viruses. Proc. Natl. Acad. Sci. U.S.A. *100,* 10890–10895.

Bose, S., Basu, M., and Banerjee, A.K. (2004). Role of nucleolin in human parainfluenza virus type 3 infection of human lung epithelial cells. J. Virol. *78,* 8146–8158.

Both, G.W., Moyer, S.A., and Banerjee, A.K. (1975). Translation and identification of the mRNA species synthesized *in vitro* by the virion-associated RNA polymerase of vesicular stomatitis virus. Proc. Natl. Acad. Sci. U.S.A. *72,* 274–278.

Bousse, T.L., Taylor, G., Krishnamurthy, S., Portner, A., Samal, S.K., and Takimoto, T. (2004). Biological significance of the second receptor binding site of Newcastle disease virus hemagglutinin-neuraminidase protein. J. Virol. *78,* 13,351–13355.

Brown, J.C. (2002). The management of croup. Br. Med. Bull. *61,* 189–202.

Brugh, M., Erickson, G.A., and Beard, C.W. (1980). Embryonated eggs compared with fragments of chorioallantois attached to egg shell for isolation of Newcastle disease virus. Avian Dis. *24,* 486–492.

Buchholz, C.J., Retzler, C., Homann, H.E., and Neubert, W.J. (1994). The carboxy-terminal domain of Sendai virus nucleocapsid protein is involved in complex formation between phosphoprotein and nucleocapsid-like particles. Virology *204,* 770–776.

Bukreyev, A., Marzi, A., Feldmann, F., Zhang, L., Yang, L., Ward, J.M., Dorward, D.W., Pickles, R.J., Murphy, B.R., Feldmann, H., *et al.* (2009). Chimeric human parainfluenza virus bearing the Ebola virus glycoprotein as the sole surface protein is immunogenic and highly protective against Ebola virus challenge. Virology *383,* 348–361.

Bukrinskaya, A.G., Klimenko, S.M., Smirnov, Y.A., and Guschin, B.V. (1968). Infective substructures of Sendai virus from infected Ehrlich ascites tumor cells. J. Virol. *2,* 752–758.

Cadd, T., Garcin, D., Tapparel, C., Itoh, M., Homma, M., Roux, L., Curran, J., and Kolakofsky, D. (1996). The Sendai paramyxovirus accessory C proteins inhibit viral genome amplification in a promoter-specific fashion. J. Virol. *70,* 5067–5074.

Canchola, J., Vargosko, A.J., Kim, H.W., Parrott, R.H., Christmas, E., Jeffries, B., and Chanock, R.M. (1964). Antigenic variation among newly isolated strains of parainfluenza type 4 virus. Am. J. Hyg. *79,* 357–364.

Canonico, P.G. (1985). Efficacy, toxicology and clinical applications of ribavirin against virulent RNA viral infections. Antiviral Res. (Suppl. 1), 75–81.

Canter, D.M., and Perrault, J. (1996). Stabilization of vesicular stomatitis virus L polymerase protein by P protein binding: a small deletion in the C-terminal domain of L abrogates binding. Virology *219,* 376–386.

Carlsen, S.R., Peluso, R.W., and Moyer, S.A. (1985). *In vitro* replication of Sendai virus wild-type and defective interfering particle genome RNAs. J. Virol. *54,* 493–500.

Castells, E., George, V.G., and Hierholzer, J.C. (1990). NCI-H292 as an alternative cell line for the isolation and propagation of the human paramyxoviruses. Arch. Virol. *115,* 277–288.

Cevik, B., Smallwood, S., and Moyer, S.A. (2003). The L-L oligomerization domain resides at the very N-terminus of the sendai virus L RNA polymerase protein. Virology *313,* 525–536.

Cevik, B., Holmes, D.E., Vrotsos, E., Feller, J.A., Smallwood, S., and Moyer, S.A. (2004). The phosphoprotein (P) and L binding sites reside in the

N-terminus of the L subunit of the measles virus RNA polymerase. Virology *327,* 297–306.

Cevik, B., Smallwood, S., and Moyer, S.A. (2007). Two N-terminal regions of the Sendai virus L RNA polymerase protein participate in oligomerization. Virology *363,* 189–197.

Chandrika, R., Horikami, S.M., Smallwood, S., and Moyer, S.A. (1995). Mutations in conserved domain I of the Sendai virus L polymerase protein uncouple transcription and replication. Virology *213,* 352–363.

Chanock, R.M. (1956). Association of a new type of cytopathogenic myxovirus with infantile croup. J. Exp. Med. *104,* 555–576.

Chanock, R.M., Parrott, R.H., Cook, K., Andrews, B.E., Bell, J.A., Reichelderfer, T., Kapikian, A.Z., Mastrota, F.M., and Huebner, R.J. (1958). Newly recognized myxoviruses from children with respiratory disease. N. Engl. J. Med. *258,* 207–213.

Chattopadhyay, D., and Banerjee, A.K. (1987). Phosphorylation within a specific domain of the phosphoprotein of vesicular stomatitis virus regulates transcription *in vitro*. Cell *49,* 407–414.

Chattopadhyay, S., and Banerjee, A.K. (2009). Phosphoprotein, P of human parainfluenza virus type 3 prevents self-association of RNA-dependent RNA polymerase, L. Virology *383,* 226–236.

Chen, J.L., Das, T., and Banerjee, A.K. (1997). Phosphorylated states of vesicular stomatitis virus P protein *in vitro* and *in vivo*. Virology *228,* 200–212.

Chen, L., Gorman, J.J., McKimm-Breschkin, J., Lawrence, L.J., Tulloch, P.A., Smith, B.J., Colman, P.M., and Lawrence, M.C. (2001). The structure of the fusion glycoprotein of Newcastle disease virus suggests a novel paradigm for the molecular mechanism of membrane fusion. Structure *9,* 255–266.

Chen, M., Ogino, T., and Banerjee, A.K. (2007). Interaction of vesicular stomatitis virus P and N proteins: identification of two overlapping domains at the N terminus of P that are involved in N^0–P complex formation and encapsidation of viral genome RNA. J. Virol. *81,* 13478–13485.

Chenik, M., Schnell, M., Conzelmann, K.K., and Blondel, D. (1998). Mapping the interacting domains between the rabies virus polymerase and phosphoprotein. J. Virol. *72,* 1925–1930.

Chiu, C.Y., Rouskin, S., Koshy, A., Urisman, A., Fischer, K., Yagi, S., Schnurr, D., Eckburg, P.B., Tompkins, L.S., Blackburn, B.G., *et al.* (2006). Microarray detection of human parainfluenzavirus 4 infection associated with respiratory failure in an immunocompetent adult. Clin. Infect. Dis. *43,* e71–76.

Choudhary, S., De, B.P., and Banerjee, A.K. (2000). Specific phosphorylated forms of glyceraldehyde 3-phosphate dehydrogenase associate with human parainfluenza virus type 3 and inhibit viral transcription *in vitro*. J. Virol. *74,* 3634–3641.

Choudhary, S.K., Malur, A.G., Huo, Y., De, B.P., and Banerjee, A.K. (2002). Characterization of the oligomerization domain of the phosphoprotein of human parainfluenza virus type 3. Virology *302,* 373–382.

Clinton, G.M., Burge, B.W., and Huang, A.S. (1979). Phosphoproteins of vesicular stomatitis virus: identity and interconversion of phosphorylated forms. Virology *99,* 84–94.

Colamonici, O., Yan, H., Domanski, P., Handa, R., Smalley, D., Mullersman, J., Witte, M., Krishnan, K., and Krolewski, J. (1994). Direct binding to and tyrosine phosphorylation of the alpha subunit of the type I interferon receptor by p135tyk2 tyrosine kinase. Mol. Cell Biol. *14,* 8133–8142.

Colonno, R.J., and Banerjee, A.K. (1976). A unique RNA species involved in initiation of vesicular stomatitis virus RNA transcription *in vitro*. Cell *8,* 197–204.

Compans, R.W., Holmes, K.V., Dales, S., and Choppin, P.W. (1966). An electron microscopic study of moderate and virulent virus-cell interactions of the parainfluenza virus SV5. Virology *30,* 411–426.

Conzelmann, K.K. (1998). Nonsegmented negative-strand RNA viruses: genetics and manipulation of viral genomes. Annu. Rev. Genet. *32,* 123–162.

Conzelmann, K.K. (2004). Reverse genetics of *mononegavirales*. Curr. Top. Microbiol. Immunol. *283,* 1–41.

Cowton, V.M., McGivern, D.R., and Fearns, R. (2006). Unravelling the complexities of respiratory syncytial virus RNA synthesis. J. Gen. Virol. *87,* 1805–1821.

Crennell, S., Takimoto, T., Portner, A., and Taylor, G. (2000). Crystal structure of the multifunctional paramyxovirus hemagglutinin-neuraminidase. Nat. Struct. Biol. *7,* 1068–1074.

Cui, S., Eisenacher, K., Kirchhofer, A., Brzozka, K., Lammens, A., Lammens, K., Fujita, T., Conzelmann, K.K., Krug, A., and Hopfner, K.P. (2008). The C-terminal regulatory domain is the RNA 5′-triphosphate sensor of RIG-I. Mol. Cell *29,* 169–179.

Curran, J.A., and Kolakofsky, D. (1991). Rescue of a Sendai virus DI genome by other parainfluenza viruses: implications for genome replication. Virology *182,* 168–176.

Curran, J., and Kolakofsky, D. (2008). Nonsegmented negative-strand RNA virus RNA synthesis *in vivo*. Virology *371,* 227–230.

Curran, J., Homann, H., Buchholz, C., Rochat, S., Neubert, W., and Kolakofsky, D. (1993). The hypervariable C-terminal tail of the Sendai paramyxovirus nucleocapsid protein is required for template function but not for RNA encapsidation. J. Virol. *67,* 4358–4364.

Curran, J., Marq, J.B., and Kolakofsky, D. (1995). An N-terminal domain of the Sendai paramyxovirus P protein acts as a chaperone for the NP protein during the nascent chain assembly step of genome replication. J. Virol. *69,* 849–855.

Das, S.C., and Pattnaik, A.K. (2004). Phosphorylation of vesicular stomatitis virus phosphoprotein P is indispensable for virus growth. J. Virol. *78,* 6420–6430.

Das, T., Gupta, A.K., Sims, P.W., Gelfand, C.A., Jentoft, J.E., and Banerjee, A.K. (1995). Role of cellular casein kinase II in the function of the phosphoprotein (P) subunit of RNA polymerase of vesicular stomatitis virus. J. Biol. Chem. *270,* 24100–24107.

De, B.P., and Banerjee, A.K. (1993). Rescue of synthetic analogs of genome RNA of human parainfluenza virus type 3. Virology *196,* 344–348.

De, B.P., and Banerjee, A.K. (1999). Involvement of actin microfilaments in the transcription/replication of human parainfluenza virus type 3: possible role of actin in other viruses. Microsc Res. Tech *47,* 114–123.

De, B.P., Lesoon, A., and Banerjee, A.K. (1991). Human parainfluenza virus type 3 transcription *in vitro*: role of cellular actin in mRNA synthesis. J. Virol. *65,* 3268–3275.

De, B.P., Burdsall, A.L., and Banerjee, A.K. (1993). Role of cellular actin in human parainfluenza virus type 3 genome transcription. J. Biol. Chem. *268,* 5703–5710.

De, B.P., Gupta, S., and Banerjee, A.K. (1995). Cellular protein kinase C isoform zeta regulates human parainfluenza virus type 3 replication. Proc. Natl. Acad. Sci. U.S.A. *92,* 5204–5208.

De, B.P., Gupta, S., Zhao, H., Drazba, J.A., and Banerjee, A.K. (1996). Specific interaction *in vitro* and *in vivo* of glyceraldehyde-3-phosphate dehydrogenase and LA protein with cis-acting RNAs of human parainfluenza virus type 3. J. Biol. Chem. *271,* 24728–24735.

De, B.P., Hoffman, M.A., Choudhary, S., Huntley, C.C., and Banerjee, A.K. (2000). Role of NH(2)- and COOH-terminal domains of the P protein of human parainfluenza virus type 3 in transcription and replication. J. Virol. *74,* 5886–5895.

Delenda, C., Hausmann, S., Garcin, D., and Kolakofsky, D. (1997). Normal cellular replication of Sendai virus without the trans-frame, nonstructural V protein. Virology *228,* 55–62.

Delenda, C., Taylor, G., Hausmann, S., Garcin, D., and Kolakofsky, D. (1998). Sendai viruses with altered P, V, and W protein expression. Virology *242,* 327–337.

Didcock, L., Young, D.F., Goodbourn, S., and Randall, R.E. (1999a). Sendai virus and simian virus 5 block activation of interferon-responsive genes: importance for virus pathogenesis. J. Virol. *73,* 3125–3133.

Didcock, L., Young, D.F., Goodbourn, S., and Randall, R.E. (1999b). The V protein of simian virus 5 inhibits interferon signalling by targeting STAT1 for proteasome-mediated degradation. J. Virol. *73,* 9928–9933.

Dillon, P.J., and Parks, G.D. (2007). Role for the phosphoprotein P subunit of the paramyxovirus polymerase in limiting induction of host cell antiviral responses. J. Virol. *81,* 11116–11127.

Dimock, K., and Collins, P.L. (1993). Rescue of synthetic analogs of genomic RNA and replicative-intermediate RNA of human parainfluenza virus type 3. J. Virol. *67,* 2772–2778.

Dowling, P.C., Giorgi, C., Roux, L., Dethlefsen, L.A., Galantowicz, M.E., Blumberg, B.M., and Kolakofsky, D. (1983). Molecular cloning of the 3′-proximal third of Sendai virus genome. Proc. Natl. Acad. Sci. U.S.A. *80,* 5213–5216.

Doyle, T. (1927). hitherto unrecorded disease of fowls due to a filter-passing virus. Comp. Pathol. Therapeut. *40,* 144–169.

Durbin, A.P., Hall, S.L., Siew, J.W., Whitehead, S.S., Collins, P.L., and Murphy, B.R. (1997). Recovery of infectious human parainfluenza virus type 3 from cDNA. Virology *235,* 323–332.

Durbin, A.P., McAuliffe, J.M., Collins, P.L., and Murphy, B.R. (1999). Mutations in the C, D, and V open reading frames of human parainfluenza virus type 3 attenuate replication in rodents and primates. Virology *261,* 319–330.

Dutch, R.E., Hagglund, R.N., Nagel, M.A., Paterson, R.G., and Lamb, R.A. (2001). Paramyxovirus fusion (F) protein: a conformational change on cleavage activation. Virology *281,* 138–150.

Egelman, E.H., Wu, S.S., Amrein, M., Portner, A., and Murti, G. (1989). The Sendai virus nucleocapsid exists in at least four different helical states. J. Virol. *63,* 2233–2243.

Emerson, S.U. (1982). Reconstitution studies detect a single polymerase entry site on the vesicular stomatitis virus genome. Cell *31,* 635–642.

Fearns, R., Peeples, M.E., and Collins, P.L. (1997). Increased expression of the N protein of respiratory syncytial virus stimulates minigenome replication but does not alter the balance between the synthesis of mRNA and antigenome. Virology *236,* 188–201.

Feller, J.A., Smallwood, S., Horikami, S.M., and Moyer, S.A. (2000). Mutations in conserved domains IV and VI of the large (L) subunit of the sendai virus RNA polymerase give a spectrum of defective RNA synthesis phenotypes. Virology *269,* 426–439.

Finch, J.T., and Gibbs, A.J. (1970). Observations on the structure of the nucleocapsids of some paramyxoviruses. J. Gen. Virol. *6,* 141–150.

Fleury, H.J., Bonnici, J.F., Babin, M., du Pasquier, P., and Samb, A. (1984). Serological evidence of human infection with the paramyxovirus Yucaipa in Senegal, West Africa. Am. J. Trop. Med. Hyg. *33,* 190–191.

Frank, A.L., Couch, R.B., Griffis, C.A., and Baxter, B.D. (1979). Comparison of different tissue cultures for isolation and quantitation of influenza and parainfluenza viruses. J. Clin. Microbiol. *10,* 32–36.

Fukuhara, N., Huang, C., Kiyotani, K., Yoshida, T., and Sakaguchi, T. (2002). Mutational analysis of the Sendai virus V protein: importance of the conserved residues for Zn binding, virus pathogenesis, and efficient RNA editing. Virology *299,* 172–181.

Gainey, M.D., Dillon, P.J., Clark, K.M., Manuse, M.J., and Parks, G.D. (2008). Paramyxovirus-induced shutoff of host and viral protein synthesis: role of the P and V proteins in limiting PKR activation. J. Virol. *82,* 828–839.

Galinski, M.S., Mink, M.A., Lambert, D.M., Wechsler, S.L., and Pons, M.W. (1986a). Molecular cloning and sequence analysis of the human parainfluenza 3 virus mRNA encoding the P and C proteins. Virology *155,* 46–60.

Galinski, M.S., Mink, M.A., Lambert, D.M., Wechsler, S.L., and Pons, M.W. (1986b). Molecular cloning and sequence analysis of the human parainfluenza 3 virus RNA encoding the nucleocapsid protein. Virology *149*, 139–151.

Galinski, M.S., Mink, M.A., and Pons, M.W. (1988). Molecular cloning and sequence analysis of the human parainfluenza 3 virus gene encoding the L protein. Virology *165*, 499–510.

Gao, J., De, B.P., and Banerjee, A.K. (1999). Human parainfluenza virus type 3 up-regulates major histocompatibility complex class I and II expression on respiratory epithelial cells: involvement of a STAT1- and CIITA-independent pathway. J. Virol. *73*, 1411–1418.

Gao, J., De, B.P., Han, Y., Choudhary, S., Ransohoff, R., and Banerjee, A.K. (2001). Human parainfluenza virus type 3 inhibits gamma interferon-induced major histocompatibility complex class II expression directly and by inducing alpha/beta interferon. J. Virol. *75*, 1124–1131.

Gao, J., De, B.P., and Banerjee, A.K. (2002). Interferon type I downregulates human parainfluenza virus type 3-induced major histocompatibility complex class II expression. Viral Immunol. *15*, 85–93.

Garcin, D., Pelet, T., Calain, P., Roux, L., Curran, J., and Kolakofsky, D. (1995). A highly recombinogenic system for the recovery of infectious Sendai paramyxovirus from cDNA: generation of a novel copy-back nondefective interfering virus. EMBO J. *14*, 6087–6094.

Garcin, D., Marq, J.B., Strahle, L., le Mercier, P., and Kolakofsky, D. (2002). All four Sendai Virus C proteins bind Stat1, but only the larger forms also induce its mono-ubiquitination and degradation. Virology *295*, 256–265.

Gerard, F.C., Ribeiro Ede, A., Jr., Leyrat, C., Ivanov, I., Blondel, D., Longhi, S., Ruigrok, R.W., and Jamin, M. (2009). Modular organization of rabies virus phosphoprotein. J. Mol. Biol. *388*, 978–996.

Gilbert, B.E., and Knight, V. (1986). Biochemistry and clinical applications of ribavirin. Antimicrob. Agents Chemother. *30*, 201–205.

Goebel, S.J., Taylor, J., Barr, B.C., Kiehn, T.E., Castro-Malaspina, H.R., Hedvat, C.V., Rush-Wilson, K.A., Kelly, C.D., Davis, S.W., Samsonoff, W.A., *et al.* (2007). Isolation of avian paramyxovirus 1 from a patient with a lethal case of pneumonia. J. Virol. *81, 12*,709–12714.

Goodbourn, S., Didcock, L., and Randall, R.E. (2000). Interferons: cell signalling, immune modulation, antiviral response and virus countermeasures. J. Gen. Virol. *81*, 2341–2364.

Goodbourn, S., and Randall, R.E. (2009). The regulation of type I interferon production by paramyxoviruses. J. Interferon Cytokine Res. *29*, 539–547.

Goswami, K.K., Cameron, K.R., Russell, W.C., Lange, L.S., and Mitchell, D.N. (1984a). Evidence for the persistence of paramyxoviruses in human bone marrows. J. Gen. Virol. *65*, 1881–1888.

Goswami, K.K., Lange, L.S., Mitchell, D.N., Cameron, K.R., and Russell, W.C. (1984b). Does simian virus 5 Infect. humans? J. Gen. Virol. *65*, 1295–1303.

Gotoh, B., Takeuchi, K., Komatsu, T., Yokoo, J., Kimura, Y., Kurotani, A., Kato, A., and Nagai, Y. (1999). Knockout of the Sendai virus C gene eliminates the viral ability to prevent the interferon-alpha/beta-mediated responses. FEBS Lett. *459*, 205–210.

Gotoh, B., Komatsu, T., Takeuchi, K., and Yokoo, J. (2002). Paramyxovirus strategies for evading the interferon response. Rev. Med. Virol. *12*, 337–357.

Grdzelishvili, V.Z., Smallwood, S., Tower, D., Hall, R.L., Hunt, D.M., and Moyer, S.A. (2005). A single amino acid change in the L-polymerase protein of vesicular stomatitis virus completely abolishes viral mRNA cap methylation. J. Virol. *79*, 7327–7337.

Grdzelishvili, V.Z., Smallwood, S., Tower, D., Hall, R.L., Hunt, D.M., and Moyer, S.A. (2006). Identification of a new region in the vesicular stomatitis virus L polymerase protein which is essential for mRNA cap methylation. Virology *350*, 394–405.

Greenberg, D.P., Walker, R.E., Lee, M.S., Reisinger, K.S., Ward, J.I., Yogev, R., Blatter, M.M., Yeh, S.H., Karron, R.A., Sangli, C., *et al.* (2005). A bovine parainfluenza virus type 3 vaccine is safe and immunogenic in early infancy. J. Infect. Dis. *191*, 1116–1122.

Grogan, C.C., and Moyer, S.A. (2001). Sendai virus wild-type and mutant C proteins show a direct correlation between L polymerase binding and inhibition of viral RNA synthesis. Virology *288*, 96–108.

Gupta, S., De, B.P., Drazba, J.A., and Banerjee, A.K. (1998). Involvement of actin microfilaments in the replication of human parainfluenza virus type 3. J. Virol. *72*, 2655–2662.

Gupta, A.K., Blondel, D., Choudhary, S., and Banerjee, A.K. (2000). The phosphoprotein of rabies virus is phosphorylated by a unique cellular protein kinase and specific isomers of protein kinase C. J. Virol. *74*, 91–98.

Haller, A.A., Miller, T., Mitiku, M., and Coelingh, K. (2000). Expression of the surface glycoproteins of human parainfluenza virus type 3 by bovine parainfluenza virus type 3, a novel attenuated virus vaccine vector. J. Virol. *74*, 11626–11635.

Hasan, M.K., Kato, A., Shioda, T., Sakai, Y., Yu, D., and Nagai, Y. (1997). Creation of an infectious recombinant Sendai virus expressing the firefly luciferase gene from the 3′ proximal first locus. J. Gen. Virol. *78*, 2813–2820.

Hasan, M.K., Kato, A., Muranaka, M., Yamaguchi, R., Sakai, Y., Hatano, I., Tashiro, M., and Nagai, Y. (2000). Versatility of the accessory C proteins of Sendai virus: contribution to virus assembly as an additional role. J. Virol. *74*, 5619–5628.

Hausmann, S., Jacques, J.P., and Kolakofsky, D. (1996). Paramyxovirus RNA editing and the requirement for hexamer genome length. RNA *2*, 1033–1045.

Hausmann, S., Garcin, D., Delenda, C., and Kolakofsky, D. (1999). The versatility of paramyxovirus RNA polymerase stuttering. J. Virol. *73*, 5568–5576.

He, B., Paterson, R.G., Ward, C.D., and Lamb, R.A. (1997). Recovery of infectious SV5 from cloned DNA and expression of a foreign gene. Virology *237*, 249–260.

He, B., Leser, G.P., Paterson, R.G., and Lamb, R.A. (1998). The paramyxovirus SV5 small hydrophobic

(SH) protein is not essential for virus growth in tissue culture cells. Virology *250,* 30–40.

He, B., Lin, G.Y., Durbin, J.E., Durbin, R.K., and Lamb, R.A. (2001). The SH integral membrane protein of the paramyxovirus simian virus 5 is required to block apoptosis in MDBK cells. J. Virol. *75,* 4068–4079.

He, B., Paterson, R.G., Stock, N., Durbin, J.E., Durbin, R.K., Goodbourn, S., Randall, R.E., and Lamb, R.A. (2002). Recovery of paramyxovirus simian virus 5 with a V protein lacking the conserved cysteine-rich domain: the multifunctional V protein blocks both interferon-beta induction and interferon signaling. Virology *303,* 15–32.

Heggeness, M.H., Scheid, A., and Choppin, P.W. (1980). Conformation of the helical nucleocapsids of paramyxoviruses and vesicular stomatitis virus: reversible coiling and uncoiling induced by changes in salt concentration. Proc. Natl. Acad. Sci. U.S.A. *77,* 2631–2635.

Heggeness, M.H., Smith, P.R., and Choppin, P.W. (1982). *In vitro* assembly of the nonglycosylated membrane protein (M) of Sendai virus. Proc. Natl. Acad. Sci. U.S.A. *79,* 6232–6236.

Henrickson, K.J. (2003). Parainfluenza viruses. Clin. Microbiol. Rev. *16,* 242–264.

Henrickson, K.J., Kuhn, S.M., and Savatski, L.L. (1994). Epidemiology and cost of infection with human parainfluenza virus types 1 and 2 in young children. Clin. Infect. Dis. *18,* 770–779.

Hercyk, N., Horikami, S.M., and Moyer, S.A. (1988). The vesicular stomatitis virus L protein possesses the mRNA methyltransferase activities. Virology *163,* 222–225.

Hermodsson, S. (1963). Inhibition of interferon by an infection with parainfluenza virus type 3 (PIV-3). Virology *20,* 333–343.

Hewitt, J.A., and Nermut, M.V. (1977). A morphological study of the M-protein of Sendai virus. J. Gen. Virol. *34,* 127–136.

Hiebert, S.W., Paterson, R.G., and Lamb, R.A. (1985a). Hemagglutinin-neuraminidase protein of the paramyxovirus simian virus 5: nucleotide sequence of the mRNA predicts an N-terminal membrane anchor. J. Virol. *54,* 1–6.

Hiebert, S.W., Paterson, R.G., and Lamb, R.A. (1985b). Identification and predicted sequence of a previously unrecognized small hydrophobic protein, SH, of the paramyxovirus simian virus 5. J. Virol. *55,* 744–751.

Hiebert, S.W., Richardson, C.D., and Lamb, R.A. (1988). Cell surface expression and orientation in membranes of the 44-amino-acid SH protein of simian virus 5. J. Virol. *62,* 2347–2357.

Hoffman, M.A., and Banerjee, A.K. (1997). An infectious clone of human parainfluenza virus type 3. J. Virol. *71,* 4272–4277.

Hoffman, M.A., and Banerjee, A.K. (2000a). Analysis of RNA secondary structure in replication of human parainfluenza virus type 3. Virology *272,* 151–158.

Hoffman, M.A., and Banerjee, A.K. (2000b). Precise mapping of the replication and transcription promoters of human parainfluenza virus type 3. Virology *269,* 201–211.

Hoffman, M.A., Thorson, L.M., Vickman, J.E., Anderson, J.S., May, N.A., and Schweitzer, M.N. (2006). Roles of human parainfluenza virus type 3 bases 13 to 78 in replication and transcription: identification of an additional replication promoter element and evidence for internal transcription initiation. J. Virol. *80,* 5388–5396.

Horikami, S.M., Curran, J., Kolakofsky, D., and Moyer, S.A. (1992). Complexes of Sendai virus NP-P and P-L proteins are required for defective interfering particle genome replication *in vitro*. J. Virol. *66,* 4901–4908.

Horikami, S.M., Smallwood, S., and Moyer, S.A. (1996). The Sendai virus V protein interacts with the NP protein to regulate viral genome RNA replication. Virology *222,* 383–390.

Horikami, S.M., Hector, R.E., Smallwood, S., and Moyer, S.A. (1997). The Sendai virus C protein binds the L polymerase protein to inhibit viral RNA synthesis. Virology *235,* 261–270.

Hornung, V., Ellegast, J., Kim, S., Brzozka, K., Jung, A., Kato, H., Poeck, H., Akira, S., Conzelmann, K.K., Schlee, M., *et al.* (2006). 5′-Triphosphate RNA is the ligand for RIG-I. Science *314,* 994–997.

Horwood, P.F., Gravel, J.L., and Mahony, T.J. (2008). Identification of two distinct bovine parainfluenza virus type 3 genotypes. J. Gen. Virol. *89,* 1643–1648.

Howe, C., Morgan, C., de Vaux St Cyr, C., Hsu, K.C., and Rose, H.M. (1967). Morphogenesis of type 2 parainfluenza virus examined by light and electron microscopy. J. Virol. *1,* 215–237.

Hsiung, G.D. (1972). Parainfluenza-5 virus. Infection of man and animal. Prog Med. Virol. *14,* 241–274.

Hsu, C.H., Re, G.G., Gupta, K.C., Portner, A., and Kingsbury, D.W. (1985). Expression of Sendai virus defective-interfering genomes with internal deletions. Virology *146,* 38–49.

Hu, C., and Gupta, K.C. (2000). Functional significance of alternate phosphorylation in Sendai virus P protein. Virology *268,* 517–532.

Hu, C.J., Kato, A., Bowman, M.C., Kiyotani, K., Yoshida, T., Moyer, S.A., Nagai, Y., and Gupta, K.C. (1999). Role of primary constitutive phosphorylation of Sendai virus P and V proteins in viral replication and pathogenesis. Virology *263,* 195–208.

Huang, C., Kiyotani, K., Fujii, Y., Fukuhara, N., Kato, A., Nagai, Y., Yoshida, T., and Sakaguchi, T. (2000). Involvement of the zinc-binding capacity of Sendai virus V protein in viral pathogenesis. J. Virol. *74,* 7834–7841.

Huang, Z., Panda, A., Elankumaran, S., Govindarajan, D., Rockemann, D.D., and Samal, S.K. (2004). The hemagglutinin-neuraminidase protein of Newcastle disease virus determines tropism and virulence. J. Virol. *78,* 4176–4184.

Hull, R.N., Minner, J.R., and Smith, J.W. (1956). New viral agents recovered from tissue cultures of monkey kidney cells. I. Origin and properties of cytopathogenic agents S.V.1, S.V.2, S.V.4, S.V.5, S.V.6, S.V.11, S.V.12 and S.V.15. Am. J. Hyg. *63,* 204–215.

Huntley, C.C., De, B.P., and Banerjee, A.K. (1997). Phosphorylation of Sendai virus phosphoprotein by cellular protein kinase C zeta. J. Biol. Chem. *272*, 16578–16584.

Huntley, C.C., De, B.P., Murray, N.R., Fields, A.P., and Banerjee, A.K. (1995). Human parainfluenza virus type 3 phosphoprotein: identification of serine 333 as the major site for PKC zeta phosphorylation. Virology *211*, 561–567.

Hwang, L.N., Englund, N., Das, T., Banerjee, A.K., and Pattnaik, A.K. (1999). Optimal replication activity of vesicular stomatitis virus RNA polymerase requires phosphorylation of a residue(s) at carboxy-terminal domain II of its accessory subunit, phosphoprotein P. J. Virol. *73*, 5613–5620.

Itoh, M., Wang, X.L., Suzuki, Y., and Homma, M. (1992). Mutation of the HANA protein of Sendai virus by passage in eggs. Virology *190*, 356–364.

Itoh, M., Isegawa, Y., Hotta, H., and Homma, M. (1997). Isolation of an avirulent mutant of Sendai virus with two amino acid mutations from a highly virulent field strain through adaptation to LLC-MK2 cells. J. Gen. Virol. *78*, 3207–3215.

Jackson, R.L., Spadafora, D., and Perrault, J. (1995). Hierarchal constitutive phosphorylation of the vesicular stomatitis virus P protein and lack of effect on P1 to P2 conversion. Virology *214*, 189–197.

Jacques, J.P., Hausmann, S., and Kolakofsky, D. (1994). Paramyxovirus mRNA editing leads to G deletions as well as insertions. EMBO J. *13*, 5496–5503.

Johnson, K.M., Chanock, R.M., Cook, M.K., and Huebner, R.J. (1960). Studies of a new human hemadsorption virus. I. Isolation, properties and characterization. Am. J. Hyg. *71*, 81–92.

Karlin, D., Ferron, F., Canard, B., and Longhi, S. (2003). Structural disorder and modular organization in *Paramyxovirinae* N and P. J. Gen. Virol. *84*, 3239–3252.

Karron, R.A., and Collins, P.L. (2007). Parainfluenza viruses. In Fields Virology, Knipe, D.M., and Howley, P.M., eds. (Lippincott Williams and Wilkins: Philadelphia, PA), pp. 1497–1526.

Kato, A., Kiyotani, K., Sakai, Y., Yoshida, T., and Nagai, Y. (1997a). The paramyxovirus, Sendai virus, V protein encodes a luxury function required for viral pathogenesis. EMBO J. *16*, 578–587.

Kato, A., Kiyotani, K., Sakai, Y., Yoshida, T., Shioda, T., and Nagai, Y. (1997b). Importance of the cysteine-rich carboxyl-terminal half of V protein for Sendai virus pathogenesis. J. Virol. *71*, 7266–7272.

Kato, H., Sato, S., Yoneyama, M., Yamamoto, M., Uematsu, S., Matsui, K., Tsujimura, T., Takeda, K., Fujita, T., Takeuchi, O., *et al.* (2005). Cell type-specific involvement of RIG-I in antiviral response. Immunity *23*, 19–28.

Kato, H., Takeuchi, O., Sato, S., Yoneyama, M., Yamamoto, M., Matsui, K., Uematsu, S., Jung, A., Kawai, T., Ishii, K.J., *et al.* (2006). Differential roles of MDA5 and RIG-I helicases in the recognition of RNA viruses. Nature *441*, 101–105.

Kawano, M., Okamoto, K., Bando, H., Kondo, K., Tsurudome, M., Komada, H., Nishio, M., and Ito, Y. (1991). Characterizations of the human parainfluenza type 2 virus gene encoding the L protein and the intergenic sequences. Nucleic Acids Res. *19*, 2739–2746.

Kawano, M., Tsurudome, M., Oki, N., Nishio, M., Komada, H., Matsumura, H., Kusagawa, S., Ohta, H., and Ito, Y. (1993). Sequence determination of the P gene of simian virus 41: presence of irregular deletions near the RNA-editing sites of paramyxoviruses. J. Gen. Virol. *74*, 911–916.

Kawano, M., Kaito, M., Kozuka, Y., Komada, H., Noda, N., Nanba, K., Tsurudome, M., Ito, M., Nishio, M., and Ito, Y. (2001). Recovery of infectious human parainfluenza type 2 virus from cDNA clones and properties of the defective virus without V-specific cysteine-rich domain. Virology *284*, 99–112.

Klein, J.D., and Collier, A.M. (1974). Pathogenesis of human parainfluenza type 3 virus infection in hamster tracheal organ culture. Infect. Immun. *10*, 883–888.

Kolakofsky, D., Pelet, T., Garcin, D., Hausmann, S., Curran, J., and Roux, L. (1998). Paramyxovirus RNA synthesis and the requirement for hexamer genome length: the rule of six revisited. J. Virol. *72*, 891–899.

Kolakofsky, D., Le Mercier, P., Iseni, F., and Garcin, D. (2004). Viral DNA polymerase scanning and the gymnastics of Sendai virus RNA synthesis. Virology *318*, 463–473.

Komada, H., Tsurudome, M., Bando, H., Nishio, M., Ueda, M., Tsumura, H., and Ito, Y. (1989). Immunological response of monkeys infected intranasally with human parainfluenza virus type 4. J. Gen. Virol. *70*, 3487–3492.

Komada, H., Bando, H., Ito, M., Ohta, H., Kawano, M., Nishio, M., Tsurudome, M., Watanabe, N., Ikemura, N., Kusagawa, S., *et al.* (1995). Sequence analyses of human parainfluenza virus type 4A and type 4B fusion proteins. J. Gen. Virol. *76*, 3205–3210.

Komada, H., Inoue, H., Yamabayashi, C., Ito, M., Kawano, M., Nishio, M., Tsurudome, M., Kozuka, Y., Noda, N., Namba, K., *et al.* (2000). Incomplete replication of human parainfluenza virus type 4 in LLC-MK2 cells and in L929 cells. Med. Microbiol. Immunol. *188*, 185–189.

Komatsu, T., Takeuchi, K., Yokoo, J., Tanaka, Y., and Gotoh, B. (2000). Sendai virus blocks alpha interferon signaling to signal transducers and activators of transcription. J. Virol. *74*, 2477–2480.

Komatsu, T., Takeuchi, K., and Gotoh, B. (2007). Bovine parainfluenza virus type 3 accessory proteins that suppress beta interferon production. Microbes Infect. *9*, 954–962.

Koyama, A.H., Irie, H., Kato, A., Nagai, Y., Adachi, A., (2003). Virus multiplication and induction of apoptosis by Sendai virus: role of the C proteins. Microbes Infect. *5*, 373–378.

Kraneveld, F. (1926). A poultry disease in the Dutch East Indies. Nederlands-Indische Bladen voor Diergeneeskunde *38*, 448–450.

Krishnamurthy, S., Huang, Z., and Samal, S.K. (2000). Recovery of a virulent strain of newcastle disease virus from cloned cDNA: expression of a foreign gene results in growth retardation and attenuation. Virology *278,* 168–182.

Kurotani, A., Kiyotani, K., Kato, A., Shioda, T., Sakai, Y., Mizumoto, K., Yoshida, T., and Nagai, Y. (1998). Sendai virus C proteins are categorically nonessential gene products but silencing their expression severely impairs viral replication and pathogenesis. Genes Cells *3,* 111–124.

Kuroya, M., and Ishida, N. (1953). Newborn virus pneumonitis (type Sendai). II. The isolation of a new virus possessing hemagglutinin activity. Yokohama Med. Bull. *4,* 217–233.

Lamb, R.A., and Jardetzky, T.S. (2007). Structural basis of viral invasion: lessons from paramyxovirus F. Curr. Opin. Struct. Biol. *17,* 427–436.

Lamb, R.A., and Parks, G.D. (2007). *Paramyxoviridae*: the viruses and their replication. In Fields Virology, Knipe, D.M., and Howley, P.M., eds. (Lippincott Williams and Wilkins: Philadelphia, PA), pp. 1449–1496.

Landry, M.L., and Ferguson, D. (2000). SimulFluor respiratory screen for rapid detection of multiple respiratory viruses in clinical specimens by immunofluorescence staining. J. Clin. Microbiol. *38,* 708–711.

Latorre, P., Kolakofsky, D., and Curran, J. (1998). Sendai virus Y proteins are initiated by a ribosomal shunt. Mol. Cell Biol. *18,* 5021–5031.

Lau, S.K., Li, K.S., Chau, K.Y., So, L.Y., Lee, R.A., Lau, Y.L., Chan, K.H., Lim, W.W., Woo, P.C., and Yuen, K.Y. (2009). Clinical and molecular epidemiology of human parainfluenza virus 4 infections in hong kong: subtype 4B as common as subtype 4A. J. Clin. Microbiol. *47,* 1549–1552.

Lawrence, M.C., Borg, N.A., Streltsov, V.A., Pilling, P.A., Epa, V.C., Varghese, J.N., McKimm-Breschkin, J.L., and Colman, P.M. (2004). Structure of the haemagglutinin–neuraminidase from human parainfluenza virus type III. J. Mol. Biol. *335,* 1343–1357.

Lawson, N.D., Stillman, E.A., Whitt, M.A., and Rose, J.K. (1995). Recombinant vesicular stomatitis viruses from DNA. Proc. Natl. Acad. Sci. U.S.A. *92,* 4477–4481.

Le Mercier, P., Garcin, D., Garcia, E., and Kolakofsky, D. (2003). Competition between the Sendai virus N mRNA start site and the genome 3′-end promoter for viral RNA polymerase. J. Virol. *77,* 9147–9155.

Lewis, V.A., Champlin, R., Englund, J., Couch, R., Goodrich, J.M., Rolston, K., Przepiorka, D., Mirza, N.Q., Yousuf, H.M., Luna, M., *et al.* (1996). Respiratory disease due to parainfluenza virus in adult bone marrow transplant recipients. Clin. Infect. Dis. *23,* 1033–1037.

Leyrer, S., Neubert, W.J., and Sedlmeier, R. (1998). Rapid and efficient recovery of Sendai virus from cDNA: factors influencing recombinant virus rescue. J. Virol. Methods *75,* 47–58.

Li, J., Fontaine-Rodriguez, E.C., and Whelan, S.P. (2005). Amino acid residues within conserved domain VI of the vesicular stomatitis virus large polymerase protein essential for mRNA cap methyltransferase activity. J. Virol. *79,* 13373–13384.

Li, J., Wang, J.T., and Whelan, S.P. (2006). A unique strategy for mRNA cap methylation used by vesicular stomatitis virus. Proc. Natl. Acad. Sci. U.S.A. *103,* 8493–8498.

Li, J., Rahmeh, A., Morelli, M., and Whelan, S.P. (2008). A conserved motif in region v of the large polymerase proteins of nonsegmented negative-sense RNA viruses that is essential for mRNA capping. J. Virol. *82,* 775–784.

Lin, G.Y., Paterson, R.G., and Lamb, R.A. (1997). The RNA binding region of the paramyxovirus SV5 V and P proteins. Virology *238,* 460–469.

Lin, G.Y., Paterson, R.G., Richardson, C.D., and Lamb, R.A. (1998). The V protein of the paramyxovirus SV5 interacts with damage-specific DNA binding protein. Virology *249,* 189–200.

Lin, Y., Bright, A.C., Rothermel, T.A., and He, B. (2003). Induction of apoptosis by paramyxovirus simian virus 5 lacking a small hydrophobic gene. J. Virol. *77,* 3371–3383.

Loney, C., Mottet-Osman, G., Roux, L., and Bhella, D. (2009). Paramyxovirus ultrastructure and genome packaging: cryo-electron tomography of sendai virus. J. Virol. *83,* 8191–8197.

Luthra, P., Sun, D., Wolfgang, M., and He, B. (2008). AKT1-dependent activation of NF-kappaB by the L protein of parainfluenza virus 5. J. Virol. *82,* 10887–10895.

Malur, A.G., Choudhary, S.K., De, B.P., and Banerjee, A.K. (2002a). Role of a highly conserved NH(2)-terminal domain of the human parainfluenza virus type 3 RNA polymerase. J. Virol. *76,* 8101–8109.

Malur, A.G., Gupta, N.K., De Bishnu, P., and Banerjee, A.K. (2002b). Analysis of the mutations in the active site of the RNA-dependent RNA polymerase of human parainfluenza virus type 3 (HPIV3). Gene Expr. *10,* 93–100.

Malur, A.G., Hoffman, M.A., and Banerjee, A.K. (2004). The human parainfluenza virus type 3 (HPIV 3) C protein inhibits viral transcription. Virus Res. *99,* 199–204.

Malur, A.G., Chattopadhyay, S., Maitra, R.K., and Banerjee, A.K. (2005). Inhibition of STAT 1 phosphorylation by human parainfluenza virus type 3 C protein. J. Virol. *79,* 7877–7882.

Mao, H., Thakur, C.S., Chattopadhyay, S., Silverman, R.H., Gudkov, A., and Banerjee, A.K. (2008). Inhibition of human parainfluenza virus type 3 infection by novel small molecules. Antiviral Res. *77,* 83–94.

Mao, H., Chattopadhyay, S., and Banerjee, A.K. (2009). N-terminally truncated C protein, CNDelta25, of human parainfluenza virus type 3 is a potent inhibitor of viral replication. Virology *394,* 143–148.

Marx, A., Torok, T.J., Holman, R.C., Clarke, M.J., and Anderson, L.J. (1997). Pediatric hospitalizations for croup (laryngotracheobronchitis): biennial increases associated with human parainfluenza virus 1 epidemics. J. Infect. Dis. *176,* 1423–1427.

Marx, A., Gary, H.E., Jr., Marston, B.J., Erdman, D.D., Breiman, R.F., Torok, T.J., Plouffe, J.F., File, T.M., Jr., and Anderson, L.J. (1999). Parainfluenza virus infection among adults hospitalized for lower respiratory tract infection. Clin. Infect. Dis. *29,* 134–140.

Masters, P.S., and Banerjee, A.K. (1988a). Complex formation with vesicular stomatitis virus phosphoprotein NS prevents binding of nucleocapsid protein N to nonspecific RNA. J. Virol. *62,* 2658–2664.

Masters, P.S., and Banerjee, A.K. (1988b). Resolution of multiple complexes of phosphoprotein NS with nucleocapsid protein N of vesicular stomatitis virus. J. Virol. *62,* 2651–2657.

Meguro, H., Bryant, J.D., Torrence, A.E., and Wright, P.F. (1979). Canine kidney cell line for isolation of respiratory viruses. J. Clin. Microbiol. *9,* 175–179.

Melchjorsen, J., Jensen, S.B., Malmgaard, L., Rasmussen, S.B., Weber, F., Bowie, A.G., Matikainen, S., and Paludan, S.R. (2005). Activation of innate defense against a paramyxovirus is mediated by RIG-I and TLR7 and TLR8 in a cell-type-specific manner. J. Virol. *79,* 12944–12951.

Merz, D.C., Scheid, A., and Choppin, P.W. (1980). Importance of antibodies to the fusion glycoprotein of paramyxoviruses in the prevention of spread of infection. J. Exp. Med. *151,* 275–288.

Merz, D.C., Prehm, P., Scheid, A., and Choppin, P.W. (1981). Inhibition of the neuraminidase of paramyxoviruses by halide ions: a possible means of modulating the two activities of the HN protein. Virology *112,* 296–305.

Miller, R.H., Pursell, A.R., Mitchell, F.E., and Johnson, K.M. (1964). A Newly Discovered Myxovirus (S.V.41) Isolated from Cell Cultures of Cynomolgus Monkey Kidney. Am. J. Hyg. *80,* 365–376.

Mori, I., Komatsu, T., Takeuchi, K., Nakakuki, K., Sudo, M., and Kimura, Y. (1995). Parainfluenza virus type 1 infects olfactory neurons and establishes long-term persistence in the nerve tissue. J. Gen. Virol. *76,* 1251–1254.

Moriuchi, H., Oshima, T., Nishimura, H., Nakamura, K., Katsushima, N., and Numazaki, Y. (1990). Human malignant melanoma cell line (HMV-II) for isolation of influenza C and parainfluenza viruses. J. Clin. Microbiol. *28,* 1147–1150.

Moscona, A. (2005). Entry of parainfluenza virus into cells as a target for interrupting childhood respiratory disease. J. Clin. Invest. *115,* 1688–1698.

Mountcastle, W.E., Compans, R.W., Lackland, H., and Choppin, P.W. (1974). Proteolytic cleavage of subunits of the nucleocapsid of the paramyxovirus simian virus 5. J. Virol. *14,* 1253–1261.

Moyer, S.A., Abraham, G., Adler, R., and Banerjee, A.K. (1975). Methylated and blocked 5′ termini in vesicular stomatitis virus *in vivo* mRNAs. Cell *5,* 59–67.

Moyer, S.A., Baker, S.C., and Lessard, J.L. (1986). Tubulin: a factor necessary for the synthesis of both Sendai virus and vesicular stomatitis virus RNAs. Proc. Natl. Acad. Sci. U.S.A. *83,* 5405–5409.

Muller, R., Poch, O., Delarue, M., Bishop, D.H., and Bouloy, M. (1994). Rift Valley fever virus L segment: correction of the sequence and possible functional role of newly identified regions conserved in RNA-dependent polymerases. J. Gen. Virol. *75,* 1345–1352.

Murphy, A.M., and Grdzelishvili, V.Z. (2009). Identification of sendai virus L protein amino acid residues affecting viral mRNA cap methylation. J. Virol. *83,* 1669–1681.

Myers, T.M., and Moyer, S.A. (1997). An amino-terminal domain of the Sendai virus nucleocapsid protein is required for template function in viral RNA synthesis. J. Virol. *71,* 918–924.

Myers, T.M., Smallwood, S., and Moyer, S.A. (1999). Identification of nucleocapsid protein residues required for Sendai virus nucleocapsid formation and genome replication. J. Gen. Virol. *80,* 1383–1391.

Nagai, Y. (1999). Paramyxovirus replication and pathogenesis. Reverse genetics transforms understanding. Rev. Med. Virol. *9,* 83–99.

Nakhaei, P., Genin, P., Civas, A., and Hiscott, J. (2009). RIG-I-like receptors: sensing and responding to RNA virus infection. Semin. Immunol. *21,* 215–222.

Neumann, G., and Kawaoka, Y. (2004). Reverse genetics systems for the generation of segmented negative-sense RNA viruses entirely from cloned cDNA. Curr. Top. Microbiol. Immunol. *283,* 43–60.

Neumann, G., Whitt, M.A., and Kawaoka, Y. (2002). A decade after the generation of a negative-sense RNA virus from cloned cDNA – what have we learned? J. Gen. Virol. *83,* 2635–2662.

Newman, J.T., Surman, S.R., Riggs, J.M., Hansen, C.T., Collins, P.L., Murphy, B.R., and Skiadopoulos, M.H. (2002). Sequence analysis of the Washington/1964 strain of human parainfluenza virus type 1 (HPIV1) and recovery and characterization of wild-type recombinant HPIV1 produced by reverse genetics. Virus Genes *24,* 77–92.

Newman, J.T., Riggs, J.M., Surman, S.R., McAuliffe, J.M., Mulaikal, T.A., Collins, P.L., Murphy, B.R., and Skiadopoulos, M.H. (2004). Generation of recombinant human parainfluenza virus type 1 vaccine candidates by importation of temperature-sensitive and attenuating mutations from heterologous paramyxoviruses. J. Virol. *78,* 2017–2028.

Nichols, W.G., Corey, L., Gooley, T., Davis, C., and Boeckh, M. (2001a). Parainfluenza virus infections after hematopoietic stem cell transplantation: risk factors, response to antiviral therapy, and effect on transplant outcome. Blood *98,* 573–578.

Nichols, W.G., Gooley, T., and Boeckh, M. (2001b). Community-acquired respiratory syncytial virus and parainfluenza virus infections after hematopoietic stem cell transplantation: the Fred Hutchinson Cancer ReseArch. Center experience. Biol. Blood Marrow Transplant. 7 (*Suppl.*), 11S–15S.

Nishio, M., Ohtsuka, J., Tsurudome, M., Nosaka, T., and Kolakofsky, D. (2008). Human parainfluenza virus type 2 V protein inhibits genome replication by binding to the L protein: possible role in promoting viral fitness. J. Virol. *82,* 6130–6138.

Novick, D., Cohen, B., and Rubinstein, M. (1994). The human interferon alpha/beta receptor: characterization and molecular cloning. Cell *77,* 391–400.

Ogawa, M., Mutsuga, N., Tsurudome, M., Kawano, M., Matsumura, H., Kusagawa, S., Komada, H., Nishio,

M., and Ito, Y. (1992). Nucleotide sequence analysis of the simian virus 41 gene encoding the large (L) protein and construction of a phylogenetic tree for the L proteins of paramyxoviruses. J. Gen. Virol. *73,* 2743–2750.

Ogino, T., and Banerjee, A.K. (2007). Unconventional mechanism of mRNA capping by the RNA-dependent RNA polymerase of vesicular stomatitis virus. Mol. Cell *25,* 85–97.

Ogino, T., and Banerjee, A.K. (2010). The HR motif in the RNA-dependent RNA polymerase L protein of Chandipura virus is required for unconventional mRNA-capping activity. J. Gen. Virol. *91,* 1311–1314.

Ogino, T., Iwama, M., Kinouchi, J., Shibagaki, Y., Tsukamoto, T., and Mizumoto, K. (1999). Involvement of a cellular glycolytic enzyme, phosphoglycerate kinase, in Sendai virus transcription. J. Biol. Chem. *274,* 35999–36008.

Ogino, T., Yamadera, T., Nonaka, T., Imajoh-Ohmi, S., and Mizumoto, K. (2001). Enolase, a cellular glycolytic enzyme, is required for efficient transcription of Sendai virus genome. Biochem. Biophys. Res. Commun. *285,* 447–455.

Ogino, T., Iwama, M., Ohsawa, Y., and Mizumoto, K. (2003). Interaction of cellular tubulin with Sendai virus M protein regulates transcription of viral genome. Biochem. Biophys. Res. Commun. *311,* 283–293.

Ogino, T., Kobayashi, M., Iwama, M., and Mizumoto, K. (2005). Sendai virus RNA-dependent RNA polymerase L protein catalyzes cap methylation of virus-specific mRNA. J. Biol. Chem. *280,* 4429–4435.

Ogino, T., Yadav, S.P., and Banerjee, A.K. (2010). Histidine-mediated RNA transfer to GDP for unique mRNA capping by vesicular stomatitis virus RNA polymerase. Proc. Natl. Acad. Sci. U.S.A. *107,* 3463–3468.

Ohgimoto, S., Bando, H., Kawano, M., Okamoto, K., Kondo, K., Tsurudome, M., Nishio, M., and Ito, Y. (1990). Sequence analysis of P gene of human parainfluenza type 2 virus: P and cysteine-rich proteins are translated by two mRNAs that differ by two nontemplated G residues. Virology *177,* 116–123.

Ohsawa, K., Yamada, A., Takeuchi, K., Watanabe, Y., Miyata, H., and Sato, H. (1998). Genetic characterization of parainfluenza virus 3 derived from guinea pigs. J. Vet. Med. Sci. *60,* 919–922.

Orstavik, I. (1981). Susceptibility of continuous lines of monkey kidney cells to influenza and parainfluenza viruses in the presence of trypsin. Acta Pathol. Microbiol. Scand. B *89,* 179–183.

Park, K.H., Huang, T., Correia, F.F., and Krystal, M. (1991). Rescue of a foreign gene by Sendai virus. Proc. Natl. Acad. Sci. U.S.A. *88,* 5537–5541.

Parks, G.D. (1994). Mapping of a region of the paramyxovirus L protein required for the formation of a stable complex with the viral phosphoprotein P. J. Virol. *68,* 4862–4872.

Parrott, R.H., Vargosko, A.J., Kimhw, Bell, J.A., and Chanock, R.M. (1962). Acute respiratory diseases of viral etiology. III. parainfluenza. Myxoviruses. Am. J. Public Health Nations Health *52,* 907–917.

Pattnaik, A.K., Hwang, L., Li, T., Englund, N., Mathur, M., Das, T., and Banerjee, A.K. (1997). Phosphorylation within the amino-terminal acidic domain I of the phosphoprotein of vesicular stomatitis virus is required for transcription but not for replication. J. Virol. *71,* 8167–8175.

Patton, J.T., Davis, N.L., and Wertz, G.W. (1984). N protein alone satisfies the requirement for protein synthesis during RNA replication of vesicular stomatitis virus. J. Virol. *49,* 303–309.

Pichlmair, A., Schulz, O., Tan, C.P., Naslund, T.I., Liljestrom, P., Weber, F., and Reis e Sousa, C. (2006). RIG-I-mediated antiviral responses to single-stranded RNA bearing 5′-phosphates. Science *314,* 997–1001.

Plattet, P., Strahle, L., le Mercier, P., Hausmann, S., Garcin, D., and Kolakofsky, D. (2007). Sendai virus RNA polymerase scanning for mRNA start sites at gene junctions. Virology *362,* 411–420.

Plumet, S., Herschke, F., Bourhis, J.M., Valentin, H., Longhi, S., and Gerlier, D. (2007). Cytosolic 5′-triphosphate ended viral leader transcript of measles virus as activator of the RIG I-mediated interferon response. PLoS One *2,* e279.

Poch, O., Blumberg, B.M., Bougueleret, L., and Tordo, N. (1990). Sequence comparison of five polymerases (L proteins) of unsegmented negative-strand RNA viruses: theoretical assignment of functional domains. J. Gen. Virol. *71,* 1153–1162.

Popa, L.M., Repanovici, R., Iliescu, R., Mihalache, O., and Cajal, N. (1985). Some physico-chemical and biological properties of egg-Sendai virus with altered haemagglutinin–neuraminidase protein. Acta Virol. *29,* 461–465.

Porotto, M., Greengard, O., Poltoratskaia, N., Horga, M.A., and Moscona, A. (2001). Human parainfluenza virus type 3 HN–receptor interaction: effect of 4-guanidino-Neu5Ac2en on a neuraminidase-deficient variant. J. Virol. *75,* 7481–7488.

Porotto, M., Murrell, M., Greengard, O., and Moscona, A. (2003). Triggering of human parainfluenza virus 3 fusion protein (F) by the hemagglutinin-neuraminidase (HN) protein: an HN mutation diminishes the rate of F activation and fusion. J. Virol. *77,* 3647–3654.

Porter, D.D., Prince, G.A., Hemming, V.G., and Porter, H.G. (1991). Pathogenesis of human parainfluenza virus 3 infection in two species of cotton rats: Sigmodon hispidus develops bronchiolitis, while Sigmodon fulviventer develops interstitial pneumonia. J. Virol. *65,* 103–111.

Prince, G.A., Ottolini, M.G., and Moscona, A. (2001). Contribution of the human parainfluenza virus type 3 HN–receptor interaction to pathogenesis *in vivo*. J. Virol. *75,* 12446–12451.

Qanungo, K.R., Shaji, D., Mathur, M., and Banerjee, A.K. (2004). Two RNA polymerase complexes from vesicular stomatitis virus-infected cells that carry out transcription and replication of genome RNA. Proc. Natl. Acad. Sci. U.S.A. *101,* 5952–5957.

Rahmeh, A.A., Li, J., Kranzusch, P.J., and Whelan, S.P. (2009). Ribose 2′-O methylation of the vesicular stomatitis virus mRNA cap precedes and facilitates subsequent guanine-N-7 methylation by the large polymerase protein. J. Virol. *83,* 11043–11050.

Randall, R.E., and Goodbourn, S. (2008). Interferons and viruses: an interplay between induction, signalling, antiviral responses and virus countermeasures. J. Gen. Virol. *89,* 1–47.

Reagan, R.L., Smith, E.J., and Brueckner, A.L. (1951). Electron micrographs of Newcastle disease virus propagated in the cave bat (Myotus lucifugus). J. Bacteriol. *61,* 37–40.

Reed, G., Jewett, P.H., Thompson, J., Tollefson, S., and Wright, P.F. (1997). Epidemiology and clinical impact of parainfluenza virus infections in otherwise healthy infants and young children <5 years old. J. Infect. Dis. *175,* 807–813.

Rehwinkel, J., Tan, C.P., Goubau, D., Schulz, O., Pichlmair, A., Bier, K., Robb, N., Vreede, F., Barclay, W., Fodor, E., *et al.* (2010). RIG-I detects viral genomic RNA during negative-strand RNA virus infection. Cell *140,* 397–408.

Rhodes, D.P., Moyer, S.A., and Banerjee, A.K. (1974). *In vitro* synthesis of methylated messenger RNA by the virion-associated RNA polymerase of vesicular stomatitis virus. Cell *3,* 327–333.

Roth, J.P., Li, J.K., and Barnard, D.L. (2010). Human parainfluenza virus type 3 (HPIV-3): construction and rescue of an infectious, recombinant virus expressing the enhanced green fluorescent protein (EGFP). Curr. Protoc. Microbiol. Chapter 15, Unit 15F 11.

Russell, C.J., Jardetzky, T.S., and Lamb, R.A. (2001). Membrane fusion machines of paramyxoviruses: capture of intermediates of fusion. EMBO J. *20,* 4024–4034.

Russell, R., Paterson, R.G., and Lamb, R.A. (1994). Studies with cross-linking reagents on the oligomeric form of the paramyxovirus fusion protein. Virology *199,* 160–168.

Sabbah, A., and Bose, S. (2009). Retinoic acid inducible gene I activates innate antiviral response against human parainfluenza virus type 3. Virol. J. *6,* 200.

Sakaguchi, T., Toyoda, T., Gotoh, B., Inocencio, N.M., Kuma, K., Miyata, T., and Nagai, Y. (1989). Newcastle disease virus evolution. I. Multiple lineages defined by sequence variability of the hemagglutinin-neuraminidase gene. Virology *169,* 260–272.

Sakaguchi, T., Kiyotani, K., Kato, A., Asakawa, M., Fujii, Y., Nagai, Y., and Yoshida, T. (1997). Phosphorylation of the Sendai virus M protein is not essential for virus replication either *in vitro* or *in vivo*. Virology *235,* 360–366.

Sakaguchi, T., Kato, A., Sugahara, F., Shimazu, Y., Inoue, M., Kiyotani, K., Nagai, Y., and Yoshida, T. (2005). AIP1/Alix is a binding partner of Sendai virus C protein and facilitates virus budding. J. Virol. *79,* 8933–8941.

Samal, S.K., and Collins, P.L. (1996). RNA replication by a respiratory syncytial virus RNA analog does not obey the rule of six and retains a nonviral trinucleotide extension at the leader end. J. Virol. *70,* 5075–5082.

Samuel, A.S., Paldurai, A., Kumar, S., Collins, P.L., and Samal, S.K. (2010). Complete genome sequence of avian paramyxovirus (APMV) serotype 5 completes the analysis of nine APMV serotypes and reveals the longest APMV genome. PLoS One *5,* e9269.

Sanchez, A., Banerjee, A.K., Furuichi, Y., and Richardson, M.A. (1986). Conserved structures among the nucleocapsid proteins of the *Paramyxoviridae*: complete nucleotide sequence of human parainfluenza virus type 3 NP mRNA. Virology *152,* 171–180.

Sanderson, C.M., Avalos, R., Kundu, A., and Nayak, D.P. (1995). Interaction of Sendai viral F, HN, and M proteins with host cytoskeletal and lipid components in Sendai virus-infected BHK cells. Virology *209,* 701–707.

Sato, M., and Wright, P.F. (2008). Current status of vaccines for parainfluenza virus infections. Pediatr. Infect. Dis. J. *27,* S123–125.

Schnell, M.J., Mebatsion, T., and Conzelmann, K.K. (1994). Infectious rabies viruses from cloned cDNA. EMBO J. *13,* 4195–4203.

Schroder, M., and Bowie, A.G. (2007). An arms race: innate antiviral responses and counteracting viral strategies. Biochem. Soc. Trans. *35,* 1512–1514.

Shima, T., Yoshimoto, G., Nonami, A., Yoshida, S., Kamezaki, K., Iwasaki, H., Takenaka, K., Miyamoto, T., Harada, N., Teshima, T., *et al.* (2008). Successful treatment of parainfluenza virus 3 pneumonia with oral ribavirin and methylprednisolone in a bone marrow transplant recipient. Int. J. Hematol. *88,* 336–340.

Shuai, K., Stark, G.R., Kerr, I.M., and Darnell, J.E., Jr. (1993). A single phosphotyrosine residue of Stat91 required for gene activation by interferon-gamma. Science *261,* 1744–1746.

Sidhu, M.S., Menonna, J.P., Cook, S.D., Dowling, P.C., and Udem, S.A. (1993). Canine distemper virus L gene: sequence and comparison with related viruses. Virology *193,* 50–65.

Skiadopoulos, M.H., Surman, S.R., Riggs, J.M., Collins, P.L., and Murphy, B.R. (2001). A chimeric human-bovine parainfluenza virus type 3 expressing measles virus hemagglutinin is attenuated for replication but is still immunogenic in rhesus monkeys. J. Virol. *75, 10,* 498–10504.

Skiadopoulos, M.H., Tatem, J.M., Surman, S.R., Mitcho, Y., Wu, S.L., Elkins, W.R., and Murphy, B.R. (2002). The recombinant chimeric human parainfluenza virus type 1 vaccine candidate, rHPIV3–1cp45, is attenuated, immunogenic, and protective in African green monkeys. Vaccine *20,* 1846–1852.

Skiadopoulos, M.H., Schmidt, A.C., Riggs, J.M., Surman, S.R., Elkins, W.R., St Claire, M., Collins, P.L., and Murphy, B.R. (2003). Determinants of the host range restriction of replication of bovine parainfluenza virus type 3 in rhesus monkeys are polygenic. J. Virol. *77,* 1141–1148.

Sleat, D.E., and Banerjee, A.K. (1993). Transcriptional activity and mutational analysis of recombinant vesicular stomatitis virus RNA polymerase. J. Virol. *67,* 1334–1339.

Smallwood, S., and Moyer, S.A. (2004). The L polymerase protein of parainfluenza virus 3 forms an oligomer and can interact with the heterologous Sendai virus L, P and C proteins. Virology *318*, 439–450.

Smallwood, S., Ryan, K.W., and Moyer, S.A. (1994). Deletion analysis defines a carboxyl-proximal region of Sendai virus P protein that binds to the polymerase L protein. Virology *202*, 154–163.

Smallwood, S., Easson, C.D., Feller, J.A., Horikami, S.M., and Moyer, S.A. (1999). Mutations in conserved domain II of the large (L) subunit of the Sendai virus RNA polymerase abolish RNA synthesis. Virology *262*, 375–383.

Smallwood, S., Cevik, B., and Moyer, S.A. (2002a). Intragenic complementation and oligomerization of the L subunit of the sendai virus RNA polymerase. Virology *304*, 235–245.

Smallwood, S., Hovel, T., Neubert, W.J., and Moyer, S.A. (2002b). Different substitutions at conserved amino acids in domains II and III in the Sendai L RNA polymerase protein inactivate viral RNA synthesis. Virology *304*, 135–145.

Strahle, L., Marq, J.B., Brini, A., Hausmann, S., Kolakofsky, D., and Garcin, D. (2007). Activation of the beta interferon promoter by unnatural Sendai virus infection requires RIG-I and is inhibited by viral C proteins. J. Virol. *81*, 12227–12237.

Stricker, R., Mottet, G., and Roux, L. (1994). The Sendai virus matrix protein appears to be recruited in the cytoplasm by the viral nucleocapsid to function in viral assembly and budding. J. Gen. Virol. *75*, 1031–1042.

Sun, M., Fuentes, S.M., Timani, K., Sun, D., Murphy, C., Lin, Y., August, A., Teng, M.N., and He, B.(2008). Akt plays a critical role in replication of nonsegmented negative-stranded RNA viruses. J. Virol. *82*, 105–114.

Sun, D., Luthra, P., Li, Z., and He, B. (2009). PLK1 down-regulates parainfluenza virus 5 gene expression. PLoS Pathog. *5*, e1000525.

Swanson, K., Wen, X., Leser, G.P., Paterson, R.G., Lamb, R.A., and Jardetzky, T.S. (2010). Structure of the Newcastle disease virus F protein in the post-fusion conformation. Virology *402*, 372–379.

Swierkosz, E.M., Erdman, D.D., Bonnot, T., Schneiderheinze, C., and Waner, J.L. (1995). Isolation and characterization of a naturally occurring parainfluenza 3 virus variant. J. Clin. Microbiol. *33*, 1839–1841.

Takeuchi, K., Komatsu, T., Kitagawa, Y., Sada, K., and Gotoh, B. (2008). Sendai virus C protein plays a role in restricting PKR activation by limiting the generation of intracellular double-stranded RNA. J. Virol. *82*, 10102–10110.

Takeuchi, O., and Akira, S. (2009). Innate immunity to virus infection. Immunol. Rev. *227*, 75–86.

Takimoto, T., and Portner, A. (2004). Molecular mechanism of paramyxovirus budding. Virus Res. *106*, 133–145.

Takimoto, T., Taylor, G.L., Crennell, S.J., Scroggs, R.A., and Portner, A. (2000). Crystallization of Newcastle disease virus hemagglutinin-neuraminidase glycoprotein. Virology *270*, 208–214.

Tang, R.S., MacPhail, M., Schickli, J.H., Kaur, J., Robinson, C.L., Lawlor, H.A., Guzzetta, J.M., Spaete, R.R., and Haller, A.A. (2004). Parainfluenza virus type 3 expressing the native or soluble fusion (F) Protein of Respiratory Syncytial Virus (RSV) confers protection from RSV infection in African green monkeys. J. Virol. *78, 11*,198–11207.

Tang, R.S., Spaete, R.R., Thompson, M.W., MacPhail, M., Guzzetta, J.M., Ryan, P.C., Reisinger, K., Chandler, P., Hilty, M., Walker, R.E., *et al.* (2008). Development of a PIV-vectored RSV vaccine: preclinical evaluation of safety, toxicity, and enhanced disease and initial clinical testing in healthy adults. Vaccine *26*, 6373–6382.

Tapparel, C., Hausmann, S., Pelet, T., Curran, J., Kolakofsky, D., and Roux, L. (1997). Inhibition of Sendai virus genome replication due to promoter-increased selectivity: a possible role for the accessory C proteins. J. Virol. *71*, 9588–9599.

Tarbouriech, N., Curran, J., Ebel, C., Ruigrok, R.W., and Burmeister, W.P. (2000a). On the domain structure and the polymerization state of the sendai virus P protein. Virology *266*, 99–109.

Tarbouriech, N., Curran, J., Ruigrok, R.W., and Burmeister, W.P. (2000b). Tetrameric coiled coil domain of Sendai virus phosphoprotein. Nat. Struct. Biol. *7*, 777–781.

Terrier, O., Rolland, J.P., Rosa-Calatrava, M., Lina, B., Thomas, D., and Moules, V. (2009). Parainfluenza virus type 5 (PIV-5) morphology revealed by cryo-electron microscopy. Virus Res. *142*, 200–203.

Thomas, S.M., Lamb, R.A., and Paterson, R.G. (1988). Two mRNAs that differ by two nontemplated nucleotides encode the amino coterminal proteins P and V of the paramyxovirus SV5. Cell *54*, 891–902.

Tsai, K.S., and Thomson, R.G. (1975). Bovine parainfluenza type 3 virus infection: virus replication in bovine embryonic cell cultures and virion separation by rate-zonal centrifugation. Infect. Immun. *11*, 770–782.

Tsurudome, M., Bando, H., Nishio, M., Iwamoto, Y., Kawano, M., Kondo, K., Komada, H., and Ito, Y. (1990). Antigenic and structural properties of a paramyxovirus simian virus 41 (SV41) reveal a close relationship with human parainfluenza type 2 virus. Virology *179*, 738–748.

Tsurudome, M., Oki, N., Higuchi, Y., Miyahara, K., Yoshimoto, M., Mutsuga, N., Kitada, S., Ogawa, M., Miyahara, Y., Okamoto, K., *et al.* (1991). Molecular relationships between human parainfluenza virus type 2, and simian viruses 41 and 5: determination of nucleoprotein gene sequences of simian viruses 41 and 5. J. Gen. Virol. 72, 2289–2292.

Tuckis, J., Smallwood, S., Feller, J.A., and Moyer, S.A. (2002). The C-terminal 88 amino acids of the Sendai virus P protein have multiple functions separable by mutation. J. Virol. *76*, 68–77.

Valle, M., and Cantell, K. (1965). The ability of Sendai virus to overcome cellular resistance to vesicular stomatitis virus. I. General characteristics of the system. Ann. Med. Exp. Biol. Fenn. *43*, 57–60.

Vidal, S., and Kolakofsky, D. (1989). Modified model for the switch from Sendai virus transcription to replication. J. Virol. *63,* 1951–1958.

Vidal, S., Curran, J., and Kolakofsky, D. (1990). A stuttering model for paramyxovirus P mRNA editing. EMBO J. *9,* 2017–2022.

Vulliemoz, D., and Roux, L. (2001). 'Rule of six': how does the Sendai virus RNA polymerase keep count? J. Virol. *75,* 4506–4518.

Vulliemoz, D., and Roux, L. (2002). Given the opportunity, the Sendai virus RNA-dependent RNA polymerase could as well enter its template internally. J. Virol. *76,* 7987–7995.

Weber, F., Wagner, V., Rasmussen, S.B., Hartmann, R., and Paludan, S.R. (2006). Double-stranded RNA is produced by positive-strand RNA viruses and DNA viruses but not in detectable amounts by negative-strand RNA viruses. J. Virol. *80,* 5059–5064.

Weinberg, G.A., Hall, C.B., Iwane, M.K., Poehling, K.A., Edwards, K.M., Griffin, M.R., Staat, M.A., Curns, A.T., Erdman, D.D., and Szilagyi, P.G. (2009). Parainfluenza virus infection of young children: estimates of the population-based burden of hospitalization. J. Pediatr. *154,* 694–699.

Wells, G., and Malur, A. (2008). Expression of human parainfluenza virus type 3 PD protein and intracellular localization in virus infected cells. Virus Genes *37,* 358–367.

Wertz, G.W. (1978). Isolation of possible replicative intermediate structures from vesicular stomatitis virus-infected cells. Virology *85,* 271–285.

Whelan, S.P. (2008). Response to 'Non-segmented negative-strand RNA virus RNA synthesis *in vivo*'. Virology *371,* 234–237.

Whelan, S.P., and Wertz, G.W. (1999). The 5′ terminal trailer region of vesicular stomatitis virus contains a position-dependent cis-acting signal for assembly of RNA into infectious particles. J. Virol. *73,* 307–315.

Whelan, S.P., and Wertz, G.W. (2002). Transcription and replication initiate at separate sites on the vesicular stomatitis virus genome. Proc. Natl. Acad. Sci. U.S.A. *99,* 9178–9183.

Whelan, S.P., Ball, L.A., Barr, J.N., and Wertz, G.T. (1995). Efficient recovery of infectious vesicular stomatitis virus entirely from cDNA clones. Proc. Natl. Acad. Sci. U.S.A. *92,* 8388–8392.

Whelan, S.P., Barr, J.N., and Wertz, G.W. (2004). Transcription and replication of nonsegmented negative-strand RNA viruses. Curr. Top. Microbiol. Immunol. *283,* 61–119.

Wiegand, M.A., Bossow, S., Schlecht, S., and Neubert, W.J. (2007). De novo synthesis of N and P proteins as a key step in Sendai virus gene expression. J. Virol. *81, 13,*835–13844.

Xiao, S., Paldurai, A., Nayak, B., Subbiah, M., Collins, P.L., and Samal, S.K. (2009). Complete genome sequence of avian paramyxovirus type 7 (strain Tennessee) and comparison with other paramyxoviruses. Virus Res. *145,* 80–91.

Yan, H., Krishnan, K., Greenlund, A.C., Gupta, S., Lim, J.T., Schreiber, R.D., Schindler, C.W., and Krolewski, J.J. (1996). Phosphorylated interferon-alpha receptor 1 subunit (IFNaR1) acts as a docking site for the latent form of the 113kDa STAT2 protein. EMBO J. *15,* 1064–1074.

Yao, Q., and Compans, R.W. (2000). Filamentous particle formation by human parainfluenza virus type 2. J. Gen. Virol. *81,* 1305–1312.

Yin, H.S., Paterson, R.G., Wen, X., Lamb, R.A., and Jardetzky, T.S. (2005). Structure of the uncleaved ectodomain of the paramyxovirus (hPIV3) fusion protein. Proc. Natl. Acad. Sci. U.S.A. *102,* 9288–9293.

Yin, H.S., Wen, X., Paterson, R.G., Lamb, R.A., and Jardetzky, T.S. (2006). Structure of the parainfluenza virus 5 F protein in its metastable, prefusion conformation. Nature *439,* 38–44.

Yoneyama, M., and Fujita, T. (2009). RNA recognition and signal transduction by RIG-I-like receptors. Immunol. Rev. *227,* 54–65.

Young, D.F., Didcock, L., Goodbourn, S., and Randall, R.E. (2000). *Paramyxoviridae* use distinct virus-specific mechanisms to circumvent the interferon response. Virology *269,* 383–390.

Yuasa, T., Bando, H., Kawano, M., Tsurudome, M., Nishio, M., Kondo, K., Komada, H., and Ito, Y. (1990). Sequence analyses of the 3′ genome end and NP gene of human parainfluenza type 2 virus: sequence variation of the gene-starting signal and the conserved 3′ end. Virology *179,* 777–784.

Zhan, X., Slobod, K.S., Krishnamurthy, S., Luque, L.E., Takimoto, T., Jones, B., Surman, S., Russell, C.J., Portner, A., and Hurwitz, J.L. (2008). Sendai virus recombinant vaccine expressing hPIV-3 HN or F elicits protective immunity and combines with a second recombinant to prevent hPIV-1, hPIV-3 and RSV infections. Vaccine *26,* 3480–3488.

Zhang, L., Bukreyev, A., Thompson, C.I., Watson, B., Peeples, M.E., Collins, P.L., and Pickles, R.J. (2005). Infection of ciliated cells by human parainfluenza virus type 3 in an *in vitro* model of human airway epithelium. J. Virol. *79,* 1113–1124.

Zhao, H., and Banerjee, A.K. (1995). Interaction between the nucleocapsid protein and the phosphoprotein of human parainfluenza virus 3. Mapping of the interacting domains using a two-hybrid system. J. Biol. Chem. *270,* 12485–12490.

Zhao, H., De, B.P., Das, T., and Banerjee, A.K. (1996). Inhibition of human parainfluenza virus-3 replication by interferon and human MxA. Virology *220,* 330–338.

Hendra and Nipah Viruses

6

Michael K. Lo, Melissa M. Coughlin and Paul A. Rota*

Abstract

Nipah and Hendra viruses comprise the genus *Henipavirus* and are highly pathogenic paramyxoviruses, which cause fatal encephalitis and respiratory disease in humans. Since their respective initial outbreaks in 1998 and 1994, they have continued to cause sporadic outbreaks resulting in fatal disease. Owing to their designation as biosafety level 4 pathogens, the level of containment required to work with live henipaviruses is available only to select laboratories around the world. This chapter provides an overview of henipaviruses in regards to their diagnosis, clinical features, and epidemiology, while highlighting recent studies which have expanded our knowledge of their molecular biology, pathogenesis, and control with novel antiviral therapeutics.

History

Hendra virus (HeV) and Nipah virus (NiV) are zoonotic viruses belonging to the genus *Henipavirus* of the family *Paramyxoviridae*. *Henipavirus* infection causes severe encephalitis, occurring in regions where the natural reservoir of the viruses is found (Eaton *et al.*, 2007; Tyler, 2009). The natural reservoir of HeV and NiV are fruit bats or flying foxes of the genus *Pteropus*. The distribution of genus *Pteropus* extends from Madagascar through India and South East Asia to Australia (Field, 2009). HeV, initially named equine *Morbillivirus*, was identified in 1994 following an outbreak of acute febrile illness in horses that resulted in 21 equine cases, two human infections and one death in Hendra, a town near Brisbane in Queensland, Australia. The first isolate of HeV was obtained from this fatal case. Subsequently, in 1995 a second outbreak in Mackay, Australia was identified to have occurred one month earlier in 1994 after a horse owner died of relapsing encephalitis (Eaton *et al.*, 2007; O'Sullivan *et al.*, 1997).

The first known human infections with NiV were detected during an outbreak of severe febrile encephalitis in peninsular Malaysia and Singapore from the fall of 1998 to the spring of 1999. A total of 276 patients with viral encephalitis due to NiV were reported in peninsular Malaysia and Singapore, mostly among adult males who were involved in pig farming or pork production activities, with 106 fatalities (case fatality rate [CFR] – 38.5%; CDC, 1999a,b; Chua *et al.*, 2000a; Parashar *et al.*, 2000). In spite of isolation of NiV from urine, saliva, and respiratory secretions from infected patients (Chua *et al.*, 2001), the primary mode of transmission to humans during this outbreak was pig-to-human (Parashar *et al.*, 2000), as the subsequent culling of over 1 million pigs lead to the eventual containment of the outbreak. Neutralizing antibodies to NiV were found primarily in *Pteropus hypomelanus* and *Pteropus vampyrus* during initial surveillance studies, but virus was not isolated (Yob *et al.*, 2001). It was not until bat urine and swabs of contaminated fruit were collected and NiV was isolated from *Pteropus hypomelanus* on Tioman Island (Chua *et al.*, 2002). Since these initial outbreaks there

*Corresponding author

have been repeated outbreaks of both viruses. HeV has been confined mainly to Queensland, Australia. However, while no additional outbreaks have been reported for NiV in Malaysia, NiV has resulted in several outbreaks in both Bangladesh and India (Chadha *et al.*, 2006; WHO, 2009a,b).

Nipah and Hendra viruses

Classification

NiV and HeV belong to the family *Paramyxoviridae*, and have non-segmented, negative-stranded (NNS) RNA genomes consisting of helical nucleocapsids encased in an envelope forming relatively spherical, pleomorphic virus particles. There are two subfamilies within the family *Paramyxoviridae*, the *Paramyxovirinae* and the *Pneumovirinae*. The subfamily *Paramyxovirinae* is divided into five genera: *Rubulavirus* (prototype, mumps virus), *Respirovirus* (prototype, human parainfluenza virus 1), *Morbillivirus* (prototype, measles virus), *Avulavirus* (prototype, Newcastle disease virus), and with the emergence of NiV and HeV, a new genus *Henipavirus* (Bellini *et al.*, 1998; Mayo, 2002).

Propagation

NiV and HeV virions adhere to host cell surfaces through the identified receptor molecule Ephrin-B2, which is expressed on neurons, smooth muscle, and endothelial cells surrounding small arteries (Bonaparte *et al.*, 2005; Negrete *et al.*, 2005). Ephrin-B3, a related molecule with less affinity for NiV G, has nonetheless also been found to serve as a functional alternative receptor for NiV, but not HeV (Negrete *et al.*, 2006).

After receptor binding by G, the F protein then mediates fusion between the viral envelope and the host cell membrane and the viral ribonucleocapsid is then released into the cytoplasm (Lamb and Parks, 2007). A recent study has indicated the ability of NiV to enter cells via macropinocytosis, which is dependent upon a single tyrosine residue in the Ephrin-B2 cytoplasmic tail (Pernet *et al.*, 2009). The physiological relevance of this finding remains to be determined. Upon ribonucleocapsid release into the host cell cytoplasm, transcription of viral mRNA begins, with the encapsidated genome serving as the template for mRNA synthesis. The polymerase complex (P and L proteins) initiates transcription at the promoter region in the 3′ leader. Each gene is transcribed, being regulated by transcriptional start and stop sites at their respective regions in the gene. The polymerase is thought to 'stutter' at the transcriptional stop signals in order to synthesize poly-A tails of the mRNA. The polymerase pauses at the gene stop signal, crosses the intergenic region, and reinitiates transcription at the next gene start signal. During this process, some of the polymerase complexes will fall off from the template. Given that there is only one promoter at the 3′ leader, these polymerase complexes can only reinitiate transcription at the 3′ end. This results in a differential gradient of transcription, in which the genes closer to the 3′ end are transcribed more often than genes closer to the 5′ end of the genome (Bellini *et al.*, 1998; Mayo, 2002).

As translation of viral mRNA occurs, viral proteins accumulate in the cell, and an unidentified mechanism triggers the polymerase to stop transcription and initiate genome replication. During the genome replication process, the polymerase complex enters the 3′ genomic promoter in a similar manner as during transcription, but ignores the mRNA transcriptional signals at gene junctions, causing the polymerase to transcribe a full-length positive-sense antigenome. A promoter in the 3′ non-coding region of the antigenome serves as the location where synthesis of new, negative-sense genomic RNA begins. Newly-made genomes can in turn, serve as templates for transcription (Lamb and Parks, 2007).

In order to assemble virus particles, the genome replication process must be terminated, newly made genomes must be properly encapsidated and regions of the cell membrane need to be prepared to accept budding nucleocapsids (Lamb and Parks, 2007). Polymerase complexes stay associated with packaged nucleocapsids and are responsible for the next round of infection. The glycoproteins are synthesized in the endoplasmic reticulum (ER) and mature through the Golgi network to the cell membrane. The NiV F glycoprotein is a unique exception to the rule, as F protein processing and maturation occurs in the endosome (Diederich *et al.*, 2005). Cytoplasmic tails of the F and G glycoproteins play a role in the interaction with the M protein, which is the protein most probably responsible for driving virus particle assembly and budding (Sanderson *et al.*, 1994; Ciancanelli and Basler, 2006; Lamb and Parks, 2007; Patch *et al.*, 2007, 2008; Ong *et al.*, 2009).

Morphology

Electron microscopy studies using both negative stain and thin section preparations depicted extracellular NiV virions as tangled collections of filamentous, helical nucleocapsids encased by the viral envelope (Chua *et al.*, 2000a; Goldsmith *et al.*, 2003). While NiV virions share common morphological characteristics with the rest of the family *Paramyxoviridae* (i.e. herringbone-like nucleocapsid structure, nucleocapsid accumulation along plasma membranes during budding), there were several distinct characteristics observed. There were unusual cytoplasmic ring-like reticular inclusions (RI) found in proximity to membranes of the endoplasmic reticulum. These inclusions were distinct from the typical nucleocapsid inclusions (NCI) found in paramyxovirus-infected cells. There were also long cytoplasmic tubules found in the periphery of NiV-infected cells, which were occasionally incorporated into virus particles. NiV virions were on average larger in diameter (500 nm) than typical paramyxoviruses (150–400 nm), with extreme variations in size (180–1900 nm; Goldsmith *et al.*, 2003). There are subtle ultrastructural differences that distinguish NiV and HeV. When viewed by negative-contrast electron microscopy, the surface projections of HeV virions have a double-fringed appearance, while those of NiV have a single fringe (Hyatt *et al.*, 2001). The localization of the cell nuclei and viral nucleocapsids within infected cells also differentiates NiV and HeV. Along with the viral nucleocapsids, the cell nuclei of NiV-infected cells localize to the cell periphery. In HeV-infected cells, both the cell nuclei and HeV nucleocapsids are either more centrally located or randomly distributed in the cytoplasm (Hyatt *et al.*, 2001). *In situ* hybridization studies using both positive and negative sense probes in NiV-infected cell culture and brain specimens taken from fatal NiV infections indicated that non-encapsidated viral mRNA is almost exclusively found in the novel RI mentioned previously, and not in the NCIs (Goldsmith *et al.*, 2003). Given that the RI complexes were found in the context of a natural infection, these results suggest the possibility that for NiV, the newly described RI structures are an integral part of the transcription and replication process.

Genome organization

The *Henipavirus* genome consists of six genes encoded from 3′ to 5′, the nucleocapsid (N), phosphoprotein (P), matrix (M), fusion (F), attachment glycoprotein (G), and large (L) polymerase, respectively. The prototypical genomes of NiV and HeV are 18,246 nucleotides (nt) and 18,234 nt long respectively, making it significantly larger than most other paramyxoviruses, with the exception of the rodent-borne J-virus and Beilong virus (Fig. 6.1; Jack *et al.*, 2005; Li *et al.*, 2006; Magoffin *et al.*, 2007b). The large genome size is due to a longer open reading frame (ORF) encoding the P protein and to the large 3′ untranslated regions (UTR) that flank each gene with the exception of the large polymerase gene (Wang *et al.*, 1998, 2000a; Yu *et al.*, 1998a,b; Harcourt *et al.*, 2000, 2001; Mayo, 2002). The organization of the *Henipavirus* genome indicates that it probably has a similar replication strategy to other paramyxoviruses, which was demonstrated by the requirement of the N, P, and L proteins for mini-genome replication (Harcourt *et al.*, 2000, 2001; Wang *et al.*, 2000b; Halpin *et al.*, 2004). The genus-specific 3′ leader and 5′ trailer regions of paramyxovirus genomes serve as promoters for both transcription and replication (Lamb and Parks, 2007). The coding sequence of each gene is flanked by UTRs that contain a gene-start (GS) sequence and a gene-end (GE) sequence. Immediately following the GE sequence is a short non-coding intergenic region that precedes the gene-start sequence of the following gene. Studies of other paramyxoviruses have shown that GE signals direct polyadenylation and termination of each mRNA, while GS signals direct initiation and capping (Lamb and Parks, 2007). The terminal 12–13 nucleotides of the genome and antigenome are complementary, which indicates their critical role as replication promoters.

Figure 6.1 Comparative genome lengths among viruses in the subfamily *Paramyxovirinae*. NiV has an exceptionally large genome compared with most paramyxoviruses by virtue of its longer non-coding regions and larger P gene (genome lengths in parentheses; see text). The solid black segments indicate leader and trailer sequences. The clear segments indicate virus gene non-coding regions, and the shaded areas indicate the respective NiV gene coding regions. MeV, measles virus; SeV, Sendai virus; MuV, mumps virus; NDV, Newcastle disease virus; J-V, J-paramyxovirus.

Viral proteins

Since their emergence, numerous studies have provided insight on the molecular characteristics of *Henipavirus* genes as well as the mechanistic details of infection, which include virus entry, protein localization and processing, genome replication, and virus particle assembly. Viruses in the subfamily *Paramyxovirinae* encode two integral membrane glycoproteins, an attachment glycoprotein (HN, H, or G) and a fusion (F) glycoprotein. Previous studies utilizing recombinant vaccinia viruses individually expressing F and G proteins of both HeV and NiV indicate that each protein induces neutralizing antibodies, and are both required to mediate cell entry (Bossart *et al.*, 2001; Tamin *et al.*, 2002).

G glycoprotein

The *Henipavirus* G proteins share an identical length of 602 amino acids, and are type II membrane proteins that form tetramers upon cell surface expression, and serve primarily in cellular receptor binding (Wang *et al.*, 2001; Levroney *et al.*, 2005). In contrast to the attachment glycoproteins of other paramyxoviruses, NiV G lacks both haemagglutinin and neuraminidase activities (Tamin *et al.*, 2002). Preliminary experiments indicated the possibility that henipaviruses utilized a protein receptor instead of a carbohydrate moiety for virus entry (Eaton *et al.*, 2004). Two independent groups using different approaches experimentally confirmed this hypothesis (Bonaparte *et al.*, 2005; Negrete *et al.*, 2005). One approach involved using a recombinant soluble immunoadhesin consisting of the ectodomain of the NiV G fused with the Fc region of human IgG1 to immunoprecipitate cell surface proteins from NiV fusion-permissive cells. Trypsin digestion and mass spectrometric analysis of the immunoprecipitated surface proteins identified only one transmembrane protein, confirmed by two independent tryptic fragments to be ephrinB2. Transfection of non-permissive cell lines with ephrinB2 rendered them permissive to both

NiV envelope-mediated fusion as well as to NiV infection, which was evaluated by using VSV-RFP viruses pseudotyped with the NiV F and G proteins (Negrete *et al.*, 2005). The second approach utilized microarray analyses of both permissive and non-permissive cell lines to NiV infection in order to identify plasma membrane proteins differentially expressed in permissive cell lines. From the pool of potential receptor-encoding genes identified, only ephrinB2 was able to rescue NiV-mediated cell fusion when transfected into the non-permissive cell line as well as to render those cells permissive to NiV infection (Bonaparte *et al.*, 2005). The expression pattern of ephrinB2 in neurons, smooth muscle, and endothelial cells correlates with the cellular tropism of NiV during infection. EphrinB2 is involved in embryogenesis, vasculogenesis and axonal guidance (Palmer and Klein, 2003; Poliakov *et al.*, 2004). Recent studies of the requirement for ephrinB2-mediated NiV infection of endothelial cell lines underscores the correlation of this receptor with observed pathogenesis in human cases (Erbar *et al.*, 2008; Thiel *et al.*, 2008). Not long after the delineation of ephrinB2 as the primary cellular receptor for NiV infection, ephrinB3 was identified as an alternative cellular receptor for NiV infection (Negrete *et al.*, 2006). A comparison of the respective binding affinities of ephrinB2 and ephrinB3 to a soluble form of NiV G showed that despite a 10-fold lower binding affinity than ephrinB2, ephrinB3 bound to NiV G at subnanomolar affinities. In addition, this study identified two critical residues shared between ephrinB2 and ephrinB3 that both mediate NiV entry and are crucial to their function as EphB receptor ligands (Negrete *et al.*, 2006).

A follow-up study by Negrete and colleagues identified and distinguished functional residues required for ephrinB2 and ephrinB3 interactions with NiV G. While NiV and HeV both utilize ephrinB2 as a cellular receptor, this study demonstrated a greater ability of NiV G to bind ephrinB3 than that of HeV G. By interchanging various homologous regions of NiV and HeV G proteins, they were able to map the determinant of ephrinB3 binding to a single residue that could confer equivalent ephrinB3 binding efficiency when incorporated into HeV G (Negrete *et al.*, 2007). The authors hypothesize that the difference in the efficiency of ephrinB3 usage between NiV and HeV may be reflected in the increased neurotropism of NiV compared with HeV in human infections. This study confirmed and provided insight relating to the results from a previous study, which initially identified several of these critical functional residues (Guillaume *et al.*, 2006a). Though another previous study of HeV G identified several different residues crucial to ephrinB2 binding, the authors conceded the possibility of those residues forming a discontinuous epitope that indirectly affected binding (Bishop *et al.*, 2007).

The crystal structures of the ephrinB2 and ephrinB3 interactions with NiV G have recently been determined (Bowden *et al.*, 2008b; Xu *et al.*, 2008). The data from both of these studies confirm the accuracy of the functional mapping performed by Negrete and colleagues, and also provide a comprehensive analysis of critical residues composing the hydrophobic binding cleft in NiV G, which interacts with high affinity to ephrinB2 and ephrinB3. Further analysis of the unbound form of NiV G protein by crystallography and mass spectrometry indicated that NiV G contains highly processed complex-type glycans with negligible amounts of oligomannose-type glycans. Interestingly, the *N*-acetylglucosamine (GlcNAc) $\beta 1 \rightarrow 2$Man terminal structures on NiV G were noted as a potential ligand for LSECtin, a C-type lectin expressed on sinusoidal endothelial cells of lymph nodes and liver (Bowden *et al.*, 2008a).

A receptor binding activation site in the stalk region of NiV G was shown to trigger fusion by the F protein (Aguilar *et al.*, 2009). This finding is consistent with observations of other paramyxovirus attachment glycoprotein interactions with their corresponding fusion proteins (Melanson and Iorio, 2006, 2004; Lamb and Parks, 2007; Corey and Iorio, 2007, 2009; Lee *et al.*, 2008). Studies of the mechanistic dynamics of *Henipavirus* fusion have recently been performed by Whitman and colleagues. Their initial study determined the optimal molar ratio of HeV F:HeV G expressed on the cell surface was 1:1, and showed that the HeV G protein has a higher turnover rate and a slower trafficking rate through the cell than for HeV F (Whitman and Dutch, 2007). Further study revealed that the HeV G undergoes an intrinsically slower process of oligomerization through the secretory pathway, suggesting that F–G interactions occur only following the initial steps of synthesis and cellular trafficking (Whitman *et al.*, 2009).

F glycoprotein

The F proteins of both HeV and NiV are 546 amino acid type I transmembrane proteins that form trimers on the cell surface, and are responsible for mediating the fusion of virus and host cell membranes during infection (Michalski *et al.*, 2000; Bossart *et al.*, 2001; Wang *et al.*, 2001; Tamin *et al.*, 2002; Levroney *et al.*, 2005). Similar to other paramyxoviruses, activation of NiV F fusion depends on specific sequences in its cytoplasmic tail (Tong *et al.*, 2002; Waning *et al.*, 2004; Aguilar *et al.*, 2007). The F proteins of paramyxoviruses are synthesized as inactive precursors called F_0, and are then typically cleaved by a cellular protease to yield the biologically active F1 and F2 subunits, which are linked by disulfide bonds (Lamb and Parks, 2007). The membrane-anchored subunit F1 contains a hydrophobic amino terminus that is highly conserved, often referred to as the fusion peptide. Paramyxovirus F proteins with a multibasic consensus sequence (R-X-R/K-R) are activated by ubiquitous host cell proteases such as furin, while F proteins with only one basic residue at the site of cleavage are activated by extracellular trypsin-like proteases upon transport of inactive F_0 precursors to the cell membrane. Viruses with multibasic sequences in their F protein usually cause systemic infections, while those with single basic cleavage sites such as Sendai virus cause localized infections (Klenk and Garten, 1994). A confounding observation was that despite the systemic nature of NiV infections, the cleavage site of the NiV F protein only contains one basic arginine residue at the cleavage site, unlike the multibasic furin-like protease consensus sequences found in most morbilliviruses, rubulaviruses, and *Pneumoviruses*. Comprehensive mutagenesis studies of the cleavage sites of both HeV and NiV F proteins demonstrated that the single arginine residue at the cleavage site was not required for F processing, which indicated that the protease responsible for processing F was neither trypsin nor any other furin-like protease (Moll *et al.*, 2004a; Craft and Dutch, 2005). Furthermore, the absence of serum proteases did not inhibit NiV in cell culture, which implicated an intracellular protease to be responsible for F protein processing (Diederich *et al.*, 2005). By tracking the route of synthesized F and G in NiV infected cells or HeV F plasmid transfected cells, it was determined that these proteins undergo endocytosis following surface expression (Meulendyke *et al.*, 2005; Vogt *et al.*, 2005). While the endocytosis of NiV G occurred at a similar rate to that of typical membrane turnover and did not accumulate inside cells, it was shown that intracellular F protein accumulated over time. When site directed mutagenesis was performed on the classical endocytosis signal (Y-X-X-φ) contained in the NiV F protein, not only was endocytosis abrogated, but the ability of F to mediate fusion when co-expressed with G was also ablated (Vogt *et al.*, 2005; Fig. 6.2). The same group determined that F processing occurred in acidified endosomes (Diederich *et al.*, 2005). Mutations performed in the same endocytosis signal of HeV F caused an increase in F processing, but these results were obscured owing to the different levels of surface expression corresponding to these particular mutations (Meulendyke *et al.*, 2005). Eventually, it was shown that the endosomal protease cathepsin L was responsible for both HeV and NiV F protein processing by using cysteine protease inhibitors and cell lines in which expression of cathepsin L was knocked down via small interfering RNA (siRNA; Pager and Dutch, 2005; Pager *et al.*, 2006).

In order to determine whether the henipaviruses utilize pH-dependent proteolysis for viral entry, Diederich and colleagues infected cells pre-treated with either an inhibitor of endosomal acidification (NH_4Cl) or a variety of endocytosis or cathepsin inhibitors with NiV (Diederich *et al.*, 2008). None of the inhibitors affected NiV infection. However, when cells were treated with either cathepsin inhibitors or NH_4Cl after NiV infection, there was a significant decrease of infectious virus released from the cells. This study demonstrated that cell-free NiV virions possessed adequate amounts of proteolytically cleaved F protein to mediate viral fusion with the host membranes,

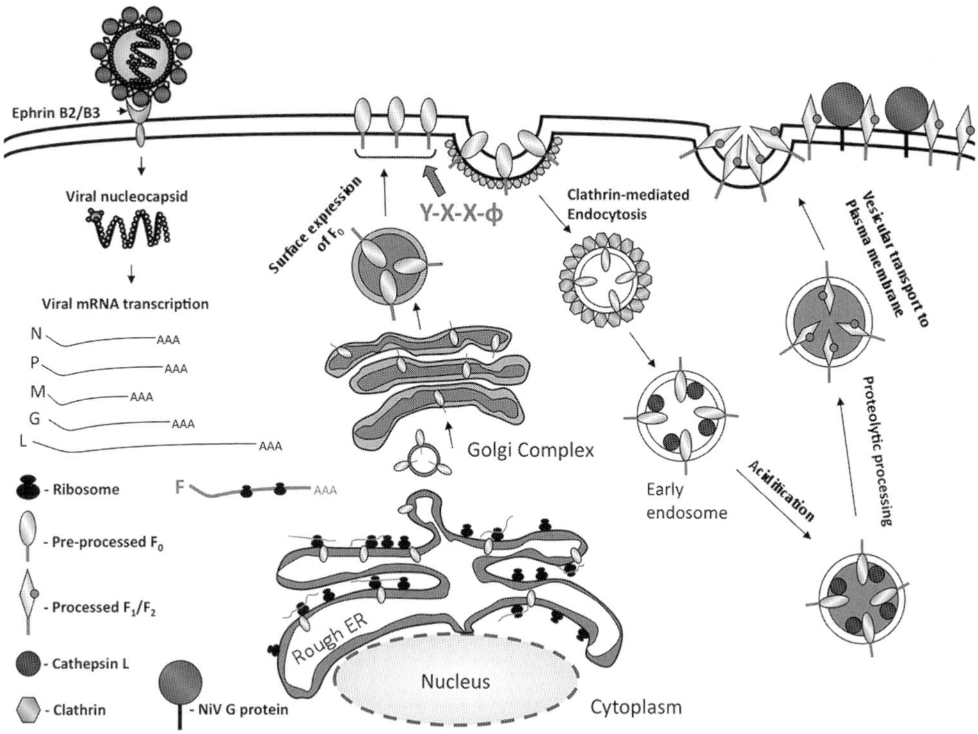

Figure 6.2 Proteolytic processing of NiV F protein. Upon ribonucleocapsid release into the host cell cytoplasm, transcription of viral mRNA begins, with the ribonucleocapsid complex serving as the template for RNA synthesis. The polymerase complex initiates transcription at the promoter region in the 3′ leader. Each gene is transcribed, being regulated by transcriptional start and stop sites at their respective regions in the gene. The NiV F gene mRNA is translated by ribosomes into F protein, which is post-translationally modified as it progresses through the cell secretory system. Pre-processed F_0 is transported to the plasma membrane, where it is endocytosed by a clathrin-dependent mechanism via its Y-X-X-ϕ motif in its C-terminus (see text). The F_0 eventually reaches the early endosome where upon its acidification, the cysteine protease cathepsin L cleaves F_0 into the F1-F2 activated form of F. The processed F is finally trafficked to the plasma membrane where it is competent to mediate cell–to-cell fusion with concurrently expressed NiV G protein.

and that the lowering of endosomal pH is only required for the maturation of new virion particles produced from an infected cell and not for viral entry (Diederich *et al.*, 2008).

Recently, Weise and colleagues determined specific tyrosine residues in the cytoplasmic domains of the NiV F and G proteins were crucial in determining their apical or basolateral localization within polarized epithelial cells (Weise *et al.*, 2010). The tyrosine residue 525 in NiV F, which is part of the endocytosis signal mentioned above, was also required for basolateral sorting. A di-tyrosine motif in NiV G was responsible for its basolateral sorting. Interestingly, it was shown that upon co-expression of NiV F and G, the localization of NiV G was contingent upon the localization of NiV F. Disruption of the basolateral targeting of the glycoproteins abrogated syncytia formation in the polarized epithelia, which the authors suggest is crucial for viral spread in respiratory epithelia (Weise *et al.*, 2010).

It has been shown for many viral glycoproteins that N-glycans are crucial structural components that facilitate proper folding and transport (Doms *et al.*, 1993; Roberts *et al.*, 1993; Segawa *et al.*, 2000; Zimmer *et al.*, 2001; McGinnes *et al.*, 2001; Tamura *et al.*, 2002). The HeV F protein contains six potential glycosylation sites, with three located in each of the F_2 and F_1 subunits (Michalski *et al.*, 2000). The NiV F protein contains five potential glycosylation sites, three located in the F_2 subunit, and 2 in the F_1 subunit. The influence of N-linked oligosaccharides on the processing and biological activity of the NiV F protein has been studied by two independent groups, yielding some similar and some contrasting results. Moll and colleagues determined that four out of the five potential sites were glycosylated (N67, N99, N414 and N464), with complex glycosylation at all, but the N464 site, which is of the high mannose variety. Site-directed mutagenesis of each glycosylation site demonstrated that both the amino acid sequence and the glycosylation of the N414 site were required for proper F protein processing and cell surface expression. It was shown that while deglycosylation negatively affected the overall cell surface expression of the F mutants, it increased the fusogenic ability of the individual N414 and N464 mutants and one N67/N99 double mutant F (Moll *et al.*, 2004b). An analogous mutagenesis study was performed on the HeV F, which yielded similar findings as Moll and colleagues, except that the N414A mutant abrogated HeV F fusion instead of causing hyperfusogenicity (Carter *et al.*, 2005). Another study on the NiV F protein done by an independent group confirmed the glycosylation pattern of NiV F, the corresponding increase in fusion ability with deglycosylation at particular sites (N414, N464), and the corresponding abrogation of cell surface expression upon combinatorial mutation of all glycosylation sites. In contrast to Moll and colleagues, however, it was determined that an individual F mutant at the N99 position significantly increased fusogenicity (Aguilar *et al.*, 2006). It is worth noting the difference in mutagenesis strategies employed by these two groups, which may explain the differences in their findings. While Moll and colleagues mutated the third residue of the glycosylation motif (N-X-S/T; underlined) to a glycine (with the exception of Fg4* in which asparagine was mutated to glutamine), Aguilar and colleagues directly mutated the glycosylated asparagine residue to glutamine (N-X-S/T; underlined). This latter study tested multiple combinations of N-glycosylation mutant F proteins and demonstrated a synergistic effect of multiple N-glycans in reducing fusion. A select panel of these mutant F proteins was used with wild-type NiV G to pseudotype VSV reporter viruses. These reporter viruses were used to evaluate the influence of F protein N-glycans on the efficiency of viral entry. The results indicated that the mutants which showed a hyperfusogenic phenotype also enhanced viral entry. This study also demonstrated that N-glycans can protect NiV from neutralizing antibodies and that reducing the glycosylation on the F protein decreases the avidity of its interaction with the G protein (Aguilar *et al.*, 2006). The progressive analyses of NiV G interactions with its cellular receptors, as well as its interaction with NiV F will facilitate the development of antiviral therapeutics which target these interactions.

M protein

The NiV and HeV M proteins have a length of 352 amino acids and like other paramyxovirus M proteins are very basic, with pIs of 9.99 and 9.44, respectively, and charges of +12 and +14 at neutral pH, respectively (Wang *et al.*, 2001). The paramyxovirus M proteins provide rigidity and structure to the virion through its interactions with the cytoplasmic tail of the F protein, the ribonucleoprotein complex, and the inner surface of the virion envelope (Lamb and Parks, 2007; Sanderson *et al.*, 1994; Tashiro *et al.*, 1993, 1996). Late domain motifs found in numerous viral matrix proteins recruit specific host factors to viral assembly sites where they facilitate virus release (Freed, 2002). There are several different late domain motifs found among negative stranded non-segmented viruses: P(T/S)AP, PPxY, $YP(x)_nL$, and more recently φ-P-x-V (Lamb and Parks, 2007). Several paramyxovirus M proteins have been shown to recruit cellular host factors to facilitate viral budding by way of these late domains (Schmitt *et al.*, 2005; Ciancanelli and Basler, 2006; Irie *et al.*, 2007; Patch *et al.*, 2008).

Ciancanelli and Basler showed that eukaryotic plasmid expression of the NiV M protein alone can yield virus-like particles (VLP), and demonstrated the first evidence of late domain-like activity in the NiV M protein by mutating the YMYL motif in the N-terminal portion of M. Mutation of the first and fourth residues of this motif severely decreased the presence of M in the culture supernatant, and skewed the subcellular localization of the M towards the nucleus. By inserting the YMYL motif into a late domain-deficient Ebola virus VP40 matrix protein, the ability of VP40 to bud and form VLPs was restored. This study demonstrated the ability of NiV M to self-associate and that the M proteins were enclosed by a lipid membrane, which protected them from proteolysis (Ciancanelli and Basler, 2006).

A following study done by Patch and colleagues identified a potentially different late domain motif YPLGVG also in the N-terminal end of NiV M. Similar to the previous study, it was shown that mutagenesis of this sequence either ablated or drastically decreased VLP budding and also skewed the NiV M subcellular localization to the nucleus. However, substitution of the Ebola virus late domain with this motif along with several surrounding residues only restored the filamentous cell morphology phenotype of cells expressing Ebola VP40 and did not result in any budded Ebola VLPs (Patch *et al.*, 2008). The authors also tested whether NiV M required VPS4A to form VLPs. VPS4A is a component of the vacuolar protein sorting (VPS) pathway that facilitates the formation of multivesicular bodies, which is a process similar to viral budding (Lamb and Parks, 2007). It was shown that co-expression of NiV M with either a wild-type or a dominant negative form of VPS4A did not affect the ability of NiV M to form VLPs. These results indicate potentially novel cellular components utilized by NiV to allow egress from an infected cell (Patch *et al.*, 2008).

Ribonucleoprotein complex

The ribonucleoprotein (RNP) complex of paramyxoviruses consists of an N protein-encapsidated genome and the viral RNA-dependent RNA polymerase complex formed by the L and P proteins (Lamb and Parks, 2007). The L protein presumably possesses all the enzymatic activities for viral RNA synthesis, while the P protein serves as a scaffold between the L and the encapsidated genome (Harty and Palese, 1995; Bankamp *et al.*, 1996; Liston *et al.*, 1997; Carsillo *et al.*, 2006; Houben *et al.*, 2007).

N protein

The N protein of henipaviruses is a 532 amino acid protein responsible for encapsidating the viral genome and also for the common herring-bone-like structures seen in electron micrographs of paramyxoviruses (Lamb and Parks, 2007). Paramyxovirus N proteins typically contain two domains, with the amino-terminal domain responsible for specific interactions with the RNA genome and other N proteins and the carboxy-terminal domain being responsible for interactions with the P protein (Egelman *et al.*, 1989). The N proteins of the subfamily *Paramyxovirinae* are thought to cover exactly six nucleotides, which results in maximal viral replication when the genome length is divisible evenly by 6. This concept, coined 'the rule of six' has been demonstrated for several paramyxoviruses, including NiV (Calain and Roux, 1993; Hausmann *et al.*, 1996; Kolakofsky *et al.*, 1998; Halpin *et al.*, 2004). As with other paramyxoviruses, NiV N has an inherent ability to assemble itself, as the expression of NiV N protein in the absence of other viral proteins results in the formation of nucleocapsid-like structures (Tan *et al.*, 2004). A detailed mapping study of the precise regions required for NiV N self-assembly determined that the minimal contiguous sequence of N required was from amino acids 30–404, while the C-terminal 128 amino acids were dispensable (Ong *et al.*, 2009).

The viral polymerase associates with the encapsidated genome by way of interaction between the P and the N proteins of the ribonucleocapsid. Studies of Sendai virus (SeV) have demonstrated the N–P interactions *in vitro* (Horikami *et al.*, 1992). Upon mixture, individually formed complexes of N-P and P-L of different paramyxoviruses have the ability to support replication of defective interfering virus *in vitro* (Errington and Emmerson, 1997; De *et al.*, 2000). Also, the SeV P protein has been shown to interact with unassembled N protein (N^0) during the nascent chain assembly step of genome replication, with P serving as a protein chaperone for N by preventing non-specific aggregation of N (Curran *et al.*, 1995). The domains responsible for the N–P interaction have been determined for many of the paramyxoviruses, including NiV and HeV (Harty and Palese, 1995; Bankamp *et al.*, 1996; Cevik *et al.*, 2004; Chan *et al.*, 2004; Houben *et al.*, 2007). It was shown that the N and the P proteins of NiV and HeV were able to form both homologous and heterologous N–P complexes. In addition, it was determined that there was more than one

P-binding site on the N protein, with one in the extreme carboxy-terminal region being sufficient to interact with the carboxy-terminal portion of the P protein (Chan *et al.*, 2004).

L protein

The L protein of paramyxoviruses is the largest and the least abundant viral protein found in an infected cell. The large size of the protein reflects the many different enzymatic functions attributed to it, including initiation, elongation and termination of both mRNA transcription (capping, methylation and polyadenylation of mRNA) and genome replication (Lamb and Parks, 2007). A comparative study done on L proteins of NNS RNA viruses lead to the identification of six similar domains shared among them (Stec *et al.*, 1991). The L protein of NiV contains all six conserved domains found in NNS viruses, but the GDNQ motif found in most NNS RNA viruses is replaced by GDNE. This motif is the proposed active site where new ribonucleotides are added to a growing strand of RNA polymerized by the L protein. The mutation of the GDNQ motif to GDNE for both rabies and rinderpest viruses resulted in significantly reduced polymerase activities (Chattopadhyay *et al.*, 2004; Schnell and Conzelmann, 1995). Site directed mutagenesis of the glutamate residue in the motif GDN<u>E</u> (underlined) of the NiV L protein was recently performed (Magoffin *et al.*, 2007a). In this study, the E residue was substituted with Q, K, P, D, A, N, I, or G residues, and the resulting mutant L proteins were tested for their function in a NiV minigenome assay, which measured green fluorescent protein expression (GFP) as an indicator of genome replication. It was determined that the GDNQ substitution did not significantly affect the function of NiV L, while the GDNK and GDNP mutants completely abolished GFP expression. The activity of the GDND, GDNA, GDNN, GDNI and GNDG L protein mutants were significantly reduced compared with the wild-type GDNE and GDNQ, retaining 2–40% of wild-type L activity. The GDNQ motif that is conserved in the L proteins among many NNS RNA viruses has been thought to be essential to the catalytic activity of the polymerase (Poch *et al.*, 1989). The ability for the NiV L protein to tolerate so many different substitutions in the GDNE motif at the E residue without completely losing its activity suggests that the amino acid residue at the E position is important for maintaining the overall protein structure rather than for its catalytic activity (Magoffin *et al.*, 2007a).

P protein

The *Henipavirus* P protein is a 707 (HeV)/709 (NiV) amino acid protein and is the only essential gene product encoded by the P gene for genome replication (Halpin *et al.*, 2004). An early study utilizing mass spectrometry described only serine phosphorylation sites of NiV P, in contrast to both serine and threonine phosphorylation sites in HeV P (Shiell *et al.*, 2003). A study of the N-P protein interaction domains showed that there were at least two independent N-binding sites in the P protein; one was at the N-terminus (amino acids 3–220), while the other was at the end of the C-terminus (amino acids 636–709; Chan *et al.*, 2004).

A common characteristic among viruses in the subfamily *Paramyxovirinae* is the ability to generate multiple proteins from the P gene that are distinct from the P protein (Lamb and Parks, 2007). These additional gene products are usually not required for virus replication *in vitro*, though they often serve as virulence factors *in vivo* (Garcin *et al.*, 1997, 1999; Kato *et al.*, 1997a,b; Kurotani *et al.*, 1998; Tober *et al.*, 1998; Valsamakis *et al.*, 1998; Patterson *et al.*, 2000; Devaux and Cattaneo, 2004; von Messling *et al.*, 2006; Devaux *et al.*, 2008; Toth *et al.*, 2009). There are a number of mechanisms by which these accessory gene products are generated.

The first mechanism is by a process of non-templated co-transcriptional editing, otherwise known as RNA editing. The viral polymerase complex incorporates non-templated guanosine residues into a specific editing site of the P gene mRNA, which creates frameshifts that result in the translation of different ORFs downstream of the editing site (Curran *et al.*, 1991). RNA editing is responsible for generating the viral mRNA transcripts of V and W, which share a common amino terminus with the P protein, but have respectively unique carboxy termini. The V transcript is created by the addition of 3n+1 guanosine residues to the editing site, effectively generating a +1 ORF distal to the editing site. The W transcript is generated by adding 3n+2 guanosine residues, which results in a +2 ORF distal to the editing site. The exception to this rule is for the genus *Rubulavirus*, in which the P protein is made when 2 guanosine residues are added, and the V protein is made when there are no insertions in the mRNA editing site (Lamb and Parks, 2007).

A second mechanism which is responsible for expressing the C proteins is a process by which host cell ribosomes read through the original P gene mRNA start codon to use an alternative downstream translation initiation sequence (Curran and Kolakofsky, 1988a,b, 1989,; Latorre *et al.*, 1998). This results in translation of a protein using a distinct overlapping ORF. Two other mechanisms include internal ribosomal binding sites and the use of non-AUG start codons (Mehdi *et al.*, 1990; Bellini *et al.*, 1998). The P gene from certain members of the subfamily *Paramyxovirinae* (SeV) can make up to seven non-structural accessory proteins (Curran and Kolakofsky, 1988a,b, 1989; Latorre *et al.*, 1998).

The P gene of henipaviruses encodes at least three non-structural proteins (C, V, and W) in addition to the P protein. The V and W proteins are generated by RNA editing, while the C protein is encoded by a second ORF that initiates 23 nt downstream of the translational initiation site for the P ORF (Harcourt *et al.*, 2000). There is also an additional accessible open reading frame within the P gene of HeV identified as the small basic protein (SB; Wang *et al.*, 2001). Experiments using transfected plasmid vectors expressing NiV P, V, and W indicated that the V and P proteins are localized in the cytoplasm, while the W protein is in the nucleus (Rodriguez *et al.*, 2002, 2004; Shaw *et al.*, 2004). However, it was not until more recently that the P gene mRNA editing frequencies of henipaviruses and the subcellular localization of the C, V, and W proteins were characterized in NiV infected cells (Lo *et al.*, 2009). In this study, the RNA editing sites for NiV Malaysia, NiV Bangladesh, and HeV were sequenced to determine the frequency and numbers of guanosine residue insertions at the editing site. For all three viruses tested, approximately two-thirds of all P gene transcripts were edited and 50% of all transcripts encoded for P, 25% for V, and 25% for W. Western blots performed against NiV infected cell lysates using specific anti-peptide sera against the unique regions of the P, V, W, and C proteins indicated robust expression levels of V and W, while the expression of C was markedly lower. The V, W, and C proteins were surprisingly also detected in significant quantities in sucrose gradient purified virions. Immunofluorescence assays performed on NiV infected cells using the aforementioned antisera showed distinct subcellular localizations of the P, C, V, and W proteins (Lo *et al.*, 2009). The P protein, while localized in the cytoplasm, concentrated near the plasma membrane, which was in contrast to the homogeneous distribution of plasmid-expressed P. The V protein was evenly distributed throughout the cytoplasm and the W protein exclusively localized to the nucleus. The C protein was distributed in the perinuclear areas in a punctuate pattern. Co-immunoprecipitation experiments from NiV infected cell lysates showed that the importin karyopherin-α3 co-precipitated with the W protein, which confirm the findings from earlier plasmid expression studies (Shaw *et al.*, 2005; Lo *et al.*, 2009). Following the publication of this study, Kulkarni and colleagues published similar results in regards to the robust expression of NiV V and W proteins owing to the high frequency of NiV P gene editing observed in NiV infected cells. This study determined that the relative number of P gene mRNA transcripts encoding P protein versus those encoding for the V and W proteins drastically decreases over the course of the first 30 h of infection (Kulkarni *et al.*, 2009).

Genetics and reverse genetics

Minigenomes of paramyxoviruses, which incorporate the leader, the GS of the N gene, a reporter gene, the GE of the L gene, and the trailer have been developed for many paramyxoviruses including NiV (Sidhu *et al.*, 1995; Hausmann *et al.*, 1996; Durbin *et al.*, 1997; Keller *et al.*, 2001; Perez *et al.*, 2003; Chattopadhyay *et al.*, 2004; Halpin *et al.*, 2004; Raha *et al.*, 2004; Rosario *et al.*, 2005; Magoffin *et al.*, 2007b; Jiang *et al.*, 2008;). An RNA polymerase I driven NiV minigenome system has also been developed subsequent to the T7 polymerase-driven system (Freiberg *et al.*, 2008). A fine mapping study of the NiV antigenomic promoter used a T7 polymerase-driven NiV minigenome system to determine the precise boundaries of the *cis*-acting signals contained therein (Walpita and Peters, 2007). The results of the study indicated a bipartite promoter structure with two critically conserved regions (nt 1–12 and 79–91). The structure of the internal antigenomic promoter element (nt 79–91), while analogous in position to its equivalent in the Sendai virus antigenomic promoter, contained a distinct 5′-$(\underline{G}NNN\underline{UG})_{14-15}(\underline{G}NNNNN)_{16}$ motif (Tapparel *et al.*, 1998; Walpita and Peters, 2007).

A study that also utilized the NiV minigenome system demonstrated that the NiV C, V, and W proteins provided in *trans* via plasmid expression inhibited NiV minigenome transcription and replication in a dose-dependent manner (Sleeman *et al.*, 2008). Interestingly, the heterologous C and V proteins of measles virus (MeV) also had the ability to inhibit NiV minigenome replication. Conversely, NiV C, V, and W proteins were able to inhibit MeV minigenome replication, although to a lesser extent than the MeV C protein. The C protein of human parainfluenza virus 3 (hPIV3) was surprisingly also shown to inhibit replication of both NiV and MeV minigenomes. The inhibitory effect of the NiV C and V proteins on NiV minigenome replication was consistent with observations for other paramyxoviruses such as Newcastle disease virus (NDV), MeV, hPIV2, hPIV3 and SeV (Grogan and Moyer, 2001; Reutter *et al.*, 2001; Bankamp *et al.*, 2005; Witko *et al.*, 2006; Nishio *et al.*, 2008).

Yoneda and colleagues were the first to develop a T7 polymerase-based full-length reverse genetic system for NiV (Yoneda *et al.*, 2006). A recombinant antigenomic cDNA clone of NiV was constructed, which incorporated a T7 promoter in front of the 5′ leader and unique restriction sites in the 3′ UTRs of each gene. By infecting CV-1 cells with a replication-deficient modified vaccinia virus expressing T7 polymerase and transfecting the cells with the cDNA clone along with helper plasmids expressing N, P, and L, syncytia were visible two days post-transfection. An additional reporter gene (EGFP) was inserted between the N and P genes to create a reporter NiV in which replication could easily be monitored by fluorescence microscopy. The growth characteristics of the recombinant viruses compared favourably with wild-type virus *in vitro*, with the EGFP-bearing NiV growing slightly slower at 24 h post-infection. When tested *in vivo* in a hamster model, wild-type NiV and the rescued NiV showed similar mortality. The recombinant reporter NiV was used to infect cell lines of various mammalian origins and required ephrinB2 for viral entry. However, mouse and rat cell lines which expressed ephrinB2 were not permissive for viral replication, which implied that there are other factors besides receptor expression that control NiV tropism (Yoneda *et al.*, 2006). Since then, an independent group developed a similar T7 polymerase-based full-length NiV reverse genetic system to study the P gene (Ciancanelli *et al.*, 2009). There are currently no known full-length HeV reverse genetic systems available.

Pathogenesis

Henipaviruses are the only zoonotic paramyxoviruses that are highly pathogenic in humans, but have a host range that spans five terrestrial mammalian orders, exceeding that of the *Morbillivirus* canine distemper virus (CDV), which infects species across three orders (Appel and Summers, 1995; Roelke-Parker *et al.*, 1996; Chua *et al.*, 2000a; Mohd *et al.*, 2000; Middleton *et al.*, 2002; Reynes *et al.*, 2005).

A post-mortem study of human NiV infection determined that a systemic multiorgan vasculitis associated with infection of endothelial cells was the main pathological feature with infection being most pronounced in the central nervous system (CNS; Wong *et al.*, 2002). In the CNS vascular endothelium, immunohistochemical analysis showed intense staining of endothelial, parenchymal, and multinucleate giant cells characteristic of paramyxovirus infections. However, evidence of endothelial infection and vasculitis was also observed in other organs, including the lung, heart, spleen, and kidney. NiV has been isolated from cerebrospinal fluid, tracheal secretions, throat swabs, nasal swabs and urine specimens of patients (Goh *et al.*, 2000; Chua *et al.*, 2001; Wong *et al.*, 2002). NiV infection can also become a chronic condition, as serious neurological disease can occur late (more than four years) following a non-encephalitic or asymptomatic infection (Tan *et al.*, 2002; Wong *et al.*, 2002; Chong and Tan, 2003).

Pathology of the first fatal human case of HeV infection revealed severe interstitial pneumonia characterized by gross lesions of congestion, haemorrhage and oedema associated with histological chronic alveolitis (Murray *et al.*, 1995b; Selvey *et al.*, 1995; Hooper *et al.*, 2001). Pathological manifestations of HeV infection in horses from the initial outbreak included dilated pulmonary lymphatics, severe pulmonary oedema and congestion (Hooper *et al.*, 1997). While the earliest cases of human HeV infection indicated more respiratory than neurological involvement, observations of HeV infected horses clearly indicate neurological involvement (Murray *et al.*, 1995b; O'Sullivan *et al.*, 1997; Westbury, 2000; Field *et al.*, 2010;). A recent study also indicated the potential for HeV to cause both acute and relapsing encephalitis in humans (Wong *et al.*, 2009).

The disease manifestations of both natural and experimental *Henipavirus* infections vary greatly between its natural chiropteran hosts and other susceptible terrestrial species. In Pteropid fruit bats, the infection is subclinical with low levels of detectable antigen, only rarely causing vasculitis in select organs (Williamson *et al.*, 1998, 2000; Hooper *et al.*, 2001; Middleton *et al.*, 2007). In contrast, among terrestrial species, systemic vasculitis is the predominant finding particularly in the pulmonary and nervous systems with virus being readily recovered from nasal secretions, urine and internal organs (Daniels *et al.*, 2001; Hooper *et al.*, 2001). Owing to their high pathogenicity and lack of therapeutics, NiV and HeV are classified as biosafety level (BSL) four agents. Given that very few laboratories have BSL-4 capabilities, data from experimental infections is limited.

Initial experimental HeV infection of guinea pigs indicated that while subcutaneous infection produced clinical signs consistent with human infections, the intradermal route did not lead to productive infection and intranasally infected animals seroconverted without clinical signs (Westbury *et al.*, 1995; Williamson *et al.*, 1998, 2001). Cats were also identified as a viable model for pathology and transmission studies (Westbury *et al.*, 1995, 1996). Results from experimental infection of horses indicated similar respiratory symptoms to those found in the equine deaths from the initial outbreak (Murray *et al.*, 1995a,b). Hamsters were infected with HeV to evaluate the efficacy of passive neutralizing antibody prophylaxis (Guillaume *et al.*, 2009).

NiV infection of golden hamsters by the intranasal or intraperitoneal routes caused a fatal neurological disease associated with severe pathological lesions in the hamster brain. Viral antigen and RNA were detected in endothelial cells, neurons, lungs, kidneys and the spleen, while virus and/or viral RNA could be recovered from most organs and urine, but not from serum (Georges-Courbot *et al.*, 2006; Guillaume *et al.*, 2006b). Cats, guinea pigs, and ferrets have also been tested as potential models for NiV infection (Weingartl *et al.*, 2005; McEachern *et al.*, 2008; Torres-Velez *et al.*, 2008; Pallister *et al.*, 2009). Most recently, squirrel monkeys and African green monkeys have been evaluated as non-human primate models of NiV infection, which are crucial for future testing of antiviral compounds and potential vaccines (Geisbert *et al.*, 2010; Marianneau *et al.*, 2010). While the clinical signs of the infected squirrel monkeys reflected those observed in human cases including neurological and respiratory involvement, vasculitis and brain abnormalities were less pronounced compared with human NiV infections (Marianneau *et al.*, 2010). In contrast, African green monkeys infected with NiV displayed severe respiratory pathology, neurological disease and generalized vasculitis, which accurately corresponded to observations of human NiV infections. Geisbert and colleagues strongly suggest that the African green monkey provides the first consistent and highly pathogenic model of NiV infection, which can serve as a valid platform for evaluating and licensing various therapeutic strategies for human use (Geisbert *et al.*, 2010).

Immune responses

In order to successfully clear a viral infection to which one has no pre-existing immunity, the innate antiviral immune response must be able to keep viral replication at levels that will not kill the host while the adaptive immune response is being primed to eventually clear the infection. There are two crucial aspects to mounting an effective innate antiviral response: (1) pathogen component recognition, and (2) the antiviral cellular response to the recognized pathogen. Various intracellular cytoplasmic and endosomal pattern recognition receptors (PRRs) such as Toll-like receptor 3 (TLR-3), TLR-7 and TLR-8, retinoic acid-inducible gene 1 (RIG-I) and melanoma differentiation-associated gene 5 (Mda-5) detect viral nucleic acids such as tri-phosphorylated RNA and single or double-stranded RNA (ssRNA, dsRNA), and in turn initiate the antiviral response (Kawai and Akira, 2008). A common and crucial aspect of the antiviral response induced by these PRRs is the production of interferons (IFNs). IFNs are soluble factors that mediate cellular antiviral responses and are classified into type I, type II, and type III IFNs. Most cells in the body produce type I IFN (IFN α/β) and type III IFN (IFN λ), while activated T lymphocytes and natural killer (NK) cells are the primary sources of type II IFN (IFNγ; Platanias, 2005; Uze and Monneron, 2007). The transcriptional activators NF-κB, ATF2-c-Jun, interferon regulatory factor 3 (IRF-3), and IRF-7 regulate the expression of type I IFNs, with IRF-3 and IRF-7 being the primary activators of type I IFN during viral infection. While RIG-I, Mda-5, and TLR-3 induce production of type I IFN (IFN-β) via IRF-3 activation (Fig. 6.3B), TLR-7 and TLR-8 induce type I (IFN-α) production via IRF-7 activation (Kawai and Akira, 2008).

When secreted type I IFNs engage the receptor subunits IFN-α receptors 1 and 2 (IFNAR 1/2) on the cell surface, Janus tyrosine kinases (JAKs, i.e. Jak1, Tyk2) associated with the cytoplasmic domain of these subunits are activated via cross-phosphorylation and phosphorylate specific tyrosine residues of the IFNARs. Signal transducer and activator of transcription (STAT) 1 and STAT2, which are transcription factors crucial to the IFN response, are then recruited to the phosphorylated IFNAR complex where they are phosphorylated by JAKs at specific tyrosine and serine residues. This phosphorylation step allows STAT1 and 2 to form heterodimers that enter the nucleus where it combines with another transcription factor, IRF-9, to form the transcriptional activator IFN-stimulated gene factor 3 (ISGF3). ISGF3 binds to promoters with IFN-stimulated response element (ISRE) sequences to induce transcription of IFN-stimulated genes (ISGs; Fig. 6.3A; Platanias, 2005). Some of these ISGs include IRF-7, 2′,5′-oligoadenylate synthetase (OAS), Rnase L, p56 and double-stranded RNA-induced protein kinase (PKR). These ISGs all contribute to the generation of an antiviral state in the cell (Platanias, 2005).

There is extensive evidence that the paramyxovirus C and V proteins can limit the induction of type I IFNs and also block IFN signalling (Lamb and Parks, 2007). Several studies utilizing eukaryotic plasmid expression systems have provided insight into possible mechanisms by which NiV C, V, and W proteins block the host antiviral response.

Initial studies of the *Henipavirus* V proteins indicated that they inhibit IFN signal transduction by sequestering STAT1 and STAT2 in high molecular weight complexes in the cytoplasm and by inhibiting STAT1 phosphorylation (Fig. 6.3A; Rodriguez *et al.*, 2002, 2003). By generating truncation mutants of NiV V, this same group identified a nuclear export sequence (NES) between amino acids 174 and 192 in the N-terminus of V that enabled it to shuttle steady-state STAT1 to the cytoplasm. Attachment of this NES to a nuclear export-deficient human immunodeficiency virus Rev protein restored its subcellular location to the cytoplasm. Through using the above-mentioned truncation mutants of NiV V, the authors were able to map the functional domains of V responsible for binding STAT1 (amino acids 100–160). Despite finding a STAT2 binding site on V (amino acids 230–271), it was shown that STAT2 binding to V required the presence of STAT1 (Rodriguez *et al.*, 2004).

Through using an IFN-sensitive NDV that contained a GFP reporter gene, Park and colleagues were able to rescue replication of NDV in IFN-treated chicken embryo fibroblasts (CEFs) when they transfected plasmids expressing NiV V, W, or C proteins. The V and W proteins rescued GFP expression in a robust manner, while the level of GFP rescued by the C protein was less pronounced. It was shown that both the V and W proteins inhibited IFN signalling by

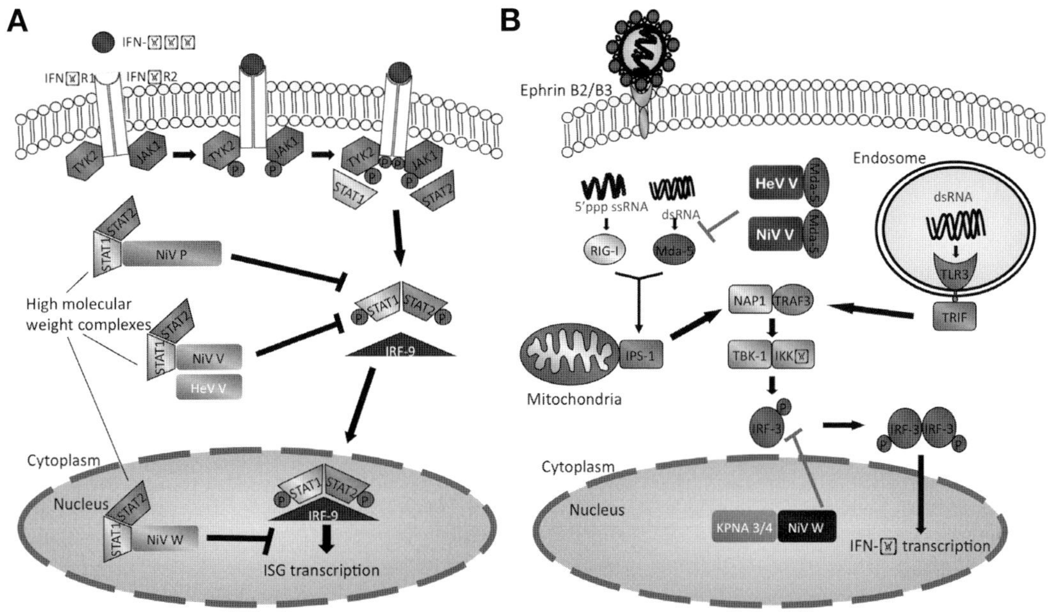

Figure 6.3 (A) When secreted type I interferons (IFNs) engage the receptor subunits IFN-α receptors 1 and 2 (IFNAR 1/2) on the surface of cells, Janus tyrosine kinases (JAKs; i.e. Jak1, Tyk2) associated with the cytoplasmic domain of these subunits become activated and phosphorylate specific tyrosine residues of IFNARs. Signal transducer and activator of transcription (STAT) 1 and STAT2, which are transcription factors crucial to the IFN response, are then recruited to the phosphorylated IFNAR complex, where they are phosphorylated by JAKs at specific tyrosine and serine residues. This phosphorylation step allows STAT1 and 2 to form heterodimers that enter the nucleus, where it combines with another transcription factor, IRF-9 to form the transcriptional activator IFN-stimulated gene factor 3 (ISGF3). ISGF3 binds to promoters with IFN-stimulated response element (ISRE) sequences to induce transcription of IFN-stimulated genes (ISGs). Some of these ISGs include IRF-7, 2′5′ oligoadenylate synthetase (OAS), Rnase L, p56 and double-stranded RNA-induced protein kinase (PKR). These ISGs all contribute to the generation of an 'antiviral state' in the cell. NiV P, V, W and HeV V proteins have individually been shown to bind STAT1 and STAT2, effectively preventing STAT1 phosphorylation in type I IFN-stimulated cells (see text). (B) The IFN induction by PRRs. TLR-3 is expressed in a variety of cells, from fibroblasts to endothelial cells to conventional dendritic cells. When TLR-3 recognizes dsRNA, a molecule called TIR domain-containing adaptor inducing IFN-β (TRIF) associates with a kinase complex of non-canonical IκB kinases (IKKs) TBK1 and IKKε that phosphorylates IRF-3. Phosphorylated IRF-3 translocates into the nucleus to activate transcription of IFN-β. RIG-I and Mda5 are two TLR-independent cytoplasmic PRRs involved in the induction of type I IFN. They are both DexD/H box RNA helicases with caspase recruitment domain (CARD)-like domains. Their helicase domains are responsible for interaction with 5′ triphosphorylated ssRNA and dsRNA, respectively; while the CARD-like domains are probably responsible for downstream signalling that eventually activates IRF3 and NFκB. These PRRs signal through a mitochondrial adaptor molecule called interferon promoter stimulator-1 (IPS-1), which in turn leads to activation of TBK1 and IKKε, resulting in IRF-3 activation. Plasmid expression of the NiV V and W proteins has been demonstrated to block IRF-3 activation (see text).

an IFN-inducible promoter reporter gene assay (Park *et al.*, 2003). This group eventually elucidated the cellular localizations of V and W by immunofluorescence, showing that V localized to the cytoplasm while W was distributed exclusively to the nucleus. In addition, the STAT1 binding domain was mapped to the 50–150 amino acid region of the N-terminus shared by the P, V, and W proteins, which corresponded with the results obtained for NiV V by Rodriguez and colleagues (Rodriguez *et al.*, 2004; Shaw *et al.*, 2004). Interestingly, in spite of the shared N-terminus, it was shown that P, V, and W had differential abilities to inhibit induction of ISRE promoters, with W having the most inhibitive capacity, then V, and then P having the least capacity for inhibition (Shaw *et al.*, 2004).

Shaw and colleagues delineated a nuclear localization signal (NLS) in the unique C-terminus of the NiV W protein, which required basic residues at amino acid positions 439, 440and 442 (Shaw *et al.*, 2005). The NLS was shown to interact with the nuclear importins karyopherin-α3 and karyopherin-α4. While NiV V and W could inhibit IRF-3 responsive promoter activation both by Sendai virus infection and cytoplasmic dsRNA, only W could inhibit IRF-3 activation via dsRNA stimulation of TLR-3 (Fig. 6.3B). An interesting observation in this study was that the addition of a non-viral NLS to the V protein resulted in the same inhibitory effect seen with W. This study also demonstrated that plasmid expression of the W protein in cells reduced levels of phosphorylated IRF-3, indicating a distinct mechanism of IFN evasion unique to W. There was a demonstrable difference in ability of the V and W proteins to inhibit the IKK-like kinases IKKε and TBK-1. Both V and W inhibited IKK –mediated activation of ISG transcription, while only W was able to inhibit TBK-1-mediated activation of ISG transcription. The authors concluded that because W localizes to the nucleus, it could inhibit both pathways leading to IRF-3 activation, the TLR-3 pathway and the virus/dsRNA pathway (Shaw *et al.*, 2005).

Through a serendipitous point mutation in the NiV V protein generated by PCR, Hagmaier and colleagues were able to show that a single amino acid change in residue 125 abolished STAT1 binding ability as well as the ability to block IFN signalling. It was also shown in this study that NiV V blocked IFN signalling in cell lines from numerous mammalian species indicating that this ability does not constrain the virus from crossing species (Hagmaier *et al.*, 2006). More recently, it has been shown that plasmid expressed NiV and HeV V proteins, along with many other paramyxovirus V proteins, bind the helicase Mda-5 via its cysteine-rich C-terminal domain and blocks activation of the IFN-β promoter by preventing the oligomerization of Mda-5 upon binding to dsRNA (Childs *et al.*, 2007, 2009; Fig. 6.3B). A novel interaction between NiV and HeV V proteins and a host cell kinase Polo-like kinase 1 (PLK1) has been demonstrated (Ludlow *et al.*, 2008). Intriguingly, the binding site for PLK1 overlaps with the STAT1 binding region. By constructing point mutants in the shared binding region, Ludlow and colleagues were able to abrogate STAT1 binding independently of PLK1 binding to NiV V protein. It was shown that ablation of the binding sites to either protein on the NiV P protein did not affect NiV minigenome replication, which indicated the possibility of attenuating NiV by disrupting host protein interactions via reverse genetics. The plasmid expression studies reviewed above have indeed shed significant light on the virus–host interactions of NiV as well as the potential mechanisms of interferon antagonism by NiV V, W, and C proteins.

It was not until the availability of a full-length NiV reverse genetic systems that detailed mutagenesis of the P gene products could be performed and tested in the context of viral infection (Yoneda *et al.*, 2006; Ciancanelli *et al.*, 2009). One study using reverse genetics determined that a NiV mutant virus with a non-functional C protein ORF drastically decreased its growth kinetics, both in IFN-β producing (HEK 293T) and non-IFN-β producing (Vero E6) cell lines. By using a STAT1–GFP fusion protein, the study also showed that wild-type NiV infection sequestered STAT1 in the nucleus of Vero cells and that a STAT1 binding mutant NiV (G121E) could not sequester STAT1 in the same manner. Immunofluorescence assays of NiV infected cells in this study demonstrated that the phosphorylated form of STAT1 did not localize to the nucleus despite IFN-β treatment (Ciancanelli *et al.*, 2009).

Most recently, an independent group characterized the infection of three distinct human primary endothelial cells and one neuronal cell line with NiV and measured their respective abilities to mount antiviral and inflammatory responses (Lo *et al.*, 2010). While the infected neuronal cell line failed to mount any detectable responses, the three primary endothelial cell types all produced functional IFN-β, which approximately corresponded inversely to the maximum viral titre generated in each cell type. Interestingly, the authors showed that the subcellular localization of the W protein was decidedly cytoplasmic in the three infected endothelial cell types while being exclusively nuclear in the neuronal cell line (Lo *et al.*, 2010). The authors proposed that this discrepancy in the localization of W between the endothelial cells and neuronal cells may explain the ability of the endothelial cells to mount an antiviral response. Two of the infected endothelial cell types (lung and lymphatic lung) also produced significantly increased levels of functional inflammatory chemokines in the cell supernatants including IL-8 (CXCL8), MCP-1 (CCL2) and IP-10 (CXCL10). The inflammatory mediators present in the supernatants from infected cells were shown to induce chemotaxis both in monocyte and T cell lines. The results of this study highlighted the importance of utilizing relevant cell types to measure immune responses against NiV and provided initial insights into the molecular mechanisms of NiV pathogenesis. Further delineation of the antiviral and inflammatory immune responses should involve deciphering which cellular PRRs are responsible for the observed responses in endothelial cells as well as the nature of other potential molecular virus–host interactions crucial to determining the fatal outcomes of infection. There has only been one study in which measuring the immune responses of NiV infected felines was a peripheral part. In this study, the authors mention that in only one of eight infected cats were there any detectable levels of cytokines (IL-2, IFN-γ, IL-10 and IL-4; McEachern *et al.*, 2008). The relative paucity of data in regards to immune responses at both the cellular and live animal model levels warrants future studies that will comprehensively define critical elements of the anti-*Henipavirus* immune response.

Clinical features

Owing to the small number of cases, HeV infection is not well characterized. Initial infection appears to occur by contact with infected horse secretions or blood via the respiratory route or through abrasions on the skin. The incubation period for HeV is estimated to be between 5 and 8 days (Murray *et al.*, 1995b; Selvey *et al.*, 1995; Playford *et al.*, 2010). Most HeV infections present initially as influenza-like illness (ILI) including fever, myalgia, and headache prior to the progression of more serious disease (Eaton *et al.*, 2007). The first case of HeV presented with ILI followed by rapid deterioration of respiratory symptoms requiring mechanical ventilation. Chest radiographs showed diffuse alveolar shadowing. The patient developed respiratory failure, renal failure and arterial thrombosis. Blood tests indicated thrombocytopenia, increased creatine phosphokinase, lactic dehydrogenase, aspartate aminotransferase, and glutamyltransferase (Murray *et al.*, 1995b; Selvey *et al.*, 1995; O'Sullivan *et al.*, 1997). The patient died on day 7 following cardiac arrest (Eaton *et al.*, 2007). Autopsy showed severe interstitial pneumonia with congestion, haemorrhage and oedema in both lungs (Selvey *et al.*, 1995). Virus was isolated by cell culture from the lung, liver, kidney and spleen demonstrating multiple organ targets following viraemia (Murray *et al.*, 1995b). Several cases of HeV infection have presented with mainly neurological symptoms in infected individuals including a fatal relapse encephalitis case in 1995 and the more recent outbreak at the Redlands veterinary clinic (Eaton *et al.*, 2007; Playford *et al.*, 2010). The relapse encephalitis case in a horse owner in Mackay initially presented with symptoms of meningitis including headache, drowsiness, vomiting, and neck stiffness. Thirteen months after his initial recovery he presented with 2 weeks of lower back pain and irritable mood which progressed to seizure. The patient's neurological symptoms continued to progress as magnetic resonance imaging (MRI) showed multifocal cortical lesions. By day 7 brain stem involvement was manifested in reduced consciousness and led to the requirement for mechanical ventilation. The patient died on day 25 (O'Sullivan *et al.*, 1997; Eaton *et al.*, 2007). The Redlands veterinarian presented with a history of ILI for 2 days, but by day 5 had developed neurological symptoms including mild confusion, ataxia and bilateral ptosis. HeV was detected in the cerebrospinal fluid (CSF) at this time and MRI showed multifocal cortical and subcortical hyperintense lesions and evidence of infarction. Over the next several days the neurological symptoms progressed to tonic-clonic seizure and the patient remained unconscious for the next 4 weeks and died on day 40 (Playford *et al.*, 2010). The second case in this outbreak also presented with several days of ILI with neurological symptoms developing on day 12 including mild confusion, dysarthria, ataxia and bilateral ptosis. Unlike the first case, however, the MRI at this time was normal and HeV could not be detected in the CSF. Symptoms progressed over the next 12 days as the MRI showed small scattered foci. The patient eventually recovered and returned home on day 37 of illness with mild ataxia (Playford *et al.*, 2010).

Most individuals infected with NiV were symptomatic within 2 weeks of exposure in the Malaysian outbreak while the Bangladesh cases had a shorter average incubation period of 9 days. Patients deteriorated rapidly after onset of symptoms, requiring hospitalization (Eaton *et al.*, 2007; Hossain *et al.*, 2008). NiV infects endothelial cells causing multiorgan vasculitis with pathology seen primarily in the CNS (Goh *et al.*, 2000; Wong *et al.*, 2002). Virus was more likely to be isolated from the CNS of fatal NiV cases (Chua *et al.*, 2000b). NiV causes acute febrile encephalitis with the most frequent symptoms being fever, headache, drowsiness, dizziness, myalgia and vomiting. During the Malaysian outbreak, fever occurred for a mean duration of 3.5 days prior to hospital admission, 10% of patients had neurological signs at presentation and 20% presented with reduced consciousness (Goh *et al.*, 2000). Patients presenting with reduced consciousness had indicators of brain stem involvement (i.e. drowsiness, areflexia, segmental myoclonus, hypertension and tachycardia; Goh *et al.*, 2000; Tan and Chua, 2008). Reduced consciousness and evidence of brain stem involvement was a poor prognostic factor. Goh and colleagues found that among the 55% of patients who experienced reduced levels of consciousness, only 15% of those patients recovered as compared with recovery in all patients with normal consciousness. The mean duration of illness from onset to death seen during the Malaysian outbreak was 10.3 days. MRI revealed multiple small lesions (2–7 mm in diameter) in the subcortical and deep white matter as the probable sites of microinfarction caused by vasculitis (Eaton *et al.*, 2007; Goh *et al.*, 2000). Laboratory tests of NiV infected individuals demonstrated thrombocytopenia, leukopenia, elevated alanine aminotransferase and aspartate aminotransferase. The CSF had increased white blood cell counts and/or protein levels (Goh *et al.*, 2000). The percentage of NiV patients presenting with respiratory disease was higher in Bangladesh (69% respiratory difficulty) than in Malaysia (25% respiratory symptoms; Luby *et al.*, 2009a; Tee *et al.*, 2009).

As seen in one HeV case, some NiV patients experienced relapse of disease or late onset encephalitis after initial infection, which occurred on average approximately 8 months after initial infection (range 9 days–22 months). Relapse and late onset encephalitis appear to have similar clinical manifestations with MRI revealing significant confluent lesions of the grey matter (Goh *et al.*, 2000; Tan *et al.*, 2002; Tyler, 2009). In addition, viral antigen was found in the brains of relapse and late onset encephalitis patients during necropsy indicating viral replication in the brains of these individuals. Unlike acute NiV encephalitis cases, relapse and late onset encephalitis cases did not show vasculitis in the CNS (Tan *et al.*, 2002; Tan and Chua, 2008; Tyler, 2009). A number of NiV infected individuals also experienced residual neurological symptoms that ranged from mild cognitive or cerebellar disabilities to more severe cognitive impairment with some remaining in a vegetative state (Goh *et al.*, 2000).

Epidemiology

HeV

The first identified outbreak of HeV, which occurred at a racehorse stable in Hendra, Queensland, in 1994 remains the largest recorded equine outbreak of HeV. The initial case in this outbreak was a pregnant mare at the stable. Within the following two weeks, 20 additional thoroughbred horses became sick with severe respiratory disease, 14 of which died. A horse trainer and stable hand that had close contact with the sick mare also developed ILI with the horse trainer succumbing to infection 6 days following onset of symptoms (Murray *et al.*, 1995b). The second human death caused by HeV occurred in a horse owner who became infected through assisting in the autopsies of two sick horses with severe respiratory illness and neurological symptoms. These horses were retrospectively diagnosed with HeV, as their infections occurred a month before the first identified outbreak (Hooper *et al.*, 1996). The man developed meningitis shortly after close contact with the sick animals and initially recovered, but succumbed 13 months later to fatal encephalitis due to suspected reactivation of latent HeV infection (Wong *et al.*, 2009).

The absence of an epidemiological link between the first two outbreaks of HeV prompted the testing of numerous wildlife species found around the stables of the two outbreak locations (Rogers *et al.*, 1996). There are four *Pteropus* species bats found in Australia, *P. alecto, P. conspicillatus, P. scapulatus* and *P. poliocephalus.* Neutralizing antibodies have been detected in all four species (Young *et al.*, 1996). The virus was isolated from *P. poliocephalus* and *P. alecto,* whose habitats are located in the eastern and southern regions of Australia or northern Australia, respectively was identical to horse and human isolates from outbreak cases (Young *et al.*, 1996; Halpin *et al.*, 2000). Horses are the only animals known to transmit HeV infection to humans. It is unclear how horses become infected from the natural reservoir hosts, but it has been postulated that they may come into contact with contaminated droppings while in pasture (Field *et al.*, 2010). The transmission from infected horses to humans is inefficient, requiring very close contact, with urine or saliva from the infected horse probably playing a role in transmission to humans. There has been no documented human-to-human transmission of HeV to date (Murray *et al.*, 1995b; McCormack *et al.*, 1999).

Since the initial identification of HeV there has been 14 subsequent outbreaks of HeV involving a total of 44 equine cases, seven human infections and four deaths, resulting in a mortality rate of 57% (ProMed-mail, 2010a; WHO, 2009a). HeV outbreaks have primarily occurred in Queensland, with the exception of one equine case in New South Wales (Fig. 6.4). All human cases of HeV have involved close contact with infected horses. In 2008, an outbreak of HeV at Redlands veterinary clinic in Queensland resulted in five equine cases and two human cases of encephalitis resulting in the death of a clinic veterinarian. The horses in this outbreak presented primarily with neurological symptoms rather than respiratory symptoms as seen in previous outbreaks. HeV isolated from this outbreak had sequence differences from a simultaneous outbreak in Proserpine, Queensland as well as from previously reported outbreaks. However, it is unclear if these differences contributed to the presentation of mainly neurological symptoms in the infected horses (Field *et al.*, 2010; Playford *et al.*, 2010). A roost of *Pteropus* bats was identified in the vicinity of the clinic and it is probable that the first equine case occurred from contact with secretions from a bat feeding in the area (Field *et al.*, 2010; Playford *et al.*, 2010). The subsequent equine cases probably occurred from either contact with the primary case in the clinic prior to onset of symptoms or with contaminated environmental surfaces as swabs of stalls housing infected horses tested PCR positive for HeV (Field *et al.*, 2010). The human cases of HeV infection during this outbreak probably contracted the disease through high-risk exposure activities including performing daily nasal cavity lavage of an infected horse 3 days before symptom onset and participating in a necroscopy of an infected horse (Playford *et al.*, 2010). Again noted in this outbreak was the necessity for close contact with nasal secretions or blood from the infected horse. This outbreak also demonstrated the possibility of transmission of HeV prior to the onset of symptoms (Field *et al.*, 2010; Playford *et al.*, 2010). The fourth death due to HeV occurred in 2009 at a horse stud near Rockhampton. This outbreak resulted in four equine cases and the death of a veterinarian who treated an ill horse subsequently diagnosed with HeV without wearing personal protective equipment (PPE; ProMed-mail, 2009). The most recent equine death due to HeV occurred in May of 2010 in Tewantin, Queensland. The horse was put down after confirmation of HeV infection. As of yet, there have been no additional equine nor human cases reported on the property in Tewantin and those in close contact with the horse were offered experimental prophylactic treatment (ProMed-mail, 2010a).

NiV

The first outbreak of NiV in Malaysia lasted nine months and began in the Kinta district of Ipoh, Perak state, as a respiratory illness and encephalitis in pigs. During this outbreak, NiV demonstrated high transmissibility among pigs

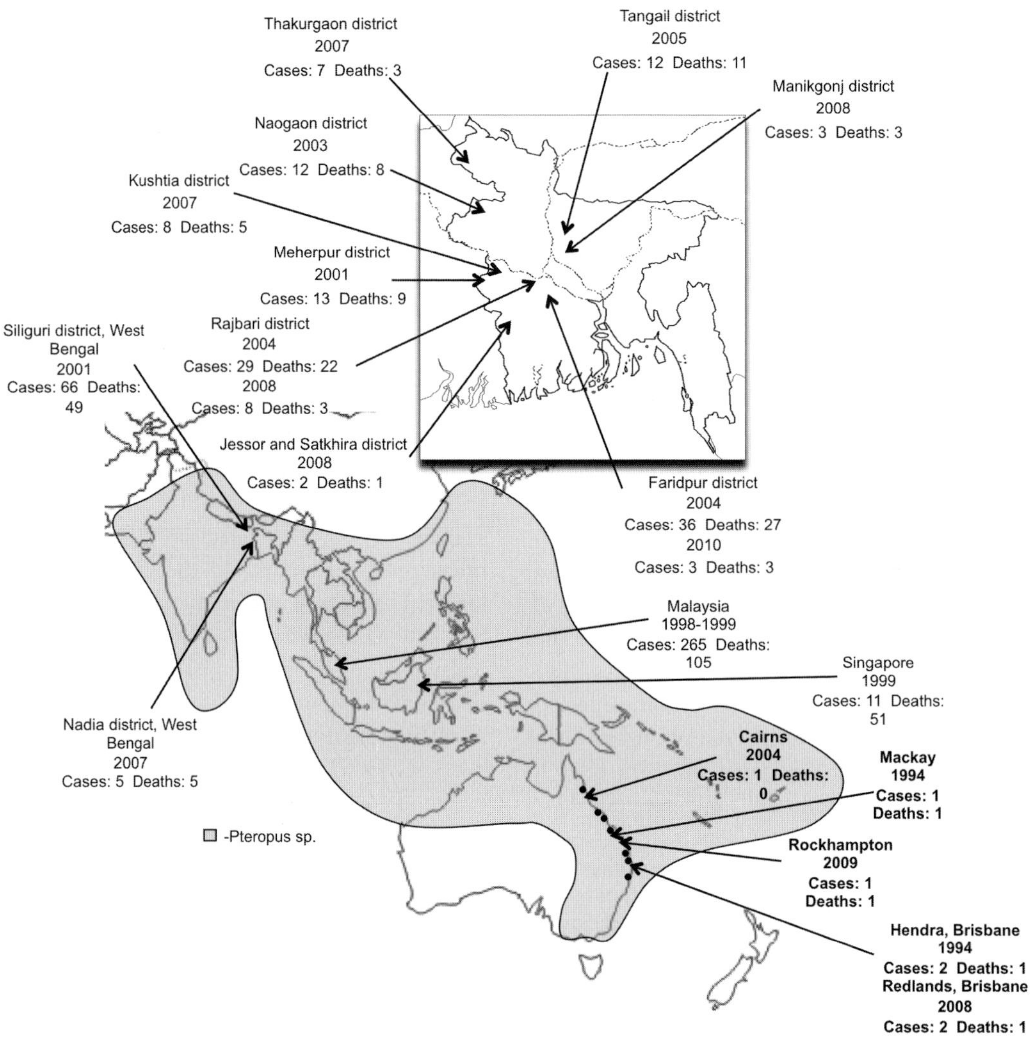

Figure 6.4 Distribution and cases of Hendra virus (HeV) and NiV infection. The shaded area depicts *Pteropus* sp. distribution in regions of reported outbreaks. Reported human cases of HeV are in bold type. The dots depict equine outbreaks with no reported human cases. Reported human cases of NiV are in standard type, inset cases occurring in Bangladesh.

via the respiratory route and subsequently pig-to-human transmission causing acute febrile encephalitis. The cases were mainly among adult males who worked in the rapidly growing pig industry in Malaysia (Eaton *et al.*, 2007; Lo and Rota, 2008; Luby *et al.*, 2009a; Tee *et al.*, 2009). The outbreak spread to several states within Malaysia as well as Singapore via the transport of infected pigs (Chua *et al.*, 2000a; Eaton *et al.*, 2007). By the summer of 1999 there were a total of 276 cases reported and 106 deaths, resulting in a 38.5% mortality rate (Eaton *et al.*, 2007; Lo and Rota, 2008; Tee *et al.*, 2009). The distribution of PPE to those in close contact with sick pigs, the prohibition of livestock transportation, and the culling of more than 1 million pigs all contributed to curtailing the outbreak (Chua *et al.*, 2000a; Luby *et al.*, 2009a).

Following the identification of NiV in Malaysia there have been repeated outbreaks reported in

Bangladesh and India from 2001–2010 (Chadha *et al.*, 2006; ProMed-mail, 2010b; WHO, 2009b; Fig. 6.4). The outbreaks in Bangladesh were the result of multiple independent introductions of NiV into the human population, as demonstrated by the greater sequence diversity of NiV isolates from Bangladesh as compared with the only Malaysian outbreak (Luby *et al.*, 2009b). There were a number of notable differences between the outbreak in Malaysia and of those reported in Bangladesh and India. The mortality rate seen in the outbreaks in Bangladesh and India range from 38%, similar to that seen in Malaysia, to as high as 100%, with an average mortality rate of 73% (Anonymous, 2010). Pigs were not the intermediate host resulting in transmission to humans in Bangladesh or India; rather index cases appeared to develop symptoms following direct contact with *Pteropus* bats. Finally, human-to-human transmission has been documented in Bangladesh and India (Chadha *et al.*, 2006; Eaton *et al.*, 2007; Anonymous, 2010).

Although there has been no evidence of NiV found in other animals or rodents in endemic areas, anti-NiV neutralizing antibodies have been detected in other bat species in Malaysia albeit within a smaller proportion of the population than seen in *Pteropus* bats (Yob *et al.*, 2001). Evidence of NiV infection has been demonstrated in fruit bats in several countries in South East Asia as well as in Africa (Anonymous, 2010; Hayman *et al.*, 2008). Initial studies determined that approximately 25% (range 17–31%) of *P. vampyrus* and *P. hypomelanus* bats found in Malaysia had neutralizing antibodies against NiV (Yob *et al.*, 2001; Luby *et al.*, 2009b). NiV has been isolated from urine and saliva specimens from *Pteropus* bats in Malaysia, Cambodia and Thailand (Luby *et al.*, 2009a b). There is a single species, *P. giganteus*, found in Bangladesh, and these bats also carry neutralizing antibodies to NiV (Luby *et al.*, 2009b). In Madagascar seropositive *Pteropus rufus* and *Eidolon dupreahum* bats have been found; additionally in Ghana *Eidolon helvum* have a 39% seroprevalence for NiV reactive antibodies (Hayman *et al.*, 2008; Iehle *et al.*, 2007). The determination of seropositive bats in Africa suggests that the range of potential NiV infections is wider than previously thought.

Three routes of transmission of NiV from bats into humans have been determined through epidemiological investigations. The first identified route of transmission was via the consumption of contaminated date palm sap (Luby *et al.*, 2009b). A study by Homaira and colleagues focused on an outbreak in the Thakurgaon district of Bangladesh in 2007 that resulted in seven NiV infections and three deaths. The study found that NiV cases were more likely to consume raw date palm sap than uninfected persons (Homaira *et al.*, 2010). The most recent deaths due to NiV encephalitis occurred in January of 2010 involving a woman and two children in Faridpur, Bangladesh with the probable route of transmission being the consumption of raw date palm juice. The second mode of transmission identified was the consumption of contaminated dropped fruit by domestic animals which subsequently infected humans or consumption of contaminated fruit directly by individuals. The third mode of transmission was direct contact with NiV infected bat secretions (Luby *et al.*, 2009a). The role of direct contact with NiV-infected bat secretions was exemplified by an outbreak in Rajbari, Bangladesh in 2004, in which there were 29 cases of NiV infection and 22 deaths, mainly among younger boys. Epidemiological investigations of the outbreak determined that tree climbing probably resulted in case patients coming into direct contact with infected bats (Tan and Chua, 2008). In Bangladesh and India, person-to-person spread has been documented. Infected individuals termed 'Nipah spreaders' in one study of human-to-human transmission occurring in Bangladesh found that these individuals transmitted NiV to an average of seven people (Luby *et al.*, 2009b). Infected individuals in Bangladesh were more likely to have respiratory symptoms and close contact while caring for sick individuals appeared to contribute to person-to-person spread (Blum *et al.*, 2009; Luby *et al.*, 2009b).

Diagnoses

Henipaviruses are internationally classified as biosafety level 4 (BSL-4) agents, thus clinical specimens must be handled with caution. Propagation of viruses from clinical specimens known to be infected with henipaviruses is not recommended without appropriate containment facilities. The Centers for Disease Control and Prevention (CDC), Atlanta, GA, USA, and the Australian Animal Health Laboratory (AAHL), Geelong, Australia, have adopted the approach that primary virus isolation from specimens of outbreaks not already proven to be either HeV or NiV takes place at BSL-3. However, if the results of cell culture suggest the presence of these agents, cultures should be transferred to BSL-4 to conform to biosafety guidelines (Daniels *et al.*, 2001).

A NiV-specific ELISA was eventually transferred to surveillance labs in Malaysia (Daniels *et al.*, 2001). For cases in which the ELISAs gave equivocal results, negative-stained CSF specimens were subjected to transmission electron microscopy to visually confirm the presence of NiV (Chow *et al.*, 2000). Immunohistochemistry was crucial in detecting NiV antigen in *ex vivo* tissues of humans, dogs and pigs (CDC, 1999a,b; Chua *et al.*, 2000a). Complete molecular characterization of both Malaysia and Bangladesh strains of NiV has enabled the development of both standard and real-time RT-PCR assays to detect viral RNA from serum, urine and CSF (AbuBakar *et al.*, 2004; Guillaume *et al.*, 2004b; Wacharapluesadee and Hemachudha, 2007). ELISAs currently in use utilize recombinant-expressed purified NiV and HeV antigens, and a number of high-affinity monoclonal antibodies have been generated for diagnostic and prophylactic purposes (Eshaghi *et al.*, 2004; Kashiwazaki *et al.*, 2004; Tan *et al.*, 2004; Tanimura *et al.*, 2004a,b; Eshaghi *et al.*, 2005a,b; Chen *et al.*, 2006; Yu *et al.*, 2006; Zhu *et al.*, 2006; Juozapaitis *et al.*, 2007; Zhu *et al.*, 2008). While serum neutralization tests (SNTs) have long been a reference standard, the next generation of SNTs will circumvent the use of live virus, and be able to differentiate between NiV and HeV specific infection (Bossart *et al.*, 2007; Chen *et al.*, 2007; Kaku *et al.*, 2009; Tamin *et al.*, 2009).

Prevention and control

The primary means of preventing human infections are the early recognition of animal disease and the use of precautions to avoid exposure. Since it is difficult to interrupt the transmission of viruses from the natural reservoir (fruit bats) to horses or pigs, early identification of infected animals and use of appropriate personal protective measures to prevent transmission are keys to reducing the risk to humans. While there are currently no vaccines, passive immunoprophylaxes, or antiviral chemoprophylaxes approved for human *Henipavirus* infections, numerous studies have significantly contributed to their respective future development as effective therapies.

The *Henipavirus* F and G proteins share high levels of amino acid sequence identity, with 83.3% identity between the G proteins, and with 88.1% identity between the F proteins (Harcourt *et al.*, 2000; Wang *et al.*, 2001). Recombinant expressed, soluble versions of the G glycoprotein (sG) from NiV (Bossart *et al.*, 2005) that were used to vaccinate cats produced high antibody titres and provided complete protection from NiV challenge (Mungall *et al.*, 2006). Purified sG retains a number of important native structural, functional and antigenic features, making it potentially suitable as a vaccine candidate, although consistent large scale expression has been problematic. Concurrent expression of NiV F, G and M proteins lead to production VLPs (Ciancanelli and Basler, 2006; Patch *et al.*, 2007). Since VLPs have been shown to generate a protective immune responses against Ebola and Marburg viruses (Swenson *et al.*, 2005), they potentially serve as a more efficient method of immunization against NiV. DNA vaccination with plasmids containing either NiV G or F genes stimulated considerable IgG responses in mice (Wang *et al.*, 2006), while canarypox virus-based vaccine vectors expressing NiV G or F (or both) protected pigs against challenge and prevented viral shedding (Weingartl *et al.*, 2006). Vaccination with these canarypox virus-based vaccines appeared to stimulate both type 1 and type 2 cytokine responses, suggesting this approach may be highly effective for the prevention of livestock infection. A similar approach using vaccinia virus vectors expressing either NiV F or G were shown to be effective in generating protective antibodies against NiV challenge in hamsters (Guillaume *et al.*, 2004a).

Several independent groups have developed neutralizing monoclonal antibodies (Mabs) against the *Henipavirus* F and G glycoproteins. Guillaume and colleagues were able to demonstrate significant levels of protection in hamsters administered with Mabs against NiV G and NiV F up to 24 h post infection. Interestingly, while the NiV G Mab provided a higher level of overall protection than the NiV F Mab, the NiV F Mab could neutralize virus while the NiV G Mab could not (Guillaume *et al.*, 2006b). The same group showed that when administered within 24 h of infection, these Mabs provided protection of hamsters against HeV infection (Guillaume *et al.*, 2009). By using the purified HeV sG protein as an antigen to screen a large naïve human phage display library of Fabs, an independent group developed human Mabs, which potently blocked *Henipavirus* F and G protein-mediated fusion as well as neutralized live virus (Zhu *et al.*, 2006). Affinity maturation by light-chain shuffling along with random mutagenesis of the heavy-chain variable domain of one particular Mab labelled 'm102' resulted in higher affinity Fab fragments. One resultant clone 'm102.4' was found to have exceptional ability to bind and neutralize both HeV and NiV (Zhu *et al.*, 2008). By treating ferrets with m102.4 10 h after being infected with NiV, Bossart and colleagues demonstrated the robust capacity of m102.4 to protect the ferrets from virus challenge (Bossart *et al.*, 2009).

Homotypic and heterotypic heptad repeat (HR) peptides designed against the fusion peptides of *Henipavirus* F proteins have been shown to inhibit viral entry and replication by disrupting the HR1–HR2 interaction, which forms the 6-helix coiled-coil bundle (Bossart *et al.*, 2002; Xu *et al.*, 2004). Intriguingly, it was shown that a HR peptide corresponding to the hPIV3 F protein carboxy-terminal (C) peptide preceding the transmembrane domain inhibited *Henipavirus* replication better than HR peptides corresponding to *Henipavirus* F proteins (Porotto *et al.*, 2006). Further investigation into the mechanism of the greater observed inhibition of the hPIV3 HRC peptide determined a correlation between the increased ability of the HR peptide to interact with the HeV F N peptide coiled-coil trimer. These results suggested that conserved interhelical packing interactions in the F protein fusion core are critical determinants of the hPIV3 HRC peptide inhibitory activity (Porotto *et al.*, 2007). Following this study, the authors showed that incorporating cholesterol molecules with the HRC peptide increase antiviral efficacy by ~2 logs. The authors suggest that the cholesterol molecules not only shield the HRC peptides from potential proteases, but target the peptides to their specific site of action at the plasma membrane (Porotto *et al.*, 2010).

In the Malaysian outbreak, ribavirin treatment was shown to reduce mortality rate of NiV infection (Chong *et al.*, 2001). The interferon inducer poly(I)–poly(C_{12}U) prevented mortality in five of six animals in a hamster model of NiV infection, while a 5-ethynyl analogue of ribavirin and several other OMP-decarboxylase inhibitors were shown to have anti-NiV activity *in vitro* (Georges-Courbot *et al.*, 2006). Several platforms have since been developed to screen for potential antiviral compounds *in vitro*. The NiV minigenome replication system has been utilized to evaluate the antiviral activity of siRNA directed against different NiV genes, as well as libraries of compounds. There have been no further studies of the potential lead compounds in animal models (Freiberg *et al.*, 2008; Mungall *et al.*, 2008). Variations of pseudotyped VLPs containing combinations of *Henipavirus* F, G, and/or M proteins have been used to screen for viral entry inhibitors (Kaku *et al.*, 2009; Tamin *et al.*, 2009; Wolf *et al.*, 2009). Aljofan and colleagues developed a high-throughput immunolabelling assay measuring viral antigen production, and have tested the suitability of several cell types for this assay (Aljofan *et al.*, 2008, 2009a). Utilizing this assay in a study to screen over 8000 compounds led to the identification of a number of lead compounds. Of the lead compounds identified, the study describes the antiviral action of three commercially available compounds as proof-of-principle for future screening of compound libraries (Aljofan *et al.*, 2009b). When this assay was used in concert with a multicycle assay developed to mimic certain stages of *Henipavirus* replication, Porotto and colleagues identified the anti-malarial drug chloroquine as a potent antiviral against henipaviruses *in vitro*. Unfortunately, chloroquine proved to be ineffective against *Henipavirus* infection in two independent studies of infected hamsters and ferrets (Freiberg *et al.*, 2009; Pallister *et al.*, 2009). In a recent collaborative study involving several groups, Wolf and colleagues identified a broad-spectrum rhodanine derivative molecule labelled 'LJ001', which not only inhibited viral entry by henipaviruses, but also by numerous other enveloped viruses including filoviruses, arenaviruses, flaviviruses, poxviruses, HIV and influenza A viruses *in vitro* (Wolf *et al.*, 2010). Although initial testing of this compound against lethal challenge of Ebola virus in mice yielded disappointing results, the outcome was explained by low initial serum concentrations of the compound and the short half life of the compound in the animal (4 h). Further development of the pharmacokinetic properties of LJ001 paired with improved dosage regimens will be crucial to comprehensively evaluate the antiviral potential of this compound against many viruses including HeV and NiV.

Perspectives

Since their emergence in Malaysia and Australia, henipaviruses have been a recurring threat to human health in South East Asia and the South Pacific region. The higher CFRs from the outbreaks in Bangladesh and India reinforce the need to improve preventative measures against NiV infection wherever possible. Expanding the surveillance and laboratory capacity for rapid diagnosis of encephalitis outbreaks is crucial to early detection and containment in at risk areas. Additional studies are needed to determine the ecological, environmental and/or epidemiological circumstances that favour transmission of NiV from its reservoir host to humans. With the advent of high throughput screening assays and a viable non-human primate animal model for human *Henipavirus* infections, it should only be a matter of time before effective antivirals are developed for human use. Knowledge regarding the molecular mechanisms of pathogenesis is still in its infancy, which warrants additional studies to decipher cellular pathways crucial to generating disease. The increasing number of studies utilizing *Henipavirus* reverse genetic systems ensures that the individual roles of viral proteins in pathogenesis will be delineated. Future studies of *Henipavirus* molecular biology and immune responses will inevitably expand from the human side to the natural host species, as Pteropid bat cell lines have recently been developed (Crameri *et al.*, 2009). Contact between the natural reservoir of henipaviruses and humans either directly or via an intermediate host is likely to continue, therefore warranting the continued study and development of preventative or post-exposure therapeutics.

References

AbuBakar, S., Chang, L.Y., Ali, A.R., Sharifah, S.H., Yusoff, K., and Zamrod, Z. (2004). Isolation and molecular identification of Nipah virus from pigs. Emerg. Infect. Dis. *10*, 2228–2230.

Aguilar, H.C., Matreyek, K.A., Filone, C.M., Hashimi, S.T., Levroney, E.L., Negrete, O.A., Bertolotti-Ciarlet, A., Choi, D.Y., McHardy, I., Fulcher, J.A., *et al.* (2006). N-glycans on Nipah virus fusion protein protect against neutralization but reduce membrane fusion and viral entry. J. Virol. *80*, 4878–4889.

Aguilar, H.C., Matreyek, K.A., Choi, D.Y., Filone, C.M., Young, S., and Lee, B. (2007). Polybasic KKR motif in the cytoplasmic tail of Nipah virus fusion protein modulates membrane fusion by inside-out signaling. J. Virol. *81*, 4520–4532.

Aguilar, H.C., Ataman, Z.A., Aspericueta, V., Fang, A.Q., Stroud, M., Negrete, O.A., Kammerer, R.A., and Lee, B. (2009). A novel receptor-induced activation site in the Nipah virus attachment glycoprotein (G) involved in triggering the fusion glycoprotein (F). J. Biol. Chem. *284*, 1628–1635.

Aljofan, M., Porotto, M., Moscona, A., and Mungall, B.A. (2008). Development and validation of a chemiluminescent immunodetection assay amenable to high throughput screening of antiviral drugs for Nipah and Hendra virus. J. Virol. Methods *149*, 12–19.

Aljofan, M., Saubern, S., Meyer, A.G., Marsh, G., Meers, J., and Mungall, B.A. (2009a). Characteristics of Nipah virus and Hendra virus replication in different cell lines and their suitability for antiviral screening. Virus Res. *142*, 92–99.

Aljofan, M., Sganga, M.L., Lo, M.K., Rootes, C.L., Porotto, M., Meyer, A.G., Saubern, S., Moscona, A., and Mungall, B.A. (2009b). Antiviral activity of gliotoxin, gentian violet and brilliant green against Nipah and Hendra virus *in vitro*. Virol. J. *6*, 187.

Anonymous (2010). Nipah virus Fact sheet. In Weekly Epidemiological Record (World Health Organization, Geneva, Switzerland), pp. 64–67.

Appel, M.J., and Summers, B.A. (1995). Pathogenicity of morbilliviruses for terrestrial carnivores. Vet. Microbiol. *44*, 187–191.

Bankamp, B., Horikami, S.M., Thompson, P.D., Huber, M., Billeter, M., and Moyer, S.A. (1996). Domains of the measles virus N protein required for binding to P protein and self-assembly. Virology *216*, 272–277.

Bankamp, B., Wilson, J., Bellini, W.J., and Rota, P.A. (2005). Identification of naturally occurring amino acid variations that affect the ability of the measles virus C protein to regulate genome replication and transcription. Virology *336,* 120–129.

Bellini, W.J., Rota, P.A., and Anderson, L.J. (1998). Paramyxoviruses. In Topley and Wilson's Microbiology and Microbial Infections, 9th edn, Mahy, B.W.J., and Collier, L., eds. (Edward Arnold: London), pp. 435–462.

Bishop, K.A., Stantchev, T.S., Hickey, A.C., Khetawat, D., Bossart, K.N., Krasnoperov, V., Gill, P., Feng, Y.R., Wang, L., Eaton, B.T., *et al.* (2007). Identification of Hendra virus G glycoprotein residues that are critical for receptor binding. J. Virol. *81,* 5893–5901.

Blum, L.S., Khan, R., Nahar, N., and Breiman, R.F. (2009). In-depth assesment of an outbreak of Nipah encephalitis with person-to-person transmission in Bangladesh: implications for prevention and control strategies. Am. J. Trop. Med. Hygiene *80,* 96–102.

Bonaparte, M.I., Dimitrov, A.S., Bossart, K.N., Crameri, G., Mungall, B.A., Bishop, K.A., Choudhry, V., Dimitrov, D.S., Wang, L.F., Eaton, B.T., *et al.* (2005). Ephrin-B2 ligand is a functional receptor for Hendra virus and Nipah virus. Proc. Natl. Acad. Sci. U.S.A. *102,* 10652–10657.

Bossart, K.N., Wang, L.F., Eaton, B.T., and Broder, C.C. (2001). Functional expression and membrane fusion tropism of the envelope glycoproteins of Hendra virus. Virology *290,* 121–135.

Bossart, K.N., Wang, L.F., Flora, M.N., Chua, K.B., Lam, S.K., Eaton, B.T., and Broder, C.C. (2002). Membrane fusion tropism and heterotypic functional activities of the Nipah virus and Hendra virus envelope glycoproteins. J. Virol. *76,* 11186–11198.

Bossart, K.N., Crameri, G., Dimitrov, A.S., Mungall, B.A., Feng, Y.R., Patch, J.R., Choudhary, A., Wang, L.F., Eaton, B.T., and Broder, C.C. (2005). Receptor binding, fusion inhibition, and induction of cross-reactive neutralizing antibodies by a soluble G glycoprotein of Hendra virus. J. Virol. *79,* 6690–6702.

Bossart, K.N., McEachern, J.A., Hickey, A.C., Choudhry, V., Dimitrov, D.S., Eaton, B.T., and Wang, L.F. (2007). Neutralization assays for differential *Henipavirus* serology using Bio-Plex protein array systems. J. Virol. Methods *142,* 29–40.

Bossart, K.N., Zhu, Z., Middleton, D., Klippel, J., Crameri, G., Bingham, J., McEachern, J.A., Green, D., Hancock, T.J., Chan, Y.P., *et al.* (2009). A neutralizing human monoclonal antibody protects against lethal disease in a new ferret model of acute nipah virus infection. PLoS Pathog. *5,* e1000642.

Bowden, T.A., Aricescu, A.R., Gilbert, R.J., Grimes, J.M., Jones, E.Y., and Stuart, D.I. (2008a). Structural basis of Nipah and Hendra virus attachment to their cell-surface receptor ephrin-B2. Nat. Struct. Mol. Biol. *15,* 567–572.

Bowden, T.A., Crispin, M., Harvey, D.J., Aricescu, A.R., Grimes, J.M., Jones, E.Y., and Stuart, D.I. (2008b). Crystal structure and carbohydrate analysis of nipah virus attachment glycoprotein: a template for antiviral and vaccine design. J. Virol. *82,* 11628–11636.

Calain, P., and Roux, L. (1993). The rule of six, a basic feature for efficient replication of Sendai virus defective interfering RNA. J. Virol. *67,* 4822–4830.

Carsillo, T., Zhang, X., Vasconcelos, D., Niewiesk, S., and Oglesbee, M. (2006). A single codon in the nucleocapsid protein C terminus contributes to *in vitro* and *in vivo* fitness of Edmonston measles virus. J. Virol. *80,* 2904–2912.

Carter, J.R., Pager, C.T., Fowler, S.D., and Dutch, R.E. (2005). Role of N-linked glycosylation of the Hendra virus fusion protein. J. Virol. *79,* 7922–7925.

CDC (1999a). Outbreak of Hendra-like virus–Malaysia and Singapore, 1998–1999. Morbidity Mortality Wkly. Rep. *48,* 265–269.

CDC (1999b). Update: outbreak of Nipah virus–Malaysia and Singapore, 1999. Morbid Mortal Wkly. Rep. *48,* 335–337.

Cevik, B., Kaesberg, J., Smallwood, S., Feller, J.A., and Moyer, S.A. (2004). Mapping the phosphoprotein binding site on Sendai virus NP protein assembled into nucleocapsids. Virology *325,* 216–224.

Chadha, M.S., Comer, J.A., Lowe, L., Rota, P.A., Rollin, P.E., Bellini, W.J., Ksiazek, T.G., and Mishra, A. (2006). Nipah virus-associated encephalitis outbreak, Siliguri, India. Emerg. Infect. Dis. *12,* 235–240.

Chan, Y.P., Koh, C.L., Lam, S.K., and Wang, L.F. (2004). Mapping of domains responsible for nucleocapsid protein–phosphoprotein interaction of henipaviruses. J. Gen. Virol. *85,* 1675–1684.

Chattopadhyay, A., Raha, T., and Shaila, M.S. (2004). Effect of single amino acid mutations in the conserved GDNQ motif of L protein of Rinderpest virus on RNA synthesis *in vitro* and *in vivo*. Virus Res. *99,* 139–145.

Chen, J.M., Yu, M., Morrissy, C., Zhao, Y.G., Meehan, G., Sun, Y.X., Wang, Q.H., Zhang, W., Wang, L.F., and Wang, Z.L. (2006). A comparative indirect ELISA for the detection of *Henipavirus* antibodies based on a recombinant nucleocapsid protein expressed in Escherichia coli. J. Virol. Methods *136,* 273–276.

Chen, J.M., Yaiw, K.C., Yu, M., Wang, L.F., Wang, Q.H., Crameri, G., and Wang, Z.L. (2007). Expression of truncated phosphoproteins of Nipah virus and Hendra virus in *Escherichia coli* for the differentiation of *Henipavirus* infections. Biotechnol. Lett. *29,* 871–875.

Childs, K., Stock, N., Ross, C., Andrejeva, J., Hilton, L., Skinner, M., Randall, R., and Goodbourn, S. (2007). Mda-5, but not RIG-I, is a common target for paramyxovirus V proteins. Virology *359,* 190–200.

Childs, K.S., Andrejeva, J., Randall, R.E., and Goodbourn, S. (2009). Mechanism of mda-5 Inhibition by paramyxovirus V proteins. J. Virol. *83,* 1465–1473.

Chong, H.T., Kamarulzaman, A., Tan, C.T., Goh, K.J., Thayaparan, T., Kunjapan, S.R., Chew, N.K., Chua, K.B., and Lam, S.K. (2001). Treatment of acute Nipah encephalitis with ribavirin. Ann. Neurol. *49,* 810–813.

Chong, H.T., and Tan, C.T. (2003). Relapsed and late-onset Nipah encephalitis, a report of three cases. Neurol. J. South East Asia *8,* 109–112.

Chow, V.T., Tambyah, P.A., Yeo, W.M., Phoon, M.C., and Howe, J. (2000). Diagnosis of nipah virus encephalitis by electron microscopy of cerebrospinal fluid. J. Clin. Virol. *19,* 143–147.

Chua, K.B., Bellini, W.J., Rota, P.A., Harcourt, B.H., Tamin, A., Lam, S.K., Ksiazek, T.G., Rollin, P.E., Zaki, S.R., Shieh, W., *et al.* (2000a). Nipah virus: a recently emergent deadly paramyxovirus. Science *288,* 1432–1435.

Chua, K.B., Lam, S.K., Tan, C.T., Hooi, P.S., Goh, K.J., Chew, N.K., Tan, K.S., Kamarulzaman, A., and Wong, K.T. (2000b). High mortality in Nipah encephalitis is associated with presence of virus in cerebrospinal fluid. Ann. Neurol. *48,* 802–805.

Chua, K.B., Lam, S.K., Goh, K.J., Hooi, P.S., Ksiazek, T.G., Kamarulzaman, A., Olson, J., and Tan, C.T. (2001). The presence of Nipah virus in respiratory secretions and urine of patients during an outbreak of Nipah virus encephalitis in Malaysia. J. Infect. *42,* 40–43.

Chua, K.B., Koh, C.L., Hooi, P.S., Wee, K.F., Khong, J.H., Chua, B.H., Chan, Y.P., Lim, M.E., and Lam, S.K. (2002). Isolation of Nipah virus from Malaysian Island flying-foxes. Microbes Infect. *4,* 145–151.

Ciancanelli, M.J., and Basler, C.F. (2006). Mutation of YMYL in the Nipah virus matrix protein abrogates budding and alters subcellular localization. J. Virol. *80, 12,*070–12078.

Ciancanelli, M.J., Volchkova, V.A., Shaw, M.L., Volchkov, V.E., and Basler, C.F. (2009). Nipah virus sequesters inactive STAT1 in the nucleus via a P gene-encoded mechanism. J. Virol. *83,* 7828–7841.

Corey, E.A., and Iorio, R.M. (2007). Mutations in the stalk of the measles virus hemagglutinin protein decrease fusion but do not interfere with virus-specific interaction with the homologous fusion protein. J. Virol. *81,* 9900–9910.

Corey, E.A., and Iorio, R.M. (2009). Measles virus attachment proteins with impaired ability to bind CD46 interact more efficiently with the homologous fusion protein. Virology *383,* 1–5.

Craft, W.W., Jr., and Dutch, R.E. (2005). Sequence motif upstream of the Hendra virus fusion protein cleavage site is not sufficient to promote efficient proteolytic processing. Virology *341,* 130–140.

Crameri, G., Todd, S., Grimley, S., McEachern, J.A., Marsh, G.A., Smith, C., Tachedjian, M., De Jong, C., Virtue, E.R., Yu, M., *et al.* (2009). Establishment, immortalisation and characterisation of pteropid bat cell lines. PLoS One *4,* e8266.

Curran, J., and Kolakofsky, D. (1988a). Ribosomal initiation from an ACG codon in the Sendai virus P/C mRNA. EMBO J. *7,* 245–251.

Curran, J., and Kolakofsky, D. (1988b). Scanning independent ribosomal initiation of the Sendai virus X protein. EMBO J. *7,* 2869–2874.

Curran, J., and Kolakofsky, D. (1989). Scanning independent ribosomal initiation of the Sendai virus Y proteins *in vitro* and *in vivo*. EMBO J. *8,* 521–526.

Curran, J., Boeck, R., and Kolakofsky, D. (1991). The Sendai virus P gene expresses both an essential protein and an inhibitor of RNA synthesis by shuffling modules via mRNA editing. EMBO J. *10,* 3079–3085.

Curran, J., Marq, J.B., and Kolakofsky, D. (1995). An N-terminal domain of the Sendai paramyxovirus P protein acts as a chaperone for the NP protein during the nascent chain assembly step of genome replication. J. Virol. *69,* 849–855.

Daniels, P., Ksiazek, T., and Eaton, B.T. (2001). Laboratory diagnosis of Nipah and Hendra virus infections. Microbes Infect. *3,* 289–295.

De, B.P., Hoffman, M.A., Choudhary, S., Huntley, C.C., and Banerjee, A.K. (2000). Role of NH(2)- and COOH-terminal domains of the P protein of human parainfluenza virus type 3 in transcription and replication. J. Virol. *74,* 5886–5895.

Devaux, P., and Cattaneo, R. (2004). Measles virus phosphoprotein gene products: conformational flexibility of the P/V protein amino-terminal domain and C protein infectivity factor function. J. Virol. *78, 11,*632–11640.

Devaux, P., Hodge, G., McChesney, M.B., and Cattaneo, R. (2008). Attenuation of V- or C-defective measles viruses: infection control by the inflammatory and interferon responses of rhesus monkeys. J. Virol. *82,* 5359–5367.

Diederich, S., Moll, M., Klenk, H.D., and Maisner, A. (2005). The nipah virus fusion protein is cleaved within the endosomal compartment. J. Biol. Chem. *280,* 29899–29903.

Diederich, S., Thiel, L., and Maisner, A. (2008). Role of endocytosis and cathepsin-mediated activation in Nipah virus entry. Virology *375,* 391–400.

Doms, R.W., Lamb, R.A., Rose, J.K., and Helenius, A. (1993). Folding and assembly of viral membrane proteins. Virology *193,* 545–562.

Durbin, A.P., Siew, J.W., Murphy, B.R., and Collins, P.L. (1997). Minimum protein requirements for transcription and RNA replication of a minigenome of human parainfluenza virus type 3 and evaluation of the rule of six. Virology *234,* 74–83.

Eaton, B.T., Wright, P.J., Wang, L.F., Sergeyev, O., Michalski, W.P., Bossart, K.N., and Broder, C.C. (2004). Henipaviruses: recent observations on regulation of transcription and the nature of the cell receptor. Arch. Virol. (Suppl.) 122–131.

Eaton, B.T., Mackenzie, J.S., and Wang, L.-F. (2007). *Paramyxoviridae*: henipaviruses, In Fields Virology, Knipe, D.M., and Howley, P., eds. (Lippincott Williams & Wilkins: Philadelphia, PA), pp1587–1600.

Egelman, E.H., Wu, S.S., Amrein, M., Portner, A., and Murti, G. (1989). The Sendai virus nucleocapsid exists in at least four different helical states. J. Virol. *63,* 2233–2243.

Erbar, S., Diederich, S., and Maisner, A. (2008). Selective receptor expression restricts Nipah virus infection of endothelial cells. Virol. J. *5,* 142.

Errington, W., and Emmerson, P.T. (1997). Assembly of recombinant Newcastle disease virus nucleocapsid protein into nucleocapsid-like structures is inhibited by the phosphoprotein. J. Gen. Virol. *78,* 2335–2339.

Eshaghi, M., Tan, W.S., Mohidin, T.B., and Yusoff, K. (2004). Nipah virus glycoprotein: production in baculovirus and application in diagnosis. Virus Res. *106*, 71–76.

Eshaghi, M., Tan, W.S., Chin, W.K., and Yusoff, K. (2005a). Purification of the extra-cellular domain of Nipah virus glycoprotein produced in *Escherichia coli* and possible application in diagnosis. J. Biotechnol. *116*, 221–226.

Eshaghi, M., Tan, W.S., Ong, S.T., and Yusoff, K. (2005b). Purification and characterization of Nipah virus nucleocapsid protein produced in insect cells. J. Clin. Microbiol. *43*, 3172–3177.

Field, H., Schaaf, K., Kung, N., Simon, C., Waltisbuhl, D., Hobert, H., Moore, F., Middleton, D., Crook, A., Smith, G., *et al.* (2010). Hendra virus outbreak with novel clinical features, Australia. Emerg. Infect. Dis. *16*, 338–340.

Field, H.E. (2009). Bats and emerging zoonoses: henipaviruses and SARS. Zoonoses Public Health *56*, 278–284.

Freed, E.O. (2002). Viral late domains. J. Virol. *76*, 4679–4687.

Freiberg, A., Dolores, L.K., Enterlein, S., and Flick, R. (2008). Establishment and characterization of plasmid-driven minigenome rescue systems for Nipah virus: RNA polymerase I- and T7-catalyzed generation of functional paramyxoviral RNA. Virology *370*, 33–44.

Freiberg, A.N., Worthy, M.N., Lee, B., and Holbrook, M.R. (2009). Combined chloroquine and ribavirin treatment does not prevent death in a hamster model of Nipah and Hendra virus infection. J. Gen. Virol. *91*, 765–772.

Garcin, D., Itoh, M., and Kolakofsky, D. (1997). A point mutation in the Sendai virus accessory C proteins attenuates virulence for mice, but not virus growth in cell culture. Virology *238*, 424–431.

Garcin, D., Latorre, P., and Kolakofsky, D. (1999). Sendai virus C proteins counteract the interferon-mediated induction of an antiviral state. J. Virol. *73*, 6559–6565.

Geisbert, T.W., Daddario-Dicaprio, K.M., Hickey, A.C., Smith, M.A., Chan, Y.P., Wang, L.F., Mattapallil, J.J., Geisbert, J.B., Bossart, K.N., and Broder, C.C. (2010). Development of an acute and highly pathogenic nonhuman primate model of nipah virus infection. PLoS One *5*, e10690.

Georges-Courbot, M.C., Contamin, H., Faure, C., Loth, P., Baize, S., Leyssen, P., Neyts, J., and Deubel, V. (2006). Poly(I)-poly(C12U) but not ribavirin prevents death in a hamster model of Nipah virus infection. Antimicrob. Agents Chemother. *50*, 1768–1772.

Goh, K.J., Tan, C.T., Chew, N.K., Tan, P.S., Kamarulzaman, A., Sarji, S.A., Wong, K.T., Abdullah, B.J., Chua, K.B., and Lam, S.K. (2000). Clinical features of Nipah virus encephalitis among pig farmers in Malaysia. N. Engl. J. Med. *342*, 1229–1235.

Goldsmith, C.S., Whistler, T., Rollin, P.E., Ksiazek, T.G., Rota, P.A., Bellini, W.J., Daszak, P., Wong, K.T., Shieh, W.J., and Zaki, S.R. (2003). Elucidation of Nipah virus morphogenesis and replication using ultrastructural and molecular approaches. Virus Res. *92*, 89–98.

Grogan, C.C., and Moyer, S.A. (2001). Sendai virus wild-type and mutant C proteins show a direct correlation between L polymerase binding and inhibition of viral RNA synthesis. Virology *288*, 96–108.

Guillaume, V., Contamin, H., Loth, P., Georges-Courbot, M.C., Lefeuvre, A., Marianneau, P., Chua, K.B., Lam, S.K., Buckland, R., Deubel, V., *et al.* (2004a). Nipah virus: vaccination and passive protection studies in a hamster model. J. Virol. *78*, 834–840.

Guillaume, V., Lefeuvre, A., Faure, C., Marianneau, P., Buckland, R., Lam, S.K., Wild, T.F., and Deubel, V. (2004b). Specific detection of Nipah virus using real-time RT-PCR (TaqMan). J. Virol. Methods *120*, 229–237.

Guillaume, V., Aslan, H., Ainouze, M., Guerbois, M., Wild, T.F., Buckland, R., and Langedijk, J.P. (2006a). Evidence of a potential receptor-binding site on the Nipah virus G protein (NiV-G): identification of globular head residues with a role in fusion promotion and their localization on an NiV-G structural model. J. Virol. *80*, 7546–7554.

Guillaume, V., Contamin, H., Loth, P., Grosjean, I., Courbot, M.C., Deubel, V., Buckland, R., and Wild, T.F. (2006b). Antibody prophylaxis and therapy against Nipah virus infection in hamsters. J. Virol. *80*, 1972–1978.

Guillaume, V., Wong, K.T., Looi, R.Y., Georges-Courbot, M.C., Barrot, L., Buckland, R., Wild, T.F., and Horvat, B. (2009). Acute Hendra virus infection: Analysis of the pathogenesis and passive antibody protection in the hamster model. Virology *387*, 459–465.

Hagmaier, K., Stock, N., Goodbourn, S., Wang, L.F., and Randall, R. (2006). A single amino acid substitution in the V protein of Nipah virus alters its ability to block interferon signalling in cells from different species. J. Gen. Virol. *87*, 3649–3653.

Halpin, K., Young, P.L., Field, H.E., and Mackenzie, J.S. (2000). Isolation of Hendra virus from pteropid bats: a natural reservoir of Hendra virus. J. Gen. Virol. *81*, 1927–1932.

Halpin, K., Bankamp, B., Harcourt, B.H., Bellini, W.J., and Rota, P.A. (2004). Nipah virus conforms to the rule of six in a minigenome replication assay. J. Gen. Virol. *85*, 701–707.

Harcourt, B.H., Tamin, A., Ksiazek, T.G., Rollin, P.E., Anderson, L.J., Bellini, W.J., and Rota, P.A. (2000). Molecular characterization of Nipah virus, a newly emergent paramyxovirus. Virology *271*, 334–349.

Harcourt, B.H., Tamin, A., Halpin, K., Ksiazek, T.G., Rollin, P.E., Bellini, W.J., and Rota, P.A. (2001). Molecular characterization of the polymerase gene and genomic termini of Nipah virus. Virology *287*, 192–201.

Harty, R.N., and Palese, P. (1995). Measles virus phosphoprotein (P) requires the NH2- and COOH-terminal domains for interactions with the nucleoprotein (N) but only the COOH terminus for interactions with itself. J. Gen. Virol. *76*, 2863–2867.

Hausmann, S., Jacques, J.P., and Kolakofsky, D. (1996). Paramyxovirus RNA editing and the requirement for hexamer genome length. RNA *2*, 1033–1045.

Hayman, D.T., Suu-Ire, R., Breed, A.C., McEachern, J.A., Wang, L., Wood, J.L., and Cunningham, A.A. (2008).

Evidence of *Henipavirus* infection in West African fruit bats. PLoS One *3*, e2739.

Homaira, N., Rahman, M., Hossain, M.J., Epstein, J.H., Sultana, R., Khan, M.S., Podder, G., Nahar, K., Ahmed, B., Gurley, E.S., *et al.* (2010). Nipah virus outbreak with person-to-person transmission in a district of Bangladesh, 2007. Epidemiol. Infect. *138*, 1630–1636.

Hooper, P.T., Gould, A.R., Russell, G.M., Kattenbelt, J.A., and Mitchell, G. (1996). The retrospective diagnosis of a second outbreak of equine *Morbillivirus* infection. Aust. Vet. J. *74*, 244–245.

Hooper, P., Zaki, S., Daniels, P., and Middleton, D. (2001). Comparative pathology of the diseases caused by Hendra and Nipah viruses. Microbes Infect. *3*, 315–322.

Hooper, P.T., Ketterer, P.J., Hyatt, A.D., and Russell, G.M. (1997). Lesions of experimental equine *Morbillivirus* pneumonia in horses. Vet. Pathol. *34*, 312–322.

Horikami, S.M., Curran, J., Kolakofsky, D., and Moyer, S.A. (1992). Complexes of Sendai virus NP-P and P-L proteins are required for defective interfering particle genome replication *in vitro*. J. Virol. *66*, 4901–4908.

Hossain, M.J., Gurley, E.S., Montgomery, J.M., Bell, M., Carroll, D.S., Hsu, V.P., Formenty, P., Croisier, A., Bertherat, E., Faiz, M.A., *et al.* (2008). Clinical presentation of nipah virus infection in Bangladesh. Clin. Infect. Dis. *46*, 977–984.

Houben, K., Marion, D., Tarbouriech, N., Ruigrok, R.W., and Blanchard, L. (2007). Interaction of the C-terminal domains of sendai virus N and P proteins: comparison of polymerase-nucleocapsid interactions within the paramyxovirus family. J. Virol. *81*, 6807–6816.

Hyatt, A.D., Zaki, S.R., Goldsmith, C.S., Wise, T.G., and Hengstberger, S.G. (2001). Ultrastructure of Hendra virus and Nipah virus within cultured cells and host animals. Microbes Infect. *3*, 297–306.

Iehle, C., Razafitrimo, G., Razainirina, J., Andriaholinirina, N., Goodman, S.M., Faure, C., Georges-Courbot, M.C., Rousset, D., and Reynes, J.M. (2007). *Henipavirus* and Tioman virus antibodies in pteropodid bats, Madagascar. Emerg. Infect. Dis. *13*, 159–161.

Irie, T., Shimazu, Y., Yoshida, T., and Sakaguchi, T. (2007). The YLDL sequence within Sendai virus M protein is critical for budding of virus-like particles and interacts with Alix/AIP1 independently of C protein. J. Virol. *81*, 2263–2273.

Jack, P.J., Boyle, D.B., Eaton, B.T., and Wang, L.F. (2005). The complete genome sequence of J. virus reveals a unique genome structure in the family *Paramyxoviridae*. J. Virol. *79, 10*,690–10700.

Jiang, Y., Liu, H., Liu, P., and Kong, X. (2008). Plasmids driven minigenome rescue system for Newcastle disease virus V4 strain. Mol. Biol. Rep. *36*,1909–1914

Juozapaitis, M., Serva, A., Zvirbliene, A., Slibinskas, R., Staniulis, J., Sasnauskas, K., Shiell, B.J., Wang, L.F., and Michalski, W.P. (2007). Generation of *Henipavirus* nucleocapsid proteins in yeast *Saccharomyces cerevisiae*. Virus Res. *124*, 95–102.

Kaku, Y., Noguchi, A., Marsh, G.A., McEachern, J.A., Okutani, A., Hotta, K., Bazartseren, B., Fukushi, S., Broder, C.C., Yamada, A., *et al.* (2009). A neutralization test for specific detection of Nipah virus antibodies using pseudotyped vesicular stomatitis virus expressing green fluorescent protein. J. Virol. Methods *160*, 7–13.

Kashiwazaki, Y., Na, Y.N., Tanimura, N., and Imada, T. (2004). A solid-phase blocking ELISA for detection of antibodies to Nipah virus. J. Virol. Methods *121*, 259–261.

Kato, A., Kiyotani, K., Sakai, Y., Yoshida, T., and Nagai, Y. (1997a). The paramyxovirus, Sendai virus, V protein encodes a luxury function required for viral pathogenesis. EMBO J. *16*, 578–587.

Kato, A., Kiyotani, K., Sakai, Y., Yoshida, T., Shioda, T., and Nagai, Y. (1997b). Importance of the cysteine-rich carboxyl-terminal half of V protein for Sendai virus pathogenesis. J. Virol. *71*, 7266–7272.

Kawai, T., and Akira, S. (2008). Toll-like receptor and RIG-I-like receptor signaling. Ann. N.Y. Acad. Sci. *1143*, 1–20.

Keller, M.A., Murphy, S.K., and Parks, G.D. (2001). RNA replication from the simian virus 5 antigenomic promoter requires three sequence-dependent elements separated by sequence-independent spacer regions. J. Virol. 75, 3993–3998.

Klenk, H.D., and Garten, W. (1994). Host cell proteases controlling virus pathogenicity. Trends Microbiol. *2*, 39–43.

Kolakofsky, D., Pelet, T., Garcin, D., Hausmann, S., Curran, J., and Roux, L. (1998). Paramyxovirus RNA synthesis and the requirement for hexamer genome length: the rule of six revisited. J. Virol. *72*, 891–899.

Kulkarni, S., Volchkova, V., Basler, C.F., Palese, P., Volchkov, V.E., and Shaw, M.L. (2009). Nipah Virus edits its P gene at high frequency to express the V and W proteins. J. Virol. *83*, 3982–3987.

Kurotani, A., Kiyotani, K., Kato, A., Shioda, T., Sakai, Y., Mizumoto, K., Yoshida, T., and Nagai, Y. (1998). Sendai virus C proteins are categorically nonessential gene products but silencing their expression severely impairs viral replication and pathogenesis. Genes Cells *3*, 111–124.

Lamb, R.A., and Parks, G.D. (2007). *Paramyxoviridae*: The viruses and their replication. In Fields Virology, Knipe, D.M., and Howley, P., eds. (Lippincott Williams & Wilkins: Philadelphia, PA), pp. 1449–1496.

Latorre, P., Kolakofsky, D., and Curran, J. (1998). Sendai virus Y proteins are initiated by a ribosomal shunt. Mol. Cell Biol. *18*, 5021–5031.

Lee, J.K., Prussia, A., Paal, T., White, L.K., Snyder, J.P., and Plemper, R.K. (2008). Functional interaction between paramyxovirus fusion and attachment proteins. J. Biol. Chem. *283, 16*,561–16572.

Levroney, E.L., Aguilar, H.C., Fulcher, J.A., Kohatsu, L., Pace, K.E., Pang, M., Gurney, K.B., Baum, L.G., and Lee, B. (2005). Novel innate immune functions for galectin-1: galectin-1 inhibits cell fusion by Nipah virus envelope glycoproteins and augments dendritic cell secretion of proinflammatory cytokines. J. Immunol. *175*, 413–420.

Li, Z., Yu, M., Zhang, H., Magoffin, D.E., Jack, P.J., Hyatt, A., Wang, H.Y., and Wang, L.F. (2006). Beilong virus, a novel paramyxovirus with the largest genome of non-segmented negative-stranded RNA viruses. Virology *346,* 219–228.

Liston, P., Batal, R., DiFlumeri, C., and Briedis, D.J. (1997). Protein interaction domains of the measles virus nucleocapsid protein (NP). Arch. Virol. *142,* 305–321.

Lo, M.K., Harcourt, B.H., Mungall, B.A., Tamin, A., Peeples, M.E., Bellini, W.J., and Rota, P.A. (2009). Determination of the *Henipavirus* phosphoprotein gene mRNA editing frequencies and detection of the C, V and W proteins of Nipah virus in virus-infected cells. J. Gen. Virol. *90,* 398–404.

Lo, M.K., Miller, D.M., Aljofan, M., Mungall, B.A., Rollin, P.E., Bellini, W.J., and Rota, P.A. (2010). Characterization of the antiviral and inflammatory responses against Nipah virus in endothelial cells and neurons. Virology *404,* 78–88.

Lo, M.K., and Rota, P.A. (2008). The emergence of Nipah virus, a highly pathogenic paramyxovirus. J. Clin. Virol. *43,* 396–400.

Luby, S.P., Gurley, E.S., and Hossain, M.J. (2009a). Transmission of Human Infections with Nipah Virus. Clin. Infect. Dis. *49,* 1743–1748.

Luby, S.P., Hossain, M.J., Gurley, E.S., Ahmed, B.-N., Banu, S., Khan, S.U., Homaira, N., Rota, P.A., Rollin, P.E., Comer, J.A., *et al.* (2009b). Recurrent zoonotic transmission of Nipah virus into humans, Bangladesh, 2001–2007. Emerg. Infect. Dis. *15,* 1229–1235.

Ludlow, L.E., Lo, M.K., Rodriguez, J.J., Rota, P.A., and Horvath, C.M. (2008). *Henipavirus* V protein association with Polo-like kinase reveals functional overlap with STAT1 binding and interferon evasion. J. Virol. *82,* 6259–6271.

Magoffin, D.E., Halpin, K., Rota, P.A., and Wang, L.F. (2007a). Effects of single amino acid substitutions at the E residue in the conserved GDNE motif of the Nipah virus polymerase (L) protein. Arch. Virol. *152,* 827–832.

Magoffin, D.E., Mackenzie, J.S., and Wang, L.F. (2007b). Genetic analysis of J-virus and Beilong virus using minireplicons. Virology *364,* 103–111.

Marianneau, P., Guillaume, V., Wong, T., Badmanathan, M., Looi, R.Y., Murri, S., Loth, P., Tordo, N., Wild, F., Horvat, B., *et al.* (2010). Experimental infection of squirrel monkeys with nipah virus. Emerg. Infect. Dis. *16,* 507–510.

Mayo, M.A. (2002). A summary of taxonomic changes recently approved by ICTV. Arch. Virol. *147,* 1655–1663.

McCormack, J.G., Allworth, A.M., Selvey, L.A., and Selleck, P.W. (1999). Transmissibility from horses to humans of a novel paramyxovirus, equine *Morbillivirus* (EMV). J. Infect. *38,* 22–23.

McEachern, J.A., Bingham, J., Crameri, G., Green, D.J., Hancock, T.J., Middleton, D., Feng, Y.R., Broder, C.C., Wang, L.F., and Bossart, K.N. (2008). A recombinant subunit vaccine formulation protects against lethal Nipah virus challenge in cats. Vaccine *26,* 3842–3852.

McGinnes, L., Sergel, T., Reitter, J., and Morrison, T. (2001). Carbohydrate modifications of the NDV fusion protein heptad repeat domains influence maturation and fusion activity. Virology *283,* 332–342.

Mehdi, H., Ono, E., and Gupta, K.C. (1990). Initiation of translation at CUG, GUG, and ACG codons in mammalian cells. Gene *91,* 173–178.

Melanson, V.R., and Iorio, R.M. (2004). Amino acid substitutions in the F-specific domain in the stalk of the newcastle disease virus HN protein modulate fusion and interfere with its interaction with the F protein. J. Virol. *78,* 13053–13061.

Melanson, V.R., and Iorio, R.M. (2006). Addition of N-glycans in the stalk of the Newcastle disease virus HN protein blocks its interaction with the F protein and prevents fusion. J. Virol. *80,* 623–633.

von Messling, V., Svitek, N., and Cattaneo, R. (2006). Receptor (SLAM [CD150]) recognition and the V protein sustain swift lymphocyte-based invasion of mucosal tissue and lymphatic organs by a *Morbillivirus.* J. Virol. *80,* 6084–6092.

Meulendyke, K.A., Wurth, M.A., McCann, R.O., and Dutch, R.E. (2005). Endocytosis plays a critical role in proteolytic processing of the Hendra virus fusion protein. J. Virol. *79,* 12643–12649.

Michalski, W.P., Crameri, G., Wang, L., Shiell, B.J., and Eaton, B. (2000). The cleavage activation and sites of glycosylation in the fusion protein of Hendra virus. Virus Res. *69,* 83–93.

Middleton, D.J., Morrissy, C.J., van der Heide, B.M., Russell, G.M., Braun, M.A., Westbury, H.A., Halpin, K., and Daniels, P.W. (2007). Experimental Nipah virus infection in pteropid bats (*Pteropus poliocephalus*). J. Comp. Pathol. *136,* 266–272.

Middleton, D.J., Westbury, H.A., Morrissy, C.J., van der Heide, B.M., Russell, G.M., Braun, M.A., and Hyatt, A.D. (2002). Experimental Nipah virus infection in pigs and cats. J. Comp. Pathol. *126,* 124–136.

Mohd, N.M.N., Gan, C.H., and Ong, B.L. (2000). Nipah virus infection of pigs in peninsular Malaysia. Rev. Sci. Tech. Off. Int. Epiz. *19,* 160–165.

Moll, M., Diederich, S., Klenk, H.D., Czub, M., and Maisner, A. (2004a). Ubiquitous activation of the Nipah virus fusion protein does not require a basic amino acid at the cleavage site. J. Virol. *78,* 9705–9712.

Moll, M., Kaufmann, A., and Maisner, A. (2004b). Influence of N-glycans on processing and biological activity of the nipah virus fusion protein. J. Virol. *78,* 7274–7278.

Mungall, B.A., Middleton, D., Crameri, G., Bingham, J., Halpin, K., Russell, G., Green, D., McEachern, J., Pritchard, L.I., Eaton, B.T., *et al.* (2006). A feline model of acute Nipah virus infection and protection with a soluble glycoprotein-based subunit vaccine. J. Virol. *80,* 12293–12302.

Mungall, B.A., Schopman, N.C., Lambeth, L.S., and Doran, T.J. (2008). Inhibition of *Henipavirus* infection by RNA interference. Antiviral Res. *80,* 324–331.

Murray, K., Rogers, R., Selvey, L., Selleck, P., Hyatt, A., Gould, A., Gleeson, L., Hooper, P., and Westbury, H. (1995a). A novel *Morbillivirus* pneumonia of horses and its transmission to humans. Emerg. Infect. Dis. *1,* 31–33.

Murray, K., Selleck, P., Hooper, P., Hyatt, A., Gould, A., Gleeson, L., Westbury, H., Hiley, L., Selvey, L., Rodwell, B., *et al.* (1995b). A *Morbillivirus* that caused fatal disease in horses and humans. Science *268,* 94–97.

Negrete, O.A., Chu, D., Aguilar, H.C., and Lee, B. (2007). Single amino acid changes in the Nipah and Hendra virus attachment glycoproteins distinguish ephrinB2 from ephrinB3 usage. J. Virol. *81,* 10804–10814.

Negrete, O.A., Levroney, E.L., Aguilar, H.C., Bertolotti-Ciarlet, A., Nazarian, R., Tajyar, S., and Lee, B. (2005). EphrinB2 is the entry receptor for Nipah virus, an emergent deadly paramyxovirus. Nature *436,* 401–405.

Negrete, O.A., Wolf, M.C., Aguilar, H.C., Enterlein, S., Wang, W., Muhlberger, E., Su, S.V., Bertolotti-Ciarlet, A., Flick, R., and Lee, B. (2006). Two key residues in ephrinB3 are critical for its use as an alternative receptor for Nipah virus. PLoS Pathog. *2,* e7.

Nishio, M., Ohtsuka, J., Tsurudome, M., Nosaka, T., and Kolakofsky, D. (2008). Human parainfluenza virus type 2 V protein inhibits genome replication by binding to the L protein: possible role in promoting viral fitness. J. Virol. *82,* 6130–6138.

O'Sullivan, J.D., Allworth, A.M., Paterson, D.L., Snow, T.M., Boots, R., Gleeson, L.J., Gould, A.R., Hyatt, A.D., and Bradfield, J. (1997). Fatal encephalitis due to novel paramyxovirus transmitted from horses. Lancet *349,* 93–95.

Ong, S.T., Yusoff, K., Kho, C.L., Abdullah, J.O., and Tan, W.S. (2009). Mutagenesis of the nucleocapsid protein of Nipah virus involved in capsid assembly. J. Gen. Virol. *90,* 392–397.

Pager, C.T., Craft, W.W., Jr., Patch, J., and Dutch, R.E. (2006). A mature and fusogenic form of the Nipah virus fusion protein requires proteolytic processing by cathepsin L. Virology *346,* 251–257.

Pager, C.T., and Dutch, R.E. (2005). Cathepsin L is involved in proteolytic processing of the Hendra virus fusion protein. J. Virol. *79, 12,*714–12720.

Pallister, J., Middleton, D., Crameri, G., Yamada, M., Klein, R., Hancock, T.J., Foord, A., Shiell, B., Michalski, W., Broder, C.C., *et al.* (2009). Chloroquine administration does not prevent Nipah virus infection and disease in ferrets. J. Virol. *83, 11,*979–11982.

Palmer, A., and Klein, R. (2003). Multiple roles of ephrins in morphogenesis, neuronal networking, and brain function. Genes Dev. *17,* 1429–1450.

Parashar, U.D., Sunn, L.M., Ong, F., Mounts, A.W., Arif, M.T., Ksiazek, T.G., Kamaluddin, M.A., Mustafa, A.N., Kaur, H., Ding, L.M., *et al.* (2000). Case–control study of risk factors for human infection with a new zoonotic paramyxovirus, Nipah virus, during a 1998–1999 outbreak of severe encephalitis in Malaysia. J. Infect. Dis. *181,* 1755–1759.

Park, M.S., Shaw, M.L., Munoz-Jordan, J., Cros, J.F., Nakaya, T., Bouvier, N., Palese, P., Garcia-Sastre, A., and Basler, C.F. (2003). Newcastle disease virus (NDV)-based assay demonstrates interferon-antagonist activity for the NDV V protein and the Nipah virus V, W, and C proteins. J. Virol. 77, 1501–1511.

Patch, J.R., Crameri, G., Wang, L.F., Eaton, B.T., and Broder, C.C. (2007). Quantitative analysis of Nipah virus proteins released as virus-like particles reveals central role for the matrix protein. Virol. J. *4,* 1.

Patch, J.R., Han, Z., McCarthy, S.E., Yan, L., Wang, L.F., Harty, R.N., and Broder, C.C. (2008). The YPLGVG sequence of the Nipah virus matrix protein is required for budding. Virol. J. *5,* 137.

Patterson, J.B., Thomas, D., Lewicki, H., Billeter, M.A., and Oldstone, M.B. (2000). V and C proteins of measles virus function as virulence factors *in vivo*. Virology *267,* 80–89.

Perez, M., Sanchez, A., Cubitt, B., Rosario, D., and de la Torre, J.C. (2003). A reverse genetics system for Borna disease virus. J. Gen. Virol. *84,* 3099–3104.

Pernet, O., Pohl, C., Ainouze, M., Kweder, H., and Buckland, R. (2009). Nipah virus entry can occur by macropinocytosis. Virology *395,* 298–311.

Platanias, L.C. (2005). Mechanisms of type-I- and type-II-interferon-mediated signalling. Nat. Rev. Immunol. *5,* 375–386.

Playford, e.g. McCall, B., Smith, G., Slinko, V., Allen, G., Smith, I., Moore, F., Taylor, C., Kung, Y.H., and Field, H. (2010). Human Hendra virus encephalitis associated with equine outbreak, Australia, 2008. Emerg. Infect. Dis. *16,* 219–223.

Poch, O., Sauvaget, I., Delarue, M., and Tordo, N. (1989). Identification of four conserved motifs among the RNA-dependent polymerase encoding elements. EMBO J. *8,* 3867–3874.

Poliakov, A., Cotrina, M., and Wilkinson, D.G. (2004). Diverse roles of eph receptors and ephrins in the regulation of cell migration and tissue assembly. Dev. Cell *7,* 465–480.

Porotto, M., Carta, P., Deng, Y., Kellogg, G.E., Whitt, M., Lu, M., Mungall, B.A., and Moscona, A. (2007). Molecular determinants of antiviral potency of paramyxovirus entry inhibitors. J. Virol. *81, 10,*567–10574.

Porotto, M., Doctor, L., Carta, P., Fornabaio, M., Greengard, O., Kellogg, G.E., and Moscona, A. (2006). Inhibition of hendra virus fusion. J. Virol. *80,* 9837–9849.

Porotto, M., Yokoyama, C., Palermo, L.M., Mungall, B., Aljofan, M., Cortese, R., Pessi, A., and Moscona, A. (2010). Viral entry inhibitors targeted to the membrane site of action. J. Virol. *84,* 6760–6768.

ProMed-mail (2009). Vet's death sparks fresh Hendra warning. http://www.promedmail.org archive no. 20090903.3098

ProMed-mail (2010a). Horse dies from Hendra virus. http://www.promedmail.org archive no. 201000522.1699

ProMed-mail (2010b). Nipah virus claims 3 lives in Faridpur. http://www.promedmail.org archive no. 20100122.0250

Raha, T., Chattopadhyay, A., and Shaila, M.S. (2004). Development of a reconstitution system for Rinderpest virus RNA synthesis *in vitro*. Virus Res. *99,* 131–138.

Reutter, G.L., Cortese-Grogan, C., Wilson, J., and Moyer, S.A. (2001). Mutations in the measles virus C protein that up regulate viral RNA synthesis. Virology *285,* 100–109.

Reynes, J.M., Counor, D., Ong, S., Faure, C., Seng, V., Molia, S., Walston, J., Georges-Courbot, M.C., Deubel, V., and Sarthou, J.L. (2005). Nipah virus in Lyle's flying foxes, Cambodia. Emerg. Infect. Dis. *11*, 1042–1047.

Roberts, P.C., Garten, W., and Klenk, H.D. (1993). Role of conserved glycosylation sites in maturation and transport of influenza A virus hemagglutinin. J. Virol. *67*, 3048–3060.

Rodriguez, J.J., Cruz, C.D., and Horvath, C.M. (2004). Identification of the nuclear export signal and STAT-binding domains of the Nipah virus V protein reveals mechanisms underlying interferon evasion. J. Virol. *78*, 5358–5367.

Rodriguez, J.J., Parisien, J.P., and Horvath, C.M. (2002). Nipah virus V protein evades alpha and gamma interferons by preventing STAT1 and STAT2 activation and nuclear accumulation. J. Virol. *76*, 11476–11483.

Rodriguez, J.J., Wang, L.F., and Horvath, C.M. (2003). Hendra virus V protein inhibits interferon signaling by preventing STAT1 and STAT2 nuclear accumulation. J. Virol. *77*, 11842–11845.

Roelke-Parker, M.E., Munson, L., Packer, C., Kock, R., Cleaveland, S., Carpenter, M., O'Brien, S.J., Pospischil, A., Hofmann-Lehmann, R., Lutz, H., *et al.* (1996). A canine distemper virus epidemic in Serengeti lions (*Panthera leo*). Nature *379*, 441–445.

Rogers, R.J., Douglas, I.C., Baldock, F.C., Glanville, R.J., Seppanen, K.T., Gleeson, L.J., Selleck, P.N., and Dunn, K.J. (1996). Investigation of a second focus of equine *Morbillivirus* infection in coastal Queensland. Aust. Vet. J. *74*, 243–244.

Rosario, D., Perez, M., and de la Torre, J.C. (2005). Functional characterization of the genomic promoter of borna disease virus (BDV): implications of 3′-terminal sequence heterogeneity for BDV persistence. J. Virol. *79*, 6544–6550.

Sanderson, C.M., Wu, H.H., and Nayak, D.P. (1994). Sendai virus M protein binds independently to either the F or the HN glycoprotein *in vivo*. J. Virol. *68*, 69–76.

Schmitt, A.P., Leser, G.P., Morita, E., Sundquist, W.I., and Lamb, R.A. (2005). Evidence for a new viral late-domain core sequence, FPIV, necessary for budding of a paramyxovirus. J. Virol. *79*, 2988–2997.

Schnell, M.J., and Conzelmann, K.K. (1995). Polymerase activity of *in vitro* mutated rabies virus L protein. Virology *214*, 522–530.

Segawa, H., Yamashita, T., Kawakita, M., and Taira, H. (2000). Functional analysis of the individual oligosaccharide chains of Sendai virus fusion protein. J. Biochem. *128*, 65–72.

Selvey, L.A., Wells, R.M., McCormack, J.G., Ansford, A.J., Murray, K., Rogers, R.J., Lavercombe, P.S., Selleck, P., and Sheridan, J.W. (1995). Infection of humans and horses by a newly described *Morbillivirus*. Med. J. Aust. *162*, 642–645.

Shaw, M.L., Cardenas, W.B., Zamarin, D., Palese, P., and Basler, C.F. (2005). Nuclear localization of the Nipah virus W protein allows for inhibition of both virus- and toll-like receptor 3-triggered signaling pathways. J. Virol. *79*, 6078–6088.

Shaw, M.L., Garcia-Sastre, A., Palese, P., and Basler, C.F. (2004). Nipah virus V and W proteins have a common STAT1-binding domain yet inhibit STAT1 activation from the cytoplasmic and nuclear compartments, respectively. J. Virol. *78*, 5633–5641.

Shiell, B.J., Gardner, D.R., Crameri, G., Eaton, B.T., and Michalski, W.P. (2003). Sites of phosphorylation of P and V proteins from Hendra and Nipah viruses: newly emerged members of *Paramyxoviridae*. Virus Res. *92*, 55–65.

Sidhu, M.S., Chan, J., Kaelin, K., Spielhofer, P., Radecke, F., Schneider, H., Masurekar, M., Dowling, P.C., Billeter, M.A., and Udem, S.A. (1995). Rescue of synthetic measles virus minireplicons: measles genomic termini direct efficient expression and propagation of a reporter gene. Virology *208*, 800–807.

Sleeman, K., Bankamp, B., Hummel, K.B., Lo, M.K., Bellini, W.J., and Rota, P.A. (2008). The C, V and W proteins of Nipah virus inhibit minigenome replication. J. Gen. Virol. *89*, 1300–1308.

Stec, D.S., Hill, M.G., 3rd., and Collins, P.L. (1991). Sequence analysis of the polymerase L gene of human respiratory syncytial virus and predicted phylogeny of nonsegmented negative-strand viruses. Virology *183*, 273–287.

Swenson, D.L., Warfield, K.L., Negley, D.L., Schmaljohn, A., Aman, M.J., and Bavari, S. (2005). Virus-like particles exhibit potential as a pan-filovirus vaccine for both Ebola and Marburg viral infections. Vaccine *23*, 3033–3042.

Tamin, A., Harcourt, B.H., Ksiazek, T.G., Rollin, P.E., Bellini, W.J., and Rota, P.A. (2002). Functional properties of the fusion and attachment glycoproteins of Nipah virus. Virology *296*, 190–200.

Tamin, A., Harcourt, B.H., Lo, M.K., Roth, J.A., Wolf, M.C., Lee, B., Weingartl, H., Audonnet, J.C., Bellini, W.J., and Rota, P.A. (2009). Development of a neutralization assay for Nipah virus using pseudotype particles. J. Virol. Methods *160*, 1–6.

Tamura, T., Yamashita, T., Segawa, H., and Taira, H. (2002). N-linked oligosaccharide chains of Sendai virus fusion protein determine the interaction with endoplasmic reticulum molecular chaperones. FEBS Lett. *513*, 153–158.

Tan, C.T., and Chua, K.B. (2008). Nipah virus encephalitis. Curr. Infect. Dis. Rep. *10*, 315–320.

Tan, C.T., Goh, K.J., Wong, K.T., Sarji, S.A., Chua, K.B., Chew, N.K., Murugasu, P., Loh, Y.L., Chong, H.T., Tan, K.S., *et al.* (2002). Relapsed and late-onset Nipah encephalitis. Ann. Neurol. *51*, 703–708.

Tan, W.S., Ong, S.T., Eshaghi, M., Foo, S.S., and Yusoff, K. (2004). Solubility, immunogenicity and physical properties of the nucleocapsid protein of Nipah virus produced in Escherichia coli. J. Med. Virol. *73*, 105–112.

Tanimura, N., Imada, T., Kashiwazaki, Y., Shahirudin, S., Sharifah, S.H., and Aziz, A.J. (2004a). Monoclonal antibody-based immunohistochemical diagnosis of Malaysian Nipah virus infection in pigs. J. Comp. Pathol. *131*, 199–206.

Tanimura, N., Imada, T., Kashiwazaki, Y., Shamusudin, S., Syed Hassan, S., Jamaluddin, A., Russell, G., and White, J. (2004b). Reactivity of anti-Nipah virus monoclonal antibodies to formalin-fixed, paraffin-embedded lung tissues from experimental Nipah and Hendra virus infections. J. Vet. Med. Sci. *66,* 1263–1266.

Tapparel, C., Maurice, D., and Roux, L. (1998). The activity of Sendai virus genomic and antigenomic promoters requires a second element past the leader template regions: a motif (GNNNNN)3 is essential for replication. J. Virol. *72,* 3117–3128.

Tashiro, M., Seto, J.T., Klenk, H.D., and Rott, R. (1993). Possible involvement of microtubule disruption in bipolar budding of a Sendai virus mutant, F1-R, in epithelial MDCK cells. J. Virol. *67,* 5902–5910.

Tashiro, M., McQueen, N.L., Seto, J.T., Klenk, H.D., and Rott, R. (1996). Involvement of the mutated M protein in altered budding polarity of a pantropic mutant, F1-R, of Sendai virus. J. Virol. *70,* 5990–5997.

Tee, K.K., Takebe, Y., and Kamarulzaman, A. (2009). Emerging and re-emerging viruses in Malaysia, 1997–2007. Int J. Infect. Dis. *13,* 307–318.

Thiel, L., Diederich, S., Erbar, S., Pfaff, D., Augustin, H.G., and Maisner, A. (2008). Ephrin-B2 expression critically influences Nipah virus infection independent of its cytoplasmic tail. Virol. J. *5,* 163.

Tober, C., Seufert, M., Schneider, H., Billeter, M.A., Johnston, I.C., Niewiesk, S., ter Meulen, V., and Schneider-Schaulies, S. (1998). Expression of measles virus V protein is associated with pathogenicity and control of viral RNA synthesis. J. Virol. *72,* 8124–8132.

Tong, S., Li, M., Vincent, A., Compans, R.W., Fritsch, E., Beier, R., Klenk, C., Ohuchi, M., and Klenk, H.D. (2002). Regulation of fusion activity by the cytoplasmic domain of a paramyxovirus F protein. Virology *301,* 322–333.

Torres-Velez, F.J., Shieh, W.J., Rollin, P.E., Morken, T., Brown, C., Ksiazek, T.G., and Zaki, S.R. (2008). Histopathologic and immunohistochemical characterization of Nipah virus infection in the guinea pig. Vet. Pathol. *45,* 576–585.

Toth, A.M., Devaux, P., Cattaneo, R., and Samuel, C.E. (2009). Protein kinase PKR mediates the apoptosis induction and growth restriction phenotypes of C protein-deficient measles virus. J. Virol. *83,* 961–968.

Tyler, K.L. (2009). Emerging viral infections of the central nervous system: Part 2. Arch. Neurol. *66,* 1065–1074.

Uze, G., and Monneron, D. (2007). IL-28 and IL-29: newcomers to the interferon family. Biochimie *89,* 729–734.

Valsamakis, A., Schneider, H., Auwaerter, P.G., Kaneshima, H., Billeter, M.A., and Griffin, D.E. (1998). Recombinant measles viruses with mutations in the C, V, or F gene have altered growth phenotypes *in vivo*. J. Virol. *72,* 7754–7761.

Vogt, C., Eickmann, M., Diederich, S., Moll, M., and Maisner, A. (2005). Endocytosis of the Nipah virus glycoproteins. J. Virol. *79,* 3865–3872.

Wacharapluesadee, S., and Hemachudha, T. (2007). Duplex nested RT-PCR for detection of Nipah virus RNA from urine specimens of bats. J. Virol. Methods *141,* 97–101.

Walpita, P., and Peters, C.J. (2007). Cis-acting elements in the antigenomic promoter of Nipah virus. J. Gen. Virol. *88,* 2542–2551.

Wang, L., Harcourt, B.H., Yu, M., Tamin, A., Rota, P.A., Bellini, W.J., and Eaton, B.T. (2001). Molecular biology of Hendra and Nipah viruses. Microbes Infect. *3,* 279–287.

Wang, L.F., Michalski, W.P., Yu, M., Pritchard, L.I., Crameri, G., Shiell, B., and Eaton, B.T. (1998). A novel P/V/C gene in a new member of the *Paramyxoviridae* family, which causes lethal infection in humans, horses, and other animals. J. Virol. *72,* 1482–1490.

Wang, L.F., Yu, M., Hansson, E., Pritchard, L.I., Shiell, B., Michalski, W.P., and Eaton, B.T. (2000a). The exceptionally large genome of hendra virus: support for creation of a new genus within the family *Paramyxoviridae*. J. Virol. *74,* 9972–9979.

Wang, L.F., Yu, M., Hansson, E., Pritchard, L.I., Shiell, B., Michalski, W.P., and Eaton, B.T. (2000b). The exceptionally large genome of Hendra virus: support for creation of a new genus within the family *Paramyxoviridae*. J. Virol. *74,* 9972–9979.

Wang, X., Ge, J., Hu, S., Wang, Q., Wen, Z., Chen, H., and Bu, Z. (2006). Efficacy of DNA immunization with F and G protein genes of Nipah virus. Ann N Y Acad Sci. *1081,* 243–245.

Waning, D.L., Russell, C.J., Jardetzky, T.S., and Lamb, R.A. (2004). Activation of a paramyxovirus fusion protein is modulated by inside-out signaling from the cytoplasmic tail. Proc. Natl. Acad. Sci. U.S.A. *101,* 9217–9222.

Weingartl, H., Czub, S., Copps, J., Berhane, Y., Middleton, D., Marszal, P., Gren, J., Smith, G., Ganske, S., Manning, L., *et al.* (2005). Invasion of the central nervous system in a porcine host by nipah virus. J. Virol. *79,* 7528–7534.

Weingartl, H.M., Berhane, Y., Caswell, J.L., Loosmore, S., Audonnet, J.C., Roth, J.A., and Czub, M. (2006). Recombinant nipah virus vaccines protect pigs against challenge. J. Virol. *80,* 7929–7938.

Weise, C., Erbar, S., Lamp, B., Vogt, C., Diederich, S., and Maisner, A. (2010). Tyrosine residues in the cytoplasmic domains affect sorting and fusion activity of the Nipah virus glycoproteins in polarized epithelial cells. J. Virol. *84,* 7634–7641.

Westbury, H.A. (2000). Hendra virus disease in horses. Rev. Sci. Tech. *19,* 151–159.

Westbury, H.A., Hooper, P.T., Selleck, P.W., and Murray, P.K. (1995). Equine *Morbillivirus* pneumonia: susceptibility of laboratory animals to the virus. Aust. Vet. J. *72,* 278–279.

Westbury, H.A., Hooper, P.T., Brouwer, S.L., and Selleck, P.W. (1996). Susceptibility of cats to equine *Morbillivirus*. Aust. Vet. J. *74,* 132–134.

Whitman, S.D., and Dutch, R.E. (2007). Surface density of the Hendra G protein modulates Hendra F protein-promoted membrane fusion: role for Hendra G protein trafficking and degradation. Virology *363,* 419–429.

Whitman, S.D., Smith, E.C., and Dutch, R.E. (2009). Differential rates of protein folding and cellular trafficking for the Hendra virus F and G proteins: implications for F–G complex formation. J. Virol. *83*, 8998–9001.

Williamson, M.M., Hooper, P.T., Selleck, P.W., Gleeson, L.J., Daniels, P.W., Westbury, H.A., and Murray, P.K. (1998). Transmission studies of Hendra virus (equine *Morbillivirus*) in fruit bats, horses and cats. Aust. Vet. J. *76*, 813–818.

Williamson, M.M., Hooper, P.T., Selleck, P.W., Westbury, H.A., and Slocombe, R.F. (2000). Experimental hendra virus infectionin pregnant guinea-pigs and fruit Bats (Pteropus poliocephalus). J. Comp. Pathol. *122*, 201–207.

Williamson, M.M., Hooper, P.T., Selleck, P.W., Westbury, H.A., and Slocombe, R.F. (2001). A guinea-pig model of Hendra virus encephalitis. J. Comp. Pathol. *124*, 273–279.

Witko, S.E., Kotash, C., Sidhu, M.S., Udem, S.A., and Parks, C.L. (2006). Inhibition of measles virus minireplicon-encoded reporter gene expression by V protein. Virology *348*, 107–119.

Wolf, M.C., Wang, Y., Freiberg, A.N., Aguilar, H.C., Holbrook, M.R., and Lee, B. (2009). A catalytically and genetically optimized beta-lactamase-matrix based assay for sensitive, specific, and higher throughput analysis of native *Henipavirus* entry characteristics. Virol. J. *6*, 119.

Wolf, M.C., Freiberg, A.N., Zhang, T., Akyol-Ataman, Z., Grock, A., Hong, P.W., Li, J., Watson, N.F., Fang, A.Q., Aguilar, H.C., *et al.* (2010). A broad-spectrum antiviral targeting entry of enveloped viruses. Proc. Natl. Acad. Sci. U.S.A. *107*, 3157–3162.

Wong, K.T., Shieh, W.J., Kumar, S., Norain, K., Abdullah, W., Guarner, J., Goldsmith, C.S., Chua, K.B., Lam, S.K., Tan, C.T., *et al.* (2002). Nipah virus infection: pathology and pathogenesis of an emerging paramyxoviral zoonosis. Am. J. Pathol. *161*, 2153–2167.

Wong, K.T., Robertson, T., Ong, B.B., Chong, J.W., Yaiw, K.C., Wang, L.F., Ansford, A.J., and Tannenberg, A. (2009). Human Hendra virus infection causes acute and relapsing encephalitis. Neuropathol. Appl. Neurobiol. *35*, 296–305.

WHO (2009a). Hendra Virus Fact Sheet (World Health Organization, Geneva, Switzerland).

WHO (2009b). Nipah Virus Fact Sheet (World Health Organization, Geneva, Switzerland).

Xu, K., Rajashankar, K.R., Chan, Y.P., Himanen, J.P., Broder, C.C., and Nikolov, D.B. (2008). Host cell recognition by the henipaviruses: crystal structures of the Nipah G attachment glycoprotein and its complex with ephrin-B3. Proc. Natl. Acad. Sci. U.S.A. *105*, 9953–9958.

Xu, Y., Gao, S., Cole, D.K., Zhu, J., Su, N., Wang, H., Gao, G.F., and Rao, Z. (2004). Basis for fusion inhibition by peptides: analysis of the heptad repeat regions of the fusion proteins from Nipah and Hendra viruses, newly emergent zoonotic paramyxoviruses. Biochem. Biophys. Res. Commun. *315*, 664–670.

Yob, J.M., Field, H., Rashdi, A.M., Morrissy, C., van der Heide, B., Rota, P., bin Adzhar, A., White, J., Daniels, P., Jamaluddin, A., *et al.* (2001). Nipah virus infection in bats (order *Chiroptera*) in peninsular Malaysia. Emerg. Infect. Dis. *7*, 439–441.

Yoneda, M., Guillaume, V., Ikeda, F., Sakuma, Y., Sato, H., Wild, T.F., and Kai, C. (2006). Establishment of a Nipah virus rescue system. Proc. Natl. Acad. Sci. U.S.A. *103*, 16508–16513.

Young, P.L., Halpin, K., Selleck, P.W., Field, H., Gravel, J.L., Kelly, M.A., and Mackenzie, J.S. (1996). Serologic evidence for the presence in Pteropus bats of a paramyxovirus related to equine *Morbillivirus*. Emerg. Infect. Dis. *2*, 239–240.

Yu, F., Khairullah, N.S., Inoue, S., Balasubramaniam, V., Berendam, S.J., Teh, L.K., Ibrahim, N.S., Abdul Rahman, S., Hassan, S.S., Hasebe, F., *et al.* (2006). Serodiagnosis using recombinant nipah virus nucleocapsid protein expressed in *Escherichia coli*. J. Clin. Microbiol. *44*, 3134–3138.

Yu, M., Hansson, E., Langedijk, J.P., Eaton, B.T., and Wang, L.F. (1998a). The attachment protein of Hendra virus has high structural similarity but limited primary sequence homology compared with viruses in the genus Paramyxovirus. Virology *251*, 227–233.

Yu, M., Hansson, E., Shiell, B., Michalski, W., Eaton, B.T., and Wang, L.F. (1998b). Sequence analysis of the Hendra virus nucleoprotein gene: comparison with other members of the subfamily *Paramyxovirinae*. J. Gen. Virol. *79*, 1775–1780.

Zhu, Z., Dimitrov, A.S., Bossart, K.N., Crameri, G., Bishop, K.A., Choudhry, V., Mungall, B.A., Feng, Y.R., Choudhary, A., Zhang, M.Y., *et al.* (2006). Potent neutralization of Hendra and Nipah viruses by human monoclonal antibodies. J. Virol. *80*, 891–899.

Zhu, Z., Bossart, K.N., Bishop, K.A., Crameri, G., Dimitrov, A.S., McEachern, J.A., Feng, Y., Middleton, D., Wang, L.F., Broder, C.C. *et al.* (2008). Exceptionally potent cross-reactive neutralization of Nipah and Hendra viruses by a human monoclonal antibody. J. Infect. Dis. *197*, 846–853.

Zimmer, G., Trotz, I., and Herrler, G. (2001). N-glycans of F protein differentially affect fusion activity of human respiratory syncytial virus. J. Virol. *75*, 4744–4751.

Measles Virus

Sibylle Schneider-Schaulies* and W. Paul Duprex

7

Abstract

Measles virus (MV) infections continue to be of high clinical relevance as they can be associated with severe disease processes such as pneumonia and central nervous system (CNS) complications, but also because they cause a generalized transient immunosuppression. Though characterized and causatively linked to MV decades ago, the pathogenesis of these diseases including the prime target cells in the respiratory tract is far from being understood. The advent of reverse genetics systems for both vaccine and wild-type viruses alongside the establishment of suitable tissue culture and animal models has helped to provide new insights into mechanisms of viral entry and tissue targeting both *in vitro* and *in vivo*. Furthermore, there is an increasing understanding of mechanisms underlying the disruption of immune functions towards secondary infections in the face of the induction of an efficient virus-specific immune response. For the latter, the interaction of MV with professional antigen-presenting cells and the consequences for T cell activation and/or inhibition, have received particular attention. The detailed knowledge of MV gene functions together with the definition of the interaction of MV with cells of the haematopoietic system is critical to improve the success of vaccination, particularly in young infants and in immunocompromised individuals, but also to use MV as a vector for targeted gene therapy.

History

Measles virus (MV) is highly contagious and causes acute measles, a disease usually contracted by children and young adults. The virus is distributed worldwide and although there are multiple genotypes, it is antigenically stable and exists as a single serotype. MV causes an acute infection in an individual, once in a lifetime, and persists in populations with a large enough birth rate to maintain transmission. There is no animal reservoir. Although monkeys are susceptible to infection, transmission from animals is not an important means of introducing the disease into a community. MV can persist for years in certain individuals, although these infections are rare and not associated with periodic shedding of infectious virus. A single infection is sufficient to confer lifelong immunity to clinical disease upon reinfection in the absence of re-exposure. Consequently, in order to remain endemic in a given community, the virus must rely on the infection of infants who are susceptible. So efficient is the process that the first known report of measles (in Egyptian hieroglyphics) failed to recognize the infectious nature of the illness, and described the typical clinical symptoms simply as a normal part of human growth and development.

In the pre-vaccine era in developed countries, the maximum incidence of measles was seen in children aged 5–9 years. Infections and epidemics

*Corresponding author

centred on elementary schools. Younger children acquired measles as secondary cases from their school-age siblings, and approximately 99% of young adults tested had been exposed to MV. From 1963, introduction of the measles vaccine has altered the age incidence and percentage of cases in different age groups. In countries with optimal vaccine coverage measles infection has shifted to the teenage group and infections tend to be due to importation of the viruses from areas of the world where the virus remains endemic. In areas with ineffective vaccine programmes, children up to four years of age are highly susceptible. In such developing countries, where measles has its greatest incidence in children under two years of age, case fatality rates range from 1% to 5% and can reach 30% in refugee and comparable settings with malnourishment, crowding and intensity of exposure mainly influencing severity and mortality (Moss and Griffin, 2006). A highly efficacious vaccine has been available for more than four decades, yet measles is still one of the leading causes of death for young children in the developing world. The Measles Initiative, a partnership launched in 2001 between the World Health Organization (WHO), the United Nations (UN), the US Centres for Disease Prevention and Control (CDC), the American Red Cross and other organizations, has developed and implemented comprehensive strategies for sustainable measles mortality reduction. As a result, measles deaths worldwide fell from estimated 733,000 in 2000 to 164,000 in 2008 (Wolfson *et al.*, 2007; http://www.measlesinitiative.org.

It was only with the isolation of the virus by Enders and Peebles in 1954 that experimentation on MV became possible. Since then the availability of tissue culture systems, monoclonal antibodies and molecular tools has permitted insights into viral structure and replication. Functional studies on MV *cis*-acting untranslated (UTR) regions and open reading frames (ORFs) have benefited from the establishment of reverse genetics systems allowing recombinant (r) viruses (rMVs) carrying stable alterations in their genomes to be generated (Radecke *et al.*, 1995; Takeda *et al.*, 2000; de Vries *et al.*, 2010). Thus, within limitations imposed by the fitness of the recombinant virus, the contribution of MV gene products and *cis*-acting sequences on the viral life-cycle in tissue culture, *ex vivo* models and experimentally infected animals can be addressed. MVs expressing fluorescent or enzymatically active reporter genes allow detection of virus infected cells *in vitro* and *in vivo* permitting tissue tropism and cell targeting to be studied with unprecedented levels of sensitivity (de Swart *et al.*, 2007). Reverse genetics also allows manipulation of the virus for applied purposes; for example, the virus is being retargeted to host cell receptors on tumour cells for use in systemic, cytoreductive regimens (Iankov *et al.*, 2007; Cattaneo *et al.*, 2008). Furthermore, rMVs expressing heterologous genes from simian immunodeficiency virus (SIV) or human immunodeficiency virus 1 (HIV-1) are being developed as multivalent vaccine vectors (Tangy and Naim, 2005; Zuniga *et al.*, 2007).

Infectious agent

Classification

MV belongs to the order *Mononegavirales* and is the type species of the *Morbillivirus* genus, one of six genera in the subfamily *Paramyxovirinae*. Morbilliviruses lack a virion-associated neuraminidase activity, cause formation of intranuclear inclusion bodies, exhibit antigenic similarities and produce similar diseases in their host species. Other members of this genus, detailed elsewhere in this book, include peste de petit ruminants virus (PPRV), which infects sheep and goats, the recently eradicated rinderpest virus (RPV), which infects cattle, canine distemper virus (CDV), which although isolated from dogs also infects a variety of wild animal species, phocine distemper virus (PDV), which infects seals and sea-lions, cetacean *Morbillivirus* (CeMV), which infects dolphins and porpoise. MV may have evolved from RPV in an environment in which cattle and humans lived in close proximity, and thus measles represents a relatively new infection of humans being regarded as 'a disease of civilization'.

Propagation

MV efficiently spreads via the aerosol route from infected individuals to initiate infection in the upper respiratory tract of susceptibles from where it targets secondary lymphatic tissues and spreads systemically by cell-associated viraemia. MV can be isolated from nasal washes, throat swabs and peripheral blood lymphocytes before antiviral immunity is established. Initially, Vero cells from the African Green monkey were used to isolate the virus from clinical samples. However, this is now known to cause tissue culture adaptation of wild-type viruses, which leads to the use of CD46 rather than CD150 as a cellular receptor (Tahara *et al.*, 2007a; see below). Therefore, B lymphoblastoid cells of human or marmoset origin, or, as recommended by WHO, Vero cells transfected to express CD150 are exclusively used for virus isolation (Ono *et al.*, 2001; http://www.cdc.gov/measles/lab-tools/vero-slam.html). When the virus is passaged in this way virulence is retained in the optimal animal model (see later). The virus spreads efficiently from cell-to-cell *in vitro* producing multinucleated syncytia or giant cell, which represent the typical cytopathic effect (CPE). Fusion occurs at neutral pH and although between 10^6 and 10^7 infectious particles are released into the supernatant, a large amount of the infectivity remains cell-associated. Typically sonication or multiple freeze–thaw rounds are used to liberate the virus from infected cells. Recombinant viruses, which express enhanced green fluorescent protein (EGFP), have been helpful in cases where cell-to-cell fusion is not evident; for example, in persistently infected cells (Ludlow *et al.*, 2005) or infected epithelial cells.

Morphology

Virions are highly pleomorphic and of average size 120–250 nm (range 300–1000 nm). They are enveloped and encapsidate the ribonucleoprotein (RNP) complex, which is the basic unit of infectivity and contains the negative-sensed genomic RNA. Two glycoproteins project as 5–8 nm-long spike-like projections (peplomers) from the 10–20 nm-thick lipid envelope. The fusion (F) glycoprotein is a type 1 protein and the haemagglutinin (H) glycoprotein is a type 2 protein, and both have short cytoplasmic tails. These cytoplasmic tails are believed to interact physically and functionally with the hydrophobic matrix (M) protein (Tahara *et al.*, 2007b), which, in turn, links the envelope to the RNP. Although not an integral membrane protein, M appears as a shell of electron dense material just below the membrane when virions are observed by electron microscopy (EM). The RNP has a density of 1.30 g/cm^3 in CsCl gradients and the genomic RNA is encapsidated fully by the nucleocapsid (N) protein to form a helical, RNase resistant structure with a length of 1 μm and a diameter of 18–21 nm. The amino-terminus of the N protein (N_{tail}) projects from the RNP and modulates processivity. Two additional proteins, the phospho- (P) and large (L) proteins associate to form the RNA-dependent RNA polymerase (RdRp), which must be incorporated into every virion for infectivity. The RNP has a herringbone structure when observed by EM and exhibits extensive conformational flexibility (Bhella *et al.*, 2004). Owing to the pleomorphic nature of the virion, the cargo space varies significantly from 3×10^5 to $>10^7$ nm^3 and can accommodate more than one RNP containing genomic or antigenomic RNA, (–) RNP or (+)RNP respectively, and many virions are functionally polyploid (Rager *et al.*, 2002). It is thought that the budding process does not actively discriminate between (–)RNPs and (+) RNPs. Defective interfering (DI) particles containing genomes with large internal deletions are common and the particle to plaque forming unit (p.f.u.) ratio is high (Afzal *et al.*, 2003). Cellular proteins, for example actin, can also be packaged into virions.

Genome organization

The single-stranded negative-sensed MV RNA genome is 15,894 nucleotides (nt) in length and, as in for most mammalian RNA viruses, shows suppression of CpG or UpA which may be important in escaping toll-like receptor (TLR) detection (Rima and McFerran, 1997). One N protein covers six nt, meaning that 2649 N proteins are required to form a (–)RNP. Thirteen N proteins make a single turn, resulting in approximately 204 helical turns/RNP. The genome obeys the 'rule of six' (Calain and Roux, 1993). Hence, the genome length must be an even multiple of six for efficient replication by the RdRp (Sidhu *et al.*, 1995). Genes are arranged linearly (3′-N-P/C/V-M-F-H-L-5′; Rima and Duprex, 2009) and it has been observed that they are located in a particular phase with respect to the positioned N protein. Whether this hexameric phasing is functionally important remains to be determined, although it is conserved among all the morbilliviruses (Rima *et al.*, 1995). The 56-nt long 3′ leader (Le) and the 40-nt long 5′ trailer (Tr) sequences at the genomic termini contain encapsidation signals and the promoters used for viral transcription and/or replication. Two sequence elements, referred to as the A-box, which comprise the first 15 nt of the Le and Tr and the B-box ($GN_5GN_5GN_5$), which is composed of the 14th, 15th and 16th hexamers of the Le, are necessary for promoter activity (Parks *et al.*, 2001; Liu *et al.*, 2006). The genome contains six transcription units (genes), which are separated by trinucleotide intergenic (Ig) sequences (5′-CUU-3′ for the first five and 5′-CGU-3′ between the H and L genes). These Ig sequences are preceded by an AU-rich gene end (GE) sequence and followed by gene start (GS) sequence (5′-AGGRNN$^{C}/_{A}$ARG$^{A}/_{T}$). A 1012 nt, GC-rich UTR, which probably folds into extensive secondary RNA structures, separates the M and F ORFs. Several ORFs have been identified within this UTR, although expression of any of the predicted proteins has never been substantiated. Reverse genetic studies (see later) demonstrated that it is dispensable for virus growth *in vitro*. The unusual high GC content and predicted secondary structures of the F 5′ UTR suggest a regulatory role, although this has been difficult to document on either the RNA or protein level (Takeda *et al.*, 2005; Rennick *et al.*, 2007). The transcription units encode at least eight proteins with the N (60 kDa), P (70 kDa), M (37 kDa), F (disulfide-linked 41 kDa F_1 and 22 kDa F_2), H (80 kDa) and L (220 kDa) being components of the virion. The second transcription unit gives rise to at least two additional proteins, the C and V proteins, which are translated from an overlapping reading frame or as a co-transcriptionally edited products (see below) of the P mRNA (Bellini *et al.*, 1985; Cattaneo *et al.*, 1989a). Accession of the V protein ORF relies on a sequence with similarity to the GE, the so-called editing site within the P gene, which instead of directing polyadenylation leads to RdRp slippage and insertion of one or more non-templated G residues. The W protein, which would result from an edited transcript containing two non-templated two G residues, has not been identified in infected cells. Over 30 complete genome sequences of representative laboratory-adapted, vaccine and wild-type viruses are available in sequence databases.

RNA and proteins

Transcription and replication: the N, P, L, C and V proteins

Upon infection of a permissive cell, the (−)RNP is liberated into the cytoplasm. Only virions containing a functional RdRp can initiate an infection and the complex acts as both transcriptase and replicase. Both processes rely on availability of the fully encapsidated RNP. The L/P complex contains the requisite catalytic activities for RdRp function although, as yet, the stoichiometry remains unknown. Two non-conserved hinge regions (H1 and H2) separate three largely conserved domains (D1-D3) within the 2183 amino acid L protein and it is very unlikely that the protein exists as a single globular entity. Although the three dimensional structure of the protein has not been determined, successful generation of a recombinant virus containing the complete ORF of EGFP within H2 region indicates that the domains are functionally important (Duprex and Rima, 2002). A canonical GDNQ motif in D2 serves as the most likely RdRp active site while D3 exhibits 2-O-methyl transferase activity (Ferron *et al.*, 2002). In addition to its interaction with the P protein, the L protein possibly homooligomerizes. The L protein also interacts with host cell factors, which modulate activities; for example, tubulin stimulates RNA synthesis *in vitro* (Cevik *et al.*, 2003). Whether such interactions serve to distinguish transcriptase from replicase activity or if it is governed by differential phosphorylation of the P protein is unknown. Phosphorylation sites (serines 86, 151 and 180 for casein kinase II and tyrosine 110), homotrimerization domains and regions which promote interactions with the N and L proteins have been mapped within the P protein sequence. The P protein carboxyl-terminal domain (PCT) contains an intrinsic disordered protein (IDP) domain, which is unfolded under physiological conditions and undergoes a disorder-to-order transition on protein interaction, a process termed induced folding (Bourhis *et al.*, 2006, 2007). Dynamic folding and unfolding of the PCT domain on interaction with N_{tail}, which is also an IDP, is thought to control RdRp processivity (Longhi, 2009). The P protein also interacts with the central structured N protein (N_{core}) domain preventing nuclear retention and self-assembly of the N protein. Typically for IDPs, N_{tail} (which contains three conserved domains, referred to as Boxes 1–3) order transits upon binding of viral (especially P protein), but also cellular interaction partners including Hsp72, which has been proposed to promote transcriptase processivity by competing with PCT binding within Box 2; thereby, destabilizing the transcriptase/template interaction (Johansson *et al.*, 2003; Zhang *et al.*, 2005).

Four phases in the dynamics of MV RNA accumulation have been described. In the first 6 h post infection (h.p.i.), transcription by the incoming RdRp is cycloheximide insensitive (Plumet *et al.*, 2005). Transcription is initiated following attachment of the RdRp to the promoters within the 3′ end of the genome. Whether or not a transcript of the 56 nt Le sequence is produced prior to the formation of a capped N mRNA, remains to be determined and it is possible that transcription could be initiated internally at the N GS sequence. The RdRp continues transcription at a calculated processivity of 3 nt/s through the first transcription unit. On encountering the GE sequence of the N gene one of a number of outcomes are possible. If the RdRp stutters or slips on the U-rich polyadenylation sequence in the template, 70–140 A residues will be added to the 3′ end of the N mRNA. Thereafter, the polymerase may either detach from the template or skip the Ig sequence and commence transcription of the P gene at the GS sequence. As for the N mRNA, an inverted G residue is added into the P mRNA and the cap is methylated. The RdRp slippage at the editing site in the P gene accounts for insertion of non-templated G residues in 30–50% of transcripts from this transcription unit. When the transcriptase detaches from the template at the gene boundary after completion of polyadenylation, transcription must be reinitiated at the 3′ end of the genomes. Consequently, a polar gradient is established where viral mRNAs produced from genes at the 3′ end of the genome are much more abundant than those that are promoter distal; thus, L-specific mRNAs are least abundant in infected cells. The slope of the transcription gradient varies with type of host cell and infection; for example, it tends to be steeper in neuronal cells and in persistent infections (Cattaneo *et al.*, 1987; Schneider-Schaulies *et al.*, 1995). As yet, mechanisms underlying transcriptional attenuation have not even been clarified in studies involving mini-genomes (Rennick *et al.*, 2007). Occasionally, GE, Ig and GS sequences are ignored and transcribed thereby giving rise to polyadenylated, read-through transcripts which can contain up to five transcription units. In the second phase of RNA synthesis (6–12 h.p.i.), newly formed RdRp complexes, produced following translation of nascent P and L mRNAs lead to an exponential increase in RNA synthesis. Transcription and replication are inextricably linked and phase three (12–24 h.p.i.) marks the transition to genome replication, which proceeds through a (+)RNP intermediate. Acting as a replicase the RdRp ignores GE, Ig, GS and editing sequences thereby, giving rise to full length encapsidated positive and negative stranded genomes, (+) and (−)RNPs. These, together with Le and Tr RNAs, are the only species which display 5′ triphosphates capable of activating innate immunity through RIG-I (Plumet *et al.*, 2007). In infected cells, Y forms of the RNP are observed by EM. RNPs have a strong promoter at the 3′ end ensuring that the ratio of (−)RNP:(+)RNP is high, somewhere in the range from 2:1 to 10:1 (Cattaneo *et al.*, 1989b; Plumet *et al.*, 2005). Replication relies on both P/L and N/P complexes, and it has been suggested that the plasticity provided by the extended conformation between the N and P IDPs are key in allowing contact between the assembly substrate (soluble, monomeric N protein complexed with P) and the replicase; thus, leading to a tripartite N/P/L complex (Longhi, 2009). Given the importance of the transcriptase/replicase, it is not surprising that expression and/or function of the L protein are being targeted *in vitro* by, for example, RNA interference or non-nucleoside inhibitors in order to exploit inhibition of the polymerase for antiviral therapy. From 24–28 h.p.i. RNA synthesis slows considerably. This could be due to an as yet unidentified mechanism, or more probably simply due to the extensive CPE in infected cells by this time point. There is no further production of viral RNA after 28 h.p.i.

In addition to host cell factors, there is evidence for modulation of the transcription/replication process by viral proteins, such as the non-structural C and V proteins (Bankamp *et al.*, 2005; Witko *et al.*, 2006; Bankamp *et al.*, 2008; Nakatsu *et al.*, 2008). Although the C protein has been shown to inhibit viral RNA synthesis in all systems tested, there is variation in the observed effect of the V protein. In two minigenome systems, the V protein has been shown to inhibit viral RNA synthesis by repressing minigenome replication through a direct V–RNA interaction (Parks *et al.*, 2006). A third system has found no such inhibition of RNA synthesis. There is currently no explanation for these conflicting observations. Late in replication, the M protein is found in association with the RNP complex and this has been interpreted as important in terminating RNA synthesis and physical engagement of the RNP as required for transport to the assembly and budding site of the virion at the plasma cell membrane (Runkler *et al.*, 2007; Suryanarayana *et al.*, 1994).

Assembly and budding: the M, F and H proteins

The viral mRNAs direct the synthesis of proteins, which are translocated to the assembly and budding sites. Glycoproteins are modified and acquire N-linked glycosylation (amino acids 200 and 215 in H, and three sites within the F_2 subunit) through the Golgi apparatus and are inserted into the plasma membrane. The ectodomain of the closely related Newcastle disease virus (NDV) is organized into a membrane proximal stalk and a membrane distal globular head region composed of a six-blade β-propeller and it was suggested that MV H structure would be similar (Crennell *et al.*, 2000). The six-bladed β-propeller fold of the head domain, which, unlike other paramyxovirus attachment proteins reveals a cube-like instead of a globular structure, has recently been confirmed by determination of the MV H crystal structure at atomic resolution, and amino acids and domains important in interaction with its different receptors were mapped (Hashiguchi *et al.*, 2007; Vongpunsawad *et al.*, 2004). Seven residues located within the globular head domain are essential for oligomerization and folding of the H glycoprotein, and a cysteine residue in the stalk regions for formation of disulfide linked dimers. In addition to mediating receptor attachment, H glycoprotein exerts a helper function in F-mediated membrane fusion at neutral pH, probably by directing the fusion peptide into the optimal distance to the target cell membrane and stabilizing the interaction. Synthesized as a precursor protein (F_0), F glycoprotein is cleaved in the Golgi apparatus by furin to yield two disulfide-linked subunits, F_1 and F_2. Glycosylation of the F_0 precursor is an essential prerequisite for cleavage, and mutations of any of glycosylation sites within F_2 affect additionally cell surface transport, stability and fusogenic activity of the F glycoprotein (Plemper *et al.*, 2003). The F_1 subunit reveals an amino terminal stretch of hydrophobic residues (the fusion peptide), and two amphipatic α-helical domains, one (HRA) is adjacent to the fusion peptide and the other (HRB) contains a leucine-zipper motif located at the amino-terminus of the transmembrane region. The fusion peptide within F_1 is masked during intracellular transport by intramolecular folding and this most likely prevents fusion of internal membranes (Doyle *et al.*, 2006). A central region mediates interaction with the H glycoprotein (Wild and Buckland, 1995).

The H and F glycoproteins associated to give a functional fusion complex on the surface of the virion in which H homooligomerizes as a dimer of dimers and F homooligomerizes into trimers. Strong intermolecular interactions form in the endoplasmic reticulum (ER) and hetero-oligomeric fusion complexes are transported through the secretory pathway to the plasma membrane (Plemper *et al.*, 2001). The strength of these interactions, which can be modulated upon interaction with the M protein, inversely correlates with the ability of the fusion complex to promote syncytium formation at the expense of particle production (Plemper *et al.*, 2002; Tahara *et al.*, 2007b). The F glycoprotein preferentially associates with membrane microdomains (lipid rafts) and as H and F form the fusion complex in the ER, F effectively recruits H glycoprotein to lipid rafts from where it would otherwise have been excluded (Vincent *et al.*, 2000; Pohl *et al.*, 2007).

The M protein condensed RNPs are recruited to the plasma membrane from the ill-defined intracytoplasmic inclusions in the perinuclear area thought to be the main sites of virus replication. Late in infection or on persistent infection, RNPs may also enter the nucleus, although the functional relevance of this is not known. Accumulation of the M protein at intracellular membranes is a prerequisite for co-transport with the RNP and subsequent virus release and budding (Runkler *et al.*, 2007; Salditt *et al.*, 2010). In polarized cells, M protein mediates the redirection of F and H glycoproteins from the basolateral to the apical surface assisting particle production (Naim *et al.*, 2000; Moll *et al.*, 2002). The M protein is highly membrane associated, although it is not known whether post-translational modifications facilitate this. About 20% of the M protein fraction co-floats with lipid raft markers, which is essential for the ability of the protein to promote the release of virus-like particles (Runkler *et al.*, 2007; Pohl *et al.*, 2007; Salditt *et al.*, 2010). Oligomerization of the M protein is required in this process confirming earlier suggestions on its aggregation into a crystalline array enabling distortion of the membrane into an outward-facing bulge, which promotes particle budding. Unlike related viruses, MV M protein does not harbour canonical late domains, which serve to recruit components of the endosomal sorting complex required for transport, identified as important in viruses which exit the cell by budding. MV particle production occurs independently of this system (Salditt *et al.*, 2010). Virion release is comparatively inefficient and particles remain highly cell associated, the basis of which is not understood although there is no evidence that tetherin is involved.

Genetics and reverse genetics

In order to generate a recombinant negative strand RNA virus it is necessary to assemble one form of the basic unit of infectivity within a permissive cell. As it is necessary to express mRNAs encoding the N, P and L proteins, typically full-length anti-genomic RNA is expressed to limit the likelihood of hybridization of genome and message. The first reverse genetic system, which was developed to generate rMV, used a fundamentally different approach to that employed to recover other members of the order *Mononegavirales*. It was based on expression of the anti-genomic RNA of a laboratory-adapted, strain Edmonston (Ed)-derived virus in 293 cells engineered to express T7 RNA polymerase and the N and P proteins (Radecke *et al.*, 1995). The resulting rMV, designated Edtag owing to the inclusion of three silent nucleotide changes in the genome, as has been used extensively by many groups across the world and it forms the foundation on which all subsequent developments were based. The use of synthetic mini-genomes was instrumental in paving the way to full-genome rescue and in understanding MV transcription and replication in general. These mini-genomes comprise reporter genes, such as chloramphenicol acetyl transferase (CAT) or EGFP flanked with the virus 3′ and 5′ promoter and other cis-acting sequences. They are typically monocistronic, but can also contain multiple reporter genes (Sidhu *et al.*, 1995; Rennick *et al.*, 2007). A replicating mini-replicon can also be accomplished by transfecting cells with *in vitro* synthesized RNA representing the mini-genome and then superinfecting the cells with replication-competent MV. This leads to the assembly of DI mini-genomic RNPs, which can be packaged into DI particles. Transcription from the GS sequence at the beginning of the reporter gene leads to the generation of capped and polyadenylated mRNAs from which the EGFP or CAT proteins are produced. Cells can be examined microscopically for fluorescence, which allows an understanding of the differences in expression between different cells or cell lysates can be assayed for acetylation of chloramphenicol which gives a population-based average. The majority of groups drive assembly of such mini-genome RNPs using four plasmids in which expression of the mini-genome and the N, P and L proteins are under the control of the T7 RNA polymerase or cytomegalovirus (CMV) immediate-early promoters. When the plasmids are introduced into eukaryotic cells by transient transfection, those containing a CMV promoter will be expressed in the nucleus. When plasmids with T7 promoters are used, the T7 RNA polymerase is supplied by co-infecting cells with the host range restricted modified vaccinia Ankara which expresses T7 RNA polymerase (MVA-T7), host range restricted modified fowlpox which expresses T7 RNA polymerase (FP-T7), replication competent vaccinia-T7 (vTF7–3) or by using BSR-T7 cells, which are stably transformed with T7 polymerase. Such approaches have also been applied to the generation of rMVs and there has been a gradual shift from the use of stably transfected cells to transiently transfected cells (Schneider *et al.*, 1997). Further developments have involved the generation of plasmids expressing N, P, L and T7 under the control of the highly efficient chicken actin promoter. Insertion of an RNA polymerase II promoter in place of the T7 promoter in the anti-genomic plasmid now makes inclusion of T7 expressing plasmid unnecessary. Although none of these approaches are technically trivial since the original system was published in 1995 they have been used to generate scores of rMVs, which have helped to shed light on many aspects of the molecular biology, cell biology and pathogenesis of MV. The original Edtag virus has limitations in terms of a number of mutations in the genome and although replication competent in transgenic mice expressing CD46 and lacking the interferon-αβ receptor (Mrkic *et al.*, 1998; Duprex *et al.*, 1999), the virus replicated poorly in macaques (Combredet *et al.*, 2003). These mutations have recently been corrected to produce an rMV which can infect macaques by the aerosol route (de Vries *et al.*, 2010). To circumvent this issue other reverse genetic systems have been developed based on the Schwarz and Moraten strains (Lorin *et al.*, 2005). Conversely, a very useful system has been generated based on a wild-type Japanese isolate IC323 (Takeda *et al.*, 2000) in which the virus is rescued in SLAM-expressing B95a cells (Tatsuo *et al.*, 2000). The recombinant virus has been shown to retain virulence in the macaque model (de Swart *et al.*, 2007) and has been used to study MV pathogenesis *in vivo* and tropism in *ex vivo* human cornea (Leonard *et al.*, 2008; Ludlow *et al.*, 2010). Additional wild-type virus based reverse genetic systems are required to ensure that observations made using IC323, which is a strain isolated in developed world, are general. As such a reverse genetic system based on a virus isolated in Khartoum, Sudan has been produced to fill this gap (Lemon *et al.*, unpublished).

Reverse genetics is typically used to generate rMVs based on vaccine, laboratory-adapted and wild-type MV genomes, which fall into seven main groups. First, viruses in which genes are replaced with those from other MV strains; second, viruses containing extra genetic material within one or more additional transcription units (ATUs); third, viruses in which the ORF has been modified at one or more nucleotides by site directed mutagenesis; fourth, viruses in which virus proteins have been extended by the insertion of short epitope tags or complete foreign proteins; fifth, viruses with modifications in the *cis*-acting UTRs; sixth, recombinants with complete genes deleted and; seventh, viruses in which the genome has been segmented. These recombinant viruses have been used to fully understand the biology of the virus and highlighted in context in the appropriate sections in this chapter. In order to illustrate the power of the reverse genetic systems, judiciously selected examples from each group are briefly covered below.

Viruses in which genes are replaced with those from other MV strains

Neurovirulence is a hallmark of MV infection and for many years rodent-brain adapted viruses, such as CAM/RB were passaged to study acute encephalitis in Lewis rats and C57 mice (see later). In these models, vaccine strains are not neurotropic, whereas CAM/RB leads to a profound infection. In order to understand the molecular basis of neurovirulence, envelope-associated genes from CAM/RB were inserted into the Edtag backbone and the H gene was shown to be the major determinant of neurovirulence (Duprex *et al.*, 1999).

Viruses containing extra genetic material within one or more ATU

Viruses that contain an ATU from which a foreign protein is expressed as a transgene have been invaluable in MV research. Technically, the most impressive virus that has been produced is one that has three ATUs; the first expressing EGFP from the promoter proximal position at the 3′ end of the genome; the second expressing β-galactosidase from an ATU positioned between the P and M genes and; the third ATU situated between the H and L genes from which CAT is expressed (Zuniga *et al.*, 2007). These genes increase the genome size by over 25% and the recombinant virus is termed a 'triporter'. Such viruses pave the way for vaccinologists who are interested in inserting tractable genes in the generation of novel vaccines.

Viruses in which the ORF has been modified at one or more nucleotides by site directed mutagenesis

Although the majority of MV tropism and pathogenesis can be understood based on the use of CD150, certain cell types, for example epithelial cell and neurons, are infected which do not express SLAM. Near saturation mutagenesis of the H glycoprotein has led to the generation of viruses which cannot use CD150 or CD46 as receptor (Vongpunsawad *et al.*, 2004). Such 'receptor-blind' will be invaluable for gene therapists who are making good progress in retargeting MV to selectively target tumour cells.

Viruses in which virus proteins have been extended by the insertion of short epitope tags or complete foreign proteins

Expression of reporter genes, especially EGFP, from ATUs has permitted cells to be identified very early after infection *in vitro*, *ex vivo* and *in vivo*. However, this approach does not localize the viral proteins directly as EGFP diffuses throughout the cell. A long term challenge in measles virology was the intractability of producing an antiserum, which would permit this detection of the L protein in infected cells and tissues. This issue was resolved upon the generation of an rMV in which the complete ORF of EGFP was inserted into the polymerase and has permitted the polymerase to be observed directly in living and fixed cells (Duprex and Rima, 2002). Interestingly, this approach slowed RdRp polymerization down from 3 nt/s to 0.6 nt/s (Plumet *et al.*, 2005) and led to the attenuation of two closely related wild-type *Morbillivirus* in disease-relevant animal models (Silin *et al.*, 2007).

Viruses with modifications in the *cis*-acting UTRs

Along with Edtag, the original rescue system reported the generation of a second recombinant virus, which had 504 nt of the GC-rich 5′ UTR of the F gene deleted (Radecke *et al.*, 1995). Interestingly this substantial deletion has no effect on the virus titre *in vitro* although a decrease in titre in a human thymus/liver implant model was reported (Valsamakis *et al.*, 1998). Similar alteration have been made in the IC323 background and this leads to a decrease in F glycoprotein levels and overall CPE (Takeda *et al.*, 2005).

Recombinant viruses with complete genes deleted

Sequencing studies examining the genomes of viruses present in the brain of subacute sclerosing panencephalitis (SSPE) patients has shown that the M protein is significantly altered. Much attention has been focused on the impact such changes have on how the virus spreads in the brain and whether or not these viruses are essentially confined to an intracellular existence. Redundancy of the M protein for the virus life-cycle is best illustrated by a recombinant virus in which the complete gene was deleted. This virus exhibited a hyper-fusogenic phenotype; virus titres were reduced approximately 250-fold compared with the parental Edtag virus and in infected animals the virus penetrated the parenchyma with greater efficiency than the equivalent M protein-expressing virus.

Viruses in which the genome has been segmented

Given that polyploidy has been observed (Rager *et al.*, 2002) and triporter viruses could be generated (Zuniga *et al.*, 2007) it was a logical extension to ask the question, 'could MV exist as a segmented virus?'. Reverse genetics was used to demonstrate that this was the case. Not only could three (–) RNPs be encapsidated and passaged but these segments could also stably maintain expression of two fluorescent proteins (EGFP and DsRed) and β-galactosidase (Takeda *et al.*, 2006). Such viruses may permit the development of novel vaccination strategies and shed light on aspects of RNP packaging, which have been difficult to investigate.

Given that recombination is not observed owing to the encapsidated nature of the MV genome, additional sequences are effectively 'locked' into the rMV. When the plethora of viruses is considered it is rather surprising that the rMVs are also very genetically stable at the level of the individual nucleotide. The error rate of the RdRp would suggest that mutations should be prevalent in non-essential genes, like for example an ATU containing EGFP. However, this does not prove to be the case and rather than what might be expected, i.e. randomization of the gene during *in vitro* or *in vivo* passage, non-fluorescent syncytia are only very rarely observed. Such fortuitous stability is another advantage for those attempting to use MV for applied purposes such as vectored vaccines or virotherapeutics.

Pathogenesis

Much of the knowledge that has been obtained in the area of pathogenesis comes from experimental infection in non-human primates or *ex vivo* studies. Precise details of the pathogenesis in humans are scant.

Animal models

One of the biggest challenges working with MV is the absence of a tractable animal model, which fully recapitulates the systemic infection observed in humans. This has held back understanding of the disease and it is interesting to note that nearly forty years after the virus was isolated from patients and cultured *in vitro* the primary target cell remains elusive, the underlying basis for immunosuppression is not yet understood and the molecular determinants of attenuation of the highly efficacious vaccine is still not known. A number of animal models have been developed and shed light on certain aspects of what is observed in patients. However, apart from rhesus and cynomolgus macaques, most tend to be uni- rather than multidimensional models of the disease in humans.

As a neurovirulent virus one of the first animal models to be developed was the intracerebral (i.c.) infection of rodents, established with a view to understanding viral neuropathogenesis. Thus, rodent brain (RB) adapted viruses, based on the CAM strain (CAM/RB) or hamster neurotropic virus (HNT), were generated by repeated passage from animal to animal. Sequencing the genome showed that the majority of nucleotide changes were in the H gene suggesting that the glycoprotein probably binds to a novel receptor on rodent neurons. The importance of the H glycoprotein was confirmed by reverse genetics, which proved that the rodent brain-adapted measles virus confers neurovirulence to the Edmonston vaccine strain (see above). Although it is important to recognize that entry into the CNS is far from natural, these viruses cause acute encephalitis which is observed in some patients. This is an important aspect of the clinical infection and it is not well understood either how the virus gains access to the CNS or how it spreads from neuron to neuron once the blood brain barrier is breached. In the future it will be important to determine if wild-type viruses can access the CNS in other models, for example the macaque. With the identification of CD46 as a cellular receptor for MV, an alternative approach to examining non-rodent brain adapted virus became feasible. Rather than adapt the virus to a rodent transgenic mice expressing CD46 were generated and backcrossed with a transgenic line in which the type I interferon receptor had been knocked-out (Ifnar^{ko}-CD46Ge). For the first time this permitted mice to be infected with vaccine or laboratory-adapted rather than brain-adapted viruses (Mrkic *et al.*, 1998). However, spread to the periphery was limited, although the animals could also be used to examine neurovirulence following i.c. infection with Edtag-based viruses expressing fluorescent proteins (Duprex *et al.*, 1999). Identification of CD150 as the cellular receptor used by wild-type viruses prompted the generation of further transgenic animals, which to a greater or lesser extent lead to more disease-relevant systemic infections (Sellin and Horvat, 2009). However, some of these models require that animals are back-crossed with Ifnar^{ko} or STAT-1-deficient animals to increase susceptibility to MV infection or enhance virus spread within the animal. Collectively, these data indicates that the intracellular milieu probably lacks one or more host factors which are necessary for key processes in the virus life-cycle such as RdRp function, RNP transport, and virion assembly.

Recognizing the challenges associated with murine models, alongside the artificial nature of the route of inoculation used in some cases and with a view to bridging the gap between them and the more disease relevant, but less tractable macaque model cotton rats (*Sigmodon hispidus*) have been developed as an alternative (Niewiesk, 2009). Although these animals are outbred, which in some ways might be considered to be an advantage and a full range of immunological tools are still in development, they do offer a number of advantages compared with transgenic and non-transgenic mice. First, it is possible to discriminate between wild-type and vaccine viruses in terms of relative levels of immunosuppression. Second, virus can be administered using the intranasal (i.n.) route and third, if enough virus is delivered in this way, a small, but significant and reproducible amount of spread from the initial sites of infection to local lymph nodes and the spleen is observed depending on the strain. Finally, this model permits the effect of maternal antibodies on disease progression and vaccine efficacy to be examined. A number of new and old world primate species have been used to model the disease. Their inherent strength lays in the fact that following intratracheal (i.t.), the full spectrum of clinical signs are observed and the virus replicates in the upper and lower respiratory tract, the gastrointestinal tract, lymphoid system and the skin (van Binnendijk *et al.*, 1994; McChesney *et al.*, 1997; Auwaerter *et al.*, 1999; Polack *et al.*, 1999). Rhesus (*Macaca mulatta*) and cynomolgus (*Macaca fascicularis*) macaques are most commonly used (Kobune *et al.*, 1996; El Mubarak *et al.*, 2007), although marmosets (*Saguinus mystax*) provide a useful extension as the virus is more neurovirulent when they are infected. Like the cotton rat, macaque infection permits the examination of existing and developmental vaccines in the presence of maternal antibodies (van Binnendijk *et al.*, 1994). Again, the fact that the animals are outbred mirrors the natural disease in humans. Combining rMVs, which express fluorescent proteins from an ATU alongside high resolution, three-dimensional reconstruction-based confocal microscopy with any of these models provides an excellent route to identifying very small numbers of cells *in* and *ex vivo*. These approaches, and more disease-relevant delivery of the virus, for example via the respiratory route, should be used with systematic immunological analyses during the prodromal, acute and recovery phase of the disease to provide unique insights not only into MV pathogenesis in particular, but also viral pathogenesis in general (de Swart, 2009). Recently, it has been shown that viruses which fail to bind to an as yet unknown epithelial receptor seem not to shed from the respiratory tract of infected monkeys (Leonard *et al.*, 2008). Identification of this receptor may help to answer the conundrum of how wild-type viruses gains access to CD150-negative cells lines, such as epithelia, endothelia and neurons, cell types, which are infected albeit at low levels in patients. It may also shed some light on aspects on the transmission of one of the most infectious human pathogens.

Host range

Measles is a human disease although it is clear that non-human primates can be infected naturally. It is interesting that free-living animals, imported into research facilities, were typically MV-antibody negative. However, close proximity with humans at a time when the virus was endemic in the developed world led to natural transmission in many primates species. Indeed, the virus can infect the great apes under this circumstance. Nonetheless, primates are most likely dead end hosts, certainly in the wild as the population size is too small to maintain transmission. Other animals cannot be infected with MV by natural routes as it is evident in the difficulties encountered in establishing small animal models.

Entry, spread and tissue tropism

A distinguishing feature is that unlike other paramyxoviruses, which use sialic acid molecules found on a range of cell surface molecules to enter a cell, MV gains entry to permissive cells using specific cell surface receptors. Identification of the first, CD46 or membrane cofactor protein (MCP) or CD46 (Dorig *et al.*, 1993; Naniche *et al.*, 1993) provides a salutary lesson to virologists who, for valid reasons, passage viruses *in vitro* through inappropriate cell lines (Buckland and Wild, 1997). Generation of the Edmonston measles vaccine utilized such approaches and these drove the virus to adapt from signalling lymphocyte activation molecule (SLAM) or CD150, which is recognized as the natural receptor used by clinical isolates (Tatsuo *et al.*, 2000; Yanagi *et al.*, 2006).

Epithelial cells do not express CD150 and are not easily infected with non-culture-adapted wild-type MV strains *in vitro* (Takeuchi *et al.*, 2003b); the same is true for neurons. It still has not been resolved whether there are specific neuronal and epithelial receptors, or indeed whether these are the same molecule. However, the identification of residues on the H glycoprotein, which affect entry into CD150-negative epithelial cells along with the generation of rMVs containing these mutations strongly suggested that MV can utilize other receptors (Takeda *et al.*, 2007; Leonard *et al.*, 2008; Tahara *et al.*, 2008). However, these remain to be formally identified. Generation of rodent-brain adapted viruses illustrates the same principle of receptor promiscuity. No naturally occurring MV strain has been isolated, which fails to use CD150 as an entry receptor and the expression of this molecule on cell types explains most aspects of viral tropism. CD150 is a member of a family of proteins, which regulate signalling in cells of the immune system. As such, it is expressed on a wide range of cells such as T- and B-cells, activated lymphocytes, macrophages and DCs, although expression is undetectable in monocytes and NK cells. Interestingly, murine CD150 cannot function as a receptor for MV, explaining, at least in part, why mouse models have been so difficult to establish (see above). As the virus binds TLR2, which upon ligation leads to CD150 induction, the virus may 'switch on' its own receptor (Bieback *et al.*, 2002). Although this remains to be shown *in vivo*, such an approach could prove to be mechanistically very attractive from a pathological perspective as a virus spreads and encounters new micro-niches within the host. Non-canonical receptors, such as DC-SIGN, which binds MV but does not lead to internalization into cells, have also been identified (de Witte *et al.*, 2006). Likewise, the NK-1 receptor may have a role in binding to the F glycoprotein (Makhortova *et al.*, 2007). This is particularly relevant in the CNS where it has been suggested, although not proven, that the virus may spread from cell-to-cell transynaptically in the absence of the release of infectious virus. Thus, MV lymphotropism, and to a lesser extent epitheliotropism, neurotropism and endotheliotropism, can and should be explained by receptor and mostly CD150 usage. Whether the conundrum of how the virus gains entry into CD150-negative cells will be resolved in the near future remains to be determined; indeed from the broader perspective of viral pathogenesis one might question its importance.

Immune responses

MV induces a marked immune activation, which usually controls acute measles efficiently. However, clearance of genomic RNA may require several months (Riddell *et al.*, 2007). Cellular immunity is crucial for recovery from measles as exemplified by the fact that morbidity and mortality is exceedingly high in children with cellular immune deficiencies agammaglobulinaemia does not impact on resolution of the disease or clearance of the virus (Griffin, 1995; Moss *et al.*, 2004; Permar *et al.*, 2003a, 2004;). Coincident with immune activation, MV induces a generalized immune suppression, which can continue for many weeks after the overt clinical symptoms have disappeared. This was recognized long before MV was isolated and is the major reason for the constantly high morbidity and mortality rate associated with acute measles, particularly in the developing world.

Immune activation

The role of innate immunity in measles is poorly understood, although in experimentally infected mice, type I interferon (IFN) has proven to be important in MV interference. However, direct evidence for its importance in human infection has not been obtained. Pointing to the importance of type I IFN in MV control, C and V non-structural proteins, which are dispensable for growth *in vitro*, are involved in evasion from type I IFN. Unlike the cytosolic V protein, C protein was found to shuttle between the nucleus and cytoplasm and associate with viral particles (Devaux and Cattaneo, 2004). Both proteins regulate the efficiency of MV replication in primary cells and IFN production or signalling (Tober *et al.*, 1998; Escoffier *et al.*, 1999; Takeuchi *et al.*, 2003a; Palosaari *et al.*, 2003; Fontana *et al.*, 2008). More recently, the P protein was found to inhibit interferon signalling as well by promoting STAT1 degradation (Devaux *et al.*, 2007). Unsurprisingly, the absence of V and C proteins affects MV virulence in animal models (Valsamakis *et al.*, 1998; Devaux *et al.*, 2008), although it is questionable if a ΔV virus based on Edtag is really functionally different from Edtag due some of the mutations in the P gene. Differential induction of type I IFN by MV strains has also been linked to the presence of DI particles in the respective preparation rather than to differences between wild-type and attenuated strains (Naniche *et al.*, 2000; Shingai *et al.*, 2007). Terminal 5′ triphosphates at ends of the genome, anti-genome and DI termini might be efficient at inducing type I IFN via detection by RIG-I (Plumet *et al.*, 2007). Whether detection by TLRs is involved in IFN-induction is unlikely as wild-type MV does not interact with TLR-4 or -9, but rather TLR-2 as shown on monocytes (Bieback *et al.*, 2002). TLR-2 independent activation of monocytes and DCs may also relate to the ability of intracellular MV to activate NFκB directly by as yet unknown mechanisms (Hummel *et al.*, 1998; Dhib-Jalbut *et al.*, 1999; Helin *et al.*, 2001). In line with the importance of these cells in induction of MV immune responses, exposure to MV promotes functional maturation of immature DC as indicated by upregulation of MHC class II and costimulatory molecules. Cytokines induced in DC as well as in monocytes by MV infection include IL-1-α/β, IL-1R antagonist, IL-6, TNF-α, type I IFN and low amounts of IL-12p35 and IL-12p40, some of which trigger DC maturation or are directly involved in polarizing ensuing T-cell responses (Schnorr *et al.*, 1997b; Klagge *et al.*, 2000; Servet-Delprat *et al.*, 2000b; Dubois *et al.*, 2001; Bieback *et al.*, 2002; Klagge *et al.*, 2004; Gringhuis *et al.*, 2007). Not surprisingly, autologous peptide loaded DC promote activation of MV-specific $CD8^+$ T cells *ex vivo* and there is no evidence that exposure of DCs to MV would abrogate priming of T-cell responses (Ndhlovu *et al.*, 2010).

$CD8^+$ T cells are important in the early control of viral replication and likely to contribute to protective immunity (Permar *et al.*, 2003a). MV-specific HLA-restricted $CD8^+$ cytotoxic T lymphocytes appear in the blood at the time of the rash, some of which develop into CD8 memory cells (Jaye *et al.*, 1998; Nanan *et al.*, 2000). MV-specific $CD4^+$ T cells are also activated in response to MV infection, and T cell epitopes are present in many MV proteins, although they are particularly abundant in the H and F glycoproteins (Oh *et al.*, 2006). Consistent with early activation of $CD8^+$ and Th1 $CD4^+$ T cells, plasma levels of IFN-γ and neopterin, soluble IL-2R, CD4 and CD8 and IL-2 initially rise, while the subsequent switch to persistent IL-4 production marks a Th2 $CD4^+$ T cell dominated response during recovery (Griffin, 1995; Griffin and Ward, 1993). On *in vitro* restimulation with MV antigens, mainly IFN-γ-producing T cells are activated in acute measles and after vaccination, while month to years later, both IFN-- and IL-4-producing cells can be detected (Gans *et al.*, 1999; Dhiman *et al.*, 2005; Ovsyannikova *et al.*, 2003a,b). Although not substantiated in humans, expansion of $CD4^+CD25^{high}FoxP3$ expressing cells has been reported in experimentally infected animals, indicating activation of regulatory T cells in measles (Sellin *et al.*, 2009). Whether these mainly act to control MV specific effector cells or contribute to immunosuppression and/or establishment of persistence has not yet been resolved. The impact of neither these parameters nor for quality, magnitude or longevity of T-cell responses has on immunogenetic factors, such as polymorphisms in HLA class I and II alleles and cytokine genes been verified (Dhiman *et al.*, 2007; Ovsyannikova *et al.*, 2006a,b).

CD4 responses shape and promote production of MV-specific antibodies. These, initially of the IgM- and then IgG1 and IgG4-isotypes. In contrast to N-specific antibodies, which are produced most abundantly and rapidly. H protein-specific Igs are neutralizing. B-cell epitopes mapped within the H glycoprotein are mainly conformational and localize to regions between amino acids 368 and 396 and in the highly conserved CD150 binding site (Ertl *et al.*, 2003; Hashiguchi, 2007). F glycoprotein specific antibodies mainly recognize linear epitopes and contribute to MV neutralization by interfering with virus-to-cell fusion at the plasma membrane (Polack *et al.*, 2000). In line with the observation that agammaglobulinaemia has no effect on the resolution of acute measles, MV clearance was not affected on depletion of $CD20^+$ lymphocytes from experimentally infected monkeys and no unusual clinical sequelae were observed (Permar *et al.*, 2004). Although

MV-specific antibodies are extensively produced in persistent infection in humans, they are also not effective at eliminating the virus from the CNS.

MV-induced immunosuppression

Measles patients are highly susceptible to opportunistic infections and lymphopenic, which may affect up to 50% of the cellular peripheral blood compartment (Albrecht *et al.*, 1980; Okada *et al.*, 2000; de Swart *et al.*, 2007). The latter is not associated with a decrease in thymic output (Permar *et al.*, 2003b) and may result from lymphocyte depletion owing to viral infection or apoptosis (Addae *et al.*, 1995; Auwaerter *et al.*, 1996; Okada *et al.*, 2000; de Swart *et al.*, 2007). This may occur particularly efficient in secondary lymphatic tissues where MV transmission from infected DCs to activated T cells would cause substantial loss of these cells (de Swart *et al.*, 2007; de Witte *et al.*, 2008). Immunosuppression, however, is observed up to weeks after the onset of the rash when the lymphocyte counts have returned to normal and MV-infected cells are, if at all, present with only low frequency (Okada *et al.*, 2000). Key features of MV-induced immunosuppression are inhibition of delayed type hypersensitivity (DTH) responses and a restricted ability of lymphocytes to proliferate in response to recall-antigens as well as polyclonal stimulation (Schneider-Schaulies and Dittmer, 2006; Schneider-Schaulies and Schneider-Schaulies, 2009). Several hypotheses have been put forth to explain this and these focus on deregulation of APC viability and function, which would be expected to significantly impact induction, efficiency and quality of T cell activation. There is evidence from murine models that development of DCs from precursors is blocked by MV and this was proposed to relate to type I IFN production and STAT2-dependent signalling (Hahm *et al.*, 2005).

The virus targets DCs *in vivo* as CD11c, MHC class II+ cells expressing MV encoded proteins were found in lymph nodes of monkeys experimentally infected (de Swart *et al.*, 2007; de Witte *et al.*, 2008) and in CD150 transgenic animals (Welstead *et al.*, 2005). As outlined above, MV is sensed by both plasma membrane and cytosolic pattern recognition receptors, which promote initial activation and presumably mobilization. Recent evidence suggests that TLR-dependent signals might be modulated on co-recognition by DC-SIGN, which has been shown to enhance MV infection of DCs (de Witte *et al.*, 2006; Gringhuis *et al.*, 2009). In immature DCs generated from precursors *in vitro*, MV promotes phenotypic and functional maturation as indicated by upregulation of MHC and costimulatory molecules as well as cytokine production (Fugier-Vivier *et al.*, 1997; Grosjean *et al.*, 1997; Kaiserlian *et al.*, 1997; Schnorr *et al.*, 1997b; Steineur *et al.*, 1998; Klagge *et al.*, 2000; Servet-Delprat *et al.*, 2000b). The extent of phenotypic maturation mostly was lower than in lipopolysaccharide (LPS)- or tumour necrosis factor α (TNFα) matured DCs, is however considerably less important for functional impairment. Rather, inefficient chemokine receptor switching might impact on recruitment of activated DCs from the periphery to secondary lymphatics (Abt *et al.*, 2009) and this may be further compromised by their reduced migratory activity on extracellular matrix (ECM) components (Shishkova *et al.*, 2007). Moreover, impairment of IL-12 production and release of IL-10 were reported in MV infected DC cultures, which expectedly would affect the quality of ensuing T-cell responses towards a Th2/Th1 phenotype which has, however, not been observed on direct co-culture with T cells (Karp *et al.*, 1996; Fugier-Vivier *et al.*, 1997; Servet-Delprat *et al.*, 2000b; Dubois *et al.*, 2001; Marie *et al.*, 2001; Hoffman *et al.*, 2003; Klagge *et al.*, 2004). Moreover, preferential induction of Th1 cells following TCR/CD46 co-ligation and promotion of Th2-biased responses by the ability of MV to abrogate TLR-dependent and –independent type I IFN production from plasmacytoid DCs were proposed (Kemper *et al.*, 2003; Schlender *et al.*, 2005). Independently of potentially dampening cellular at the expense of regulatory and humoral immunity, MV-infected DCs may affect viability of conjugating T cells upon induction of TRAIL-mediated apoptosis or transfer of virus, which efficiently replicates in activated T cells which may, in turn, also act to deplete DCs in a CD95-dependent manner (Fugier-Vivier *et al.*, 1997; Servet-Delprat *et al.*, 2000; a de Vidalain *et al.*, 2000; Swart *et al.*, 2007; de Witte *et al.*, 2008). However, these scenarios require formation of stable interfaces between MV-infected DCs and conjugating T cells which, as revealed by live imaging experiments, are rarely formed *in vitro* and this most likely explains the inefficient viral transfer from DCs to non-activated T cells (de Witte *et al.*, 2008) and the inability of MV-infected DCs to promote allogenic T-cell expansion in MLRs (Schneider-Schaulies *et al.*, 2003; Servet-Delprat *et al.*, 2003; Shishkova *et al.*, 2007).

Although the mechanisms underlying formation of unstable immune synapses (IS) by MV have not been resolved as yet, viral proteins expressed on the DC surface may actively contribute to T-cell silencing (Kerdiles *et al.*, 2006; Schneider-Schaulies and Dittmer, 2006). Studies addressing their role in interfering with stimulated T-cell expansion have mainly been performed *in vitro*. There, the N protein, cell surface display by an as yet unidentified cellular pathway has been postulated, which reportedly blocked stimulated expansion of transformed lymphoid cell lines and activated T cells via N_{tail} while interacting with FcgII receptors on B cells to block Ig synthesis via N_{core} (Laine *et al.*, 2005). Signalling by the MV glycoprotein fusion complex, as expressed on viral particles and infected cells also including DCs to T cells, has been investigated in more detail both *in vitro* and *in vivo* (Schlender *et al.*, 1996; Niewiesk *et al.*, 1997; Niewiesk *et al.*, 1999; Weidmann *et al.*, 2000; Avota *et al.*, 2001). As a consequence of non-fusogenic interaction with the glycoprotein fusion complex, T-cells arrest in the G1-phase of the cell cycle, yet do not undergo apoptosis (Schnorr *et al.*, 1997a; Engelking *et al.*, 1999; Niewiesk *et al.*, 1999). The receptor structure is unknown, yet apparently signals to target activation of the phosphatidyl-inositol-3-/Akt kinase (PI3/AktK) pathway, which is key to transmission of signals essential in cell survival, especially in lymphoid cells, and cell cycle entry (Avota *et al.*, 2001, 2004, 2006; Schneider-Schaulies *et al.*, 2002). In addition to promote activation of proteins directly involved in cell cycle progression such as cyclin-dependent kinases, PI3K essentially controls actin cytoskeletal dynamics, which is required for most parameters in T-cell activation such as motility, mobility, morphing (as relevant for migration and formation of interface with DCs) and receptor segregation into functional clusters (Vicente-Manzanares and Sanchez-Madrid, 2000; Wulfing *et al.*, 2003; Billadeau and Burkhardt, 2006). Not surprisingly, MV interaction interfered with the ability of T cells to reorganize their actin cytoskeleton on encounter with ECM components or stimulatory interfaces yet also caused loss of actin based protrusions indicating that the glycoprotein fusion complex directly signals to these cells (Muller *et al.*, 2006). A recent study provided evidence that these involve activation of sphingomyelinases causing accumulation of membrane ceramides and thereby ceramide-enriched platforms known to regulate lateral segregation of receptors and signalling components (Gulbins *et al.*, 2004; Zhang *et al.*, 2008; Gassert *et al.*, 2009).

Clinical features

Acute measles

MV first gains entry into the body through the upper respiratory tract or conjunctiva. Although the identity of the primary target cells remains elusive, professional antigen presenting cells such as tissue resident macrophages or DCs rather than epithelial cells may be first targets of infection, which acquire virus by CD150-dependent, DC-SIGN enhanced infection (de Swart, 2007, 2009; de Witte *et al.*, 2008). APCs *cis-* or *trans-* infected with MV most likely mediate transport of MV to secondary lymphatics where MV replicates at a very high level in experimentally infected macaques with B cells being most efficiently infected (de Swart *et al.*, 2007). From here the virus spreads to the rest of the reticuloendothelial system and respiratory tract through infected lymphocytes (primary viraemia). Giant cells, containing inclusion bodies (Warthin-Finkeldy cells), are formed in lymphoid tissues, but are rarely observed in tracheal and bronchial epithelial cells. Less than 2% of peripheral B and T cells harbour MV in the macaque model (de Swart *et al.*, 2007). In humans, MV can be isolated from patient PBMCs in the early stages of infection and this is greatly assisted if the cells are mitogenically stimulated. About five days following initial infection, MV spreads from compartments supporting extensive replication to the skin and viscera, kidney and bladder (secondary viraemia). Giant cells within infected tissues are also characterized by lymphoid hyperplasia and inflammatory infiltrates. Patients enter a 2–4 days prodromal phase and 10–11 days following the initial infection display symptoms, such as fever, malaise, sneezing, rhinitis, conjunctivitis, cough, occasionally associated with a transitory urticarial or macular rash. At this time, giant cells are present in the sputum, nasopharyngeal secretions and urinary sediment cells, virus is present in blood and secretions and the patient is highly infectious. During this period, Koplik's spots, the pathognomonic exanthema of measles, appear on the buccal and lower labial mucosa opposite the lower molars. The distinctive rash appears about 14 days after exposure and starts behind the ears and on the forehead from where it spreads to involve face, neck, trunk, and upper and lower extremities. Histologically, the rash is characterized by vascular congestion, oedema, epithelial necrosis and round cell infiltrates. Virus antigen is absent from skin lesions, but rather concentrated near blood vessels and in the endothelial cells of the dermal capillaries. The rash fades after 3–4 days and fever, conjunctivitis and respiratory symptoms subside. Antibody titres rise and restrict virus shedding and the patient shows rapid improvement. Modified measles can occur in partially immunized individuals and this follows a similar course with very reduced symptomatology. Atypical measles, a severe manifestation of acute measles, were seen after wild-type MV infection of recipients of a formalin-inactivated vaccine, which was therefore withdrawn from the market in 1969. The pathogenesis of the atypical measles involves a strong anamnestic humoral response marked by low avidity, complement-fixing and non-protective antibodies and deposition of immune complexes in the lung and a pronounced eosinophilia (Polack *et al.*, 1999). In immunocompromised individuals, especially those with cellular immune deficiencies, severe, often fatal complications, such as giant cell pneumonia or measles inclusion body encephalitis (MIBE), which is established within a few months after acute measles and shares certain features with SSPE (see below).

Complications of measles infection

Complications of acute measles in immunocompetent hosts are relatively rare and result mainly from opportunistic secondary infection of necrotic surfaces, such as those in the respiratory tract. On rare occasions, myocarditis, pericarditis, hepatitis, appendicitis, mesenteric lymphadenitis and ileocolitis occurs. MV has been suggested to relate to the pathogenesis of a variety of diseases also including chronic active hepatitis, multiple sclerosis and osteitis deformans and otosclerosis, yet for neither of those pathogenic significance has been substantiated (Rima and Duprex, 2006).

Severe CNS complications are acute post-infection measles encephalitis (APME) and SSPE with no precipitating factors known. APME is characterized by resurgence of fever, headache, seizures, cerebellar ataxia and coma which occurs during acute measles at a frequency of about 1 per 1000 to 5000 cases and can be fatal in up to 20 to 40%. In those who recover, lasting neurological sequelae, such as selective brain damage with retardation, recurrent convulsive seizures, hemi and paraplegia can occur. Histologically, demyelination, perivascular cuffing, gliosis, fat-laden macrophages near the blood vessel walls and, in some cases, inclusion bodies in brain cells can be seen. Neither MV nor MV-specific antibodies can usually be detected in APME brain tissue or cerebrospinal fluid (CSF), respectively; lending support to the hypothesis that myelin breakdown promoted by virus-induced autoimmune reactions is centrally involved and, though not experimentally proven, may involve molecular mimicry or a deregulation of autoreactive cells occurring secondary to viral infections.

With an incidence of 1 case per 10^4 cases of acute measles (Takasu *et al.*, 2003; Bellini *et al.*, 2005), SSPE is a rare, fatal, slowly progressing degenerative CNS disease. It generally manifests in children and young adults six to eight years following acute measles, however, it can also occur after up to 20–30 years. Predisposing factors remain unknown, yet do not include unusual features of the acute infection. Both genetic factors, as documented by occurrence of SSPE in siblings (Vieker *et al.*, 2003; Tuxhorn, 2004) or high incidence in Papua New Guinea (Takasu *et al.*, 2003) and immaturity of the immune system upon primary infection, may contribute since half of SSPE patients have contracted measles before the age of 2 years.

The course of SSPE is variable. The onset is usually marked by generalized intellectual deterioration or psychological disturbance followed by neurological or motor dysfunctions, such as dyspraxia, generalized convulsions, aphasia, visual disturbances or mild, repetitive simultaneous myoclonic jerks. In 75% of cases, the invasion of the retina by the virus leads to a chorioretinitis, often affecting the macular area followed by blindness. Finally, the disease proceeds to progressive cerebral degeneration leading to coma and death. Remissions are common, yet SSPE inevitably leads to death.

Histologically, a diffuse encephalitis affecting both the white and grey matter is seen, characterized by perivascular cuffing, diffuse lymphocytic infiltration, glial cells proliferation, intranuclear inclusion bodies (Cowdry bodies) in fibrous astrocytes, neurones and oligodendrocytes, and the absence of giant cell formation. EEG changes typically consist of periodic high-amplitude slow wave complexes (Radermecker), which are bilateral, usually synchronous and symmetrical and synchronous with myoclonic jerks recurring at 3.5–20 s intervals. Usually, two or more delta waves appearing bi- or polyphasic are seen, yet the EEG pattern can vary depending on the patient or disease course and even disappear on disease progression. Hyperimmunity against MV proteins in serum and CSF is a hallmark of SSPE. CSF antibodies are IgG, IgA and IgD isotypes, and in this compartment, M-specific antibodies are usually absent and the immune response is oligoclonally restricted reflecting local antibody production from plasma cells having invaded the CNS from the periphery.

Although the role of MV in the aetiology of SSPE is well documented, questions concerning mode of brain invasion and initial establishment of persistence are unresolved. The presence of large amounts of viral proteins in the absence of detectable giant cell formation together with the absence of free virus in the CSF and the difficulty to isolate MV from brain tissue led to the hypotheses that (1) infection may spread slowly strictly in a cell-associated manner and not involve particle formation, and (2) viral gene expression might be defective supporting a persistent rather than a lytic course of the infection. In both SSPE and MIBE, only N and P proteins are routinely detected in brain sections, while transcriptional and translational alterations affect the expression of envelope protein specific genes. Mutations are found all over the MV genome, yet alterations are most frequent in the M gene, followed by F, H, P and N genes the L gene is most conserved (Billeter *et al.*, 1994). Both point mutations most probably accumulating as a result of the error rate of the RdRp, or clustered transitions are seen, which are thought to result from the activity of a cellular enzyme complex, now referred to as ADAR. As a result, translation from viral mRNAs is abolished or leads to the synthesis of truncated or unstable MV proteins, which may well account for lack of infectious particle production in SSPE. Viruses were rarely isolated post mortem by co-cultivation (referred to as SSPE isolates), some of which proved to be contaminations of ordinary laboratory MV strains on sequence analysis and it is still unclear whether the 'true SSPE-isolates' represent subpopulations of replication/maturation competent viruses or revertants that were selected by isolation. Factors involved in the establishment of persistence by a non-defective MV in brain tissue poorly understood, yet may involve cell type specific restriction of MV gene expression, such as the type I IFN-inducible MxA protein, which acts to slow down viral replication (Schneider-Schaulies *et al.*, 1995; Schneider-Schaulies and ter Meulen, 1999). In addition, the ADAR activity actively modifying the primary sequence of viral RNAs has been demonstrated in these cells. Whether single nucleotide polymorphisms identified in the promoter regions of HLA, IL-4 and MxA genes found in SSPE contribute to establishment or maintenance of persistence is unknown (Inoue *et al.*, 2002; Torisu *et al.*, 2004; Pipo-Deveza *et al.*, 2006).

Immunological features associated with establishment of SSPE are unknown. Transient immunosuppression and immaturity of the immune system on primary exposure are thought to contribute, yet this has not been attributed to specific cellular compartments (Oldstone *et al.*, 2005). In mice, CD4, CD8 and B cells in conjunction control primary MV CNS infection (Tishon *et al.*, 2006), yet it is unknown if defective responses of specific compartments favour persistence. However, as there is no direct evidence for defective cell-mediated immunity in the control of MV infection in SSPE, where neither T-cell frequencies, proliferation and effector functions including interleukin synthesis are not detectably impaired. Humoral hyperimmunity is defining for SSPE, yet unable to control, and has even been proposed to rather support CNS disease antibodies. They may act to clear antigen from brain cell surfaces by internalization, but also directly regulate MV gene expression by activating signalling pathways upon ligation of the H protein by neutralizing antibodies (a process referred to as 'antibody-induced antigenic modulation'; Liebert *et al.*, 1990; Schneider-Schaulies *et al.*, 1992).

Epidemiology

In the northern hemisphere, measles incidence tends to rise in winter and spring when lowered relative humidity would favour aerosol transmission to the upper respiratory tract, the nose and, possibly, the conjunctivae. The spread of measles illustrates the principles of epidemiology. Populations smaller than 500,000 do not have birth rates supplying the number of susceptibles required for the maintenance of the virus, which was completely eliminated from isolated groups until reintroduction from outside. Then, the disease tends to be more serious because all age groups are susceptible to the infection. Mortality from uncomplicated measles was historically as high as 30% and currently may be as low as 1/1000 in industrialized countries, 5–10% in endemic areas in sub-Saharan Africa and may reach 25% in refugee camps or other situations of overcrowding (Moss, 2007; Moss and Griffin, 2006). These figures can also be achieved on re-introduction of MV into populations not exposed to it for long periods, such as Fiji Islands in 1875 or in Greenland in 1951 where epidemics with fatality rates of up to 20–25% occurred. In spite of the existence of genotype-specific neutralizing antibodies (Finsterbusch *et al.*, 2009; de Swart *et al.*, 2009), MV is monotypic and infection by any strain confers protective immunity. As detailed below, MV genotypes are defined, which, when active, accumulate mutations continuously. As yet, there is no evidence for the existence of MV genotypes differing in pathogenic potential such as lymphotropism and neurovirulence, nor antigenic drift in wild-type MV strains that may impair the protective effect of the current vaccine.

Molecular epidemiological data have proven highly important in evaluating the efficiency of mass vaccination campaigns as they allow track imported MVs. For instance, in 1995, in the United States 60% of measles reported were either directly imported or were found to be directly linked to an imported case (Rota *et al.*, 2002).

Sequence analysis of MVs have enabled us to establish their relationship and assignment to lineage groups, which are referred to as 'clades' (designated A–H) containing different genotypes. Genotyping is based on the carboxy-terminal 151 amino acids of the N protein, where up to 10.6% divergence in sequence between unrelated strains can occur. In addition, the sequence of the H gene is available for most recent isolates to assess for potential antigenic variants. While some MV clades are more or less extinct (i.e. have not been isolated for at least 15 years), other clades are co-circulating (see http://www.cdc.gov/ncidod/dvrd/revb/measles/index.htm for actual information). The heterogeneity of genetically related recent isolates, which are drifting on a genetic level mirrors the activity within a clade, and molecular surveillance, analysed in conjunction with standard epidemiological data, supported documentation of interruption of transmission of endemic measles. In addition, molecular epidemiology has allowed clear association of wild-type, but not vaccine MV with SSPE. In the late 1980s and beginning 1990s of the last century, wild-type MV of a known genotype was re-imported into the US causing about 50,000 cases. About 10 years later, seven cases of SSPE were noted, which could clearly be assigned to the wild-type MV having caused the acute infection indicating that SSPE is caused by circulating wild-type viruses and develops more frequently than thought (Takasu *et al.*, 2003; Bellini *et al.*, 2005).

Diagnosis

Acute measles can be diagnosed on the basis of the distinctive symptoms, yet the diagnostic acumen of many physicians to recognize acute measles has decreased with implementation of vaccine programmes. Since vaccination itself has led to the emergence of unusual presentations and milder forms of measles and more patients are immunocompromised, the need for laboratory diagnosis/confirmation of MV has increased. Microscopy has been widely used to detect multinucleate giant cells with inclusion bodies (Cowdry bodies) in the nasopharyngeal secretions (NPS) or the MV proteins by indirect immunofluorescence (IIF) in cells shed in nasal secretions or urine (from 2 days before to up to 5 days after the appearance of the rash). Virus isolation on specimen collected within 0 to 5 days of rash onset is not part of routine diagnosis, yet encouraged particularly in outbreak settings where there is a requirement for molecular epidemiological data. Urinary sediment cells and mitogen-activated peripheral blood lymphocytes are generally used as sources for isolation, and, as indicated above, Epstein–Barr virus transformed marmoset B lymphocyte cell line (B95a) (Kobune *et al.*, 1990) or transgenic Vero cells stably expressing CD150 are now exclusively used. For diagnostic purposes, RT-PCR-based detection of viral RNA (N- and/or F-gene specific primers are the most commonly used) is best applied in specific instances, such as suspected infection of the immunosuppressed, developing pneumonia without a rash, or unexplained encephalitis.

Nevertheless, diagnosis of measles is still mostly based on surrogate markers haemagglutination inhibition (HI), plaque-reduction neutralization (PRN) and, now, most widely employed enzyme-linked immunoassay (EIA), have been used extensively, which measure measles-specific IgM as well as IgG responses. As recommended laboratory method by the WHO Global Measles Laboratory Network for confirmation of measles diagnosis, serum-based capture and indirect IgM EIAs are now commercially available for use by public health, clinical and commercial laboratories (Ratnam *et al.*, 2000). Less invasive, oral fluid samples have been used for detection of measles-specific IgM and IgG antibodies, although specimens have heightened background reactivity in commercial assays (Helfand *et al.*, 1996). Alternatively, serum eluted from blood spots can be collected onto filter paper and can be used in commercially available measles EIA assays without the loss of sensitivity and specificity (Helfand *et al.*, 2001; de Swart *et al.*, 2001). This method can also be used for molecular characterization of measles virus genome, but cannot be used for virus isolation (Helfand *et al.*, 2001; de Swart *et al.*, 2001).

The diagnosis of SSPE formerly relied on brain biopsies inspected for inclusion bodies or MV antigens by IF. The characteristic intrathecal antibody synthesis allowed determination of measles antibody titres in the CSF, which, usually sufficient, may be confirmed for MV-specific heterogeneity by isoelectric focusing. Virus isolation from SSPE brain tissue often fails, yet RT-PCR amplification is possible from post mortem tissue, yet not from CSF samples, and has proven very informative for confirming the identification of circulating wild-type MV as causative agent of the disease. As also detailed above, specific EEG patterns (Radermecker complexes) essentially signify SSPE.

Prevention and control

Vaccination is the only effective way to prevent measles infections. The PAHO region of the WHO and US have effectively eliminated measles using live-attenuated measles vaccine by providing a two-dose vaccine schedule (de Quadros *et al.*, 2004), where the first dose is administered at 12–15 months of age, and the second dose any time between 4 weeks later and school entry. As by announcement of the Measles Initiative (a partnership committed to reducing measles deaths globally launched in 2001), measles deaths worldwide fell by 78% between 2000 and 2008. Reduction of measles mortality by vaccination campaigns by 90% has been achieved ahead of schedule already in 2008 in most regions except for South-East Asia with delayed implementation of large-scale vaccine campaigns largely accounting for the lack of progress. Yet, measles cases and also fatalities may quickly resurge if vaccination efforts are not sustained (Measles Initiative, 2010).

In developing countries where measles is still endemic, factors such as the presence of maternal antibody, underlying infectious diseases, and malnutrition profoundly interfere and confound vaccine efficacy in most vulnerable age groups. Moreover, influxes of migrating refugees from other areas have resulted in the reintroduction of measles into populations that have undergone mass vaccination efforts. Live-attenuated measles-containing vaccines have been successfully used to control and eliminate measles in industrialized countries, the experience in Africa has led some to consider development of alternative vaccines, such as high-titre vaccines, DNA-based vaccines with or without adjuvants, thermostable dried powder vaccines and more. Except for the high-titre vaccines, which are no longer recommended by the WHO following their association with adverse outcomes (Aaby *et al.*, 1993), none of the other alternatives have been clinically tested. In spite of the effectiveness of the vaccine currently in use, which is at least equally effective when administered by the aerosol rather than subcutaneous route (de Swart *et al.*, 2006), new candidate measles vaccines are being exploited in terms of efficacy (particularly in the presence of maternal antibodies), but also other aspects such as thermostability, standardization, production costs and application mode in animal models. Candidates being evaluated include expression of MV antigens in the context of live viral (like pox and adenovirus) or bacterial (like *Shigella flexneri* and *Salmonella typhimurium*) vectors, subunit or synthetic vaccines (like iscoms, biodegradable particles, lipopeptides) and nucleic acid vaccines (plasmid DNA). In addition, recombinant MVs varying heterologous viral antigens are tested as vectors for multivaccine strategies, yet again so far only in animal studies.

All WHO measles vaccines that are currently in use are derived from Edmonston Enders A or B seed stocks. Side effects were common on administration of the first attenuated vaccine strain, the Edmonston B strain, and therefore, further attenuated and less reactogenic strains (Schwarz and Moraten) were developed on further passage in chick embryo at lowered temperature. Both vaccines produced a 95% seroconversion rate, although antibody titres induced by the Schwarz strain declined more rapidly than those produced in response to the Moraten strain, but remained at protective levels (Clements and Cutts, 1995; Katz, 2009). The Edmonston Zagreb strain, generated from the Edmonston B vaccine strain by passage in human diploid cells produced even higher seroconversion rates than the Schwarz vaccine as did the AIK-C vaccine, produced in Japan, and originating from the Edmonston A virus.

Primarily IgG1 MV-specific antibodies are efficiently induced to protective levels on administration of the vaccine after waning of the maternal antibodies, and though their levels and half-lives may be inferior to those induced by natural infection, they are still detectable in most individuals 30 years after immunization in the absence of boosting infections (Dine *et al.*, 2004). As for acute measles, both MV-specific CD4 and CD8 T cells are induced and activated by the vaccine, although CTL-responses after restimulation *in vitro* are considerably lower. Live measles vaccine can associate with transient lymphopenia, loss of DTH skin test responses to recall antigen and decreased *in vitro* proliferative responses to mitogens (Hussey *et al.*, 1996).

The vaccine has an excellent record of safety. It is only on rare instances that a transient rash or low-grade fever (or, very rarely, febrile seizures) are observed in otherwise asymptomatic recipients, which by far are outweighed by the benefits of the protective effects. CNS conditions, such as encephalitis and encephalopathy, have been reported with a frequency of less than one per million doses administered an incidence, which is lower than that of encephalitis of unknown aetiology. Hypersensitivity reactions following the administration of live measles vaccine are rare, yet a history of anaphylactic reactions to eggs, gelatin or neomycin following ingestion contradicts administration of the vaccine (Duclos and Ward, 1998). This also refers to individuals immunocompromised by haematopoietic malignancies or therapy with alkylating agents, antimetabolites, radiation, or large doses of corticosteroids. Asymptomatic HIV-infection, however, is not a contra-indication and the vaccine may also be given to symptomatic HIV-infected children, since measles in these children can be severe. The vaccine has not been associated with severe or unusual adverse events in these children (Moss *et al.*, 2003). Exposed symptomatic HIV-infected (as well as other immunocompromised) should receive high doses of measles Ig regardless of their previous vaccination status.

As indicated above, measles vaccination has efficiently reduced case numbers and measles mortality. This also applies to complications such as AME and SSPE (Duclos and Ward, 1998). The measles outbreak in the USA in 1989–1990 affecting 55,000 cases (Atkinson *et al.*, 1992) reflected the highly contagious nature of the virus and established, owing to an inappreciably low vaccine coverage among preschool aged children, vaccine failure in a low percentage of school-aged children who received only a single dose of vaccine. The two-dose vaccination schedule issued as a consequence is only one strategy used to ensure high vaccine coverage in the quest for measles elimination. PAHO has used live-attenuated MV in a strategy known as keep-up, catch-up, follow-up. This strategy combines routine vaccination (keep-up) with mass vaccination campaigns (catch-up), and 4-year follow-up campaigns (follow-up; de Quadros *et al.*, 1996). When used in conjunction with the two opportunities for administration, the current live-attenuated measles vaccine has proven to be efficacious, effective and safe. In high-risk areas, initial vaccination can be administered at six months of age to children, though efficacy is reduced at ages younger than 12 months, and thus a second dose of vaccine given before the determined average age of onset of natural measles for the region/country is critical to insure protection for those children.

Perspectives

Significant strides have been made in reducing the global impact of MV by the Measles Initiative, which was established in 2001. Just over a decade ago MV infection was the leading cause of death from a vaccine preventable disease, now mortality rates have been reduced from over 1 million per annum to under 200,000 and the burden of disease continues to diminish. This is a tremendous success for such a highly transmissible virus. Huge swathes of the globe are now MV free where the virus is no longer endemic and vaccine coverage is increasing in many parts of the developing world. Nonetheless, importations occur into regions from where the virus has been eliminated and it is imperative to ensure that the vaccine uptake remains high in places where the virus is no longer perceived to be a problem. Given that over 95% of susceptibles must be vaccinated to break transmission complacency and unfounded and unsubstantiated claims that vaccination is unsafe provide the public health organizations with tremendous logistical challenges in their goal of eradication. Whether this is practicable remains to be seen and given the significant challenges for poliovirus, which has been much more difficult to eradicate than smallpox, many within the global health community consider whether eradication in the short to medium term might be very difficult. Delivery still presents a problem given that the vaccine must be distributed through the cold chain, reconstituted and administered within 8 h to ensure that potency is retained. Vaccine vial monitors help to ensure that excursions from the cold chain are detected and thermally-compromised vaccines are discarded rather than administered. It is encouraging to see funders like the Bill and Melinda Gates Foundation under the Grand Challenges in Global Health (GCGH) Initiative, Foundation for the National Institutes of Health and the WHO support studies, which aim to enhance the thermostability of vaccines, both existing and developmental, as this is a real issue in the delivery of some of the most complex biopharmaceuticals to be globally distributed. Steps have been made to increase both solution and lyophilized vaccine using high-throughput formulation approaches (Schlehuber *et al.*, unpublished). In addition, novel routes of vaccine delivery are being actively pursued with a view to needle-free administration. Thus, WHO support and wet aerosol programme and GCGH are pursuing spray-dried powders, which would be inherently more thermostable. It is clear that even if the virus were to be eradicated, vaccination should not be discontinued immediately, if at all. The impact on an infection in a huge pool of highly susceptible humans who are *Morbillivirus* antibody naïve and thereby lack any protection to MV by another CD150 utilizing virus, such as CDV, which is much more neurotropic would be immense. Moreover, in a world where bioterrorism is now a reality and virus genomes can be synthesized *de novo*, serious thought needs to be given to discontinuation of vaccination. Understating the barriers which restrict cross species infection is probably one of the most important challenges, not just for MV, but in general. Reverse genetic approaches are being harnessed with a view to producing novel vectored vaccines based on MV. This is due to the excellent safety record of the vaccine, the virus ability to retain very large insertions stably and the lack of recombination, which keeps the transgene locked within the genome. It will be exciting in years to come to see if those industrial scientists can overcome the regulatory hurdles and harness the virus to treat new diseases (Billeter *et al.*, 2009). Another beneficial prospect would be if the academic scientists involved in retargeting, shielding and arming the virus to infect and eliminate tumour cells are successful (Russell and Peng, 2009). Solid progress has been made in that regard; human trials are under way using GMP-grade viruses. From a molecular perspective, now that the H glycoprotein structure is resolved at the atomic level, it would be valuable from a therapeutic perspective if other proteins, particularly those which make up the RdRp, could also be purified, crystallized and resolved. This could open up the door to the generation of novel antivirals, possibly with cross-specificity for other members of the order *Mononegavirales*, something which has proved elusive to date.

References

Aaby, P., Andersen, M., Sodemann, M., Jakobsen, M., Gomes, J., and Fernandes, M. (1993). Reduced childhood mortality after standard measles vaccination at 4–8 months compared with 9–11 months of age. BMJ *307*, 1308–1311.

Abt, M., Gassert, E., and Schneider-Schaulies, S. (2009). Measles virus modulates chemokine release and chemotactic responses of dendritic cells. J. Gen. Virol. *90*, 909–914.

Addae, M.M., Komada, Y., Zhang, X.L., and Sakurai, M. (1995). Immunological unresponsiveness and apoptotic cell death of T cells in measles virus infection. Acta Paediatr. Jpn. *37*, 308–314.

Afzal, M.A., Osterhaus, A.D., Cosby, S.L., Jin, L., Beeler, J., Takeuchi, K., and Kawashima, H. (2003). Comparative evaluation of measles virus-specific RT-PCR methods through an international collaborative study. J. Med. Virol. *70*, 171–176.

Albrecht, P., Lorenz, D., Klutch, M.J., Vickers, J.H., and Ennis, F.A. (1980). Fatal measles infection in marmosets pathogenesis and prophylaxis. Infect. Immun. *27*, 969–978.

Atkinson, W.L., Orenstein, W.A., and Krugman, S. (1992). The resurgence of measles in the United States, 1989–1990. Annu. Rev. Med. *43*, 451–463.

Auwaerter, P.G., Kaneshima, H., McCune, J.M., Wiegand, G., and Griffin, D.E. (1996). Measles virus infection of thymic epithelium in the SCID-hu mouse leads to thymocyte apoptosis. J. Virol. *70*, 3734–3740.

Auwaerter, P.G., Rota, P.A., Elkins, W.R., Adams, R.J., DeLozier, T., Shi, Y., Bellini, W.J., Murphy, B.R., and Griffin, D.E. (1999). Measles virus infection in rhesus macaques: altered immune responses and comparison of the virulence of six different virus strains. J. Infect. Dis. *180*, 950–958.

Avota, E., Avots, A., Niewiesk, S., Kane, L.P., Bommhardt, U., ter Meulen, V., and Schneider-Schaulies, S. (2001). Disruption of Akt kinase activation is important for immunosuppression induced by measles virus. Nat. Med. *7*, 725–731.

Avota, E., Muller, N., Klett, M., and Schneider-Schaulies, S. (2004). Measles virus interacts with and alters signal transduction in T-cell lipid rafts. J. Virol. *78*, 9552–9559.

Avota, E., Harms, H., and Schneider-Schaulies, S. (2006). Measles virus induces expression of SIP110, a constitutively membrane clustered lipid phosphatase, which inhibits T cell proliferation. Cell Microbiol. *8*, 1826–1839.

Bankamp, B., Wilson, J., Bellini, W.J., and Rota, P.A. (2005). Identification of naturally occuring amino acid variations that affect the ability of the measles virus C protein to regulate genome replication and transcription. Virology *336*, 120–129.

Bankamp, B., Hodge, G., McChesney, M.B., Bellini, W.J., and Rota, P.A. (2008). Genetic changes that affect the virulence of measles virus in the macaque model. Virology *373*, 39–50.

Bellini, W.J., Englund, G., Rozenblatt, S., Arnheiter, H., and Richardson, C.D. (1985). Measles virus P gene codes for two proteins. J. Virol. *53*, 908–919.

Bellini, W.J., Rota, J.S., Lowe, L.E., Katz, R.S., Dyken, P.R., Zaki, S.R., Shieh, W.J., and Rota, P.A. (2005). Subacute sclerosing panencephalitis: more cases of this fatal disease are prevented by measles immunization than was previously recognized. J. Infect. Dis. *192*, 1686–1693.

Bhella, D., Ralph, A., and Yeo, R.P. (2004). Conformational flexibility in recombinant measles virus nucleocapsids visualised by cryo-negative stain electron microscopy and real-space helical reconstruction. J. Mol. Biol. *340*, 319–331.

Bieback, K., Lien, E., Klagge, I.M., Avota, E., Schneider-Schaulies, J., Duprex, W.P., Wagner, H., Kirschning, C.J., Ter Meulen, V., and Schneider-Schaulies, S. (2002). Hemagglutinin protein of wild-type measles virus activates toll-like receptor 2 signaling. J. Virol. *76*, 8729–8736.

Billadeau, D.D., and Burkhardt, J.K. (2006). Regulation of cytoskeletal dynamics at the immune synapse: new stars join the actin troupe. Traffic *7,* 1451–1460.

Billeter, M.A., Cattaneo, R., Spielhofer, P., Kaelin, K., Huber, M., Schmid, A., Baczko, K., and ter Meulen, V. (1994). Generation and properties of measles virus mutations typically associated with subacute sclerosing panencephalitis. Ann. N.Y. Acad. Sci. *724,* 367–377.

Billeter, M.A., Naim, H.Y., and Udem, S.A. (2009). Reverse genetics of measles virus and resulting multivalent recombinant vaccines: applications of recombinant measles viruses. Curr. Top. Microbiol. Immunol. *329,* 129–162.

van Binnendijk, R.S., van der Heijden, R.W., van Amerongen, G., UytdeHaag, F.G., and Osterhaus, A.D. (1994). Viral replication and development of specific immunity in macaques after infection with different measles virus strains. J. Infect. Dis. *170,* 443–448.

Bourhis, J.M., Canard, B., and Longhi, S. (2006). Structural disorder within the replicative complex of measles virus: functional implications. Virology *344,* 94–110.

Bourhis, J.M., Canard, B., and Longhi, S. (2007). Predicting protein disorder and induced folding: from theoretical principles to practical applications. Curr. Protein Pept. Sci. *8,* 135–149.

Buckland, R., and Wild, T.F. (1997). Is CD46 the cellular receptor for measles virus? Virus Res. *48,* 1–9.

Calain, P., and Roux, L. (1993). The rule of six, a basic feature for efficient replication of Sendai virus defective interfering RNA. J. Virol. *67,* 4822–4830.

Cattaneo, R., Rebmann, G., Schmid, A., Baczko, K., ter Meulen, V., and Billeter, M.A. (1987). Altered transcription of a defective measles virus genome derived from a diseased human brain. Embo J. *6,* 681–688.

Cattaneo, R., Kaelin, K., Baczko, K., and Billeter, M.A. (1989a). Measles virus editing provides an additional cysteine-rich protein. Cell *56,* 759–764.

Cattaneo, R., Schmid, A., Spielhofer, P., Kaelin, K., Baczko, K., ter Meulen, V., Pardowitz, J., Flanagan, S., Rima, B.K., Udem, S.A., *et al.* (1989b). Mutated and hypermutated genes of persistent measles viruses which caused lethal human brain diseases. Virology *173,* 415–425.

Cattaneo, R., Miest, T., Shashkova, E.V., and Barry, M.A. (2008). ReprogramMed. viruses as cancer therapeutics: targeted, arMed. and shielded. Nat. Rev. Microbiol. *6,* 529–540.

Cevik, B., Smallwood, S., and Moyer, S.A. (2003). The L-L oligomerization domain resides at the very N-terminus of the sendai virus L RNA polymerase protein. Virology *313,* 525–536.

Clements, C.J., and Cutts, F.T. (1995). The epidemiology of measles: thirty years of vaccination. Curr. Top. Microbiol. Immunol. *191,* 13–33.

Combredet, C., Labrousse, V., Mollet, L., Lorin, C., Delebecque, F., Hurtrel, B., McClure, H., Feinberg, M.B., Brahic, M., and Tangy, F. (2003). A molecularly cloned Schwarz strain of measles virus vaccine induces strong immune responses in macaques and transgenic mice. J. Virol. *77,* 11546–11554.

Crennell, S., Takimoto, T., Portner, A., and Taylor, G. (2000). Crystal structure of the multifunctional paramyxovirus hemagglutinin-neuraminidase. Nat. Struct. Biol. *7,* 1068–1074.

Devaux, P., and Cattaneo, R. (2004). Measles virus phosphoprotein gene products: conformational flexibility of the P/V protein amino-terminal domain and C protein infectivity factor function. J. Virol. *78,* 11632–11640.

Devaux, P., von Messling, V., Songsungthong, W., Springfeld, C., and Cattaneo, R. (2007). Tyrosine 110 in the measles virus phosphoprotein is required to block STAT1 phosphorylation. Virology *360,* 72–83.

Devaux, P., Hodge, G., McChesney, M.B., and Cattaneo, R. (2008). Attenuation of V- or C-defective measles viruses: infection control by the inflammatory and interferon responses of rhesus monkeys. J. Virol. *82,* 5359–5367.

Dhib-Jalbut, S., Xia, J., Rangaviggula, H., Fang, Y.Y., and Lee, T. (1999). Failure of measles virus to activate nuclear factor-kappa B in neuronal cells: implications on the immune response to viral infections in the central nervous system. J. Immunol. *162,* 4024–4029.

Dhiman, N., Ovsyannikova, I.G., Oberg, A.L., Grill, D.E., Jacobson, R.M., and Poland, G.A. (2005). Immune activation at effector and gene expression levels after measles vaccination in healthy individuals: a pilot study. Hum Immunol. *66,* 1125–1136.

Dhiman, N., Ovsyannikova, I.G., Cunningham, J.M., Vierkant, R.A., Kennedy, R.B., Pankratz, V.S., Poland, G.A., and Jacobson, R.M. (2007). Associations between measles vaccine immunity and single-nucleotide polymorphisms in cytokine and cytokine receptor genes. J. Infect. Dis. *195,* 21–29.

Dine, M.S., Hutchins, S.S., Thomas, A., Williams, I., Bellini, W.J., and Redd, S.C. (2004). Persistence of vaccine-induced antibody to measles 26–33 years after vaccination. J. Infect. Dis. *189 (Suppl. 1),* S123–130.

Dorig, R.E., Marcil, A., Chopra, A., and Richardson, C.D. (1993). The human CD46 molecule is a receptor for measles virus (Edmonston strain). Cell *75,* 295–305.

Doyle, J., Prussia, A., White, L.K., Sun, A., Liotta, D.C., Snyder, J.P., Compans, R.W., and Plemper, R.K. (2006). Two domains that control prefusion stability and transport competence of the measles virus fusion protein. J. Virol. *80,* 1524–1536.

Dubois, B., Lamy, P.J., Chemin, K., Lachaux, A., and Kaiserlian, D. (2001). Measles virus exploits dendritic cells to suppress $CD4^+$ T-cell proliferation via expression of surface viral glycoproteins independently of T-cell trans-infection. Cell Immunol. *214,* 173–183.

Duclos, P., and Ward, B.J. (1998). Measles vaccines: a review of adverse events. Drug Saf. *19,* 435–454.

Duprex, W.P., and Rima, B.K. (2002). Using green fluorescent protein to monitor measles virus cell-to-cell spread by time-lapse confocal microscopy. Methods Mol. Biol. *183,* 297–307.

Duprex, W.P., Duffy, I., McQuaid, S., Hamill, L., Cosby, S.L., Billeter, M.A., Schneider-Schaulies, J., ter Meulen, V., and Rima, B.K. (1999). The H gene of rodent brain-adapted measles virus confers neurovirulence to the Edmonston vaccine strain. J. Virol. *73*, 6916–6922.

El Mubarak, H.S., Yuksel, S., van Amerongen, G., Mulder, P.G., Mukhtar, M.M., Osterhaus, A.D., and de Swart, R.L. (2007). Infection of cynomolgus macaques (Macaca fascicularis) and rhesus macaques (Macaca mulatta) with different wild-type measles viruses. J. Gen. Virol. *88*, 2028–2034.

Engelking, O., Fedorov, L.M., Lilischkis, R., ter Meulen, V., and Schneider-Schaulies, S. (1999). Measles virus-induced immunosuppression *in vitro* is associated with deregulation of G1 cell cycle control proteins. J. Gen. Virol. *80*, 1599–1608.

Ertl, O.T., Wenz, D.C., Bouche, F.B., Berbers, G.A., and Muller, C.P. (2003). Immunodominant domains of the Measles virus hemagglutinin protein eliciting a neutralizing human B cell response. Arch. Virol. *148*, 2195–2206.

Escoffier, C., Manie, S., Vincent, S., Muller, C.P., Billeter, M., and Gerlier, D. (1999). Nonstructural C protein is required for efficient measles virus replication in human peripheral blood cells. J. Virol. *73*, 1695–1698.

Ferron, F., Longhi, S., Henrissat, B., and Canard, B. (2002). Viral RNA-polymerases – a predicted 2′-O-ribose methyltransferase domain shared by all *Mononegavirales*. Trends Biochem. Sci. *27*, 222–224.

Finsterbusch, T., Wolbert, A., Deitemeier, I., Meyer, K., Mosquera, M.M., Mankertz, A., and Santibanez, S. (2009). Measles viruses of genotype H1 evade recognition by vaccine-induced neutralizing antibodies targeting the linear haemagglutinin noose epitope. J. Gen. Virol. *90*, 2739–2745.

Fontana, J.M., Bankamp, B., Bellini, W.J., and Rota, P.A. (2008). Regulation of interferon signaling by the C and V proteins from attenuated and wild-type strains of measles virus. Virology *374*, 71–81.

Fugier-Vivier, I., Servet-Delprat, C., Rivailler, P., Rissoan, M.C., Liu, Y.J., and Rabourdin-Combe, C. (1997). Measles virus suppresses cell-mediated immunity by interfering with the survival and functions of dendritic and T cells. J. Exp. Med. *186*, 813–823.

Gans, H.A., Maldonado, Y., Yasukawa, L.L., Beeler, J., Audet, S., Rinki, M.M., DeHovitz, R., and Arvin, A.M. (1999). IL-12, IFN-gamma, and T cell proliferation to measles in immunized infants. J. Immunol. *162*, 5569–5575.

Gassert, E., Avota, E., Harms, H., Krohne, G., Gulbins, E., and Schneider-Schaulies, S. (2009). Induction of membrane ceramides: a novel strategy to interfere with T lymphocyte cytoskeletal reorganisation in viral immunosuppression. PLoS Pathog. *5*, e1000623.

Griffin, D.E. (1995). Immune responses during measles virus infection. Curr. Top. Microbiol. Immunol. *191*, 117–134.

Griffin, D.E., and Ward, B.J. (1993). Differential CD4 T cell activation in measles. J. Infect. Dis. *168*, 275–281.

Gringhuis, S.I., den Dunnen, J., Litjens, M., van Het Hof, B., van Kooyk, Y., and Geijtenbeek, T.B. (2007). C-type lectin DC-SIGN modulates Toll-like receptor signaling via Raf-1 kinase-dependent acetylation of transcription factor NF-kappaB. Immunity *26*, 605–616.

Gringhuis, S.I., den Dunnen, J., Litjens, M., van der Vlist, M., and Geijtenbeek, T.B. (2009). Carbohydrate-specific signaling through the DC-SIGN signalosome tailors immunity to Mycobacterium tuberculosis, HIV-1 and Helicobacter pylori. Nat. Immunol. *10*, 1081–1088.

Grosjean, I., Caux, C., Bella, C., Berger, I., Wild, F., Banchereau, J., and Kaiserlian, D. (1997). Measles virus infects human dendritic cells and blocks their allostimulatory properties for CD4$^+$ T cells. J. Exp. Med. *186*, 801–812.

Gulbins, E., Dreschers, S., Wilker, B., and Grassme, H. (2004). Ceramide, membrane rafts and infections. J. Mol. Med. *82*, 357–363.

Hahm, B., Trifilo, M.J., Zuniga, E.I., and Oldstone, M.B. (2005). Viruses evade the immune system through type I interferon-mediated STAT2-dependent, but STAT1-independent, signaling. Immunity *22*, 247–257.

Hashiguchi, T., Kajikawa, M., Maita, N., Takeda, M., Kuroki, K., Sasaki, K., Kohda, D., Yanagi, Y., and Maenaka, K. (2007). Crystal structure of measles virus hemagglutinin provides insight into effective vaccines. Proc. Natl. Acad. Sci. U.S.A. *104, 19*,535–19540.

Helfand, R.F., Kebede, S., Alexander, J.P., Jr., Alemu, W., Heath, J.L., Gary, H.E., Jr., Anderson, L.J., Beyene, H., and Bellini, W.J. (1996). Comparative detection of measles-specific IgM in oral fluid and serum from children by an antibody-capture IgM EIA. J. Infect. Dis. *173*, 1470–1474.

Helfand, R.F., Keyserling, H.L., Williams, I., Murray, A., Mei, J., Moscatiello, C., Icenogle, J., and Bellini, W.J. (2001). Comparative detection of measles and rubella IgM and IgG derived from filter paper blood and serum samples. J. Med. Virol. *65*, 751–757.

Helin, E., Vainionpaa, R., Hyypia, T., Julkunen, I., and Matikainen, S. (2001). Measles virus activates NF-kappa B and STAT transcription factors and production of IFN-alpha/beta and IL-6 in the human lung epithelial cell line A549. Virology *290*, 1–10.

Hoffman, S.J., Polack, F.P., Hauer, D.A., and Griffin, D.E. (2003). Measles virus infection of rhesus macaques affects neutrophil expression of IL-12 and IL-10. Viral Immunol. *16*, 369–379.

Hummel, K.B., Bellini, W.J., and Offermann, M.K. (1998). Strain-specific differences in LFA-1 induction on measles virus-infected monocytes and adhesion and viral transmission to endothelial cells. J. Virol. *72*, 8403–8407.

Hussey, G.D., Goddard, E.A., Hughes, J., Ryon, J.J., Kerran, M., Carelse, E., Strebel, P.M., Markowitz, L.E., Moodie, J., Barron, P., *et al.* (1996). The effect of Edmonston-Zagreb and Schwarz measles vaccines on immune response in infants. J. Infect. Dis. *173*, 1320–1326.

Iankov, I.D., Blechacz, B., Liu, C., Schmeckpeper, J.D., Tarara, J.E., Federspiel, M.J., Caplice, N., and Russell, S.J. (2007). Infected cell carriers: a new strategy for systemic delivery of oncolytic measles viruses in cancer virotherapy. Mol. Ther. *15*, 114–122.

Inoue, T., Kira, R., Nakao, F., Ihara, K., Bassuny, W.M., Kusuhara, K., Nihei, K., Takeshita, K., and Hara, T. (2002). Contribution of the interleukin 4 gene to susceptibility to subacute sclerosing panencephalitis. Arch. Neurol. *59,* 822–827.

Jaye, A., Magnusen, A.F., Sadiq, A.D., Corrah, T., and Whittle, H.C. (1998). *Ex vivo* analysis of cytotoxic T lymphocytes to measles antigens during infection and after vaccination in Gambian children. J. Clin. Invest. *102,* 1969–1977.

Johansson, K., Bourhis, J.M., Campanacci, V., Cambillau, C., Canard, B., and Longhi, S. (2003). Crystal structure of the measles virus phosphoprotein domain responsible for the induced folding of the C-terminal domain of the nucleoprotein. J. Biol. Chem. *278,* *44,*567–44573.

Kaiserlian, D., Grosjean, I., and Caux, C. (1997). Infection of human dendritic cells by measles virus induces immune suppression. Adv. Exp. Med. Biol. *417,* 421–423.

Karp, C.L., Wysocka, M., Wahl, L.M., Ahearn, J.M., Cuomo, P.J., Sherry, B., Trinchieri, G., and Griffin, D.E. (1996). Mechanism of suppression of cell-mediated immunity by measles virus. Science *273,* 228–231.

Katz, S.L. (2009). John F. Enders and measles virus vaccine–a reminiscence. Curr. Top. Microbiol. Immunol. *329,* 3–11.

Kemper, C., Chan, A.C., Green, J.M., Brett, K.A., Murphy, K.M., and Atkinson, J.P. (2003). Activation of human $CD4^+$ cells with CD3 and CD46 induces a T-regulatory cell 1 phenotype. Nature *421,* 388–392.

Kerdiles, Y.M., Sellin, C.I., Druelle, J., and Horvat, B. (2006). Immunosuppression caused by measles virus: role of viral proteins. Rev. Med. Virol. *16,* 49–63.

Klagge, I.M., ter Meulen, V., and Schneider-Schaulies, S. (2000). Measles virus-induced promotion of dendritic cell maturation by soluble mediators does not overcome the immunosuppressive activity of viral glycoproteins on the cell surface. Eur J. Immunol. *30,* 2741–2750.

Klagge, I.M., Abt, M., Fries, B., and Schneider-Schaulies, S. (2004). Impact of measles virus dendritic-cell infection on Th-cell polarization *in vitro*. J. Gen. Virol. *85,* 3239–3247.

Kobune, F., Sakata, H., and Sugiura, A. (1990). Marmoset lymphoblastoid cells as a sensitive host for isolation of measles virus. J. Virol. *64,* 700–705.

Kobune, F., Takahashi, H., Terao, K., Ohkawa, T., Ami, Y., Suzaki, Y., Nagata, N., Sakata, H., Yamanouchi, K., and Kai, C. (1996). Nonhuman primate models of measles. Lab. Anim. Sci. *46,* 315–320.

Laine, D., Bourhis, J.M., Longhi, S., Flacher, M., Cassard, L., Canard, B., Sautes-Fridman, C., Rabourdin-Combe, C., and Valentin, H. (2005). Measles virus nucleoprotein induces cell-proliferation arrest and apoptosis through NTAIL-NR and NCORE–FcgammaRIIB1 interactions, respectively. J. Gen. Virol. *86,* 1771–1784.

Leonard, V.H., Sinn, P.L., Hodge, G., Miest, T., Devaux, P., Oezguen, N., Braun, W., McCray, P.B., Jr., McChesney, M.B., and Cattaneo, R. (2008). Measles virus blind to its epithelial cell receptor remains virulent in rhesus monkeys but cannot cross the airway epithelium and is not shed. J. Clin. Invest. *118,* 2448–2458.

Liebert, U.G., Schneider-Schaulies, S., Baczko, K., and ter Meulen, V. (1990). Antibody-induced restriction of viral gene expression in measles encephalitis in rats. J. Virol. *64,* 706–713.

Liu, X., Bankamp, B., Xu, W., Bellini, W.J., and Rota, P.A. (2006). The genomic termini of wild-type and vaccine strains of measles virus. Virus Res. *122,* 78–84.

Longhi, S. (2009). Nucleocapsid structure and function. Curr. Top. Microbiol. Immunol. *329,* 103–128.

Lorin, C., Delebecque, F., Labrousse, V., Da Silva, L., Lemonnier, F., Brahic, M., and Tangy, F. (2005). A recombinant live attenuated measles vaccine vector primes effective HLA-A0201-restricted cytotoxic T lymphocytes and broadly neutralizing antibodies against HIV-1 conserved epitopes. Vaccine *23,* 4463–4472.

Ludlow, M., McQuaid, S., Cosby, S.L., Cattaneo, R., Rima, B.K., and Duprex, W.P. (2005). Measles virus superinfection immunity and receptor redistribution in persistently infected NT2 cells. J. Gen. Virol. *86,* 2291–2303.

Ludlow, M., Rennick, L.J., Sarlang, S., Skibinski, G., McQuaid, S., Moore, T., de Swart, R.L., and Duprex, W.P. (2010). Wild-type measles virus infection of primary epithelial cells occurs via the basolateral surface without syncytium formation or release of infectious virus. J. Gen. Virol. *91,* 971–979.

McChesney, M.B., Miller, C.J., Rota, P.A., Zhu, Y.D., Antipa, L., Lerche, N.W., Ahmed, R., and Bellini, W.J. (1997). Experimental measles. I. Pathogenesis in the normal and the immunized host. Virology *233,* 74–84.

Makhortova, N.R., Askovich, P., Patterson, C.E., Gechman, L.A., Gerard, N.P., and Rall, G.F. (2007). Neurokinin-1 enables measles virus trans-synaptic spread in neurons. Virology *362,* 235–244.

Marie, J.C., Kehren, J., Trescol-Biemont, M.C., Evlashev, A., Valentin, H., Walzer, T., Tedone, R., Loveland, B., Nicolas, J.F., Rabourdin-Combe, C., *et al.* (2001). Mechanism of measles virus-induced suppression of inflammatory immune responses. Immunity *14,* 69–79.

Moll, M., Klenk, H.D., and Maisner, A. (2002). Importance of the cytoplasmic tails of the measles virus glycoproteins for fusogenic activity and the generation of recombinant measles viruses. J. Virol. *76,* 7174–7186.

Moss, W.J. (2007). Measles still has a devastating impact in unvaccinated populations. PLoS Med. *4,* e24.

Moss, W.J., and Griffin, D.E. (2006). Global measles elimination. Nat. Rev. Microbiol. *4,* 900–908.

Moss, W.J., Clements, C.J., and Halsey, N.A. (2003). Immunization of children at risk of infection with human immunodeficiency virus. Bull. World Health Organ. *81,* 61–70.

Moss, W.J., Ota, M.O., and Griffin, D.E. (2004). Measles: immune suppression and immune responses. Int. J. Biochem. Cell Biol. *36,* 1380–1385.

Mrkic, B., Pavlovic, J., Rulicke, T., Volpe, P., Buchholz, C.J., Hourcade, D., Atkinson, J.P., Aguzzi, A., and Cattaneo, R. (1998). Measles virus spread and pathogenesis in genetically modified mice. J. Virol. *72,* 7420–7427.

Muller, N., Avota, E., Schneider-Schaulies, J., Harms, H., Krohne, G., and Schneider-Schaulies, S. (2006). Measles virus contact with T cells impedes cytoskeletal remodeling associated with spreading, polarization, and CD3 clustering. Traffic *7,* 849–858.

Naim, H.Y., Ehler, E., and Billeter, M.A. (2000). Measles virus matrix protein specifies apical virus release and glycoprotein sorting in epithelial cells. EMBO J. *19,* 3576–3585.

Nakatsu, Y., Takeda, M., Ohno, S., Shirogane, Y., Iwasaki, M., and Yanagi, Y. (2008). Measles virus circumvents the host interferon response by different actions of the C and V proteins. J. Virol. *82,* 8296–8306.

Nanan, R., Rauch, A., Kampgen, E., Niewiesk, S., and Kreth, H.W. (2000). A novel sensitive approach for frequency analysis of measles virus-specific memory T-lymphocytes in healthy adults with a childhood history of natural measles. J. Gen. Virol. *81,* 1313–1319.

Naniche, D., Varior-Krishnan, G., Cervoni, F., Wild, T.F., Rossi, B., Rabourdin-Combe, C., and Gerlier, D. (1993). Human membrane cofactor protein (CD46) acts as a cellular receptor for measles virus. J. Virol. *67,* 6025–6032.

Naniche, D., Yeh, A., Eto, D., Manchester, M., Friedman, R.M., and Oldstone, M.B. (2000). Evasion of host defenses by measles virus: wild-type measles virus infection interferes with induction of Alpha/Beta interferon production. J. Virol. *74,* 7478–7484.

Ndhlovu, Z.M., Oelke, M., Schneck, J.P., and Griffin, D.E. (2010). Dynamic regulation of functionally distinct virus-specific T cells. Proc. Natl. Acad. Sci. U.S.A. *107,* 3669–3674.

Niewiesk, S. (2009). Current animal models: cotton rat animal model. Curr. Top. Microbiol. Immunol. *330,* 89–110.

Niewiesk, S., Eisenhuth, I., Fooks, A., Clegg, J.C., Schnorr, J.J., Schneider-Schaulies, S., and ter Meulen, V. (1997). Measles virus-induced immune suppression in the cotton rat (Sigmodon hispidus) model depends on viral glycoproteins. J. Virol. *71,* 7214–7219.

Niewiesk, S., Ohnimus, H., Schnorr, J.J., Gotzelmann, M., Schneider-Schaulies, S., Jassoy, C., and ter Meulen, V. (1999). Measles virus-induced immunosuppression in cotton rats is associated with cell cycle retardation in uninfected lymphocytes. J. Gen. Virol. *80,* 2023–2029.

Oh, S., Stegman, B., Pendleton, C.D., Ota, M.O., Pan, C.H., Griffin, D.E., Burke, D.S., and Berzofsky, J.A. (2006). Protective immunity provided by HLA-A2 epitopes for fusion and hemagglutinin proteins of measles virus. Virology *352,* 390–399.

Okada, H., Kobune, F., Sato, T.A., Kohama, T., Takeuchi, Y., Abe, T., Takayama, N., Tsuchiya, T., and Tashiro, M. (2000). Extensive lymphopenia due to apoptosis of uninfected lymphocytes in acute measles patients. Arch. Virol. *145,* 905–920.

Oldstone, M.B., Dales, S., Tishon, A., Lewicki, H., and Martin, L. (2005). A role for dual viral hits in causation of subacute sclerosing panencephalitis. J. Exp. Med. *202,* 1185–1190.

Ono, N., Tatsuo, H., Hidaka, Y., Aoki, T., Minagawa, H., and Yanagi, Y. (2001). Measles viruses on throat swabs from measles patients use signaling lymphocytic activation molecule (CDw150) but not CD46 as a cellular receptor. J. Virol. *75,* 4399–4401.

Ovsyannikova, I.G., Dhiman, N., Jacobson, R.M., Vierkant, R.A., and Poland, G.A. (2003a). Frequency of measles virus-specific $CD4^+$ and $CD8^+$ T cells in subjects seronegative or highly seropositive for measles vaccine. Clin. Diagn. Lab. Immunol. *10,* 411–416.

Ovsyannikova, I.G., Reid, K.C., Jacobson, R.M., Oberg, A.L., Klee, G.G., and Poland, G.A. (2003b). Cytokine production patterns and antibody response to measles vaccine. Vaccine *21,* 3946–3953.

Ovsyannikova, I.G., Ryan, J.E., Jacobson, R.M., Vierkant, R.A., Pankratz, V.S., and Poland, G.A. (2006a). Human leukocyte antigen and interleukin 2, 10 and 12p40 cytokine responses to measles: is there evidence of the HLA effect? Cytokine *36,* 173–179.

Ovsyannikova, I.G., Vierkant, R.A., and Poland, G.A. (2006b). Importance of HLA-DQ and HLA-DP polymorphisms in cytokine responses to naturally processed HLA-DR-derived measles virus peptides. Vaccine *24,* 5381–5389.

Palosaari, H., Parisien, J.P., Rodriguez, J.J., Ulane, C.M., and Horvath, C.M. (2003). STAT protein interference and suppression of cytokine signal transduction by measles virus V protein. J. Virol. *77,* 7635–7644.

Parks, C.L., Lerch, R.A., Walpita, P., Wang, H.P., Sidhu, M.S., and Udem, S.A. (2001). Analysis of the noncoding regions of measles virus strains in the Edmonston vaccine lineage. J. Virol. *75,* 921–933.

Parks, C.L., Witko, S.E., Kotash, C., Lin, S.L., Sidhu, M.S., and Udem, S.A. (2006). Role of V protein RNA binding in inhibition of measles virus minigenome replication. Virology *348,* 96–106.

Permar, S.R., Klumpp, S.A., Mansfield, K.G., Kim, W.K., Gorgone, D.A., Lifton, M.A., Williams, K.C., Schmitz, J.E., Reimann, K.A., Axthelm, M.K., *et al.* (2003a). Role of CD8(+) lymphocytes in control and clearance of measles virus infection of rhesus monkeys. J. Virol. *77,* 4396–4400.

Permar, S.R., Moss, W.J., Ryon, J.J., Douek, D.C., Monze, M., and Griffin, D.E. (2003b). Increased thymic output during acute measles virus infection. J. Virol. *77,* 7872–7879.

Permar, S.R., Klumpp, S.A., Mansfield, K.G., Carville, A.A., Gorgone, D.A., Lifton, M.A., Schmitz, J.E., Reimann, K.A., Polack, F.P., Griffin, D.E., *et al.* (2004). Limited contribution of humoral immunity to the clearance of measles viremia in rhesus monkeys. J. Infect. Dis. *190,* 998–1005.

Pipo-Deveza, J.R., Kusuhara, K., Silao, C.L., Lukban, M.B., Salonga, A.M., Sanchez, B.C., Kira, R., Takemoto, M., Torisu, H., and Hara, T. (2006). Analysis of MxA, IL-4, and IRF-1 genes in Filipino patients with subacute sclerosing panencephalitis. Neuropediatrics *37,* 222–228.

Plemper, R.K., Hammond, A.L., and Cattaneo, R. (2001). Measles virus envelope glycoproteins hetero-oligomerize in the endoplasmic reticulum. J. Biol. Chem. *276, 44*,239–44246.

Plemper, R.K., Hammond, A.L., Gerlier, D., Fielding, A.K., and Cattaneo, R. (2002). Strength of envelope protein interaction modulates cytopathicity of measles virus. J. Virol. *76,* 5051–5061.

Plemper, R.K., Lakdawala, A.S., Gernert, K.M., Snyder, J.P., and Compans, R.W. (2003). Structural features of paramyxovirus F protein required for fusion initiation. Biochemistry *42,* 6645–6655.

Plumet, S., Duprex, W.P., and Gerlier, D. (2005). Dynamics of viral RNA synthesis during measles virus infection. J. Virol. *79,* 6900–6908.

Plumet, S., Herschke, F., Bourhis, J.M., Valentin, H., Longhi, S., and Gerlier, D. (2007). Cytosolic 5′-triphosphate ended viral leader transcript of measles virus as activator of the RIG I-mediated interferon response. PLoS One *2,* e279.

Pohl, C., Duprex, W.P., Krohne, G., Rima, B.K., and Schneider-Schaulies, S. (2007). Measles virus M and F proteins associate with detergent-resistant membrane fractions and promote formation of virus-like particles. J. Gen. Virol. *88,* 1243–1250.

Polack, F.P., Auwaerter, P.G., Lee, S.H., Nousari, H.C., Valsamakis, A., Leiferman, K.M., Diwan, A., Adams, R.J., and Griffin, D.E. (1999). Production of atypical measles in rhesus macaques: evidence for disease mediated by immune complex formation and eosinophils in the presence of fusion-inhibiting antibody. Nat. Med. *5,* 629–634.

Polack, F.P., Lee, S.H., Permar, S., Manyara, E., Nousari, H.G., Jeng, Y., Mustafa, F., Valsamakis, A., Adams, R.J., Robinson, H.L., *et al.* (2000). Successful DNA immunization against measles: neutralizing antibody against either the hemagglutinin or fusion glycoprotein protects rhesus macaques without evidence of atypical measles. Nat. Med. *6,* 776–781.

de Quadros, C.A., Olive, J.M., Hersh, B.S., Strassburg, M.A., Henderson, D.A., Brandling-Bennett, D., and Alleyne, G.A. (1996). Measles elimination in the Americas. Evolving strategies. JAMA *275,* 224–229.

de Quadros, C.A., Izurieta, H., Venczel, L., and Carrasco, P. (2004). Measles eradication in the Americas: progress to date. J. Infect. Dis. *189 (Suppl. 1),* S227–235.

Radecke, F., Spielhofer, P., Schneider, H., Kaelin, K., Huber, M., Dotsch, C., Christiansen, G., and Billeter, M.A. (1995). Rescue of measles viruses from cloned DNA. Embo J. *14,* 5773–5784.

Rager, M., Vongpunsawad, S., Duprex, W.P., and Cattaneo, R. (2002). Polyploid measles virus with hexameric genome length. EMBO J. *21,* 2364–2372.

Ratnam, S., Tipples, G., Head, C., Fauvel, M., Fearon, M., and Ward, B.J. (2000). Performance of indirect immunoglobulin M (IgM) serology tests and IgM capture assays for laboratory diagnosis of measles. J. Clin. Microbiol. *38,* 99–104.

Rennick, L.J., Duprex, W.P., and Rima, B.K. (2007). Measles virus minigenomes encoding two autofluorescent proteins reveal cell-to-cell variation in reporter expression dependent on viral sequences between the transcription units. J. Gen. Virol. *88,* 2710–2718.

Riddell, M.A., Moss, W.J., Hauer, D., Monze, M., and Griffin, D.E. (2007). Slow clearance of measles virus RNA after acute infection. J. Clin. Virol. *39,* 312–317.

Rima, B.K., and Duprex, W.P. (2006). Morbilliviruses and human disease. J. Pathol. *208,* 199–214.

Rima, B.K., and Duprex, W.P. (2009). The measles virus replication cycle. Curr. Top. Microbiol. Immunol. *329,* 77–102.

Rima, B.K., and McFerran, N.V. (1997). Dinucleotide and stop codon frequencies in single-stranded RNA viruses. J. Gen. Virol. *78,* 2859–2870.

Rima, B.K., Earle, J.A., Baczko, K., Rota, P.A., and Bellini, W.J. (1995). Measles virus strain variations. Curr. Top. Microbiol. Immunol. *191,* 65–83.

Rota, P.A., Liffick, S.L., Rota, J.S., Katz, R.S., Redd, S., Papania, M., and Bellini, W.J. (2002). Molecular epidemiology of measles viruses in the United States, 1997–2001. Emerg. Infect. Dis. *8,* 902–908.

Runkler, N., Pohl, C., Schneider-Schaulies, S., Klenk, H.D., and Maisner, A. (2007). Measles virus nucleocapsid transport to the plasma membrane requires stable expression and surface accumulation of the viral matrix protein. Cell Microbiol. *9,* 1203–1214.

Russell, S.J., and Peng, K.W. (2009). Measles virus for cancer therapy. Curr. Top. Microbiol. Immunol. *330,* 213–241.

Salditt, A., Koethe, S., Pohl, C., Harms, H., Kolesnikova, L., Becker, S., and Schneider-Schaulies, S. Measles virus M protein driven particle production does not involve the ESCRT system. J. Gen. Virol.

Schlender, J., Schnorr, J.J., Spielhoffer, P., Cathomen, T., Cattaneo, R., Billeter, M.A., ter Meulen, V., and Schneider-Schaulies, S. (1996). Interaction of measles virus glycoproteins with the surface of uninfected peripheral blood lymphocytes induces immunosuppression *in vitro*. Proc. Natl. Acad. Sci. U.S.A. *93, 13*,194–13199.

Schlender, J., Hornung, V., Finke, S., Gunthner-Biller, M., Marozin, S., Brzozka, K., Moghim, S., Endres, S., Hartmann, G., and Conzelmann, K.K. (2005). Inhibition of toll-like receptor 7- and 9-mediated alpha/beta interferon production in human plasmacytoid dendritic cells by respiratory syncytial virus and measles virus. J. Virol. *79,* 5507–5515.

Schneider, H., Spielhofer, P., Kaelin, K., Dotsch, C., Radecke, F., Sutter, G., and Billeter, M.A. (1997). Rescue of measles virus using a replication-deficient vaccinia-T7 vector. J. Virol. Methods *64,* 57–64.

Schneider-Schaulies, S., and ter Meulen, V. (1999). Pathogenic aspects of measles virus infections. Arch. Virol. *15,* 139–158.

Schneider-Schaulies, S., and Dittmer, U. (2006). Silencing T cells or T-cell silencing: concepts in virus-induced immunosuppression. J. Gen. Virol. *87,* 1423–1438.

Schneider-Schaulies, S., and Schneider-Schaulies, J. (2009). Measles virus-induced immunosuppression. Curr. Top. Microbiol. Immunol. *330,* 243–269.

Schneider-Schaulies, S., Liebert, U.G., Segev, Y., Rager-Zisman, B., Wolfson, M., and ter Meulen, V. (1992). Antibody-dependent transcriptional regulation of measles virus in persistently infected neural cells. J. Virol. *66,* 5534–5541.

Schneider-Schaulies, S., Schneider-Schaulies, J., Dunster, L.M., and ter Meulen, V. (1995). Measles virus gene expression in neural cells. Curr. Top. Microbiol. Immunol. *191,* 101–116.

Schneider-Schaulies, S., Klagge, I.M., and ter Meulen, V. (2003). Dendritic cells and measles virus infection. Curr. Top. Microbiol. Immunol. *276,* 77–101.

Schneider-Schaulies, S., Bieback, K., Avota, E., Klagge, I., and ter Meulen, V. (2002). Regulation of gene expression in lymphocytes and antigen-presenting cells by measles virus: consequences for immunomodulation. J. Mol. Med. *80,* 73–85.

Schnorr, J.J., Seufert, M., Schlender, J., Borst, J., Johnston, I.C., ter Meulen, V., and Schneider-Schaulies, S. (1997a). Cell cycle arrest rather than apoptosis is associated with measles virus contact-mediated immunosuppression *in vitro*. J. Gen. Virol. *78,* 3217–3226.

Schnorr, J.J., Xanthakos, S., Keikavoussi, P., Kampgen, E., ter Meulen, V., and Schneider-Schaulies, S. (1997b). Induction of maturation of human blood dendritic cell precursors by measles virus is associated with immunosuppression. Proc. Natl. Acad. Sci. U.S.A. *94,* 5326–5331.

Sellin, C.I., and Horvat, B. (2009). Current animal models: transgenic animal models for the study of measles pathogenesis. Curr. Top. Microbiol. Immunol. *330,* 111–127.

Sellin, C.I., Jegou, J.F., Renneson, J., Druelle, J., Wild, T.F., Marie, J.C., and Horvat, B. (2009). Interplay between virus-specific effector response and Foxp3 regulatory T cells in measles virus immunopathogenesis. PLoS One *4,* e4948.

Servet-Delprat, C., Vidalain, P.O., Azocar, O., Le Deist, F., Fischer, A., and Rabourdin-Combe, C. (2000a). Consequences of Fas-mediated human dendritic cell apoptosis induced by measles virus. J. Virol. *74,* 4387–4393.

Servet-Delprat, C., Vidalain, P.O., Bausinger, H., Manie, S., Le Deist, F., Azocar, O., Hanau, D., Fischer, A., and Rabourdin-Combe, C. (2000b). Measles virus induces abnormal differentiation of CD40 ligand-activated human dendritic cells. J. Immunol. *164,* 1753–1760.

Servet-Delprat, C., Vidalain, P.O., Valentin, H., and Rabourdin-Combe, C. (2003). Measles virus and dendritic cell functions: how specific response cohabits with immunosuppression. Curr. Top. Microbiol. Immunol. *276,* 103–123.

Shingai, M., Ebihara, T., Begum, N.A., Kato, A., Honma, T., Matsumoto, K., Saito, H., Ogura, H., Matsumoto, M., and Seya, T. (2007). Differential type I IFN-inducing abilities of wild-type versus vaccine strains of measles virus. J. Immunol. *179,* 6123–6133.

Shishkova, Y., Harms, H., Krohne, G., Avota, E., and Schneider-Schaulies, S. (2007). Immune synapses forMed. with measles virus-infected dendritic cells are unstable and fail to sustain T cell activation. Cell Microbiol. *9,* 1974–1986.

Sidhu, M.S., Chan, J., Kaelin, K., Spielhofer, P., Radecke, F., Schneider, H., Masurekar, M., Dowling, P.C., Billeter, M.A., and Udem, S.A. (1995). Rescue of synthetic measles virus minireplicons: measles genomic termini direct efficient expression and propagation of a reporter gene. Virology *208,* 800–807.

Silin, D., Lyubomska, O., Ludlow, M., Duprex, W.P., and Rima, B.K. (2007). Development of a challenge-protective vaccine concept by modification of the viral RNA-dependent RNA polymerase of canine distemper virus. J. Virol. *81,* 13649–13658.

Steineur, M.P., Grosjean, I., Bella, C., and Kaiserlian, D. (1998). Langerhans cells are susceptible to measles virus infection and actively suppress T cell proliferation. Eur J. Dermatol. *8,* 413–420.

Suryanarayana, K., Baczko, K., ter Meulen, V., and Wagner, R.R. (1994). Transcription inhibition and other properties of matrix proteins expressed by M genes cloned from measles viruses and diseased human brain tissue. J. Virol. *68,* 1532–1543.

de Swart, R.L. (2009). Measles studies in the macaque model. Curr. Top. Microbiol. Immunol. *330,* 55–72.

de Swart, R.L., Nur, Y., Abdallah, A., Kruining, H., El Mubarak, H.S., Ibrahim, S.A., Van Den Hoogen, B., Groen, J., and Osterhaus, A.D. (2001). Combination of reverse transcriptase PCR analysis and immunoglobulin M detection on filter paper blood samples allows diagnostic and epidemiological studies of measles. J. Clin. Microbiol. *39,* 270–273.

de Swart, R.L., Kuiken, T., Fernandez-de Castro, J., Papania, M.J., Bennett, J.V., Valdespino, J.L., Minor, P., Witham, C.L., Yuksel, S., Vos, H., *et al.* (2006). Aerosol measles vaccination in macaques: preclinical studies of immune responses and safety. Vaccine *24,* 6424–6436.

de Swart, R.L., Ludlow, M., de Witte, L., Yanagi, Y., van Amerongen, G., McQuaid, S., Yuksel, S., Geijtenbeek, T.B., Duprex, W.P., and Osterhaus, A.D. (2007). Predominant infection of CD150+ lymphocytes and dendritic cells during measles virus infection of macaques. PLoS Pathog. *3,* e178.

de Swart, R.L., Yuksel, S., Langerijs, C.N., Muller, C.P., and Osterhaus, A.D. (2009). Depletion of measles virus glycoprotein-specific antibodies from human sera reveals genotype-specific neutralizing antibodies. J. Gen. Virol. *90,* 2982–2989.

Tahara, M., Takeda, M., Seki, F., Hashiguchi, T., and Yanagi, Y. (2007a). Multiple amino acid substitutions in hemagglutinin are necessary for wild-type measles virus to acquire the ability to use receptor CD46 efficiently. J. Virol. *81,* 2564–2572.

Tahara, M., Takeda, M., and Yanagi, Y. (2007b). Altered interaction of the matrix protein with the cytoplasmic tail of hemagglutinin modulates measles virus growth by affecting virus assembly and cell–cell fusion. J. Virol. *81,* 6827–6836.

Tahara, M., Takeda, M., Shirogane, Y., Hashiguchi, T., Ohno, S., and Yanagi, Y. (2008). Measles virus infects both polarized epithelial and immune cells by using distinctive receptor-binding sites on its hemagglutinin. J. Virol. *82,* 4630–4637.

Takasu, T., Mgone, J.M., Mgone, C.S., Miki, K., Komase, K., Namae, H., Saito, Y., Kokubun, Y., Nishimura, T., Kawanishi, R., *et al.* (2003). A continuing high incidence of subacute sclerosing panencephalitis (SSPE) in the Eastern Highlands of Papua New Guinea. Epidemiol. Infect. *131,* 887–898.

Takeda, M., Takeuchi, K., Miyajima, N., Kobune, F., Ami, Y., Nagata, N., Suzaki, Y., Nagai, Y., and Tashiro, M. (2000). Recovery of pathogenic measles virus from cloned cDNA. J. Virol. *74,* 6643–6647.

Takeda, M., Ohno, S., Seki, F., Nakatsu, Y., Tahara, M., and Yanagi, Y. (2005). Long untranslated regions of the measles virus M and F genes control virus replication and cytopathogenicity. J. Virol. *79,* 14346–14354.

Takeda, M., Nakatsu, Y., Ohno, S., Seki, F., Tahara, M., Hashiguchi, T., and Yanagi, Y. (2006). Generation of measles virus with a segmented RNA genome. J. Virol. *80,* 4242–4248.

Takeda, M., Tahara, M., Hashiguchi, T., Sato, T.A., Jinnouchi, F., Ueki, S., Ohno, S., and Yanagi, Y. (2007). A human lung carcinoma cell line supports efficient measles virus growth and syncytium formation via a SLAM- and CD46-independent mechanism. J. Virol. *81,* 12091–12096.

Takeuchi, K., Kadota, S.I., Takeda, M., Miyajima, N., and Nagata, K. (2003a). Measles virus V protein blocks interferon (IFN)-alpha/beta but not IFN-gamma signaling by inhibiting STAT1 and STAT2 phosphorylation. FEBS Lett. *545,* 177–182.

Takeuchi, K., Miyajima, N., Nagata, N., Takeda, M., and Tashiro, M. (2003b). Wild-type measles virus induces large syncytium formation in primary human small airway epithelial cells by a SLAM(CD150)-independent mechanism. Virus Res. *94,* 11–16.

Tangy, F., and Naim, H.Y. (2005). Live attenuated measles vaccine as a potential multivalent pediatric vaccination vector. Viral Immunol. *18,* 317–326.

Tatsuo, H., Ono, N., Tanaka, K., and Yanagi, Y. (2000). SLAM (CDw150) is a cellular receptor for measles virus. Nature *406,* 893–897.

Tishon, A., Lewicki, H., Andaya, A., McGavern, D., Martin, L., and Oldstone, M.B. (2006). CD4 T cell control primary measles virus infection of the CNS: regulation is dependent on combined activity with either CD8 T cells or with B cells: CD4, CD8 or B cells alone are ineffective. Virology *347,* 234–245.

Tober, C., Seufert, M., Schneider, H., Billeter, M.A., Johnston, I.C., Niewiesk, S., ter Meulen, V., and Schneider-Schaulies, S. (1998). Expression of measles virus V protein is associated with pathogenicity and control of viral RNA synthesis. J. Virol. *72,* 8124–8132.

Torisu, H., Kusuhara, K., Kira, R., Bassuny, W.M., Sakai, Y., Sanefuji, M., Takemoto, M., and Hara, T. (2004). Functional MxA promoter polymorphism associated with subacute sclerosing panencephalitis. Neurology *62,* 457–460.

Tuxhorn, i.e. (2004). Familial subacute sclerosing panencephalitis in two siblings. Pediatr. Neurol. *31,* 291–294.

Valsamakis, A., Schneider, H., Auwaerter, P.G., Kaneshima, H., Billeter, M.A., and Griffin, D.E. (1998). Recombinant measles viruses with mutations in the C, V, or F gene have altered growth phenotypes *in vivo*. J. Virol. *72,* 7754–7761.

Vicente-Manzanares, M., and Sanchez-Madrid, F. (2000). Cell polarization: a comparative cell biology and immunological view. Dev. Immunol. *7,* 51–65.

Vidalain, P.O., Azocar, O., Lamouille, B., Astier, A., Rabourdin-Combe, C., and Servet-Delprat, C. (2000). Measles virus induces functional TRAIL production by human dendritic cells. J. Virol. *74,* 556–559.

Vieker, S., Schmitt, J.J., Behrens, C., Weissbrich, B., and Hartmann, H. (2003). Subacute sclerosing panencephalitis in two brothers. Neuropediatrics *34,* 326–329.

Vincent, S., Gerlier, D., and Manie, S.N. (2000). Measles virus assembly within membrane rafts. J. Virol. *74,* 9911–9915.

Vongpunsawad, S., Oezgun, N., Braun, W., and Cattaneo, R. (2004). Selectively receptor-blind measles viruses: Identification of residues necessary for SLAM- or CD46-induced fusion and their localization on a new hemagglutinin structural model. J. Virol. *78,* 302–313.

de Vries, R.D., Lemon, K., Ludlow, M., McQuaid, S., Yuksel, S., van Amerongen, G., Rennick, L.J., Rima, B.K., Osterhaus, A.D., de Swart, R.L., *et al.* (2010). *In vivo* tropism of attenuated and pathogenic measles virus expressing green fluorescent protein in macaques. J. Virol. *84,* 4714–4724.

Weidmann, A., Fischer, C., Ohgimoto, S., Ruth, C., ter Meulen, V., and Schneider-Schaulies, S. (2000). Measles virus-induced immunosuppression *in vitro* is independent of complex glycosylation of viral glycoproteins and of hemifusion. J. Virol. *74,* 7548–7553.

Welstead, G.G., Iorio, C., Draker, R., Bayani, J., Squire, J., Vongpunsawad, S., Cattaneo, R., and Richardson, C.D. (2005). Measles virus replication in lymphatic cells and organs of CD150 (SLAM) transgenic mice. Proc. Natl. Acad. Sci. U.S.A. *102,* 16415–16420.

Wild, T.F., and Buckland, R. (1995). Functional aspects of envelope-associated measles virus proteins. Curr. Top. Microbiol. Immunol. *191,* 51–64.

de Witte, L., Abt, M., Schneider-Schaulies, S., van Kooyk, Y., and Geijtenbeek, T.B. (2006). Measles virus targets DC-SIGN to enhance dendritic cell infection. J. Virol. *80,* 3477–3486.

de Witte, L., de Vries, R.D., van der Vlist, M., Yuksel, S., Litjens, M., de Swart, R.L., and Geijtenbeek, T.B. (2008). DC-SIGN and CD150 have distinct roles in transmission of measles virus from dendritic cells to T-lymphocytes. PLoS Pathog. *4,* e1000049.

Wolfson, L.J., Strebel, P.M., Gacic-Dobo, M., Hoekstra, E.J., McFarland, J.W., and Hersh, B.S. (2007). Has the 2005 measles mortality reduction goal been achieved? A natural history modelling study. Lancet *369,* 191–200.

Wulfing, C., Purtic, B., Klem, J., and Schatzle, J.D. (2003). Stepwise cytoskeletal polarization as a series of checkpoints in innate but not adaptive cytolytic killing. Proc. Natl. Acad. Sci. U.S.A. *100,* 7767–7772.

Yanagi, Y., Takeda, M., and Ohno, S. (2006). Measles virus: cellular receptors, tropism and pathogenesis. J. Gen. Virol. *87*, 2767–2779.

Zhang, X., Bourhis, J.M., Longhi, S., Carsillo, T., Buccellato, M., Morin, B., Canard, B., and Oglesbee, M. (2005). Hsp72 recognizes a P binding motif in the measles virus N protein C-terminus. Virology *337*, 162–174.

Zhang, Y., Li, X., Becker, K.A., and Gulbins, E. (2008). Ceramide-enriched membrane domains-Structure and function. Biochim. Biophys. Acta.

Zuniga, A., Wang, Z., Liniger, M., Hangartner, L., Caballero, M., Pavlovic, J., Wild, P., Viret, J.F., Glueck, R., Billeter, M.A., *et al.* (2007). Attenuated measles virus as a vaccine vector. Vaccine *25*, 2974–2983.

Canine Distemper Virus

8

Bevan Sawatsky, Sébastien Delpeut and Veronika von Messling*

Abstract

Canine distemper virus (CDV), a member of the genus *Morbillivirus*, causes one of the most devastating infectious diseases in carnivores. It is highly contagious and transmitted by aerosol or contact with body fluids from infected animals. Immune cells expressing the *Morbillivirus* receptor signalling lymphocyte activation molecule (SLAM, CD150) are the initial CDV targets, and first infected peripheral blood mononuclear cells are detected after two days. Infection of epithelia coincides with the onset of clinical signs, and is frequently followed by neuroinvasion and neurological sequelae. Disease severity can range from mild and short-lived to lethal, depending on strain and sensitivity of the host species. Therefore, the commercial live-attenuated vaccines, even though they are safe and efficacious in dogs, have to be carefully evaluated for each species. The characterization of genetic virulence determinants has revealed the importance of efficient entry and viral interference with innate immune activation, which constitute starting points for the development of the specifically attenuated vaccines.

History

Canine distemper virus (CDV) is an important pathogen of wild and domesticated carnivores. The disease is characterized by skin rash, fever, gastrointestinal and respiratory signs, and a profound immunosuppression as well as by frequent neurological complications. Canine distemper disease was first described by Henri Carré in 1905 (Carré, 1905), and its viral aetiology was proven in 1926 (Laidlaw and Dunkin, 1926). The disease is highly contagious and was endemic in domestic dogs until live-attenuated vaccines became available in the 1950s (Martella *et al.*, 2007). Since then, CDV-infected dogs are rarely seen in countries with high vaccine coverage, and wild carnivores have become the main reservoir. In contrast, in regions with low vaccine coverage, infected domestic dogs are frequently the source for outbreaks in susceptible wildlife populations.

Infectious agent

Classification

CDV is classified in the order *Mononegavirales*, in the genus *Morbillivirus* of the family *Paramyxoviridae*. Other members of this genus include Measles virus (MeV), Rinderpest virus (RPV), Peste-des-petits-ruminants virus (PPRV), Phocine distemper virus (PDV), and several morbilliviruses that infect marine mammals. Even though there is only one CDV serotype, isolates fall into phylogenetically distinct groups according to geographic distribution. In addition to the vaccine strains, the North American, European, African, Asian, and Arctic clades are currently recognized (McCarthy *et al.*, 2007; Kapil *et al.*, 2008).

Propagation

CDV is transmitted by aerosol droplets and contact with body fluids from infected animals. In animals, the first cells to be infected are probably

*Corresponding author

dendritic cells in the respiratory tract, which in turn transmit virus to circulating T and B cells in the lymph nodes (De Swart *et al.*, 2007; Griffin, 2007). The virus then disseminates rapidly throughout the lymphatic system, with first infected peripheral blood mononuclear cells (PBMCs) being detected as early as 2 days after infection. Spread to epithelia and in some cases to the central nervous system (CNS) coincides with the onset of clinical signs. The virus subsequently shed in mucosal secretions and can also be found in urine and probably in faeces (von Messling *et al.*, 2006). In cultured cells, virus propagation occurs primarily by glycoprotein-mediated fusion between infected and non-infected cells and subsequent cell-to-cell spread (Rankin *et al.*, 1972). Depending on the cell type and virus strain used, widespread formation of giant multinucleated cells, so called syncytia, can be observed as the infection progresses (Cosby *et al.*, 1981; von Messling *et al.*, 2001).

Morphology

The CDV particle is pleomorphic and enveloped with a diameter of approximately 100–200 nm. Glycoprotein spikes can sometimes be discerned on high-magnification electron micrographs. The nucleocapsid structure of CDV is similar to that of other paramyxoviruses. The paramyxovirus nucleocapsid protein associates very tightly to the viral nucleic acid, which is seen in electron micrographs as a 'herringbone' structure (Finch and Gibbs, 1970).

Genome organization

CDV is an enveloped, single-stranded RNA virus of negative polarity with a total genome length of 15,690 nucleotides. The CDV genome is flanked by a 3′ leader sequence and a 5′ trailer sequence. These sequences are necessary for genome and anti-genome production during viral replication. CDV contains six genes that encode eight proteins. All genes are separated by a CUU intergenic sequence and have their own 3′ and 5′ untranslated regions (UTRs). The nucleocapsid (N) gene is located immediately upstream of the leader sequence, followed by the phosphoprotein (P), matrix (M), fusion (F), haemagglutinin (H), and RNA-dependent RNA polymerase (L) genes. While all CDV isolates sequenced to date have the same genome length, there is approximately 10% sequence difference between the most widely divergent strains. Even though differences are distributed throughout the genome, the N, P, M, and L proteins are more highly conserved (Bellini *et al.*, 1986; Sidhu *et al.*, 1993), while the H protein and the region between the M and F genes are more variable (Martella *et al.*, 2006; Anderson and von Messling, 2008).

RNA and proteins

The N protein

Of the eight mRNAs produced from the six transcriptional units of the CDV genome, the N gene is located at the 3′ extremity of the genome. Its ORF encodes a 523 amino acid protein with a molecular weight of approximately 58 kDa (Masuda *et al.*, 2006). The N protein is produced in the largest abundance and encapsidates both genomic and anti-genomic viral RNA. The association between the N protein and viral nucleic acid is very strong and is thought to form the template with which the viral polymerase associates during transcription of viral mRNAs and genome replication (Lamb and Parks, 2007). Even though high levels of N protein-specific antibodies are produced after vaccination or infection, prophylactic injection of monoclonal antibodies against the N protein in animals does not protect against subsequent challenge (Hirayama *et al.*, 1991). Very little difference has been noted between wild-type and vaccine isolates in the N gene coding sequence and the UTRs, and there is over 95% amino acid sequence identity between N proteins of wild-type and vaccine strains (Stettler and Zurbriggen, 1995). Some more recent Asian isolates have lower levels of amino acid sequence identity, but are still antigenically identical to vaccine strain N proteins (Yoshida *et al.*, 1998).

The P, V, and C proteins

The P gene produces three distinct mRNAs. The P gene mRNA encodes the phosphoprotein, which is 507 amino acids in length with a predicted molecular weight of 54 kDa. The apparent molecular weight by SDS-PAGE is usually higher (~72 kDa), probably owing to post-translational phosphorylation (Sugai *et al.*, 2009). Based on studies with other paramyxovirus P proteins (Smallwood *et al.*, 1994), the CDV P protein likely interacts with the L protein to form the transcriptase complex via its C-terminal domain. The N-terminal domain of P is associated with the soluble, monomeric N protein that is not associated with viral nucleic acid (Sugai *et al.*, 2009).

The V protein mRNA is produced by non-templated insertion of a G nucleotide residue in the codon for glycine-231 by the viral polymerase complex, also called RNA editing, leading to a frame shift (Sugai *et al.*, 2009). The P and V proteins, therefore, have identical N-terminal amino acid sequences, but unique C-terminal domains. The final V ORF encodes a 299 amino acid protein with a predicted molecular weight of 33 kDa. Similar to MeV, the CDV V protein has been identified as main immune interference protein. The tyrosine-110 and neighbouring residues in the P/V shared N-terminal region are important for the inhibition of STAT1 nuclear translocation, while the C-terminal region, which forms a unique zinc finger motif conserved among paramyxoviruses (Li *et al.*, 2006), is involved in interference with STAT2 and mda5 signalling (Röthlisberger *et al.*, 2010).

The C mRNA is produced from the P gene via transcription of an alternative open reading frame. The C protein consists of 174 amino acids and has a predicted molecular weight of 20 kDa. Little is known about the role of the CDV C protein, but it may have a role in particle assembly and release analogous to the MeV C protein (Devaux *et al.*, 2004). Even though a C-deficient virus remained lethal in ferrets, the lack of C further increased attenuation of a V-deficient virus, indicating that the C protein contributes to virulence (von Messling *et al.*, 2006).

The M protein

The M ORF encodes a 335 amino acid protein with a molecular weight of 38 kDa. The M protein is targeted to the perimembrane compartment and interacts with the inner leaflet of the plasma membrane (Bellini *et al.*, 1986). It drives budding of virus particles and is thought to interact on the one hand with the N protein and on the other hand with the cytoplasmic tails of the viral envelope glycoproteins to ensure incorporation of the viral genome in new particles (Sanderson *et al.*, 1993). The C-terminus of the CDV M protein interacts with the glycoproteins that are expressed on the cell surface and inhibits fusion (Wiener *et al.*, 2007). It has been proposed that the M and N proteins act synergistically to limit cellular damage during infection by virulent CDV strains (Wiener *et al.*, 2007). Virus release is significantly impaired in M deleted MeVs (Cathomen *et al.*, 1998) and this is likely to be the case for CDV as well.

The F protein

The F protein is a type I integral membrane protein with an exceptionally long N-terminal signal peptide consisting of up to 135 amino acids (Anderson and von Messling, 2008; von Messling and Cattaneo, 2002). The 527-amino acid F_0 protein is translated and post-translationally modified in the endoplasmic reticulum (ER) and the Golgi apparatus, where it acquires its N-glycans. The F protein contains four to five potential N-linked glycosylation sites, of which three are used (Cherpillod *et al.*, 1999; von Messling and Cattaneo, 2003). These N-glycans are important for ensuring correct cell surface expression and fusion activity (von Messling and Cattaneo, 2003). The F_0 protein is then cleaved at the multibasic cleavage site upstream of the hydrophobic fusion peptide by furin or other subtilisin-like proteases in the trans–Golgi network into its disulfide bond-linked F1 and F2 subunits (Lamb and Parks, 2007). This results in the expression on the cell surface of mature membrane-anchored F protein trimers that hetero-oligomerize with H protein tetramers (Plemper *et al.*, 2001).

The H protein

The H protein is a 607 amino acid type II integral membrane protein that mediates attachment to the cellular receptors, thereby playing a critical role in its pathogenesis (Stern *et al.*, 1995; von Messling *et al.*, 2001, 2005, 2006). It has up to eight potential N-glycosylation sites of which up to five conserved sites are used (Sawatsky and von Messling, 2010). As with all morbilliviruses, the signalling lymphocyte activation molecule (SLAM; CD150; Tatsuo *et al.*, 2000, 2001; Hsu *et al.*, 2001) acts as a CDV receptor on immune cells, but the entry mechanisms and molecules involved in the infection of epithelial cells and cells of the CNS remain to be identified. Upon receptor binding, the H protein undergoes a conformational change, which in turn triggers the pH-independent membrane fusion activity of the F protein. By analogy to other paramyxovirus attachment proteins, CDV H is expressed on the cell surface as disulfide-linked dimer (Plemper *et al.*, 2000). Each dimer further associates non-covalently with another dimer, forming H protein tetramers, though the presence of dimeric attachment proteins is sufficient to trigger fusion (Plemper *et al.*, 2000, 2001). Most neutralizing antibodies are directed against CDV H, and in turn, the H gene displays the most genetic diversity (Bolt *et al.*, 1997; Haas *et al.*, 1997). However, vaccine-induced antibodies still efficiently neutralize recent wild type isolates, indicating that the principal antigenic epitopes remain conserved (Sawatsky and von Messling, 2010).

The L protein

With a site of 2184 amino acids, the L protein is the largest protein encoded by the CDV genome. By virtue of its position near the 5′ terminus of the genome, it is produced in the smallest quantity in the infected cell. The CDV L protein shares a high degree of amino acid sequence identity not only among different strains, but also with the homologous MeV protein (Sidhu *et al.*, 1993; McIlhatton *et al.*, 1997), and the conserved *Mononegavirales* L protein domains identified by Poch *et al.* (1990) are also found in CDV L (Sidhu *et al.*, 1993). Together with P, the L protein forms a transcriptase complex, which transcribes the individual viral mRNAs using the negative-sense genomic RNA as a template (Horikami *et al.*, 1994; Gao and Lenard, 1995). During later stages of infection, the L-P protein complex also interacts with the N protein to drive replication of full-length anti-genomic intermediate RNA and genomic RNA for incorporation into virus particles (Gupta *et al.*, 2003).

Transcription and replication

Most of what is known about CDV transcription and replication is derived from comparisons with other paramyxoviruses (Plumet *et al.*, 2005). The paramyxovirus transcription and replication occurs in the cytoplasm. The viral RNA polymerase enters at the 3′ end of the genome and first synthesizes leader RNA, which is neither capped nor polyadenylated. The viral genes are then transcribed in a sequential manner. The transcription of each gene begins at the 3′ transcriptional start signal and ends at the 5′ transcriptional stop signal (Li and Pattnaik, 1999; Whelan and Wertz, 1999; Whelan and Wertz, 2002). All mRNAs are capped and polyadenylated by the viral RNA polymerase (Stillman and Whitt, 1999; Ogino *et al.*, 2005; Ogino and Banerjee, 2007; Wang *et al.*, 2007). Since viral transcription occurs in the cytoplasm, viral mRNAs are directly accessible to ribosomes for translation. The viral RNA polymerase is thought to pause at the intergenic function before either disassociating from the genomic RNA template, or re-initiating at the next gene transcriptional start signal (Kolakofsky *et al.*, 2004; Plumet *et al.*, 2005). This mechanism of transcriptional start-stop generates an mRNA gradient, whereby genes near the 3′ terminus of the genome are transcribed with greater frequency than those near the 5′ terminus (Cattaneo *et al.*, 1987; Homann *et al.*, 1990; Wright *et al.*, 2005). The intracellular N protein levels are thought to determine the activity of the viral polymerase (Plumet *et al.*, 2005). Early in the infection, under conditions of low N protein expression, the polymerase complex functions predominantly as transcriptase, producing viral mRNAs (Gao and Lenard, 1995; Gao *et al.*, 1996), but later in the infection, when N protein levels are high, it switches to act as replicase synthesizing anti-genomic and genomic transcripts (Gupta *et al.*, 2003). This has been suggested to be the major mechanism by which mononegaviruses regulate the generation of full-length genomic and anti-genomic RNA.

The genome length of members of the subfamily *Paramyxovirinae* follows the 'rule of six', indicating that the genome size must be an exact multiple of six to support efficient replication (Calain and Roux, 1993; Durbin *et al.*, 1997; Kolakofsky *et al.*, 1998; Peeters *et al.*, 2000; Halpin *et al.*, 2004). Although this has not been explicitly demonstrated for CDV, all genomes sequenced so far have hexameric genome lengths.

Genetics and reverse genetics

Several reverse genetics systems have been developed to recover infectious CDV from plasmid DNA. Thus far, full-length cDNA genomes have been generated for the CDV strains Onderstepoort (Gassen *et al.*, 2000; von Messling *et al.*, 2001; Parks *et al.*, 2002), 5804P (von Messling *et al.*, 2003, 2004), A75/17 (Plattet *et al.*, 2004; Rudd *et al.*, 2006), and Snyder Hill (Rudd *et al.*, 2010). The genomic plasmids contain a T7 RNA polymerase promoter to drive initial transcription of anti-genomic RNA. Expression plasmids encoding either CDV or MeV N, P, and L genes produce the proteins needed to form the RNP complex that then transcribes and replicates the genome. The source of T7 RNA polymerase can vary; earlier recovery systems used a recombinant modified vaccinia virus strain Ankara expressing T7 RNA polymerase or the modified vaccinia vector vTF7–3 (Gassen *et al.*, 2000; von Messling *et al.*, 2001), while later protocols are based on co-transfection of a T7 polymerase expression plasmid (Anderson and von Messling, 2008) or the use of T7 polymerase-expressing cell lines (Plattet *et al.*, 2004). More recently, cellular RNA polymerase II promoters have begun to replace T7 RNA polymerase promoters for the production of full-length viral RNA from plasmids (Martin *et al.*, 2006). These systems no longer require the addition of an exogenous polymerase, such as T7 polymerase, but instead use host cell polymerases to transcribe the genomic plasmid.

To generate recombinant viruses, cell lines such as 293, HeLa, or B95a cells are transfected with the genomic plasmid and defined ratios of the accessory N, P, L, and T7 RNA polymerase expression plasmids. After 2–3 days in culture, these transfected cells are overlaid onto Vero cells that stably express the canine SLAM receptor. These cultures are then monitored for syncytium formation, which are then transferred onto fresh cells expressing SLAM to grow virus stocks (von Messling *et al.*, 2003; Fujita *et al.*, 2007). The enhanced green fluorescent protein (eGFP) has been introduced as an additional transcription unit at various positions in the genomes of the above CDV strains (von Messling *et al.*, 2004), which facilitates the detection of infected cells during the initial virus recovery and *in vivo* (Rudd *et al.*, 2006; Bonami *et al.*, 2007).

Pathogenesis

Host range

Unlike other members of the genus *Morbillivirus*, CDV naturally infects a broad range of terrestrial and marine carnivores (Harder and Osterhaus, 1997). In countries with low vaccine coverage, domestic dogs are the main reservoir and CDV outbreaks in wild carnivores are thought to originate from contact between domestic and wild animals (Cleaveland *et al.*, 2000; Gowtage-Sequeira *et al.*, 2009). Indigenous wild carnivores that live in close proximity to urban areas, such as foxes and raccoons, are considered the primary source of CDV outbreaks in dogs in the developed world (Bolt *et al.*, 1997; Kapil *et al.*, 2008). Similar modes of transmission have also been suggested for marine mammals, which in addition are susceptible to species-specific morbilliviruses that are closely related to, but genetically distinct from CDV (Barrett *et al.*, 2004; Butina *et al.*, 2010).

Experimental intracerebral inoculation with CDV of mice, guinea pigs, and some non-human primates can result in encephalitis or encephalomyelitis (Yamanouchi *et al.*, 1977; Yoshikawa *et al.*, 1981; Bernard *et al.*, 1983; Nagata *et al.*, 1990), but other routes of inoculation generally do not result in disease. However, a CDV outbreak was recently reported in hand-feeding rhesus monkeys in China (Sun *et al.*, 2010), and an earlier outbreak in captive Japanese monkeys was associated with one case of fatal viral encephalitis (Yoshikawa *et al.*, 1989). The involvement of CDV in the aetiology of multiple sclerosis and Paget's disease has been suggested (Cook *et al.*, 1979; Madden *et al.*, 1981; Mee *et al.*, 1998), but there is so far no experimental evidence to support this hypothesis (Stephenson *et al.*, 1980; Haile *et al.*, 1982).

Entry, spread, and tissue tropism

The primary *Morbillivirus* receptor, SLAM (Tatsuo *et al.*, 2000; Hsu *et al.*, 2001), is expressed on activated lymphocytes and certain subsets of dendritic cells (Cocks *et al.*, 1995), but is not present in epithelial or neuronal tissues. Thus far, no other receptors for CDV have been identified, so the molecular mechanism by which CDV is able to invade epithelial and neuronal tissues remains undefined. It has been suggested, however, that heparin sulfate in the extracellular matrix might be a factor in virus attachment and entry (Fujita *et al.*, 2007).

Resident immune cells in the upper respiratory tract mucosa are the initial sites of infection following inhalation of contaminated aerosol droplets. These cells then migrate to local lymph nodes, where the virus spreads to T and B lymphocytes, which in turn disseminate the infection throughout the lymphatic system. The onset of clinical signs coincides with spread to epithelial cells, and at the peak of infection CDV is found in respiratory, gastro-intestinal, urinary, endocrine, lymphoid, vascular, and in some cases CNS tissues (Appel, 1970; Frisk *et al.*, 1999). Infection of the CNS may result in neurological clinical signs during the acute disease phase or several weeks to months after recovery (Higgins *et al.*, 1989; Muller *et al.*, 1995; Stettler *et al.*, 1997). In addition to the haematogenous route, either through infection of endothelial cells in the CNS and subsequent release of the virus into the brain, infection of the choroid plexus, or trafficking into the CNS of infected PBMCs (Summers *et al.*, 1978, 1979; Higgins *et al.*, 1982), CDV also enters the CNS anterogradely via the olfactory bulb. Anatomically, olfactory neurons project axons to the brain through the olfactory nerve and enable direct entry into the CNS. The close proximity of dendrites of mature olfactory receptor neurons and highly infected epithelial cells in the olfactory mucosa probably facilitates CNS entry through this pathway (Rudd *et al.*, 2006).

CDV neuropathology

In naturally infected dogs, CDV CNS infection can present as polioencephalitis (Nesseler *et al.*, 1999) or demyelinating leucoencephalomyelitis (Wisniewski *et al.*, 1972), depending on the presence of lesions in the grey or white matter. Several different forms of polioencephalitis have been described, all of which occur infrequently. Among those, old dog encephalitis is characterized by predominant lesions in the grey matter of the cerebral hemisphere and brain stem (Lincoln *et al.*, 1971), while post-vaccine encephalitis results in the appearance of CNS lesions in the cerebral cortex and brain stem 1–3 weeks after vaccination with a modified-live vaccine (Hartley, 1974; Bestetti *et al.*, 1978). In inclusion body encephalitis, which is characterized by high levels of viral mRNA and reduced or absent viral protein translation, lesions are often restricted in the diencephalon, mesencephalon, and medulla oblongata, but there may also be lesions in the cerebral cortex (Nesseler *et al.*, 1997, 1999).

The early stage of demyelinating leukoencephalomyelitis is associated with expression of viral proteins and RNA in the lesions, moderate upregulation of MHC class II expression, and little or no infiltration of $CD8^+$ cytotoxic T cells (Alldinger *et al.*, 1996; Tipold *et al.*, 1999; Wunschmann *et al.*, 1999). The observed demyelination may be the result of degeneration of infected oligodendrocytes (Graber *et al.*, 1995; Zurbriggen *et al.*, 1998; Schobesberger *et al.*, 2002), antibody-mediated (Krakowka *et al.*, 1973; Vandevelde *et al.*, 1982), or due to microglia activation (Stein *et al.*, 2004), but the exact mechanisms remain to be determined. The chronic stage is characterized by virus clearance from the affected areas and a strong inflammatory immune response (Zurbriggen *et al.*, 1995; Alldinger *et al.*, 1996; Wunschmann *et al.*, 1999; Vandevelde and Zurbriggen, 2005). Although CDV is effectively cleared from these inflammatory lesions in the white matter (Bollo *et al.*, 1986), it remains detectable in other areas of the CNS (Vandevelde and Zurbriggen, 2005), which may be due to long-distance spread via astrocytic processes (Wyss-Fluehmann *et al.*, 2010) or along neuronal signalling pathways.

Virulence factors

While specific genetic virulence factors remain to be identified, pathogenesis studies using specifically modified recombinant viruses have elucidated the contribution of different viral proteins or sequence motifs. The exchange of envelope and replication complex genes between closely related attenuated and virulent strains revealed that multiple genes were involved in virulence, since both chimeric viruses still caused disease, but were no longer lethal (von Messling *et al.*, 2003). Since the duration and severity of infection-induced immunosuppression correlates with disease outcome (Bonami *et al.*, 2007; Svitek and von Messling, 2007), a virus lacking the accessory V protein, which is the primary immune interference protein (Devaux *et al.*, 2007; Nakatsu *et al.*, 2008), was generated. In addition to a complete lack of immunosuppression, the virus induced a robust immune response and rapid virus clearance and caused only mild clinical signs (von Messling *et al.*, 2006).

Subsequent investigations have focused on the contribution of receptor binding and entry. Recombinant viruses with altered receptor tropism revealed that SLAM binding is essential for a lethal disease outcome (von Messling *et al.*, 2005). A more recent study demonstrated the importance of H protein N-glycosylation for virulence, even though the underlying mechanism remains to be determined (Sawatsky and von Messling, 2010). A systematic analysis of the unusually long untranslated region between the M and F protein revealed it that it tightly controls F protein mRNA synthesis and is essential for lethal disease (Anderson and von Messling, 2008).

Animal models

Among the natural CDV hosts, dogs, ferrets, and to a lesser degree mink, are standard laboratory animals. Dogs were used for most of the initial pathogenesis studies in the context of CDV vaccine development. However, considerable variability in disease severity was observed within groups infected with the same virus, ranging from asymptomatic or mild and short-lived to lethal (Appel *et al.*, 1982). In contrast, ferrets and mink are more sensitive to the virus and wild-type strains are often lethal (von Messling *et al.*, 2003; Nielsen *et al.*, 2009a). This higher reproducibility of clinical signs and the fact that certain strains are reliably neuroinvasive, have resulted in an increased use of ferrets to study pathogenesis and neuroinvasion mechanisms (Axthelm and Krakowka, 1987; Vandevelde and Zurbriggen, 2005; Rudd *et al.*, 2006, 2010; Pillet *et al.*, 2009). Small rodents, including hamsters and mice, have also been used to investigate neurodissemination and evaluate the efficacy of experimental vaccine. However, infection and replication in these models is limited, usually requiring intracerebral inoculation, which limits their use for pathogenesis studies (Cosby *et al.*, 1981; Allen *et al.*, 1987). Since CDV infection in dogs or ferrets closely reproduces the course and clinical signs of MeV in patients (Imagawa, 1968), these models are also used to gain insight in general *Morbillivirus* pathogenesis mechanisms (Kauffman *et al.*, 1982b; von Messling *et al.*, 2003; Pillet *et al.*, 2009).

Immune responses

Despite the severe and long-lasting immunosuppression, which is a hallmark of all *Morbillivirus* infections (von Pirquet, 1908), survivors ultimately mount a robust immune response that leads to virus clearance and confers lifelong protection. Even though the exact mechanism of this apparent paradox remains unclear, there is increasing understanding of the immune response kinetics involved. During the lymphatic phase of infection, CDV efficiently inhibits innate immune activation, and while this inhibition is maintained throughout the disease in lethal cases, IFN and cytokine mRNA induction is detected as early as three days after infection in survivors (Krakowka *et al.*, 1987; Grone *et al.*, 1998; Svitek and von Messling, 2007).

The first detection of CDV-specific antibodies coincides with the onset of rash and clinical signs. Neutralizing antibody titres above 80, which are considered protective, are reached two to three weeks later (Svitek and von Messling, 2007). The adaptive immune response elicited by either wild type infection or a live-attenuated vaccine probably persists life-long and provides near-sterile protection from re-infection (Schultz, 2006). The kinetics of the cellular immune response are less well characterized, but early studies reported that the level of lymphocyte-mediated cytotoxicity is strain dependent. Similar to the humoral response, the cytotoxic T-cell response is very low or absent in dogs that succumb to the disease and strong in survivors (Shek *et al.*, 1980; Appel *et al.*, 1982). A recent study in mink infected with a wild type strain confirms this correlation between the adaptive immune response and virus clearance, since an increase in lymphocytes expressing the pro-inflammatory cytokines IFN-γ and TNF-α was first detected two weeks after infection, which coincided with the appearance of neutralizing antibodies and the decrease of infected PBMCs (Nielsen *et al.*, 2009b).

There is strong circumstantial evidence for a link between a reduced or delayed host response and CDV-associated chronic CNS disease (Bonami *et al.*, 2007; Beineke *et al.*, 2009). Infection kinetics in ferrets indicate that neuroinvasion occurs only after massive infection of immune and epithelial tissues, and may be a consequence of inefficient virus clearance (Rudd *et al.*, 2006). During the acute disease phase, the presence of CDV in the CNS triggers a robust local immune response, including activation of microglia, astrocytosis, and expression of pro-inflammatory cytokines, and probably provides sufficient antigenic stimulus to induce humoral and cellular responses (Yamanouchi *et al.*, 1979; Alldinger *et al.*, 1996; Rudd *et al.*, 2010). While this response may result in virus clearance in some cases, there is also a correlation between microglial activation, recruitment of cytotoxic T cells to the CNS, and acute demyelination (Larsen *et al.*, 1989; Tipold *et al.*, 1999; Stein *et al.*, 2004). Even though antimyelin antibodies are probably not involved in the initiation of demyelination (Vandevelde *et al.*, 1981, 1982), their presence in chronic lesions indicates an important role in the progression of demyelination (Wisniewski *et al.*, 1972; Cammer *et al.*, 1978; Kristensson *et al.*, 1979; Griot *et al.*, 1989a,b).

Clinical features

During the early prodromal disease phase there are no outward signs of disease, although some animals may experience a slight fever peak at days 3–5 post infection (Metzler *et al.*, 1980; Appel *et al.*, 1982; von Messling *et al.*, 2003). The first clinical signs of wild type infections are observed at 6–10 days after infection, starting with small red spots and ulcers on lips and eye lids and chin rash developing shortly thereafter. Infected animals also experience sustained fever in excess of 39°C. The rash becomes progressively worse, affecting large areas of skin and mucosal tissues. Macroscopic examination of ferrets infected with eGFP-expressing recombinant viruses revealed massive infection in many mucosal areas such as the eyes, ear, nose, mouth, genitals, and anus (von Messling *et al.*, 2004). As the disease progresses, gastrointestinal and respiratory signs are frequently observed and the animals become increasingly lethargic and dehydrated. Depending on the strain and species involved, neurological complications may occur (Confer *et al.*, 1975; Summers *et al.*, 1979, 1984; Appel *et al.*, 1982; Appel and Summers, 1995). In addition, the severe immunosuppression increases susceptibility to secondary infections (Krakowka *et al.*, 1975; Appel *et al.*, 1982), which in turn may contribute to the clinical disease.

T and B lymphocytes are the first targets of CDV infection *in vivo* (von Messling *et al.*, 2004). Infected PBMCs can be detected as early as 2 days post infection and infection levels can reach peak levels of more than 80% within 7–10 days. While animals that succumb to the disease are unable to clear the virus, viral loads in survivors gradually decrease (Metzler *et al.*, 1980; Appel *et al.*, 1982; Sawatsky and von Messling, 2010). Infected animals show a precipitous decline in white blood cell count, with lymphocyte and granulocyte populations being equally affected (Pillet and von Messling, 2009). In mild cases, the white blood cell count quickly reaches pre-infection levels, while severe infections are characterized by prolonged leukopenia, even after the virus has been cleared and the animal has recovered from clinical disease (McCullough *et al.*, 1974; Krakowka, 1975, 1982; Appel *et al.*, 1982; Kauffman *et al.*, 1982a). CDV also inhibits the proliferative activity of lymphocytes with the extent of inhibition correlating with disease severity (Appel *et al.*, 1982; Kauffman *et al.*, 1982a; von Messling *et al.*, 2003).

Epidemiology

CDV is found throughout the world, and owing to its rather broad host range, it is endemic on most continents (Mori *et al.*, 1994; Norris *et al.*, 2006; Calderon *et al.*, 2007; Dales Nava *et al.*, 2008; Kapil *et al.*, 2008; Ferreyra *et al.*, 2009). Despite this broad distribution, there is still only one circulating serotype, and even the most divergent isolates still share ~90% homology with older strains (Haas *et al.*, 1997; Cherpillod *et al.*, 1999), highlighting the overall stability of CDV on the nucleotide and protein level.

CDV isolates can be phylogenetically grouped according to their geographic origin of isolation (McCarthy *et al.*, 2007). The oldest available isolates, Snyder Hill and Onderstepoort, were isolated in the 1950s in New York and South Africa, respectively, and were extensively used for vaccine development (Martella *et al.*, 2007). Both viruses are phylogenetically quite close, differing only by 1–2% in their H proteins at the amino acid level, and constitute the America-1 phylogenetic group (Pardo *et al.*, 2005; Martella *et al.*, 2007). The America-2 group of viruses are more distantly related to America-1 viruses and comprise isolates from North American wildlife and zoo animals. Arctic-like viruses are typically found in the northern parts of Canada, China, and Siberia (Maes *et al.*, 2003), although related viruses have recently been found in dogs in Italy and Hungary (Martella *et al.*, 2006; Demeter *et al.*, 2007). This may not necessarily be a reflection of virus migration into Europe from the Arctic, but rather a diversification of the H protein sequence in wild populations of European animals. Other European isolates group together, indicating that there is a common pool of circulating viruses in this region (Haas *et al.*, 1996; Demeter *et al.*, 2010).

Some of the more diverse CDV strains are found in the Asia-1 and Asia-2 phylogenetic groups. An additional potential N-glycosylation site at N584 in the primary H protein sequence of some of these Asian viruses has been suggested as cause of suspected vaccine failure by shielding antigenic sites of the vaccine strain (Iwatsuki *et al.*, 1997; Mochizuki *et al.*, 1999). However, *in vitro* studies have recently shown that this extra N-glycosylation site is not used (Sawatsky and von Messling, 2010), and there is no evidence as of yet that these changes in the H protein primary sequence result in a genuine evolution in the antigenic structure of the virus.

Diagnosis

Before molecular diagnostic tools became available, CDV diagnosis was based on the detection of viral antigen in smears or impressions of ocular or oral mucosa by fluorescently-labelled antibodies (Fairchild *et al.*, 1967, 1971). Alternatively, presence of IgM antibodies against CDV or a more than fourfold increase in neutralizing antibodies was considered strongly indicative of a recent CDV infection (Blixenkrone-Moller *et al.*, 1993; Soma *et al.*, 2001). Furthermore, antibody detection in the cerebrospinal fluid can be used to confirm involvement of CDV in neurological disease (Rima *et al.*, 1987; Johnson *et al.*, 1988).

First CDV sequence information became available in 1985 (Rozenblatt *et al.*, 1985), and the development of PCR-based diagnostic assays for CDV followed shortly thereafter (Haas *et al.*, 1990; Shin *et al.*, 1995). In addition, several enzyme-linked immunosorbent assays for the detection of antibodies or antigen in different body fluids have been developed (Bernard *et al.*, 1982; Gemma *et al.*, 1995; von Messling *et al.*, 1999; Soma *et al.*, 2003; Latha *et al.*, 2007), some of which are commercially available.

Prevention and control

Since there are no CDV-specific antiviral drugs available, existing treatment regimen aim at counteracting the clinical signs through administration of anti-nausea and anti-diarrhoea drugs and rehydration. In addition, antibiotics are frequently used to prevent or treat secondary bacterial infections. Anti-CDV activity of RNA-dependent RNA polymerase inhibitors such as ribavirin and 5-ethynyl-1-β-ribofuranosylimidazole-carboxamide (EICAR) has been demonstrated *in vitro* (Elia *et al.*, 2008), but their therapeutic potential remains to be evaluated.

The first live-attenuated CDV vaccines were developed in the 1950s and have a good safety in domestic dogs (Croghan, 1966; Keeble, 1962; Prydie, 1968; Rockborn, 1959). The traditional vaccination protocol involves two to three immunizations within the first 6 months of life to assure a robust induction of an immune response as pre-existing maternal antibodies decline (Krakowka *et al.*, 1978; Winters, 1981), followed by annual repeats. The ability to induce a protective immune response in the presence of maternal antibodies has been shown for different candidate vaccines based on the expression of CDV proteins either in viral vectors or from DNA plasmids (Welter *et al.*, 2000; Griot *et al.*, 2004; Pardo *et al.*, 2007), and a canarypox-based recombinant vaccine has received regulatory approval in North America.

While the available live-attenuated vaccines have an excellent safety record in dogs, vaccination-associated disease and deaths have occurred in domestic ferrets and several wild carnivore species that are more sensitive to the virus (Carpenter *et al.*, 1976). The efficacy of several of the above-mentioned candidate vaccines has been demonstrated in ferrets, and annual vaccination with the canarypox-based vaccine is currently recommended for domestic ferrets and constitutes an attractive option for exotic carnivores living in captivity. However, the need for several initial and regular repeat vaccinations to induce and maintain protective immunity does not make this approach feasible for wildlife. Specifically attenuated recombinant CDVs are attenuated and immunogenic in ferrets and may be promising candidates for wild life vaccines (von Messling *et al.*, 2005; Rouxel *et al.*, 2009).

Perspectives

Despite the availability of an efficient vaccine with good safety profile in dogs, the broad host range of CDV makes its eradication unlikely. There is thus a continued need for the development of safe vaccines for species that develop disease after immunization with the current live-attenuated vaccines. In addition, the apparent adaptability of CDV to new hosts (Harder and Osterhaus, 1997) makes this virus a potential for entering the human population after MeV eradication. At the same time, the genetic and clinical similarities between these two viruses also make the study of CDV in its natural host an attractive model for the characterization of MeV pathogenesis mechanisms.

References

Alldinger, S., Wunschmann, A., Baumgartner, W., Voss, C., and Kremmer, E. (1996). Up-regulation of major histocompatibility complex class II antigen expression in the central nervous system of dogs with spontaneous canine distemper virus encephalitis. Acta Neuropathol. *92*, 273–280.

Allen, I.V., Cosby, S.L., Kirk, J., Martin, S.J., and Dinsmore, S. (1987). Neuropathology and neurovirulence of canine distemper virus plaque isolates in the hamster. Neuropathol. Appl. Neurobiol. *13*, 349–369.

Anderson, D.E., and von Messling, V. (2008). The region between the canine distemper virus M and F genes modulates virulence by controlling fusion protein expression. J. Virol. *82*, 10510–10518.

Appel, M.J. (1970). Distemper pathogenesis in dogs. J. Am. Vet. Med. Assoc. *156*, 1681–1684.

Appel, M.J.G., and Summers, B.A. (1995). Pathogenicity of morbilliviruses for terrestrial carnivores. Vet. Microbiol. *44*, 187–191.

Appel, M.J.G., Shek, W.R., and Summers, B.A. (1982). Lymphocyte-mediated immune cytotoxicity in dogs infected with virulent canine distemper virus. Infect. Immun. *37*, 592–600.

Axthelm, M.K., and Krakowka, S. (1987). Canine distemper virus: the early blood–brain barrier lesion. Acta Neuropathol. *75*, 27–33.

Barrett, T., Wohlsein, P., Bidewell, C.A., and Rowell, S.F. (2004). Canine distemper virus in a Californian sea lion (*Zalophus californianus*). Vet. Rec. *154*, 334–336.

Beineke, A., Puff, C., Seehusen, F., and Baumgärtner, W. (2009). Pathogenesis and immunopathology of systemic and nervous canine distemper. Vet. Immunol. Immunopathol. *127*, 1–18.

Bellini, W.J., Englund, G., Richardson, C.D., Rozenblatt, S., and Lazzarini, R.A. (1986). Matrix genes of measles virus and canine distemper virus: cloning, nucleotide sequences, and deduced amino acid sequences. J. Virol. *58*, 408–416.

Bernard, A., Wild, T.F., and Tripier, M.F. (1983). Canine Distemper infection in mice: characterization of a neuro-adapted virus strain and its long-term evolution in the mouse. J. Gen. Virol. *64*, 1571–1579.

Bernard, S.L., Shen, D.T., and Gorham, J.R. (1982). Antigen requirements and specificity of enzyme-linked immunosorbent assay for detection of canine IgG against canine distemper viral antigens. Am. J. Vet. Res. *43*, 2266–2269.

Bestetti, G., Fatzer, R., and Frankhauser, R. (1978). Encephalitis following vaccination against distemper and infectious hepatitis in the dog. An optical and ultrastructural study. Acta Neuropathol. *43*, 69–75.

Blixenkrone-Moller, M., Svansson, V., Have, P., Orvell, C., Appel, M., Pedersen, I.R., Dietz, H.H., and Henriksen, P. (1993). Studies on manifestations of canine distemper virus infection in an urban dog population. Vet. Microbiol. *37*, 163–173.

Bollo, E., Zurbriggen, A., Vandevelde, M., and Fankhauser, R. (1986). Canine distemper virus clearance in chronic inflammatory demyelination. Acta Neuropathol. *72*, 69–73.

Bolt, G., Jensen, T.D., Gottschalck, E., Arctander, P., Appel, M.J.G., Buckland, R., and Blixenkrone-Møller, M. (1997). Genetic diversity of the attachment (H) protein gene of current field isolates of canine distemper virus. J. Gen. Virol. *78*, 367–372.

Bonami, F., Rudd, P.A., and von Messling, V. (2007). Disease duration determines canine distemper virus neurovirulence. J. Virol. *81,* 12066–12070.

Butina, T.V., Denikina, N.N., and Belikov, S.I. (2010). Canine distemper virus diversity in Lake Baikal seal (*Phoca sibirica*) population. Vet. Microbiol. *144,* 192–197.

Calain, P., and Roux, L. (1993). The rule of six, a basic feature for efficient replication of Sendai virus defective interfering RNA. J. Virol. *67,* 4822–4830.

Calderon, M.G., Remorini, P., Periolo, O., Iglesias, M., Mattion, N., and La Torre, J. (2007). Detection by RT-PCR and genetic characterization of canine distemper virus from vaccinated and non-vaccinated dogs in Argentina. Vet. Microbiol. *125,* 341–349.

Cammer, W., Bloom, B.R., Norton, W.T., and Gordon, S. (1978). Degradation of basic protein in myelin by neutral proteases secreted by stimulated macrophages: a possible mechanism of inflammatory demyelination. Proc. Natl. Acad. Sci. U.S.A. *75,* 1554–1558.

Carpenter, J.W., Appel, M.J., Erickson, R.C., and Novilla, M.N. (1976). Fatal vaccine-induced canine distemper virus infection in black-footed ferrets. J. Am. Vet. Med. Assoc. *169,* 961–964.

Carré, H. (1905). Sur la maladie des jeunes chiens. Comptes rendus de l'Academie des Sciences *140,* 1489–1491.

Cathomen, T., Mrkic, B., Spehner, D., Drillien, R., Naef, R., Pavlovic, J., Aguzzi, A., Billeter, M.A., and Cattaneo, R. (1998). A matrix-less measles virus is infectious and elicits extensive cell fusion: consequences for propagation in the brain. EMBO J. *17,* 3899–3908.

Cattaneo, R., Rebmann, G., Schmid, A., Baczko, K., ter Meulen, V., and Billeter, M.A. (1987). Altered transcription of a defective measles virus genome derived from a diseased human brain. EMBO J. *6,* 681–688.

Cherpillod, P., Beck, K., Zurbriggen, A., and Wittek, R. (1999). Sequence Analysis and Expression of the Attachment and Fusion Proteins of Canine Distemper Virus Wild-Type Strain A75/17. J. Virol. *73,* 2263–2269.

Cleaveland, S., Appel, M.J.G., Chalmers, W.S.K., Chillingworth, C., Kaare, M., and Dye, C. (2000). Serological and demographic evidence for domestic dogs as a source of canine distemper virus infection for Serengeti wildlife. Vet. Microbiol. *72,* 217–227.

Cocks, B.G., Chang, C.-C.J., Carballido, J.M., Yssel, H., de Vries, J.E., and Aversa, G. (1995). A novel receptor involved in T-cell activation. Nature *376,* 260–263.

Confer, A.W., Kahn, D.E., Koestner, A., and Krakowka, S. (1975). Biological properties of a canine distemper virus isolate associated with demyelinating encephalomyelitis. Infect. Immun. *11,* 835–844.

Cook, S.D., Dowling, P.C., and Russell, W.C. (1979). Neutralizing antibodies to canine distemper and measles virus in multiple sclerosis. J. Neurol. Sci. *41,* 61–70.

Cosby, S.L., Lyons, C., Fitzgerald, S.P., Martin, S.J., Pressdee, S., and Allen, I.V. (1981). The isolation of large and small plaque canine distemper viruses which differ in their neurovirulence for hamsters. J. Gen. Virol. *52,* 345–353.

Croghan, D.L. (1966). Attenuated live-virus canine distemper vaccines. J. Am. Vet. Med. Assoc. *149,* 662–665.

Dales Nava, A.F., Cullen, L., Jr., Sana, D.A., Nardi, M.S., Ramos Filho, J.D., Lima, T.F., Abreu, K.C., and Ferreira, F. (2008). First evidence of canine distemper in Brazilian free-ranging felids. EcoHealth *5,* 513–518.

De Swart, R.L., Ludlow, M., de Witte, L., Yanagi, Y., van Amerongen, G., McQuaid, S., Yüksel, S., Geijtenbeek, T.B.H., Duprex, W.P., and Osterhaus, A.D.M.E. (2007). Predominant infection of CD150$^+$ lymphocytes and dendritic cells during measles virus infection of macaques. PLoS Pathog. *3,* e178.

Demeter, Z., Lakatos, B., Palade, E.A., Kozma, T., Forgách, P., and Rusvai, M. (2007). Genetic diversity of Hungarian canine distemper virus strains. Vet. Microbiol. *122,* 258–269.

Demeter, Z., Palade, E.A., Hornyák, Á., and Rusvai, M. (2010). Controversial results of the genetic analysis of a canine distemper vaccine strain. Vet. Microbiol. *142,* 420–426.

Devaux, P., von Messling, V., Songsungthong, W., Springfeld, C., and Cattaneo, R. (2007). Tyrosine 110 in the measles virus phosphoprotein is required to block STAT1 phosphorylation. Virology *360,* 72–83.

Durbin, A.P., Siew, J.W., Murphy, B.R., and Collins, P.L. (1997). Minimum protein requirements for transcription and RNA replication of a minigenome of human parainfluenza virus type 3 and evaluation of the rule of six. Virology *234,* 74–83.

Elia, G., Belloli, C., Cirone, F., Lucente, M.S., Caruso, M., Martella, V., Decaro, N., Buonavoglia, C., and Ormas, P. (2008). *In vitro* efficacy of ribavirin against canine distemper virus. Antiviral. Res. *77,* 108–113.

Fairchild, G.A., Wyman, M., and Donovan, E.F. (1967). Fluorescent antibody technique as a diagnostic test for canine distemper infection: detection of viral antigen in epithelial tissues of experimentally infected dogs. Am. J. Vet. Res. *28,* 761–768.

Fairchild, G.A., Steinberg, S.A., and Cohen, D. (1971). The fluorescent antibody test as a diagnostic test for canine distemper in naturally infected dogs. Cornell Vet. *61,* 214–223.

Ferreyra, H., Calderón, M.G., Marticorena, D., Marull, C., and Leonardo, B.C. (2009). Canine Distemper Infection in Crab-eating Fox (Cerdocyon thous) from Argentina. J. Wildl. Dis. *45,* 1158–1162.

Finch, J.T., and Gibbs, A.J. (1970). Observations on the structure of the nucleocapsids of some paramyxoviruses. J. Gen. Virol. *6,* 141–150.

Frisk, A.L., Konig, M., Moritz, A., and Baumgartner, W. (1999). Detection of canine distemper virus nucleoprotein RNA by reverse transcription-PCR using serum, whole blood, and cerebrospinal fluid from dogs with distemper. J. Clin. Microbiol. *37,* 3634–3643.

Fujita, K., Miura, R., Yoneda, M., Shimizu, F., Sato, H., Muto, Y., Endo, Y., Tsukiyama-Kohara, K., and Kai,

C. (2007). Host range and receptor utilization of canine distemper virus analyzed by recombinant viruses: Involvement of heparin-like molecule in CDV infection. Virology *359,* 324–335.

Gao, Y., and Lenard, J. (1995). Cooperative binding of multimeric phosphoprotein (P) of vesicular stomatitis virus to polymerase (L) and template: pathways of assembly. J. Virol. *69,* 7718–7723.

Gao, Y., Greenfield, N.J., Cleverley, D.Z., and Lenard, J. (1996). The transcriptional form of the phosphoprotein of vesicular stomatitis virus is a trimer: structure and stability. Biochemistry *35,* 14569–14573.

Gassen, U., Collins, F.M., Duprex, W.P., and Rima, B.K. (2000). Establishment of a rescue system for canine distemper virus. J. Virol. *74,* 10737–10744.

Gemma, T., Miyashita, N., Shin, Y.S., Okita, M., Mori, T., Iwatsuki, K., Mikami, T., and Kai, C. (1995). Serological survey of canine distemper virus infection using enzyme-linked immunosorbent assay. J. Vet. Med. Sci. *57,* 761–763.

Gowtage-Sequeira, S., Banyard, A.C., Barrett, T., Buczkowski, H., Funk, S.M., and Cleaveland, S. (2009). Epidemiology, Pathology, and Genetic Analysis of a Canine Distemper Epidemic in Namibia. J. Wildl. Dis. *45,* 1008–1020.

Graber, H.U., Muller, C.F., Vandevelde, M., and Zurbriggen, A. (1995). Restricted infection with canine distemper virus leads to down-regulation of myelin gene transcription in cultured oligodendrocytes. Acta Neuropathol. *90,* 312–318.

Griffin, D.E. (2007). Measles Virus. In Fields Virology, 5th edn, Knipe, D.M., Howley, P.M., Griffin, D.E., Lamb, R.A., Martin, M.A., Roizman, B., and Straus, S.E., eds. (Lippincott Williams & Wilkins, Philadelphia, PA), pp. 1551–1586.

Griot, C., Burge, T., Vandevelde, M., and Peterhans, E. (1989a). Antibody-induced generation of reactive oxygen radicals by brain macrophages in canine distemper encephalitis: a mechanism for bystander demyelination. Acta Neuropathol. *78,* 396–403.

Griot, C., Burge, T., Vandevelde, M., and Peterhans, E. (1989b). Bystander demyelination through antibody induced macrophage activation in canine distemper virus infection. Schweiz Arch. Neurol. Psychiatr. *140,* 39–41.

Griot, C., Moser, C., Cherpillod, P., Bruckner, L., Wittek, R., Zurbriggen, A., and Zurbriggen, R. (2004). Early DNA vaccination of puppies against canine distemper in the presence of maternally derived immunity. Vaccine *22,* 650–654.

Grone, A., Frisk, A.L., and Baumgartner, W. (1998). Cytokine mRNA expression in whole blood samples from dogs with natural canine distemper virus infection. Vet. Immunol. Immunopathol. *65,* 11–27.

Gupta, A.K., Shaji, D., and Banerjee, A.K. (2003). Identification of a novel tripartite complex involved in replication of vesicular stomatitis virus genome RNA. J. Virol. *77,* 732–738.

Haas, L., Barrett, T., Harder, T., and Bostock, C.J. (1990). Detection of phocine distemper virus using the polymerase chain reaction. Dtsch Tierarztl Wochenschr. *97,* 93–95.

Haas, L., Hofer, H., East, M., Wohlsein, P., Liess, B., and Barrett, T. (1996). Canine distemper virus infection in Serengeti spotted hyenas. Vet. Microbiol. *49,* 147–152.

Haas, L., Martens, W., Greiser-Wilke, I., Mamaev, L., Butina, T., Maack, D., and Barrett, T. (1997). Analysis of the haemagglutinin gene of current wild-type canine distemper virus isolates from Germany. Virus Res. *48,* 165–171.

Haile, R., Smith, P., Read, D., Nassim, D., Warlow, C., and Russell, W.C. (1982). A study of measles virus and canine distemper virus antibodies, and of childhood infections in multiple sclerosis patients and controls. J. Neurol. Sci. *56,* 1–10.

Halpin, K., Bankamp, B., Harcourt, B.H., Bellini, W.J., and Rota, P.A. (2004). Nipah virus conforms to the rule of six in a minigenome replication assay. J. Gen. Virol. *85,* 701–707.

Harder, T.C., and Osterhaus, A.D.M.E. (1997). Canine distemper virus – a *Morbillivirus* in search of new hosts? Trends Microbiol. *5,* 120–124.

Hartley, W.J. (1974). A post-vaccinal inclusion body encephalitis in dogs. Vet. Pathol. *11,* 301–312.

Higgins, R.J., Krakowka, S.G., Metzler, A.E., and Koestner, A. (1982). Experimental canine distemper encephalomyelitis in neonatal gnotobiotic dogs. A sequential ultrastructural study. Acta Neuropathol. *57,* 287–295.

Higgins, R.J., Child, G., and Vandevelde, M. (1989). Chronic relapsing demyelinating encephalomyelitis associated with persistent spontaneous canine distemper virus infection. Acta Neuropathol. *77,* 441–444.

Hirayama, N., Senda, M., Nakashima, N., Takagi, M., Sugiyama, M., Yoshikawa, Y., and Yamanouchi, K. (1991). Protective effects of monoclonal antibodies against lethal canine distemper virus infection in mice. J. Gen. Virol. *72,* 2827–2830.

Homann, H.E., Hofschneider, P.H., and Neubert, W.J. (1990). Sendai virus gene expression in lytically and persistently infected cells. Virology *177,* 131–140.

Horikami, S.M., Smallwood, S., Bankamp, B., and Moyer, S.A. (1994). An amino-proximal domain of the L protein binds to the P protein in the measles virus RNA polymerase complex. Virology *205,* 540–545.

Hsu, E.C., Iorio, C., Sarangi, F., Khine, A.A., and Richardson, C.D. (2001). CDw150(SLAM) Is a receptor for a lymphotropic strain of measles virus and may account for the immunosuppressive properties of this virus. Virology *279,* 9–21.

Imagawa, D.T. (1968). Relationships among measles, canine distemper and rinderpest viruses. Prog. Med. Virol. *10,* 160–193.

Iwatsuki, K., Miyashita, N., Yoshida, E., Gemma, T., Shin, Y.-S., Mori, T., Hirayama, N., Kai, C., and Mikami, T. (1997). Molecular and phylogenetic analyses of the haemagglutinin (H) proteins of field isolates of canine distemper virus from naturally infected dogs. J. Gen. Virol. *78,* 373–380.

Johnson, G.C., Fenner, W.R., and Krakowka, S. (1988). Production of immunoglobulin G and increased antiviral antibody in cerebrospinal fluid of dogs with delayed-onset canine distemper viral encephalitis. J. Neuroimmunol. *17,* 237–251.

Kapil, S., Allison, R.W., Johnston, L., III, Murray, B.L., Holland, S., Meinkoth, J., and Johnson, B. (2008). Canine Distemper virus strains circulating among North American dogs. Clin.. Vaccine Immunol. *15,* 707–712.

Kauffman, C.A., Bergman, A.G., and O'Connor, R.P. (1982a). Distemper virus infection in ferrets: an animal model of measles-induced immunosuppression. Clin. Exp. Immunol. *47,* 617–625.

Kauffman, C.A., Bergman, A.G., and O'Connor, R.P. (1982b). Distemper virus infection in ferrets: an animal model of measles-induced immunosuppression. Clin. Exp. Immunol. *47,* 617–625.

Keeble, S.A. (1962). Attenuated canine distemper vaccine. Proc. R. Soc. Med. *55,* 847–849.

Kolakofsky, D., Pelet, T., Garcin, D., Hausmann, S., Curran, J., and Roux, L. (1998). Paramyxovirus RNA Synthesis and the Requirement for Hexamer Genome Length: the Rule of Six Revisited. J. Virol. *72,* 891–899.

Kolakofsky, D., Le Mercier, P., Iseni, F., and Garcin, D. (2004). Viral RNA polymerase scanning and the gymnastics of Sendai virus RNA synthesis. Virology *318,* 463–473.

Krakowka, S. (1982). Mechanisms of *in vitro* immunosuppression in canine distemper virus infection. J. Clin. Lab. Immunol. *8,* 187–196.

Krakowka, S., McCullough, B., Koestner, A., and Olsen, R. (1973). Myelin-specific autoantibodies associated with central nervous system demyelination in canine distemper virus infection. Infect. Immun. *8,* 819–827.

Krakowka, S., Cockerell, G., and Koestner, A. (1975). Effects of canine distemper virus infection on lymphoid function *in vitro* and *in vivo*. Infect. Immun. *11,* 1069–1078.

Krakowka, S., Long, D., and Koestner, A. (1978). Influence of transplacentally acquired antibody on neonatal susceptibility to canine distemper virus in gnotobiotic dogs. J. Infect. Dis. *137,* 605–608.

Krakowka, S., Ringler, S.S., Lewis, M., Olsen, R.G., and Axthelm, M.K. (1987). Immunosuppression by canine distemper virus: modulation of *in vitro* immunoglobulin synthesis, interleukin release and prostaglandin E2 production. Vet. Immunol. Immunopathol. *15,* 181–201.

Kristensson, K., Thormar, H., and Wisniewski, H.M. (1979). Myelin lesions in the rabbit eye model as a bystander effect of herpes simplex and visna virus sensitization. Acta Neuropathol. *48,* 215–217.

Laidlaw, P.P., and Dunkin, G.W. (1926). Studies in dog-distemper. III. The nature of the virus. J. Comp Pathol. Ther. *39,* 222–230.

Lamb, R.A., and Parks, G.D. (2007). *Paramyxoviridae*: The viruses and their replication. In Fields Virology, 5th edn, Knipe, D.M., Howley, P.M., Griffin, D.E., Lamb, R.A., Martin, M.A., Roizman, B., and Straus, S.E., eds. (Lippincott Williams & Wilkins: Philadelphia, PA).

Larsen, C.G., Anderson, A.O., Appella, E., Oppenheim, J.J., and Matsushima, K. (1989). The neutrophil-activating protein (NAP-1) is also chemotactic for T lymphocytes. Science *243,* 1464–1466.

Latha, D., Geetha, M., Ramadass, P., and Narayanan, R.B. (2007). Development of recombinant nucleocapsid protein based IgM-ELISA for the early detection of distemper infection in dogs. Vet. Immunol. Immunopathol. *119,* 278–286.

Li, T., and Pattnaik, A.K. (1999). Overlapping signals for transcription and replication at the 3′ terminus of the vesicular stomatitis virus genome. J. Virol. *73,* 444–452.

Li, T., Chen, X., Garbutt, K.C., Zhou, P., and Zheng, N. (2006). Structure of DDB1 in complex with a paramyxovirus V protein: Viral hijack of a propeller cluster in ubiquitin ligase. Cell *124,* 105–117.

Lincoln, S.D., Gorham, J.R., Ott, R.L., and Hegreberg, G.A. (1971). Etiologic studies of old dog encephalitis. I. Demonstration of canine distemper viral antigen in the brain in two cases. Vet. Pathol. *8,* 1–8.

McCarthy, A.J., Shaw, M.-A., and Goodman, S.J. (2007). Pathogen evolution and disease emergence in carnivores. Proc. Roy. Soc. B. *274,* 3165–3174.

McCullough, B., Krakowka, S., and Koestner, A. (1974). Experimental canine distemper virus-induced lymphoid depletion. Am. J. Pathol. *74,* 155–170.

McIlhatton, M.A., Curran, M.D., and Rima, B.K. (1997). Nucleotide sequence analysis of the large (L) genes of phocine distemper virus and canine distemper virus (corrected sequence). J. Gen. Virol. *78,* 571–576.

Madden, D.L., Wallen, W.C., Houff, S.A., Shekarchi, I.C., Leinikki, P.O., Castellano, G.A., and Sever, J.L. (1981). Measles and canine distemper antibody. Presence in sera from patients with multiple sclerosis and matched control subjects. Arch. Neurol. *38,* 13–15.

Maes, R.K., Wise, A.G., Fitzgerald, S.D., Ramudo, A., Kline, J., Vilnis, A., and Benson, C. (2003). A canine distemper outbreak in Alaska: diagnosis and strain characterization using sequence analysis. J. Vet. Diagn Invest *15,* 213–220.

Martella, V., Cirone, F., Elia, G., Lorusso, E., Decaro, N., Campolo, M., Desario, C., Lucente, M.S., Bellacicco, A.L., Blixenkrone-Møller, M., *et al.* (2006). Heterogeneity within the hemagglutinin genes of canine distemper virus (CDV) strains detected in Italy. Vet. Microbiol. *116,* 301–309.

Martella, V., Elia, G., Lucente, M.S., Decaro, N., Lorusso, E., Banyai, K., Blixenkrone-Møller, M., Lan, N.T., Yamaguchi, R., Cirone, F., *et al.* (2007). Genotyping canine distemper virus (CDV) by a hemi-nested multiplex PCR provides a rapid approach for investigation of CDV outbreaks. Vet. Microbiol. *122,* 32–42.

Martin, A., Staeheli, P., and Schneider, U. (2006). RNA polymerase II-controlled expression of antigenomic RNA enhances the rescue efficacies of two different members of the *mononegavirales* independently of the site of viral genome replication. J. Virol. *80,* 5708–5715.

Masuda, M., Sato, H., Kamata, H., Katsuo, T., Takenaka, A., Miura, R., Yoneda, M., Tsukiyama-Kohara, K.,

Mizumoto, K., and Kai, C. (2006). Characterization of monoclonal antibodies directed against the canine distemper virus nucleocapsid protein. Comp Immunol. Microbiol. Infect. Dis *29,* 157–165.

Mee, A.P., Dixon, J.A., Hoyland, J.A., Davies, M., Selby, P.L., and Mawer, E.B. (1998). Detection of canine distemper virus in 100% of Paget's disease samples by *in situ*-reverse transcriptase-polymerase chain reaction. Bone *23,* 171–175.

von Messling, V., and Cattaneo, R. (2002). Amino-terminal precursor sequence modulates canine distemper virus fusion protein function. J. Virol. *76,* 4172–4180.

von Messling, V., and Cattaneo, R. (2003). N-linked glycans with similar location in the fusion protein head modulate paramyxovirus fusion. J. Virol. *77,* 10202–10212.

von Messling, V., Harder, T.C., Moennig, V., Rautenberg, P., Nolte, I., and Haas, L. (1999). Rapid and sensitive detection of immunoglobulin M (IgM) and IgG antibodies against canine distemper virus by a new recombinant nucleocapsid protein-based enzyme-linked immunosorbent assay. J. Clin. Microbiol. *37,* 1049–1056.

von Messling, V., Zimmer, G., Herrler, G., Haas, L., and Cattaneo, R. (2001). The hemagglutinin of canine distemper virus determines tropism and cytopathogenicity. J. Virol. *75,* 6418–6427.

von Messling, V., Springfeld, C., Devaux, P., and Cattaneo, R. (2003). A ferret model of canine distemper virus virulence and immunosuppression. J. Virol. *77,* 12579–12591.

von Messling, V., Milosevic, D., and Cattaneo, R. (2004). Tropism illuminated: Lymphocyte-based pathways blazed by lethal *Morbillivirus* through the host immune system. Proc. Natl. Acad. Sci. U.S.A. *101,* 14216–14221.

von Messling, V., Oezguen, N., Zheng, Q., Vongpunsawad, S., Braun, W., and Cattaneo, R. (2005). Nearby Clusters of Hemagglutinin Residues Sustain SLAM-Dependent Canine Distemper Virus Entry in Peripheral Blood Mononuclear Cells. J. Virol. *79,* 5857–5862.

von Messling, V., Svitek, N., and Cattaneo, R. (2006). Receptor (SLAM [CD150]) Recognition and the V Protein sustain swift lymphocyte-based invasion of mucosal tissue and lymphatic organs by a *Morbillivirus.* J. Virol. *80,* 6084–6092.

Metzler, A.E., Higgins, R.J., Krakowka, S., and Koestner, A. (1980). Virulence of tissue culture-propagated canine distemper virus. Infect. Immun. *29,* 940–944.

Mochizuki, M., Hashimoto, M., Hagiwara, S., Yoshida, Y., Ishiguro, and Shinryo (1999). Genotypes of canine distemper virus determined by analysis of the hemagglutinin genes of recent isolates from dogs in Japan. J. Clin. Microbiol. *37,* 2936–2942.

Mori, T., Shin, Y.-S., Okita, M., Hirayama, N., Miyashita, N., Gemma, T., Kai, C., and Mikami, T. (1994). The biological characterization of field isolates of canine distemper virus from Japan. J. Gen. Virol. *75,* 2403–2408.

Muller, C.F., Fatzer, R.S., Beck, K., Vandevelde, M., and Zurbriggen, A. (1995). Studies on canine distemper virus persistence in the central nervous system. Acta Neuropathol. *89,* 438–445.

Nagata, T., Ochikubo, F., Yoshikawa, Y., and Yamanouchi, K. (1990). Encephalitis induced by a canine distemper virus in squirrel monkeys. J. Med. Primatol. *19,* 137–149.

Nakatsu, Y., Takeda, M., Ohno, S., Shirogane, Y., Iwasaki, M., and Yanagi, Y. (2008). Measles virus circumvents the host interferon response by different actions of the C and V Proteins. J. Virol. *82,* 8296–8306.

Nesseler, A., Baumgartner, W., Gaedke, K., and Zurbriggen, A. (1997). Abundant expression of viral nucleoprotein mRNA and restricted translation of the corresponding viral protein in inclusion body polioencephalitis of canine distemper. J. Comp. Pathol. *116,* 291–301.

Nesseler, A., Baumgärtner, W., Zurbriggen, A., and Örvell, C. (1999). Restricted virus protein translation in canine distemper virus inclusion body polioencephalitis. Vet. Microbiol. *69,* 23–28.

Nielsen, L., Søgaard, M., Jensen, T.H., Andersen, M.K., Aasted, B., and Blixenkrone-Møller, M. (2009a). Lymphotropism and host responses during acute wild-type canine distemper virus infections in a highly susceptible natural host. J. Gen. Virol. *90,* 2157–2165.

Nielsen, L., Søgaard, M., Karlskov-Mortensen, P., Jensen, T.H., Jensen, T.D., Aasted, B., and Blixenkrone-Møller, M. (2009b). Humoral and cell-mediated immune responses in DNA immunized mink challenged with wild-type canine distemper virus. Vaccine *27,* 4791–4797.

Norris, J.M., Krockenberger, M.B., Baird, A.A., and Knudsen, G. (2006). Canine distemper: re-emergence of an old enemy. Aust. Vet. J. *84,* 362–363.

Ogino, T., and Banerjee, A.K. (2007). Unconventional Mechanism of mRNA Capping by the RNA-Dependent RNA Polymerase of Vesicular Stomatitis Virus. Mol. Cell *25,* 85–97.

Ogino, T., Kobayashi, M., Iwama, M., and Mizumoto, K. (2005). Sendai virus RNA-dependent RNA polymerase L protein catalyzes cap methylation of Virus-specific mRNA. J. Biol. Chem. *280,* 4429–4435.

Pardo, I.D.R., Johnson, G.C., and Kleiboeker, S.B. (2005). Phylogenetic characterization of canine distemper viruses detected in naturally infected dogs in North America. J. Clin. Microbiol. *43,* 5009–5017.

Pardo, M.C., Tanner, P., Bauman, J., Silver, K., and Fischer, L. (2007). Immunization of puppies in the presence of maternally derived antibodies against canine distemper virus. J. Comp Pathol. *137* (*Suppl. 1*), S72–75.

Parks, C.L., Wang, H.-P., Kovacs, G.R., Vasilakis, N., Kowalski, J., Nowak, R.M., Lerch, R.A., Walpita, P., Sidhu, M.S., and Udem, S.A. (2002). Expression of a foreign gene by recombinant canine distemper virus recovered from cloned DNAs. Virus Res. *83,* 131–147.

Peeters, B.P.H., Gruijthuijsen, Y.K., de Leeuw, O.S., and Gielkens, A.L.J. (2000). Genome replication of Newcastle disease virus: involvement of the rule of six. Arch. Virol. *145,* 1829–1845.

Pillet, S., and von Messling, V. (2009). Canine distemper virus selectively inhibits apoptosis progression in infected immune cells. J. Virol. *83,* 6279–6287.

Pillet, S., Svitek, N., and von Messling, V. (2009). Ferrets as a model for *Morbillivirus* pathogenesis, complications, and vaccines. Curr. Top. Microbiol. Immunol. *330,* 73–88.

von Pirquet, C.E. (1908). Das Verhalten der kutanin Tuberkulinreaktion während der Masern. Dtsch Med. Wochenschr. *34,* 1297–1300.Plattet, P., Zweifel, C., Wiederkehr, C., Belloy, L., Cherpillod, P., Zurbriggen, A., and Wittek, R. (2004). Recovery of a persistent Canine distemper virus expressing the enhanced green fluorescent protein from cloned cDNA. Virus Res. *101,* 147–153.

Plemper, R.K., Hammond, A.L., and Cattaneo, R. (2000). Characterization of a region of the measles virus hemagglutinin sufficient for its dimerization. J. Virol. *74,* 6485–6493.

Plemper, R.K., Hammond, A.L., and Cattaneo, R. (2001). Measles virus envelope glycoproteins hetero-oligomerize in the endoplasmic reticulum. J. Biol. Chem. *276,* 44239–44246.

Plumet, S., Duprex, W.P., and Gerlier, D. (2005). Dynamics of viral RNA synthesis during measles virus infection. J. Virol. *79,* 6900–6908.

Poch, O., Blumberg, B.M., Bougueleret, L., and Tordo, N. (1990). Sequence comparison of five polymerases (L proteins) of unsegmented negative-strand RNA viruses: theoretical assignment of functional domains. J. Gen. Virol. *71,* 1153–1162.

Prydie, J. (1968). Studies with an attenuated canine distemper vaccine derived from chick embryo cultures. Res. Vet. Sci. *9,* 443–452.

Rankin, A.M., Fisher, L.E., and Bussell, R.H. (1972). Cell fusion by canine distemper virus-infected cells. J. Virol. *10,* 1179–1183.

Rima, B.K., Baczko, K., Imagawa, D.T., and ter Meulen, V. (1987). Humoral immune response in dogs with old dog encephalitis and chromic distemper meningo-encephalitis. J. Gen. Virol. *68,* 1723–1735.

Rockborn, G. (1959). An attenuated strain of canine distemper virus in tissue culture. Nature *184* (*Suppl. 11*), 822.

Röthlisberger, A., Wiener, D., Schweizer, M., Peterhans, E., Zurbriggen, A., and Plattet, P. (2010). Two domains of the V protein of virulent canine distemper virus selectively inhibit STAT1 and STAT2 nuclear import. J. Virol. *84,* 6328–6343.

Rouxel, R.N., Svitek, N., and von Messling, V. (2009). A chimeric measles virus with canine distemper envelope protects ferrets from lethal distemper challenge. Vaccine *27,* 4961–4966.

Rozenblatt, S., Eizenberg, O., Englund, G., and Bellini, W.J. (1985). Cloning and characterization of DNA complementary to the canine distemper virus mRNA encoding matrix, phosphoprotein, and nucleocapsid protein. J. Virol. *53,* 691–694.

Rudd, P.A., Cattaneo, R., and von Messling, V. (2006). Canine distemper virus uses both the anterograde and the hematogenous pathway for neuroinvasion. J. Virol. *80,* 9361–9370.

Rudd, P.A., Bastien-Hamel, L.-É., and von Messling, V. (2010). Acute canine distemper encephalitis is associated with rapid neuronal loss and local immune activation. J. Gen. Virol. *91,* 980–989.

Sanderson, C.M., McQueen, N.L., and Nayak, D.P. (1993). Sendai virus assembly: M protein binds to viral glycoproteins in transit through the secretory pathway. J. Virol. *67,* 651–663.

Sawatsky, B., and von Messling, V. (2010). Canine distemper viruses expressing a hemagglutinin without N-glycans lose virulence but retain immunosuppression. J. Virol. *84,* 2753–2761.

Schobesberger, M., Zurbriggen, A., Doherr, M.G., Weissenbock, H., Vandevelde, M., Lassmann, H., and Griot, C. (2002). Demyelination precedes oligodendrocyte loss in canine distemper virus-induced encephalitis. Acta Neuropathol. *103,* 11–19.

Schultz, R.D. (2006). Duration of immunity for canine and feline vaccines: A review. Vet. Microbiol. *117,* 75–79.

Shek, W.R., Schultz, R.D., and Appel, M.J. (1980). Natural and immune cytolysis of canine distemper virus-infected target cells. Infect. Immun. *28,* 724–734.

Shin, Y., Mori, T., Okita, M., Gemma, T., Kai, C., and Mikami, T. (1995). Detection of canine distemper virus nucleocapsid protein gene in canine peripheral blood mononuclear cells by RT-PCR. J. Vet. Med. Sci. *57,* 439–445.

Sidhu, M.S., Menonna, J.P., Cook, S.D., Dowling, P.C., and Udem, S.A. (1993). Canine distemper virus L gene: Sequence and comparison with related viruses. Virology *193,* 50–65.

Smallwood, S., Ryan, K.W., and Moyer, S.A. (1994). Deletion Analysis Defines a Carboxyl-Proximal Region of Sendai Virus P Protein That Binds to the Polymerase L Protein. Virology *202,* 154–163.

Soma, T., Ishii, H., Hara, M., Yamamoto, S., Yoshida, T., Kinoshita, T., and Nomura, K. (2001). Comparison of immunoperoxidase plaque staining and neutralizing tests for canine distemper virus. Vet. Res. Commun. *25,* 311–325.

Soma, T., Ishii, H., Hara, M., Ohe, K., Hagimori, I., Ishikawa, Y., and Taneno, A. (2003). Detection of canine distemper virus antigen in canine serum and its application to diagnosis. Vet. Rec. *153,* 499–501.

Stein, V.M., Czub, M., Schreiner, N., Moore, P.F., Vandevelde, M., Zurbriggen, A., and Tipold, A. (2004). Microglial cell activation in demyelinating canine distemper lesions. J. Neuroimmunol. *153,* 122–131.

Stephenson, J.R., ter Meulen, V., and Kiessling, W. (1980). SeArch. for canine-distemper-virus antibodies in multiple sclerosis. A detailed virological evaluation. Lancet *2,* 772–775.

Stern, L.B.-L., Greenberg, M., Gershoni, J.M., and Rozenblatt, S. (1995). The hemagglutinin envelope protein of canine distemper virus (CDV) confers cell tropism as illustrated by CDV and measles virus complementation analysis. J. Virol. *69,* 1661–1668.

Stettler, M., and Zurbriggen, A. (1995). Nucleotide and deduced amino acid sequences of the nucleocapsid protein of the virulent A75/17-CDV strain of canine distemper virus. Vet. Microbiol. *44,* 211–217.

Stettler, M., Beck, K., Wagner, A., Vandevelde, M., and Zurbriggen, A. (1997). Determinants of persistence in canine distemper viruses. Vet. Microbiol. *57,* 83–93.

Stillman, E.A., and Whitt, M.A. (1999). Transcript initiation and 5′-end modifications are separable events during vesicular stomatitis virus transcription. J. Virol. *73,* 7199–7209.

Sugai, A., Kooriyama, T., Sato, H., Yoneda, M., and Kai, C. (2009). Epitope mapping of canine distemper virus phosphoprotein by monoclonal antibodies. Microbiol. Immunol. *53,* 667–674.

Summers, B.A., Greisen, H.A., and Appel, M.J. (1978). Possible initiation of viral encephalomyelitis in dogs by migrating lymphocytes infected with distemper virus. Lancet *2,* 187–189.

Summers, B.A., Greisen, H.A., and Appel, M.J. (1979). Early events in canine distemper demyelinating encephalomyelitis. Acta Neuropathol. *46,* 1–10.

Summers, B.A., Greisen, H.A., and Appel, M.J.G. (1984). Canine distemper encephalomyelitis: Variation with virus strain. J. Comp. Pathol. *94,* 65–75.

Sun, Z., Li, A., Ye, H., Shi, Y., Hu, Z., and Zeng, L. (2010). Natural infection with canine distemper virus in hand-feeding Rhesus monkeys in China. Vet. Microbiol. *141,* 374–378.

Svitek, N., and von Messling, V. (2007). Early cytokine mRNA expression profiles predict *Morbillivirus* disease outcome in ferrets. Virology *362,* 404–410.

Tatsuo, H., Ono, N., Tanaka, K., and Yanagi, Y. (2000). SLAM (CDw150) is a cellular receptor for measles virus. Nature *406,* 893–897.

Tatsuo, H., Ono, N., and Yanagi, Y. (2001). morbilliviruses use signaling lymphocyte activation molecules (CD150) as cellular receptors. J. Virol. *75,* 5842–5850.

Tipold, A., Moore, P., Zurbriggen, A., Burgener, I., Barben, G., and Vandevelde, M. (1999). Early T-cell response in the central nervous system in canine distemper virus infection. Acta Neuropathol. *97,* 45–56.

Vandevelde, M., and Zurbriggen, A. (2005). Demyelination in canine distemper virus infection: a review. Acta Neuropathol. *109,* 56–68.

Vandevelde, M., Fankhauser, R., Kristensen, F., and Kristensen, B. (1981). Immunoglobulins in demyelinating lesions in canine distemper encephalitis. An immunohistological study. Acta Neuropathol. *54,* 31–41.

Vandevelde, M., Kristensen, F., Kristensen, B., Steck, A.J., and Kihm, U. (1982). Immunological and pathological findings in demyelinating encephalitis associated with canine distemper virus infection. Acta Neuropathol. *56,* 1–8.

Wang, J.T., McElvain, L.E., and Whelan, S.P.J. (2007). Vesicular stomatitis virus mRNA capping machinery requires specific cis-acting signals in the RNA. J. Virol. *81,* 11499–11506.

Welter, J., Taylor, J., Tartaglia, J., Paoletti, E., and Stephensen, C.B. (2000). Vaccination against canine distemper virus infection in infant ferrets with and without maternal antibody protection, using recombinant attenuated poxvirus vaccines. J. Virol. *74,* 6358–6367.

Whelan, S.P.J., and Wertz, G.W. (1999). Regulation of RNA synthesis by the genomic termini of vesicular stomatitis virus: identification of distinct sequences Essential for transcription but not replication. J. Virol. *73,* 297–306.

Whelan, S.P.J., and Wertz, G.W. (2002). Transcription and replication initiate at separate sites on the vesicular stomatitis virus genome. Proc. Natl. Acad. Sci. U.S.A. *99,* 9178–9183.

Wiener, D., Plattet, P., Cherpillod, P., Zipperle, L., Doherr, M.G., Vandevelde, M., and Zurbriggen, A. (2007). Synergistic inhibition in cell–cell fusion mediated by the matrix and nucleocapsid protein of canine distemper virus. Virus Res. *129,* 145–154.

Winters, W.D. (1981). Time dependent decreases of maternal canine virus antibodies in newborn pups. Vet. Rec. *108,* 295–299.

Wisniewski, H., Raine, C.S., and Kay, W.J. (1972). Observations on viral demyelinating encephalomyelitis. Canine distemper. Lab. Invest. *26,* 589–599.

Wright, P.J., Crameri, G., and Eaton, B.T. (2005). RNA synthesis during infection by Hendra virus: an examination by quantitative real-time PCR of RNA accumulation, the effect of ribavirin and the attenuation of transcription. Arch. Virol. *150,* 521–532.

Wunschmann, A., Alldinger, S., Kremmer, E., and Baumgartner, W. (1999). Identification of $CD4^+$ and $CD8^+$ T cell subsets and B cells in the brain of dogs with spontaneous acute, subacute-, and chronic-demyelinating distemper encephalitis. Vet. Immunol. Immunopathol. *67,* 101–116.

Wyss-Fluehmann, G., Zurbriggen, A., Vandevelde, M., and Plattet, P. (2010). Canine distemper virus persistence in demyelinating encephalitis by swift intracellular cell-to-cell spread in astrocytes is controlled by the viral attachment protein. Acta Neuropathol. *119,* 617–630.

Yamanouchi, K., Yoshikawa, Y., Sato, T.A., Katow, S., Kobune, F., Kobune, K., Uchida, N., and Shishido, A. (1977). Encephalomyelitis induced by canine distemper virus in non-human primates. Jpn. J. Med. Sci. Biol. *30,* 241–257.

Yamanouchi, K., Sato, T.A., Kobune, F., and Shishido, A. (1979). Antibody responses in the cerebrospinal fulid of cynomolgus monkeys after intracerebral inoculation with paramyxoviruses. Infect. Immun. *23,* 185–191.

Yoshida, E., Iwatsuki, K., Miyashita, N., Gemma, T., Kai, C., and Mikami, T. (1998). Molecular analysis of the nucleocapsid protein of recent isolates of canine distemper virus in Japan. Vet. Microbiol. *59,* 237–244.

Yoshikawa, Y., Yamanouchi, K., Okada, K., Taniguchi, H., Nagashima, K., Nagai, Y., and Akiyama, K. (1981). Potentiation of neurovirulence of canine distemper virus in guinea pigs by sensitization with neural antigen. Jpn. J. Exp. Med. *51,* 299–307.

Yoshikawa, Y., Ochikubo, F., Matsubara, Y., Tsuruoka, H., Ishii, M., Shirota, K., Nomura, Y., Sugiyama, M., and Yamanouchi, K. (1989). Natural infection with canine distemper virus in a Japanese monkey (Macaca fuscata). Vet. Microbiol. *20,* 193–205.

Zurbriggen, A., Graber, H.U., and Vandevelde, M. (1995). Selective spread and reduced virus release leads to canine distemper virus persistence in the nervous system. Vet. Microbiol. *44,* 281–288.

Zurbriggen, A., Schmid, I., Graber, H.U., and Vandevelde, M. (1998). Oligodendroglial pathology in canine distemper. Acta Neuropathol. *95,* 71–77.

Rinderpest and Peste des Petits Ruminants Viruses

9

Michael D. Baron

Abstract

In terms of their impact on livestock, and therefore on human well-being and development, *Rinderpest virus* (RPV) and *Peste des petits ruminants virus* (PPRV) are two of the most important members of the Paramyxovirus family. RPV, the cause of the most feared of all cattle diseases, now appears to have been eradicated; however, over the past 20 years, PPRV has increased its global distribution through most of sub-Saharan and North Africa, the Middle East, the Indian subcontinent and eastwards into Tibet. While most scientific effort has been focused on developing means of control of the diseases, effort in a small number of laboratories has thrown light on the detailed molecular biology of the viruses, providing information on those areas in which they are the same as, or differ from, other members of the same genus. Such findings have highlighted how important it is to understand the way that these viruses are restricted in the range of organisms in which they will cause disease, an understanding that will become increasingly important with the success in eradicating these diseases on a local and global level.

History of the diseases

Rinderpest

A highly infectious and often fatal disease of cattle, oxen and buffalo, rinderpest (RP) acquired the sobriquets 'plague' or 'murrain' because of the devastating effects it had on populations of livestock animals. There is a strong argument for saying that RP has been the most important of all veterinary diseases. Descriptions of what is probably RP disease have been found in an Indian document on cattle diseases dating from the first millennium BC (Leclainche, 1955). References to great plagues or sicknesses in cattle are found in Roman writings from the fifth century BC through the fourth century AD, although it is not always clear whether the disease referred to is rinderpest or anthrax. The disease appears to have originated, and been maintained, in Asia; some of its earliest recorded incursions into Europe probably came with Hun invasions in the fourth and fifth centuries and its association with Russia can be seen in one of the names for the disease in seventeenth century Britain, 'the Steppe murrain', referring to the arrival of the disease with cattle imported from eastern Russia. Outbreaks of the disease have had historically significant effects, weakening the Roman empire as RP advanced in front of the Huns coming from the east and influencing the invasion plans of Charlemagne in the early ninth century. In more recent times, the Great African Pandemic that followed the introduction of RP to that continent in the 1880s severely affected the indigenous population, possibly facilitating the colonization of East Africa (Pastoret *et al.*, 2006). On the positive side, the establishment of the first European veterinary schools, in France in 1762 and Britain in 1791, has been ascribed to the need to improve the handling of the regular incursions of the disease (Rweyemamu *et al.*, 2006).

By the end of the eighteenth century, RP was more generally recognized as distinct from anthrax or foot and mouth disease (see Fig. 9.1

Figure 9.1 (A) Illustrative eighteenth century lithograph from the Netherlands of a contemporary outbreak of rinderpest; (B) photograph taken in early 1890s during the Great African rinderpest pandemic; (C) campaign poster for the global rinderpest eradication campaign (GREP).

for a contemporary warning about the disease from eighteenth century Netherlands), its nature as a contagion, introduced into an area through the movement of infected animals, was accepted and systems were being developed to control outbreaks. RP was introduced to the African continent in the late nineteenth century, probably in cattle being imported from Aden or Bombay into what is now Eritrea in 1887. The resultant pandemic spread through the whole of sub-Saharan Africa, reaching the Cape in about 1897. Contemporary estimates (Edington, 1899) were that 80–90% of cattle and buffalo succumbed, as well as large numbers of giraffe, wildebeest and different kinds of antelope and other wild ungulates (Fig. 9.1). A number of internationally eminent microbiologists, including Robert Koch, spent some time in South Africa working on preventative measures, independent groups proposing the treatment of naïve cattle with a combination of immune serum and infectious material, the only means of protection until the development of various killed or live attenuated vaccines (Taylor *et al.*, 2006). While slaughter and movement control essentially eliminated rinderpest from Europe by the early twentieth century, and prevented all outbreaks of the disease in Australia or the Americas bar one incursion into each continent, it took the development of mass-produced vaccines and more systematic epidemiology to control the disease in Africa and the Indian sub-continent. These efforts culminated in 1992 with the Global Rinderpest Eradication Programme (GREP; Fig. 9.1), jointly sponsored by the Food and Agriculture Organization of the United Nations (FAO) and the Office International des Epizooties (OIE), the World Organization for Animal Health. The aim of this project was, working through regional and national bodies, to eliminate rinderpest entirely from the world, recognizing that all previous purely local efforts at control had suffered from reincursions of the disease from elsewhere. The effectiveness of GREP can be seen in the decrease in the distribution of the disease during the 1990s (Fig. 9.2) to the point where there were only a few potential foci of disease left at the turn of the millennium, and these in areas where political upheaval made monitoring difficult.

A chapter such as this cannot do justice to the impact of RP on human populations and scientific thought through history, or to the many dedicated scientists, doctors, veterinarians and livestock keepers who have fought the disease over the centuries. I shall instead refer the interested reader to the comprehensive work '*Cattle Plague: A History*' (Spinage, 2003) as well as a slightly more recent book on both rinderpest and peste des petits ruminants (Barrett *et al.*, 2006). What can be added to both those works is the final success of the FAO/OIE 'Rinderpest 2010' GREP project. As of the time of writing, no definitive outbreak of RP has occurred anywhere in the world since 2001. Most countries with any historic evidence of rinderpest disease have been accredited as rinderpest free by the OIE, a process that should be complete by 2011. Planning is already under way

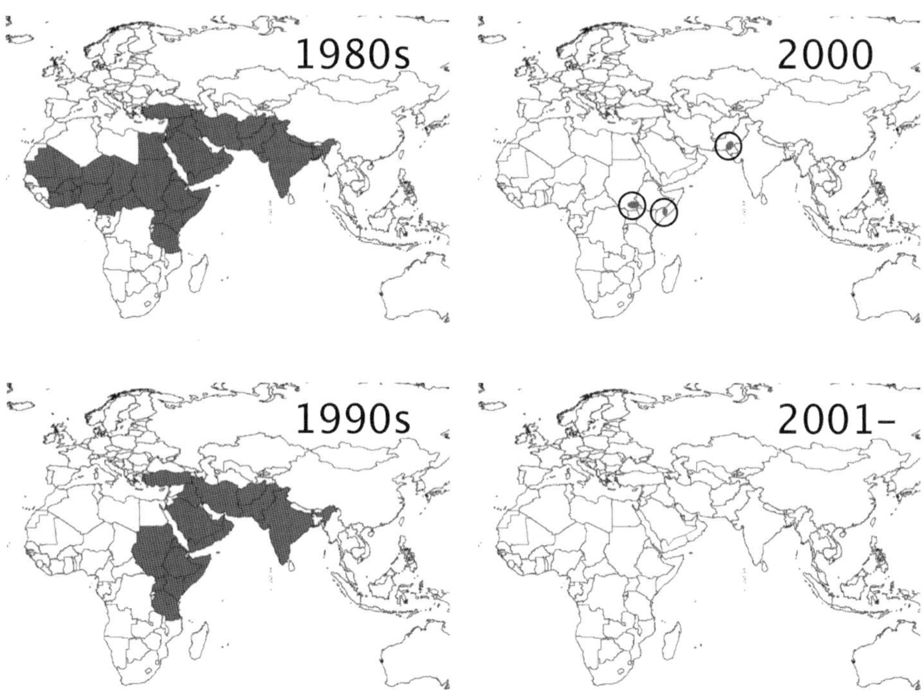

Figure 9.2 The eradication of rinderpest. Maps showing the countries reporting rinderpest outbreaks in the 1980s and 1990s, the three remaining suspect foci that existed in 2000 (circled), and the status since the last confirmed outbreak in 2001.

for the situation when the only rinderpest that exists is inside laboratories and vaccine stockpiles. This is only the second disease ever completely eradicated, and the first veterinary one, and must be recognized as a great achievement by the many hundreds of people who contributed to GREP.

Peste des petits ruminants

The history of peste des petits ruminants (PPR) is less clear than that of RP, but provides us with some important questions of epidemiology and host-specificity. In contrast to RP, PPR has only been recognized as a disease relatively recently. A sickness of sheep and goats that does not spread to in-contact cattle, PPR was first identified in the early 1940s in Ivory Coast and other countries in West Africa, when it was considered a strain of RP that had become fully adapted to small ruminants (Gargadennec and Lalanne, 1942). The PPR virus (PPRV) was identified as a morbillivirus distinct from the virus that causes RP (RPV) only 40 years later (Gibbs *et al.*, 1979). Since its first discovery, PPR has apparently spread through the whole of sub-Saharan Africa, the Middle East, Turkey and the Indian sub-continent. In the last few years, first outbreaks of PPR have occurred in China (Tibet), in East Africa (Kenya, Uganda and Tanzania), and in North Africa (Morocco and Tunisia; Fig. 9.3). PPR, therefore, appears to be an emerging disease, expanding into the world as RP has been eliminated.

What is not clear is whether this 'emergence' is entirely due to the spread of a new disease, possibly enabled by the removal of RP, or if much of the 'emergence' of PPR is due to its recognition as a distinct disease now that this kind of illness in sheep/goats can no longer be ascribed to RP. RP-like illnesses in sheep and goats have been recorded for centuries, as has the apparent transfer of RP from cattle to sheep, though only when the animals are in close contact (Spinage, 2003). Modern experiments have shown that certain strains of the virus that causes RP (RPV) can be transmitted from cattle to sheep and goats, as well as from sheep to sheep (again only when in close contact), and can cause disease in goats, if not sheep (Anderson *et al.*, 1990a). The development of tests that discriminate between anti-RPV and anti-PPRV antibodies (Anderson and McKay, 1994) has allowed the acquisition of serological evidence to show that RPV can and does infect sheep/goats under normal conditions, apparently subclinically (or causing only mild or inapparent disease; Anderson *et al.*, 1990b; Wambura *et al.*, 1999). Sheep exposed to RPV are immune to PPRV (Taylor, 1979). Circulating RP may therefore have suppressed PPR spread. However, the symptoms of PPR in sheep and goats are very similar to those of RP in cattle (see below) and the conditions under which RP would be expected to be spread (movement of livestock for trade or during war) would also favour the spread of PPR. Co-circulation of RPV and PPRV may have occurred without the latter being explicitly recognized. The discovery of PPRV in India happened at multiple locations apparently simultaneously (Shaila *et al.*, 1989, 1996), suggesting that the virus may have been present for some time before

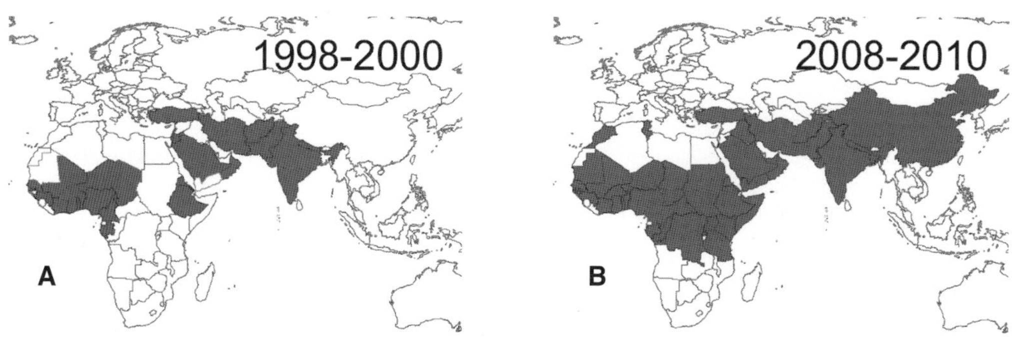

Figure 9.3 The recent spread of *Peste de petit ruminants virus* (PPRV). Maps showing countries reporting outbreaks of peste de petit ruminants (PPR) disease in (A) 1998–2000 and (B) 2008–2010.

its discovery. Examination of the genetic lineages of PPRV and RPV provides evidence suggesting that PPRV may have also originated in Asia (see below). Without analysis of appropriate historical samples, it will not be possible to determine how long PPRV has actually been circulating in sheep and goat populations, although this is not to deny the real spread of the disease in the last 10 years, since it has appeared recently in countries that have been free of RP for decades.

Description and classification of the viruses

RPV and PPRV are morbilliviruses, closely related to measles virus (MV), canine and phocine (seal) distemper viruses (CDV and PDV). Other morbillivirusesare known which infect porpoise, dolphins and whales, and are sometimes referred to as separate viruses, the porpoise, dolphin and cetacean morbilliviruses (PMV, DMV, CMV). Fig. 9.4 shows a phylogenetic tree of the morbilliviruses based on the nucleocapsid (N) protein gene sequence of each virus; as can be seen, the porpoise and dolphin viruses are as similar as individual strains of RPV and it is arguable whether PMV, DMV and CMV are separate viruses or strains of a single Cetacean morbillivirus (CMV). The morbilliviruses form one genus within the sub-family *Paramyxovirinae* of the family *Paramyxoviridae*. As with all the paramyxoviruses, the morbilliviruses have a single-segmented RNA genome of negative sense. The virions are pleiomorphic enveloped structures; the size range of RPV virions can be between 200–700 nm (Plowright *et al.*, 1962; Tajima and Ushijima, 1971). The membrane envelope is derived from the host cell during the budding of progeny virions (Walsh *et al.*, 2000). A negative stained picture of a RPV virion is shown in Fig. 9.5, illustrating the characteristic 'herringbone' appearance of the nucleocapsid structures found in the virion. This nucleocapsid consists of the viral genome entirely wrapped by multiple

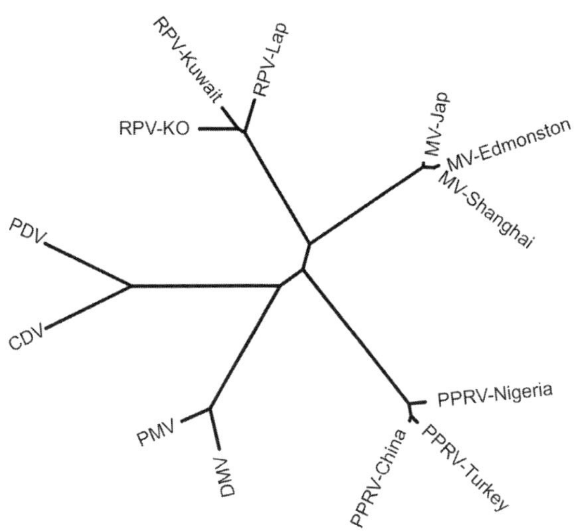

Figure 9.4 Relationships between the morbilliviruses. Phylogenetic tree based on the N gene sequences of the indicated viruses. The evolutionary history was inferred using the neighbour-joining method (Saitou and Nei, 1987). The evolutionary distances (number of base substitutions per site) were computed using the maximum composite likelihood method (Tamura *et al.*, 2004) and the tree was drawn with branch lengths determined by the evolutionary distances used to infer the phylogenetic tree. The rate variation among sites was modelled with a gamma distribution (shape parameter = 1) and all positions containing gaps and missing data were eliminated from the dataset. There were a total of 1649 positions in the final dataset. Phylogenetic analyses were conducted using the program MEGA4 (Tamura *et al.*, 2007). The accession numbers of the sequences used were: RPV-KO, NC_006296; RPV-Kuwait, Z34262; RPV-Lap, E06018; MV-Jap, NC_001498; MV-Edmonston, K01711; MV-Shanghai, EU435017; CDV, NC_001921; PDV, X75717; DMV, NC_005283; PMV, AY949833; PPRV-China, FJ905304; PPRV-Turkey, NC_006383; PPRV-Nigeria, X74443.

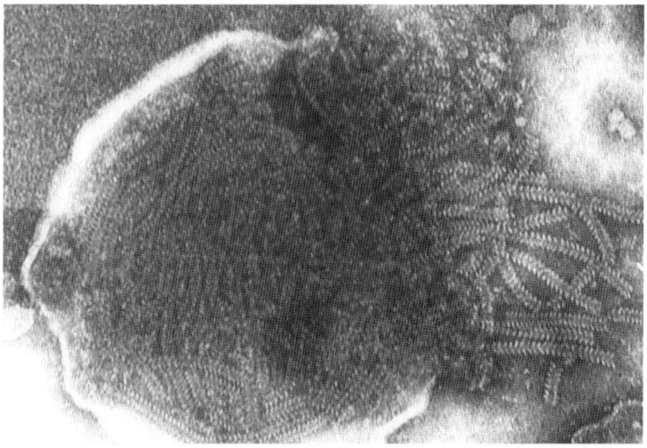

Figure 9.5 Negative stained electron micrograph of a rinderpest virion. With thanks to R. Kaushik (University of Massachusetts Medical School) and M.S. Shaila (Indian Institute of Science).

copies of the viral nucleocapsid (N) protein, the helical packing of which in the nucleocapsid gives rise to the appearance in electron micrographs. A characteristic of these nucleocapsids is that the nucleic acid of the genome is resistant to digestion with endonucleases such as micrococcal nuclease. Given a genome length of 15,882 or 15,948 for RPV and PPRV, respectively (Baron and Barrett, 1995; Bailey *et al.*, 2005) and steric and genetic considerations, which suggest that each N protein associates with six genome nucleotides (see below), each genome must associate with approximately 2650 copies of N. Analysis of the pitch and diameter of the nucleocapsid helix (Bhella *et al.*, 2004) suggests that each turn of the helix involves slightly over 13 copies of the N protein, and therefore a genome would involve just over 200 turns of the nucleocapsid helix. As can be seen from such images as those in Fig. 9.5, an individual paramyxovirus virion can hold much more than a single copy of the virus genome in the form of encapsidated RNA (Loney *et al.*, 2009), a property that has allowed recombinant measles viruses to be created where essentially two full copies of the genome have to be maintained in each infectious unit (Rager *et al.*, 2002).

Propagation

When first isolated as an infectious agent at the end of the nineteenth century, RPV was maintained by passage in cattle, there being no other way to prepare the infectious material. The history of the attempts to grow the virus in other animals and in tissue culture centres on the aim of microbiologists of the time to prepare material for use as either inactivated or live attenuated vaccines. Although both RPV and PPRV cause disease in only a limited range of hosts, epidemiological evidence shows that, just as RPV can infect sheep and goats subclinically, PPRV can infect cattle without causing disease (Anderson and McKay, 1994). This ability of the viruses to grow in other hosts has been used several times to select for strains that, adapting to high growth (and often virulence) in a non-bovine host, become attenuated in the cattle for which protection is being sought. This work is extensively chronicled in Spinage (2003) and in briefer form by Scott (1964) and Taylor *et al.* (2006). Both Schein (1926) and Edwards (1930) independently showed that serial passage of RPV in goats resulted in a virus attenuated in cattle, although this attenuation was by no means complete; after 625 passages the resultant virus still caused 17% mortality in calves (Krishnamurti, 1945). Variations on this goat-adapted vaccine were nevertheless used extensively in the 1930s and 1940s, sometimes accompanied with immune serum to prevent sickness in the immunized animals.

During the same period, perhaps driven by the comparative rarity of sheep and goats in Japan, Nakamura and coworkers used rabbits as the attenuating host. They found that, though RPV was essentially innocuous in rabbits when first injected, the pathogenicity of the virus in the new host increased with increasing numbers of passages (Nakamura *et al.*, 1938). After 350 passages, an isolate was prepared that was sufficiently attenuated in Mongolian cattle to be used in RP control programmes in the 1940s (Isogai, 1944). This lapinized strain was further passaged in rabbits or adapted by other workers to pigs, goats, sheep or eggs in order to improve either the attenuation of the virus or the production of vaccine (a single rabbit only producing 600 doses of vaccine). Direct adaptation of virulent RPV to eggs was carried out in Canada (Shope *et al.*, 1946) and later in Kenya (Taylor *et al.*, 2006). The lapinized-avianized RPV was further passaged in rabbits to provide a vaccine that was attenuated sufficiently for use in Korean and Japanese cattle, which had always been particularly susceptible to RP (Scott, 1964).

First attempts at tissue culture growth of RPV were in the mid 1950s using strains previously adapted to goats, rabbits or eggs and growing the viruses either on primary cells homologous to the animals to which they had been adapted, or on primary bovine cells. Curiously, none of these attempts were successful and the first tissue culture-grown RPV was achieved with the direct adaptation of the virulent Kabete O strain of the virus to grow in bovine kidney (BK) cells (Plowright and Ferris, 1959a). Cytopathic effect was not seen until the fifth passage in BK cells, but thereafter the effect of the virus on the cultured cells was easy to observe. Interestingly, the virus became more pathogenic over the first 10 passages (Plowright and Ferris, 1959b), perhaps owing to the elimination of defective interfering (DI) particles. Unfortunately, those early isolates were not preserved; it would have been most interesting to study the differences that led to the increase in virulence. From the sixteenth passage on, the virus began to lose virulence; from the 21st passage, only temperature reactions were seen and none of the lesions and diarrhoea characteristic of RP disease. Virus of the 70th, 90th and 120th passage was fully attenuated, and virus of the 90th passage was stably attenuated through seven back passages in cattle (Plowright and Ferris, 1962).

Subsequently, RPV has been propagated in a large number of cell types of both ruminant and primate origin. The original BK-adapted virus readily grew in Vero cells (Mirchamsy *et al.*, 1970), and has been found to grow in a wide variety of primate and bovine cells in the author's own lab. All kinds of RPV (wild-type or tissue culture adapted) were found to grow in bovine T cells immortalized by the intracellular parasite *Theileria* (Rossiter *et al.*, 1992), and the Epstein–Barr virus-transformed tamarin monocyte-derived line B95a was shown to be a good host for RPV (Kobune *et al.*, 1991) just as it is for MV (Kobune *et al.*, 1990). While these last two cell types have proven to be excellent for propagating RPV, the presence of *Theileria* or Epstein–Barr virus, respectively, makes them unsuitable for growing virus that will subsequently be used to infect animals. The discovery that the highly conserved host cell membrane protein CD150, also known as signalling lymphocyte activation molecule (SLAM), is the normal receptor for all kinds of morbilliviruses (Tatsuo *et al.*, 2000, 2001; Baron, 2005) prompted the creation of modified tissue culture lines expressing SLAM; these lines have proven the most useful for propagating wild type RPV isolates. We have not observed significant differences in the sensitivity of cells expressing human, canine or bovine SLAM to infection with RPV (unpublished observations), so any of these cell lines could be used. Tissue culture-adapted virus appears to use heparan sulfate as its receptor (Baron, 2005), the nearly ubiquitous distribution of this protein explaining why the virus is able to enter so many cell types from so many species. Interestingly, not all wild-type RPV isolates require SLAM for growth; the highly virulent Saudi/81 strain (causes 100% mortality in cattle) infects non-SLAM expressing cells such as primary bovine skin cells or Vero cells directly from the infected animal, causing cytopathic effect and growing to high titre without apparent need for adaptation, at least at the level of the viral haemagglutinin (H) protein responsible for receptor binding. The H gene sequences of virus grown on SLAM-expressing and not-expressing cells were the same (unpublished observations), indicating that the ability to use heparan sulfate as a receptor may be inherent in some wild RPV isolates. In other cases, passage in tissue culture may be required to select for improved binding to this alternative receptor (as is the case for the virulent

Kabete O strain (Baron, 2005)), but the basic ability appears to be there already in the structure of the viral proteins.

Despite their frequent use for isolating different types of virus, BHK-21 cells have proven to be one of the few types of cells that have been very poor hosts for RPV, even for tissue culture-adapted strains. There is not an absolute block on hamster cells, since RPV will grow in CHO cells (Baron, 2005), and the reason for the failure of RPV to grow in BHK cells remains to be established.

Interestingly, just as BK cells were the first successful host for RPV, primary lamb kidney cells (LK) have proven to be one of the best for isolating PPRV (Gilbert and Monnier, 1962; Taylor and Abegunde, 1979). However, because of the ease of use of Vero cells, many isolates of PPRV have been first propagated in these cells (Hamdy *et al.*, 1976). Either LK cells or Veros are recommended for PPRV isolation (OIE, 2008). More recently, it has become preferable to use one of the Vero-SLAM lines available. There is some evidence that Vero-canine SLAM may be more sensitive for isolating PPRV (i.e. PPRV has a preference for canine SLAM over human or bovine SLAM; Barrett, T., unpublished observations), but any of these cells could be used. Since PPRV infects Vero-SLAM more effectively than plain Veros, and causes syncytia formation in Vero-SLAMs but not in Vero cells, it is currently assumed that SLAM is the receptor for PPRV as well, although this has not yet been rigorously proven.

Neither RPV nor PPRV form plaques in the normally accepted sense of causing cell lysis or rapid cell death in tissue culture, so titration is usually carried out by measuring the 50% tissue culture infectious dose ($TCID_{50}$; Anderson *et al.*, 1996). Regions of infected cells can be observed in the light microscope by the cytopathic effect (cell rounding or distortion, syncytium formation). Apoptosis has not been observed in tissue culture up to 4 days post infection, in contrast to cells infected with MV, where we have observed apoptosis within 24 h post infection. These observations correspond with the activation of transcription of apoptosis-related genes seen in MV-infected cells and the absence of such activation in cells infected with RPV (Nanda *et al.*, 2009).

Genome organization

RPV and PPRV require genomes that are a multiple of six bases in length for efficient replication (Baron and Barrett, 1995; Bailey *et al.*, 2005), an observation termed 'the rule of six' when it was first observed for *Sendai virus* (Calain and Roux, 1993), and since found to be true for most members of the subfamily *Paramyxovirinae*. There are six genes, or transcription units, with promoter sequences (that is to say, binding sites for the viral RNA-dependent RNA polymerase (RdRP)) only at the 3′ ends of the genome and antigenome. The virus genes encode the nucleocapsid (N) protein, the phosphoprotein (P), the matrix protein (M), the fusion (F) and H membrane glycoproteins and the large (L) protein, which is the viral polymerase. The P gene also encodes the three non-structural proteins V, W and C. The arrangement of transcriptional control elements and protein coding sequences in RPV and PPRV is illustrated in Fig. 9.6 and compared with the genomes of MV and CDV. As can be seen, RPV and PPRV have very similar genomes, with major differences in the length of coding or non-coding sequences only being seen in the untranslated regions (UTRs) at the end of the M gene and the start of the F gene, and two single base differences in the lengths of the 3′ UTRs at the end of the F and H genes. The +6 and –6 differences in, respectively, the open reading frame (ORF) and 3′ UTR of the P gene are because the PPRV P ORF is two codons longer, and the 3′ UTR is correspondingly six bases shorter than that of RPV making the total P gene the same length for the two viruses.

There is a relatively conserved motif at the junction between genes ($A^{/G}TT^{/AC}ANA^{/G}A^{/C}AAAACTTAGGA^{/G}$), with a similar sequence at the junction of genome promoter (GP) and N gene and that of L gene and Anti-Genome Promoter (AGP). The sequence prior to the CTT motif marks the end of one gene and the polyadenylation site, while the sequence after the CTT is the 5′ end of the next gene. The CTT is transcribed only in full-length antigenome RNAs and does not appear in normal viral mRNAs. It is assumed that the viral RdRP, when in mRNA transcription mode, recognizes these sequences in some way to initiate mRNA transcription (including capping) at the start of the transcription unit,

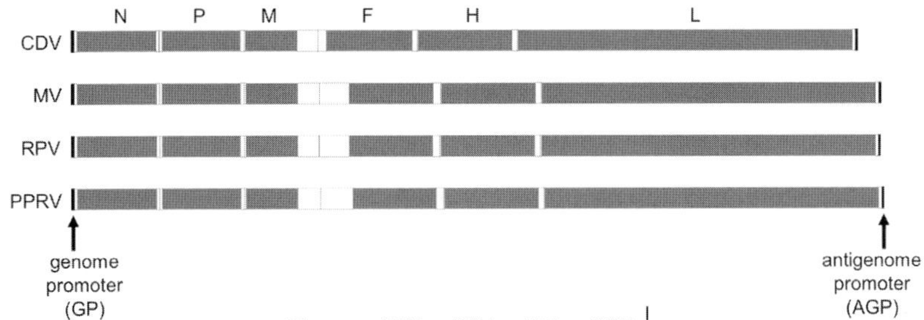

Gene:	Virus: region	PDV	RPV	MV	CDV
GP	*promoter*	55	55	55	55
N	*5'UTR*	52	52	52	52
	ORF	1578	1578	1578	1572
	3'UTR	62	62	62	62
P	*5'UTR*	59	59	59	59
	ORF	1530	1524	1524	1524
	3'UTR	69	75	75	75
M	*5'UTR*	32	32	32	32
	ORF	1008	1008	1008	1008
	3'UTR	446	423	429	410
F	*5'UTR*	634	589	583	167
	ORF	1641	1641	1653	1677
	3'UTR	139	140	140	105
H	*5'UTR*	20	20	20	20
	ORF	1830	1830	1830	1824
	3'UTR	110	111	111	105
L	*5'UTR*	22	22	22	22
	ORF	6552	6552	6552	6555
	3'UTR	72	72	72	68
AGP	*promoter*	37	37	37	38

Figure 9.6 Gene layout of morbilliviruses. The distribution of protein coding sequence (grey), promoters (black) and UTRs (white) in the genomes of rinderpest virus (RPV), peste de petit ruminants virus (PPRV), measles virus (MV) and canine distemper virus (CDV) are shown. The actual length of each coding and non-coding region is given in detail in the accompanying table.

and to terminate transcription at the appropriate point, where the viral RdRP adds a poly(A) tail to the mRNA. The sequences usually described as the GP and AGP are quite short, 52 and 37 bases respectively. However, these are not the minimum sequences necessary for transcription and replication. Studies with RPV and PPRV minigenomes (essentially genomes with almost all viral sequences apart from the promoters removed and replaced with a single transcription unit containing the coding sequence for a reporter gene) showed that approximately 100 bases at each end of the genome are essential for minigenome function (Mioulet *et al.*, 2001; Bailey *et al.*, 2007). Taken with data from other paramyxoviruses (Tapparel *et al.*, 1998; Murphy and Parks, 1999; Hoffman and Banerjee, 2000; Walpita, 2004), it is clear that there is another sequence element essential for transcription that lies within the first and last genes, the so-called 'B-Box' (Blumberg *et al.*, 1991) or 'CRII' (Murphy and Parks, 1999). This motif occupies a region 79–96 nucleotides from the 3′ end of the genome/antigenome. Given that morbillivirus nucleocapsids have slightly over 13 N proteins per turn of the helix (Bhella *et al.*, 2004), and probably six bases per N, each turn is about 78 nucleotides, so this motif would lie on the same side of the nucleocapsid to the highly conserved 18–20 base sequence at the 3′ end of the GP or AGP and presumably represents an extended binding site for the viral RdRP.

While most of the morbillivirus genes show a very high level of conservation in length of coding and non-coding regions (Fig. 9.6), for the F genes things are not so clear. Simple examination of the F gene sequences of the morbilliviruses would suggest that the open reading frame begins anywhere from 86 to 634 bases from the start of the gene. In addition, in some cases, the first AUG in the gene transcript is the start of a short, possibly unconserved ORF and it is the second or third AUG that is the start of the F protein ORF (Fig. 9.7). Comparisons of multiple strains of the same virus and comparison of RPV, PPRV and other morbilliviruses shows that the hypothetical amino-terminus of the F proteins is highly variable, both in sequence and in length, up to the putative signal sequence of this class 1 membrane-anchored protein (Fig. 9.7). In some cases there is a start codon in a suitable place to give rise to a more typical membrane protein, with a normal length signal sequence followed by a signal protease cleavage point (Fig. 9.7), though the alternative start codons are not conserved. Removal of the viral sequences upstream of this shorter ORF has been observed to improve protein expression in transfected cells (Hasel *et al.*, 1987; Evans *et al.*, 1990). The effects of the removal of these long untranslated sequences on virus replication have not been consistent; in RPV there was no effect on virus (Barret and Baron, unpublished observations), while MV replication was greatly altered by deleting either the 3′UTR at the end of the M gene or the 5′UTR at the start of

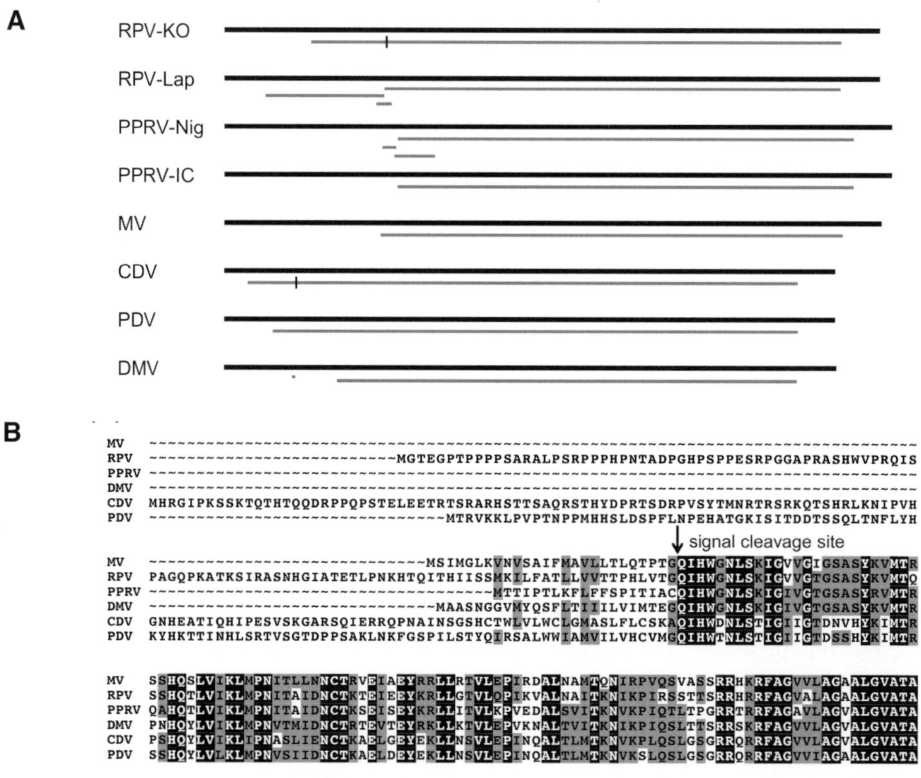

Figure 9.7 Morbillivirus F proteins. (A) Open reading frames in morbillivirus F genes. Shown are open reading frame (ORF) maps for two strains each of rinderpest virus (RPV) and peste de petit ruminants virus (PPRV), as well as examples of measles virus (MV), canine distemper virus (CDV), phocine distemper virus (PDV) and dolphin morbillivirus (DMV). Alternative start translation points in the main ORF are marked (|). (B)The amino-terminal sequences of the F proteins of MV (EU435017), RPV (X98291), PPRV (EU267273), DMV (NC_005283), CDV (GU138403) and PDV (D10371) are shown, including the first 105 amino acids downstream of the signal sequence. The protein sequences were translated from the first AUG in the mRNA that is in-frame with the F protein sequence, not necessarily the first AUG in the mRNA. The putative signalase cleavage site is marked.

the F gene (Takeda *et al.*, 2005). A highly attenuated strain of RPV was found to have a mutation in the F gene that removed the downstream start codon, meaning that F translation had to start at the upstream codon (Evans *et al.*, 1994); this virus grew poorly in cell culture as well as in the host. Finally, we have found that we could replace the downstream F ORF of RPV with that of PPRV, ignoring the upstream sequence, and generate a fully viable virus (Das *et al.*, 2000). Clearly much of this protein sequence upstream of the signal sequence is disposable, as might be expected for a peptide that is expected to be cleaved from the mature protein during synthesis. What is not clear is whether this extra section of protein is synthesized during viral replication, or whether F protein synthesis begins at one of the downstream start codons.

The exact role of this region of the genome between the end of the M protein ORF and the ORF for the known functional part of the F protein is not clear. The unusually long 3′ UTR of the M gene just prior to the F gene, coupled with (often) long 5′ UTR prior to the F protein coding sequence, means there could be a block of more than 1 kb of untranslated sequence in the middle of the virus genome. It has been suggested that this long UTR regulates the expression of F and hence viral virulence (Takeda *et al.*, 2005; Anderson and von Messling, 2008). For several of the morbilliviruses, this region has a very high GC content (Fig. 9.8), and it might form local stem-loop or pseudoknot structures that have an effect on transcription or translation. However, no clear differences have been observed between the viruses with high GC content in this region (RPV, PPRV, MV) and those where the GC content here is no different to that found in the rest of the genome (CDV, PDV, DMV). In the case of RPV, we found that both sections of UTR could be removed from the vaccine strain without affecting viral growth or expression of the F protein (Barrett and Baron, unpublished).

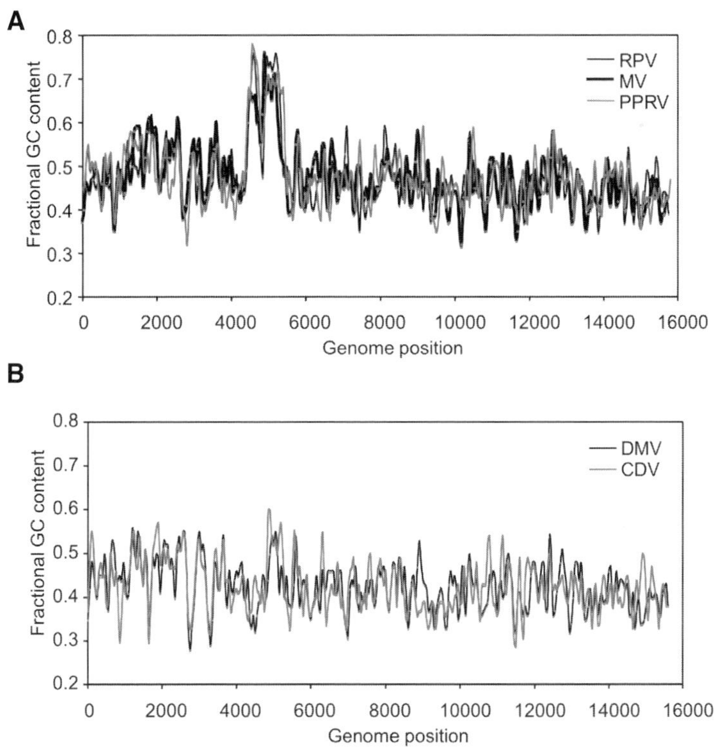

Figure 9.8 Plots of GC content for *Rinderpest virus* (RPV), *Peste de petit ruminants virus* (PPRV), *Dolphin Morbillivirus* (DMV), *Measles virus* (MV), and *Canine distemper virus* (CDV) genomes. The fractional GC content was calculated for a window size of 100 bases and a step size of 50 bases. For clarity, RPV, MV and PPRV are shown in (A) and DMV and CDV in (B).

If we accept that the main translation start point is not always the first available start codon, there must be some mechanism in some cases to cause the ribosome to skip to the start codon actually used. The F gene 5′ UTR has no internal ribosome entry site (IRES) function (Evans *et al.*, 1990), but it might still be acting to force the ribosome to skip sequences, in the same way that *Sendai virus* forces translation to start internal to the P gene to generate some of the C proteins generated by that virus (Latorre *et al.*, 1998).

Viral RNA transcription and replication

RNA synthesis

The viral RdRP has to function in two different modes, replication mode and mRNA transcription mode. When in mRNA transcription mode, the polymerase recognizes the gene start and gene end signals, and initiates the synthesis of separate capped and polyadenylated mRNAs from each gene. When the RdRP is in replication mode, these signals are not recognized and transcription continues through the length of the genome (or anti-genome). We have very little data as to the differences between mRNA transcription mode and replication mode. An early hypothesis was that replication mode required sufficient N protein to enable the co-transcriptional encapsidation that is characteristic of these viruses (Kingsbury, 1974; Lamb and Kolakofsky, 1996). Although little work has been done on this subject in the case of the morbilliviruses, the balance between the two types of transcription in another paramyxovirus, *Respiratory syncytial virus* (RSV) was found to be unaffected by the level of N protein (Fearns *et al.*, 1997). Studies on MV and RPV showed that transcription begins at the 3′ end of the genome even in mRNA transcription mode (Horikami and Moyer, 1991; Ghosh *et al.*, 1996). Our own studies have shown that the RPV C protein is essential for efficient genome transcript synthesis (Baron and Baron, unpublished), while phosphorylation of the RPV P protein has also been found to be required for replication mode transcription (Kaushik and Shaila, 2004; Raha *et al.*, 2004b; Saikia *et al.*, 2008). Elegant studies on vesicular stomatitis virus (VSV), another negative strand RNA virus (though from the rhabdoviruses rather than the paramyxoviruses), have purified distinct viral RdRP complexes associated with replication and mRNA transcription modes (Qanungo *et al.*, 2004), and there are sufficient common areas between these two larger groups of viruses to allow us to hypothesize that similar distinct complexes probably exist for the paramyxoviruses. At the same time, there are enough differences between the two virus families that the way these complexes are formed, their composition and the way they operate in the two modes need to be determined for the paramyxoviruses as well.

The two transcription modes presumably have different promoter requirements. The GP must support transcription initiation in both modes, while the AGP need only support initiation of replicative mode transcription. Although there is data to support this hypothesis in other paramyxoviruses, notably the extensively studied respirovirus *Sendai virus*, the differences in these promoters and the definition of the exact sequence requirements for genome or mRNA transcription have not been studied in morbilliviruses.

One consequence of the single promoter for RdRP entry and transcription initiation is that gene transcription (mRNA synthesis) must always occur in the same order, beginning with the N gene. At the end of each gene, the polymerase has to add a poly(A) tail and then initiate the next mRNA. If this happened with 100% efficiency each time, there would be the same amount of each virus message. However, based on studies of MV, the relative levels of the viral mRNAs are not the same, but form a gradient related to the position of the gene in the genome. There are higher levels of mRNAs from genes at the 3′ end of the genome (closest to the promoter) than from those more distal (Cattaneo *et al.*, 1987a; Schneider-Schaulies *et al.*, 1989). This led to the deduction that, at each gene end, there is a specific probability that the polymerase will detach and fail to initiate at the next gene. The frequency of reinitiation at each gene junction varies (Rennick *et al.*, 2007) and, by analogy with data on other paramyxoviruses, the probability of reinitiation varies with the sequence at each gene junction (Rassa and Parks, 1998; He and Lamb, 1999; Kato *et al.*, 1999); this may represent one way of modulating the profile of the transcription gradient to give finer control of the relative levels of each viral protein.

Early studies on the transcription products of morbilliviruses showed that, in addition to full-length genomes and the expected mRNAs, other transcripts with the characteristics of mRNAs were observed (Barrett and Underwood, 1985; Hirayama *et al.*, 1985; Yoshikawa *et al.*, 1986). These were deduced to be read-through transcripts, or bicistronic mRNAs, which are, in essence, errors in mRNA transcription where the viral RdRP fails to recognize a gene end signal and produces a mRNA containing the sequences of two successive genes; only the upstream gene is translated (Hasel *et al.*, 1987; Wong and Hirano, 1987).

RNA editing

Co-transcriptional editing of mRNA transcripts from the P gene occurs in RPV and PPRV, as is found for all the morbilliviruses (Cattaneo *et al.*, 1989a; Blixenkrone-Möller *et al.*, 1992; Baron *et al.*, 1993; Haas *et al.*, 1995; Mahapatra *et al.*, 2003). This mechanism, first identified for MV (Cattaneo *et al.*, 1989a), inserts one or more extra G residues into P gene mRNA transcripts at a defined editing site roughly half way along the P protein ORF (Fig. 9.9). This editing process appears to be a result of the viral RdRP stuttering at the editing site in a similar way to the mechanism proposed for the addition of poly(A) tails to viral mRNAs, which is by repetitively transcribing the short run of Ts at the end of each gene (Vidal *et al.*, 1990a,b; Hausmann *et al.*, 1999). The extra nucleotides inserted at the editing site result in a frame shift when the mRNA is being translated, giving rise to different proteins depending on how many nucleotides are inserted: the P protein (no insertions), the V protein (one nucleotide insertion) or the W protein (insertion of two nucleotides). Insertion of three nucleotides gives rise to a P protein once again (albeit now with an extra glycine amino acid). All three proteins share the first 231 amino acids of the P protein sequence, followed by different carboxy-terminal sequences (see below). Studies in cell culture and samples from infected animals showed that the ratio of P:V:W-type transcripts varied slightly with the strain of RPV, but not with the time of infection or the tissue type (Haas *et al.*, 1995), suggesting it is purely a property of normal function of the viral RdRP in mRNA transcription mode. The different roles of the P, V and W proteins are discussed in the following section.

Viral protein structure and function

Although RPV and PPRV have only six genes, they produce nine different proteins. By virtue of the co-transcriptional 'editing' in the P gene and by use of two alternative reading frames in P gene transcripts, four different proteins can be produced from the P gene, a remarkable example of efficient use of genetic material (and highlighting again the significance of the large, apparently unused, sequence between the M and F protein coding sequences discussed above). The viral proteins can be divided into roughly three functional groups, those associated with the nucleocapsid core (N, P, L), those associated with the membrane envelope (M, F, H) and the non-structural proteins (C, V, W). Most of our basic knowledge of the structure and functions of morbillivirus proteins has come from studies on MV, not surprisingly given its importance as a

```
     ..TCAGAGAAACCCATTAAAAAGGGCACAGACGTGAAATCGAC..
P     ..  E  K  P  I  K  K  G  T  D  V  K  S  ..
     ..TCAGAGAAACCCATTAAAAAGGGgCACAGACGTGAAATCGAC..
V     ..  E  K  P  I  K  K  G  H  R  R  E  I  ..
     ..TCAGAGAAACCCATTAAAAAGGGggCACAGACGTGAAATCGAC..
W     ..  E  K  P  I  K  K  G  A  Q  T  *
```

Figure 9.9 The editing site in the P gene of *Rinderpest virus* (RPV) (underlined), showing the reading frames used for no (P protein), one (V protein) or two (W protein) G insertions at the editing site. The stop codon in the W protein is represented by *.

human pathogen, and the restrictions on handling livestock pathogens such as RPV and PPRV. The close similarity in the sequences of most of the *morbillivirus* proteins means that it is reasonable to assume fairly close similarity in structure and function. I will consider here the *morbillivirus* proteins as a group, highlighting where we have become aware of virus-specific details.

The nucleocapsid protein (N)

The N protein is the product of the most transcribed gene and is the most common viral protein. It accumulates to high levels in infected cells (see, e.g. Sweetman *et al.*, 2001). The protein is 525 amino acids (RPV, PPRV or MV) or 523 amino acids (CDV, PDV, CMV) in length. The protein interacts directly with other N proteins and with the P protein (Huber *et al.*, 1991; Bankamp *et al.*, 1996). The amino-terminal 398 aa are highly conserved across all morbilliviruses (Fig. 9.10), and appear to form the helical core of the nucleocapsid, since this section of N, expressed alone, assembles into nucleocapsid-like structures (Bankamp *et al.*, 1996). It is necessary to express the first 375 aa of MV N to give a protein that will self-assemble (Kingston *et al.*, 2004a); internal deletions have prevented self assembly (Bankamp *et al.*, 1996). There appear to be two kinds of N–P interaction: that giving rise to stable, approximately equimolar, N–P complexes, and requiring extended regions of the N protein core as well as the carboxy-terminus (Bankamp *et al.*, 1996), and the interaction typical of the polymerase, where the P protein, in association with the L protein, is binding to N protein oligomers (Kingston *et al.*, 2004a,b). In the latter case, with much of the core sequence buried in the nucleocapsid oligomer, P–N interaction appears to occur only at the carboxy-terminus of N, an intrinsically disordered region that is constrained to a helix on binding to a triple helix bundle formed in the P protein carboxy-terminus (Johansson *et al.*, 2003). Although various structural studies have described this N–P interaction as involving multiple interaction sites (Longhi *et al.*, 2003; Bourhis *et al.*, 2005), or a single, short site (Kingston *et al.*, 2004a,b; Yegambaram and Kingston, 2010), at the carboxy-terminus of N, the general structure of the interactions are fairly clear.

Preparative expression of N protein in bacteria (Warnes *et al.*, 1995) or insect cells (Fooks *et al.*, 1993; Kingston *et al.*, 2004a; Schoehn *et al.*, 2004) leads to the assembly of pseudo nucleocapsids which contain host cell RNA (Warnes *et al.*, 1995; Schoehn *et al.*, 2004). The carboxy-terminus of the N protein can be removed by trypsinisation from natural (Giraudon *et al.*, 1988) or artificial nucleocapsids (Schoehn *et al.*, 2004), leaving an intact nucleocapsid-like structure composed of oligomers of the core N sequence. Full nucleocapsids are normally flexible, with some variability in pitch (Bhella *et al.*, 2004); removal of the carboxy-terminal end by trypsinization leads to a much more rigid structure of uniform pitch, approximately 13.4 N proteins/turn of the helix (Bhella *et al.*, 2004; Schoehn *et al.*, 2004).

The carboxy-terminus of the N protein also binds to the host cell protein hsp72 (Zhang *et al.*, 2005), a dominant member of the hsp70 family of heat shock proteins; this interaction appears to be essential for viral RNA transcription (Zhang *et al.*, 2002). This role of hsp72 in part explains the observations of the increased replication of MV and other morbilliviruses in cells after a heat shock or other cellular stress (Oglesbee *et al.*, 1993, 1996; Vasconcelos *et al.*, 1998; Parks *et al.*, 1999). These observations have led to potential improvements in measles virus-based oncolytic virotherapy by upregulating hsp70 family proteins through blocking hsp90 function (Liu *et al.*, 2008).

Interaction of the MV N with the host cell signalling protein interferon regulatory factor 3 (IRF-3) has been reported (tenOever *et al.*, 2002), though further studies have failed to show direct interaction between these two proteins under conditions where N and P interact well, suggesting that this interaction is complex and occurs only in the presence of other host cell components (Colombo *et al.*, 2009).

The phosphoprotein (P)

The P protein was first identified as a highly phosphorylated, nucleocapsid-associated protein (Robbins and Bussell, 1979). Although the *morbillivirus* P proteins have similar sizes (506–509 aa), with predicted molecular weights of approximately 55 kDa, they all migrate anomalously

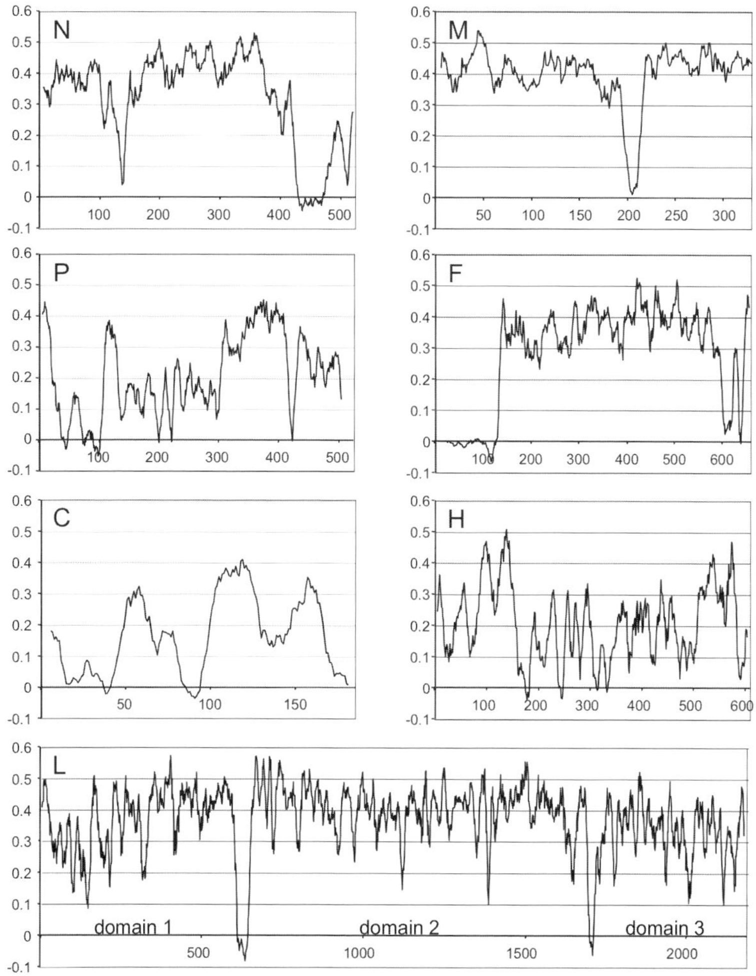

Figure 9.10 Plots of sequence conservation in morbillivirus proteins. The degree of sequence conservation at each position was calculated using *plotcon* from the EMBOSS suite. The EBLOSUM62–12 comparison matrix was used with a sliding window of 11 amino acids. The three L protein domains referred to in the text are labelled.

slowly on SDS-PAGE gels (Bellini *et al.*, 1985; Diallo *et al.*, 1987). This is thought to be due to phosphorylation of primarily serine, and some threonine residues, of which the P protein sequence is particularly rich. P is the second most common morbilliviral protein and appears to play two major roles in the virus life cycle, as a structural subunit of the viral RdRP, enabling the progress of the polymerase along the genome through its interaction with the N proteins and as a chaperone in the assembly of nucleocapsids. Therefore, the P protein binds strongly to both N and L proteins as well as forming an oligomer of itself. Cross-linking studies on the RPV P protein have suggested that it forms a tetramer (Rahaman *et al.*, 2004).

Coexpression of the P protein with N limits the self-assembly of the N protein and alters the density of the resulting nucleocapsid-like filaments (Spehner *et al.*, 1997). The stable interaction of P with N monomers requires domains at both the carboxy and amino ends of P (Harty and Palese, 1995; Shaji and Shaila, 1999), although there is some evidence that just the N-binding site in the amino-terminal half of P is sufficient for transient interaction with monomeric N (Tober *et al.*,

1998). It is thought that the P protein regulates the assembly of N during viral replication to limit the spontaneous oligomerization of N monomers and keep them available for assembly into proper nucleocapsids, as has been demonstrated for *Sendai virus* (Curran *et al.*, 1995). Interestingly, interaction of P with nucleocapsids (N–RNA complexes) appears to require only the carboxy-terminal approximately 50 amino acids of P (Johansson *et al.*, 2003; Kingston *et al.*, 2004a). This domain forms a triple helix coiled-coil structure which binds to the carboxy-terminal tail of the N protein (Johansson *et al.*, 2003; Kingston *et al.*, 2004a).

The amino-terminal half of the P protein is less conserved in its sequence than the carboxy-terminal half and little is known of its function other than its potential role as a chaperone in nucleocapsid assembly. Because of the position of the editing site, this part of the P protein is common to P, V and W proteins. In addition, the coding sequence for the viral C protein is contained within this region of the P/V/W ORF in the +1 reading frame and it is possible that requirements for virus-specific C protein sequences have led to greater differences in the amino acid sequences encoded by the P/V/W ORF in the different morbilliviruses.

Binding of P to L has been less well characterized. The L protein appears to be stabilized to some extent by the presence of P (Horikami *et al.*, 1994; Chattopadhyay and Shaila, 2004), and P–L complexes have to be formed during coexpression to be effective (Raha *et al.*, 2004a). Although the P protein binds to a region in the amino-terminal 500 amino acids of L and the same domain is responsible for L-L dimerization, different residues are involved in the two processes (Cevik *et al.*, 2004). The P protein appears to act as the 'legs and feet' of the full polymerase in progressing along the nucleocapsid (Kingston *et al.*, 2004b; Yegambaram and Kingston, 2010). The P protein of RPV has been suggested as binding to the transcript of the GP, the so-called 'leader' RNA (Raha *et al.*, 2004b), an interaction which requires the same phosphorylation of the P protein as replicative RNA transcription (Kaushik and Shaila, 2004). The details of these processes remain to be determined.

The large protein (L)

The L protein is the enzymatically functional subunit of the viral RdRP. It is a very large protein in morbilliviruses, 2183 amino acids long with an estimated molecular weight of 250 kDa. Enzymatically, it is responsible for the essential RNA-dependent RNA polymerase activity of the virus and all the motifs associated with such polymerases (Tordo *et al.*, 1988; Poch *et al.*, 1989, 1990) are found in the morbillivirus L proteins (Blumberg *et al.*, 1988; Baron and Barrett, 1995; Muthuchelvan *et al.*, 2005). In addition, a motif associated with capping has been identified in all the *Mononegavirales* (Ferron *et al.*, 2002), so it is likely that the morbillivirus L protein also caps viral mRNAs. The addition of poly(A) tails to viral mRNAs appears to be the result of 'stuttering' or repeated copying by the polymerase of a short stretch of T residues at the end of each gene. Other activities may be associated with the L protein, given its size, but these have not yet been identified.

The sequence of the morbillivirus L proteins is very highly conserved over the whole length of the protein, except for two short regions of very low conservation (Fig. 9.10) which have been termed 'hinge' regions (McIlhatton *et al.*, 1997), thought to be flexible connecting regions between functional domains. Insertion of the coding sequence for green fluorescent protein (GFP) while maintaining polymerase function has been possible at the second of these 'hinges', but not the first (Duprex *et al.*, 2002), showing that there are physical constraints on the flexibility or relative positioning of domains 1 and 2, but much greater flexibility allowed between domains 2 and 3. Recombinant RPV with such an inserted GFP grew normally in tissue culture, but was attenuated *in vivo* (Brown *et al.*, 2005b). The major polymerase-associated sequence motifs are distributed in all three domains, suggesting that the protein functions as a single complex unit. Although the L protein homooligomerizes (Cevik *et al.*, 2004), whether or to what extent the individual L proteins in the functional oligomer work together is unknown.

As mentioned above, L–L interaction and L–P interaction both map to the amino-terminus of the protein. We have also found that the non-structural C protein binds L (Sweetman *et al.*, 2001), binding specifically to domain 2 (Baron and Baron, unpublished data). The C protein of morbilliviruses has been shown to have a number of effects on viral RNA transcription; this is discussed below under *C protein*. Our own studies have also found that RPV L also binds to the host cell protein striatin (Sleeman and Baron, 2005), a component of protein kinase C (PKC) complexes. The role of striatin or PKC in viral RNA transcription is not known. Anti-striatin antibodies had a slight stimulating effect on RNA synthesis in a cell-free viral transcription system (Baron and Baron, unpublished).

The nucleoprotein core (N-P-L)

Replication of and/or transcription from viral genomic RNA requires all three proteins (Sidhu *et al.*, 1995; Baron and Barrett, 1997), while transcription from isolated or synthetic nucleocapsids requires coexpressed L+P and N+P complexes (Raha *et al.*, 2004a). Interestingly, the N, P and L proteins of the morbilliviruses, although they have highly conserved sequences in general, and particularly in the defined domains involved in N-P and P–L interactions, do not function well when mixed in heterologous sets for the transcription of minigenomes (Brown *et al.*, 2005a), suggesting that the interactions between the core proteins are complex and that sufficient virus-specific sequence changes have taken place that these viral proteins cannot work together well unless they are in a set of proteins that have co-evolved. However, despite this limitation, it would appear that viral RNA transcription is not a limiting step in viral growth, since a chimeric virus in which the N protein of RPV was replaced by the N protein of PPRV was not only viable, it grew at the same rate in tissue culture as the parental RPV strain used (Parida *et al.*, 2007). The transcription complex also appears to need host proteins, since purified RPV nucleocapsids show defective RNA synthesis, requiring the addition of whole cytoplasmic extract or purified host cell proteins, either tubulin (Moyer *et al.*, 1990) or microtubule-associated proteins (MAPs; Baron and Baron, unpublished) as well as hsp70 family proteins (Oglesbee *et al.*, 1996), for efficient, full-length RNA synthesis.

The matrix protein (M)

The exact role of the morbillivirus M protein has not been entirely elucidated, but it appears to play a complex role in the assembly of the virion, particularly virus budding. The M protein appears to associate directly with membranes without the presence of the viral glycoproteins, though it requires full viral infection to be transported to and accumulate at the normal budding site, the apical surface of the cell (Manie *et al.*, 2000; Riedl *et al.*, 2002). The M protein is found in detergent resistant membrane fractions (DRMs; Manie *et al.*, 2000; Vincent *et al.*, 2000), and expression of the M protein alone can lead to budding of virus-like membrane vesicles (Pohl *et al.*, 2007). At the same time there is clearly interaction with the viral glycoproteins, because sorting of F and H to the apical surface of the host cell requires the presence of the M protein (Naim *et al.*, 2000). This interaction of the M protein with the viral glycoproteins has a significant role in virus budding and subsequent fusion, so that a chimaeric RPV with the glycoproteins of PPRV instead of those of RPV was viable, but grew relatively slowly (Das *et al.*, 2000); growth was improved by replacing the M protein also with that of PPRV (Mahapatra *et al.*, 2006). M predominantly seems to interact with the cytoplasmic tail of the F glycoprotein (Cathomen *et al.*, 1998a; Naim *et al.*, 2000; Moll *et al.*, 2002), although mutations that strengthen interactions with the cytoplasmic tail of the H protein have been found on adaptation of MV to Vero cell culture (Tahara *et al.*, 2005, 2007), simultaneously improving budding from Vero cells and decreasing infection through the normal virus receptor (SLAM). These findings indicate a complex set of interactions at the inner surface of the host cell membranes leading to budding. The M protein is currently considered to be the main driving force for the budding process, although generation of virus-like particles has also been seen when just the F protein is expressed (Pohl *et al.*, 2007). A recombinant MV without an M protein gene grew, but was severely defective in budding, spreading mostly by cell–cell fusion (Cathomen *et al.*, 1998a).

While no similar disease has been reported in RPV or PPRV infections, a rare fatal complication of MV infection has been ascribed to mutations in the M protein. Subacute sclerosing panencephalitis (SSPE) can take many years to appear in the victim, and appears to be a result of the virus establishing a persistent infection in the brain; these persistent infections are associated with high levels of mutation in the M gene and decreased expression of M protein and viral glycoproteins (Liebert *et al.*, 1986; Cattaneo *et al.*, 1987a,b, 1988, 1989b; Wong *et al.*, 1989) as well as changes in M protein function (Hirano *et al.*, 1992, 1993; Suryanarayana *et al.*, 1994).

The M protein also binds to the N protein and inhibits RNA transcription from viral nucleocapsids (Suryanarayana *et al.*, 1994; Reuter *et al.*, 2006; Iwasaki *et al.*, 2009) as well as enabling nucleocapsids to associate with host cell membranes (Runkler *et al.*, 2007; Iwasaki *et al.*, 2009). It thus provides a bridge between the viral envelope (glycoproteins F and H) and the ribonucleoprotein core (N, P, L). The nucleocapsids may be transported to the membranes along actin microfilaments, since disassembly of microfilaments using cytochalasin B inhibits MV budding (Stallcup *et al.*, 1983), while association of complete nucleocapsids with actin has been reported (Moyer *et al.*, 1990). The M protein of morbilliviruses, unlike the orthologous VP40 of filoviruses, does not utilize the endosomal sorting complex required for transport (ESCRT) pathway in its budding process (Salditt *et al.*, 2010).

The fusion protein (F)

As might be expected from the name of this protein, it is responsible for the fusion of the viral envelope with the plasma membrane of the infected cell, allowing the release of the nucleocapsid into the host cell cytoplasm. Expression of the viral proteins on the cell surface of infected cells leads to fusion with neighbouring cells, forming polynucleated syncytia, and allowing the virus nucleocapsid to be passed into new cells without necessarily forming a virion. Most researchers have found that effective fusion requires coexpression of the F and H proteins on the same cell (Wild *et al.*, 1991; Heminway *et al.*, 1994), although there is some evidence for the PPRV F protein being able to induce cell–cell fusion on its own (Seth and Shaila, 2001a). Expression of the M protein as well as the two glycoproteins limits cell–cell fusion (Cathomen *et al.*, 1998a; Tahara *et al.*, 2007), while alterations to the cytoplasmic tails of the MV glycoproteins can increase cell–cell fusion, presumably by preventing normal interaction with the M protein (Cathomen *et al.*, 1998b).

The F protein itself is synthesized as a type 1 membrane protein, which is glycosylated at several positions along its length. As described under *Genome Organization*, the actual start translation point in virus infection is not always clear, but assuming use of the first start codon in frame with the F protein ORF, these proteins have leader peptides of variable length and unconserved sequence (Fig. 9.7) ending in all cases except PPRV with classical (von Heijne, 1983) hydrophobic signal sequences and signal peptide cleavage sites. In the case of PPRV, there is a relatively short sequence upstream of the start of the conserved F protein and it would seem that this remains attached to the rest of the protein, as this sequence does not fit the requirements of a signalase cleavage site (von Heijne, 1983). From the cleavage point on, the morbillivirus F protein sequences are highly conserved (Buckland *et al.*, 1987; Evans *et al.*, 1994; Meyer and Diallo, 1995). The protein downstream of the signalase cleavage site, F_0, is processed to the mature protein, which exists as a disulfide-linked F_1–F_2 heterodimer following post-translational cleavage (Sato *et al.*, 1988). Separation of the (amino-terminal) F_2 peptide by this cleavage exposes the hydrophobic fusion domain at the amino-terminus of the larger F_1 peptide, which also includes the membrane anchor and cytoplasmic tail of the protein at its carboxy-terminus. The cleavage site in morbillivirus F proteins has the pattern R-R-X-(R/K)-R-*cut*, which resembles the furin consensus R-X-X-R. Inhibition of cellular furin (Watanabe *et al.*, 1995) or mutation of the last arginine in the motif to leucine (Alkhatib *et al.*, 1994a) inhibits syncytium formation (cell–cell fusion) and release of infectious virus, suggesting that the cleavage of F_0 is normally performed by furin and showing that this cleavage is essential for the fusion functionality of the protein.

There are three highly conserved N-linked glycosylation sites, all in the F_2 domain (Meyer and Diallo, 1995), and all are required for proper folding of the protein and transport to the cell surface

(Alkhatib *et al.*, 1994b; Hu *et al.*, 1995; Bolt *et al.*, 1999). The F protein is further modified by attachment of palmitoyl residues, primarily to a cysteine residue in the transmembrane domain (Caballero *et al.*, 1998). This modification is essential for fusion activity and may be directing the protein to the correct type of membrane, as the F protein, like the M protein, is found associated with DRMs (Pohl *et al.*, 2007). The mature protein exists as a homotrimer (Zhu *et al.*, 2002; Rahaman *et al.*, 2003), based on crystal structure of expressed sections of the membrane-proximal core of the protein, or as a homotetramer (Malvoisin and Wild, 1993), based on cross-linking of the mature protein. A leucine zipper motif, located proximal to the membrane anchor, is required for fusion activity, but not for homooligomerization of the F protein (Buckland *et al.*, 1992).

The haemagglutinin protein (H)

Despite its name, the H proteins of most of the morbilliviruses do not act as haemagglutinins, there being no suitable virus receptor on red blood cells. An exception (and the partial origin of the protein's name) is the H protein of tissue-culture adapted MV, which will agglutinate rhesus monkey, but not human, red cells. The monkey erythrocytes express CD46, the protein that MV uses as a receptor after adaptation to tissue culture cells (Ono *et al.*, 2001). An early assay for anti-measles antibodies was the ability of sera to inhibit this haemagglutination.

An interesting difference between the morbillivirus H protein and the corresponding attachment proteins of other paramyxoviruses (usually called HN) is the lack of that letter 'N' (for neuraminidase). Sialic acid forms, or is an important part of, the host cell receptor for many of the viruses of the *Respirovirus* and *Rubulavirus* genera of paramyxoviruses, so these viruses have maintained the neuraminidase function to enable release of progeny virus from infected cells. Neuraminidase activity has not generally been observed among the morbilliviruses, presumably because they have adapted to use direct protein–protein interaction with their common receptor (SLAM) in binding to the target cell. Sequence analysis has identified the residual signature of a neuraminidase in morbillivirus H proteins (Langedijk *et al.*, 1997) as well as low-level enzyme activity in the expressed H proteins of RPV and PPRV (Langedijk *et al.*, 1997; Seth and Shaila, 2001b).

The morbillivirus H proteins are type 2 membrane proteins, with a typical signal-anchor sequence near the amino-terminus, leaving 34 amino-terminal amino acids in the cytoplasm. The extracellular domain is thought to form a membrane proximal extended stalk-like structure with the carboxy-terminal half of the protein forming the globular 'head' of the protein that contains the receptor binding sites. The mature protein forms cysteine-linked homodimers (Hu *et al.*, 1994; Plemper *et al.*, 2000) in which the disulfide links are in the stalk. These dimers themselves appear to dimerize to give tetrameric H proteins (Plemper *et al.*, 2000). The globular head of MV H protein, minus its stalk, has recently been crystalized both in isolation (Colf *et al.*, 2007) and associated with one of its receptors, CD46 (Santiago *et al.*, 2010). The structure shows similarities to known neuraminidases, including the deep pocket that normally accommodates the glycan chain, but without the requisite active sites, thus confirming the earlier speculation that the morbilliviruses are derived from an ancestor with an HN protein, but have lost this functionality during their evolution. The fact that PPRV has retained the most neuraminidase activity suggests that it may be closer to the putative ancestral morbillivirus than any of the other known members of the genus.

Unlike their F proteins, the glycosylation patterns of morbillivirus H proteins vary with the virus. Glycosylation is required for correct folding of the H protein and transport from the endoplasmic reticulum (E.R.) to the cell surface. The H and F protein oligomers interact physically and functionally: adding an E.R. retention signal to the MV F protein leads to retention of H protein in the E.R. as well (Plemper *et al.*, 2001); mutation of the H protein cytoplasmic tail decreased H–F interactions and led to a concomitant increase in infectivity and cytopathic effect (Plemper *et al.*, 2002); various mutations in the H protein stalk have been shown to affect cell–cell fusion (Corey and Iorio, 2007). Heterologous combinations of F and H proteins from different viruses show variation in their functional compatibility, so that PPRV H and F proteins only functioned as an homologous pair (Das *et al.*, 2000), while CDV and MV H proteins could be exchanged (von Messling *et al.*, 2001) as could RPV and CDV H proteins (Brown *et al.*, 2005a).

The non-structural proteins V and W

The V and W proteins are the products of co-transcriptional editing of P gene mRNA transcripts, as described above. Judging by the frequency of double G insertions during editing, W-type transcripts are a very minor (1–2%) component of the total P gene output in RPV (Baron *et al.*, 1993) or PPRV (Mahapatra *et al.*, 2003), and W proteins are, therefore, probably very minor components of the infected cell. Whether they have a distinct role or are simply accidental by-products of the editing mechanism that produces the V protein is not yet known. The V protein, on the other hand, is produced from about a third of P gene transcripts and is an important factor in both viral replication and the ability of the virus to block the host innate immune response.

The production of a V protein (or, properly, co-transcriptional editing) seems to have evolved in an ancestor of the whole *Paramyxovirinae* sub-family (Jordan *et al.*, 2000), since editing is found in almost all the members of this group of viruses, with the exception of the human parainfluenza virus type 1 (hPIV1) (Matsuoka *et al.*, 1991). The major part of the protein is, of course, the same as the amino-terminal half of the P protein, and is relatively poorly conserved among morbilliviruses. The V-specific sequence downstream of the editing site shows a Cys-rich motif that is highly conserved across all the paramyxovirus V proteins (Fig. 9.11). This motif has been shown to bind two zinc ions (Liston and Briedis, 1994). The V protein is not essential for virus growth in cell culture, since recombinant RPV and MV have been made that do not express V and these viruses are viable, although they show changes in RNA transcription (Schneider *et al.*, 1997; Baron and Barrett, 2000). Specifically, preventing V expression in RPV gave rise to a virus with increased mRNA expression and increased tendency to form syncytia (Baron and Barrett, 2000); such constructs grew at the normal rate in multistep growth curves, but to lower final titres than V+ virus when preparing stocks. MV that did not express V showed a similar increase in

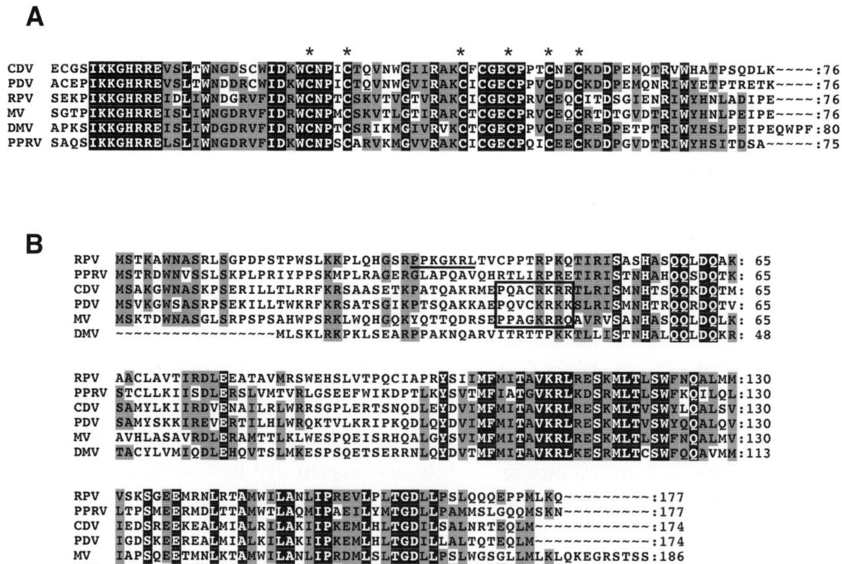

Figure 9.11 (A) Alignment of morbillivirus V protein-specific domains, showing the conserved cysteine residues (*) that form the zinc-binding site; (B) Alignment of morbillivirus C proteins. The position of the putative NLS sequence in measles virus (MV), canine distemper virus (CDV) and phocine distemper virus (PDV) is outlined, and that found in rinderpest virus (RPV) is underlined.

syncytium formation and viral mRNA expression (Schneider *et al.*, 1997; Tober *et al.*, 1998). Correspondingly, expression of excess V protein inhibits viral RNA transcription in either the whole virus (Tober *et al.*, 1998) or minigenome system (Parks *et al.*, 2006; Witko *et al.*, 2006).

The V protein is found uniformly distributed in infected cells (Sweetman *et al.*, 2001) and does not specifically associate with the viral N/P/L proteins, which exist in large complexes in the infected cell, although direct interaction of RPV V and L proteins have been reported (Sweetman *et al.*, 2001). It is possible that the V protein binds transiently to monomeric N (N^0; Tober *et al.*, 1998), which helps to prevent improper assembly of N during infection. The absence of V might limit the availability of N for replicative transcription, biasing the processes in the infected cells towards relatively increased mRNA synthesis, which is what is observed. However, this hypothesis implies a functional difference between V–N^0 and P–N^0 dimers, since preventing the expression of V by blocking editing will lead to a concomitant increase in the expression of the P protein, which can also bind N^0.

The other major role of the V protein is in helping the virus overcome the host's innate immune defences. The first of these to come into play in most virus infections is the type 1 interferon (IFN α/β) response, triggered through cytoplasmic and/or extracytoplasmic pathogen-associated molecular pattern (PAMP) receptors. The expression of IFN leads, in both an autocrine and paracrine manner, to the induction of a large number of host cell proteins that can inhibit viral replication, up to and including the suicide (through apoptosis) of the infected cell. Type 1 IFN also stimulates further innate and adaptive immune responses, including the expression of type 2 IFN (IFN γ). These host cell defences and the ways that paramyxoviruses set out to block them have been extensively reviewed recently (see Fontana *et al.*, 2008b; Goodbourn and Randall, 2009), so I shall highlight here what we know about the mechanisms employed by RPV and PPRV. Importantly, the way in which the morbilliviruses interfere with the induction and action of type 1 and type 2 IFNs appears to be one of the few areas where significant differences are seen between the morbilliviruses.

The V protein of RPV blocks both type 1 and type 2 IFN signalling in infected cells, preventing the phosphorylation of both STAT1 and STAT2 (Nanda and Baron, 2006). STAT1 is bound by the V protein to a sequence in the P/V/W shared domain, and all three proteins can be shown to co-immunoprecipitate this signalling protein (Nanda and Baron, 2006). Interestingly, although all three proteins bind STAT1 equally, only the V protein is efficient in blocking STAT1 and STAT2 phosphorylation (Nanda and Baron, 2006) and only the V protein could prevent cells entering the IFN-induced anti-viral state (Chinnikannan, Nanda and Baron, unpublished), indicating that this binding of STAT1 is not the key element in blocking IFN signalling. Although our initial studies failed to show any interaction of the V protein with STAT2 (Nanda and Baron, 2006), we have since found that this is largely because we were using the V protein of the vaccine strain of RPV, which binds STAT2 only weakly, while the V protein of a virulent RPV isolate binds STAT2 well (Chinnikannan and Baron, unpublished). The V protein of PPRV similarly binds both STAT1 and STAT2 and blocks both type 1 and type 2 IFN signalling (Chinnikannan and Baron, unpublished). There is agreement that the V protein of MV blocks type 1 IFN signalling (Palosaari *et al.*, 2003; Takeuchi *et al.*, 2003; Yokota *et al.*, 2003; Ohno *et al.*, 2004; Caignard *et al.*, 2007; Fontana *et al.*, 2008a; Ramachandran *et al.*, 2008), but not about whether it blocks type 2 IFN signalling (Palosaari *et al.*, 2003; Takeuchi *et al.*, 2003; Ramachandran *et al.*, 2008). The MV V protein has also been found to bind STAT1 and STAT2; binding of STAT1 by the MV V protein has also been mapped to the amino-terminal part of V, the sequence shared with the P protein, while binding of STAT2 maps to the V protein-specific domain (Ramachandran *et al.*, 2008; Caignard *et al.*, 2009). Some studies have also identified a direct interaction of MV V with components of the type 1 IFN receptor or its associated Janus kinases (Yokota *et al.*, 2003; Caignard *et al.*, 2007), while others have ascribed the function of MV V to the ability to bind STAT1 and STAT2, either preventing their phosphorylation (Takeuchi *et al.*,

2003; Caignard *et al.*, 2007; Devaux *et al.*, 2007) or allowing phosphorylation of these proteins but retaining them in the cytoplasm (Palosaari *et al.*, 2003). It is probable that the differences in reported activities of MV V proteins derive from differences between different isolates of MV, but the importance of the different activities of the V protein *in vivo* will be harder to establish. The V protein of CDV has been reported as blocking nuclear import of STAT1 and STAT2 without affecting their phosphorylation (Rothlisberger *et al.*, 2010). In a direct comparison, we have found that V proteins of RPV, PPRV, CDV and MV all block type 1 IFN signalling, regardless of the ability to bind STAT1 or STAT2, but the ability to block type 2 IFN signalling correlates with the ability to coimmunoprecipitate STAT1.

The V proteins of the rubulavirus *Parainfluenza virus type* 5 (PIV5; originally called *Simian virus* 5, or SV5) was the first to be shown to have a significant effect on the induction of type 1 IFNs as opposed to blocking the action of this cytokine (He *et al.*, 2002; Poole *et al.*, 2002). This block on IFN induction was due to the binding of the host cell DExD/H box helicase known as mda-5 (melanoma differentiation-associated protein 5; Andrejeva *et al.*, 2004). This protein acts as a cytoplasmic pattern recognition receptor (PRR), binding dsRNAs (Kato *et al.*, 2006; Pichlmair *et al.*, 2009) and inducing, though a complex pathway, the activation of transcription from the IFN-β promoter. Binding of mda-5 by the PIV5 V protein was shown to inhibit IFN induction and these findings were extended to MV as well as other paramyxoviruses (Childs *et al.*, 2007). However, the RPV V has no effect on IFN induction, that role being played by the RPV C protein (Boxer *et al.*, 2009). We have found that, while we also see co-immunoprecipitation of mda-5 with PIV5 and MV V proteins, we don't see this with the V proteins of RPV, PPRV or CDV (Boxer, Chinnikannan and Baron, unpublished). These data suggest that some viruses have evolved the ability to bind mda-5 as it is particularly important in the defence systems of their specific hosts, while other viruses, with other hosts, have not found this to be necessary for effective growth and spread. One area that is still under investigation is why paramyxoviruses should block the activity of mda-5 in the first place, since these viruses do not produce detectable dsRNA during infection (Weber *et al.*, 2006) and an alternative PRR, the product of retinoic acid inducible gene 1 (RIG-I), has been found to be essential for IFN responses to MV and other paramyxoviruses (Hornung *et al.*, 2006; Pichlmair *et al.*, 2006; Plumet *et al.*, 2007; Hausmann *et al.*, 2008; Loo *et al.*, 2008; Rehwinkel *et al.*, 2010). There is recent evidence that both RIG-I and mda-5 proteins are important in responses to MV infection, although the former is at least quantitatively more important (Ikegame *et al.*, 2010), and the mda-5 protein appears to be particularly important for normal IFN responses to MV infection *in vivo* (Gitlin *et al.*, 2010).

The V proteins of morbilliviruses may have yet more ways of inhibiting the induction of IFN in infected cells. It has been shown that MV V acts as an alternative or decoy substrate for IκB kinase α (IKKα), an important constituent of the kinase complex that phosphorylates interferon regulatory factor 7 (IRF-7) in plasmacytoid dendritic cells (pDCs) in response to signalling through Toll-like receptors (TLRs; Pfaller and Conzelmann, 2008). Binding of MV V to IKKα decreased the activation of IFN expression in response to TLR signalling. The same authors observed a direct interaction of MV V with IRF-7, the critical transcription factor in IFN induction in pDCs. Such an activity might be particularly important in terms of the infected host (as opposed to the individual infected cell), since pDCs are the major source of type 1 IFN during *in vivo* infection.

The non-structural protein C

The C proteins are the least conserved among all the morbillivirus proteins (Fig. 9.10). As with the V protein, the C protein appears to play at least two roles, one in viral RNA synthesis and one antagonizing the host IFN response. When expressed from plasmids, the RPV C protein accumulates almost entirely in the nucleus (Boxer *et al.*, 2009) as does that of MV (Nishie *et al.*, 2007), although no conserved nuclear localization signal (NLS) can be found in the C protein sequences. A 'pat7' type NLS (P, X_2, $(K/R)_{3/4}$) identified in MV C protein (Nishie *et al.*, 2007) is found in the C proteins of CDV and PDV, but not RPV, PPRV or DMV (Fig. 9.11); similarly, the RPV C contains a possible NLS at positions 32–38 (Fig. 9.11), but this is also not a conserved motif across the morbilliviruses. A conserved positively charged motif is found in the short highly conserved domain near the centre of the C protein sequence (Fig. 9.11), which might form a non-standard NLS. It may be that nuclear localization is an evolutionarily recent function for some of the *Morbillivirus* C proteins and different viruses have acquired this capability through different mutational events. In infected cells, the RPV C (Sweetman *et al.*, 2001) and the MV C are found both in the nucleus and associated with the N/P/L complexes in the cytoplasm.

The RPV C protein was found to bind to the L protein (Sweetman *et al.*, 2001) and RPV that does not express the C protein was viable in tissue culture, but grew more slowly and showed overall decreased viral mRNA synthesis (Baron and Barrett, 2000). Further studies have shown that C binds to the central domain of L and, in a cell-free viral RNA synthesis system, revealed two effects of the C protein on viral RNA transcription. Firstly, as in cells infected with the C- RPV mutant, nucleocapsids from such cells show a decreased rate of mRNA transcription; they also showed a greatly decreased replicative transcription relative to viral mRNA transcription (Baron, J., and Baron, M.D., unpublished). The exact balance of C protein may be important, since we found that expressing extra C protein in infected cells also decreased viral mRNA synthesis while increasing viral genome synthesis. In this respect RPV C and MV C appear to differ slightly, since MV C has been reported to decrease all types of viral RNA transcription (Bankamp *et al.*, 2005).

As mentioned above, the RPV C protein is a strong inhibitor of type 1 IFN induction (Boxer *et al.*, 2009), as are all the morbillivirus C proteins we have tested (RPV, CDV, PPRV and MV). Studies on MV have shown that the absence of the C protein greatly reduces viral growth in IFN-producing cells (Escoffier *et al.*, 1999) or those with active IFN-induced genes (Devaux and Cattaneo, 2004; Toth *et al.*, 2009), or in experimental animals (Patterson *et al.*, 2000; Takeuchi *et al.*, 2005; Devaux *et al.*, 2008). Recent work has shown a clear role for the MV C protein in inhibiting IFN induction (Nakatsu *et al.*, 2006; McAllister *et al.*, 2010), although at least one group has ascribed this to an indirect effect of MV C on IFN induction through an effect on viral RNA transcription (Nakatsu *et al.*, 2008). Since the C protein has roles in both viral transcription and resistance to the innate immune response, it will take further work to distinguish the exact role played by each function.

The nature of disease

As mentioned in a previous section, RPV and PPRV cause similar diseases in their respective hosts. I shall outline here the nature of RP as a disease and then highlight those areas where PPR has unique features. The classical acute form of RP is a severe disease, with high morbidity and causing high mortality; however, strains of the virus that caused only a mild disease in cattle became dominant in areas where the virus was persisting towards the end of the RPV eradication campaign. This appeared to be a specific adaptation to domestic livestock, as those same strains appeared to cause a severe disease in wildlife (Kock *et al.*, 1999; Wohlsein and Saliki, 2006). There is variation in the inherent resistance of different breeds of cattle to RPV and of sheep and goats to PPRV, as well as age-related resistance to PPRV in sheep and goats, with younger animals showing higher mortality rates.

Clinical features of RPV

Full clinical rinderpest is seen in all strains of *Bos taurus* (unhumped cattle), *Bubalus bubalis* (buffalo) and *Bos grunniens* (yaks); the humped cattle, or zebu (*Bos indicus*), was historically considered as showing slightly less severe clinical disease. RPV affects animals of all ages and mortality rates may exceed 90% in cases where the virus enters a naïve population. The last reported such outbreak was in Pakistan in the 1990s (Rossiter *et al.*, 1998). Where the virus was endemic, the normal pattern was for the frequency of mild or even inapparent infections to increase (Wohlsein and Saliki, 2006), making detection problematic in the closing stages of the RPV eradication campaign. The changes in virulence were not due to selection for more resistant hosts, as the virus strains proved to be attenuated even when tested in known highly susceptible hosts (Plowright, 1963; Wamwayi *et al.*, 1995). The nature of pathogenesis in such mild strains was not a change in tissue tropism, but a change in quantity of virus replication and tissue destruction (Wohlsein *et al.*, 1993).

The classic acute form of the disease can be divided into five phases (Plowright, 1965).

1 An incubation period, lasting 3–9 days depending on the strain of virus and the route of infection. In some experiments, this incubation period could last up to 2 weeks following contact infection.
2 The prodromal phase, lasting 2–5 days, and marked by a sudden rise in body temperature, peaking at 40.0–41.5°C. Animals become depressed and anorexic, restless and with shallow respiration. Milk yield is reduced, although the animals often increase water intake. The mucous membranes of eyes, nose and mouth become congested and there is a lot of discharge from the eyes and nose. There is a marked and rapid loss of circulating white blood cells (leucopenia) at this time. The white cell count will remain low (about 10% of normal) and only return to normal during the convalescent phase.
3 The mucosal phase begins with the appearance of the first necrotic lesions in the epithelial tissue lining the mouth and vagina. These begin as very small grey/white spots that spread and enlarge, first in the gums and the inside of the lower lip, and can be seen on the top and sides of the tongue and in other parts of the buccal mucosa. The papillae in the sides of the mouth become congested and then necrotic. The necrotic areas throughout the mouth and gums readily erode, becoming covered with a yellow or off-white mass of cellular debris and encrusted with exudate from the inflamed tissues. The erosions can spread into the pharynx and to the underside of the tongue. The animals salivate profusely, drink enormously but cease to eat, probably owing to the discomfort of chewing and swallowing. The breath becomes fetid.
4 The diarrhoeal phase begins 1–3 days later, often accompanied, or just preceded, by a sudden drop in core body temperature to below normal. The diarrhoea is highly liquid, shooting and profuse, with blood and mucous and other debris from the gut mucosa. The necrosis of the mucosal epithelium may spread to the nasal passages, the vagina, vulva or prepuce. By this stage the animal is severely apathetic, with laboured breathing and unwillingness to move. The back arches and the animals become dehydrated and emaciated and may eventually become prostrate, unable even to stand. Death ensues usually between 5 and 10 days after the onset of disease, except for the few animals that recover.
5 The last phase, convalescence, takes several weeks. Any animals pregnant at the time of infection will abort. The white cell count slowly returns to normal and the oral lesions heal over a period of 2–3 weeks. This transient loss of white cells, and the generalized immunosuppression that can go on for even longer (see below), means that the animal is susceptible to activation of latent or chronic infections (e.g. with parasites) or to secondary infection by other pathogens.

The mild strains have a longer incubation period (>2 weeks) and the symptoms seen are confined to a slight rise in temperature, congestion in, and discharges from, the eyes and nose and only very slight or no lesions in the mouth. Diarrhoea will be mild if any, and not all infected animals will show all signs. Morbidity is usually much lower than for the virulent disease and mortality almost zero. RPV has also been reported as causing a rare peracute form of the disease in which death occurs 1–2 days after the first symptoms are seen. The underlying cause of this form of the disease is unclear.

In addition to cattle and buffalo, disease has also been reported in a number of other wild ungulates, including giraffe, lesser kudu, eland and impala (Kock *et al.*, 1999). In India, the disease has been reported in axis deer, and wild cattle (gaur, mithun and banteng; Wohlsein and Saliki, 2006). Most reports of experimental rinderpest infection of small ruminants (sheep, goats) found a subclinical infection, possibly with transient fever and slight discharges from eyes or nose. However, Asian isolates of RPV used to infect local breeds of sheep and goats elicited a rinderpest-like disease (Anderson *et al.*, 1990a), albeit still with low mortality. Infection of pigs has been reported (Ramani *et al.*, 1974; Govindarajan *et al.*, 1996), as has the identification of the virus in warthogs in Africa (Shanthikumar and Atilola, 1990). European pigs are less susceptible than Asian varieties (Govindarajan *et al.*, 1996); the disease in Asian pigs is similar to a mild rinderpest in cattle, with pyrexia, nasal and ocular discharges, buccal erosions, anorexia, depression and diarrhoea. The diarrhoea in pigs can be prolonged and lead to death.

Despite all these alternative hosts, the epidemiology of the disease shows that it is primarily a disease of cattle. Vaccination of domestic bovid livestock has always led to control and eventual elimination of the disease from wildlife as well as livestock and it seems that wild ungulates, small ruminants or pigs act as sentinels for the presence of the virus in the local cattle populations, rather than alternative reservoirs for the virus.

Rinderpest pathology

The morphological changes to tissues in rinderpest-infected animals have been described in great detail elsewhere (Khera, 1958; Wohlsein *et al.*, 1993; Wohlsein *et al.*, 1995; Spinage, 2003; Wohlsein and Saliki, 2006). In summary, the pathological findings follow and extend the clinical signs, in that the animals show extensive damage to mucous membranes of the digestive tract and to lymphoid organs, and the number and extent of the lesions are correlated with the severity of the disease. Immunohistological examination shows that the virus is lympho- and epitheliotropic, that these tissues contain large amounts of viral protein, and that the amount of protein and degree of spread are related to the virulence of the virus (Wohlsein *et al.*, 1995). Unlike MV and CDV in their respective hosts, RPV has not been found associated with neural tissues in infected animals. Necrotic lesions and haemorrhage can be found anywhere in the gastrointestinal tract, though the fore stomachs and small intestine are less affected than the abomasum and large intestine. A characteristic of RPV and PPRV infection is the so-called 'zebra striping' or 'tiger striping' on the walls of the caecum and colon, caused by the congestion of the blood vessels along the ridges of the mucosal folds in the walls of this part of the intestine.

Although the major clinical effects of the virus are due to the pathogenesis in the mucosal membranes, the primary target of the virus are the lymphoid cells. As mentioned above, the specific receptor for RPV and PPRV is SLAM/CD150, a protein primarily expressed on the surface of activated B and T cells, activated macrophages and mature dendritic cells (Cocks *et al.*, 1995). In addition to the leucopenia observed at the same time as the pyrexia in infected animals, extensive damage to lymphoid organs are seen in acute disease. The Peyer's patches in the walls of the small intestine are depleted of lymphocytes. Lymph nodes are swollen in early stages, but appear collapsed in animals that die after a prolonged or severe disease. Histologically, so called 'giant cells' are common in lymphoid follicles of lymph nodes, spleen, tonsils and the mucosal lymphoid tissues of the intestine; these giant cells are actually syncytia formed when cells expressing the viral glycoproteins on their surface fuse with neighbouring cells expressing the viral receptor. As disease progresses, these tissues show progressive lymphocytic depletion.

Clinical features and pathology of PPR

Sheep and goats infected with PPRV show a similar, if slightly less severe, clinical picture to cattle infected with RPV. As in rinderpest, there is a rare peracute form of the disease causing death 4–6 days after the onset of fever. The more normal acute form is very like the description of rinderpest above, except for the involvement of the lungs in PPRV infection. Many animals with PPR show abnormally rapid or laboured breathing, and a productive cough. In post-mortem examination, the main differences are again in the existence of significant lung pathology in PPRV infection, with patches of congestion in the lung tissue and signs of pneumonia. The picture in the lungs can be confused by the presence of secondary bacterial infections. Histological examination of lung tissues show marked infiltration of alveolar macrophages and proliferation of pneumocytes; syncytia are also seen in the lung, which is not true of RPV infections. Further details on the pathology of PPR disease can be found in Wohlsein and Saliki, 2006 and references therein (Wohlsein and Saliki, 2006).

European sheep and goats most commonly show a mild subacute form of disease, with low level pyrexia, congestion of the eyes and nose, but little discharge or diarrhoea. This is particularly true of experimental infections in highly secure units, which are concomitantly free of most secondary infectious agents. The contribution of such secondary infections to the pathology of PPRV remains to be determined. Like RPV, the virus is highly lymphotropic, causing profound leucopenia and a general immunosuppression.

The mortality rates vary enormously (anywhere from 10% to 90%) depending on the species infected, their ages and the prevalence of secondary infectious agents. Goats have been reported as more susceptible than sheep (Nanda *et al.*, 1996), but this has not been confirmed in other outbreaks. West African goats have been found to be more susceptible than European varieties, and, within the former group, the dwarf varieties were most susceptible to disease (Couacy-Hymann *et al.*, 2007). Cattle and pigs can be infected with the virus, but this is only detected by the presence of specific antibodies while the animals show no clinical signs (Nawathe and Taylor, 1979; Anderson and McKay, 1994). Wildlife is again a complicating factor, with infection reported in Dorcas gazelle, Thompson's gazelle, gemsbok and ibex (Wohlsein and Saliki, 2006).

Virus entry

The mode of entry of RPV and PPRV into the host is a matter of some speculation. The virus is excreted from the infected animals in tears, saliva, faeces and in the ocular and nasal discharges common in disease. Transmission to new hosts is assumed to be through contact with such material and its introduction to the hosts' mucosal membranes, since there is no evidence for these viruses being able to penetrate skin. This means that the first contact of the virus with the host will be with epithelial cells of the mucosal tissues. Although the virus is found in both lymphoid and epithelial tissues during infection, many wild type RPV isolates, like wild-type MV isolates, show a strong dependence on the presence of SLAM/CD150 for cell entry (Tatsuo *et al.*, 2000, 2001; Baron, 2005). SLAM is not expressed on epithelial cells and wild type viruses show relatively poor entry to such cells. In addition, experiments with MV have shown that it has preferential entry to the basolateral side of epithelial cells in an intact epithelium (Sinn *et al.*, 2002; Ludlow *et al.*, 2010), not the apical side with which it would first come into contact. A mutant MV that cannot infect epithelial cells is infectious via the oro-nasal route *in vivo*, but is not excreted (Leonard *et al.*, 2008), showing that initial infection is not of the epithelial cells, but that infection of such cells is necessary for excretion of the virus and spread to a new host. One possibility is that the viruses rely on mucosal dendritic cells or other lymphatic cells in the respiratory airway or gut for entry (Ludlow *et al.*, 2009). These cells have processes that penetrate the epithelial layer, presenting part of their cell surface to the gut lumen. These processes allow the cells to sample the antigenic pool in the gut, taking up material which can be put into the normal antigen presenting mechanism and thereby eliciting an early immune response to a potential invading pathogen. These cells may be the first targets of RPV and other morbilliviruses. Once these cells have encountered a pathogen, they begin to mature and migrate naturally to the lymphoid centres. By this mechanism, RPV or PPRV-infected cells could be transported to where they would encounter a large number of suitable host cells in which to replicate. The progeny virus would enter the blood where it would have access to the basolateral surface of epithelia. This scenario explains much of what we observe in the progress of infection in RPV and PPRV disease, but remains to be proved absolutely.

Immune responses

As mentioned above under *V protein* and *C protein*, RPV and PPRV both have mechanisms to control the initial innate immune response. Virulent RPV infection of bovine dendritic cells elicits a very limited interferon response with few changes observed in the transcription of host cell genes (Nanda *et al.*, 2009), while the tissue culture-adapted vaccine strain of RPV induced higher levels of interferon (Nanda *et al.*, 2009). The C proteins of these two strains showed correspondingly high and low abilities to block the induction of transcription from the IFN-β promoter (Boxer *et al.*, 2009). There is also an inverse correlation between virulence and interferon induction in RPV infection *in vivo* (Hussain *et al.*, 1982) as well as a direct correlation between virulence and virus replication and spread (Wohlsein *et al.*, 1995). These findings suggest that the ability to inhibit the induction of type 1 interferons is associated with greater replication and spread of the virus inside the host and consequent greater pathogenicity. A similar correlation between virulence and low interferon induction has been suggested for wild-type MV strains (Naniche *et al.*, 2000), although the degree of inhibition of interferon induction may not be comparable between the two viruses, since wild-type MV induces significantly more IFN than wild-type RPV (Nanda *et al.*, 2009).

The relative importance of humoral (antibody) and cell-mediated (cytotoxic T cell) responses in recovery from infection with RPV or PPRV is not clear. Animals that recover from infection (including infection with the attenuated vaccine strains of RPV or PPRV) have high levels of circulating neutralizing antibody as well as antigen-specific proliferating CD4$^+$ T cells (Dhar *et al.*, 1995; Lund *et al.*, 2000), so either or both responses may be involved in the clearance of virus. Historically, an early form of vaccination was to inject cattle simultaneously with serum from recovered animals and with infectious material (Taylor *et al.*, 2006), so serum antibody on its own can be protective. The nature of such antibody, or the way such antibody is induced, seems to be important; the various forms of vaccine developed against rinderpest did not always show a correlation between development of neutralizing antibody and protection (see below). Evidence for the importance of the humoral antibody response comes from experiments with a rabbit-adapted strain of RPV, which causes a mild but clear disease in rabbits; clearance of virus and recovery from infection in this model was unaffected by thymectomy and anti-thymocyte antiserum, both of which treatments reduced cell-mediated immune responses (Kobune *et al.*, 1976), suggesting that antibody-mediated protection was sufficient. Mononuclear cells, which would include macrophages and dendritic cells, were also shown to be important for recovery from infection in the rabbit model (Yamanouchi, 1980). So far no system in which only cell-mediated immune responses were present has been tested in RPV or PPRV infections.

The defensive immune response of the host to RPV or PPRV infection is, in any event, obscured to some extent by the generalized immunosuppression common to infections with any of the morbilliviruses (reviewed in Schneider-Schaulies and Schneider-Schaulies, 2008). This is not just the marked lymphodepletion seen in severe disease, since morbillivirus-induced immunosuppression persists after the recovery of peripheral blood white cell count. Peripheral blood leucocytes from animals infected with virulent RPV show reduced response to mitogens and to both RPV and other, unrelated, antigens (Lund *et al.*, 2000), although the vaccine strain of RPV only had a transient immunosuppressive effect (Heaney *et al.*, 2005). Similar results were seen in PPRV-infected goats (Rajak *et al.*, 2005) as well as in rabbits infected with the lapinized strain of RPV (Yamanouchi *et al.*, 1974a,b). Cells from naïve animals (cattle or goats) are inhibited from proliferation by exposure to UV-inactivated RPV or PPRV, or to cells expressing both the RPV glycoproteins (Heaney *et al.*, 2002). This proliferation inhibition is not a general contact-mediated growth arrest, but is specific for lymphoid cells (Heaney *et al.*, 2002).

Diagnosis

For most of its history, diagnosis of rinderpest has had to be based on the clinical signs of disease (traditionally this was 'The three Ds – Death, Discharge, Diarrhoea'), with the only method of confirmation the infection of naïve animals with tissue homogenates from a suspected case. The first laboratory tests were based on variations of the 'precipitin' (agar gel immunodiffusion) test (White, 1958), which depended on the quality of the antiserum used, but which was robust and needed little in the way of equipment. It was thus appropriate for most of the countries in which it needed to be used. The reproducibility of this test improved with the development of standard techniques for the production of hyperimmune sera in rabbits. At the time of the discovery of PPRV, reliable differentiation of RPV and PPRV was dependent on virus isolation in tissue culture and reinfection of animals, as polyclonal antisera (e.g. from animals that had recovered from disease, or generated in rabbits) showed strong cross-reactivity between the two viruses in precipitin assays, indirect ELISAs, or virus neutralization tests. This antigenic similarity can be made use of, as it leads to cross-protection between all the morbilliviruses. This enabled the use of the Plowright RPV vaccine in many parts of the world in the 1970s and 1980s to help control PPR outbreaks, since the RPV vaccine effectively protected sheep and goats from PPRV (Taylor, 1979; Mariner *et al.*, 1993).

The introduction of monoclonal antibodies (mAbs) made it possible not only to have unlimited amounts of standardized reagents, but to have reagents with complete specificity for one or the other virus. These antibody tests have subsequently been complemented with variations on the polymerase chain reaction (PCR), a process that again provides the capacity for complete discrimination between RPV and PPRV as well as subsequently providing information about the epidemiological origin of the virus through determining the sequence of the PCR product.

MAb-based tests are primarily ELISAs, either *immunocapture* (icELISA) or *competitive* (cELISA). These two tests are complementary, the former being a test for the presence of viral antigen (in samples from acutely diseased animals) and the latter a test for the presence of specific antibodies in sera (useful for blood samples from animals that have recovered from disease, and for serum surveillance for virus prevalence as well as determination of efficiency of vaccination programmes). In each case the specificity of the mAbs used is critical.

In an icELISA, viral antigen is captured in a 96-well plate by a polyclonal antiserum or a first mAb bound to the plate. The captured antigen is then detected with a second mAb, usually coupled to horseradish peroxidase (HRP) or biotin for subsequent detection. The specificity is ensured by that of the second mAb. The first icELISA was used by Saliki (1994) for identifying PPRV and used a polyclonal capture antibody. A subsequent combined test using mAbs for both capture and detection was developed by Libeau and co-workers (Libeau and Lefevre, 1990; Libeau *et al.*, 1994). This test showed complete discrimination of RPV and PPRV and good sensitivity. It also works on samples that have been stored at room temperature (i.e. when the virus infectivity would have been destroyed), and this is now considered a definitive diagnostic test by the OIE. It has the advantages of simplicity and rapidity and is still widely used for PPR diagnosis.

An extension of the principles of the icELISA can be found in the pen-side test developed for RPV diagnosis towards the end of the eradication campaign (Bruning *et al.*, 1999; Wambura *et al.*, 2000). These were based on the chromatographic strip test technology ('Clearview™') most widely used in human pregnancy tests. In the RPV test, a liquid test sample (e.g. ocular or nasal discharge material from a suspect animal) is placed in the sample window of the device where it is absorbed into a pad of material connected to a nitrocellulose membrane strip. In the pad, any viral antigen is bound by virus-specific mAb coupled to blue latex beads. These beads flow along a nitrocellulose membrane as liquid from the sample is wicked along the nitrocellulose. As they flow, they pass over two strips of immobilized mAbs. The first is a second virus-specific mAb, which holds any beads that have already bound virus antigen. A blue band here indicates a positive test. The second strip has a generic anti-mouse IgG mAb where the remaining beads are captured, providing a signal that the test has functioned correctly. The sensitivity of the RPV test was approximately 1000 $TCID_{50}$; field tests showed it to be more sensitive than icELISA, but less sensitive than PCR, for individual animals. Such a device has benefits in greatly speeding the diagnosis of infected premises and hence helping in the control of disease spread; a similar test for PPRV is currently under development.

Diagnosis of samples in reference laboratories is now based almost entirely on PCR, after copying the viral genome to cDNA using reverse transcriptase (RT-PCR). Early RT-PCR-based methods suffered from the problems of sequence variability in RNA viruses, as the more conserved a target sequence the more likely there was to be cross-reaction with another morbillivirus. It was usual to use several primer pairs, based on conserved sequences in several genes, for diagnosis (Forsyth and Barrett, 1995). The availability of PCR-derived material allowed the determination of sections of genome sequence and hence sequence-based epidemiology (see below), an important tool in the analysis of geographical movement of both RPV and PPRV. Gel-based PCR has since been replaced with real-time PCR owing to the increased sensitivity this technology supplies. Detection of around 10 copies of genome is common. Real-time PCR tests for PPRV (Bao *et al.*, 2008; Kwiatek *et al.*, 2010) have been developed based on the N protein sequence. Both of these tests showed clear and specific detection of PPRV from multiple geographically and temporally disparate isolates, without detection of RPV, although one did cross-react with MV. The possibility that a strain of PPRV will appear that has a base change that prevents detection remains and it may be necessary to run both tests to be sure of negative results.

For animals that recover from disease, viral antigen or genome cannot usually be detected in samples more than a week after the initial rise in temperature, and determination of whether animals have been infected with RPV or PPRV requires analysis of the presence or absence of virus-specific antibodies in the serum. For the detection of antibodies to virus, the most common test is the cELISA. In this case, viral antigen is bound to a 96-well plate and incubated with the serum sample to be tested followed by incubation with a virus-specific mAb. The amount of bound mAb is then quantitated by incubating with a HRP-coupled anti-mouse IgG followed by a colorimetric enzyme reaction. Virus specific antibodies in the serum sample inhibit mAb binding and so reduce the colour produced. The assay has the advantage that the same assay works for serum samples from any species. All kinds of livestock as well as various related wildlife species can be screened for virus-specific antibodies using a single assay kit. The test depends on finding a mAb against an epitope specific for one virus or the other, so that polyclonal anti-PPRV sera are relatively ineffective in displacing the specific anti-RPV mAb and vice versa. Specific cELISAs for RPV (Anderson *et al.*, 1990c) and PPRV (Anderson and McKay, 1994; Libeau *et al.*, 1995) have been developed and used in the field.

Control in the field

Virus transmission is essentially by contact with infected animals, or fresh secretions or faeces from infected animals. The viruses do not survive for any extended period in normal environments, being susceptible to heat, UV light, dehydration, and high and low pH. Transfer of the viruses across any distance is through the movement of livestock, usually through migration or trade. Historically, rinderpest was a disease associated with war, as cattle were moved as people fled during military campaigns, or were moved with armies as sources of food for the soldiers. In current times, introduction of disease into new areas is almost always associated with the trade of livestock, although the spread of PPR may be being assisted by transmission through wildlife, linking otherwise separate areas of livestock grazing. The first level of control is therefore the restriction of movement of animals from endemic areas, with rigorous quarantine and surveillance procedures if a total ban is not practical.

Control in endemic areas, leading ultimately to the eradication of RPV, has been based on homologous attenuated virus vaccines. The early history of RPV vaccine development has already been described (see *Propagation*). The tissue culture-adapted 'Plowright' vaccine was used throughout Africa, the Middle East and the Indian sub-continent for the various control and eradication programmes. The Plowright vaccine has never been recorded as causing disease or reverting to virulence in use; the immune response generated appears to be protective for the life of the animal (up to 11 years) (Plowright and Taylor, 1967; Rweyemamu *et al.*, 1974; Bansal and Joshi, 1979; Plowright, 1984), although another trial found protection lasted less than 3 years (Provost *et al.*, 1969)). Some animals with low levels of neutralizing antibody showed transient viraemia after challenge, but no clinical signs were seen. Various improvements and optimizations of freeze-drying procedures (Mariner *et al.*, 1990a) meant that the vaccine could be distributed even into areas where no cold-chain existed (Mariner *et al.*, 1990b). The use of the vaccine, coupled with extensive serum surveillance to establish the true coverage of the vaccination programmes (and subsequently to validate the absence of RPV from national herds once vaccination had been stopped (a requirement for the certification of a country as 'rinderpest-free')), were among the key elements that led to the success of GREP.

Similar tissue culture-adapted viruses were developed in the then Soviet Union in the 1960s and 70s, but the first (derived from the Japanese lapinized virus) was withdrawn owing to adverse reactions, while the second, derived from an Iranian isolate of RPV, was suspected of reverting to virulence on several occasions (Taylor *et al.*, 2006).

A number of alternative RPV vaccines were developed that, although they were never used in the field, provided useful information about immune responses that could be protective against morbillivirus infections as well as about the immune responses in cattle. Recombinant vaccinia viruses expressing one or both of the surface glycoproteins of RPV were made by several groups (Yilma *et al.*, 1988; Belsham *et al.*, 1989; Tsukiyama *et al.*, 1989; Giavedoni *et al.*, 1991; Yamanouchi *et al.*, 1993). Animals vaccinated with these constructs were protected from virulent challenge; while they did not always show RPV neutralizing antibody in a standard assay, such a test only detects some of the anti-RPV humoral antibody that may have been involved in protection. These recombinants did not provide as long-lasting protection as the attenuated homologous vaccine, with clinical signs and even disease being seen at challenge after two or three years (Ohishi *et al.*, 2000), although a double recombinant expressing both F and H proteins gave complete protection up to 16 months after vaccination (Verardi *et al.*, 2002). Interestingly, while all of the recombinant vaccinia viruses expressing a surface glycoprotein were protective, a similar construct expressing the RPV N protein was not (Ohishi *et al.*, 1999). In this case, T-cells from infected animals proliferated in response to the RPV N protein and anti-N antibodies were found, but not neutralizing antibodies, and vaccinated animals were not protected from virulent RPV challenge. These observations suggest that (a) the surface glycoproteins are the main target for protective responses and (b) humoral antibody binding to the surface of the virus, or the surface of virus-infected cells, may be the main protective response.

An interesting variant on the recombinant vaccinia was the use of a vaccine strain of the sheep pathogen capripox virus, known in cattle as lumpy skin disease virus (LSDV). Like vaccinia, this is a pox virus, and recombinant versions were made which expressed either the H or F glycoproteins of RPV (Romero *et al.*, 1993, 1994a,b). These double vaccines protected animals against both LSDV and RPV although, like the vaccinia recombinants, the protection against rinderpest was of limited duration (Ngichabe *et al.*, 2002).

Individual viral glycoproteins were also prepared in insect cells using baculovirus; on their own these proteins were not effective vaccines (Bassiri *et al.*, 1993), although the RPV H protein did elicit a protective immune response in cattle if first packaged into an ISCOM (Kamata *et al.*, 2001). None of these vaccines was ever used in the field, partly because of safety concerns over the use of vaccinia in areas where there was widespread HIV-induced immunosuppression, and partly because of the excellent track record of the Plowright vaccine (any new vaccine had to provide very significant advantages to justify using it instead of the existing effective vaccine). The main driving force for alternative vaccines was the desire for a vaccine which would allow differentiation between vaccinated and previously infected animals based on the serum antibody profile (something which cannot be done using the homologous attenuated RPV vaccine), a so-called DIVA vaccine (differentiation of infected from vaccinated animals). The major humoral antibody responses in RPV infection or vaccination are to the N and H proteins, which is why those have been the targets of the diagnostic ELISAs in production. Animals treated with a recombinant pox virus vaccine that gave antibodies to the H protein could be distinguished from infected animals by the absence of anti-N antibodies.

The development of reverse genetics for RPV allowed the creation of a number of modified versions of the vaccine strain (Das *et al.*, 2000; Walsh *et al.*, 2000; Mahapatra *et al.*, 2006; Parida *et al.*, 2007) that were potential DIVA vaccines. These recombinant RPVs either expressed extra antigenic proteins or replaced RPV proteins with their PPRV equivalent (RPV/PPRV chimeras). These chimeric viruses have the advantage of eliciting a unique combination of anti-RPV and anti-PPRV antibodies, but these developments came too late to be used in RPV control.

An attenuated tissue culture-adapted strain of PPRV was derived by Diallo and Taylor (1989) and has been used extensively in Africa and the Middle East to try to suppress outbreaks. At least three similar PPRV vaccine candidates were developed in India (Saravanan *et al.*, 2010). While all these vaccines are effective, a coordinated or systematic approach to eradication of PPRV has not been established. Double vaccines similar to the RPV-lumpy skin disease recombinant construct have been made in which the PPRV glycoproteins are expressed from recombinant capripox virus (Diallo *et al.*, 2002; Berhe *et al.*, 2003; Chen *et al.*, 2010). As in the case of the RPV–LSDV constructs, these vaccines could be used as DIVA vaccines, and may have an important use in an organized control programme. The use of chimeric recombinant PPRV might be possible, although no system for making recombinant PPRV has been published. It will almost certainly not be possible to use RPV-PPRV chimeras, as the use of any virus containing RPV sequences (other than emergency stocks of RPV vaccine) will be restricted to high-containment laboratories once RPV eradication is formally announced and we enter a post-rinderpest world.

Genetics and reverse genetics

Reverse genetics, the production of live virus from a cDNA copy of the virus genome, was relatively easy to establish for RPV (Baron and Barrett, 1997) using similar methods as were used for the first negative strand virus 'rescue', rabies virus (Schnell *et al.*, 1994). Many of the applications of this technique have already been described above. Curiously, up to now the same procedures have failed to produce a rescue system for PPRV, although a number of RPV-PPRV chimeric viruses have been produced based on the RPV genome backbone (Das *et al.*, 2000; Mahapatra *et al.*, 2006; Parida *et al.*, 2007), and a functional PPRV minigenome replication system has been produced (Bailey *et al.*, 2007). We have used gene swaps between the vaccine strain of RPV and the virulent strain from which it was derived to show that the stability of the vaccine is due to the large number and wide distribution of attenuating mutations in the vaccine strain (Banyard *et al.*, 2005; Baron *et al.*, 2005). Similar work using the rabbit model of RPV, in which the normal RPV vaccine strain is also attenuated, have shown that virulence could be restored by replacing the H and P genes with those from the pathogenic lapinized RPV (Yoneda *et al.*, 2002, 2004). Gene swaps between tissue-culture-adapted and wild-type virus helped confirm the natural and cell-culture receptors for RPV (Baron, 2005), while knockout viruses lacking the expression of one or other non-structural protein and P gene swaps between virulent and non-virulent RPV strains have also been helpful in determining the functions of the V and C proteins (Baron and Barrett, 2000; Nanda and Baron, 2006; Boxer *et al.*, 2009).

The genetic analysis of RPV and PPRV isolates has largely focused on phylogenetic analysis based on short stretches of sequence. The early systems for diagnosis of RPV or PPRV using PCR generated products of length around 300bp, partly because this was about the length that could be reliably sequenced using normal gel-based sequencing and reading from the PCR primers. Although it is now easy to generate much longer stretches of sequence data, the phylogenetic comparison with older isolates will still be based only on the ~300 bases of data we have for those isolates. We will continue to use these shorter sections until we have time to go back and resequence all the old viruses.

There are no serotypes of RPV and PPRV, but there are clear lineages of both viruses based on their geographic source (Chamberlain *et al.*, 1993; Shaila *et al.*, 1996; Dhar *et al.*, 2002). Illustrative phylogenetic trees (unrooted) are shown in Fig. 9.12. These data are based on specific 322 nucleotide sections of the viral F genes for each virus; Libeau and co-workers use a similar section of the N gene for lineage determination in PPRV with essentially the same results for most virus isolates (Kwiatek *et al.*, 2010). The utility of lineage identification lies in the information it provides on the probable origin of the virus causing a new outbreak. Discovery that an outbreak of RPV in wildlife in Kenya was caused by a virus of a lineage (African lineage 2) previously thought to have died out (Barrett *et al.*, 1998) led to searches for previously hidden endemic sources of virus in the region, utilizing local livestock keepers to acquire primary epidemiological data (Mariner and Roeder, 2003). This, in turn, led to a greater realization of the need for better epidemiological intelligence and targeted vaccination based on that information (Mukhopadhyay *et al.*, 1999). Knowledge of the distribution of the two African lineages led to mathematical models of their transmission (Mariner *et al.*, 2005). More recently, discovery of RPV in samples from Iran, long after RPV was thought to have been eradicated there, was found not to be a concern when the samples were shown to be most closely related to the vaccine strain (compare *Iran/09* with *Kabete(vaccine)* in Fig. 9.12A), and were therefore laboratory contaminants. PPRV lineage identification proved that the outbreak in Morocco in 2007, the first time the virus had been seen in North Africa, was a lineage 4 virus, and therefore had not come from West Africa, where all the viruses are lineage 3, but from the Middle East or Turkey, where lineage 4 PPRV strains circulate (Fig. 9.12B).

Origins of PPRV

The classical course of the diseases caused by RPV and PPRV in their respective hosts are very

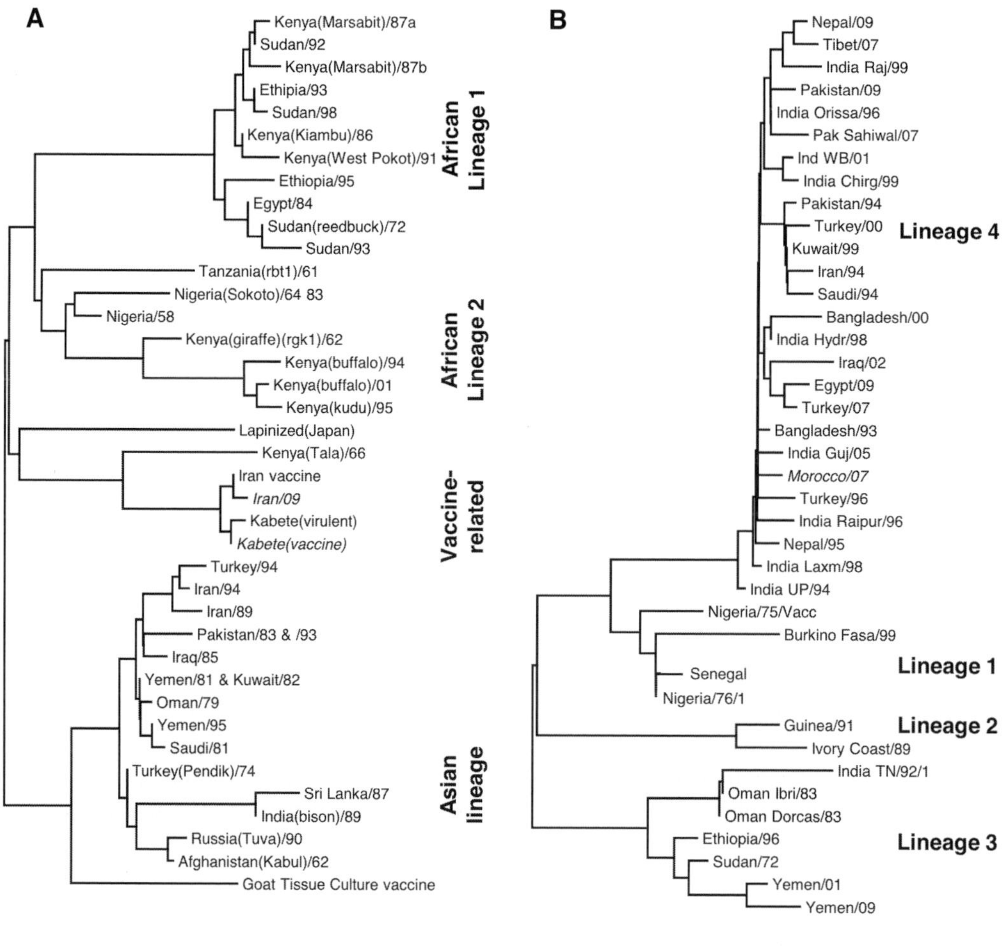

Figure 9.12 Example phylogenetic trees based on the F gene sequences of (A) *Rinderpest virus* (RPV) and (B) *Peste de petit ruminants virus* (PPRV), illustrating the lineages for these two viruses. Phylogenetic trees were derived using the program MEGA4 as described in the legend for Fig. 9.4.

similar, with similar clinical signs and mortality rates. Because of this similarity, it is likely that, in the past, many cases of PPR were ascribed to RPV since diagnostic tests that discriminated between the two viruses only became available in the last 20 years. There are historical records of outbreaks of 'rinderpest' in sheep or goats that strangely did not cause disease in local cattle (outbreaks that occurred at times before the identification of PPR as a separate disease). This confusion has meant that we do not truly know the origin of PPRV, nor how long it has been circulating in different parts of the world. The phylogenetic tree of the morbilliviruses (Fig. 9.5) shows that both RPV and PPRV have diverged a similar distance from their most recent common ancestor; assuming that the sequence drift rate is similar for both viruses (i.e. both mutate according to the same evolutionary 'clock'), PPRV has been in circulation for as long as RPV. Examination of the detailed genome sequences of the two viruses highlights another reason for considering that PPRV has been circulating for at least as long as RPV. All the morbilliviruses have to have genomes that are a multiple of six bases long (see above under *Genome Organization*), and hence any gain or loss of sequence has to be a multiple of six. The only differences in length between RPV and MV, viruses thought to have diverged within the last 1300 years (Furuse *et al.*, 2010) are indeed multiples of six (Fig. 9.6). The total difference in length of RPV and PPRV genomes is, as expected, a multiple of six (66 bases). However, those 66 bases are made up of four differences in gene length at different points in the genome (Fig. 9.6), none of which are a multiple of six. The large losses of sequence in RPV relative to PPRV in the M gene 3′ UTR and F gene 5′ UTR are balanced by single base changes in the lengths of the F and H gene 3′ UTRs. It is not, of course, possible to say which virus is the older, and whether length differences are due to sequence gain or loss or a combination of both, but the pattern of sequence difference suggests a number of such events on the path between RPV and PPRV, some of which must have involved either simultaneous loss and gain in different regions or transition through a state where the virus genome was not a multiple-of-six in length. This would require a compensating mutation to restore efficient growth and there would be strong selection for such a mutation. Minigenome studies have shown that PPRV does not absolutely require a multiple of six for its genome length, though it is very much more efficient in replication when this state exists (Bailey *et al.*, 2007).

As to the origin of PPRV, comparison of the phylogenetic trees of RPV and PPRV shows an important similarity, in that each has a single lineage encompassing all of Asia and the Middle East (RPV Lineage 3, PPRV Lineage 4), with two or three lineages covering Africa. Historically, it is known that RPV originated (had its main focus) in eastern parts of Asia (Spinage, 2003), and the African lineages have been assumed to originate in separate introductions of the virus to that continent followed by evolution from these genetic choke points. If PPRV originated in West Africa, where it was originally identified, and was then introduced into different parts of Asia in the last 100 years or so, it is unlikely that the Asian lineage would be as homogeneous as it is. In fact, the lineage pattern of PPRV suggests that this virus was circulating and becoming homogeneous in Asia for centuries, and has only recently been introduced into Africa, where several African lineages have developed.

Perspectives

Eradication of rinderpest – lessons learned

It is difficult to imagine, for those of us brought up in the second half of the twentieth century, the impact that smallpox had on human society until the advent of reliable vaccination. In the same way, with no serious outbreaks of rinderpest for more than 30 years, and much of the virus circulating even then being the mild strains, we probably find it difficult to conceive of the impact the disease had on society. For this reason, it is important that the eradication of rinderpest (the formal announcement of which is expected at around the time of publication of this book) is celebrated for the achievement and the benefit that it is, the second viral disease permanently (we hope) removed from the world. It is important also that the experience and knowledge of those who were actively involved in the eradication process is captured and passed on so that it can be applied to other such programmes. In committing themselves not just to control of the disease, but to its eradication, the FAO Expert Consultation meeting held in 1992 determined that the factors favouring the possibility of eradication were: an excellent vaccine; the availability of simple and effective diagnostic tests; a short infectious period (no persistence or carrier state); transmission only by close contact; a restricted geographical distribution; an economic incentive for local and national participation and for compliance with the programme (Rweyemamu *et al.*, 2006). Against those factors had to be set the possibility of disruption by civil strife, or the weakness of local veterinary services, since the GREP was above all a programme, co-ordinated by FAO and OIE, but carried out with local resources. In contrast, smallpox eradication was a campaign, with very strong, centrally funded, international control (Fenner, 1982). In carrying out the programme, the importance of epidemiologically targeted vaccination, extensive and continued surveillance to ensure the continued absence of disease and the preparation of proper contingency plans for re-emergence or re-introduction of the virus were found to be important, as was a time-limited programme which put pressure on countries to establish clearly that they were free of disease by a certain time (2010). This last point was important, as it required countries to take the risk of ceasing vaccination, as this was the only way to be sure that all disease was gone and it was not simply being contained or hidden by large scale vaccination.

If we consider the situation with PPRV, it is clear that the eradication of this disease is as achievable as it was for RPV. There are good diagnostics, as good a vaccine and the disease characteristics are nearly identical. The major difference for some time has been the economic incentive: sheep and goats are of much less individual value than cattle, and have a shorter working life, which means a faster turnover of animals, and a consequent need for more vaccine and people to administer it in the field. The ever advancing spread of PPRV has made the economic imperative for coordinated action stronger, and a recent document from FAO stated as its conclusion 'There are good reasons to believe that the eradication of PPR is an achievable goal like rinderpest' (FAO Animal Production and Health Division, 2009). Control programmes and eventual eradication would be improved by the development of a DIVA vaccine and associated test, which would allow serum surveillance for disease even in areas undergoing vaccination to control the spread of the disease as well as enabling specific screening for vaccination as part of trade controls. Work in the near future is being focused on developing and testing these tools.

Host specificity of pathogenesis

One area of study that has become increasingly important with the eradication of RPV, the hoped-for eradication of MV and the possible eradication of PPRV, is the factor(s) determining species specificity of disease. We know that these viruses all use the same receptor and have similar cell and tissue specificity within the hosts in which they cause disease. There is some indication that there are slight preferences of viruses for the SLAM receptor from their normal host (Tatsuo *et al.*, 2001), but this would not be sufficient to restrict the growth of an RNA virus since such viruses have little or no error correction associated with their RdRPs. Consequently, they have a high mutation rate and hence a high capability to scan all available mutations that might provide an improved ability to enter cells. In any event, as described above under 'Propagation', wild-type strains of RPV have a variable, but intrinsic, ability to enter all kinds of cells using heparan sulfate as a receptor. Most of the morbilliviruses have shown very little specificity for the cell types they will grow in, so that RPV grows well in bovine cells, human cells, monkey cells, hamster cells, etc., and this is reflected in the observation that several of the morbilliviruses have been shown to replicate in hosts other than the ones with which they are associated, and do so enough to elicit an immune response, but not enough to cause disease (e.g. Warren *et al.*, 1960; DeLay *et al.*, 1965; Provost *et al.*, 1967; Yamanouchi *et al.*, 1969; Norrby and Appel, 1980). After several blind passages, RPV was transformed from a cattle-virulent virus with no effect in rabbits to one that causes a severe disease in rabbits, but is highly attenuated in cattle (Nakamura *et al.*, 1938; Nakamura *et al.*, 1955), showing that there is no absolute barrier to the virus moving into a new species, no completely missing gene, as has been found to restrict the host range of some DNA viruses (e.g. Brooks *et al.*, 1995).

The eradication of RPV, and possibly MV and PPRV, and cessation of vaccination will lead to the respective host populations eventually having little or no acquired immunity to this group of viruses, while other related viruses such as CDV remain in circulation. There is concern that the loss of herd immunity may leave a window of opportunity for other animal morbilliviruses to enter the human population and cause a zoonosis, or to enter the cattle population and cause a socially and economically devastating pandemic.

The factors governing host-range restriction of these (and other related) viruses are areas of considerable importance. Proper risk assessment of the changing immune status of livestock and humans requires a thorough understanding of the mechanism of infection and pathogenesis of these viruses, and their interactions with their respective host cells, to enable reasonable estimation of the likelihood of various outcomes. It is clear that host organisms have evolved a number of systems to combat the invasion of pathogens. Many of these systems are only poorly understood, particularly outside the human and mouse field. The one area where we have found very clear differences between the morbilliviruses is in the mechanisms with which they control innate immune responses (see above under 'V protein' and 'C protein'), and there is some evidence that this may have an effect on the actual transcriptional response to infection with different viruses (Nanda *et al.*, 2009). We need to know to what extent the barrier to trans-species jumps is purely within the virus (in which case there is a high risk of mutations arising which would allow the virus to adapt to a new host) or if that barrier is the result of real differences in the defence systems of the different hosts, so that adaptation of a virus to the defence systems of one host limits the likelihood of the virus evolving out of what is in effect a local optimum in the viable mutation space. It is clear that we need to continue our efforts to understand not only the viruses themselves, but also the innate and adaptive immune systems of livestock species and the way that they differ from our own.

References

Alkhatib, G., Roder, J., Richardson, C., Briedis, D., Weinberg, R., Smith, D., Taylor, J., Paoletti, E., and Shen, S.H. (1994a). Characterization of a cleavage mutant of the measles virus fusion protein defective in syncytium formation. J. Virol. *68*, 6770–6774.

Alkhatib, G., Shen, S.H., Briedis, D., Richardson, C., Massie, B., Weinberg, R., Smith, D., Taylor, J., Paoletti, E., and Roder, J. (1994b). Functional analysis of N-linked glycosylation mutants of the measles virus fusion protein synthesized by recombinant vaccinia virus vectors. J. Virol. *68*, 1522–1531.

Anderson, D.E., and von Messling, V. (2008). Region between the canine distemper virus M and F genes modulates virulence by controlling fusion protein expression. J. Virol. *82, 10*,510–10518.

Anderson, E.C., Hassan, A., Barrett, T., and Anderson, J. (1990a). Observations on the pathogenicity for sheep and goats and the transmissibility of the strain of virus isolated during the Rinderpest outbreak in Sri-Lanka in 1987. Vet. Microbiol. *21*, 309–318.

Anderson, E.C., Jago, M., Mlengeya, T., Timms, C., Payne, A., and Hirji, K. (1990b). A serological survey of rinderpest antibody in wildlife and sheep and goats in northern Tanzania. Epidemiol Infect. *105*, 203–214.

Anderson, J., and McKay, J.A. (1994). The detection of antibodies against peste des petits ruminants virus in cattle, sheep and goats and the possible implications to rinderpest control programmes. Epidemiol. Infect. *112*, 225–231.

Anderson, J., Barreet, T., and and Scott, G.R. (1997). Manual on the Diagnosis of Rinderpest. 2nd edn. (FAO, Rome).

Anderson, J., McKay, J.A., and Butcher, R.N. (1990c). The use of monoclonal antibodies in competitive ELISA for the detection of antibodies to rinderpest and peste des petits ruminants viruses. Paper presented at: Seromonitoring of Rinderpest throughout Africa: Phase One, Proceedings of the Final ReseArch. Coordination Meeting of the IAEA Rinderpest Control Projects, Cote d'Ivoire, 19–23 November 1990 IAEA-TECDOC-623 (Vienna, International Atomic Energy Agency).

Andrejeva, J., Childs, K.S., Young, D.F., Carlos, T.S., Stock, N., Goodbourn, S., and Randall, R.E. (2004). The V proteins of paramyxoviruses bind the IFN-inducible RNA helicase, mda-5, and inhibit its activation of the IFN-beta promoter. Proc. Natl. Acad. Sci. U.S.A. *101*, 17264–17269.

Bailey, D., Banyard, A., Dash, P., Ozkul, A., and Barrett, T. (2005). Full genome sequence of peste des petits ruminants virus, a member of the *Morbillivirus* genus. Virus Res. *110*, 119–124.

Bailey, D., Chard, L.S., Dash, P., Barrett, T., and Banyard, A.C. (2007). Reverse genetics for peste-des-petits-ruminants virus (PPRV): promoter and protein specificities. Virus Res. *126,* 250–255.

Bankamp, B., Horikami, S.M., Thompson, P.D., Huber, M., Billeter, M., and Moyer, S.A. (1996). Domains of the measles virus N protein required for binding to P protein and self-assembly. Virology *216,* 272–277.

Bankamp, B., Wilson, J., Bellini, W.J., and Rota, P.A. (2005). Identification of naturally occurring amino acid variations that affect the ability of the measles virus C protein to regulate genome replication and transcription. Virology *336,* 120–129.

Bansal, R.P., and Joshi, R.C. (1979). Immunogenicity of tissue-culture rinderpest vaccine. Indian J. Anim. Sci. *49,* 260–265.

Banyard, A.C., Baron, M.D., and Barrett, T. (2005). A role for virus promoters in determining the pathogenesis of Rinderpest virus in cattle. J. Gen. Virol. *86,* 1083–1092.

Bao, J., Li, L., Wang, Z., Barrett, T., Suo, L., Zhao, W., Liu, Y., Liu, C., and Li, J. (2008). Development of one-step real-time RT-PCR assay for detection and quantitation of peste des petits ruminants virus. J. Virol. Methods *148,* 232–236.

Baron, M.D. (2005). Wild-type rinderpest virus uses SLAM (CD150) as its receptor. J. Gen. Virol. *86,* 1753–1757.

Baron, M.D., and Barrett, T. (1995). Sequencing and analysis of the nucleocapsid (N) and polymerase (L) genes and the terminal extragenic domains of the vaccine strain of rinderpest virus. J. Gen. Virol. *76,* 593–602.

Baron, M.D., and Barrett, T. (1997). Rescue of rinderpest virus from cloned cDNA. J. Virol. *71,* 1265–1271.

Baron, M.D., and Barrett, T. (2000). Rinderpest viruses lacking the C and V proteins show specific defects in growth and transcription of viral RNAs. J. Virol. *74,* 2603–2611.

Baron, M.D., Shaila, M.S., and Barrett, T. (1993). Cloning and sequence analysis of the phosphoprotein gene of rinderpest virus. J. Gen. Virol. *74,* 299–304.

Baron, M.D., Banyard, A.C., Parida, S., and Barrett, T. (2005). The Plowright vaccine strain of rinderpest virus has attenuating mutations in most genes. J. Gen. Virol. *86,* 1093–1101.

Barrett, T., and Underwood, B. (1985). Comparison of messenger RNAs induced in cells infected with each member of the *Morbillivirus* group. Virology *145,* 195–199.

Barrett, T., Forsyth, M.A., Inui, K., Wamwayi, H.M., Kock, R., Wambua, J., Mwanzia, J., and Rossiter, P.B. (1998). Rediscovery of the second African lineage of rinderpest virus: its epidemiological significance. Vet. Rec. *142,* 669–671.

Barrett, T., Pastoret, P.-P., and Taylor, W.P., eds. (2006). Rinderpest and Peste des Petits Ruminants. Virus Plagues of Large and Small Ruminants (Academic Press: London, Burlington, San Diego).

Bassiri, M., Ahmad, S., Giavedoni, L., Jones, L., Saliki, J.T., Mebus, C., and Yilma, T. (1993). Immunological responses of mice and cattle to baculovirus-expressed F and H proteins of rinderpest virus: lack of protection in the presence of neutralizing antibody. J. Virol. *67,* 1255–1261.

Bellini, W.J., Englund, G., Rozenblatt, S., Arnheiter, H., and Richardson, C.D. (1985). Measles virus P gene codes for two proteins. J. Virol. *53,* 908–919.

Belsham, G.J., Anderson, E.C., Murray, P.K., Anderson, J., and Barrett, T. (1989). Immune response and protection of cattle and pigs generated by a vaccinia virus recombinant expressing the F protein of rinderpest virus. Vet. Rec. *124,* 655–658.

Berhe, G., Minet, C., Le Goff, C., Barrett, T., Ngangnou, A., Grillet, C., G., L., Fleming, M., Black, D.N., and Diallo, A. (2003). Development of a dual recombinant vaccine to protect small ruminants against peste-des-petits-ruminant virus and capripoxvirus infections. J. Virol. *77,* 1571–1577.

Bhella, D., Ralph, A., and Yeo, R.P. (2004). Conformational flexibility in recombinant measles virus nucleocapsids visualised by cryo-negative stain electron microscopy and real-space helical reconstruction. J. Mol. Biol. *340,* 319–331.

Blixenkrone-Möller, M., Sharma, B., Varsanyi, T.M., Hu, A., Norrby, E., and Kovamees, J. (1992). Sequence analysis of the genes encoding the nucleocapsid protein and phosphoprotein (P) of phocid distemper virus, and editing of the P gene transcript. J. Gen. Virol. *73,* 885–893.

Blumberg, B.M., Crowley, J.C., Silverman, J.I., Mennona, J., Cook, S.D., and Dowling, P.C. (1988). Measles virus L protein evidences elements of ancestral RNA polymerase. Virology *164,* 487–497.

Blumberg, B.M., Chan, J., and Udem, S. (1991). Function of Paramyxovirus 3′ and 5′ end sequences in theory and practice. In The Paramyxoviruses, Kinsbury, D.W., ed. (Plenum Press: New York and London), pp. 235–247.

Bolt, G., Pedersen, I.R., and Blixenkrone-Moller, M. (1999). Processing of N-linked oligosaccharides on the measles virus glycoproteins: importance for antigenicity and for production of infectious virus particles. Virus Res. *61,* 43–51.

Bourhis, J.M., Receveur-Brechot, V., Oglesbee, M., Zhang, X., Buccellato, M., Darbon, H., Canard, B., Finet, S., and Longhi, S. (2005). The intrinsically disordered C-terminal domain of the measles virus nucleoprotein interacts with the C-terminal domain of the phosphoprotein via two distinct sites and remains predominantly unfolded. Protein Sci. *14,* 1975–1992.

Boxer, E.L., Nanda, S.K., and Baron, M.D. (2009). The rinderpest virus non-structural C protein blocks the induction of type 1 interferon. Virology *385,* 134–142.

Brooks, M.A., Ali, A.N., Turner, P.C., and Moyer, R.W. (1995). A rabbitpox virus serpin gene controls host range by inhibiting apoptosis in restrictive cells. J. Virol. *69,* 7688–7698.

Brown, D.D., Collins, F.M., Duprex, W.P., Baron, M.D., Barrett, T., and Rima, B.K. (2005a). 'Rescue' of mini-genomic constructs and viruses by combinations of *Morbillivirus* N, P and L proteins. J. Gen. Virol. *86,* 1077–1081.

Brown, D.D., Rima, B.K., Allen, I.V., Baron, M.D., Banyard, A.C., Barrett, T., and Duprex, W.P. (2005b). Rational attenuation of a *Morbillivirus* by modulating the activity of the RNA-dependent RNA polymerase. J. Virol. *79,* 14330–14338.

Bruning, A., Bellamy, K., Talbot, D., and Anderson, J. (1999). A rapid chromatographic strip test for the pen-side diagnosis of rinderpest virus. J. Virol. Methods *81,* 143–154.

Buckland, R., Gerald, C., Barker, R., and Wild, T.F. (1987). Fusion glycoprotein of measles virus: nucleotide sequence of the gene and comparison with other paramyxoviruses. J. Gen. Virol. *68,* 1695–1703.

Buckland, R., Malvoisin, E., Beauverger, P., and Wild, F. (1992). A leucine zipper structure present in the measles virus fusion protein is not required for its tetramerization but is essential for fusion. J. Gen. Virol. *73,* 1703–1707.

Caballero, M., Carabana, J., Ortego, J., Fernandez-Munoz, R., and Celma, M.L. (1998). Measles virus fusion protein is palmitoylated on transmembrane-intracytoplasmic cysteine residues which participate in cell fusion. J. Virol. *72,* 8198–8204.

Caignard, G., Guerbois, M., Labernardiere, J.L., Jacob, Y., Jones, L.M., Wild, F., Tangy, F., and Vidalain, P.O. (2007). Measles virus V protein blocks Jak1-mediated phosphorylation of STAT1 to escape IFN-alpha/beta signaling. Virology *368,* 351–362.

Caignard, G., Bourai, M., Jacob, Y., Tangy, F., and Vidalain, P.O. (2009). Inhibition of IFN-alpha/beta signaling by two discrete peptides within measles virus V protein that specifically bind STAT1 and STAT2. Virology *383,* 112–120.

Calain, P., and Roux, L. (1993). The rule of six, a basic feature for efficient replication of Sendai virus defective interfering RNA. J. Virol. *67,* 4822–4830.

Cathomen, T., Mrkic, B., Spehener, D., Drillien, R., Naef, R., Pavlovic, J., Aguzzi, A., Billeter, M.A., and Cattaneo, R. (1998a). A matrix-less measles virus is infectious and elicits extensive cell fusion: consequences for propagation in the brain. EMBO J. *17,* 3899.

Cathomen, T., Naim, H.Y., and Cattaneo, R. (1998b). Measles viruses with altered envelope protein cytoplasmic tails gain cell fusion competence. J. Virol. *72,* 1224–1234.

Cattaneo, R., Rebmann, G., Baczko, K., ter Meulen, V., and Billiter, M.A. (1987a). Altered ratios of measles virus transcripts in diseased human brains. Virology *160,* 523–526.

Cattaneo, R., Rebmann, G., Schmid, A., Bazko, K., Meulen, V.T., and Billeter, M. (1987b). Altered transcription of a defective measles virus genome derived from a diseased human brain. EMBO J. *6,* 681–688.

Cattaneo, R., Schmid, A., Eschle, D., Baczko, K., ter Meulen, V., and Billeter, M.A. (1988). Biased hypermutation and other genetic changes in defective measles viruses in human brain infections. Cell *55,* 255–265.

Cattaneo, R., Kaelin, K., Baczko, K., and Billeter, M.A. (1989a). Measles virus editing provides an additional cysteine-rich protein. Cell *56,* 759–764.

Cattaneo, R., Schmid, A., Spielhofer, P., Kaelin, K., Baczko, K., ter Meulen, V., Pardowitz, J., Flanagan, S., Rima, B.K., Udem, S.A., *et al.* (1989b). Mutated and hypermutated genes of persistent measles viruses which caused lethal human brain disease. Virology *173,* 415–425.

Cevik, B., Holmes, D.E., Vrotsos, E., Feller, J.A., Smallwood, S., and Moyer, S.A. (2004). The phosphoprotein (P) and L binding sites reside in the N-terminus of the L subunit of the measles virus RNA polymerase. Virology *327,* 297–306.

Chamberlain, R.W., Wamwayi, H.M., Hockley, E., Shaila, M.S., Goatley, L., Knowles, N.J., and Barrett, T. (1993). Evidence for different lineages of rinderpest virus reflecting their geographic isolation. J. Gen. Virol. *74,* 2775–2780.

Chattopadhyay, A., and Shaila, M.S. (2004). Rinderpest virus RNA polymerase subunits: mapping of mutual interacting domains on the large protein L and phosphoprotein *p*. Virus Genes *28,* 169–178.

Chen, W., Hu, S., Qu, L., Hu, Q., Zhang, Q., Zhi, H., Huang, K., and Bu, Z. (2010). A goat poxvirus-vectored peste-des-petits-ruminants vaccine induces long-lasting neutralization antibody to high levels in goats and sheep. Vaccine.

Childs, K., Stock, N., Ross, C., Andrejeva, J., Hilton, L., Skinner, M., Randall, R., and Goodbourn, S. (2007). Mda-5, but not RIG-I, is a common target for paramyxovirus V proteins. Virology *359,* 190–200.

Cocks, B.G., Chang, C.C., Carballido, J.M., Yssel, H., de Vries, J.E., and Aversa, G. (1995). A novel receptor involved in T-cell activation. Nature *376,* 260–263.

Colf, L.A., Juo, Z.S., and Garcia, K.C. (2007). Structure of the measles virus hemagglutinin. Nat. Struct. Mol. Biol. *14,* 1227–1228.

Colombo, M., Bourhis, J.M., Chamontin, C., Soriano, C., Villet, S., Costanzo, S., Couturier, M., Belle, V., Fournel, A., Darbon, H., *et al.* (2009). The interaction between the measles virus nucleoprotein and the Interferon Regulator Factor 3 relies on a specific cellular environment. Virol. J. *6,* 59.

Corey, E.A., and Iorio, R.M. (2007). Mutations in the stalk of the measles virus hemagglutinin protein decrease fusion but do not interfere with virus-specific interaction with the homologous fusion protein. J. Virol. *81,* 9900–9910.

Couacy-Hymann, E., Bodjo, C., Danho, T., Libeau, G., and Diallo, A. (2007). Evaluation of the virulence of some strains of peste-des-petits-ruminants virus (PPRV) in experimentally infected West African dwarf goats. Vet. J. *173,* 178–183.

Curran, J., Marq, J.B., and Kolakofsky, D. (1995). An N-terminal domain of the Sendai paramyxovirus P protein acts as a chaperone for the NP protein during the nascent chain assembly step of genome replication. J. Virol. *69,* 849–855.

Das, S.C., Baron, M.D., and Barrett, T. (2000). Recovery and characterization of a chimeric rinderpest virus with the glycoproteins of peste-des-petits-ruminants virus: homologous F and H proteins are required for virus viability. J. Virol. *74,* 9039–9047.

DeLay, P.D., Stone, S.S., Karzon, D.T., Katz, S., and Enders, J. (1965). Clinical and immune response of alien hosts to inoculation with measles, rinderpest, and canine distemper viruses. Am. J. Vet. Res. *26*, 1359–1373.

Devaux, P., and Cattaneo, R. (2004). Measles virus phosphoprotein gene products: conformational flexibility of the P/V protein amino-terminal domain and C protein infectivity factor function. J. Virol. *78*, 11632–11640.

Devaux, P., von Messling, V., Songsungthong, W., Springfeld, C., and Cattaneo, R. (2007). Tyrosine 110 in the measles virus phosphoprotein is required to block STAT1 phosphorylation. Virology *360*, 72–83.

Devaux, P., Hodge, G., McChesney, M.B., and Cattaneo, R. (2008). Attenuation of V- or C-defective measles viruses: infection control by the inflammatory and interferon responses of rhesus monkeys. J. Virol. *82*, 5359–5367.

Dhar, P., Joshi, R.C., and Bandyopadhyay, S.K. (1995). Humoral antibody response in animals infected with virulent rinderpest virus. Trop. Anim. Health Prod. *27*, 26–30.

Dhar, P., Sreenivasa, B.P., Barrett, T., Corteyn, M., Singh, R.P., and Bandyopadhyay, S.K. (2002). Recent epidemiology of peste des petits ruminants virus (PPRV). Vet. Microbiol. *88*, 153–159.

Diallo, A., Barrett, T., Lefevre, P.C., and Taylor W., P. (1987). Comparison of proteins induced in cells infected with rinderpest and peste des petits ruminants viruses. J. Gen. Virol. *68*, 2033–2038.

Diallo, A., Taylor, W.P., Lefevre, P.C., and Provost, A. (1989). Atténuation d'une souche de virus de la peste des petits ruminants: candidat pour un vaccin homologue. Rev. Elev. Med. Vet. Pays. Trop. *42*, 311–317.

Diallo, A., Minet, C., Berhe, G., Goff, C.L., Black, D.N., Fleming, M., Barrett, T., Grillet, C., and Libeau, G. (2002). Goat immune response to capripox vaccine expressing the hemagglutinin protein of peste des petits ruminants. Ann. N .Y. Acad. Sci. *969*, 88–91.

Duprex, W.P., Collins, F.M., and Rima, B.K. (2002). Modulating the function of the measles virus RNA-dependent RNA polymerase by insertion of green fluorescent protein into the open reading frame. J. Virol. *76*, 7322–7328.

Edington, A. (1899). A retrospect of the rinderpest campaign in South Africa. Lancet *i*, 357–359.

Edwards, J.T. (1930). The problem of rinderpest in India. Imp. Inst. Ag. Res. Pusa. Bull. *199*, 1–16.

Escoffier, C., Manié, S., Vincent, S., Muller, C.P., Billeter, M.A., and Gerlier, D. (1999). Nonstructural C protein is required for efficient measles virus replication in human peripheral blood cells. J. Virol. *73*, 1695–1698.

Evans, S.A., Belsham, G.J., and Barrett, T. (1990). The role of the 5′ nontranslated regions of the fusion protein mRNAs of canine distemper virus and rinderpest virus. Virology *177*, 317–323.

Evans, S.A., Baron, M.D., Chamberlain, R.W., Goatley, L., and Barrett, T. (1994). Nucleotide sequence comparisons of the fusion protein gene from virulent and attenuated strains of rinderpest virus. J. Gen. Virol. *75*, 3611–3617.

FAO Animal Production and Health Division (2009). Peste des Petits Ruminants (PPR). A Challenge for Small Ruminant Production (FAO: Rome).

Fearns, R., Peeples, M.E., and Collins, P.L. (1997). Increased expression of the N protein of respiratory syncytial virus stimulates minigenome replication but does not alter the balance between the synthesis of mRNA antigenome. Virology *236*, 188–201.

Fenner, F. (1982). A successful eradication campaign. Global eradication of smallpox. Rev. Infect. Dis. *4*, 916–930.

Ferron, F., Longhi, S., Henrissat, B., and Canard, B. (2002). Viral RNA-polymerases – a predicted 2′-O-ribose methyltransferase domain shared by all *Mononegavirales*. Trends Biochem. Sci. *27*, 222–224.

Fontana, J.M., Bankamp, B., Bellini, W.J., and Rota, P.A. (2008a). Regulation of interferon signaling by the C and V proteins from attenuated and wild-type strains of measles virus. Virology *374*, 71–81.

Fontana, J.M., Bankamp, B., and Rota, P.A. (2008b). Inhibition of interferon induction and signaling by paramyxoviruses. Immunol. Rev. *225*, 46–67.

Fooks, A.R., Stephenson, J.R., Warnes, A., Dowsett, A.B., Rima, B.K., and Wilkinson, G. (1993). Measles virus nucleocapsid protein expressed in insect cells assembles into nucleocapsid-like structures. J. Gen. Virol. *74*, 1439–1444.

Forsyth M., A., and Barrett, T. (1995). Evaluation of polymerase chain reaction for the detection and characterisation of rinderpest and peste des petits ruminants viruses for epidmiological studies. Virus Res. 151–163.

Furuse, Y., Suzuki, A., and Oshitani, H. (2010). Origin of measles virus: divergence from rinderpest virus between the 11th and 12th centuries. Virol. J. *7*, 52.

Gargadennec, L., and Lalanne, A. (1942). La peste des petits ruminants. Bull. Services Zoo Tech. Epizzoties Afrique Occidentale Francaise *5*, 16–21.

Ghosh, A., Nayak, R., and Shaila, M.S. (1996). Synthesis of leader RNA and editing of P mRNA during transcription by rinderpest virus. Virus Res. *41*, 69–76.

Giavedoni, L., Jones, L., Mebus, C., and Yilma, T. (1991). A vaccinia virus double recombinant expressing the F and H genes of rinderpest virus protects cattle against rinderpest and causes no pock lesions. Proc. Natl. Acad. Sci. U.S.A. *88*, 8011–8015.

Gibbs, E.P., Taylor, W.P., Lawman, M.J., and Bryant, J. (1979). Classification of peste des petits ruminants virus as the fourth member of the genus *Morbillivirus*. Intervirology *11*, 268–274.

Gilbert, Y., and Monnier, J. (1962). Adaptation du virus de la PPR aux cultures cellulaire. Rev. Elev. Med. Vet. Pays Trop. *15*, 321–335.

Giraudon, P., Jacquier, M.F., and Wild, T.F. (1988). Antigenic analysis of African measles virus field isolates: identification and localisation of one conserved and two variable epitope sites on the NP protein. Virus Res. *10*, 137–152.

Gitlin, L., Benoit, L., Song, C., Cella, M., Gilfillan, S., Holtzman, M.J., and Colonna, M. (2010). Melanoma differentiation-associated gene 5 (MDA5) is involved in the innate immune response to *Paramyxoviridae* infection *in vivo*. PLoS Pathog. *6,* e1000734.

Goodbourn, S., and Randall, R.E. (2009). The regulation of type I interferon production by paramyxoviruses. J. Interferon Cytokine Res. *29,* 539–547.

Govindarajan, R., Muralimanohar, B., Koteeswaran, A., Venugopalan, A.T., Varalakshmi, P., Shaila, M.S., and Ramachandran, S. (1996). Occurrence of rinderpest in European pigs in India. Vet. Rec. *139,* 473.

Haas, L., Baron, M.D., Liess, B., and Barrett, T. (1995). Editing of *Morbillivirus* P gene transcripts in infected animals. Vet. Microbiol. *44,* 299–306.

Hamdy, F.M., Dardiri, A.H., Nduaka, O., Breese, S.S., Jr., and Ihemelandu, E.C. (1976). Etiology of the stomatitis pneumoenteritis complex in Nigerian dwarf goats. Can. J. Comp. Med. *40,* 276–284.

Harty, R.N., and Palese, P. (1995). Measles virus phosphoprotein (P) requires the NH2- and COOH-terminal domains for interactions with the nucleoprotein (N) but only the COOH terminus for interactions with itself. J. Gen. Virol. *76,* 2863–2867.

Hasel, K.W., Day, S., Millward, S., Richardson, C.D., Bellini, W.J., and Greer, P.A. (1987). Characterization of cloned measles virus mRNAs by *in vitro* transcription, translation, and immunoprecipitation. Intervirology *28,* 26–39.

Hausmann, S., Garcin, D., Delenda, C., and Kolakofsky, D. (1999). The versatility of paramyxovirus RNA polymerase stuttering. J. Virol. *73,* 5568–5576.

Hausmann, S., Marq, J.B., Tapparel, C., Kolakofsky, D., and Garcin, D. (2008). RIG-I and dsRNA-induced IFNbeta activation. PLoS One *3,* e3965.

He, B., and Lamb, R.A. (1999). Effect of inserting paramyxovirus simian virus 5 gene junctions at the HN/L gene junction: analysis of accumulation of mRNAs transcribed from rescued viable viruses. J. Virol. *73,* 6228–6234.

He, B., Paterson, R.G., Stock, N., Durbin, J.E., Durbin, R.K., Goodbourn, S., Randall, R.E., and Lamb, R.A. (2002). Recovery of paramyxovirus simian virus 5 with a V protein lacking the conserved cysteine-rich domain: the multifunctional V protein blocks both interferon-beta induction and interferon signaling. Virology *303,* 15–32.

Heaney, J., Barrett, T., and Cosby, S.L. (2002). Inhibition of *in vitro* leukocyte proliferation by morbilliviruses. J. Virol. *76,* 3579–3584.

Heaney, J., Cosby, S.L., and Barrett, T. (2005). Inhibition of host peripheral blood mononuclear cell proliferation *ex vivo* by Rinderpest virus. J. Gen. Virol. *86,* 3349–3355.

von Heijne, G. (1983). Patterns of amino acids near signal-sequence cleavage sites. Eur J. Biochem *133,* 17–21.

Heminway, B.R., Yu, Y., and Galinski, M.S. (1994). Paramyxovirus mediated cell fusion requires coexpression of both the fusion and hemagglutinin-neuraminidase glycoproteins. Virus Res. *31,* 1–16.

Hirano, A., Wang, A.H., Gombart, A.F., and Wong, T.C. (1992). The matrix proteins of neurovirulent subacute sclerosing panencephalitis virus and its acute measles virus progenitor are functionally different. Proc. Natl. Acad. Sci. U.S.A. *89,* 8745–8749.

Hirano, A., Ayata, M., Wang, A.H., and Wong, T.C. (1993). Functional analysis of matrix proteins expressed from cloned genes of measles virus variants that cause subacute sclerosing panencephalitis reveals a common defect in nucleocapsid binding. J. Virol. *67,* 1848–1853.

Hirayama, N., Senda, M., Yamamoto, H., Yoshikawa, Y., and Yamanouchi, K. (1985). Isolation and characterization of canine distemper virus-specific RNA. Microbiol. Immunol. *29,* 47–54.

Hoffman, M.A., and Banerjee, A.K. (2000). Precise mapping of the replication and transcription promoters of human parainfluenza virus type 3. Virology *269,* 201–211.

Horikami, S.M., and Moyer, S.A. (1991). Synthesis of leader RNA and editing of the P messenger-RNA during transcription by purified measles virus. J. Virol. *65,* 5342–5347.

Horikami, S.M., Smallwood, S., Bankamp, B., and Moyer, S.A. (1994). An amino-proximal domain of the L protein binds to the P protein in the measles virus RNA polymerase complex. Virology *205,* 540–545.

Hornung, V., Ellegast, J., Kim, S., Brzozka, K., Jung, A., Kato, H., Poeck, H., Akira, S., Conzelmann, K.K., Schlee, M., *et al.* (2006). 5′-Triphosphate RNA is the ligand for RIG-I. Science *314,* 994–997.

Hu, A.Z., Kovamees, J., and Norrby, E. (1994). Intracellular processing and antigenic maturation of measles virus hemagglutinin protein. Arch. Virol. *136,* 239–253.

Hu, A.Z., Cathomen, T., Cattaneo, R., and Norrby, E. (1995). Influence of N-linked oligosaccharide chains on the processing, cell surface expression and function of the measles virus fusion protein. J. Gen. Virol. *76,* 705–710.

Huber, M., Cattaneo, R., Spielhofer, P., Örvell, C., Norrby, E., Messerli, M., Perriard, J.-C., and Billeter, M.A. (1991). Measles virus phosphoprotein retains the nucleocapsid protein in the cytoplasm. Virology *185,* 299–308.

Hussain, S.F., Akhtar, A.S., Rweyemamu, M.M., Kaminjolo, J.S., and Mugera, G.M. (1982). Studies on viral interference induced by rinderpest virus: 3. Interferon induction by attenuated and virulent strains of rinderpest virus in tissue culture. Bull. Anim. Health Prod. Afr. *30,* 1–6.

Ikegame, S., Takeda, M., Ohno, S., Nakatsu, Y., Nakanishi, Y., and Yanagi, Y. (2010). Both RIG-I and MDA5 RNA helicases contribute to the induction of alpha/beta interferon in measles virus-infected human cells. J. Virol. *84,* 372–379.

Isogai, S. (1944). On the rabbit virus inoculation as an active immunization methodagainst rinderpest for Mogolian cattle. Jpn. J. Vet. Sci. *6,* 388–390.

Iwasaki, M., Takeda, M., Shirogane, Y., Nakatsu, Y., Nakamura, T., and Yanagi, Y. (2009). The matrix protein of measles virus regulates viral RNA synthesis and assembly by interacting with the nucleocapsid protein. J. Virol. *83,* 10374–10383.

Johansson, K., Bourhis, J.M., Campanacci, V., Cambillau, C., Canard, B., and Longhi, S. (2003). Crystal structure of the measles virus phosphoprotein domain responsible for the induced folding of the C-terminal domain of the nucleoprotein. J. Biol. Chem. *278,* 44,567–44573.

Jordan, I.K., Sutter I.V., B.A., and McClure, M.A. (2000). Molecular evolution of the *Paramyxoviridae* and *rhabdoviridae* multiple-protein-encoding P gene. Mol. Biol. Evol *17,* 75–86.

Kamata, H., Ohishi, K., Hulskotte, E., Osterhaus, A.D., Inui, K., Shaila, M.S., Yamanouchi, K., and Barrett, T. (2001). Rinderpest virus (RPV) ISCOM vaccine induces protection in cattle against virulent RPV challenge. Vaccine *19,* 3355–3359.

Kato, A., Kiyotani, K., Hasan, M.K., Shioda, T., Sakai, Y., Yoshida, T., and Nagai, Y. (1999). Sendai virus gene start signals are not equivalent in reinitiation capacity: moderation at the fusion protein gene. J. Virol. *73,* 9237–9246.

Kato, H., Takeuchi, O., Sato, S., Yoneyama, M., Yamamoto, M., Matsui, K., Uematsu, S., Jung, A., Kawai, T., Ishii, K.J., *et al.* (2006). Differential roles of MDA5 and RIG-I helicases in the recognition of RNA viruses. Nature *441,* 101–105.

Kaushik, R., and Shaila, M.S. (2004). Cellular casein kinase II-mediated phosphorylation of rinderpest virus P protein is a prerequisite for its role in replication/transcription of the genome. J. Gen. Virol. *85,* 687–691.

Khera, K.S. (1958). Histological study on rinderpest: I Pathogenesis of rinderpest virus in lymphatic glands; II Histological lesions of the digestive tract; III. Lesions in various organs. Rev. Elev. Med. Vet. Pays. Trop. *11,* 399–420.

Kingsbury, D.W. (1974). The molecular biology of paramyxoviruses. Med. Microbiol. Immunol. *160,* 73–83.

Kingston, R.L., Baase, W.A., and Gay, L.S. (2004a). Characterization of nucleocapsid binding by the measles virus and mumps virus phosphoproteins. J. Virol. *78,* 8630–8640.

Kingston, R.L., Hamel, D.J., Gay, L.S., Dahlquist, F.W., and Matthews, B.W. (2004b). Structural basis for the attachment of a paramyxoviral polymerase to its template. Proc. Natl. Acad. Sci. U.S.A. *101,* 8301–8306.

Kobune, F., Chino, F., and Yamanouchi, K. (1976). Studies on recovery mechanism from rinderpest virus infection in rabbits. I. Effect of anti-thymocyte serum and thymectomy. Jpn J. Med. Sci. Biol. *29,* 265–275.

Kobune, F., Sakata, H., and Sugiura, A. (1990). Marmoset lymphoblastoid cells as a sensitive host for isolation of measles virus. J. Virol. *64,* 700–705.

Kobune, F., Sakata, H., Sugiyama, M., and Sugiura, A. (1991). B95a, a marmoset lymphoblastoid cell line, as a sensitive host for rinderpest virus. J. Gen. Virol. *72,* 687–692.

Kock, R.A., Wambua, J.M., Mwanzia, J., Wamwayi, H., Ndungu, E.K., Barrett, T., Kock, N.D., and Rossiter, P.B. (1999). Rinderpest epidemic in wild ruminants in Kenya 1993–97. Vet. Rec. *145,* 275–283.

Krishnamurti, R. (1945). A comparative study of Bull-Goat-Virus, with a brief survey of the work done with rinderpest goat virus in Madras. Indian J. Vet. Sci. Anim. Husb. Calcutta *15,* 247–252.Kwiatek, O., Keita, D., Gil, P., Fernandez-Pinero, J., Jimenez Clavero, M.A., Albina, E., and Libeau, G. (2010). Quantitative one-step real-time RT-PCR for the fast detection of the four genotypes of PPRV. J. Virol. Methods *165,* 168–177.

Lamb, R.A., and Kolakofsky, D. (1996). *Paramyxoviridae*: The viruses and their replication. In Field's Virology, Fields, B.N., Knipe, D.M., Howley, P.M., Chanock, R.M., Melnick, J.L., Monath, T.P., Roizman, B., and Straus, S.E., eds. (Raven Press: Philidelphia, PA), pp. 1177–1204.

Langedijk, J.P., Daus, F.J., and van Oirschot, J.T. (1997). Sequence and structure alignment of *Paramyxoviridae* attachment proteins and discovery of enzymatic activity for a *Morbillivirus* hemagglutinin. J. Virol. *71,* 6155–6167.

Latorre, P., Kolakofsky, D., and Curran, J. (1998). Sendai virus Y proteins are initiated by a ribosomal shunt. Mol. Cell Biol. *18,* 5021–5031.

Leclainche, E. (1955). Histoire illustrée de la médecine vétérinaire, vol 1 (Albin Michel: Paris).

Leonard, V.H., Sinn, P.L., Hodge, G., Miest, T., Devaux, P., Oezguen, N., Braun, W., McCray, P.B., Jr., McChesney, M.B., and Cattaneo, R. (2008). Measles virus blind to its epithelial cell receptor remains virulent in rhesus monkeys but cannot cross the airway epithelium and is not shed. J. Clin. Invest *118,* 2448–2458.

Libeau, G., and Lefevre, P.C. (1990). Comparison of rinderpest and peste des petits ruminants viruses using anti-nucleoprotein monoclonal antibodies. Vet. Microbiol. *25,* 1–16.

Libeau, G., Diallo, A., Colas, F., and Guerre, L. (1994). Rapid differential diagnosis of rinderpest and peste des petits ruminants using an immunocapture ELISA. Vet. Rec. *134,* 300–304.

Libeau, G., Prehaud, C., Lancelot, R., Colas, F., Guerre, L., Bishop, D.H., and Diallo, A. (1995). Development of a competitive ELISA for detecting antibodies to the peste des petits ruminants virus using a recombinant nucleoprotein. Res. Vet. Sci. *58,* 50–55.

Liebert, U.G., Baczko, K., Budka, H., and ter Meulen, V. (1986). Restricted expression of measles virus proteins in brains from cases of subacute sclerosing panencephalitis. J. Gen. Virol. *67,* 2435–2444.

Liston, P., and Briedis, D.J. (1994). Measles virus V protein binds zinc. Virology *198,* 399–404.

Liu, C., Erlichman, C., McDonald, C.J., Ingle, J.N., Zollman, P., Iankov, I., Russell, S.J., and Galanis, E. (2008). Heat shock protein inhibitors increase the efficacy of measles virotherapy. Gene Ther. *15,* 1024–1034.

Loney, C., Mottet-Osman, G., Roux, L., and Bhella, D. (2009). Paramyxovirus ultrastructure and genome packaging: cryo-electron tomography of sendai virus. J. Virol. *83,* 8191–8197.

Longhi, S., Receveur-Brechot, V., Karlin, D., Johansson, K., Darbon, H., Bhella, D., Yeo, R., Finet, S., and Canard, B. (2003). The C-terminal domain of the measles virus nucleoprotein is intrinsically disordered and folds upon binding to the C-terminal moiety of the phosphoprotein. J. Biol. Chem. *278,* 18638–18648.

Loo, Y.M., Fornek, J., Crochet, N., Bajwa, G., Perwitasari, O., Martinez-Sobrido, L., Akira, S., Gill, M.A., Garcia-Sastre, A., Katze, M.G., *et al.* (2008). Distinct RIG-I and MDA5 signaling by RNA viruses in innate immunity. J. Virol. *82,* 335–345.

Ludlow, M., Allen, I., and Schneider-Schaulies, J. (2009). Systemic spread of measles virus: overcoming the epithelial and endothelial barriers. Thromb. Haemost. *102,* 1050–1056.

Ludlow, M., Rennick, L.J., Sarlang, S., Skibinski, G., McQuaid, S., Moore, T., de Swart, R.L., and Duprex, W.P. (2010). Wild-type measles virus infection of primary epithelial cells occurs via the basolateral surface without syncytium formation or release of infectious virus. J. Gen. Virol. *91,* 971–979.

Lund, B.T., Tiwari, A., Galbraith, S., Baron, M.D., Morrison, W.I., and Barrett, T. (2000). Vaccination of cattle with attenuated rinderpest virus stimulates CD4(+) T-cell responses with broad viral antigen specificity. J. Gen. Virol. *81,* 2137–2146.

McAllister, C.S., Toth, A.M., Zhang, P., Devaux, P., Cattaneo, R., and Samuel, C.E. (2010). Mechanisms of protein kinase PKR-mediated amplification of beta interferon induction by C protein-deficient measles virus. J. Virol. *84,* 380–386.

McIlhatton, M.A., Curran, M.D., and Rima, B.K. (1997). Nucleotide sequence analysis of the large (L) genes of phocine distemper virus and canine distemper virus (corrected sequence). J. Gen. Virol. *78,* 571–576.

Mahapatra, M., Parida, S., Egziabher, B.G., Diallo, A., and Barrett, T. (2003). Sequence analysis of the phosphoprotein gene of peste des petits ruminants (PPR) virus: editing of the gene transcript. Virus Res. *96,* 85–98.

Mahapatra, M., Parida, S., Baron, M.D., and Barrett, T. (2006). Matrix protein and glycoproteins F and H of Peste-des-petits-ruminants virus function better as a homologous complex. J. Gen. Virol. *87,* 2021–2029.

Malvoisin, E., and Wild, T.F. (1993). Measles virus glycoproteins: studies on the structure and interaction of the haemagglutinin and fusion proteins. J. Gen. Virol. *74,* 2365–2372.

Manie, S.N., de Breyne, S., Vincent, S., and Gerlier, D. (2000). Measles virus structural components are enriched into lipid raft microdomains: a potential cellular location for virus assembly. J. Virol. *74,* 305–311.

Mariner, J.C., and Roeder, P.L. (2003). Use of participatory epidemiology in studies of the persistence of lineage 2 rinderpest virus in East Africa. Vet. Rec. *152,* 641–647.

Mariner, J.C., House, J.A., Sollod, A.E., Stem, C., van den Ende, M., and Mebus, C.A. (1990a). Comparison of the effect of various chemical stabilizers and lyophilization cycles on the thermostability of a Vero cell-adapted rinderpest vaccine. Vet. Microbiol. *21,* 195–209.

Mariner, J.C., van den Ende, M.C., House, J.A., Mebus, C.A., Salifou, S., and Stem, C. (1990b). The serological response to a thermostable Vero cell-adapted rinderpest vaccine under field conditions in Niger. Vet. Microbiol. *22,* 119–127.

Mariner, J.C., House, J.A., Mebus, C.A., and van den Ende, M.C. (1993). The use of thermostable Vero cell-adapted rinderpest vaccine as a heterologous vaccine against peste des petits ruminants. Res. Vet. Sci. *54,* 212–216.

Mariner, J.C., McDermott, J., Heesterbeek, J.A., Catley, A., and Roeder, P. (2005). A model of lineage-1 and lineage-2 rinderpest virus transmission in pastoral areas of East Africa. Prev. Vet. Med. *69,* 245–263.

Matsuoka, Y., Curran, J., Pelet, T., Kolakofsky, D., Ray, R., and Compans, R.W. (1991). The P gene of human parainfluenza virus type 1 encodes P and C proteins but not a cysteine-rich V protein. J. Virol. *65,* 3406–3410.

von Messling, V., Zimmer, G., Herrler, G., Haas, L., and Cattaneo, R. (2001). The hemagglutinin of canine distemper virus determines tropism and cytopathogenicity. J. Virol. *75,* 6418–6427.

Meyer, G., and Diallo, A. (1995). The nucleotide sequence of the fusion protein gene of the peste des petits ruminants virus – the long untranslated region in the 5′-end of the F-protein gene of morbilliviruses seems to be specific to each virus. Virus Res. *37,* 23–35.

Mioulet, V., Barrett, T., and Baron, M.D. (2001). Scanning mutagenesis identifies critical residues in the rinderpest virus genome promoter. J. Gen. Virol. *82,* 2905–2911.

Mirchamsy, H., Shafyi, A., and Bahrami, S. (1970). Use of Vero cells for titration of rinderpest virus and its neutralizing antibody. Appl. Microbiol. *19,* 545–546.

Moll, M., Klenk, H.D., and Maisner, A. (2002). Importance of the cytoplasmic tails of the measles virus glycoproteins for fusogenic activity and the generation of recombinant measles viruses. J. Virol. *76,* 7174–7186.

Moyer, S.A., Baker, S.C., and Horikami, S.M. (1990). Host cell proteins required for measles virus reproduction. J. Gen. Virol. *71,* 775–783.

Mukhopadhyay, A.K., Taylor, W.P., and Roeder, P.L. (1999). Rinderpest: a case study of animal health emergency management. Rev. Sci. Tech. *18,* 164–178.

Murphy, S.K., and Parks, G.D. (1999). RNA replication for the paramyxovirus simian virus 5 requires an internal repeated (CGNNNN) sequence motif. J. Virol. *73,* 805–809.

Muthuchelvan, D., Sanyal, A., Singh, R.P., Hemadri, D., Sen, A., Sreenivasa, B.P., Singh, R.K., and Bandyopadhyay, S.K. (2005). Comparative sequence analysis of the large polymerase protein (L) gene of peste-des-petits ruminants (PPR) vaccine virus of Indian origin. Arch. Virol. *150,* 2467–2481.

Naim, H.Y., Ehler, E., and Billeter, M.A. (2000). Measles virus matrix protein specifies apical virus release and glycoprotein sorting in epithelial cells. EMBO J. *19,* 3576–3585.

Nakamura, J., Wagatuma, S., and Fukusho, K. (1938). On the experimental infection with rinderpest virus in rabbits I. Some fundamental experiments. J. Jpn Soc. Vet. Sci. *17,* 185–204.

Nakamura, J., Kishi, S., Kiuchi, J., and Reisinger, R. (1955). An investigation of antibody response in cattle vaccinated with the rabbit-passaged LA rinderpest virus in Korea. Am. J. Vet. Res. *16,* 71–75.

Nakatsu, Y., Takeda, M., Ohno, S., Koga, R., and Yanagi, Y. (2006). Translational inhibition and increased interferon induction in cells infected with C protein-deficient measles virus. J. Virol. *80,* 11861–11867.

Nakatsu, Y., Takeda, M., Ohno, S., Shirogane, Y., Iwasaki, M., and Yanagi, Y. (2008). Measles virus circumvents the host interferon response by different actions of the C and V proteins. J. Virol. *82,* 8296–8306.

Nanda, S.K., and Baron, M.D. (2006). Rinderpest virus blocks type I and type II interferon action: role of structural and nonstructural proteins. J. Virol. *80,* 7555–7568.

Nanda, Y.P., Chatterjee, A., Purohit, A.K., Diallo, A., Innui, K., Sharma, R.N., Libeau, G., Thevasagayam, J.A., Bruning, A., Kitching, R.P., *et al.* (1996). The isolation of peste des petits ruminants virus from northern India. Vet. Microbiol. *51,* 207–216.

Nanda, S.K., Baron, J., Royall, E., Robinson, L., Falciani, F., and Baron, M.D. (2009). Infection of bovine dendritic cells by rinderpest or measles viruses induces different changes in host transcription. Virology *395,* 223–231.

Naniche, D., Yeh, A., Eto, D., Manchester, M., Friedman, R.M., and Oldstone, M.B.A. (2000). Evasion of host defenses by measles virus: wild-type measles virus infection interferes with the induction of alpha/beta interferon production. J. Virol. *74,* 7478–7484.

Nawathe, D.R., and Taylor, W.P. (1979). Experimental infection of domestic pigs with the virus of peste des petits ruminants. Trop. Anim. Health Prod. *11,* 120–122.

Ngichabe, C.K., Wamwayi, H.M., Ndungu, E.K., Mirangi, P.K., Bostock, C.J., Black, D.N., and Barrett, T. (2002). Long term immunity in African cattle vaccinated with a recombinant capripox-rinderpest virus vaccine. Epidemiol. Infect. *128,* 343–349.

Nishie, T., Nagata, K., and Takeuchi, K. (2007). The C protein of wild-type measles virus has the ability to shuttle between the nucleus and the cytoplasm. Microbes Infect. Institut. Pasteur *9,* 344–354.

Norrby, E., and Appel, M.J. (1980). Humoral immunity to canine distemper after immunization of dogs with inactivated and live measles virus. Arch. Virol. *66,* 169–177.

tenOever, B.R., Servant, M.J., Grandvaux, N., Lin, R., and Hiscott, J. (2002). Recognition of the measles virus nucleocapsid as a mechanism of IRF-3 activation. J. Virol. *76,* 3659–3669.

Oglesbee, M.J., Kenney, H., Kenney, T., and Krakowka, S. (1993). Enhanced production of *Morbillivirus* gene-specific RNAs following induction of the cellular stress response in stable persistent infection. Virology *192,* 556–567.

Oglesbee, M.J., Liu, Z., Kenney, H., and Brooks, C.L. (1996). The highly inducible member of the 70kDa family of heat shock proteins increases canine distemper virus polymerase activity. J. Gen. Virol. *77,* 2125–2135.

Ohishi, K., Inui, K., Yamanouchi, K., and Barrett, T. (1999). Cell-mediated immune responses in cattle vaccinated with a vaccinia recombinant expressing the nucleocapsid (N) protein of rinderpest virus (RPV). J. Gen. Virol. *80,* 1627–1634.

Ohishi, K., Inui, K., Barrett, T., and Yamanouchi, K. (2000). Long-term protective immunity to rinderpest in cattle following a single vaccination with a recombinant vaccinia virus expressing the virus haemagglutinin protein. J. Gen. Virol. *81,* 1439–1446.

Ohno, S., Ono, N., Takeda, M., Takeuchi, K., and Yanagi, Y. (2004). Dissection of measles virus V protein in relation to its ability to block alpha/beta interferon signal transduction. J. Gen. Virol. *85,* 2991–2999.

OIE (2008). Manual of Diagnostic Tests and Vaccines for Terrestrial Animals, Vol *2,* 6th edn. (Office International des Epizooties, Paris).

Ono, N., Tatsuo, H., Hidaka, Y., Aoki, T., Minagawa, H., and Yanagi, Y. (2001). Measles viruses on throat swabs from measles patients use signaling lymphocytic activation molecule (CDw150) but not CD46 as a cellular receptor. J. Virol. *75,* 4399–4401.

Palosaari, H., Parisien, J.P., Rodriguez, J.J., Ulane, C.M., and Horvath, C.M. (2003). STAT protein interference and suppression of cytokine signal transduction by measles virus V protein. J. Virol. *77,* 7635–7644.

Parida, S., Mahapatra, M., Kumar, S., Das, S.C., Baron, M.D., Anderson, J., and Barrett, T. (2007). Rescue of a chimeric rinderpest virus with the nucleocapsid protein derived from peste-des-petits-ruminants virus: use as a marker vaccine. J. Gen. Virol. *88,* 2019–2027.

Parks, C.L., Lerch, R.A., Walpita, P., Sidhu, M.S., and Udem, S.A. (1999). Enhanced measles virus cDNA rescue and gene expression after heat shock. J. Virol. *73,* 3560–3566.

Parks, C.L., Witko, S.E., Kotash, C., Lin, S.L., Sidhu, M.S., and Udem, S.A. (2006). Role of V protein RNA binding in inhibition of measles virus minigenome replication. Virology *348,* 96–106.

Pastoret, P.-P., Yamanouchi, K., Mueller-Doblies, U., Rweyemamu, M.M., Horzinek, M., and Barrett, T. (2006). Rinderpest – an old and worldwide story: history to c. 1902. In Rinderpest and Peste des Petits Ruminants, T. Barrett, P.-P. Pastoret, and W.P. Taylor, eds. (Academic Press: London, Burlington, San Diego).

Patterson, J.B., Thomas, D., Lewicki, H., Billeter, M.A., and Oldstone, M.B. (2000). V and C proteins of measles virus function as virulence factors *in vivo*. Virology *267,* 80–89.

Pfaller, C.K., and Conzelmann, K.K. (2008). Measles virus V protein is a decoy substrate for IkappaB kinase alpha and prevents Toll-like receptor 7/9-mediated interferon induction. J. Virol. *82,* 12365–12373.

Pichlmair, A., Schulz, O., Tan, C.P., Naslund, T.I., Liljestrom, P., Weber, F., and Reis e Sousa, C. (2006). RIG-I-mediated antiviral responses to single-stranded RNA bearing 5′-phosphates. Science *314,* 997–1001.

Pichlmair, A., Schulz, O., Tan, C.P., Rehwinkel, J., Kato, H., Takeuchi, O., Akira, S., Way, M., Schiavo, G., and Reis e Sousa, C. (2009). Activation of MDA5 requires higher-order RNA structures generated during virus infection. J. Virol. *83,* 10761–10769.

Plemper, R.K., Hammond, A.L., and Cattaneo, R. (2000). Characterization of a region of the measles virus hemagglutinin sufficient for its dimerization. J. Virol. *74,* 6485–6493.

Plemper, R.K., Hammond, A.L., and Cattaneo, R. (2001). Measles virus envelope glycoproteins hetero-oligomerize in the endoplasmic reticulum. J. Biol. Chem. *276,* 44239–44246.

Plemper, R.K., Hammond, A.L., Gerlier, D., Fielding, A.K., and Cattaneo, R. (2002). Strength of envelope protein interaction modulates cytopathicity of measles virus. J. Virol. *76,* 5051–5061.

Plowright, W. (1963). Some properties of strains of rinderpest virus recently isolated in E. Africa. Res. Vet. Sci. *4,* 96–108.

Plowright, W. (1965). Symposium: The smallest stowaways. 3. Rinderpest. Vet. Rec. *77,* 1431–1437.

Plowright, W. (1984). The duration of immunity in cattle following inoculation of rinderpest cell culture vaccine. J. Hyg (Lond.) *92,* 285–296.

Plowright, W., and Ferris, R.D. (1959a). Studies with rinderpest virus in tissue culture. I. Growth and cytopathogenicity. J. Comp. Pathol. *69,* 152–172.

Plowright, W., and Ferris, R.D. (1959b). Studies with rinderpest virus in tissue culture. II. Pathogenicity for cattle of culture-passaged virus. J. Comp. Pathol. *69,* 173–184.

Plowright, W., and Ferris, R.D. (1962). Studies with rinderpest virus in tissue culture. The use of attenuated culture virus as a vaccine for cattle. Res. Vet. Sci. *3,* 172–182.

Plowright, W., and Taylor, W.P. (1967). Long-term studies of the immunity in East African Cattle following inoculation with rinderpest culture vaccine. Res. Vet. Sci. *8,* 118–128.

Plowright, W., Cruickshank, J.G., and Waterson, A.P. (1962). The morphology of rinderpest virus. Virology *17,* 118–122.

Plumet, S., Herschke, F., Bourhis, J.M., Valentin, H., Longhi, S., and Gerlier, D. (2007). Cytosolic 5′-triphosphate ended viral leader transcript of measles virus as activator of the RIG I-mediated interferon response. PLoS ONE *2,* e279.

Poch, O., Sauvaget, I., Delarue, M., and Tordo, N. (1989). Identification of four conserved motifs among the RNA-dependent polymerase encoding elements. EMBO J. *8,* 3867–3874.

Poch, O., Blumberg, B.M., Bougueleret, L., and Tordo, N. (1990). Sequence comparison of five polymerases (L proteins) of unsegmented negative-strand RNA viruses: theoretical assignment of functional domains. J. Gen. Virol. *71,* 1153–1162.

Pohl, C., Duprex, W.P., Krohne, G., Rima, B.K., and Schneider-Schaulies, S. (2007). Measles virus M and F proteins associate with detergent-resistant membrane fractions and promote formation of virus-like particles. J. Gen. Virol. *88,* 1243–1250.

Poole, E., He, B., Lamb, R.A., Randall, R.E., and Goodbourn, S. (2002). The V proteins of simian virus 5 and other paramyxoviruses inhibit induction of interferon-beta. Virology *303,* 33–46.

Provost, A., Borredon, C., and Maurice, Y. (1967). [Antibovipestic protection conferred to cattle by a measles virus adapted to bovine cells]. C.R. Acad. Sci. Hebd. Seances Acad. Sci. D. *264,* 2961–2964.

Provost, A., Maurice, Y., and Borredon, C. (1969). [Clinical and immunologic behavior during rinderpest contamination of cattle vaccinated several years ago against rinderpest with cell culture vaccines]. Rev. Elev. Med. Vet. Pays Trop. *22,* 453–464.

Qanungo, K.R., Shaji, D., Mathur, M., and Banerjee, A.K. (2004). Two RNA polymerase complexes from vesicular stomatitis virus-infected cells that carry out transcription and replication of genome RNA. Proc. Natl. Acad. Sci. U.S.A. *101,* 5952–5957.

Rager, M., Vongpunsawad, S., Duprex, W.P., and Cattaneo, R. (2002). Polyploid measles virus with hexameric genome length. EMBO J. *21,* 2364–2372.

Raha, T., Chattopadhyay, A., and Shaila, M.S. (2004a). Development of a reconstitution system for Rinderpest virus RNA synthesis *in vitro*. Virus Res. *99,* 131–138.

Raha, T., Kaushik, R., and Shaila, M.S. (2004b). Phosphoprotein P of Rinderpest virus binds to plus sense leader RNA: regulation by phosphorylation. Virus Res. *104,* 191–200.

Rahaman, A., Srinivasan, N., Shamala, N., and Shaila, M.S. (2003). The fusion core complex of the peste des petits ruminants virus is a six-helix bundle assembly. Biochemistry (Mosc.) *42,* 922–931.

Rahaman, A., Srinivasan, N., Shamala, N., and Shaila, M.S. (2004). Phosphoprotein of the rinderpest virus forms a tetramer through a coiled coil region important for biological function. A structural insight. J. Biol. Chem. *279, 23,*606–23614.

Rajak, K.K., Sreenivasa, B.P., Hosamani, M., Singh, R.P., Singh, S.K., Singh, R.K., and Bandyopadhyay, S.K. (2005). Experimental studies on immunosuppressive effects of peste des petits ruminants (PPR) virus in goats. Comp Immunol. Microbiol. Infect. Dis. *28,* 287–296.

Ramachandran, A., Parisien, J.P., and Horvath, C.M. (2008). STAT2 is a primary target for measles virus V protein-mediated alpha/beta interferon signaling inhibition. J. Virol. *82,* 8330–8338.

Ramani, K., Charles, Y.S., Srinivas, R.P., Narayanaswamy, M., and Ramachandran, S. (1974). Isolation of rinderpest virus from an outbreak in domestic pigs in Karnataka. Indian Vet. J. *51,* 36–41.

Rassa, J.C., and Parks, G.D. (1998). Molecular basis for naturally occurring elevated readthrough transcription across the M–F junction of the paramyxovirus SV5. Virology *247,* 274–286.

Rehwinkel, J., Tan, C.P., Goubau, D., Schulz, O., Pichlmair, A., Bier, K., Robb, N., Vreede, F., Barclay, W., Fodor, E., *et al.* (2010). RIG-I detects viral genomic RNA during negative-strand RNA virus infection. Cell *140,* 397–408.

Rennick, L.J., Duprex, W.P., and Rima, B.K. (2007). Measles virus minigenomes encoding two autofluorescent proteins reveal cell-to-cell variation in reporter expression dependent on viral sequences between the transcription units. J. Gen. Virol. *88,* 2710–2718.

Reuter, T., Weissbrich, B., Schneider-Schaulies, S., and Schneider-Schaulies, J. (2006). RNA interference with measles virus N, P, and L mRNAs efficiently prevents and with matrix protein mRNA enhances viral transcription. J. Virol. *80,* 5951–5957.

Riedl, P., Moll, M., Klenk, H.D., and Maisner, A. (2002). Measles virus matrix protein is not cotransported with the viral glycoproteins but requires virus infection for efficient surface targeting. Virus Res. *83,* 1–12.

Robbins, S.J., and Bussell, R.H. (1979). Structural phosphoproteins associated with purified measles virions and cytoplasmic nucleocapsids. Intervirology *12,* 96–102.

Romero, C.H., Barrett, T., Evans, S.A., Kitching, R.P., Gershon, P.D., Bostock, C., and Black, D.N. (1993). Single capripoxvirus recombinant vaccine for the protection of cattle against rinderpest and lumpy skin disease. Vaccine *11,* 737–742.

Romero, C.H., Barrett, T., Chamberlain, R.W., Kitching, R.P., Fleming, M., and Black, D.N. (1994a). Recombinant capripoxvirus expressing the hemagglutinin protein gene of rinderpest virus: protection of cattle against rinderpest and lumpy skin disease viruses. Virology *204,* 425–429.

Romero, C.H., Barrett, T., Kitching, R.P., Carn, V.M., and Black, D.N. (1994b). Protection of cattle against rinderpest and lumpy skin disease with a recombinant capripoxvirus expressing the fusion protein gene of rinderpest virus. Vet. Rec. *135,* 152–154.

Rossiter, P.B., Herniman, K.A., and Wamwayi, H.M. (1992). Improved isolation of rinderpest virus in transforMed. bovine T lymphoblast cell lines. Res. Vet. Sci. *53,* 11–18.

Rossiter, P.B., Hussain, M., Raja, R.H., Moghul, W., Khan, Z., and Broadbent, D.W. (1998). Cattle plague in Shangri-La: observations on a severe outbreak of rinderpest in northern Pakistan 1994–1995. Vet. Rec. *143,* 39–42.

Rothlisberger, A., Wiener, D., Schweizer, M., Peterhans, E., Zurbriggen, A., and Plattet, P. (2010). Two domains of the V protein of virulent canine distemper virus selectively inhibit STAT1 and STAT2 nuclear import. J. Virol. Epub April 2010.

Runkler, N., Pohl, C., Schneider-Schaulies, S., Klenk, H.D., and Maisner, A. (2007). Measles virus nucleocapsid transport to the plasma membrane requires stable expression and surface accumulation of the viral matrix protein. Cell Microbiol. *9,* 1203–1214.

Rweyemamu, M.M., Reid, H.W., and Okuna, N. (1974). Observations on the behaviour of rinderpest virus in immune animals challenged intransally. Bull. Epizoot Dis. Afr. *22,* 1–9.

Rweyemamu, M.M., Roeder, P.L., and Taylor, W.P. (2006). Towards the global eradication of rinderpest. In Rinderpest and Peste des Petits Ruminants, Barrett, T., Pastoret, P.-P., and Taylor, W.P., eds. (Academic Press: London, Burlington, San Diego).

Saikia, P., Gopinath, M., and Shaila, M.S. (2008). Phosphorylation status of the phosphoprotein P of rinderpest virus modulates transcription and replication of the genome. Arch. Virol. *153,* 615–626.

Saitou, N., and Nei, M. (1987). The neighbor-joining method: a new method for reconstructing phylogenetic trees. Mol. Biol. Evol *4,* 406–425.

Salditt, A., Koethe, S., Pohl, C., Harms, H., Kolesnikova, L., Becker, S., and Schneider-Schaulies, S. (2010). Measles virus M protein driven particle production does not involve the ESCRT system. J. Gen Virol. *91,* 1464–1472.

Saliki, J.T., House, J.A., Mebus, C.A., and Dubovi, E.J. (1994). Comparison of monoclonal antibody-based sandwich enzyme-linked immunosorbent assay and virus isolation for detection of peste des petits ruminants virus in goat tissues and secretions. J. Clin. Microbiol. *32,* 1349–1353.

Santiago, C., Celma, M.L., Stehle, T., and Casasnovas, J.M. (2010). Structure of the measles virus hemagglutinin bound to the CD46 receptor. Nat. Struct. Mol. Biol. *17,* 124–129.

Saravanan, P., Sen, A., Balamurugan, V., Rajak, K.K., Bhanuprakash, V., Palaniswami, K.S., Nachimuthu, K., Thangavelu, A., Dhinakarraj, G., Hegde, R., *et al.* (2010). Comparative efficacy of peste des petits ruminants (PPR) vaccines. Biologicals *38,* 479–485.

Sato, T.A., Kohama, T., and Sugiura, A. (1988). Intracellular processing of measles virus fusion protein. Arch. Virol. *98,* 39–50.

Schein, H. (1926). Expériences sur la peste bovine. Bull. Soc. Path. Exot. *19,* 915–928.

Schneider-Schaulies, J., and Schneider-Schaulies, S. (2008). Receptor interactions, tropism, and mechanisms involved in *Morbillivirus*-induced immunomodulation. Adv. Virus Res. *71,* 173–205.

Schneider-Schaulies, S., Liebert, U.G., Baczko, K., Cattaneo, R., Billeter, M., and ter Meulen, V. (1989). Restriction of measles virus gene-expression in acute and subacute encephalitis of Lewis rats. Virology *171,* 525–534.

Schneider, H., Kaelin, K., and Billeter, M.A. (1997). Recombinant measles viruses defective for RNA editing and V protein synthesis are viable in cultured cells. Virology *227,* 314–322.

Schnell, M.J., Mebatsion, T., and Conzelmann, K.-K. (1994). Infectious rabies virus from cloned cDNA. EMBO J. *13,* 4195–4203.

Schoehn, G., Mavrakis, M., Albertini, A., Wade, R., Hoenger, A., and Ruigrok, R.W. (2004). The 12 A structure of trypsin-treated measles virus N-RNA. J. Mol. Biol. *339,* 301–312.

Scott, G.R. (1964). Rinderpest. Adv. Vet. Sci. *9,* 113–224.

Seth, S., and Shaila, M.S. (2001a). The fusion protein of Peste des petits ruminants virus mediates biological fusion in the absence of hemagglutinin-neuraminidase protein. Virology *289,* 86–94.

Seth, S., and Shaila, M.S. (2001b). The hemagglutinin-neuraminidase protein of peste des petits ruminants virus is biologically active when transiently expressed in mammalian cells. Virus Res. *75,* 169–177.

Shaila, M.S., Purushothaman, V., Bhavasar, D., Venugopal, K., and Venkatesan, R.A. (1989). Peste des petits ruminants of sheep in India. Vet. Rec. *125,* 602.

Shaila, M.S., Shamaki, D., Forsyth, M.A., Diallo, A., Goatley, L., Kitching, R.P., and Barrett, T. (1996). Geographic distribution and epidemiology of peste des petits ruminants virus. Virus Res. *43,* 149–153.

Shaji, D., and Shaila, M.S. (1999). Domains of rinderpest virus phosphoprotein involved in interaction with itself and the nucleocapsid protein. Virology *258,* 415–424.

Shanthikumar, S.R., and Atilola, M.A. (1990). Outbreaks of rinderpest in wild and domestic animals in Nigeria. Vet. Rec. *126,* 306–307.

Shope, R.E., Griffith, H.J., and Jenkins, D.L. (1946). The cultivation of rinderpest virus in developing hen's egg. Am. J. Vet. Res. *7 (Part II),* 135–141.

Sidhu, M.S., Chan, J., Kaelin, K., Spielhofer, P., Radecke, F., Schneider, H., Masurekar, M., Dowling, P.C., Billeter, M.A., and Udem, S.A. (1995). Rescue of synthetic measles virus minireplicons – measles genomic termini direct efficient expression and propagation of a reporter gene. Virology *208,* 800–807.

Sinn, P.L., Williams, G., Vongpunsawad, S., Cattaneo, R., and McCray, P.B., Jr. (2002). Measles virus preferentially transduces the basolateral surface of well-differentiated human airway epithelia. J. Virol. *76,* 2403–2409.

Sleeman, K., and Baron, M.D. (2005). The polymerase (L) protein of rinderpest virus interacts with the host cell protein striatin. Virology *332,* 225–234.

Spehner, D., Drillien, R., and Howley, P.M. (1997). The assembly of the measles virus nucleoprotein into nucleocapsid-like particles is modulated by the phosphoprotein. Virology *232,* 260–268.

Spinage, C.A. (2003). Cattle Plague: A History (Kluwer Academic/Plenum: New York, Boston, Dordrecht).

Stallcup, K.C., Raine, C.S., and Fields, B.N. (1983). Cytochalasin B inhibits the maturation of measles virus. Virology *124,* 59–74.

Suryanarayana, K., Baczko, K., ter Meulen, V., and Wagner, R.R. (1994). Transcription inhibition and other properties of matrix proteins expressed by M genes cloned from measles viruses and diseased human brain tissue. J. Virol. *68,* 1532–1543.

Sweetman, D.A., Miskin, J., and Baron, M.D. (2001). Rinderpest virus C and V proteins interact with the major (L) component of the viral polymerase. Virology *281,* 193–204.

Tahara, M., Takeda, M., and Yanagi, Y. (2005). Contributions of matrix and large protein genes of the measles virus edmonston strain to growth in cultured cells as revealed by recombinant viruses. J. Virol. *79,* 15218–15225.

Tahara, M., Takeda, M., and Yanagi, Y. (2007). Altered interaction of the matrix protein with the cytoplasmic tail of hemagglutinin modulates measles virus growth by affecting virus assembly and cell–cell fusion. J. Virol. *81,* 6827–6836.

Tajima, M., and Ushijima, T. (1971). The pathogenesis of rinderpest in the lymph nodes of cattle. Light and electron microscopic studies. Am. J. Pathol. *62,* 221–235.

Takeda, M., Ohno, S., Seki, F., Nakatsu, Y., Tahara, M., and Yanagi, Y. (2005). Long untranslated regions of the measles virus M and F genes control virus replication and cytopathogenicity. J. Virol. *79,* 14346–14354.

Takeuchi, K., Kadota, S.I., Takeda, M., Miyajima, N., and Nagata, K. (2003). Measles virus V protein blocks interferon (IFN)-alpha/beta but not IFN-gamma signaling by inhibiting STAT1 and STAT2 phosphorylation. FEBS Lett. *545,* 177–182.

Takeuchi, K., Takeda, M., Miyajima, N., Ami, Y., Nagata, N., Suzaki, Y., Shahnewaz, J., Kadota, S., and Nagata, K. (2005). Stringent requirement for the C protein of wild-type measles virus for growth both *in vitro* and in macaques. J. Virol. *79,* 7838–7844.

Tamura, K., Nei, M., and Kumar, S. (2004). Prospects for inferring very large phylogenies by using the neighbor-joining method. Proc. Natl. Acad. Sci. U.S.A. *101,* 11030–11035.

Tamura, K., Dudley, J., Nei, M., and Kumar, S. (2007). MEGA4: Molecular Evolutionary Genetics Analysis (MEGA) software version 4.0. Mol. Biol. Evol *24,* 1596–1599.

Tapparel, C., Maurice, D., and Roux, L. (1998). The activity of Sendai virus genomic and antigenomic promoters requires a second element past the leader template regions: a motif (GNNNNN)(3) is essential for replication. J. Virol. *72,* 3117–3128.

Tatsuo, H., Ono, N., Tanaka, K., and Yanagi, Y. (2000). SLAM (CDw150) is a cellular receptor for measles virus. Nature *406,* 893–897.

Tatsuo, H., Ono, N., and Yanagi, Y. (2001). morbilliviruses use signaling lymphocyte activation molecules (CD150) as cellular receptors. J. Virol. *75,* 5842–5850.

Taylor, W.P. (1979). Protection of goats against peste des petit ruminants with attenuated rinderpest virus. Res. Vet. Med. *27,* 321–324.

Taylor, W.P., and Abegunde, A. (1979). The isolation of peste des petits ruminants virus from Nigerian sheep and goats. Res. Vet. Sci. *26,* 94–96.

Taylor, W.P., Roeder, P.L., and Rweyemamu, M.M. (2006). History of vaccines and vaccination. In Rinderpest and Peste des Petits Ruminants, Barrett, T., Pastoret, P.-P., and Taylor, W.P., eds. (Academic Press: London, Burlington, San Diego).

Tober, C., Seufert, M., Schneider, H., Billeter, M.A., Johnston, I.C., Niewiesk, S., ter Meulen, V., and Schneider-Schaulies, S. (1998). Expression of measles virus V protein is associated with pathogenicity and control of viral RNA synthesis. J. Virol. *72,* 8124–8132.

Tordo, N., Poch, O., Ermine, A., Keith, G., and Rougeon, F. (1988). Completion of the rabies virus genome sequence determination: highly conserved domains among the L (polymerase) proteins of unsegmented negative-strand RNA viruses. Virology *165,* 565–576.

Toth, A.M., Devaux, P., Cattaneo, R., and Samuel, C.E. (2009). Protein kinase PKR mediates the apoptosis induction and growth restriction phenotypes of C protein-deficient measles virus. J. Virol. *83,* 961–968.

Tsukiyama, K., Yoshikawa, Y., Kamata, H., Imaoka, K., Asano, K., Funahashi, S., Maruyama, T., Shida, H., Sugimoto, M., and Yamanouchi, K. (1989). Development of heat-stable recombinant rinderpest vaccine. Arch. Virol. *107,* 225–235.

Vasconcelos, D., Norrby, E., and Oglesbee, M. (1998). The cellular stress response increases measles virus-induced cytopathic effect. J. Gen. Virol. *79,* 1769–1773.

Verardi, P.H., Aziz, F.H., Ahmad, S., Jones, L.A., Beyene, B., Ngotho, R.N., Wamwayi, H.M., Yesus, M.G., Egziabher, B.G., and Yilma, T.D. (2002). Long-term sterilizing immunity to rinderpest in cattle vaccinated with a recombinant vaccinia virus expressing high levels of the fusion and hemagglutinin glycoproteins. J. Virol. *76,* 484–491.

Vidal, S., Curran, J., and Kolakofsky, D. (1990a). Editing of the Sendai virus P/C mRNA by G insertion occurs during mRNA synthesis via a virus-encoded activity. J. Virol. *64,* 239–246.

Vidal, S., Curran, J., and Kolakofsky, D. (1990b). A stuttering model for paramyxovirus P mRNA editing. EMBO J. *9,* 2017–2022.

Vincent, S., Gerlier, D., and Manie, S.N. (2000). Measles virus assembly within membrane rafts. J. Virol. *74,* 9911–9915.

Walpita, P. (2004). An internal element of the measles virus antigenome promoter modulates replication efficiency. Virus Res. *100,* 199–211.

Walsh, E.P., Baron, M.D., Rennie, L.F., Monaghan, P., Anderson, J., and Barrett, T. (2000). Recombinant rinderpest vaccines expressing membrane-anchored proteins as genetic markers: evidence of exclusion of marker protein from the virus envelope. J. Virol. *74,* 10165–10175.

Wambura, P.N., Mollel, J.O., Moshy, D.W., Mbise, A.N., and Taylor, W.P. (1999). Rinderpest antibody detected in sheep and goats before an outbreak of rinderpest reported in cattle in northern Tanzania. Trop. Anim. Health Prod. *31,* 9–14.

Wambura, P.N., Moshy, D.W., Mbise, A.N., Mollel, G.O., Taylor, W.P., Anderson, J., and Bruning, A. (2000). Diagnosis of rinderpest in Tanzania by a rapid chromatographic strip-test. Trop. Anim. Health Prod. *32,* 141–145.

Wamwayi, H.M., Fleming, M., and Barrett, T. (1995). Characterisation of African isolates of rinderpest virus. Vet. Microbiol. *44,* 151–163.

Warnes, A., Fooks, A.R., Dowsett, A.B., Wilkinson, G.W.G., and Stephenson, J.R. (1995). Expression of the measles virus nucleoprotein gene in *Escherichia coli* and assembly of nucleocapsid-like structures. Gene *160,* 173–178.

Warren, J., Nadel, M.K., Slater, E., and Millian, S.J. (1960). The canine distemper-measles complex. I. Immune response of dogs to canine distemper and measles viruses. Am. J. Vet. Res. *21,* 111–119.

Watanabe, M., Hirano, A., Stenglein, S., Nelson, J., Thomas, G., and Wong, T.C. (1995). Engineered serine protease inhibitor prevents furin-catalyzed activation of the fusion glycoprotein and production of infectious measles virus. J. Virol. *69,* 3206–3210.

Weber, F., Wagner, V., Rasmussen, S.B., Hartmann, R., and Paludan, S.R. (2006). Double-stranded RNA is produced by positive-strand RNA viruses and DNA viruses but not in detectable amounts by negative-strand RNA viruses. J. Virol. *80,* 5059–5064.

White, G. (1958). A specific diffusible antigen of rinderpest virus demonstrated by the agar double-diffusion precipitation reaction. Nature *181,* 1409.

Wild, T.F., Malvoisin, E., and Buckland, R. (1991). Measles virus: both the haemagglutinin and fusion glycoproteins are required for fusion. J. Gen. Virol. *72,* 439–442.

Witko, S.E., Kotash, C., Sidhu, M.S., Udem, S.A., and Parks, C.L. (2006). Inhibition of measles virus minireplicon-encoded reporter gene expression by V protein. Virology *348,* 107–119.

Wohlsein, P., and Saliki, J. (2006). Rinderpest and peste des petits ruminants – the diseases: clinical signs and pathology. In Rinderpest and Peste des Petits Ruminants, Barrett, T., Pastoret, P.-P., and Taylor, W.P., eds. (Academic Press: London, Burlington, San Diego), pp. 68–85.

Wohlsein, P., Trautwein, G., Harder, T.C., Liess, B., and Barrett, T. (1993). Viral antigen distribution in organs of cattle experimentally infected with rinderpest virus. Vet. Pathol. *30,* 544–554.

Wohlsein, P., Wamwayi, H.M., Trautwein, G., Pohlenz, J., Liess, B., and Barrett, T. (1995). Pathomorphological and immunohistological findings in cattle experimentally infected with rinderpest virus isolates of different pathogenicity. Vet. Microbiol. *44,* 141–149.

Wong, T.C., and Hirano, A. (1987). Structure and function of bicistronic RNA encoding the phosphoprotein and matrix protein of measles virus. J. Virol. *61,* 584–589.

Wong, T.C., Ayata, M., Hirano, A., Yoshikawa, Y., Tsuruoka, H., and Yamanouchi, K. (1989). Generalized and localized biased hypermutation affecting the matrix gene of a measles virus strain that causes subacute sclerosing panencephalitis. J. Virol. *63,* 5464–5468.

Yamanouchi, K. (1980). Comparative aspects of pathogenicity of measles, canine distemper, and rinderpest viruses. Jpn J. Med. Sci. Biol. *33,* 41–66.

Yamanouchi, K., Fukuda, A., Kobune, F., Hayami, M., and Shishido, A. (1969). Serologic response in monkeys inoculated with rinderpest and measles viruses. Am. J. Vet. Res. *30,* 1831–1835.

Yamanouchi, K., Chino, F., Kobune, F., Fukuda, A., and Yoshikawa, Y. (1974a). Pathogenesis of rinderpest virus infection in rabbits I. Clinical signs, immune response, histological changes, and virus growth patterns. Infect. Immun. *9,* 199–205.

Yamanouchi, K., Fukuda, A., Kobune, F., Yoshikawa, Y., and Chino, F. (1974b). Pathogenesis of rinderpest virus infection in rabbits. II. Effect of rinderpest virus on the immune functions of rabbits. Infect. Immun. *9,* 206–211.

Yamanouchi, K., Inui, K., Sugimoto, M., Asano, K., Nishimaki, F., Kitching, R.P., Takamatsu, H., and Barrett, T. (1993). Immunisation of cattle with a recombinant vaccinia vector expressing the haemagglutinin gene of rinderpest virus. Vet. Rec. *132,* 152–156.

Yegambaram, K., and Kingston, R.L. (2010). The feet of the measles virus polymerase bind the viral nucleocapsid protein at a single site. Protein Sci. *19,* 893–899.

Yilma, T., Hsu, D., Jones, L., Owens, S., Grubman, M., Mebus, C., Yamanaka, M., and Dale, B. (1988). Protection of cattle against rinderpest with vaccinia virus recombinants expressing the HA or F gene. Science *242,* 1058–1061.

Yokota, S., Saito, H., Kubota, T., Yokosawa, N., Amano, K., and Fujii, N. (2003). Measles virus suppresses interferon-alpha signaling pathway: suppression of Jak1 phosphorylation and association of viral accessory proteins, C and V, with interferon-alpha receptor complex. Virology *306,* 135–146.

Yoneda, M., Bandyopadhyay, S.K., Shiotani, M., Fujita, K., Nuntaprasert, A., Miura, R., Baron, M.D., Barrett, T., and Kai, C. (2002). Rinderpest virus H protein: role in determining host range in rabbits. J. Gen. Virol. *83,* 1457–1463.

Yoneda, M., Miura, R., Barrett, T., Tsukiyama-Kohara, K., and Kai, C. (2004). Rinderpest virus phosphoprotein gene is a major determinant of species-specific pathogenicity. J. Virol. *78,* 6676–6681.

Yoshikawa, Y., Mizumoto, K., and Yamanouchi, K. (1986). Characterization of messenger RNAs of measles virus. J. Gen. Virol. *67,* 2807–2812.

Zhang, X., Glendening, C., Linke, H., Parks, C.L., Brooks, C., Udem, S.A., and Oglesbee, M. (2002). Identification and characterization of a regulatory domain on the carboxyl terminus of the measles virus nucleocapsid protein. J. Virol. *76,* 8737–8746.

Zhang, X., Bourhis, J.M., Longhi, S., Carsillo, T., Buccellato, M., Morin, B., Canard, B., and Oglesbee, M. (2005). Hsp72 recognizes a P binding motif in the measles virus N protein C-terminus. Virology *337,* 162–174.

Zhu, J., Zhang, C.W.-H., Qi, Y., Tien, P., and Gao, G.F. (2002). The fusion protein core of measles virus forms stable coiled-coil trimer. Biochem. Biophys. Res. Commun. *299,* 897–902.

Human Respiratory Syncytial Virus

10

Peter L. Collins

Abstract

Human respiratory syncytial virus (RSV) is a ubiquitous pathogen that infects essentially everyone worldwide during infancy and early childhood and is a leading cause of paediatric hospitalization for respiratory disease. RSV also is a frequent cause of less severe disease in healthy adults and is an important cause of morbidity and mortality in the elderly and in severely immunosuppressed individuals. RSV is an enveloped non-segmented negative strand RNA virus classified in the *Paramyxovridae* family, and its genome organization is one of the more complex of this family. The genome includes: two separate genes encoding type I and type III interferon (IFN) antagonists (NS1 and NS2); a gene (M2) with two open reading frames encoding novel proteins (M2-1 and M2-2) involved in RNA synthesis; and a glycoprotein G that has a number of unusual features, including high sequence variability, heavy glycosylation, cytokine mimicry, and a shed form that helps the virus evade neutralizing antibodies. RSV is able to efficiently infect and cause disease in very young infants, with the peak of hospitalization at 2–3 months of age, despite the presence of maternally derived virus-neutralizing serum antibodies. RSV has a single serotype but is able to re-infect symptomatically throughout life without the need for significant antigenic change, although immunity from prior infection reduces disease. It is widely thought that re-infection is due to an ability of RSV to inhibit or subvert the host immune response, but this remains largely speculative. The development of an effective vaccine or specific antiviral therapy against RSV is considered a high priority, but these goals remain unfulfilled. RSV is notable for a historic vaccine failure: a formalin-inactivated RSV vaccine that was evaluated in infants and children in the 1960s was poorly protective and paradoxically primed for enhanced RSV disease upon subsequent natural RSV infection. However, RSV also is notable because of the development of a successful strategy for passive immunoprophylaxis of infants at high risk for serious RSV disease using an RSV-neutralizing monoclonal antibody (MAb).

History

Human respiratory syncytial virus (RSV) was first isolated in 1956 from a laboratory chimpanzee with disease resembling the common cold (Morris *et al.*, 1956). Soon after, the same virus was isolated from children with croup and pneumonia (Chanock *et al.*, 1957). Epidemiological studies quickly showed that RSV is a ubiquitous paediatric pathogen of worldwide distribution. It is considered to be the most important viral agent of serious respiratory tract disease in infants and children worldwide. Subsequent studies showed that RSV symptomatically infects individuals of all ages, and can cause severe disease and substantial mortality in severely immunocompromised individuals and the elderly. RSV was dubbed 'syncytial' because it induces syncytia in cell culture. However, this turns out to be somewhat of a misnomer, because other paramyxoviruses also readily produce syncytia in cell culture, and syncytia usually are not a prominent feature of RSV pathology *in vivo*.

RSV has a bovine relative, bovine RSV (BRSV), which is highly related. It is tempting to speculate that human RSV began as a zoonotic infection of humans with BRSV (or the converse), perhaps as a consequence of increased species interaction due to the domestication of bovines. RSV exists as a single serotype with two antigenic subgroups A and B. The divergence of the two subgroups has been estimated to have occurred approximately 350 years ago (Zlateva *et al.*, 2005).

Infectious agent

Classification

The classification of RSV is illustrated by comparison with other selected members of family *Paramyxoviridae* (order *Mononegavirales*) by alignment of gene maps (Fig. 10.1) and phylogenetic analysis of complete genome sequences (Fig. 10.2). *Paramyxoviridae* has two subfamilies, *Pneumovirinae* and *Paramyxovirinae*. RSV is a member of subfamily *Pneumovirinae*, which has two genera: *Pneumovirus*, comprising RSV and its animal relatives BRSV and pneumonia virus of mice (PVM); and *Metapneumovirus*, comprising human *Metapneumovirus* (HMPV) and avian *Metapneumovirus* (AMPV). The other subfamily, *Paramyxoviridae*, has five genera containing viruses of animals and birds (these are shown in Fig. 10.1) and two newly created genera containing viruses of fish and reptiles, which respectively are Genus *Aquaparamyxovirus*, represented by Atlantic salmon paramyxovirus (ASPV), and *Ferlavirus*, represented by Fer-de-Lance paramyxovirus (FDLV; these are included in Fig. 10.2).

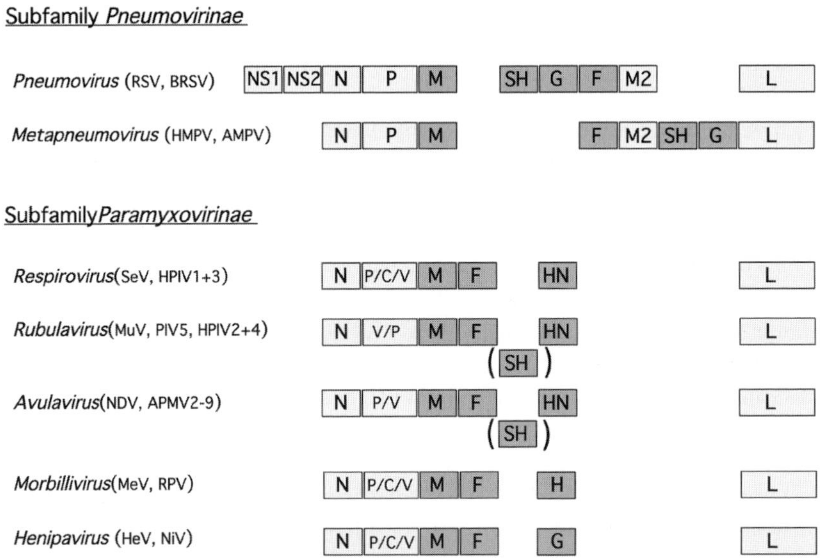

Figure 10.1 Comparison of the gene order of respiratory syncytial virus (RSV) with those of selected members of family *Paramyxoviridae*. The genes are shown in their 3′–5′ order in genomic RNA, which is the order of transcription. Subfamily and genera names are listed to the left, together with representative viral species in parentheses. Genes are not to scale and have been arrayed so that homologous genes are aligned vertically as much as possible, with gaps introduced to maximize the alignments. Genes encoding membrane-associated proteins have dark shading. The SH gene is shown offset and in parentheses for genera *Rubulavirus* and *Avulavirus* because it is present only in some members. BRSV, bovine RSV; HMPV and AMPV, human and avian metapneumovirus, respectively; SeV, Sendai virus; HPIV, human parainfluenza virus (present as serotypes 1–4); MuV, mumps virus; PIV5, parainfluenza virus 5, previously known as simian virus 5; NDV, Newcastle disease virus; APMV, avian paramyxovirus (present as serotypes 1–9, with NDV being serotype 1); MeV, measles virus; RPV, rinderpest virus; HeV, Hendra virus; NiV, Nipah virus. Proteins shown that have no direct homologue in RSV: C, non-structural accessory protein; V, cysteine-rich accessory protein; HN, haemagglutinin–neuraminidase glycoprotein.

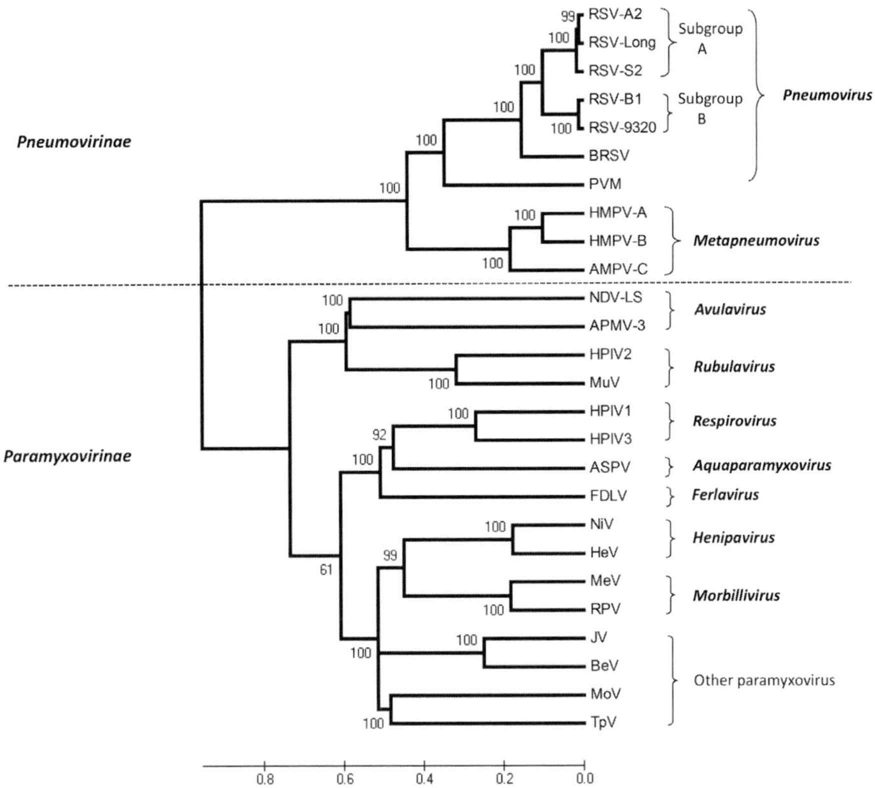

Figure 10.2 Phylogenetic analysis based on the complete nucleotide sequences of the genomes of respiratory synctial virus (RSV) and other selected members of *Paramyxoviridae*. The scale at the bottom indicates evolutionary distance as the number of nucleotide substitutions per site. The analysis is based on the neighbour-joining method (Saitou and Nei, 1987) and was performed with MEGA4 (Tamura *et al*., 2007). The numbers at branch points indicate the percentage in which the associated taxa clustered together in the bootstrap test (500 replicates). Virus names additional to those in Fig. 10.1: PVM, pneumonia virus of mice; HMPV-A and -B, HMPV antigenic subgroups A and B; AMPV-C, avian metapneumovirus subgroup C; NDV-LS, NDV LaSota strain; ASPV, Atlantic salmon paramyxovirus; FDLV, Fer-de-lance virus; JV, J-virus; BeV, Beilong virus; MoV, Mossman virus; TpV, Tupaia virus. The analysis was kindly provided by Drs. Sachin Kumar and Siba Samal, University of Maryland at College Park.

This subfamily also contains additional members not yet classified into genera, some of which are shown (Fig. 10.2, 'other paramyxoviruses').

RSV has one of the more complex genetic maps in *Paramyxoviridae* (Fig. 10.1). All members of the family share a set of common genes encoding five common proteins: the nucleoprotein N, phosphoprotein P, matrix protein M, fusion glycoprotein F, and large polymerase protein L. All members of the family also have a gene encoding an attachment protein whose structure and name can differ depending on the subfamily or genus: it is called the haemagglutinin–neuraminidase HN, haemagglutinin H, or glycoprotein G.

Other genes/proteins can be present depending on the subfamily, genus, or species (Fig. 10.1). RSV and all members of its subfamily have an M2 gene encoding two proteins, M2-1 and M2-2, from overlapping ORFs. RSV and other members of its genus have two genes encoding non-structural proteins NS1 and NS2. RSV and all members of its subfamily encode a third surface protein, the small hydrophobic SH protein. An SH protein also is encoded by certain members of Genus *Rubulavirus* and *Avulavirus*. Most members of *Paramyxovirinae* have C and/or V accessory proteins encoded by additional open reading frames in the P/V mRNA. Table 10.1 shows the per cent amino acid sequence identity between RSV subgroup A versus RSV subgroup B, BRSV, PVM, HMPV subgroup A, and AMPV subgroup A.

Propagation

RSV replicates most efficiently in immortalized cell lines derived from human epithelial cells. However, RSV can infect and replicate to a significant extent in cell lines representing various tissues from other hosts, including humans, monkeys, bovines, hamsters, and mice. HEp-2 generally is the most permissive cell line: these human epithelial-like cells originally were from an epithelial carcinoma of the larynx, but appear to have been overgrown and replaced during the passage history by contaminating HeLa cells, which are epithelial-like cells from a cervical adenocarcinoma (Masters, 2002). African green monkey Vero cells, which do not produce type I IFN because of gene deletions, also are commonly used to propagate RSV, but are not more efficient that HEp-2 cells. Virus titration usually is by plaque titration; immunostaining is often used to enhance visualization of plaques. Fresh clinical isolates of RSV may undergo some sort of adaptation to cell culture (Marsh *et al.*, 2007), but this is poorly understood. However, passaged laboratory strains retain their virulence for chimpanzees and humans (Karron *et al.*, 1997b; Whitehead *et al.*, 1998b).

RSV replicates slowly in cell culture, with the peak of virus production in a single-step growth curve occurring between approximately 20 and 40 h post-infection. Ninety per cent of infectious progeny RSV remains cell-associated, apparently reflecting a deficiency in budding *in vitro*. Sonication or freeze–thawing typically is employed to dislodge the attached virus. The final yield is not very high, usually in the range of 10^6 to 5 X $10^{7\,PFU}$/ml, with considerable contamination by cellular debris.

Virion

The RSV particle produced in cell culture has two forms, a spherical particle of approximately 150–350 nm in diameter, and a filamentous form that often predominates and is 60–200 nm in diameter and up to 10 μm in length (Norrby *et al.*, 1970; Bachi and Howe, 1973; Roberts *et al.*, 1995; Jeffree *et al.*, 2003; Fig. 10.3). It is not known which form(s) occur *in vivo*. More than 95% of the infectious particles in a typical virus preparation made *in vitro* do not pass through a 0.45 μm filter, indicating that most of the infectious particles are large filaments or aggregates (Paccaud and Jacquier, 1970; Roberts *et al.*, 1995). When exposed to UV irradiation, viral infectivity is lost with single-hit kinetics. This indicates that, even though the infectious particles can be large filaments, the vast majority contains a single functional genome (Dickens *et al.*, 1984).

The virion has a lipid envelope that is derived from the host cell plasma membrane during budding (Fig. 10.3). The envelope contains three virally encoded transmembrane surface glycoproteins, namely glycoprotein G and the fusion F and small hydrophobic SH glycoproteins (Fig. 10.4). These species form homooligomers that are visualized as short (11–20 nm) surface

Table 10.1 Per cent amino acid sequence identity between the proteins of respiratory syncytial virus (RSV) subgroup A (RSV-A) and the indicated members of subfamily *Pneumovirinae*[1]

Viruses compared		Per cent amino acid sequence identity for the indicated protein										
		NS1	NS2	N	P	M	SH	G	F	M2-1	M2-2	L
RSV-A versu	RSV-B	87	92	96	91	91	76	53	89	92	72	93
	BRSV	69	84	93	81	89	38	30	81	80	42	84
	PVM	16	20	60	33	42	23	12	43	43	10	53
	HMPV-A	–[2]	–[2]	42	35	38	23	15	33	36	17	45
	AMPV-A	–[2]	–[2]	41	32	38	19	16	35	37	12	43

[1]Viruses are listed in order of decreasing relatedness to human respiratory syncytial virus (HRSV) A. Respiratory syncytial virus (RSV)-B is RSV subgroup B; human metapneumovirus (HMPV) A and avian metapneumovirus (AMPV) A are subgroup A of HMPV and AMPV. PVM, pneumonia virus of mice; BRSV, bovine RSV.

[2]This virus does not have this gene.

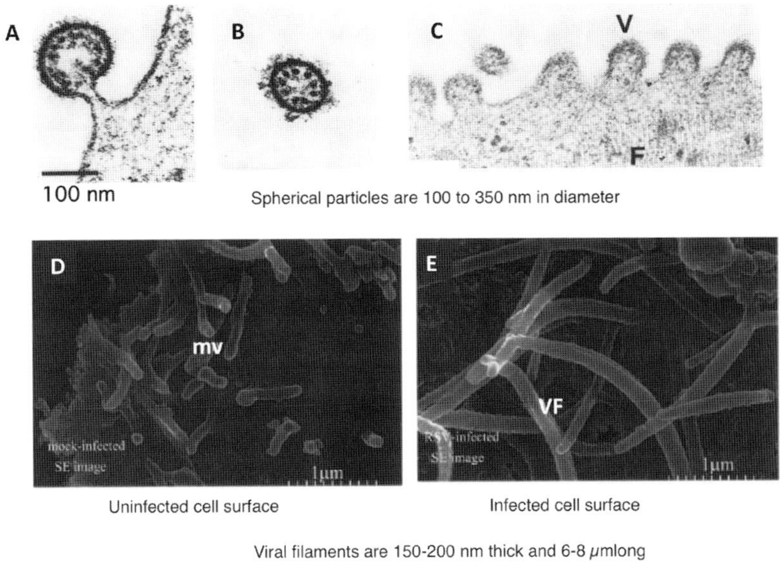

Figure 10.3 Electron micrographs of respiratory syncytial virus (RSV) particles and cell-associated viral structures. A–C show negatively stained electron micrographs of budding (A and C) and free (B) RSV particles. In C, V indicates a budding virion, and F indicates filamentous intracellular structures thought to be nucleocapsids (Kalica *et al*., 1973). D and E are field emission scanning electron micrographs of the surface of uninfected (D) and RSV-infected (E) cells (Jeffree *et al*., 2003). E shows surface viral filaments (VF) on the cell surface that are thought to yield viral filamentous particles, and D shows microvilli (mv) found on the surface of uninfected cells.

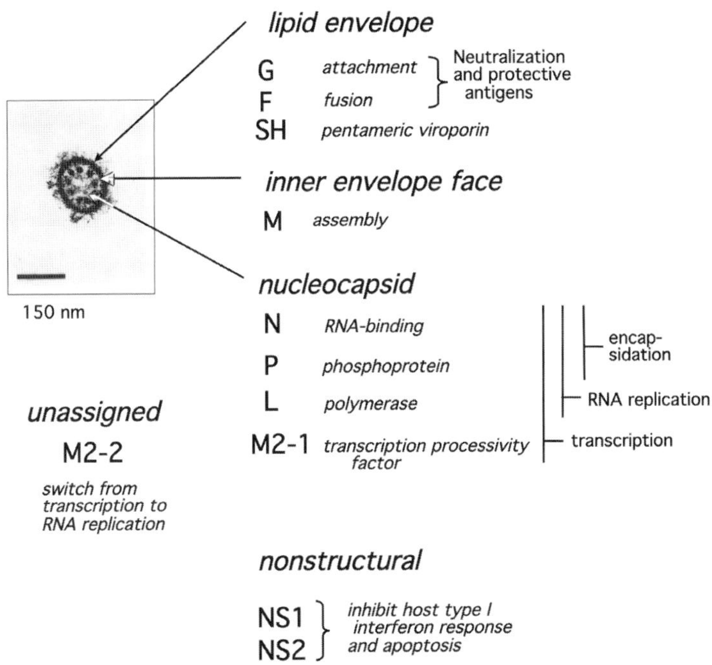

Figure 10.4 The proteins of respiratory syncytial virus (RSV): their major functions and locations in the virus particle.

spikes. The matrix M protein is a non-glycosylated species that is thought to line the inner surface of the envelope.

Contained within the lipid envelope is the helical viral nucleocapsid, consisting of the viral RNA tightly bound by the nucleoprotein N (Fig. 10.4). This appears in electron micrographs as a herringbone structure that is characteristic of *Paramyxoviridae*. However, the RSV nucleocapsid is narrower than those of prototype members of *Paramyxovirinae* (12–16 nm compared with 17–20 nm) and has a steeper pitch (Norrby *et al.*, 1970; Bachi and Howe, 1973; Bhella *et al.*, 2002). The phosphoprotein P, polymerase L, and transcription factor M2-1 proteins also are associated with the nucleocapsid and are packaged in the virion. The NS1 and NS2 proteins are detected only in trace amounts in virion preparations, which probably reflects contamination owing to the difficulty of purifying virions. They are considered provisionally to be non-structural. The status of M2-2 as a potential virion component is unknown.

RSV can readily lose infectivity during handling, presumably reflecting particle instability. This can be partly overcome by agents such as sugars that reduce aggregation and improve thermal stability (Ausar *et al.*, 2007). When recombinant RSV was engineered to contain a foreign attachment protein from baculovirus in place of the RSV glycoproteins, the stability of infectivity was improved (Sastre *et al.*, 2007). This suggests that the lability of the particle may reside in one or more of the glycoproteins. The long filamentous shape of the particle may also contribute to fragility and loss of infectivity.

Genome organization, transcription, mRNAs, and replication

Organization

The genome of RSV is a single negative sense strand of RNA that is approximately 15.2 kb in length: the map of strain A2 is shown in Fig. 10.5A. To date, six RSV strains have been sequenced completely. Four of these are from subgroup A, including the prototype A2 strain (15,222 nucleotides in length; GenBank accession number M74568), the prototypic Long strain (15,226 nucleotides; AY911262), strain S2 (15,191 nucleotides; NC_001803), and strain 19 (also called line 19; 15,191 nucleotides; FJ614813). Two strains of subgroup B have been completely sequenced: strains B1 (15,225 nucleotides; NC_001781) and 9320 (15225 nucleotides; AY353550). Extensive sequence information also is available for the subgroup B strain 18,537 (Johnson *et al.*, 1987b).

The RSV genome contains 10 genes in the order 3′ NS1-NS2-N-P-M-SH-G-F-M2-L (Fig. 10.5A). As with all *Mononegavirales*, the first gene is preceded by a short extragenic leader region and the last gene is followed by a short extragenic trailer region: in the case of strain A2, the leader and trailer regions are 44 and 155 nucleotides in length, respectively. The first 26 nucleotides at the 3′ end of the RSV genome are complementary at 21 positions with the first 24 nucleotides at the 3′ end of the antigenome, after which there is no further significant relatedness (Fig. 10.6A). This conserved sequence contains elements of the genomic and antigenomic promoters, as discussed further below (Cowton *et al.*, 2006; Fearns *et al.*, 2000).

Transcription and mRNAs

RSV transcription follows the general *Mononegavirales* model (Cowton *et al.*, 2006). The transcribing polymerase enters the genome at the 3′ end and transcribes the genes in their 3′ to 5′ order into individual mRNAs by a start-stop-restart mechanism in which the polymerase remains template-bound between genes (Dickens *et al.*, 1984). The polymerase is guided by short gene-start (GS) and gene-end (GE) transcription signals at the gene junctions (Kuo *et al.*, 1996b; Fig. 10.5B).

Mini-replicon studies indicated that the transcription was highly sensitive to substitutions at positions 3, 5, 8, 9, 10, and 11 at the beginning of the leader region (Fearns *et al.*, 2002; Fig. 10.6A). In addition, transcription required the GS signal of the first gene and was enhanced when its distance from the 3′ end was approximately that of the wild-type situation. The efficiency of transcription was enhanced by the presence of the 9-nucleotide sequence immediately preceding the GS signal. It remains unclear whether the transcriptase (i) begins synthesis at the exact 3′

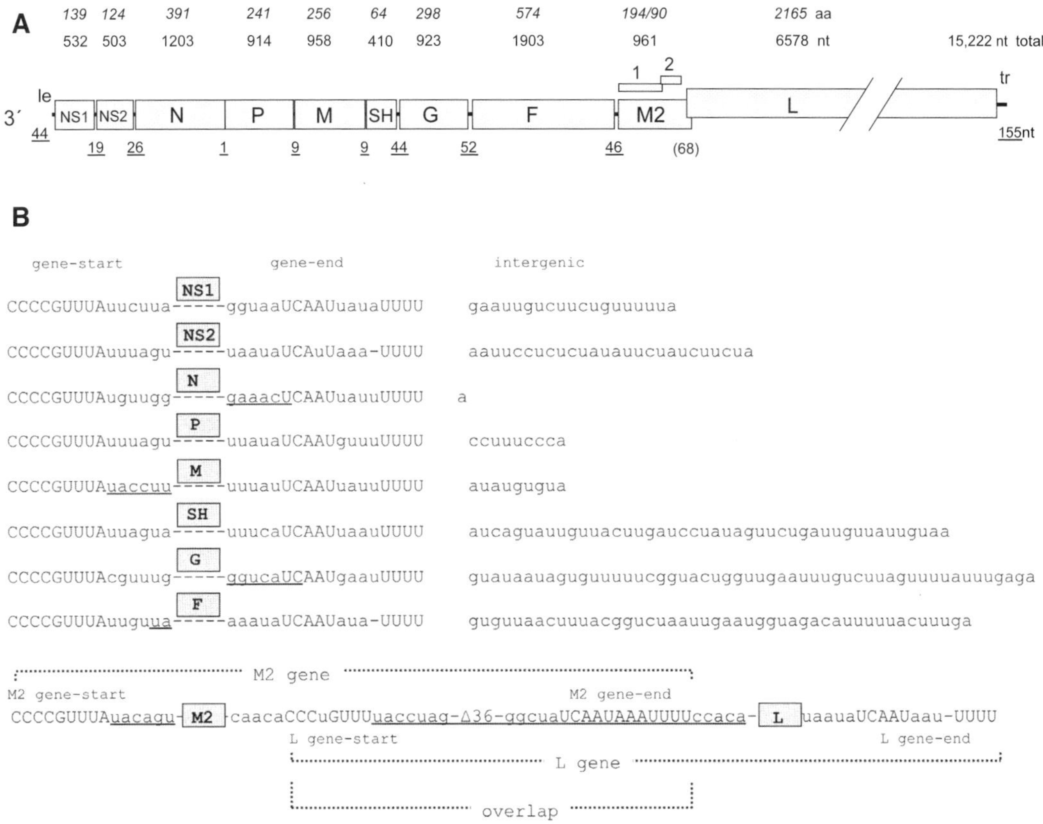

Figure 10.5 The genome of respiratory syncytial virus (RSV) strain A2. (A) Map of the genome of RSV drawn to scale and shown 3′–5′, which is the direction of transcription. The first row of numbers above the map indicates the nucleotide lengths of the genes, and the second row of numbers (italicized) indicates the amino acid lengths of the unmodified encoded protein(s). The underlined numbers under the map indicate the nucleotide lengths of the non-transcribed leader (le), trailer (tr) and intergenic regions, and the number in parentheses is the nucleotide length of the overlap between the M2 and L genes. The spacing of the two open reading frames (ORFs) present in the M2 mRNA is illustrated above the gene. (B) Gene-start, gene-end, and intergenic sequences (all sequences are negative sense). The first eight genes in the RSV gene order are represented by the first eight lines of sequence. Each line shows the first 15 and last 18 nucleotides of the indicated gene, with the main body of the gene indicated by the shaded box, followed by the downstream intergenic sequence. The bottom line depicts the M2 and L genes, showing the first 15 nucleotides and last 18 nucleotides of each gene as well as the 68-nucleotide overlap between the genes (part of the overlap was deleted – indicated as Δ36 – in order to be accommodated in the figure). For each of the genes, the conserved residues in the gene-start and gene-end sequences are capitalized. Sequences involved in ORFs are underlined.

end followed by release of a short leader transcript and re-initiation at the first GS signal, or (ii) enters the polymerase at the 3′ end followed by scanning without synthesis to the first GS signal, or (iii) coordinately recognizes the sequence at the 3′end and the first GS signal and initiates synthesis at the GS signal.

The GS signal is a 9-nucleotide sequence that is exactly conserved among the first nine RSV genes (Kuo *et al.*, 1997; Fig. 10.5B) with an occasional substitution in some strains. The GS signal of the tenth gene, L, has two nt differences, but these were not associated with any functional difference when assayed in a mini-replicon system. Synthesis of the mRNA begins at the first position of the GS signal. Interestingly, studies with a mini-replicon showed that the first nt of the encoded mRNA often was the usual G residue even when

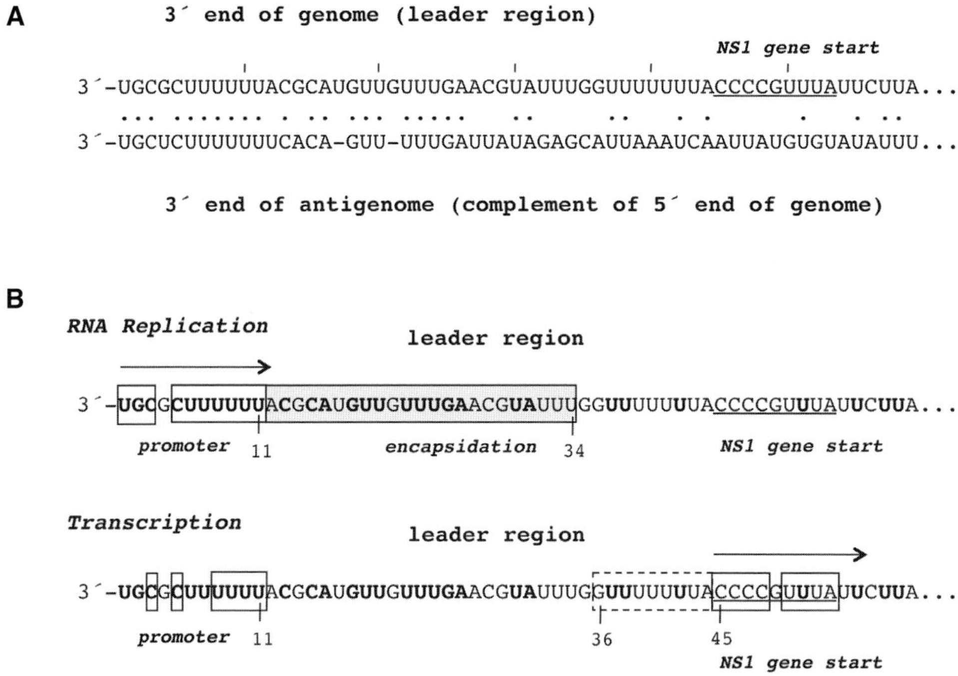

Figure 10.6 Genome termini and the genomic promoter of respiratory syncytial virus (RSV) strain A2. (A) Alignment of the 3′ ends of the genome (top) and antigenome (bottom), the latter being the complement of the 5′ end of the genome. Sequences are in negative sense, and sequence identities are indicated by dots. (B) Nucleotide positions in the leader region that are important for RNA replication (top) and transcription (bottom). Sequences important for initiation of RNA replication or transcription are indicated by open boxes (Fearns *et al.*, 2002). A region that contains an apparent encapsidation domain necessary to produce full-length replication products is indicated with a shaded box (McGivern *et al.*, 2005). A region that increases the efficiency of transcription is indicated with a hatched box (McGivern *et al.*, 2005). Sequences are in negative-sense and nucleotides that are identical between the 3′ ends of genomic and antigenomic RNA are in bold.

the template had been changed so that the first position of the GS signal was no longer the usual C (Kuo *et al.*, 1996b). This suggests that the transcribing polymerase can select the initiating nucleotide independent of the template, although this independence was not complete.

The GE signal is a 12–13 nucleotide sequence that is less well conserved than the GS signal (Kuo *et al.*, 1997; Fig. 10.5B). The GE signal directs the polymerase to execute polyadenylation by re-iterative copying on a tract of 4–7 U residues at the downstream end of the signal. This is followed by termination of synthesis and release of the mRNA. During sequential transcription, the polymerase occasionally fails to terminate at any given GE signal and continues to transcribe until it encounters a GE signal of a downstream gene. This creates read-through transcripts representing two or more adjacent genes. These read-through transcripts in aggregate comprise approximately 10% of the total mRNA. However, the amount of read-through varies considerably among the various gene junctions reflecting differences in GE efficiency associated with sequence differences. Variability in GE efficiency and sequence also can occur between RSV isolates (Moudy *et al.*, 2003).

The first nine genes are separated by short intergenic sequences (IGS) that, from one gene junction to the next, vary in length from 1–58 nucleotides for the strains sequenced to date (1–53 nucleotides for strain A2, Fig. 10.5). These are not transcribed except in the case of read-through transcription. There are no obvious conserved sequences or structural features between the various IGS, which are among the most divergent genome regions between strains and subgroups (Johnson and Collins, 1988). The naturally occurring variations in sequence and length, or even deletion of the IGS altogether, generally had little effect on transcription in a mini-genome system (Kuo *et al.*, 1996a), although there were some cases where one or more IGS nucleotide assignments flanking a GS or GE signal appeared to affect the efficiency of the signal (Moudy *et al.*, 2004). Increasing the length of an IGS in recombinant virus beyond the maximum length found in nature modestly reduced the efficiency of virus replication (Bukreyev *et al.*, 2000). Thus, the IGS appear to be non-specific, non-conserved spacers.

The GS signal of the last gene, the L gene, is located within its upstream neighbour, the M2 gene: specifically, it is located 68 nucleotides from the end of the M2 gene (Collins *et al.*, 1987; Fig. 10.5B). Thus, the two genes overlap by 68 nucleotides. Overlapping genes are not found in any other members of *Paramxoviridae*, but do exist elsewhere in *Mononegavirales*, such as in Ebola virus. This arrangement seemed inconsistent with sequential transcription. However, studies with a mini-replicon containing this M2/L overlap in the context of marker genes showed that polymerases that complete transcription at the M2 GE signal efficiently gained access to the upstream L GS signal (Fearns and Collins, 1999a). This appeared to occur by retrograde scanning, and indeed the polymerase appears to be able to scan in both directions. It may be that this back-and-forth scanning occurs at each gene junction following termination at the GE signal and prior to re-initiation at the next GS signal. Another seeming complication was that, owing to the overlap, polymerases that initiate at the L GS signal must traverse the M2 GE signal during synthesis of the L mRNA and thus presumably would prematurely terminate. Direct analysis confirmed this was the case: 90% of L gene transcripts were found to terminate at the M2 GE signal, producing a truncated L mRNA that is polyadenylated (Collins *et al.*, 1987). It is not known whether this small mRNA has any further significance. Therefore, the synthesis of full-length L mRNA depends on the occasional failure of the polymerase to terminate at the M2 GE signal.

Premature termination of nascent L transcripts at the M2 GE signal might have been expected to severely down-regulate the synthesis of full-length L mRNA. The relative molar amount of L mRNA indeed is much lower than that of the other mRNAs (Collins and Wertz, 1983); however, this also is true across *Paramyxoviridae*. In a recent study, a foreign gene was inserted into recombinant RSV downstream of the L gene, and its efficiency of expression was found to be only 4-fold less than that of the F gene, located three positions upstream (Kwilas *et al.*, 2010). This indicates that there was not a severe drop-off in transcription across the L gene, implying that the low abundance of L mRNA is due to some other factor such as stability. This indicates that the high frequency of termination of nascent L transcripts at the M2 GE signal presumably is compensated for by efficient retrograde re-initiation.

The RSV mRNAs contain a methylated 5′ cap structure: m7G[5′]ppp[5′]Gp (Barik, 1993). When capping was inhibited with a novel inhibitor being developed as a potential anti-RSV drug, transcription produced uncapped abortive mRNAs of ~45–50 nt (Liuzzi *et al.*, 2005). This suggests that capping is an essential step for efficient mRNA elongation, which also appears to be the case for vesicular stomatitis virus (family *Rhabdoviridae*, order *Mononegavirales*).

Each of the 10 RSV mRNAs has a single major ORF except for the M2 mRNA, which contains two ORFs encoding two separate proteins, M2-1 and M2-2. These two ORFs overlap slightly: the M2-1 ORF (in strain A2) comprises nucleotides 10–594 of the mRNA, and the M2-2 ORF is downstream and comprises nucleotides 563–835. Expression of the downstream M2-2 ORF is mediated by ribosomes that exit the M2-1 ORF and reinitiate at the M2-2 ORF, a process of coupled translation (Gould and Easton, 2005). Interestingly, the efficiency of coupled translation was dependent on the presence of several hundred nucleotides of upstream M2-1 coding sequence, a sequence that included a potential folded structure in the mRNA (Gould and Easton, 2005). RSV and the other members of its subfamily do not engage in RNA editing, which most members of subfamily *Paramyxovirinae* use to shift the reading frame of the P/V mRNA to alternatively encode the P or V proteins (Kolakofsky *et al.*, 2005).

RNA Replication

For RSV and for *Mononegavirales* in general, the replicating polymerase begins copying at the first nucleotide at the 3′ end of the genome and, ignoring GS and GE signals, produces a complete, exact positive-sense copy of the viral genome called the antigenome. The antigenome serves in turn as the template for producing progeny genomes. The synthesis of full-length antigenomes and genomes depends on their concurrent encapsidation as nascent RNAs. RSV and other members of its subfamily do not appear to have any requirement with regard to the precise length of the genome, and thus differ from members of subfamily *Paramyxovirinae*, for which the nucleotide length of the genome must be an even multiple of six ('rule of six') for efficient RNA replication to occur (Samal and Collins, 1996; Kolakofsky *et al.*, 2005).

In a mini-replicon system, RNA replication was highly sensitive to substitutions at positions 1–3 and 5–11 at the beginning of leader region: this included the positions important in transcription plus additional positions (Fig. 10.6B). The first 15 nucleotides were fully sufficient for efficient initiation of antigenome synthesis, but nucleotides 16–34 were required in addition for encapsidation, which in turn was essential for elongation to yield full-length antigenomes (McGivern *et al.*, 2005).

RNA replication (and transcription) can tolerate short extensions at the 3′end but were substantially inhibited by extensions of six or more nucleotides (Collins *et al.*, 1991; Cowton and Fearns, 2005). Thus, the conserved promoter sequence needs to be near, but not exactly at, the 3′ end. When the genome contained an extension, initiation of RNA synthesis occurred predominantly at the first nucleotide of the conserved promoter, and only to a small extent at the extended 3′ end, indicating that the conserved promoter sequence dictates the position of initiation. However when the template was modified so that the 3′-terminal U residue was substituted or deleted, the replication product initiated with the usual A residue. This implies that the RSV replicase selects the initiating nucleotide independent of the template (Noton *et al.*, 2010), resembling the situation noted above for transcription (Kuo *et al.*, 1996b).

Proteins

RSV encodes 11 proteins (Figs 4 and 7). F, G and SH are transmembrane surface glycoproteins; M is an internal membrane-associated matrix protein; N, P, M2-1, and L are components of the nucleocapsid/polymerase complex; M2-2 is a low-abundance protein; and NS1 and NS2 are presumptive non-structural species.

Fusion F glycoprotein

The RSV F protein (Fig. 10.7) and L proteins are the RSV species that most closely resemble their counterparts in *Paramyxovirinae*. The per cent amino acid sequence identity between the F proteins of RSV and the *Paramyxovirinae* is low (~20%), but relatedness is unambiguous and there is a strong similarity with regard to overall size and locations of hydrophobic domains, heptad repeats, cysteine residues, and the major cleavage activation site. Like other *Paramyxoviridae* F proteins, RSV F directs viral penetration by fusion between the viral and target cell membranes. Later in infection, F protein expressed on the cell surface can mediate fusion with neighbouring cells. As noted later (see Replicative cycle), recent findings indicate that the RSV F protein also plays a major role in viral attachment by binding to cellular nucleolin (F. Tayyari, D. Marchant, W. Duan, T. Moraes, P. Mastrangelo, and R. G. Hegele, presented at the Seventh International RSV Symposium, Rotterdam, 2010).

As is typical for *Paramyxoviridae*, the F protein is a type I membrane glycoprotein with a cleaved N-terminal signal peptide and a membrane anchor near the C-terminus (Fig. 10.7). F is synthesized as an inactive F_0 precursor that is activated by cleavage by a host protease to yield two disulfide-linked subunits: NH_2-F_2-S-S-F_1-COOH. The N-terminus of the F1 subunit that is created by cleavage contains a third hydrophobic domain, called the fusion peptide, which is inserted into the target membrane during fusion. The F1 subunit also contains two areas of heptad repeats, one downstream of the fusion peptide and the other upstream of the membrane anchor. These domains associate during fusion, creating a conformational shift that drives fusion of the viral and cellular membranes. Cys-69 and Cys-212 in the RSV F protein may participate in the disulfide bonding

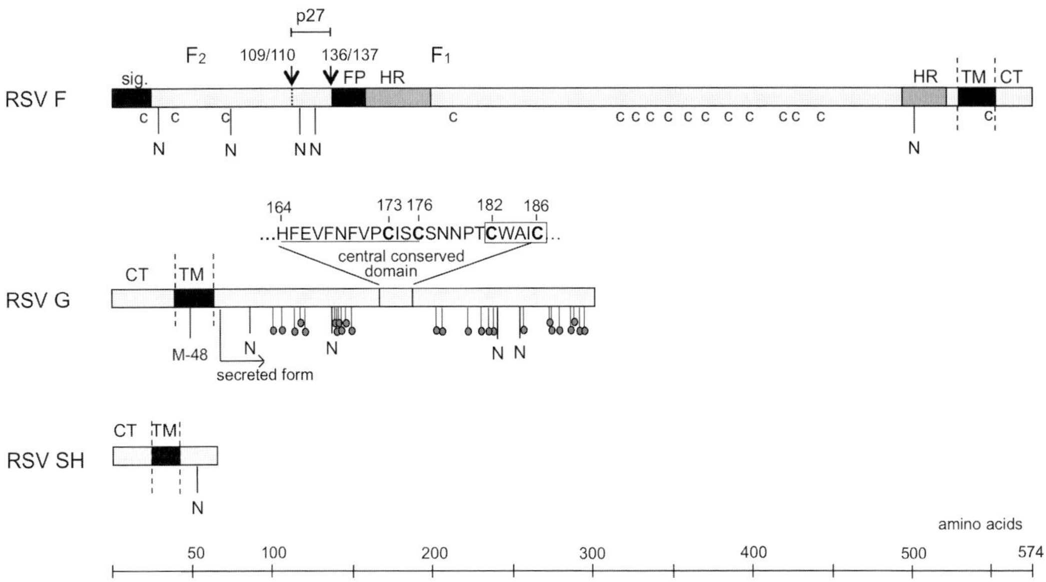

Figure 10.7 Diagrams of the primary sequences of the F, G, and SH glycoproteins of respiratory syncytial virus (RSV) strain A2, approximately to scale. Structural features: black boxes, hydrophobic domains; TM, transmembrane anchor; CT, cytoplasmic tail; downward-facing stalks with N, potential acceptor sites for N-linked carbohydrate. Annotations specific to the F protein: sig., cleaved signal peptide; FP, fusion peptide; grey boxes, heptad repeats (HR); c, cysteine residues; downward-facing arrows, cleavage activation sites with amino acid coordinates indicated; p27, peptide released by cleavage activation. Annotations specific to the G protein: downward-facing stalks with shaded circles, predicted sites for O-linked sugars; the sequence above G shows the conserved central domain with the highly conserved segment underlined and amino acid positions numbered relative to the complete sequence; cysteine residues in the cysteine noose are in bold, and the CX3C fractalkine domain is boxed.

that links the F1 and F2 subunits, because they correspond in sequence alignments to the two cysteine residues known to link the two subunits of the Sendai virus F protein. Molecular modelling suggests that there may be a second inter-subunit disulfide bond unique to RSV involving Cys-37 and Cys-439, which are two cysteine residues that do not seem to have counterparts in *Paramyxovirinae* (Day *et al.*, 2006).

RSV F protein can direct efficient fusion independent of other viral proteins. For example, when RSV F was expressed by a recombinant vesicular stomatitis virus, it was active both in viral penetration to initiate infection and in the subsequent formation of syncytia (Kahn *et al.*, 1999). This differs from the situation for most members of *Paramyxovirinae*, for which efficient fusion requires interaction with the homologous attachment protein.

The F proteins of RSV and BRSV are unique in *Paramyxoviridae* in having two, rather than one, cleavage sites (Gonzalez-Reyes *et al.*, 2001; Fig. 10.7). The site that is immediately upstream of the fusion peptide (with cleavage between residues 136/137 in human RSV strain A2) is the one that corresponds to other *Paramyxoviridae*. This site in RSV contains six tandem Arg and Lys residues (Lys-Lys-Arg-Lys-Arg-Arg↓). The second, novel site is located 27 amino acids upstream (with cleavage between residues 109/110) and has the sequence Arg-Ala-Arg-Arg↓. Thus, both sites contain the consensus motif for the ubiquitous cellular endoprotease furin found in the trans Golgi network (Arg-X-Arg/Lys-Arg↓). RSV F appears to be readily cleaved during intracellular processing, although some of the F protein packaged in virions remained uncleaved at site 109/100. Cleavage of RSV F does not appear to be limiting or to be a significant determinant of RSV tropism or pathogenesis, in contrast to the well-known models of NDV and avian influenza virus.

Cleavage at both sites of RSV or BRSV produces a short peptide whose unprocessed length is 27 amino acids (p27). For BRSV, this peptide contains a tachykinin sequence motif and is processed and released from BRSV-infected cells to yield a virokinin that induced smooth muscle contraction *in vitro*, which is a property of tachykinins (Zimmer *et al.*, 2003). Tachykinins also have pro-inflammatory and immunomodulatory activities. Recombinant BRSV in which most of p27 was deleted, or in which the upstream cleavage site was mutated, replicated as efficiently as the wild-type parent in calves, but induced less pulmonary inflammation and a somewhat lower titre of serum neutralizing antibody (Valarcher *et al.*, 2006). Thus, p27 of BRSV has the potential to augment viral disease both by effects on smooth muscle and pulmonary inflammation, and might also stimulate the host adaptive immune response. However, the sequence of p27 of human RSV does not resemble a tachykinin, and human RSV p27 did not have tachykinin-like properties *in vitro* (Zimmer *et al.*, 2003).

The RSV F_0 protein contains 5–6 potential acceptor sites for N-linked carbohydrate, depending on the strain. In the case of the Long strain, three of the sites remain in the mature, cleaved F protein (the rest are in p27) and each was shown to contain a carbohydrate side chain: Asn-27 and Asn-70 in the F_2 subunit and Asn-500 in the F_1 subunit (Zimmer *et al.*, 2001). These N-linked side chains are processed during intracellular maturation to a complex form. The presence of carbohydrate side chains at these sites was not important for cleavage or cell-surface transport, but did influence fusion; in particular, glycosylation at Asn-500 was essential for efficient fusion (Zimmer *et al.*, 2001).

The RSV F protein assembles into a homotrimer. The 3-D structure of RSV F has not been determined by direct analysis, but a model was generated based on structures that were solved for the F proteins of NDV and HPIV3 (Smith *et al.*, 2002).

Glycoprotein G

The RSV G protein (Fig. 10.7) plays a role in viral attachment that is important for efficient viral replication *in vivo*, and G also influences host immune responses. The G protein has no apparent relatedness by sequence or structure to the attachment HN, H, and G proteins of *Paramyxovirinae*. RSV G is 282–319 amino acids in length, depending on the strain, and thus is only one-half the size of the attachment proteins of *Paramyxovirinae*. RSV G is a type II transmembrane glycoprotein with a hydrophobic signal/anchor located near the N-terminus (amino acids 38–63 in strain A2), and with the C-terminal two-thirds of the molecular oriented extracellularly. This type II organization is one feature that RSV G does share with the attachment glycoproteins of *Paramyxovirinae*.

A secreted version of G is produced by translational initiation at the second ATG in the ORF (codon 48), which lies within the signal/anchor sequence (Hendricks *et al.*, 1988; Roberts *et al.*, 1994). This truncated form is then trimmed by proteolysis, removing the remainder of the signal/anchor and creating a new N-terminus at Asn-66 for the final secreted form. The secreted form constitutes approximately 20% of the total G protein synthesized in RSV-infected cell culture. However, because this truncated form is rapidly secreted, it constitutes 80% of the G protein that is released into the medium by 24 h post-infection, with the remainder of the released G protein being the full-length form present in virus particles (Hendricks *et al.*, 1988).

G is modified during synthesis at the rough endoplasmic reticulum by the addition of an estimated four N-linked carbohydrate side chains to yield an intermediate of ~45 kDa (for strain A2; Wertz *et al.*, 1989; Collins and Mottet, 1992). The side chains are later processed into a complex form. In the endoplasmic reticulum, G assembles into homooligomers that probably are trimers or tetramers. O-linked sugars are subsequently added in the trans-Golgi compartment or network (Wertz *et al.*, 1989; Collins and Mottet, 1992). In the case of strain A2, mature G migrates in gel electrophoresis as a broad diffuse band of 80–90 kDa. Analysis of a recombinant F/G fusion protein expressed in insect cells by a recombinant baculovirus indicated that G contains approximately 24–25 O-linked side chains (Wathen *et al.*, 1991). The presence of a sheath of host-specified sugars might shield the G protein from immune recognition, and might interfere with antigen processing and presentation. The secreted form of G is processed in a similar fashion to the membrane-anchored form except that it appears to be secreted as a monomer, and no antigenic differences were detected when the secreted and membrane-bound forms were analysed with an extensive panel of MAbs (Escribano-Romero *et al.*, 2004).

The ectodomain of RSV G consists of two 'mucin-like' domains separated by a short, conserved domain described below. Like mucin, these domains have a high content of serine, threonine and proline residues, are heavily glycosylated, and are thought to have a structure that is mostly extended and non-globular (Johnson *et al.*, 1987b). Of all the RSV proteins, these mucin-like domains have the greatest divergence between strains (Johnson *et al.*, 1987b). The presence of more than 75 serine and threonine residues in the mucin-like domains as potential acceptor sites for O-linked sugars, compared with the estimate of 24–25 O-linked side chains, suggests that there may be heterogeneity in site usage, which could provide additional antigenic heterogeneity. The significance of this mucin-like character remains unknown. Cellular mucins form gel-like secretions that serve as lubricants and protective barriers on mucosal surfaces, and one possibility is that the G protein may facilitate viral movement through mucin-containing respiratory secretions.

There also is evidence that the G protein may contain more carbohydrate than previously appreciated. This is based on analysis of RSV grown in an *in vitro* model of human airway epithelium (HAE; Kwilas *et al.*, 2009). HAE cultures are prepared from primary cells isolated from human airway tissue that are seeded on a filter support *in vitro* and differentiate under an air–liquid interface into pseudostratified mucociliary tissue that closely resembles the authentic airway epithelium (Zhang *et al.*, 2002). The G protein in virions produced in HAE cultures had an Mr of 180 kDa. This suggests that G protein synthesized in HAE cells, the most authentic *in vitro* model for the airway epithelium, contains substantially more carbohydrate than that produced in immortalized cell lines (Kwilas *et al.*, 2009).

The two 'mucin' domains are separated by a central conserved domain that includes 13 residues (positions 164–176 in strain A2) that are completely conserved among RSV strains, as well as an overlapping segment containing four closely spaced cysteine residues (positions 173, 176, 182, and 186) that also are completely conserved (Fig. 10.7; Johnson *et al.*, 1987b). The secondary structure of this region is predicted to be a tight turn, and disulfide linkages occur between Cys-173 and Cys-186, and between Cys-176 and Cys-182 to create a cysteine noose (Gorman *et al.*, 1997). This central domain can be deleted from recombinant RSV with little effect on virus replication in cell culture or in mice, and thus the function of the highly conserved 13-amino acid segment remains unknown (Teng and Collins, 2002). The cysteine-rich segment contains a CX3C motif involving Cys-182 and Cys-186 that is embedded in a region of limited sequence relatedness with the CX3C chemokine called fractalkine (Tripp *et al.*, 2001). A peptide containing this sequence mimicked the leukocyte chemoattractant activity of fractalkine in an *in vitro* assay.

The central conserved domain described above is followed in the G protein primary structure by a heparin-binding domain (amino acids 184–198) that was identified in binding studies using synthetic peptides spanning the G ectodomain (Feldman *et al.*, 1999). Heparin is a glycosaminoglycan (GAG), which are long unbranched chains of repeating disaccharide subunits that are part of the extracellular matrix and play a role in RSV attachment, as noted later (Replicative cycle). However, the heparin-binding domain in G probably is not important in attachment, since it could be deleted with little effect on RSV infection and replication *in vitro* or in mice (Teng and Collins, 2002).

The C-terminal region of the G protein appears to be susceptible to cleavage in cell culture, yielding shorter forms including a predominant species of 55 kDa (Kwilas *et al.*, 2009). The relative proportion of this cleaved form varied among different cell lines and was the most pronounced in Vero cells, which are used for vaccine manufacture. While the virus remained infectious in the immortalized cell lines in which it was grown, it bound less efficiently to GAGs and also had reduced infectivity in HAE cultures. This indicates that the C-terminal region of G is important for GAG binding and for infection in cultures thought to closely resemble authentic respiratory epithelium.

The G gene can be deleted from RSV with little effect on the ability of the virus to replicate in Vero cells *in vitro*. However, in HEp-2 cells, the absence of G appeared to reduce the efficiency of virus binding, fusion, assembly, and budding (Teng and Collins, 1998; Techaarpornkul *et al.*, 2001, 2002; Teng *et al.*, 2001). In addition, the ΔG virus was highly restricted for replication in mice (Teng *et al.*, 2001), and a candidate vaccine virus in which G had been spontaneously deleted was over-attenuated in humans (Karron *et al.*, 1997a). Thus, the G protein is dispensable in cell culture under some conditions, but is needed for efficient replication *in vivo*.

Small hydrophobic SH protein

The RSV SH protein is a short (64 amino acids for strain A2) transmembrane protein that, like the G protein, has a hydrophobic signal/anchor sequence near the N-terminus with the C-terminus oriented extracellularly (Collins and Mottet, 1993; Fig. 10.7). Most of the SH protein accumulates without glycosylation, but there is also a form that contains the addition of a single N-linked side chain as well as a form in which the side chain is modified by the further addition of heterogeneous amounts of polylactosaminoglycan (Olmsted and Collins, 1989; Anderson *et al.*, 1992). In addition, there is a 42-amino acid form (also present in the membrane) produced by translational initiation at a second methionyl residue, as well as glycosylated versions of this shorter form. The significance of these multiple forms is unknown. SH assembles into homooligomers that appear to be pentamers (Collins and Mottet, 1993; Kochva *et al.*, 2003; Gan *et al.*, 2008).

Expression of SH in bacteria resulted in its incorporation into the membrane, which increased its permeability to low-molecular-weight compounds (Perez *et al.*, 1997). Incorporation of partial or full-length SH protein into artificial membranes *in vitro* resulted in cation-selective channel-like activity (Gan *et al.*, 2008; Carter *et al.*, 2010). Visualization by electron microscopy detected both pentameric and hexameric pore-like structures (Carter *et al.*, 2010). Thus, the SH protein has properties of a viroporin, which typically are small proteins that modify membrane permeability and can have roles in budding and apoptosis.

The SH gene can be deleted from recombinant virus with no reduction in replication in cell culture (Bukreyev *et al.*, 1997). Virus lacking the SH protein induced more apoptosis (Fuentes *et al.*, 2007), consistent with the general hypothesis that paramyxovirus SH proteins play a role in inhibiting apoptosis (Wilson *et al.*, 2006). SH also appeared to inhibit signalling from tumour necrosis factor (TNF) α, although this did not appear to be responsible for the effect on apoptosis (Fuentes *et al.*, 2007). *In vivo*, the ΔSH virus was slightly restricted for replication in mice and in chimpanzees (Bukreyev *et al.*, 1997; Whitehead *et al.*, 1999a). However, in clinical trials, the comparison of two attenuated strains that differed only in the presence or absence of the SH protein did not reveal any difference in infection or replication (Karron *et al.*, 2005). Thus, the function and impact of the SH protein seem unclear.

Matrix M protein

The RSV M protein is a non-glycosylated internal virion component. RSV M is smaller than its *Paramyxovirinae* counterparts (256 amino acids vs. approximately 335–375 amino acids) and lacks significant sequence relatedness with these counterparts. RSV M accumulates independently at the plasmid membrane and interacts with F and other factors, possibly including the cytoskeleton, to form stable structures in preparation for particle morphogenesis (Henderson *et al.*, 2002). M is essential for forming virus-like particles (Teng and Collins, 1998). M also associates with nucleocapsids, and its removal from nucleocapsids in a cell-free system resulted in enhanced transcription (Ghildyal *et al.*, 2003). Thus, RSV M appears to play two roles typical for *Mononegavirales*: M plays a central role in organizing virion components at the plasma membrane, and M may silence the nucleocapsid in preparation for packaging into the virus particle.

A crystal structure determined for the RSV M protein showed that it is a monomer that is organized into compact N- terminal (residues 1 to 126) and C-terminal (residues 140–255) domains joined by a 13-residue linker (Money *et al.*, 2009). The surface of the M protein was found to contain a large positively-charged area that extends across the two domains and the linker and may mediate association with negatively-charged membrane as well as with nucleocapsids (Money *et al.*, 2009).

Nucleocapsid/polymerase proteins N, P, M2-1, and L

The N, P, and L proteins of RSV appear to be close functional analogues of their counterparts in other members of *Paramyxoviridae*. The N protein binds tightly along the entire length of genomic and antigenomic RNA to form an RNAse-resistant helical nucleocapsid that serves as the template for replication and transcription. The P protein is a multifunctional adapter that helps mediate interactions between components of the nucleocapsid and polymerase complex. The L protein contains the polymerase catalytic domains. In contrast, the M2-1 protein is a novel transcription factor that lacks a counterpart outside of *Pneumovirinae*.

The 391- amino acid RSV N protein is shorter than its counterparts in *Paramyxovirinae* (approximately 490–555 amino acids). It does not share significant sequence relatedness with *Paramyxovirinae* except for several conserved segments within residues 254 to 360 in the C-terminal region (Barr *et al.*, 1991) that include four α-helices also found in other members of *Mononegavirales* (Tawar *et al.*, 2009). RSV N protein produced in bacteria was recovered in decamer rings bound to bacterial RNA, and a crystal structure was obtained (Tawar *et al.*, 2009). Each decamer represents one turn of the helical nucleocapsid. Each N monomer interacts with an extended chain of seven nucleotides. Each N monomer consists of N-terminal and C-terminal domains separated by a hinge, and the RNA groove is located at the domain junction. Adjacent monomers were oriented in the same direction and loosely connected. Of the seven nucleotides associated with each N monomer, nucleotides 2–4 are oriented into the groove while the other four nucleotides face outward. It was suggested that passage of the polymerase may induce a transient hinge movement that makes the three buried nucleotides flip out, resulting in a row of 11 nucleotides (three from the upstream monomer, seven from the middle monomer, and one from the downstream monomer) available for reading (Tawar *et al.*, 2009).

The RSV P protein (241 amino acids) is shorter than its *Paramyxovirinae* counterparts (approximately 390–605 amino acids) and lacks sequence relatedness. However, it is thought to have the same general array of functions. Like its counterparts in *Paramyxovirinae*, RSV P operates as a stable homotetramer formed through a multimerization domain, which for RSV involves residues 120–150 in the middle of the molecule (Asenjo and Villanueva, 2000; Castagne *et al.*, 2004; Llorente *et al.*, 2006). P binds to soluble N protein – probably through the P N-terminal domain – and delivers it to nascent genomes/antigenomes during RNA replication. P thus confers binding specificity to the N monomer and prevents both premature association with RNA and self-aggregation (Castagne *et al.*, 2004). P also binds to the L (Khattar *et al.*, 2001) and M2-1 (Asenjo *et al.*, 2006) proteins and helps mediate their interactions with the nucleocapsid. Interaction between the P and N proteins in the nucleocapsid appears to involve the C-terminal segments of P and N, the latter of which protrudes from the decamer ring (Tawar *et al.*, 2009). P also may contribute to conformational changes that help the polymerase access the RNA template (Castagne *et al.*, 2004).

P is the major phosphorylated RSV protein and contains phosphate at more than 10–12 sites, with different sites exhibiting differing rates of turnover: the C-terminal domain contains low-turnover phosphates and accounts for most of the total phosphate, the middle domain contains intermediate-turnover phosphates, and the N-terminal domain contains high-turnover phosphates (Navarro *et al.*, 1991; Asenjo *et al.*, 2005). Phosphorylation is not necessary for tetramerization (Castagne *et al.*, 2004), but affects a number of other activities. Mutation of five major serine acceptor sites (including the two low turnover sites at 232 and 237 and three of the intermediate sites at 116, 117, and 119) reduced phosphorylation by 80–98% (Villanueva *et al.*, 2000; Lu *et al.*, 2002). These phosphates were dispensable for transcription and RNA replication by a mini-replicon (Villanueva *et al.*, 2000), and were dispensable for replication of recombinant RSV *in vitro*, but had effects on N–P interaction and also resulted in a budding defect that attenuated the virus (Lu *et al.*, 2002). Phosphorylation at another site, Ser-54, has been described to inhibit interaction of P and M and appears to be essential for dissociating the M protein from the nucleocapsid during uncoating to initiate infection (Asenjo *et al.*, 2008). Phosphorylation at yet another site, T108, appears to have a very high turnover rate and prevents interaction between the P and M2-1 proteins, controlling the association of M2-1 with the polymerase complex (Asenjo *et al.*, 2006). However, another study suggested that substitution of non-acceptor residues in the same region of P affected P–M2-1 interactions (Mason *et al.*, 2003), indicating that the effect might be a consequence of the amino acid substitution rather than the difference in phosphorylation. In a cell-free system, transcription by nucleocapsids that were reconstituted with bacterially expressed P protein that completely lacked phosphate produced only abortive RNA products, which may mean that P phosphorylation is necessary to convert newly initiated polymerase into a stable complex (Dupuy *et al.*, 1999), or may mean that phosphorylation facilitates access to nucleotides buried in the RNA groove of the N protein (Tawar *et al.*, 2009). These observations suggest that dynamic phosphorylation of P, probably involving a subset of sites and a small percentage of the total phosphate content, influences a number of viral activities.

The 194-amino acid RSV M2-1 protein is encoded by the upstream ORF in the M2 mRNA. The M2-1 protein is a transcription factor essential for viral viability (Collins *et al.*, 1995, 1996, 1999). M2-1 promotes transcription processivity; in its absence, the viral polymerase terminates prematurely and non-specifically within several hundred nucleotides and downstream genes are not significantly transcribed (Collins *et al.*, 1996; Fearns and Collins, 1999b). M2-1 also decreases the efficiency of termination at the transcription gene-end (GE) signals – possibly a reflection of the same processivity activity – resulting in increased production of read-through mRNAs (Hardy and Wertz, 1998).

M2-1 appears to form a homotetramer mediated by an oligomerization domain at residues 32 to 63 (Tran *et al.*, 2009). M2-1 contains a cysteine–histidine zinc finger motif near its amino-terminus (residues 7 to 25) that is essential for its activity (Hardy and Wertz, 2000). Gel electrophoresis of M2-1 from RSV-infected cells revealed at least two electrophoretically distinct species, with the more slowly migrating one being phosphorylated (Hardy and Wertz, 2000). The M2-1 protein binds RNA non-specifically (Cuesta *et al.*, 2000; Tran *et al.*, 2009). In an *in vitro* assay, M2-1 bound RNA or the P protein in a competitive fashion that involved residues 59–177 and was independent of the phosphorylation status of M2-1 (Tran *et al.*, 2009). This suggests that the P protein may deliver the M2-1 protein to the RNA template.

The 2165-amino acid RSV L protein is very similar in length to its other *Paramyxoviridae* counterparts and is thought to be a close functional counterpart. RSV L has low but unambiguous sequence relatedness to its counterparts in *Paramyxovirinae*. The L protein has six discrete segments containing conserved residues that appear to include polymerase motifs (Stec *et al.*, 1991). A putative nucleotide-binding domain involved in capping mRNA was identified in the central region of the L protein based on sequence analysis of a RSV mutant selected for resistance to a novel capping inhibitor (Liuzzi *et al.*, 2005).

M2-2 protein

The RSV M2-2 protein is encoded by the downstream ORF in the M2 mRNA, which slightly overlaps the upstream M2-1 ORF (Collins *et al.*, 1990a). The M2-2 ORF of strain A2 has three potential AUG start sites at codons 1, 3, and 7; mutagenesis indicated that the first two may be utilized (Cheng *et al.*, 2005). Therefore, the M2-2 ORF overlaps with the M2-1 ORF by 32 or 26 nucleotides, and encodes an M2-2 protein of 90 or 87 amino acids. M2-2 is expressed intracellularly at a low level (Ahmadian *et al.*, 1999).

Recombinant RSV in which the M2-2 ORF has been silenced grows more slowly *in vitro* than wild-type RSV, although it eventually reached a similar titre (Bermingham and Collins, 1999; Jin *et al.*, 2000). In cells infected with this ΔM2-2 virus, the accumulation of mRNA was increased while that of the genome and antigenome was decreased. The accumulation of viral proteins also was increased and syncytia formed more rapidly and extensively, probably because of increased expression of the F protein. Conversely, overexpression of M2-2 during RSV infection in cell culture was achieved by complementation with a transfected plasmid or using recombinant RSV in which the M2-2 coding sequence was uncoupled from that of M2-1 and expressed as a separate gene (Cheng *et al.*, 2005). Overexpression of M2-2 inhibited RSV replication. Expression of M2-2 also was inhibitory to RSV RNA synthesis by a mini-replicon (Collins *et al.*, 1996). These observations indicate that M2-2 plays a role in shifting RNA synthesis from transcription to RNA replication, that M2-2 can be inhibitory to RNA synthesis, and that the level of M2-2 is critical. Replication of the ΔM2-2 virus in chimpanzees was reduced approximately 500- to 1000-fold, thus identifying M2-2 as a non-essential accessory protein (Bermingham and Collins, 1999; Teng *et al.*, 2000).

Non-structural proteins NS1 and NS2

The NS1 and NS2 proteins (139 and 124 amino acids, respectively) are non-structural accessory proteins that interfere with host responses. Wild-type RSV is a poor inducer of type I IFN, but deletion of NS1 and/or NS2 from recombinant RSV resulted in a large increase in type I IFN expression in infected human epithelial cell cultures and macrophages (Spann *et al.*, 2004). This suggested that each of these proteins acts as an antagonist of IFN induction, an observation previously made with BRSV (Schlender *et al.*, 2000). The magnitude of the effect was greater for NS1 than for NS2, but was greatest for both proteins together (Spann *et al.*, 2004). Expression of either protein during RSV infection inhibited the activation and nuclear translocation of transcription factor IRF3, which is necessary for type I IFN induction (Spann *et al.*, 2004). Coimmunoprecipitation studies showed that NS1 and NS2 form a heterodimer (Swedan *et al.*, 2009), which may enhance their activity but apparently is not essential. Another study provided evidence of several uncharacterized complexes involving NS1 and NS2, and showed that monomeric NS2 was unstable with a half-life of 30 min (Evans *et al.*, 1996).

Interestingly, NS1 and NS2 appear to hit multiple targets in the signal transduction pathway leading to the activation of IRF3. Tagged NS2 protein expressed from a transfected plasmid or from recombinant RSV was shown to bind to RIG-I (Ling *et al.*, 2009), the cytoplasmic pattern recognition receptor that detects 3′-triphosphorylated RNA and initiates signal transduction to activate IRF3 (Ling *et al.*, 2009). The binding of NS2 to RIG-I blocks signal transduction by preventing the interaction between RIG-I and the outer mitochondrial membrane-associated adaptor MAVS involved in signal transduction. In other studies, the expression of either NS1 or NS2 from transfected plasmids resulted in a decrease in TRAF3 (Swedan *et al.*, 2009), which is downstream of MAVS in the signal transduction pathway leading to IRF3 activation. NS1 had the greater effect. NS1 also decreased the level of IKKε, a protein kinase that activates IRF3, whereas NS2 had no effect (Swedan *et al.*, 2009).

Wild-type RSV also strongly suppresses type I IFN-mediated signal transduction from the type I IFN receptor through the JAK/STAT pathway (Ramaswamy *et al.*, 2004). This suppression is mediated by proteosomal degradation of the STAT2 signal transduction factor (Ramaswamy *et al.*, 2004). STAT2 indeed was shown to be ubiquitinated in RSV-infected cells, consistent with proteosomal degradation. In addition, the NS1 protein was shown to bind in an *in vitro* assay to E3 ligase complex components (Elliott *et al.*, 2007). The E3 ligase complex is the component of the ubiquitination apparatus that binds to the target protein and thus confers specificity, which suggests that NS1 may be important in targeting STAT2. If this is the case, then the RSV NS1 protein might play a role similar to that described for the *Rubulavirus* V protein, linking the ubiquitination apparatus to the target STAT2 protein. When expressed from transfected cDNAs, both NS1 and NS2 mediated STAT2 degradation with the effect being greatest when the two proteins were expressed together (Elliott *et al.*, 2007). However, three other studies provided evidence that NS2 plays the major role in STAT2 degradation, whereas NS1 enhanced the activity of NS2 when expressed together but had no independent effect (Lo *et al.*, 2005; Ramaswamy *et al.*, 2006; Swedan *et al.*, 2009). An E3 ligase involvement by NS1 also presumably would not account for the degradation of TRAF3 or IKK noted above, since that appeared to involve non-proteosomal mechanisms (Swedan *et al.*, 2009). While some details of the mechanisms remain to be clarified, the overall picture is that the RSV NS proteins inhibit type I IFN induction and signalling in multiple ways.

RSV had no effect on signal transduction by type II IFN (INFγ) in epithelial cells (Ramaswamy *et al.*, 2004). This is consistent with the observation that RSV targets STAT2 and spares STAT1, since signalling from the INFγ receptor involves STAT1 homodimers and not STAT2. However, RSV does inhibit the induction of type III IFN (λ1, λ2, and λ3; also known as IL-29, IL28A, and IL28B, respectively) in human epithelial cells and macrophages (Spann *et al.*, 2004). The induction pathway of III IFN induction is not fully understood, but appears to have general similarities to that of type I IFN (Donnelly and Kotenko, 2010; Iversen and Paludan, 2010). Consistent with that premise, the observed effects of the RSV NS1 and NS2 proteins on type III and type I IFN were similar (Spann *et al.*, 2004). The effects of RSV on type III IFN-mediated signalling have not been investigated, but probably are similar to what is observed for type I IFN. This is because, while type III and type I INF use different receptors, they both signal through the JAK/STAT pathway involving STAT1/STAT2 dimers (Donnelly and Kotenko, 2010), and thus both should be susceptible to STAT2 degradation targeted by the RSV NS proteins. Type III IFN is made by a variety of cells, but its receptor seems to be more restricted than that of type I IFN and is expressed mainly on epithelial cells, which would make it very relevant for pulmonary infection (Donnelly and Kotenko, 2010).

An initial report showed that deletion of NS2 from recombinant RSV decreased its ability to induce activation of transcription factor NF-κB (Spann *et al.*, 2005). Subsequently, siRNA-mediated knock-down of the expression of either NS1 or NS2 in RSV-infected cells was found to reduce activation NF-κB and AKT signalling, with the effect of speeding the onset of apoptosis (Bitko *et al.*, 2007). Thus, the two NS proteins have the effect of activating pro-survival pathways, thereby prolonging the life of the cell and increasing the viral yield (Bitko *et al.*, 2007). This cell survival effect also operates in Vero cells – which lack the ability to make type I IFN – and thus appears to be independent of type I IFN.

In a mini-replicon system complemented by plasmid-expressed proteins, co-expression of NS1 was strongly inhibitory to transcription and RNA replication, affecting both the genomic and antigenomic promoters (Atreya *et al.*, 1998). NS2 also was somewhat inhibitory to both processes, but only at high levels of expressed protein. These effects remain to be further defined. It's worth noting that the V and C accessory proteins of rubulaviruses and respiroviruses also have been shown to have inhibitory effects on viral RNA synthesis.

As might be expected from the results described above, RSV lacking NS1 and/or NS2 replicates to reduced titre in cultured cells competent for type I IFN, as well as in seronegative chimpanzees, with the attenuation effect increasing in the order ΔNS2<ΔNS1≤ΔNS1+NS2 (Teng and Collins, 1999; Whitehead *et al.*, 1999a; Teng *et al.*, 2000; Spann *et al.*, 2004).

Replicative cycle

Efficient infection of cell lines *in vitro* by RSV involves binding to cellular GAGs, especially heparan sulfate and chondroitin sulfate B. The G and F proteins each appear to mediate GAG-binding (Hallak *et al.*, 2000; Techaarpornkul *et al.*, 2002). In addition, a number of other potential receptor molecules have been tentatively identified, including intracellular adhesion molecule (ICAM)–1 (Behera *et al.*, 2001), RhoA (Pastey *et al.*, 1999), the CX3CR1 fractalkine receptor (Tripp *et al.*, 2001), and annexin II (Malhotra *et al.*, 2003), but none of these appears to be a convincing candidate. More recently, efficient RSV infection *in vitro* and in the mouse model was shown to depend on binding to the cellular protein nucleolin (F. Tayyari, D. Marchant, T. J. Moraes, W. Duan, P. Mastrangelo, and R. G. Hegele, presented at the Seventh International RSV Symposium, Rotterdam, 2010). Somewhat unexpectedly, binding to nucleolin is mediated by the F protein rather than by the G protein. It may be that RSV attachment is a two-step process, involving an initial, possibly low-affinity binding to GAGs followed by binding to nucleolin. There also is evidence that nucleolin can serve as an attachment co-factor for HPIV3.

In polarized cells such as HAE cells, RSV infection is specific to the apical surface. Viral entry occurs by fusion of the viral envelope with the cell plasma membrane (Srinivasakumar *et al.*, 1991; Kahn *et al.*, 1999) or with endosomal membranes following clathrin-mediated endocytosis (Kolokoltsov *et al.*, 2007). In cell culture, the initiation of fusion by attached virions appeared to be a slow step as visualized by video microscopy but, once started, the process was rapid (Bachi, 1988). Genome transcription and replication occur in the cytoplasm and the virus can grow in enucleated cells and in the presence of actinomycin D, indicating a lack of essential nuclear involvement.

RSV mRNAs and proteins can first be detected intracellularly at 4 to 6 h after infection and reach a peak accumulation by 15–20 h. The release of progeny virus begins by 10–12 h post infection, reaches a peak after 24 h, and continues until the cells deteriorate by 30–48 h. Transcription and RNA replication are concurrent processes. As noted, the low-abundance M2-2 protein accumulates during infection and shifts the balance of RNA synthesis from transcription to RNA replication (Bermingham and Collins, 1999).

Several factors influence the relative levels of expression of the various viral genes. As is typical for *Mononegavirales*, a fraction of the polymerase disengages and irreversibly leaves the template at each gene junction during sequential transcription, conferring a polar gradient to gene expression. This gradient of expression is not very steep, with the exception of the L mRNA (Collins and Wertz, 1983; Krempl *et al.*, 2002). However, as already noted, the low abundance of L mRNA appears to be due to a post-transcriptional effect, presumably mRNA stability, rather than polymerase fall-off (Kwilas *et al.*, 2010). The relative levels of expression of the viral proteins also are affected by differences in the production of read-through mRNAs at the various gene junctions (Harmon and Wertz, 2002; Moudy *et al.*, 2004). This is because only the first ORF in a read-through mRNA would be accessed efficiently by ribosomes and translated (Kozak, 2005). Thus, the production of a read-through mRNA across a gene junction has the effect of reducing the amount of translatable mRNA for any downstream gene(s) represented in the transcript. Read-through has another effect: by reading across a junction, the polymerase is spared the possibility of disengagement and fall-off that can occur at each junction. Thus, while read-through may decrease the amount of translatable mRNA for downstream genes represented in a given read-through mRNA, it has the effect of delivering more polymerase to genes that are further downstream. Through this interplay of effects, differences in the efficiency of the GE signals between genes and between strains can modify gene expression at the levels of both transcription and translation. However, it is not known whether these GE-specific effects are of sufficient magnitude to have much effect on virus replication.

In experiments with mini-replicons, increasing the level of the M2-1 protein altered the transcriptional gradient (Fearns and Collins, 1999b). However, examination of viral mRNAs during RSV infection *in vitro* provided no comparable evidence that the increase in the intracellular levels of the viral proteins with time was associated with any effect on the gradient of transcription (Fearns and Collins, 1999b). In the absence of the M2-1 protein, a mini-replicon modelling the upstream end of the genome transcribed the NS1 and NS2 genes alone, suggesting a possible mechanism whereby reduction in M2-1 might facilitate persistent infection by down-regulating expression of most of the viral genes while maintaining expression of the NS1 and NS2 host defence antagonists (Fearns and Collins, 1999b). In other mini-replicon experiments, altering the level of expression of the N and P proteins did not affect the balance between RNA replication and transcription, implying that the availability of protein for encapsidation do not control this balance (Fearns *et al.*, 1997).

Virus assembly and budding in polarized cells occur at the apical surface (Roberts *et al.*, 1995; Zhang *et al.*, 2002). Video microscopy of RSV-infected cells showed that budding occurs within circumscribed regions on the cell surface and appeared visually to be the reverse of fusion (Bachi, 1988). RSV virion assembly occurs in localized, virus-modified lipid rafts involving all three viral surface proteins and the M protein (Henderson *et al.*, 2002; Jeffree *et al.*, 2003; McCurdy and Graham, 2003; McDonald *et al.*, 2004; Yeo *et al.*, 2009). The minimum viral protein requirements for the formation of virus-like particles are the F, M, N, and P proteins (Teng and Collins, 1998), and expression of these proteins induced the formation of viral filaments (Utley *et al.*, 2008).

As has been observed for other members of *Paramyxoviridae*, RSV utilizes the host cytoskeleton in its replicative cycle. Viral substructures are associated with polymerized actin throughout the infectious cycle (Jeffree *et al.*, 2007). Actin is packaged in the virion and was required, in combination with profilin (an actin-binding protein involved in restructuring actin polymers), for efficient RNA synthesis by intracellular nucleocapsids in a cell-free reaction (Burke *et al.*, 2000). In the infected cell, chemical agents that interfere with polymerized actin or tubulin affected every step of the replicative cycle except transcription and RNA replication, and in particular resulted in a 1000-fold reduction in assembly and release (Kallewaard *et al.*, 2005). RSV appears to hijack cellular apical recycling endosomes (ACE) for budding, a pathway that is distinct from that described for a number of other enveloped RNA viruses (Brock *et al.*, 2003; Utley *et al.*, 2008).

Antigens

Post-infection sera from experimental animals or humans contain antibodies that recognize a number of RSV proteins (Ward *et al.*, 1983). However, the F and G proteins are the only viral neutralization antigens, with the third surface glycoprotein, SH, appearing to play no significant role (Connors *et al.*, 1991). F and G are also the major protective antigens. This was demonstrated in experiments in which cotton rats or mice were immunized with vaccinia virus recombinants individually expressing each of the viral proteins except L, and subsequently challenged with RSV (Olmsted *et al.*, 1986; Connors *et al.*, 1991). F and G each conferred long lasting protection, although F appeared to be more immunogenic and protective than G (Olmsted *et al.*, 1986; Connors *et al.*, 1991), possibly reflecting the unusual structure and heavy glycosylation of G as already noted. The N and M2-1 proteins also provided protective immunity that waned within weeks and was shown in the case of M2-1 to be mediated by $CD8^+$ cytotoxic T lymphocytes (CTL; Connors *et al.*, 1991; Kulkarni *et al.*, 1995). The antigens involved in CTL responses would be expected to vary in different genetic backgrounds and host species.

A large fraction of MAbs specific to the F protein neutralize infectivity and a number inhibit fusion as measured by syncytium formation in cell culture. Analysis of viral mutants selected to be resistant to individual MAbs mapped five major antigenic sites that are located in the F1 subunit (Crowe *et al.*, 1998; Lopez *et al.*, 1998). In some cases, mapping has been extended by visualization of F–antibody complexes by electron microscopy (Calder *et al.*, 2000) or by crystallographic analysis (McLellan *et al.*, 2010), and have been mapped on a predicted three dimensional model of the F protein (Smith *et al.*, 2002). Operationally, most of the epitopes of F appear to be conformation-dependent, since most monoclonal and polyclonal antibodies raised against native F react poorly with denatured protein, such as in Western blots, and synthetic peptides have proven to be inefficient in inducing neutralizing antibodies or protective immunity. This is typical of globular viral surface glycoproteins.

In contrast, G has a very different antigenic nature. Essentially all G-specific MAbs bind efficiently to denatured G, such as in Western blots, and many bind efficiently to synthetic peptides, suggesting that reactivity does not depend on a complex folded structure. Also, MAbs against the G protein have only marginal effects in neutralizing RSV infectivity *in vitro* when tested individually. Neutralization depends on mixtures of different MAbs, or the addition of secondary antibodies or complement, presumably to increase steric hindrance (Martinez *et al.*, 1997; Martinez and Melero, 1998). G-specific MAbs fall into three categories: (i) a small number react with viruses from both antigenic subgroups and thus recognize 'conserved' epitopes; (ii) a larger number react broadly within but not between subgroups and are 'group-specific'; and (iii) a yet larger number react with some strains but not others within a subgroup and are 'strain-specific' (Martinez *et al.*, 1997; Melero *et al.*, 1997). The conserved and group-specific MAbs typically do not depend on glycosylation of G for binding, and their epitopes are located in the central conserved region of the G. In contrast, strain-specific MAbs often react only with glycosylated G, and their epitopes are found mostly in the mucin-like domain in the C-terminal third of the molecule, although at least one mapped instead to the upstream mucin-like domain (Martinez *et al.*, 1997; Melero *et al.*, 1997). Human post-infection sera have been shown to have reactivity with the central conserved region of G as well as the non-conserved C-terminal region (Cane, 1997; Murata *et al.*, 2010). Post-infection human serum antibodies to the G protein were found to be enriched for the IgG2 subclass, consistent with G being recognized in part as a polysaccharide antigen (Wagner *et al.*, 1987b).

Sequence analysis of viral mutants selected for resistance to individual G MAbs (with neutralization augmented with secondary antibodies or complement) revealed four types of changes: (i) single amino acid substitutions, (i) biased hypermutation thought to be introduced by cellular adenosine deaminase, (iii) frameshifts generated by the insertion or deletion of individual adenosine residues, which have the effect of altering/deleting the C-terminal third of G, and (iv) premature stop codons that delete part of the C-terminal third of G (Martinez *et al.*, 1997; Melero *et al.*, 1997). The drastic mutations involved in the last three categories presumably were tolerated because G is dispensable *in vitro*, as already noted. However, they sometimes are observed in nature. For example, in one clinical isolate, the G reading frame appeared to shift and then shift back, such that seven consecutive amino acids appeared to be encoded by an alternative frame, and in other isolates there was heterogeneity among the strains in the placement of the G termination codon (Sullender *et al.*, 1991). Also, RSV isolated from an immunocompromised child receiving anti-RSV antibody treatment contained a premature stop codon that deleted 42 amino acids from the C-terminus (Lazar *et al.*, 2006).

Antigenic subgroups and diversity

RSV has a single serotype with two antigenic subgroups A and B. The two subgroups exhibit a 3- to 4-fold reciprocal difference in neutralization by polyclonal post-infection sera. Studies with MAbs demonstrated extensive antigenic difference in the G protein with substantially less difference in the other viral proteins (Anderson *et al.*, 1985; Mufson *et al.*, 1985). Analysis of glycoprotein-specific responses with post-infection sera from cotton rats and human infants using purified F and G glycoproteins in enzyme-linked immunoadsorbent assays (ELISA) showed that F has 50% antigenic relatedness between subgroups compared with 1–7% relatedness for G. Consistent with this, the F protein expressed from a vaccinia virus recombinant was equally protective in cotton rats against either subgroup, whereas the G protein was 13-fold less protective against the heterologous subgroup (Johnson *et al.*, 1987a). This indicates that F is responsible for most of the cross-subgroup neutralization and protection.

The two subgroups have a genome-wide nucleotide sequence identity of approximately 81%. The individual proteins vary considerably in their level of subgroup divergence (Table 10.1). The G, SH, and M2-2 proteins are the most divergent species, whereas the NS2, N, P, M, M2-1, and L proteins are all more than 90% identical between subgroups. Cross-subgroup divergence is greatest for the ectodomains of SH and G, which have only 50% and 44% amino acid sequence identity, respectively, between the two subgroups. The genome-wide nature of the sequence divergence between the subgroups indicates that they represent divergent lines of evolution rather than variants that differ at only a few major antigenic sites.

There also is considerable sequence divergence with each subgroup. For example, the G protein of subgroup B has as little as 80% amino acid sequence identity among strains analysed to date. The regions that are most divergent between the subgroups are also the most divergent within each subgroup. Distinct lineages or clades have been identified within each subgroup, based largely on the sequence of the G protein.

The RSV F protein appears to be relatively stable antigenically, consistent with its high level of sequence conservation between and within the two subgroups. Analysis of RSV strains recovered from geographically diverse regions over 30 years using F-specific murine MAbs representing 16 separate neutralization epitopes in four antigenic sites showed that seven epitopes were conserved in all but one of the strains (Beeler and van Wyke Coelingh, 1989). An additional unique epitope defined by an RSV-neutralizing Fab fragment selected from a phage-display library also was conserved in every one of 19 isolates recovered over a 30-year period (Crowe *et al.*, 1998).

In contrast, analysis of viral isolates obtained over successive epidemics or over several decades provided evidence suggestive of progressive amino acid changes in G associated with differences in reactivity with MAbs. In one study, analysis of strains isolated over 47 years suggested that positive selection occurred at 13 codons in the G ectodomain, a number of which are involved in epitopes previously identified in mutants selected for resistance to individual MAbs (Zlateva *et al.*, 2004). A second study of strains isolated over 38 years provided evidence of progressive amino acid changes at an average rate of 0.25% per year that were paralleled by changes in reactivity with MAbs (Cane and Pringle, 1995). The large mucin-like domains in G are more divergent at the amino acid level than at the nucleotide level, suggesting that there is selective immune pressure for amino acid change (Johnson *et al.*, 1987b). The F protein probably is subject to the same selective pressure, but presumably is less able to accept amino acid substitutions owing to constraints from its folded structure and functional requirements. However, the antigenic changes in G appear to be slow and may have a modest effect on total antigenicity. Analysis of BRSV isolates, including those from countries in which vaccination was in wide use, provided evidence of modest progressive changes not only in G, but also in N and F (Valarcher *et al.*, 2000).

Infection of experimental animals

A number of animal species can be infected in the respiratory tract by intranasal instillation of RSV. These include mice, cotton rats, ferrets, guinea pigs, hamsters, marmosets, lambs, and a number of species of non-human primates including chimpanzees. The chimpanzee alone approaches the permissiveness of the natural human host. Chimpanzees are readily infected, support moderate to high levels of virus replication, and develop disease resembling the common cold (Whitehead *et al.*, 1998b). This is exemplified by the original isolation of RSV from symptomatic chimpanzees, which presumably acquired the virus from their human handlers. However, chimpanzees are not readily available and are not convenient for experimentation. African green monkeys also support moderate levels of virus replication – although no disease – and may be the next-best non-human primate alternative. Mice (usually BALB/c) and cotton rats are the most widely used experimental animals, with cotton rats supporting a 10-fold higher level of RSV replication compared with the BALB/c mouse. However, neither animal is very permissive: infection requires a large non-physiological viral inoculum, virus replication occurs at a low to moderate level and is quickly cleared, disease is inapparent, and the immune response in the non-natural host may be exaggerated. RSV administered into cotton rats by intramuscular injection undergoes an abortive replicative cycle (Prince *et al.*, 1979).

Animal relatives

RSV has two animal relatives, namely the closely related BRSV, and the more distantly related PVM. BRSV is a respiratory pathogen of bovines and is one of the agents of shipping fever, an economically important disease of livestock (Paccaud and Jacquier, 1970). The BRSV gene map is identical to that of RSV with respect to the number of genes, the number and types of encoded proteins, and the gene order including the overlap between the M2 and L genes (Buchholz *et al.*, 1999; Yunus *et al.*, 2001). The genome of BRSV strain A51908 (NC_001989 and AF295543) contains 15,140 nucleotides and has 73% nucleotide identity with that of RSV. The level of amino acid sequence identity ranges from 30% (G protein) to 93% (N protein; Table 10.1). BRSV can infect non-human primates, but is highly restricted (Buchholz *et al.*, 2000). Like human RSV, BRSV has two antigenic subgroups, although the extent of amino acid sequence divergence in the G protein was only 15% and thus was much less than for the human RSV subgroups (Furze *et al.*, 1997).

Ovine and caprine RSVs also have been described, but may not be distinct species. The G protein of ovine RSV has 60–62% amino acid sequence identity to that of BRSV, compared with 21–29% identity with human RSV (Mallipeddi and Samal, 1993). Sequence information is not available for caprine RSV, although an RNAse mismatch analysis was consistent with the expectation that it is more closely related to BRSV than to human RSV. These findings suggest that ovine and bovine RSV may represent different branches of a ruminant RSV with an extent of divergence resembling that of the two human RSV subgroups.

PVM is a respiratory pathogen that readily infects mice and other rodents and can be highly virulent. It was first identified during experiments involving serial passage of lung homogenates of mice that were originally inoculated with clinical material in an effort to identify possible new human pathogens (Horsfall, 1939). However, the natural host of PVM is unclear: the virus occasionally appears in colonies of laboratory mice, but serological studies have not found clear evidence of widespread infection in wild rodents. The PVM gene map is similar to that of RSV and BRSV. One difference is that the M2 and L genes of PVM do not overlap. A second difference is that the PVM P gene contains a second ORF that encodes a predicted 137-amino acid product that represents a novel, twelfth PVM protein (Barr *et al.*, 1994). The complete nucleotide sequence of PVM strain 15 (AY729016) is 14,886 nucleotides in length and has 52% identity with that of human RSV. The level of amino acid sequence identity ranges from 10% (M2-2 protein) to 60% (N protein; Table 10.1).

In their respective hosts, BRSV and PVM have the potential to serve as models for RSV. One limitation for BRSV is that virus grown in cell culture often has reduced virulence *in vivo*. Also, bovines are outbred, lack immunological reagents, and are too large to be convenient as experimental animals. PVM is highly virulent and lethal in mice and thus differs from RSV, which is fatal only in a small percentage of cases, but could serve as a model for severe disease.

Genetics and reverse genetics

Mutation rate and recombination

RSV has a high rate of nucleotide substitution, as is typical for RNA viruses. RSV mutants selected to be resistant to neutralization by individual F-specific MAbs were recovered at frequencies of 10^{-4} to 10^{-6} (Beeler and van Wyke Coelingh, 1989). This would be an underestimate of the true mutation rate, since not all substitutions result in viable virus or permit escape from neutralization. Another study estimated that the rate of nucleotide substitution in natural isolates collected over 47 years was $10^{-2.7}$ per position per year (Zlateva *et al.*, 2004): this was based on the G gene, which is more tolerant of substitutions.

Members of *Mononegavirales* frequently engage in non-homologous recombination. In non-homologous recombination, the viral polymerase disengages from one genome or antigenome template and, with the nascent RNA still attached, reinitiates copying elsewhere on the same template, or re-initiates on a second template. Sequence relatedness appears to play no role in the choice of the re-initiation site. The most prominent example of this type of recombination is the generation of defective interfering (DI) genomes, which within *Mononegavirales* have been studied mostly with vesicular stomatitis virus and Sendai virus. DI genomes contain large internal deletions that eliminate multiple genes. DI genomes thus depend on co-infecting standard virus for the proteins necessary for replication. They typically compete with and inhibit standard virus. For RSV, DI particles generated in cell culture have been identified on the basis of differences in buoyant density in sucrose gradients and the ability to inhibit standard virus replication, although no DI genomes were defined molecularly (Valdovinos and Gomez, 2003). Co-passage *in vitro* of DI RSV and standard RSV resulted in persistent infections that were relatively non-cytopathic and were stably maintained for >3 yr. Non-homologous recombination also can result in smaller deletions that leave the virus capable of autonomous replication. For example, one study described a passage series that contained progressively larger deletions in the G and SH genes: these deletions could be tolerated because the G and SH proteins were not required for efficient growth under the conditions used in the study (Karron *et al.*, 1997a). There also is evidence suggestive of non-homologous recombination in nature. For example, the difference in the order of the SH, G, F, and M2 genes between RSV (SH-G-F-M2) and HMPV (F-M2-SH-G) is suggestive of recombination during evolution from a presumed common ancestor, and the same is suggested by the lack of NS1 and NS2 genes in HMPV.

Homologous recombination, which is guided by sequence relatedness, is thought to be very infrequent for *Mononegavirales* (Collins *et al.*, 2008). The mechanism by which this occurs is unclear, but it may involve template switching in which the polymerase (once again) disengages from the template with the nascent RNA still attached. However, in this case, base-pairing by the nascent RNA guides the polymerase to re-initiate at a homologous region of the new template. Thus, this would tend to be limited to highly related RNAs. Some studies have described genomes in nature, including RSV, that may be mosaic molecules arising from homologous recombination (Zheng *et al.*, 1999).

The ability of RSV to recombine to form replication-competent mosaic viruses was evaluated directly by co-infecting cells *in vitro* with two mutant viruses, one in which the NS1 and NS2 genes were deleted, and one in which the G gene was deleted and a set of silent markers was inserted into the L gene (Spann *et al.*, 2003). Each of these mutant viruses grew substantially less efficiently than wild-type virus in the HEp-2 cells used in this co-infection, and thus any recombinant that regained a complete set of genes would regain replication fitness, amplify, and be readily identified by plaque size. In the experiments, only a single putative mosaic virus was recovered. Unexpectedly, the virus did not contain the full set of genes, lacking most of SH and G. There was evidence of one non-homologous crossover and one homologous crossover. This supports the idea that both types of recombination indeed can occur, but that the generation of mosaic virus is not frequent.

Reverse genetics

As for *Mononegavirales* in general, RSV can be produced entirely from cloned cDNA, a process called reverse genetics. This involves transfection of cultured cells with (i) a plasmid expressing a full-length RSV antigenome, and (ii) support plasmids expressing the proteins of the nucleocapsid and polymerase complex, namely the N, P, M2-1 and L proteins (Collins *et al.*, 1995, 1999; Fig. 10.8A). Plasmid expression is mediated by T7 RNA polymerase, which can be supplied by a vaccinia virus recombinant, by a constitutively expressing cell line, or by a eukaryotic expression plasmid. The expressed RSV antigenome and support proteins assemble into nucleocapsids that launch a productive infection. The expression of the positive-sense antigenome rather than the negative-sense genome avoids hybridization with the positive-sense RNAs expressed from the support plasmids. One difference for RSV compared with *Paramyxovirinae* is that expression of the M2-1 protein is required in addition to N, P, and L (Collins *et al.*, 1999). Reverse genetics provides a method for introducing predetermined changes into infectious virus via the recombinant DNA intermediate, with applications in basic research as well as for developing live vaccines.

A second type of reverse genetics system that has been used extensively for RSV is based on cDNA-encoded mini-replicons modelled after DI RNA (Fig. 10.8B). They contain the two ends of the genome or antigenome, but most or all of the viral genes have been deleted and replaced by one or more marker genes, such as chloramphenicol acetyl transferase, luciferase, or green fluorescent protein, under the control of RSV GS and GE signals. The mini-replicon is complemented by viral proteins expressed from co-transfected support plasmids, which can provide whatever mix of RSV proteins is desired (Collins *et al.*, 1996). Depending on the support proteins, this system can execute genome encapsidation, transcription, RNA replication, and particle morphogenesis (Collins *et al.*, 1996; Teng and Collins, 1998; Fearns and Collins, 1999a; McGivern *et al.*, 2005; Noton *et al.*, 2010). Owing to its small size and simplicity, the mini-replicon approach is ideal for detailed structure-function studies of *cis*-acting RNA signals or *trans*-acting viral proteins. Since the viral proteins are supplied *in trans* rather than by the mini-replicon, the system can be used to study mutations that have drastic effects on gene expression or template replication and could not be recovered and studied in complete recombinant virus.

Epidemiology

Infection of infants and young children

RSV is the most important cause of severe acute viral lower respiratory tract infection (LRI) in

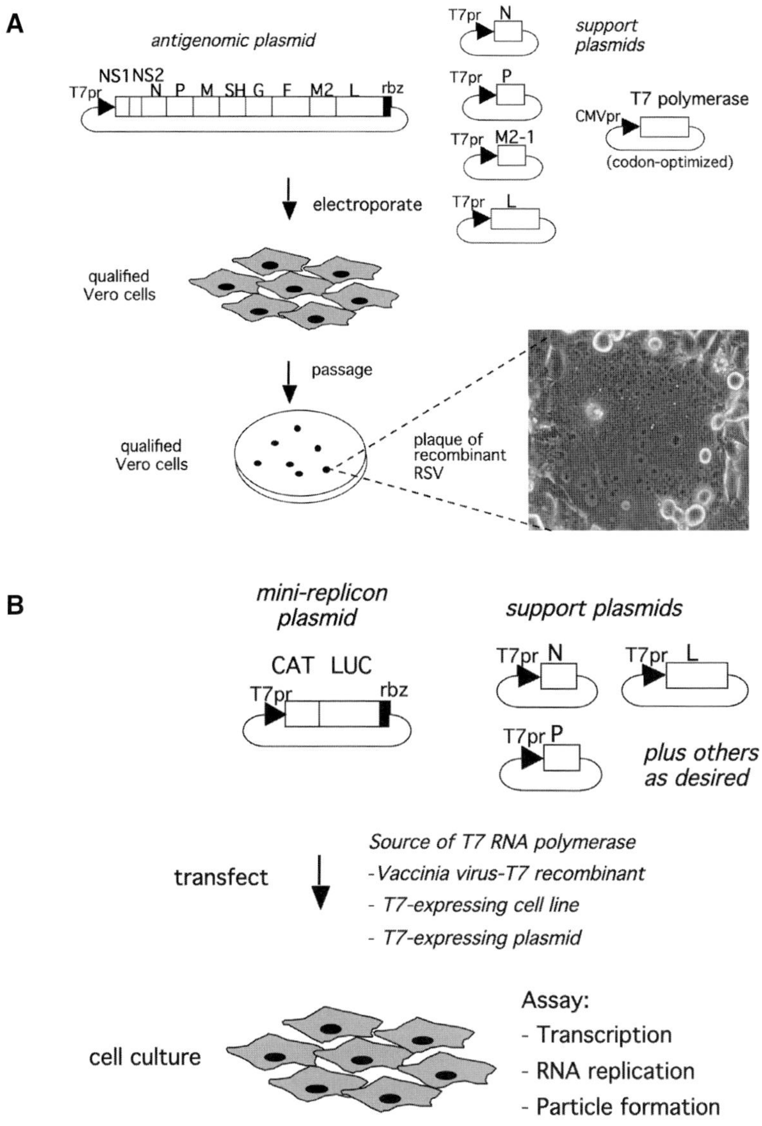

Figure 10.8 Reverse genetic systems. (A) Recovery of complete infectious virus; the photomicrograph of infected cell culture shows a viral plaque. This specifically shows conditions for recovering infectious virus that would be suitable for preparing attentuated vaccine strains for human use, including the expression of T7 RNA polymerase from a transfected plasmid and the use of Vero cells qualified for human products. (B) Helper-dependent mini-replicon system, illustrated with a mini-genome containing chloramphenicol acetyl transferase (CAT) and luciferase (LUC) marker genes. T7pr, T7 promoter; rbz, self-cleaving ribozyme.

infants and young children worldwide (note: LRI is usually defined as involving the respiratory tract below the larynx; severe LRI is usually defined as involving hospitalization; infants are defined as individuals of age one year or less). In a 13-year surveillance study in Washington, DC, RSV was detected in 23.3% of hospitalizations for respiratory tract disease in infants and young children, compared with values of 11.5%, 6%, 3.2%, and 5.2% for HPIV3, HPIV1, HPIV2, and influenza A virus, respectively (Murphy *et al.*, 1988b). Essentially the same value for RSV prevalence was confirmed in a more recent study of paediatric hospitalization using RT-PCR for detection

(Henrickson *et al.*, 2004). Similarly, in a prospective study of LRI in children followed from birth in Tucson, RSV was the virus that was identified most frequently (39.5% of all episodes), followed by HPIV3 (9%; Holberg *et al.*, 1991). RSV infection is much more likely to result in paediatric LRI compared with influenza or HPIV1, 2, or 3 (Fisher *et al.*, 1997). Rhinoviruses also cause a substantial burden of upper and lower tract disease in infants and young children, but are less frequently associated with hospitalization (Kusel *et al.*, 2006).

RSV infects and causes disease earlier in infancy than other common respiratory tract viruses (Fig. 10.9). HMPV, HPIV3 and influenza A virus also can infect early in life, but they do not infect as early and as frequently as RSV (Iwane *et al.*, 2004). Hospitalization for paediatric RSV disease is most frequent between 6 weeks and six months of life, with a peak incidence at 2–3 months of life (Fig. 10.10). In a prospective study in which children were followed from birth, 69% of the children were infected with RSV during the first year of life, and nearly everyone was infected by age the age of two years (Glezen *et al.*, 1986). There can be differences in the attack rate from year to year depending on the severity of the epidemic, but usually more than 50% of infants are infected during the first year of life (Kim *et al.*, 1973; Henderson *et al.*, 1979). Primary infection is usually symptomatic, and 20–40% of primary infections result in LRI (Kapikian *et al.*, 1961; Glezen *et al.*, 1986).

RSV also stands out in being able to re-infect symptomatically in subsequent RSV seasons without need of antigenic change. For example, in the prospective study noted above, 47% and 45% of the children were re-infected during the second and third years of life, respectively (Glezen *et al.*, 1986). Thus, children often are infected two or more times with RSV during the first few years of life. One study found that the severity of RSV disease was similar for the first and second infections early in life, but for subsequent infections there was a considerable reduction in disease severity reflecting increasing protective immunity (Henderson *et al.*, 1979). Re-infection at all ages is thought to usually be symptomatic, but severe RSV disease usually is limited to young infants and to adults at high risk owing to underlying disease or frail old age (see below). Other viruses such as HMPV and HPIV3 also can re-infect without need for significant antigen change, but this occurs less frequently and has less overall clinical impact compared with RSV.

In the United States alone, there are an estimated 73,400 to 126,300 annual hospitalizations for RSV bronchiolitis and pneumonia in infants under one year of age (Shay *et al.*, 1999). The impact of RSV also is considerable beyond infancy (i.e. beyond the first year of life). A large

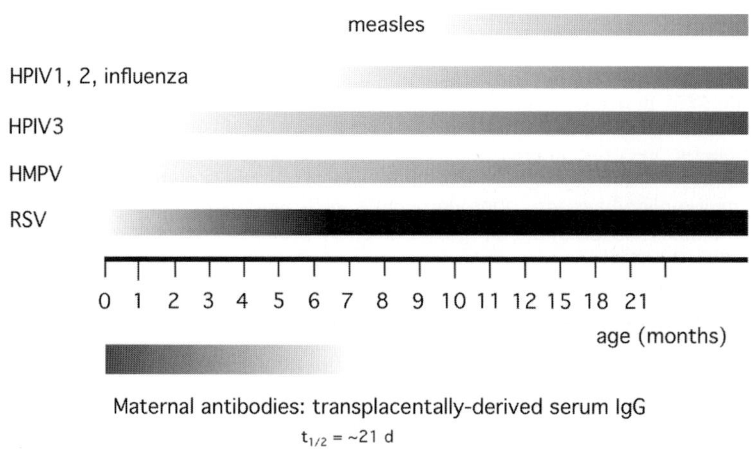

Figure 10.9 Idealized depiction of the timing in life of the acquisition of susceptibility to the indicated viruses, and of the decay of maternal serum antibodies. This illustrates that respiratory syncytial virus (RSV) infects more efficiently early in life compared with other common respiratory viruses or measles virus.

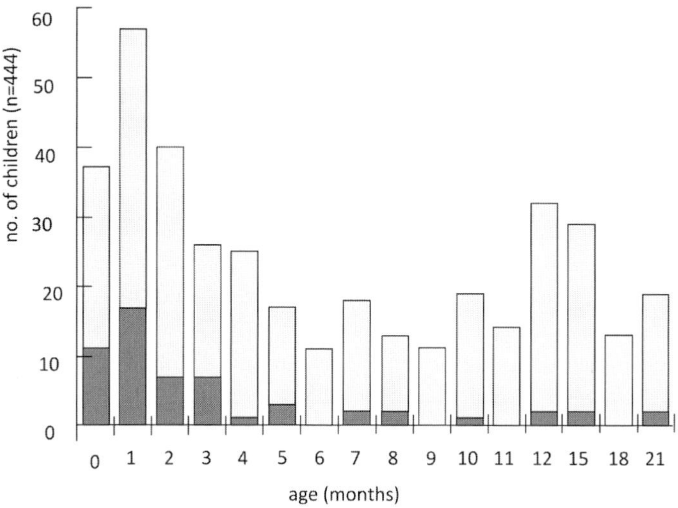

Figure 10.10 Age at time of hospitalization for respiratory syncytial virus (RSV) disease at the Johns Hopkins Hospital during 1993–96. The filled bars indicate more severe disease. Adapted from (Karron *et al.*, 1999).

prospective surveillance study estimated that, in the United States, 2.1 million children under five years of age require medical attention each year owing to RSV infection (Hall *et al.*, 2009). Seventy-eight per cent of these were over one year of age.

Paediatric deaths due to RSV are uncommon in developed countries. For example, in the United States, the paediatric mortality rate was estimated to be 5.3 and 0.9 per 100,000 per year for individuals under one year of age and one to four years of age, respectively (i.e. totalling approximately 300; Thompson *et al.*, 2003). The mortality rate appears to have dropped over time: in 1985, RSV was estimated by the Institute of Medicine, National Academy of Sciences USA, to be responsible for 4500 paediatric deaths annually in the United States. This decrease probably reflects improved supportive care.

Worldwide, RSV was estimated to cause 33.8 million new cases of LRI in children under five years of age in 2005, accounting for 22% of all LRI in that age group (Hall, 2010; Nair *et al.*, 2010). This resulted in an estimated 3.4 million hospitalizations and 66,000–199,000 deaths due to RSV-associated LRI. The majority of deaths (99%) occurred in developing countries, which can only be partially explained by the observation that the majority (90%) of children under five live in these countries. These estimates probably are under-estimates, since RSV testing was incomplete and community LRI undercounted.

Risk factors for paediatric RSV infection include factors that increase exposure, such as day care attendance, siblings under five years of age, and admission to the hospital during RSV season. Risk factors for severe disease include low titres of maternal antibodies, lack of previous RSV infection, premature birth (<36 months gestation, and especially <28 months), young age, chronic lung disease, congenital heart disease, immunodeficiency, asthma or a family history of asthma, poverty, exposure to tobacco smoke, and male gender. Infants are the most at risk for severe disease, with the risk being greatest for those less than three months of age, but also being high for those between 3–15 months (Fisher *et al.*, 1997). In one study, the estimated number of RSV hospitalizations per 1000 during the first year of life was 388 for infants with chronic lung disease, 92 for those with congenital heart disease, 66 for those born at 29 to <33 weeks, and 30 for term infants with no underlying disease (Boyce *et al.*, 2000). Fatality rates for children born prematurely or having underlying pulmonary or cardiac disease were similarly elevated (Welliver *et al.*, 2010). However, it is important to note that more than 50% of RSV hospitalizations occur in previously healthy full term infants (Boyce *et al.*, 2000). Another study noted that, among infants admitted to the intensive care unit for RSV bronchiolitis in Israel, 84% were born after 32 weeks gestation and lacked other known risk factors (Prais *et al.*, 2003).

Host genetic factors also play a role in susceptibility to paediatric RSV disease. Very high rates of hospitalization can occur in certain populations, such as rates of 5.3–24.6% reported for Alaskan natives less than one year of age, suggesting that at least some of this increase is due to a genetic predisposition (Karron *et al.*, 1999). In addition, a recent study found an increased concordance of severe paediatric RSV infection in identical versus fraternal twins (Thomsen *et al.*, 2008). A number of studies have described positive or negative associations of severe paediatric RSV disease with polymorphisms in a variety of host genes, notably ones involved in innate immunity and the Th2 subclass of $CD4^+$ T lymphocytes. These include genes for surfactant proteins, TLR4, IL-4, IL-8, IL-9, IL-10, IL-13, IL-18, RANTES (CCL5), CX3CR1, vitamin D receptor, nitric oxide synthetase, transcription factor Jun, IFNα5, and other innate immunity genes (Janssen *et al.*, 2007; Miyairi and DeVincenzo, 2008; Siezen *et al.*, 2009; Lofgren *et al.*, 2010; Schuurhof *et al.*, 2010). These gene-association studies remain preliminary and are sometimes contradictory, but they suggest that a number of genes contribute to resistance or susceptibility to RSV disease.

Infection of adults

RSV is gaining recognition as a major cause of disease in adults, for whom the re-infection rate is estimated at 5–10% per year (Falsey, 2007; Falsey *et al.*, 2005). In a recent study in the UK using RT-PCR to identify pathogens in adults 45 years and older with medically attended respiratory disease, RSV was identified in 10–22% (depending on the year) of cases compared with 13–42% with influenza A virus. Re-infection is even more frequent in adults exposed to the virus: during a typical RSV season, 25–50% of health care workers are re-infected, and family members of sick children are readily re-infected. RSV is considered to be second to influenza as a cause of medically significant respiratory tract disease in adulthood.

RSV can cause significant disease even in healthy adults. In one prospective study of healthy adults aged 18–60 years, 84% of documented RSV infections were symptomatic, with 74% involving only upper respiratory tract infection (URI) and 26% involving mild LRT (Hall *et al.*, 2001). Hospitalization of healthy, non-elderly adults for RSV disease is rare. However, severe disease – which can necessitate mechanical ventilation and can even result in death – can occur as a result of RSV infection of young, previously healthy adults (Shapiro and Jean, 2001). It is likely that adults become susceptible to re-infection as titres of secretory and serum RSV-neutralizing antibodies from previous exposure decline with time (Mills *et al.*, 1971; Hall *et al.*, 1991; Falsey and Walsh, 1998).

RSV is a particularly important cause of morbidity and mortality in the elderly (>65 years of age) and in adults with underlying pulmonary or cardiac disease or who are severely immunocompromised, especially with T-cell deficiencies (Whimbey *et al.*, 1996; Murata and Falsey, 2007). The mortality rate associated with severe RSV disease in adults with profound immunosuppression due to leukaemia or haematopoietic stem cell transplant can be as high as 80–100% (Whimbey *et al.*, 1996). RSV is estimated to cause on average 17,358 deaths annually in the USA, with 78% of these deaths in adults over the age of 65 years(Thompson *et al.*, 2003). Thus, in developed countries, deaths due to RSV are much more frequent in the elderly than in the paediatric population, whereas in less affluent countries the paediatric burden is greater.

Epidemics

RSV is highly infectious and easily spread, especially in settings of close contact such as day care facilities, hospitals, and families. Spread typically occurs by person-to-person contact, by large droplets, or by contact with fomites (i.e. contaminated objects) followed by self-inoculation. RSV thus differs from influenza virus, for which airborne spread is more common (Hall, 1982, 2007). Although RSV generally is labile, it is recoverable for hours from fomites, and for more than 1 h from rubber gloves (Hall, 1982). RSV infection can be reduced by infection control measures such as hand hygiene including gloving, washing, and hand sanitizers, limiting the sharing of toys, and reducing self-inoculation by avoiding touching one's face.

In temperate climates, RSV occurs in yearly epidemics that peak in the winter and early spring (Fig. 10.11), whereas in warmer climates RSV activity is continuous through the year with peaks in summer and early autumn (Halstead and Jenkins, 1998; Yusuf *et al.*, 2007). Globally, RSV activity is greatest during periods of moderate humidity and either cool (2–6°C) or warm (24–30°C) temperature (Yusuf *et al.*, 2007).

Five to seven genotypes representing both antigenic subgroups typically circulate during the same season in a given location, with the pattern of local strains gradually changing in successive years (Peret *et al.*, 1998, 2000). There typically are one or two dominant local genotypes that are replaced in dominance in successive years. In addition, there can be shifts in the predominance of subgroup A versus B occurring in one- or two-year cycles (Waris, 1991). The patterns of genotypes frequently are distinct between different geographic regions in the same season (Peret *et al.*, 2000). Thus, RSV epidemics do not involve wide regional spread of predominant new strains, as is the case with influenza A virus. Rather, epidemics appear to involve local endemic strains as well new genotypes introduced from other regions. The appearance in 1998 of a new genotype of subgroup B, which is called BA and contains a 60-nucleotide duplication in the coding region of the G gene, provided the opportunity to characterize the global spread and evolution of an individual genotype (Trento *et al.*, 2010). This confirmed the shifting pattern of local dominance and replacement, the emergence of new variants, and the ability of an RSV genotype to disseminate worldwide.

Pathogenesis

Humans are the only natural hosts for RSV, although non-human primates and possibly other experimental animals can be infected from their human handlers. In cell culture, the virus has a wider host range, as already noted.

Cell culture

In vitro, RSV-infected cells develop large electron-dense cytoplasmic inclusion bodies of up to

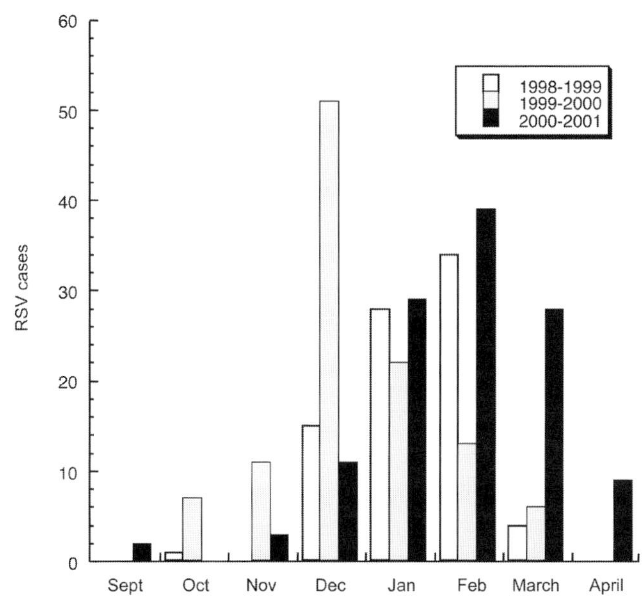

Figure 10.11 Variation in respiratory syncytial virus (RSV) epidemics during three successive years at the Johns Hopkins Hospital, based on paediatric hospitalizations for RSV disease. Unpublished data kindly provided by Dr. Ruth Karron, Johns Hopkins Bloomberg School of Public Health.

several microns in diameter that include the N, P, and M2-1 proteins and presumably contain active nucleocapsids (Garcia *et al.*, 1993). Infected cells also develop long surface filaments that bear viral antigen and probably give rise to filamentous virus particles (Fig. 10.3; Santangelo and Bao, 2007). RSV infection has modest inhibitory effects on cellular DNA and RNA synthesis and little effect on overall protein synthesis (Levine *et al.*, 1977). RSV infection activates protein kinase R (PKR), which ordinarily phosphorylates eukaryotic translation factor 2a and thereby inhibits cellular translation as part of antiviral defence (Groskreutz *et al.*, 2010). However, the RSV N protein binds to PKR and blocks its interaction with the translation factor, thereby preventing inhibition of translation and presumably favouring virus replication (Groskreutz *et al.*, 2010). Stress granules may be another potential host defence mechanism to restrict virus replication, but the formation of stress granules is blocked by RSV RNA derived from the trailer region (Hanley *et al.*, 2010). Apoptosis of RSV-infected cells develops slowly and is inhibited by the NS1, NS2, and SH proteins, as already noted. In typical monolayer cultures, syncytia appear during infection and result in cell death and destruction of the culture, but this prominent *in vitro* effect is much less frequent *in vivo*, as noted below.

The characteristics of infection are quite different in human adenoid or organized HAE cultures (Tristram *et al.*, 1998; Zhang *et al.*, 2002; Wright *et al.*, 2005). RSV appeared to preferentially infect ciliated cells, and infection was limited to the apical surface and did not spread to underlying cells (Fig. 10.12). Infection was maintained for several weeks, with infected cells being shed and replaced with little visible alteration of the microscopic appearance of the tissue (Zhang *et al.*, 2002; Wright *et al.*, 2005). Cilial beating usually, but not always, became impaired during infection. Syncytia were not observed, probably because the F glycoprotein was expressed on the apical surface and did not contact adjacent cells. The results were similar whether the virus was an extensively passaged laboratory strain or a recent clinical isolate (Wright *et al.*, 2005). RSV contrasted sharply with influenza A virus examined in parallel, which quickly spread to underlying cells and was rapidly destructive (Zhang *et al.*, 2002; Wright *et al.*, 2005).

Infection of the host

In the human host, the incubation period of RSV from exposure to illness is 3–5 days, based on observations of outbreaks in closed populations as well as experimental infections in volunteers (Kapikian *et al.*, 1961; Karron *et al.*, 1997b;

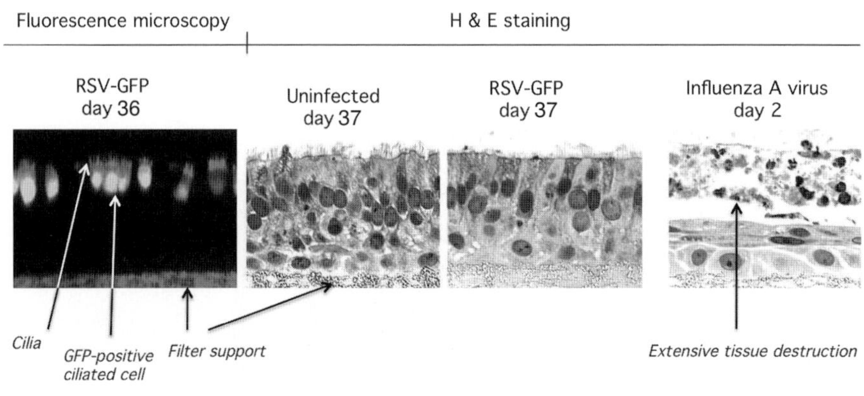

Figure 10.12 Infection of an *in vitro* model of human airway epithelium (HAE) with respiratory syncytial virus (RSV) and influenza A virus, shown in cross-section. The left hand panel is a fluorescence photomicrograph of cells 36 days following infection with RSV expressing the green fluorescent protein (GFP), also stained with antibody specific to cilia. The three panels to the right show an uninfected culture compared with cultures 37 days following infection with RSV and two days following infection with influenza A virus (Zhang *et al.*, 2002). In all of the panels, the cells were counterstained with haematoxylin and eosin.

Devincenzo *et al.*, 2010b; Fig. 10.13). The virus replicates initially in the nasopharynx. Lower respiratory tract disease signs usually appear 1–3 days following the onset of rhinorrhoea, presumably reflecting viral spread to the bronchi and bronchioles. Infants hospitalized for RSV disease shed infectious virus in nasal secretions with typical titres of approximately 10^5 50%-tissue-culture-infectious-doses ($TCID_{50}$) per ml, although it is common to observe infants shedding virus at 10- to 100-fold higher titres (Hall *et al.*, 1975; Wright *et al.*, 2002; DeVincenzo *et al.*, 2005). Following hospitalization, shedding of infectious virus typically continues for approximately a week, but in some individuals it can persist for several weeks, especially in young infants. With the use of quantitative RT-PCR instead of conventional culture, detection of RSV in secretions continues for even longer, suggesting that virus shedding persists but is neutralized by antibodies made in response to the infection (DeVincenzo *et al.*, 2010a; Lukens *et al.*, 2010). Viral titres in the lower respiratory tract are not known, although histopathology of fatal RSV cases provided evidence that viral replication in the alveoli does occur and can be abundant in cases of pneumonia (Gardner *et al.*, 1970; Johnson *et al.*, 2007; Welliver *et al.*, 2007). Clinical recovery usually coincides with the termination of shedding detected by conventional culture, although some infants continue to shed for weeks following recovery.

Histopathological findings from fatal RSV infections indicate that RSV infection of the respiratory tract is limited to the superficial cells, consistent with the findings in HAE cultures (Aherne *et al.*, 1970; Gardner *et al.*, 1970; Johnson *et al.*, 2007; Welliver *et al.*, 2007). There is necrosis and occasional proliferation of the bronchiolar epithelium and destruction of ciliated cells, but basal cells are spared. In one study, tissue damage was judged to be greater with RSV than with influenza virus (Welliver *et al.*, 2007). Ciliated epithelial cells rarely reappear before two weeks post-infection, and complete restoration requires 4–8 weeks (Hall, 2001), similar to the duration of post-infection altered lung function and airway reactivity in adults (Hall *et al.*, 1978b). Syncytia are sometimes observed, but are not prominent. However, syncytium formation and giant cell pneumonia are hallmarks of infection in individuals with extreme T cell deficiency. Inflammatory infiltrates were centred on the bronchial and pulmonary arterioles and included activated monocytes, double negative T cells, alveolar macrophages, $CD8^+$ T cells, and neutrophils. Peribronchiolar infiltrates included lymphocytes, plasma cells, and macrophages. Neutrophils were present between vascular structures and small

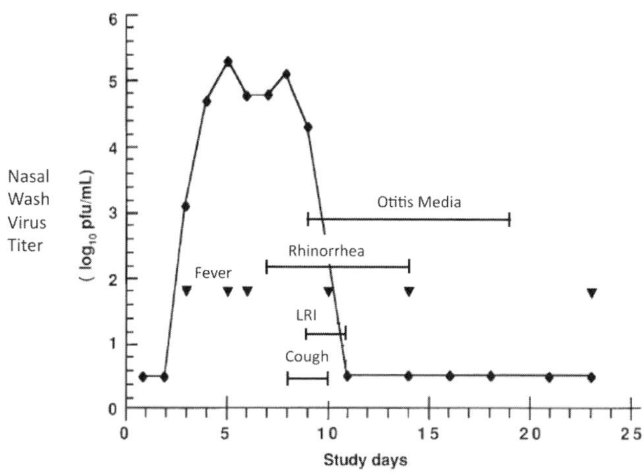

Figure 10.13 Time course of viral shedding and disease signs following infection of a seronegative vaccine with $10^{4 PFU}$ of an experimental live attenuated RSV vaccine called cpts248/955. Adapted from (Karrone *et al.*, 1997b). This is thought to exemplify a primary RSV infection with moderate disease.

airways. Most inflammatory cells were concentrated submuscular to the mucosa, but many cells also entered the airway epithelium and lumen (Johnson *et al.*, 2007). Oedema of submucosal tissue together with the presence in the airways of excessive mucus, immune cells, and sloughed epithelial cells contribute to obstruction the bronchioles and alveoli, causing either collapse or overinflation of distal portions of the airway. In cases of pneumonia, the interalveolar walls thicken as a result of mononuclear cell infiltration, and the alveolar spaces may fill with fluid. RSV antigen and tissue damage tends to be patchy rather than diffuse. Information is lacking on pathogenic changes in milder cases of RSV disease.

RSV infection is largely limited to the respiratory tract. Infectious virus has not been recovered from the blood of RSV-infected humans, although antigen and/or nucleic acid can frequently be detected in the blood of RSV-infected infants (Rohwedder *et al.*, 1998). Spread of RSV outside of the respiratory tract was detected in a small proportion of immunosuppressed cotton rats, with infectious virus cultivated from the liver and kidneys (Johnson *et al.*, 1982). RSV also has been cultured from the myocardium of an infected infant with combined immunodeficiency as well as from a liver biopsy from an infected immunocompetent infant. Viral RNA was detected in cerebral spinal fluid of an infected immunocompetent infant (Fishaut *et al.*, 1980; Eisenhut, 2006). This suggests that RSV has the potential to spread beyond the respiratory tract in some individuals, but usually is restricted by host immunity. A number of different extrapulmonary disease manifestations have been noted in some infants with severe RSV disease (Eisenhut, 2006), which deserves further study. RSV does not appear to replicate when administered subcutaneously to infants and children or intramuscularly to cotton rats (Buynak *et al.*, 1978).

RSV is usually considered to be an acute, self-limiting infection. The virus is rarely recovered in the absence of respiratory disease. However, there is some evidence of persistence. Infectious virus has been recovered from guinea pigs 60–100 days post infection (Dakhama *et al.*, 1997), and was recovered sporadically from mice 100 days post infection upon T cell depletion (Schwarze *et al.*, 2004). BRSV RNA and proteins have been detected in bovine pulmonary lymph nodes 71 days following infection, and there was indirect evidence for *in vitro* infectivity from isolated bovine B lymphocytes (Valarcher *et al.*, 2001). With human RSV, there is evidence that the virus might persist in some individuals with stable chronic obstructive pulmonary disease (Sikkel *et al.*, 2008), but another study suggested its presence was linked to acute symptomatic infection (Falsey *et al.*, 2006a). Thus, whether RSV can persist in humans as either a reservoir for epidemics or as a significant feature of RSV disease remains unclear.

A number of studies have indicated an association between RSV LRI in infancy and lingering abnormalities in pulmonary function, including reduced oxygenation, deficits in pulmonary function tests suggestive of peripheral airway obstruction, and airway reactivity including coughing and wheezing (Hall *et al.*, 1984; Long *et al.*, 1995; Noble *et al.*, 1997; Stein *et al.*, 1999). These effects could be detected for at least as long as the first 11 years of life; in one study they subsided by age 13 (Stein *et al.*, 1999), whereas in another study effects were still detected at the age of 18 (Sigurs *et al.*, 2010). Rates of recurrent respiratory tract abnormalities in infants previously hospitalized for RSV disease are high: 82% at 2 years, 69% at 3 years, and 30% at 6–8 years following hospitalization (Long *et al.*, 1995). Most of these studies did not observe a link between these lingering respiratory abnormalities and allergic sensitization and the development of persistent asthma, but other studies have suggested such a link (Sigurs *et al.*, 2010; Sigurs *et al.*, 2005), and this remains an area of controversy. A recent study showed that previous passive immunoprophylaxis using the RSV-neutralizing MAb palivizumab (described in 'Prevention and control') decreased the relative risk of recurrent wheezing by 80% in non-atopic (i.e. no predisposition to allergic disease) children, but had no effect in children with a family history of atopy (Simoes *et al.*, 2010). This suggests that RSV LRI indeed increases the risk of wheezing in non-atopic children but has no discernible effect in atopic individuals, perhaps because the risk already is high. It is not clear whether the sequelae that follow RSV LRI were caused *de novo* by the viral infection, or whether these individuals had an underlying disposition – such as underlying anatomical or functional abnormalities or a tendency to airway reactivity – that existed before infection and indeed may have made them vulnerable to RSV LRI. Several studies suggest that the latter often is the case (Young *et al.*, 1995; Castro-Rodriguez *et al.*, 1999; Thomsen *et al.*, 2009).

Pathogenic mechanisms

There is a broad heterogeneity of RSV disease in primary infection, encompassing upper respiratory tract symptoms, lower respiratory tract symptoms of varied severity with and without wheeze, death in rare cases, airway reactivity that can last through childhood, and possible allergic sensitization. The available information indicates that this diversity is not due to viral strain differences. A few reports have suggested that subgroup A or a particular genotype of subgroup A is associated with greater clinical disease severity, but the majority of studies have not found such a link (Brandenburg *et al.*, 2000b; Martinello *et al.*, 2002; Smyth *et al.*, 2002). Recently, an RSV strain called line 19 was shown to induce increased lung IL-13 expression and mucus secretion in BALB/c mice, an effect that mapped to the F protein (Moore *et al.*, 2009). While this suggests that some heterogeneity in strain virulence may exist, at the present there is no evidence that this is of sufficient magnitude to be relevant clinically. Instead, the heterogeneity of RSV disease probably reflects host differences such as age, maternal antibody titre, and the many other risk factors already noted.

RSV is a highly infectious virus. When introduced into any setting, it rapidly infects most susceptible individuals. Its ability to re-infect and be shed by RSV-experienced individuals facilitates viral spread. These attributes certainly contribute to its high attack rate and epidemic nature.

There is a continuing controversy as to what proportion of RSV disease is due to direct viral damage, and what proportion is due to the host immune response. It seems likely that both effects play a role. A number of studies have noted a positive correlation between viral load and disease severity based on the results from prospective studies of natural infection and experimental infection of wild-type virus and attenuated vaccine candidates (DeVincenzo *et al.*, 2005, 2010a; Karron *et al.*, 1997b), although there also are contradictory data (Hall *et al.*, 1975, 1976; Wright *et al.*, 2002). In any event, this does not necessarily indicate a major role for direct viral damage, since the magnitude of the immune response probably increases with increasing viral load. Studies to attenuate RSV to make live vaccines have not identified any specific viral virulence factors: every mutation or gene deletion that attenuates the virus appears to do so by reducing viral load.

RSV directly damages the airway epithelium by infection, necrosis, and cilial damage. In experimental infection in adults, the onset of disease coincided with the onset of viral shedding and the timing of peak viral load coincided with peak disease severity (DeVincenzo *et al.*, 2010a). This close coincidence suggested an importance of direct viral effects in RSV disease. However, in an experimental infection causing LRI in a seronegative child – an unusual situation that occurred as a result of an incompletely attenuated experimental vaccine – disease seemed to overlap and trail viral replication rather than being closely coincident (Fig. 10.13). RSV can cause very severe disease in immunocompromised individuals, as noted, and causes severe pulmonary pathology in immunosuppressed cotton rats (Wong *et al.*, 1985). This indicates that an intact adaptive immune system is not necessary for severe RSV disease.

On the other hand, the results with HAE cultures noted above suggest that RSV is not a rapidly destructive virus and does not efficiently penetrate beyond the superficial layer of cells. In RSV-infected cotton rats, treatment with RSV-neutralizing antibodies rapidly reduced pulmonary titres but had little effect on lung histopathology. In contrast, treatment with anti-inflammatory glucocorticoids reduced lung histopathology even though clearance of the virus was slowed, suggesting that the host immune response played the major role in the observed disease in this model (Prince *et al.*, 2000). In the mouse model, depletion of the $CD4^+$ or $CD8^+$ T lymphocyte populations resulted in markedly prolonged viral replication and yet disease was ablated (Graham *et al.*, 1991b). Treatment of infected humans with virus-neutralizing antibodies or antiviral drugs that effect a reduction in viral titres does not reduce disease (see Treatment, below section), suggesting that factors in addition to direct viral damage are important to disease. These findings suggest that the host response can make a substantial contribution to RSV disease.

There are three major models for immune-mediated contributions to RSV disease, which are not mutually exclusive. These are (i) an overly robust CD8$^+$ CTL response, resulting in excessive tissue damage; (ii) an overly robust CD4$^+$ Th2 response, providing a response that is not optimally effective against intracellular pathogens and is associated with excessive mucus production and airway reactivity, among other effects; and (iii) an overly robust innate immune response, resulting in heightened inflammation due to an excessive pulmonary influx of immune cells. These are discussed in the following section on immune responses.

Additional pathogenic mechanisms also have been described. As noted, the G protein contains a fractalkine-like CX3C domain. Fractalkine, like G, is produced in membrane bound and secreted forms. These function as chemoattractant and cell adhesion molecules to mediate the influx of CX3CR1-bearing leukocytes, which include subsets of NK cells and CD4$^+$ and CD8$^+$ T cells. In one study, infection of mice with an RSV mutant in which the CX3C motif was ablated was associated with a substantial increase in pulmonary NK cells and CD4$^+$ and CD8$^+$ T cells compared with wild-type RSV (Harcourt *et al.*, 2006), implying that fractalkine mimicry has the effect of reducing the cellular response to infection. The central conserved domain of G, which includes the CX3C domain, has also been shown to strongly down-regulate the NF-κB-mediated inflammatory responses in human monocytes in response to agonists of TLR2, TLR4, and TLR9 (Polack *et al.*, 2005). This indicated that G can suppress the host innate immune response.

The secreted form of the G protein helps RSV escape antibody-mediated neutralization (Bukreyev *et al.*, 2008). This was demonstrated by comparing wild-type recombinant RSV with a version containing an amino acid substitution that removes the start codon for the secreted form. In the absence of secreted G, RSV was more sensitive to neutralization by G-specific antibodies *in vitro*, indicating that secreted G acts as an antigen decoy. Also, in mice that received passive antibodies prior to infection, RSV that lacked G was more sensitive to restriction by both G- and F-specific antibodies. This effect required the presence of inflammatory cells in the lung and was Fc receptor-dependent (Bukreyev *et al.*, 2008). These cells may normally restrict RSV replication by uptake and destruction of antibody–antigen complexes as well as destruction of virus-infected cells by antibody-dependent cell-mediated cytotoxicity. The ability of secreted G to reduce these cell-mediated events might be an effect of fractalkine mimicry and the reduced influx of CX3CR1$^+$ leucocytes.

The G protein also can influence cytokine production by T lymphocytes in the mouse model, and provides an example of how an isolated antigen can induce a biased response. Immunization of BALB/c mice with a vaccinia virus recombinant expressing the G protein induces a Th2-biased response that leads to pulmonary eosinophilia and increased lung pathology following RSV challenge (Openshaw *et al.*, 1992). This was found to be the result of a Vβ14-restricted clonal response to a single epitope in G (contained in amino acids 185–198; Varga *et al.*, 2000). An important factor in this response was the failure of G to induce CD8$^+$ T cells in this inbred strain, which otherwise down-regulate the Th2-biased response (Olson *et al.*, 2008; Stevens *et al.*, 2009). This illustrates how a biased immune response can occur in a particular genetic background and can prime for disease. This model of priming with G protein and challenging with RSV is commonly used to investigate RSV disease. However, this model is somewhat artificial since disease depends on priming in the absence of a CD8$^+$ T-cell response, depends on this clonal CD4$^+$ T-cell response, and is observed only following challenge. In contrast, authentic RSV disease is most severe in RSV-naïve individuals, occurs in the context of a CD8$^+$ T-cell response, and does not depend on this clonal CD4$^+$ response.

RSV infection of epithelial cells induces a rapid inhibition of Na+ transport, resulting in apical fluid accumulation (Kunzelmann *et al.*, 2007). This might be a host mechanism to dilute and remove irritants, but also might contribute to excess fluid and virus spread. This effect was not unique to RSV, and was observed with parainfluenza and influenza viruses as well as with bacterial pathogens.

The ability of RSV to infect so early in life probably contributes to its virulence. Young infants are less able to tolerate acute disease in general. They also are less able to restrict and clear infection and prevent re-infection because of immunological immaturity (see Immunity, next section). The diameter of the lumen of the bronchioles is smaller with young age and small body size, making these small airways particularly susceptible to obstruction by oedema, secretions, and immune and exfoliated cells (Hogg *et al.*, 1970). Early infancy is a time of considerable lung growth and development, and the lungs of infants born prematurely are incompletely developed (Gern *et al.*, 2005). It can be speculated that lungs that are undergoing growth and development may be more susceptible to disease, and that severe infection may have long-term consequences for lung function. As one example, infants with severe RSV infection express increased levels of neurotrophic factors and receptors (Tortorolo *et al.*, 2005). Studies in a rodent model suggest that this can lead to remodelling of neural networks that innervate the respiratory mucosa, contributing to recurrent airway inflammation and hyperreactivity (Piedimonte *et al.*, 2004). Children with a history of severe RSV bronchiolitis early in life have a higher index of obstructive sleep apnoea, which is associated with increased expression of neurotrophic factors and receptors (Snow *et al.*, 2009).

Immune responses

RSV is an acute infection in which the innate and adaptive host immune responses restrict virus replication and mediate clearance and recovery. Young infants are partially protected by maternal serum IgG antibodies. Infection induces mucosal and systemic virus-neutralizing antibody responses as well as cell-mediated responses. Secretory and serum antibodies play the primary role in protecting against re-infection. The immune response to RSV has broad similarities with the responses to other respiratory viruses such as the influenza and parainfluenza viruses. However, there is ongoing controversy whether aspects of the response to RSV play a major role in RSV disease, and whether RSV mediates substantial inhibition of host immune responses.

Innate and inflammatory responses

The first line of host defence includes the physical barriers of the glycocalyx of the superficial epithelial cells, the secreted mucus layer, and ciliary sweeping. Pulmonary surfactants, which are lipoprotein complexes produced by type II alveolar cells, play a role in restricting RSV. Surfactant protein A (SP-A), considered a C-type lectin, has been shown to bind to the RSV F protein, neutralize viral infectivity *in vitro*, and promote opsonization of RSV by monocytes and macrophages *in vitro* (Ghildyal *et al.*, 1999; Barr *et al.*, 2000). Similarly, SP-D binds both F and G and enhances uptake of RSV by alveolar macrophages and neutrophils *in vitro* (LeVine *et al.*, 2004). Mice with targeted disruption of SP-A, SP-C, or SP-D exhibited increased pulmonary RSV titres and disease, which was ameliorated by topical administration of the missing surfactant (LeVine *et al.*, 1999; LeVine *et al.*, 2004; Glasser *et al.*, 2009). Genetic polymorphisms in SP-A, SP-B, and SP-D have been associated with severe paediatric RSV disease (Puthothu *et al.*, 2007; Miyairi and DeVincenzo, 2008; El Saleeby *et al.*, 2010), and infants with severe RSV disease were found to have reduced surfactant concentration and function (Kerr and Paton, 1999).

Receptors that recognize viral molecular patterns play a major role in initiating host responses to RSV. These include Toll-like receptors (TLRs) present on the plasma membrane and endosomes (Klein Klouwenberg *et al.*, 2009). The RSV F protein stimulates TLR4 on leukocytes independent of viral replication, leading to activation of the NFκB pathway to produce cytokines such IL-6 and IL-8 and activation of IRF3 to induce type I IFN (Kurt-Jones *et al.*, 2000). A significant role for TLR4 signalling in host defence is indicated by the observation that TLR4-deficient mice are less efficient than normal mice in resolving RSV infection (Klein Klouwenberg *et al.*, 2009; Kurt-Jones *et al.*, 2000). In addition, a number of studies have linked severe paediatric RSV disease with genetic polymorphisms in TLR4 reported to reduce its activation by lipopolysaccharide agonist. This is consistent with a significant protective role in host defence, although other studies did not detect this association (Klein Klouwenberg *et al.*, 2009). RSV has also been shown to activate signalling from the TLR2/TLR6 heterodimer complex on the surface of leukocytes, resulting in the expression of pro-inflammatory cytokines, neutrophil influx, and dendritic cell activation in mice (Murawski *et al.*, 2009). Following RSV challenge, mice lacking TLR2 or TLR6 had higher titres of RSV compared with normal mice (Fuentes *et al.*, 2007). RSV also activates TLR3 in epithelial cells, resulting in the production of inflammatory cytokine and chemokines (Rudd *et al.*, 2005). Studies in TLR3-knock-out mice indicated that TLR3 may not make a significant contribution to restricting RSV replication, but helps prevent IL-13 up-regulation and increased mucus production (Rudd *et al.*, 2006).

Intracellular synthesis of RSV RNA is detected by the cytoplasmic sensor RIG-I, which can initiate signalling through both the IRF3 and NFκB pathways (Yoboua *et al.*, 2010). Plasmacytoid DCs (pDCs) are a major source of IFNα/β, and are stimulated by RSV both through a cytoplasmic sensor (presumably RIG-I) and via TLR7/8, which detect viral single-stranded RNA or its degradation products (Schlender *et al.*, 2005). RSV can also inhibit signalling through these TLR- and RIG-I- mediated pathways in pDCs. We have already noted that the RSV NS proteins block signalling to IRF3 and NFκB at several points. RSV infection also causes degradation of MAVS, which is immediately downstream of RIG-I (Yoboua *et al.*, 2010). TLRs also can potentially contribute to pathogenesis: viral infection increases the expression of TLR3 and 4, promoting sensitivity both to RSV and to possible co-infecting viruses or bacteria.

Cultured airway epithelial cells produce a broad array of pro-inflammatory cytokines in response to RSV infection, including IL-8/CXCL8, RANTES/CCL5, MIP-1α/CCL3, MCP-1/CCL2, IP-10/CXCL10, and IL-6 (Becker and Soukup, 1999; Mellow *et al.*, 2004; Miller *et al.*, 2004). Up-regulation was dependent on virus replication (Miller *et al.*, 2004). Macrophages and neutrophils exposed to RSV *in vitro* expressed some of the same cytokines, such as IL-8, as well as additional species, such as IL-1, IL-10, and TNFα in the case of macrophages. Many of these same cytokines have been found in secretions for the upper and lower respiratory tract of infants with RSV bronchiolitis, including IL-8, RANTES, MIP-1α, MCP-1, IP-10, and IL-6 (Bonville *et al.*, 1999; Harrison *et al.*, 1999).

The expression of pro-inflammatory cytokines results in recruitment of innate and adaptive immune cells to the infected lung. The total number of cells recovered in washes from the upper and lower respiratory tract of infants was several fold higher in cases of severe RSV disease compared with uninfected controls (Everard *et al.*, 1994; McNamara *et al.*, 2003; Heidema *et al.*, 2007). Neutrophils were by far the most abundant immune cells present in the airway, comprising 76–85% of recovered cells in the lower respiratory tract and 93% in the upper, with lymphocytes and mononuclear cells being present at ≤10% each and eosinophils at <1% (Everard *et al.*, 1994; McNamara *et al.*, 2003). As already noted, histopathological analysis of fatal RSV cases indicated the presence in the lung parenchyma of monocytes, alveolar macrophages, lymphocytes, plasma cells, and neutrophils (Johnson *et al.*, 2007). Incoming immune cells express cytokines and other pro-inflammatory molecules such as cysteinyl leukotrienes that can heighten the inflammatory response. Granulocytes release peptides and enzymes that can be antimicrobial or mediate cytotoxicity, such as neutrophil elastase and eosinophil-associated ribonucleases (Borregaard *et al.*, 2007; Rosenberg and Domachowske, 2001). Neutrophils can adhere to infected cells via cell surface adhesion molecules that are up-regulated by viral infection (Stark *et al.*, 1996; Wang *et al.*, 1998). Increased levels of eosinophil cationic protein and neutrophil elastase have been observed in respiratory washes from infants with severe RSV disease, indicating the presence and activation of these granulocytes (Garofalo *et al.*, 1992; Abu-Harb *et al.*, 1999). One study suggested that the influx of inflammatory cells into the airways was less in pre-term infants compared with term infants (McNamara *et al.*, 2003), which might reflect a reduced response due to immunological immaturity (Adkins *et al.*, 2004; Levy, 2007).

Inflammation plays a protective role in restricting RSV replication, but also probably contributes to disease. This is exemplified by the study in cotton rats noted above in which anti-inflammatory treatment with a glucocorticosteroid dramatically reduced pulmonary pathology, but resulted in delayed clearance of the virus (Prince *et al.*, 2000). Mice that are hypereosinophilic due to increased IL-5 expression had increased viral clearance, indicating a protective role of eosinophils (Phipps *et al.*, 2007). Conversely, the amount of mRNA for IL-8 – a cytokine that promotes neutrophil chemotaxis and survival – contained in nasal aspirates of children with bronchiolitis correlated with the severity of disease (Smyth *et al.*, 2002), and an IL-8 haplotype characterized by increased IL-8 gene transcription was associated with increased susceptibility to RSV disease (Hacking *et al.*, 2004). Also, bronchoalveolar washes from infants with severe RSV disease had increased levels of mRNA for IL-9, which is secreted by neutrophils and promotes eosinophilia, chemokine expression by respiratory epithelial cells, bronchial hyperresponsiveness, and increased mucus production (McNamara *et al.*, 2004). In infants with severe RSV disease, the kinetics of neutrophil influx into the airways closely followed the peak of clinical symptoms and viral load (Lukens *et al.*, 2010). These findings suggest that neutrophilia in particular has the potential to contribute to RSV disease as well to viral clearance.

IFNα/β were detected less frequently and at lower levels in nasal secretions from infants, young children, and adults infected with RSV, compared with influenza and parainfluenza viruses (Hall *et al.*, 1978a, 1981; McIntosh, 1978). RSV infection and replication are more resistant to prophylaxis or treatment with IFNα compared with influenza and parainfluenza viruses (Sung *et al.*, 1993; Gao *et al.*, 2010). The low production and low efficacy probably reflect the multiple ways in which RSV inhibits type I IFN induction and signalling. In mice with a targeted gene knock-out of STAT1, there was little or no increase in virus replication compared with control mice, but there was increased lung pathology attributed to impaired stimulation of Th1 $CD4^+$ T lymphocytes (Durbin *et al.*, 2002). Thus, while RSV does not appear to be efficiently restricted *in vivo* by type I IFN, these cytokines appear play a role in modulating pathogenesis.

Antibodies

Secretory and serum antibodies play an important role in restricting RSV infection and in preventing re-infection. In young infants, the decrease in virus shedding during infection corresponds with the appearance of secretory IgA antibodies (McIntosh *et al.*, 1978). The clinical success of RSV immunoprophylaxis using passively-administered neutralizing antibodies confirms the protective role of serum antibodies (see Prevention and Control section). Experimental infections in adults showed that resistance to re-infection correlated with the titre of RSV-specific secretory or serum antibodies (Mills *et al.*, 1971; Hall *et al.*, 1991), and there are similar findings for natural infection (Walsh and Falsey, 2004b).

The RSV-naïve young infant typically is not RSV-seronegative, but rather possesses maternally derived serum IgG antibodies including ones specific to RSV (Crowe, 2001). Maternal antibodies are acquired by transcytosis across the placenta beginning at ~28 weeks of gestation and continuing until the time of birth. Thus, infants who are born prematurely have lower titres than if they were full term, which is one factor in their greater susceptibility to RSV infection and disease. The IgG1 subclass is preferentially transferred, but otherwise the titre and specificity of maternally derived IgG antibodies in the neonate are similar to those of the mother. Thereafter, these antibodies decay with a half-life of approximately 21 days. Breast milk may also contribute protective antibodies, although this has not been clearly documented and probably is a minor effect.

Epidemiological studies suggest that RSV disease in infants born with high levels of maternal RSV-neutralizing antibodies was milder and occurred at an older age than in infants with lower antibody levels (Glezen *et al.*, 1981). Studies in cotton rats (Prince *et al.*, 1985) as well as the clinical experience with RSV-neutralizing antibodies in prophylaxis (see Prevention and Control section) confirm that RSV-neutralizing serum antibodies are protective. However, serum

IgG antibodies gain access to the respiratory tract primarily by the relatively inefficient process of passive transudation. Because of this, there is a gradient in titre of approximately 1:350 between circulation and the lumen of the respiratory tract (Wagner *et al.*, 1987a). Thus, relatively high serum titres are needed to confer substantial protection in the respiratory tract. For example, passive transfer studies in the cotton rat showed that serum neutralizing titres of ≥1:380 conferred solid protection of the lower respiratory tract against virus replication, but titres of ≤1:100 did not protect (Prince *et al.*, 1985), and the clinical experience with RSV-neutralizing antibodies in immunoprophylaxis is similar. Furthermore, in the cotton rat experiments, even high serum antibody titres were inefficient in protecting the upper respiratory tract against RSV infection. Thus, protection conferred by serum antibodies obtained by maternal transfer or prior infection depends on a high titre of RSV-neutralizing antibodies and is particularly inefficient in the upper respiratory tract. In the case of maternal antibodies, protection wanes during the first weeks of life as maternal antibodies decay.

Both serum and secretory antibodies are made in response to RSV infection in young infants. In hospitalized infants, virus-specific serum IgA and IgG responses were detected 10 days following the onset of illness, and peak titres were achieved at 3 to 4 weeks (Welliver *et al.*, 1980). The magnitude of the serum and secretory responses in young infants are reduced compared with older individuals. For example, in a study of infants and young children undergoing a primary RSV infection, the titres of serum antibodies that bound to the F or G glycoprotein or that neutralized RSV were 8- to 10-fold lower in individuals of 4–8 months compared with those of 9–21 months of age (Murphy *et al.*, 1986b). The frequency of the secretory IgA response to RSV in infected infants was directly related to age, and these antibodies often were non-neutralizing *in vitro* (Kaul *et al.*, 1981; McIntosh *et al.*, 1978). In addition, antibody responses in the young infant are short-lived compared with older individuals, such that post-infection increases in RSV-specific IgA and IgG responses were absent or at very low titre 1 year later (Welliver *et al.*, 1980). Thus, secretory and serum antibody responses are reduced in infants of around 8 months of age compared with older individuals with respect to the (i) magnitude, (ii) quality and (iii) longevity of the response. The reduced nature of the antibody response to RSV in young infants is at least partially responsible for the frequent re-infections observed during the first years of life.

The reduced antibody responses to RSV early in life probably are due in part to immunological immaturity (Adkins *et al.*, 2004; Levy, 2007). Immunological immaturity has been characterized with a variety of antigens and is generally known to have broad effects on reducing innate, antibody, and cellular responses, although these effects are not absolute and adult-like responses have been elicited under some circumstances. Examination of the human neonatal B cell response to RSV showed that neonates exhibited a bias in antibody gene repertoire and a dramatically reduced frequency of somatic mutations compared with individuals older than three months, which probably contributes to the poorly neutralizing nature of the response in the young infant (Williams *et al.*, 2009).

A second factor in the reduced antibody response early in life is suppression by pathogen-specific maternal antibodies, an effect that is not unique to RSV (Crowe, 2001; Siegrist, 2003). Immunosuppression by RSV-specific maternal or passive serum antibodies has been documented in natural infection (Murphy *et al.*, 1986a), in experimental animals (Murphy *et al.*, 1988a, 1991), and in the setting of clinical evaluation of live RSV vaccines in infants, where the titre of maternal antibody had little effect on virus replication, but suppressed the serum and secretory antibody response (Wright *et al.*, 2000). In mice, passive RSV-specific serum antibodies suppressed both serum and secretory antibody responses but did not suppress the cell-mediated response or priming for a secondary antibody response (Crowe *et al.*, 2001). Conversely, in a selected study group of infants <6 months of age in whom the titre of RSV-specific serum antibodies was low at the time of natural infection, the post-infection increases in RSV-neutralizing serum antibody titres were indistinguishable from those of individuals aged 6–24 months (Shinoff *et al.*, 2008).

This indicates that young infants are capable of mounting a substantial antibody response when maternal antibodies are not present to exert their suppressive role. Thus, maternal antibodies are beneficial in providing some disease sparing early in life, but also are detrimental because they suppress antibody responses.

Secretory IgA antibodies are thought to be particularly efficient in preventing RSV infection. The secretory antibody response is not long-lasting following primary infection, but as individuals grow older and are re-infected this results in a higher and more sustained response (Kaul *et al.*, 1981). The serum antibody response is long-lived (except very early in life, as noted), and peak titres of antibodies detected by ELISA were reached following the second infection in life (Wagner *et al.*, 1989). In adults, natural infection induced an approximately 8-fold rise in serum RSV-neutralizing antibodies, which subsequently declined 4-fold during the subsequent year (Falsey *et al.*, 2006b). Titres in uninfected control subjects did not decline during this period suggesting that there may be a relatively stable set-point, whereas post-infection rises are more short-lived. In comparing elderly versus young adults with natural RSV infection, pre-infection titres of RSV-binding and RSV-neutralizing serum antibodies were comparable for the two groups, whereas post-infection increases were greater in the elderly (Walsh and Falsey, 2004a). This suggests that the greater susceptibility of the elderly to RSV cannot be attributed to reduced serum antibody levels and responses.

T lymphocytes

Experience with immunosuppressed patients indicates the importance of cell-mediated immunity in restricting RSV and other enveloped respiratory viruses (Fishaut *et al.*, 1980; Hall *et al.*, 1986). Immunosuppressed children have difficulty clearing the virus and can shed for months. Severely immunocompromised adults, such as haematopoietic stem cell transplant recipients or individuals with leukaemia, are readily infected and have a high incidence of serious disease and death.

In mice, depletion of the $CD4^+$ and $CD8^+$ T-lymphocyte subsets individually or together showed that both are important in clearing a primary infection (Graham *et al.*, 1991b). In contrast, depletion of B lymphocytes showed that RSV-specific antibodies are not required for clearance of a primary infection in mice, although they reduced disease and were important for restricting secondary infection (Graham *et al.*, 1991a). While treatment with RSV-neutralizing antibodies can greatly reduce viral replication in RSV-infected mice, it was inefficient in eliminating the virus, illustrating the importance of cellular immunity (Ottolini *et al.*, 1999). As already noted (see Antigens section), immunization of BALB/c mice with recombinant vaccinia viruses expressing RSV antigens that induced $CD8^+$ T cells, but not neutralizing antibodies, provided resistance to re-infection that waned within several weeks (Connors *et al.*, 1991; Kulkarni *et al.*, 1993). This suggests that, while $CD8^+$ CTL are important in clearing infection, this effect is not augmented by prior infection or vaccination, at least in this animal model.

In the depletion experiments in the mouse model noted above, $CD8^+$ CTL cells made the greater contribution to disease during primary infection (Graham *et al.*, 1991b). Adoptive transfer of RSV-specific CTL into RSV-infected mice cleared the virus, but caused enhanced disease and even death (Cannon *et al.*, 1988). A similar effect of increased disease was observed following transfer of donor lymphocytes to a severely immunocompromised child with a long-term RSV infection (El Saleeby *et al.*, 2004). These observations suggested that virus-specific $CD8^+$ CTLs have the potential to make a substantial contribution to RSV disease. However, this remains controversial. Robust $CD8^+$ T-cell responses have been documented in the peripheral blood of infants and children during primary and secondary RSV infections, and these cells appeared to subsequently migrate to the airways (Heidema *et al.*, 2007, 2008). Infants with severe RSV disease have increased numbers of T lymphocytes in the airways, with a greater abundance of the $CD8^+$ versus $CD4^+$ subset (Everard *et al.*, 1994; McNamara *et al.*, 2003; Heidema *et al.*, 2008). RSV-infected children had an increased proportion of airway $CD8^+$ T cells and decreased proportion of $CD4^+$ T cells compared with rhinovirus-infected children (Everard *et al.*,

1994; McNamara *et al.*, 2003; Heidema *et al.*, 2008). However, despite these increases, CD8+ T cells were not highly abundant either in fluid recovered from the airways of infants with severe RSV disease (Everard *et al.*, 1994; McNamara *et al.*, 2003; Heidema *et al.*, 2008) or in pulmonary tissue from infants that experienced severe or fatal RSV infection (Johnson *et al.*, 2007; Welliver *et al.*, 2007). In RSV-infected infants and children, the appearance of virus-specific CD8+ T cells in peripheral blood – which precedes their migration to the airways – peaked during recovery 9 to 15 days following the onset of symptoms (Heidema *et al.*, 2007; Lukens *et al.*, 2010). Thus, the CD8+ T-cell response did not correlate with the timing of disease. In addition, the magnitude of the systemic virus-specific CD8+ T-cell response did not correlate with disease severity (Heidema *et al.*, 2007). These findings suggest that CD8+ T cells are not a major determinant of RSV disease.

Increased stimulation of the Th2 subset of CD4+ T lymphocytes has been suggested to be a factor in RSV disease. Th2-biased responses have been implicated in the disease enhancement associated with a formalin-inactivated RSV vaccine evaluated in infants and children in the 1960s (Prevention and Treatment). In addition, the clinical picture of severe RSV disease, including airway plugging, wheezing, and long-lasting effects on lung function, has some similarity with asthma, which involves Th2-biased responses. Also, immune responses in the very young have a reduced Th1 component, which is protective for the fetus prior to birth, but in early infancy can contribute to a Th2 bias (Culley *et al.*, 2002; Adkins *et al.*, 2004). A Th2-biased response could contribute to RSV disease by promoting mucus production, promoting class switching to IgE, down-regulating CD8+ CTL and thereby interfering with clearance, promoting airway hypersensitivity, and promoting eosinophil recruitment and survival that may have antiviral or pathogenic effects. An allergic-like Th2-biased response also could contribute to long-term effects such as airway hyper-reactivity.

Peripheral blood mononuclear cells from infants hospitalized for RSV disease have been shown in some studies to have Th2-biased responses when stimulated *in vitro*, including the signature Th2 cytokine IL-4 (Roman *et al.*, 1997; Bendelja *et al.*, 2000). Two studies demonstrated a positive association between RSV disease and a genetic polymorphism in the IL-4 gene that increases gene expression (Miyairi and DeVincenzo, 2008). Also, as already noted, eosinophil degranulation products have been detected in secretions from infants with severe RSV disease. Th2-biased cytokine responses detected in respiratory washes from infants infected with RSV were significantly higher than from influenza virus-infected infants (Sung *et al.*, 2001). Conversely, another study of peripheral mononuclear cells from infants hospitalized for RSV disease found responses dominated by the Th1 cytokine IFNγ (Brandenburg *et al.*, 2000a). Other studies found heterogeneity in responses, with some infants having Th1-biased responses and others Th2 (Mobbs *et al.*, 2002; Lee *et al.*, 2007). Th2-biased responses were detected in RSV-infected infants and children with a history of asthma or allergy, suggesting an involvement specific to individuals predisposed to Th2 responses (Kim *et al.*, 2003; Legg *et al.*, 2003). Another study showed that the levels of Th2 cytokines in respiratory secretions of RSV-infected infants were greater for those ≤3 months of age compared with those >3 months of age (Kristjansson *et al.*, 2005). Furthermore, responses to influenza and parainfluenza viruses in the younger infants were similarly Th2-biased. This suggests that the Th2 responses were not unique to RSV, and instead may reflect the Th2 bias of early infancy noted above. However, since RSV has a much higher attack rate in very young infants this might be particularly significant for RSV. These studies suggest that Th2 responses to RSV infection are not uniformly dominant and thus may not be an essential element of RSV disease, but may be important in some individuals.

The Th2 cytokines IL-4 and IL-13 promote isotype class switching to IgE. Receptors on mast cells and basophils bind IgE that, upon contact with antigen, induces degranulation and the release of mediators including histamine and leukotrienes. These can mediate neural, vascular, and muscular responses including rhinorrhoea, cough, and wheeze. Increased RSV-specific IgE and histamine were detected in respiratory secretions from RSV-infected infants with wheezing and were less frequent with infection without wheezing (Welliver *et al.*, 1981). In addition, the levels of virus-specific IgE and released histamine correlated with the degree of hypoxia. However, other groups have not found convincing evidence of RSV-specific IgE (De Alarcon *et al.*, 2001).

Inhibition versus avoidance of host immunity

The finding that RSV can re-infect symptomatically throughout life without need for antigenic change is widely taken as evidence that the immune response to this virus is particularly incomplete or short-lived, presumably reflecting direct viral inhibition of immune responses. RSV indeed does have mechanisms that inhibit host immune responses, such as the IFN antagonists and the chemokine mimicry of the G protein already mentioned. As another example, dendritic cell maturation in response to RSV is dependent in large part on type I IFN, and is reduced by the NS1 protein in particular (Munir *et al.*, 2008). There also are reports of reduced activation and altered polarization of $CD4^+$ T lymphocytes *in vitro* in response to RSV-exposed human dendritic cells (Rothoeft *et al.*, 2007), although other results suggest that RSV does not differ significantly from other common respiratory viruses in this regard (Le Nouen *et al.*, 2009). Thus, while RSV does have mechanisms that inhibit host immunity, this probably is true to some extent for most or all viruses in general. It is not clear whether RSV is unusually effective in inhibiting or altering host immunity, resulting in inefficient protective immunity.

RSV infection of experimental animals provides the opportunity to evaluate immune responses in a host that is immunologically mature, free of maternal antibody, and RSV-naive. Under these conditions, the observed responses are robust, protective, and long-lasting. This suggests that the virus is not inherently highly immunosuppressive, at least in these models.

The same experiment cannot be done for humans, given the universal early exposure to the virus. Thus, as noted, the initial infection occurs early in life when responses usually are weak and short-lived owing to immunological immaturity and antibody-mediated suppression. These host effects, rather than viral inhibition, may account for the high rate of re-infection during the first years of life. When maternal antibodies are low, the humoral response to RSV in young infants is robust (as noted), suggesting that the virus is not inherently immunosuppressive in young infants.

It has been speculated that early exposure to RSV might have a long-term effect that blunts immunity throughout life. This might be hypothesized to be a host effect, perhaps a consequence of exposure to the virus in the presence of immunosuppressive antibodies and an immature immune system. Viral factors might also be involved. For example, the ability of RSV to strongly suppress the type I and type III IFN responses may interplay with the immature immune system to have long-term effects on RSV-specific immunity. However, this remains speculative.

Symptomatic re-infection without antigenic change is not unique to RSV, but also occurs to a somewhat lesser extent with the HPIVs and HMPV and even sometimes with influenza A virus (Sonoguchi *et al.*, 1986). An important factor is the inefficiency of host immunity in protecting the respiratory tract, owing to the short-term nature of secretory antibodies and the inefficiency with which serum or maternal antibodies access the respiratory lumen, as already noted. It also is noteworthy that re-infection usually is restricted to URI or mild LRI. This indicates that host immunity does substantially restrict the virus, and thus is substantially functional.

While strong inhibition of the host IFN response probably facilitates re-infection, it may be that RSV does not otherwise inhibit host immunity more efficiently than other common respiratory viruses. Instead, it may partially evade it. The high infectivity of the virus would facilitate infection. As noted, the secreted form of the G protein reduces antibody-mediated neutralization and clearance. Clonal analysis of the human antibody response to the F protein showed that half of the repertoire was against non-native protein, suggesting that unfolded F may divert and reduce the immune response to the native form (Sakurai *et al.*, 1999). The presence of two antigenic subgroups, and the divergent, heavily glycosylated nature of the G protein also would facilitate re-infection. Restriction of RSV infection to the superficial cells of the respiratory lumen, and the non-destructive and non-invasive nature of infection, would limit exposure of the virus and viral antigens to the immune system. RSV does not efficiently infect human dendritic cells *in vitro*, which may result in reduced stimulation of maturation, and also would limit the intracellular synthesis of viral proteins and their presentation on major histocompatibility class I molecules to stimulate CD8$^+$ T lymphocytes (Le Nouen *et al.*, 2009). These various features may help account for the ability of RSV to re-infect throughout life without significant antigenic change. They may also explain why RSV can infect earlier and more efficiently in life than other common respiratory viruses.

Clinical features

In otherwise healthy infants and young children, RSV infection usually causes URI including congestion, rhinorrhoea, and sometimes fever. Asymptomatic infection is uncommon (Fisher *et al.*, 1997). In 20–40% of primary infections, the infection progresses to LRI (Fisher *et al.*, 1997; Kim *et al.*, 1973), involving the respiratory tract below the larynx with disease signs including wheezing, coughing, increased respiratory rate, shortness of breath, rhonchi (snoring-like sounds due to secretions in airways), and rales (inspiratory crackling sounds due to the opening of inflamed small airways). Rhinorrhoea can be profuse. Incidence of lower respiratory tract involvement is higher in young infants (Fisher *et al.*, 1997). In most cases, uneventful recovery occurs after an illness of 7 to 12 days.

Severe RSV LRI usually involves bronchiolitis and pneumonia, which may represent a continuum of increasing severity of disease. Most children who are hospitalized are hypoxemic. In a 13-year surveillance study of young children hospitalized for respiratory disease in Washington, DC, RSV was more frequently associated with bronchiolitis (43%) and pneumonia (25%), than with bronchitis (11%) and croup (10%; Murphy *et al.*, 1988b). In a prospective outpatient study of infants and children, of those diagnosed with RSV LRI, 67% had bronchiolitis and 23% had pneumonia (Fisher *et al.*, 1997).

RSV is the most common virus associated with otitis media, followed closely by rhinovirus (Alper *et al.*, 2009). Otitis media linked to an acute viral infection may often be bacterial in nature, but the frequent detection of infectious virus or viral macromolecules in middle ear fluid suggests that virus might also be a direct causative agent (Alper *et al.*, 2009; Yano *et al.*, 2009). Bacterial infection of the airways in infants hospitalized for RSV disease has been reported to be rare (1.2%) (Hall *et al.*, 1988), although a recent study in Japan detected a much higher incidence (44%; Hishiki *et al.*, 2010). Use of a pneumococcal conjugate vaccine in South Africa resulted in a substantial decrease in pneumonia associated with seven respiratory viruses including RSV, suggesting that bacteria can play an important role in pneumonia associated with RSV as well as with other common respiratory viruses (Madhi and Klugman, 2004).

RSV and rhinoviruses are the viruses most frequently associated with asthma exacerbations: in one recent study of hospitalized children, they are associated with, respectively, 40% and 24.5% of cases (Maffey *et al.*, 2010). As already noted, severe RSV LRI in infancy frequently is followed by long-term airway reactivity and deficits in pulmonary function (Hall *et al.*, 1984; Long *et al.*, 1995; Noble *et al.*, 1997; Stein *et al.*, 1999), and has been controversially linked to allergic sensitization and asthma (Sigurs *et al.*, 2005, 2010; see Pathogenesis section).

In a large prospective study in healthy adults, more than three-fourths of documented RSV infections were symptomatic, with three-fourths of the cases experiencing URI and the remainder with mild LRI (Hall *et al.*, 2001). RSV disease in healthy adults is thought to be more severe than the common cold, with similarities to influenza (Hashem and Hall, 2003). In adults, RSV and influenza have overlapping spectra of disease signs: sore throat, and non-productive cough were equally frequent with both viruses, whereas nasal congestion and productive cough were more frequent with RSV and fever and headache were more frequent with influenza (Hall *et al.*, 2001). Duration of disease was greater with RSV, but work absence was greater with influenza. The duration of RSV shedding ranged from 1–17 days, with an average of 3.9 days (Hall *et al.*, 2001). In a study in healthy adults, RSV infection and disease were followed by 8 weeks of increased airway resistance and increased airway reactivity, indicative of protracted effects following the acute infection (Hall *et al.*, 1978b).

Diagnosis

RSV diagnosis by physical examination alone is not definitive because many respiratory illnesses share a broad spectrum of disease signs. RSV may be suspected based on profuse rhinorrhoea, wheezing and other airway sounds, the age of the individual, and the time of year and presence of RSV in the community. Airway sounds are evaluated, blood oxygen deficits are measured by pulse oximetry, and a chest x-ray can be used to detect areas of alveolar collapse (atelectasis) and lung consolidation indicative of pneumonia.

Definitive diagnosis depends on laboratory tests. The gold standard is isolation of virus in tissue culture from nasal swabs or nasal washes. However, the sensitivity of conventional cell culture is lower than with other techniques and the assay is slower (Johnston and Siegel, 1990). Also, this method is less sensitive for adults, who usually shed infectious virus less abundantly than infants and children (Falsey, 2007). The labile nature of RSV requires that the samples be kept cold without freeze–thawing and promptly inoculated into permissive cell cultures. Virus replication can be assessed by the development of cytopathic effect over a period of 3–7 days and by immunostaining or immunofluorescence. Shell vial centrifugation cultures, in which the sample is briefly centrifuged onto a cell monolayer, provides an increase in sensitivity and speed with cultures being read by immunostaining or immunofluorescence 24–48 h post inoculation (Johnston and Siegel, 1990).

More rapid diagnosis can be made by the detection of viral antigen in nasal swabs or washes. In the direct immunofluorescence assay (DFA), exfoliated cells and other debris are concentrated from secretions from centrifugation, reacted with virus-specific fluorescence-labelled antibodies, and visualized by microscopy. This has the advantage that a knowledgeable technician can confirm that the signal has the pattern expected for infected cells, but the method is time-consuming and requires expertise and a fluorescence microscope. Alternatively, antigen in secretions can be detected by antigen-capture enzyme-linked immunoassay (EIA), in which the sample is incubated with immobilized virus-specific antibodies to capture viral antigen, which is then detected with a second, enzyme-linked antibody. The sensitivities of DFA and EIA are similar (Johnston and Siegel, 1990). EIA is available in rapid commercial kits. Another rapid antigen test available in commercial kits is based on lateral flow immunochromatography (Liao *et al.*, 2009; Ginocchio *et al.*, 2010). The test sample of secretions is solubilized, mixed with a coloured gold-labelled RSV MAb, and applied to a test membrane on which antibody–antigen complexes migrate laterally and are captured by a line of immobilized RSV-specific antibody to produce a visible band. The sensitivity of these rapid commercial tests, which produce results in less than 30 minutes, may be lower than that of DFA, and negative results should be confirmed by DFA. In addition, antigen-based tests in general often have reduced sensitivity in adults because of low levels of viral shedding (Falsey, 2007).

RT-PCR detection of viral RNA in nasal swabs or washes provides excellent sensitivity in both paediatric and adult populations because reduced viral shedding is less of an issue. RT-PCR can be used in multiplex formats capable of detecting multiple pathogens. RSV RT-PCR assays are available in commercial kits, although the procedure is somewhat time-consuming (3–6 h) and requires technical expertise and specialized equipment (Liao *et al.*, 2009; Manji *et al.*, 2009).

Serological methods are used less frequently and provide only retrospective information, since antibody rises are detected by analysis of sera collected 2–4 weeks post infection. Typically, serum antibodies are detected by ELISA or microneutralization assay, with four-fold or greater increases considered indicative of infection.

Rapid detection of RSV is desirable to guide the use of appropriate infection-control measures as well as to guide supportive care and antibiotic use, and will facilitate the timely use of antiviral agents once these are developed.

Treatment

Most infants and children with mild RSV lower respiratory tract disease can be managed as outpatients. Treatment of more severe LRI can be guided by measurement of oxygenation, monitoring respiratory rate and the trend in respiratory rate, and evaluation of dehydration. Severe respiratory distress, hypoxia and dehydration are among the indications for hospitalization. Treatment requires considerable supportive care: mechanical removal of secretions, proper positioning of the infant, administration of humidified oxygen, intravenous fluids, and, in the most severe cases, respiratory assistance with mechanical ventilation. Antibiotic therapy does not improve the outcome in viral pneumonia and does not alter the risk of bacterial complication of viral pneumonia as a superinfection. The youngest infants with RSV pneumonia may require monitoring for the risk of apnoea, although apnoea usually presents near the beginning of the illness. Improvements in supportive care clearly have made a major impact on mortality from severe RSV disease, resulting in the current overall low mortality rate.

Corticosteroids have been used in therapy for RSV bronchiolitis since the 1960s, based in part on their effectiveness in the treatment of acute asthma and on the premise that much of RSV disease is immune-mediated. However, a meta-analysis of placebo-controlled studies indicated that systemic glucocorticoids on their own provided no improvement in oxygenation, respiratory rate, or length of hospitalization (Patel *et al.*, 2004), and a large controlled study indicated that inhaled glucocorticoids provided no improvement in wheezing following RSV bronchiolitis (Ermers *et al.*, 2009). Interestingly, studies in weanling rats suggest that a major element of RSV-induced wheezing involves virus-induced neuro-immune dysregulation that is resistant to the anti-inflammatory effects of steroids (Mohtasham *et al.*, 2007). A cautionary finding is that, in cotton rats, treatment with corticosteroids alone reduced host immune resistance and prolonged virus shedding (Prince *et al.*, 2000).

Drugs that have been used successfully to treat reversible airways smooth muscle constriction in asthma are also used with severe RSV infection. Hospitalized infants in the United States frequently are treated with inhaled bronchodilators, but systematic reviews indicate that they provide only modest short-term improvement (Gadomski and Bhasale, 2006; Levin *et al.*, 2008). This suggests that the increased airway resistance in infant viral bronchiolitis is not predominantly due to reversible smooth muscle constriction, but rather is a consequence of obstruction with mucus, immune cells, and sloughed epithelial cells.

Ribavirin, a nucleoside analogue, is a long-standing broad spectrum antiviral compound whose mode of action remains unclear (Sidwell and Barnard, 2006). It exhibits potent activity against RSV in cell culture and experimental animals. Ribavirin was approved in 1986 in the U.S. for use in the treatment of RSV infection (Ventre and Randolph, 2004). Initial small trials indicated a modest beneficial effect on both virus shedding and clinical illness. Follow-up studies indicated that the drug did not significantly reduce mortality, respiratory deterioration, or length of hospitalization, although there were trends towards benefit (Randolph and Wang, 1996). Small studies indicated a possible improvement in post-bronchiolitis reactive airway disease (Rodriguez *et al.*, 1999). The American Academy of Pediatrics recommends that ribavirin 'may be considered' for RSV-infected patients. The typical aerosol mode of delivery is lengthy and inconvenient, and there is concern for side effects including anaemia and teratogenicity. Currently, most centres do not routinely treat otherwise healthy infants with RSV disease with ribavirin, but the drug is sometimes used in severely immunocompromised individuals in whom viral replication is not easily controlled.

RSV disease in young infants can be reduced by prophylaxis with RSV-neutralizing antibodies (see Prevention and Control section). Therefore, these antibodies also have been evaluated for therapy of established infection. However, while treatment with neutralizing antibodies reduced virus shedding both in experimental animals and in infants and young children, there was no clear benefit to the treatment whether given systemically (Malley *et al.*, 1998; Rodriguez *et al.*, 1997a,b) or by aerosol (Rimensberger *et al.*, 1996). However, antibodies sometimes are used therapeutically in severely immunocompromised individuals (Lazar *et al.*, 2006).

Small molecule inhibitors have been developed that target various RSV proteins (Sidwell and

Barnard, 2006; Olszewska and Openshaw, 2009; Empey *et al.*, 2010). For example, a compound specific to the N protein (RSV604) is presently in phase 2 clinical trials (Chapman *et al.*, 2007). A number of inhibitors of the F protein have been identified, but in most cases development has been suspended: a single compound (BTA-9881) is in clinical trials. Compounds that appear to target the G or L proteins are also in pre-clinical studies (Liuzzi *et al.*, 2005; Sudo *et al.*, 2005; Sidwell and Barnard, 2006).

Antisense strategies are also being explored. Conventional antisense oligonucleotides probably have little potential on their own, but they also have been chemically linked to the 2–5A molecule of the host IFN system (Leaman *et al.*, 2002). This serves to recruit the cellular antiviral nuclease RNAse L to cleave the hybridized RNA. An 2–5A-antisense molecule that targets the conserved GS signals in the viral genome reduced nasal RSV replication up to 10,000-fold in African green monkeys when the drug was administered intranasally shortly after infection and during the following 5 days (Leaman *et al.*, 2002), but further development has not been reported. More recently, small interfering RNAs (siRNAs) have been developed against RSV. In particular, an siRNA targeting the N gene (ALN-RSV01) was shown to reduce RSV replication in mice 100- to 1000-fold when administered intranasally 1 to 2 days following infection. This compound was shown to be safe in a phase 1 study and was evaluated in a phase 2 challenge study in 85 healthy adults (DeVincenzo *et al.*, 2010a). The drug was administered daily by nasal spray for 2 days before and 3 days after RSV challenge. This resulted in a 38% decrease in the number of individuals infected, and modest but insignificant decreases in viral load in those who were infected (DeVincenzo *et al.*, 2010a). It will be important to evaluate efficacy in a situation where the drug is administered after, rather than beginning before, infection.

Combination therapies also are being developed, involving either the combination of two different antiviral drugs to achieve a higher level of inhibition of viral replication, or the combination of one or more antiviral drugs with an anti-inflammatory drug to restrict replication and reduce immune-mediated disease. This latter combination strategy effectively resolved viral replication and disease in the cotton rat model (Prince *et al.*, 2000). In particular, early and aggressive combined therapy, with ribavirin, corticosteroids, and antibodies, is being explored for severely immunocompromised adults (Liu *et al.*, 2010; Whimbey *et al.*, 1995). In the more usual setting of RSV infection of otherwise healthy individuals, antiviral therapy of an acute, rapid infection such as RSV is challenging because by the time the viral agent is identified, it may be too late to control disease solely by inhibiting virus replication.

Prevention and control

As noted, RSV infection can be reduced by hand hygiene and other common sense infection control measures.

Passive immunoprophylaxis

Infants at high risk for severe RSV disease can be substantially protected by passive antibody prophylaxis (Meissner and Long, 2003; Cardenas *et al.*, 2005; Mejias and Ramilo, 2008). The first product developed for this use consisted of total serum immunoglobulin for intravenous administration (RSV-IVIG), which was purified from human donor sera that had been screened for high RSV-neutralizing activity. This product (RespiGam) was licensed in 1996 for use in infants and young children at high risk of RSV disease owing to prematurity or chronic lung disease. Monthly administrations of RSV-IVIG during the RSV season reduced the frequency of hospitalization for RSV disease by 55% and reduced the number of days spent in intensive care by 97%. RSV-IVIG prophylaxis inconveniently involves a 4-h intravenous administration; the large volume precludes its use for individuals with haemodynamically significant chronic heart disease. RSV-IVIG can also interfere with some childhood live vaccines owing to the presence of antibodies against multiple common human pathogens, and has the theoretical concern of contamination with adventitious agents.

To address these limitations, an RSV-neutralizing MAb, palivizumab (Synagis), was developed by MedImmune and was licensed in 1998 for use in high-risk paediatric patients. Palivizumab is based on a murine MAb called 1129 that recognizes a highly conserved epitope on the F glycoprotein and efficiently neutralizes both RSV subgroups (Beeler and van Wyke Coelingh, 1989). This MAb was humanized by recombinantly transferring its complementarity-determining regions (~5% of the molecule) onto a human IgG1 backbone (~95% of the molecule). Palivizumab is 50- to 100-fold more effective on a weight basis than RSV-IVG in neutralizing RSV, and thus can be administered in a much smaller volume by intramuscular injection. Palivizumab is administered monthly in up to five doses during the RSV season. Indications for use include: (i) chronic lung or heart disease and age <24 months at the start of RSV season; (ii) birth at ≤28 months gestation and age ≤12 months at the start of RSV season; (iii) birth at 29–32 months gestation and age ≤6 months at the start of RSV season; or (iv) birth at 32–35 months gestation, age ≤3 months at the start of RSV season, plus the risk factor of day care use or sibling <5 years of age (2009; Meissner and Long, 2003). The efficacy of palivizumab is similar to that of RSV-IVIG. There is continued discussion of the cost-effectiveness of palivizumab owing to its expense (Prescott *et al.*, 2010), but antibodies remain the only effective prophylaxis available for RSV disease other than infection control.

Subsequently, palivizumab has been modified by *in vitro* affinity maturation. This resulted in a derivative called motavizumab that binds the RSV F protein 70-fold better, exhibits a 20-fold increase in neutralization activity *in vitro*, reduces RSV replication in treated cotton rats up to 100-fold lower levels, and was effective in the upper as well as the lower respiratory tract of cotton rats (Wu *et al.*, 2007). Motavizumab differs from palivizumab at 13 amino acid positions, seven of which individually increase its affinity for the F protein (Wu *et al.*, 2007). Binding of motavizumab to a peptide containing its epitope (residues 254–277 of the F protein) was visualized by crystallographic analysis, and provided evidence that the antibody may bind to a transitional form of the F protein (McLellan *et al.*, 2010). In a phase 3 trial comparing motavizumab to palivizumab, the incidence of medically-attended outpatient LRI and hospitalization was reduced 50% and 26%, respectively, and those infants who were hospitalized had less severe disease (Gill and Welliver, 2009; Carbonell-Estrany *et al.*, 2010). However, in a 2 June 2010 meeting, the FDA Antiviral Drugs Advisory Committee recommended rejection of this drug application because of concerns that included an apparent increased incidence of hypersensitivity reactions and anti-drug antibodies. MedImmune subsequently suspended development of motavizumab for RSV prophylaxis, although it is still being considered for other applications such as RSV therapy.

In further work, the Fc domain of motavizumab has been engineered to increase its affinity for the neonatal Fc receptor (Dall'Acqua *et al.*, 2006). The so-called neonatal Fc receptor (which is a misnomer because it is active throughout life) is responsible for transport of maternal antibodies across the placenta to the fetus. It also binds to internalized IgG destined for degradation and recycles it to circulation, thus increasing its serum half-life. Consistent with this, the engineered motavizumab exhibited a nearly 4-fold increase in serum half-life in cynomolgus monkeys and up to a 4-fold increase in lung availability (Dall'Acqua *et al.*, 2006). This drug presently is being evaluated for safety in adults.

In other studies, fully human MAbs have been developed against RSV. One approach has been to develop monoclonal IgG or IgA Fab fragments by constructing recombinant combinatorial libraries copied from donor bone marrow RNA, followed by affinity selection against immobilized RSV F antigen (Barbas *et al.*, 1992; Moreno de Alboran *et al.*, 1995). Another approach has been to prepare and screen libraries of transformed B cells, which resulted in the identification of high affinity human MAbs specific to the central conserved domain of G (Collarini *et al.*, 2009).

Vaccines and vaccine-associated disease enhancement

The need for an RSV vaccine is greatest for young infants and the elderly. Live or vectored vaccines are the preferred strategy for the young infant because they are not associated with disease enhancement (see below section). Live vaccines may be too restricted in RSV-experienced older individuals because of the presence of neutralizing antibodies (Gonzalez *et al.*, 2000). However, protein-based vaccines are safe in RSV-experienced individuals and can be developed for that population.

Disease enhancement

An experimental paediatric RSV vaccine evaluated in the 1960s had the paradoxical effect of enhancing RSV disease upon subsequent natural exposure. This vaccine consisted of tissue culture-growth RSV that was inactivated with formalin, concentrated by centrifugation, and administered with an alum adjuvant to infants and young children by three successive intramuscular injections. This formalin-inactivated vaccine (FI-RSV) was poorly protective and, upon subsequent natural infection, vaccinees had a greatly increased frequency and severity of disease, with 80% of the vaccinees requiring hospitalization compared with 5% in a control group. There were two fatalities attributed to the enhanced disease, and autopsy provided evidence of extensive pulmonary RSV replication, inflammation, and bacterial infection. In the ensuing years, this phenomenon has been extensively studied, relying in large part on mouse and cotton rat models. In rodents, disease enhancement typically is monitored by histopathology, since overt disease is rare (Prince *et al.*, 2001). Results with these rodent models must be viewed critically because of the use of the non-natural poorly-permissive host.

Retrospective studies showed that the immune response to FI-RSV differs markedly from that to RSV infection. The serum antibodies induced by FI-RSV in the original vaccinees as well as in experimental animals bound efficiently to RSV antigen in ELISA but neutralized infectivity inefficiently. Studies in experimental animals showed that FI-RSV did not induce a significant CTL response. Because it was administered intramuscularly, the FI-RSV vaccine probably did not induce appreciable local immunity in the respiratory tract, which is important in restricting RSV replication (Collins *et al.*, 1990b; Matsuoka *et al.*, 2002). Peripheral blood lymphocytes from the original vaccinees exhibited an exaggerated proliferative response to RSV antigen *in vitro*, suggestive of a heightened $CD4^+$ T-cell response to FI-RSV compared with natural infection (Kim *et al.*, 1976). Consistent with this, subsequent studies in experimental animals showed that FI-RSV induces a disproportionately increased stimulation of the Th2 subset of $CD4^+$ T cells compared with RSV infection.

This atypical response was not directly pathological, but was poorly protective and primed for a pathological secondary response upon subsequent RSV infection. A number of factors have been implicated in disease enhancement.

1. The poor protective efficacy of the vaccine permitted subsequent RSV infection and the expression of viral antigens necessary to trigger the secondary response (Murphy *et al.*, 1990).
2. The poor neutralizing activity of post-vaccination sera probably resulted from denaturation of neutralization epitopes during formalin treatment and handling. In addition, it also appears to reflect deficient antibody affinity maturation (Delgado *et al.*, 2009).
3. Antibody–antigen complexes and complement fixation appear to contribute to disease enhancement: the formation of these complexes would be favoured by viral replication in the presence of antibodies that do not neutralize infectivity, but efficiently bind viral antigen (Melendi *et al.*, 2007).
4. The Th2-biased response appears to mediate much of the disease enhancement, because the effect was not observed when FI-RSV-immunized rodents were treated with antibodies to deplete either $CD4^+$ T cells or selected Th2 cytokines prior to RSV challenge (Connors *et al.*, 1992).
5. The Th2-biased response was favoured by the lack of stimulation of NK cells and $CD8^+$ CTL, which have been shown to normally have a regulatory role that down-regulates Th2 responses to RSV antigens (Hussell and Openshaw, 1998; Olson and Varga, 2007).
6. The formalin treatment created carbonyl groups that enhanced the Th2 response (Moghaddam *et al.*, 2006).

Thus, the major elements of disease

enhancement include an inadequate protective response, an altered antibody response, and a heightened Th2 response. Similar phenomena of vaccine-related disease enhancement have been described in clinical use of formalin-inactivated measles virus vaccine (Polack, 2007) and in experimental animals receiving formalin-inactivated HPIV3 (Ottolini *et al.*, 2000) and HMPV (de Swart *et al.*, 2007) vaccines, and thus may be a common phenomenon in *Paramyxoviridae*.

Disease enhancement has also been observed in rodents that have been immunized with purified RSV F and G glycoproteins as experimental vaccines (Murphy *et al.*, 1990). This suggests that it may be a common property of killed virus and subunit vaccines. In contrast, disease enhancement is not observed with natural infection and re-infection, or in most cases with viral or DNA vectors expressing RSV antigens. This distinction between killed-virus/subunit vaccines versus live/vectored vaccines probably reflects the greater efficiency of the latter in broadly stimulating innate and adaptive immunity, including stimulation of TLRs, NK cells, and $CD8^+$ CTL important in regulating RSV-specific immune responses. Also, expression of RSV antigen *in situ* from virus or a vector avoids the issue of antigen denaturation.

Vaccines

Because of disease enhancement, present strategies for a paediatric RSV vaccine have focused on live attenuated vaccines, which have been confirmed to be free of disease enhancement in RSV-naive infants (Wright *et al.*, 2007). The intranasal route of administration is surprisingly immunogenic in chimpanzees and protects both the upper and lower respiratory tract. Also, the intranasal route has been shown to partly evade the immunosuppressive effects of serum antibodies (Crowe *et al.*, 1995). Challenges to a live RSV vaccine strategy include the inherent genetic instability of RNA viruses, the poor growth and physical lability of RSV, and the reduction in immunogenicity that accompanies attenuation.

Beginning in the 1960s, a series of live attenuated RSV strains was developed by classic biological methods, including passage at increasing low, sub-optimal temperature (Friedewald *et al.*, 1968) and chemical mutagenesis followed by selection for temperature-sensitive mutants (Crowe *et al.*, 1994). None of the viruses produced by these methods had a satisfactory level of attenuation in 1- to 2-month old, RSV-naïve infants, although one virus (called *cpts*248/404) appeared to be safe and immunogenic in an older age group, namely seronegative infants and children six months of age and older (Wright *et al.*, 2000). Further efforts to obtain satisfactory derivatives by biological means were unsuccessful, and all subsequent vaccine candidates have been made by reverse genetics.

Reverse genetics provides a number of advantages for developing live vaccines (Collins and Murphy, 2005).

1 Producing virus from cDNA provides a virus with a short and well-defined passage history, which is important for safety and regulatory reasons (Surman *et al.*, 2007).
2 The method can be used to identify mutations – typically point mutations – in existing attenuated derivatives, such as *cpts*248/404 mentioned above (Whitehead *et al.*, 1998a).
3 Reverse genetics can be used to make new types of attenuating mutations not seen in nature, such as by identifying and deleting non-essential accessory proteins including the NS1, NS2, SH, and M2-2 proteins (Bermingham and Collins, 1999; Teng *et al.*, 2000; Whitehead *et al.*, 1999a).
4 Once an attenuating mutation is identified in one virus, it often can be 'transferred' to a second virus using sequence alignments to identify the appropriate site in the second virus, thus expediting the development of a menu of attenuating mutations (Newman *et al.*, 2004).
5 The genetic stability of amino acid point mutations can be increased by choosing codons that are less likely to revert to wild-type-like phenotypes (McAuliffe *et al.*, 2004). Another expedient is to delete the codon altogether (Bartlett *et al.*, 2006), reducing the likelihood of reversion.
6 Host range elements can provide attenuation. For example, RSV, HPIV3, and HMPV have animal counterparts (BRSV, BPIV3, and AMPV, respectively) that are restricted for

replication in primates because of the host range difference. Genes from the animal virus can be used to replace the corresponding genes in the related human virus, with the result that the human virus becomes more restricted in primates (Buchholz *et al.*, 2000; Skiadopoulos *et al.*, 2003).

7 Once a menu of attenuating mutations is identified, they can be introduced in desired combinations to make a series of vaccine candidates with varying levels of attenuation (Whitehead *et al.*, 1999b).

8 A variety of other manipulations are possible, such as moving glycoprotein genes to be promoter proximal to increase expression (Krempl *et al.*, 2002).

To date, five different attenuated RSV strains produced by reverse genetics have been evaluated in clinical trials (Karron *et al.*, 2005; Wright *et al.*, 2006). One of these, a virus called rA2cp248/404/1030ΔSH (Fig. 10.14), appeared to be satisfactorily attenuated in 1- to 2-month old infants and was moderately immunogenic and highly protective against a second vaccine dose (Karron *et al.*, 2005). This virus is a recombinant version of the previously mentioned cpts248/404 virus that was further modified by the addition of another point mutation in the L protein (1030 mutation) and deletion of the SH gene. Thus, this virus contains five attenuating elements: (i) a set of five amino acid substitutions in the N, F, and L proteins that together recreate the phenotype of a cold-passaged (cp) virus called cp-RSV (Whitehead *et al.*, 1998b); (ii, iii, and iv) three independent temperature-sensitivity point mutations, namely two amino acid substitutions in the L protein (called the 248 and 1030 mutations) and a nucleotide mutation in the gene-start signal of the M2 gene (called the 404 mutation; Whitehead *et al.*, 1999b); and (v) deletion of the SH gene (Whitehead *et al.*, 1999a). The virus is strongly temperature sensitive owing to the 248, 404, and 1030 mutations, and does not form plaques above 35° C; this would provide restricted replication *in vivo*, especially in the warmer lower

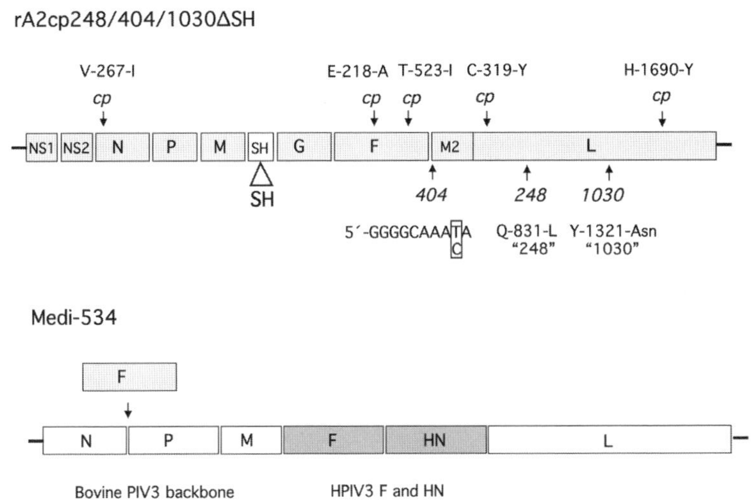

Figure 10.14 Two live-attenuated respiratory syncytial virus (RSV) candidate vaccines presently in clinical trials, both made by reverse genetics. The rA2cp248/404/1030ΔSH virus (top) contains five cold-passaged (cp) mutations in the N, F, and L proteins, two amino acid substitutions (248 and 1030) in L and one nucleotide substitution in the M2 GS signal (404) that independently confer temperature-sensitivity, and deletion of the SH gene (Whitehead *et al.*, 1999b; Karron *et al.*, 2005). The Medi-534 virus (bottom) is a bivalent vaccine against RSV and human parainfluenza virus (HPIV) 3. It consists of bovine parainfluenza virus (BPIV) 3 in which the F and HN genes have been replaced by their counterparts from HPIV3 and the coding sequence for the RSV F gene was placed under the control of HPIV3 transcription signals and inserted as an added gene between the N and P genes (Schmidt *et al.*, 2002; Tang *et al.*, 2008; Gomez *et al.*, 2009).

respiratory tract. Analysis of vaccine virus recovered from vaccinees provided evidence of genetic instability; one third of the isolates had a 1–2°C shift in temperature-sensitivity, and some isolates were shown to have lost either the 248 or 1030 mutation. However, since these viruses retained the other four attenuating elements and remained highly attenuated, this degree of instability might be acceptable. This virus presently is in expanded phase 1 and phase 2 studies to measure safety, stability, immunogenicity, and protective efficacy against natural RSV infection during the subsequent RSV season. Efforts to increase the genetic stability of this virus are under way (Luongo *et al.*, 2009).

Other live attenuated RSV derivatives are being prepared for clinical trials. One of these viruses is attenuated by the deletion of the coding sequence for the M2-2 protein. This ΔM2-2 virus exhibits reduced RNA replication and increased gene transcription and antigen production, which may provide for increased immunogenicity (Bermingham and Collins, 1999). Another virus is attenuated by deletion of the NS1 gene encoding the major RSV IFN antagonist (Teng *et al.*, 2000). The increased expression of IFN associated with this virus has the potential to increase innate and adaptive immune responses to the vaccine (Valarcher *et al.*, 2003).

Another type of live attenuated RSV vaccine presently in clinical trials involves using an attenuated version of HPIV3 as a vector to express the RSV F and/or G proteins from added genes (Fig. 10.14; Schmidt *et al.*, 2002). This provides a bivalent intranasal vaccine against these two important paediatric pathogens. Importantly, HPIV3 replicates more efficiently and is more stable physically than RSV. This should facilitate vaccine manufacture and use. The HPIV3 vaccine candidate that presently is being used as a vector consists of the bovine PIV3 backbone in which the F and HN genes have been replaced with their counterparts from HPIV3, thus providing host range attenuation from bovine PIV3 combined with the major protective antigens from HPIV3. The construct presently in clinical trials expresses the F protein from a gene added between the N and P genes (Gomez *et al.*, 2009; Tang *et al.*, 2008). An analogous approach is being pursued using Sendai virus as vector (Jones *et al.*, 2009). Sendai virus is murine relative of HPIV1 that has sufficient antigenic cross-reactivity that intranasal immunization with Sendai virus protects African green monkeys against HPIV1 challenge. Sendai virus was safe and immunogenic in adults as a potential vaccine against HPIV1, and has been engineered to express RSV antigens as a potential vaccine against HPIV1 and RSV (Jones *et al.*, 2009).

As noted, killed virus or protein-based RSV vaccines are presently considered inappropriate for paediatric use because of the potential for disease enhancement. However, priming for enhanced disease appears to be limited to the initial immunization in life and does not seem to occur when non-live vaccines are used in RSV-experienced individuals. This has been demonstrated in experimental animals (Waris *et al.*, 1997) and also is supported by studies of RSV protein vaccines in RSV-experienced individuals, as described below. Thus, a protein-based vaccine could be safely used to boost immunity in older children and adults. It is also possible that a protein vaccine might be developed that is safe for paediatric use, perhaps because of improvements in antigen quality or the use of improved adjuvants such as ones that stimulate TLRs (Boukhvalova *et al.*, 2006). However, given the limitations of current animal models of enhanced disease, it may be difficult to obtain the convincing safety data that would be needed to initiate clinical studies in RSV-naïve young infants.

A number of studies have been performed with a vaccine based on purified F protein (PFP, called PFP-1, -2, or –3 depending on differences in preparation) isolated from RSV-infected cells. This vaccine was given by intramuscular infection with alum adjuvant and was evaluated extensively in adults, in children with or without underlying disease, and in the elderly. PFP was well tolerated and, at best, was moderately immunogenic (Piedra *et al.*, 2003, 1996). The increases in RSV-neutralizing antibody responses and protection against infection appeared to largely wane within 1 year, suggesting that yearly re-immunization would be necessary (Tristram *et al.*, 1994). PFP-2 was also evaluated for maternal immunization with the goal of increasing the titre of RSV-neutralizing maternally derived antibodies in the young infant and thus providing greater protection during the first months of life. This strategy has been shown to be safe and effective for influenza virus (Zaman *et al.*, 2008). However, in an initial study, only modest increases in RSV-specific antibodies were observed in vaccine recipients and their infants (Munoz *et al.*, 2003).

Another experimental protein-based vaccine, called BBG2Na, consists of a bacterially expressed fragment of the G protein, including its central conserved domain, fused to the albumin-binding domain of the streptococcal G protein. The vaccine had promising properties when evaluated in mice. However, in RSV-naive infant macaques it was associated with a low level of pulmonary eosinophilia in some animals upon challenge, which was suggestive of disease enhancement and would make it unsuitable for RSV-naïve recipients (de Waal *et al.*, 2004). The vaccine initially appeared to be safe in adults, but induced low levels of RSV neutralizing antibodies (Power *et al.*, 2001). Development was subsequently suspended owing to incidents of hypersensitivity in some vaccine recipients, although these results have not been published.

More recently, a vaccine based on the RSV F, G, and M proteins purified from RSV-infected cells was evaluated in an elderly population in conjunction with the inactivated seasonal influenza virus vaccine, and was moderately immunogenic with or without an alum adjuvant (Falsey *et al.*, 2008). Other approaches to a subunit vaccine are being pursued by various researchers, including developing improved methods of purifying antigenically-intact F protein, or recombinantly modifying the F protein to obtain a more stable, immunogenic antigen (Baudoux, 2009).

Numerous additional approaches to an RSV vaccine continue to be evaluated in pre-clinical studies. These include a number of protein-based vaccines, including synthetic peptides, engineered multiepitope vaccines, recombinantly-derived antigen expressed by recombinant baculoviruses or produced in plants, antigen fused genetically to carrier proteins such as cholera toxin B subunit, immune-stimulating complexes containing RSV antigens, liposome-encapsidated RSV antigen, and the use of adjuvants such as CpG oligonucleotides or Neisseria species outer membrane proteins, or biopolymer nanoparticles. Other approaches are based on recombinant vectors, including DNA vectors, helper-dependent or helper-independent adenovirus, replication-competent or replication-defective vesicular stomatitis virus, vaccinia virus MVA, rhinovirus, Newcastle disease virus, *Mycobacterium bovis* BCG, *Salmonella typhimurium*, and alphavirus replicons. These novel approaches often are found to be immunogenic and protective against RSV challenge in rodent models. However, foreign antigens often are highly immunogenic in rodents, and the poor permissiveness of these models for RSV replication provides overly optimistic evaluations of protective efficacy.

The challenges to making an RSV vaccine are considerable. The vaccine must be safe and well-tolerated in young infants, and immunogenic in the presence of maternal antibodies. In addition, a paediatric RSV vaccine should (i) induce local respiratory tract immunity as well as systemic immunity; (ii) induce secretory and serum antibodies that efficiently neutralize infectivity; (iii) be free of disease enhancement; and (iv) have significant effectiveness following a single dose, since the early age of RSV disease does not provide time for prime-boost strategies.

Perspectives

RSV is among the most common and widespread human viruses, since it infects everyone worldwide multiple times during life. It causes a considerable burden of morbidity and mortality, especially at both ends of life. Although the importance of RSV in human disease has been known for more than 50 years, we lack effective antiviral drugs or vaccines. The successes that have been achieved in controlling or preventing RSV disease involve improvements in supportive care and the development of antibody immunoprophylaxis, although the latter is expensive and limited to high-risk individuals. The development of an RSV vaccine is impeded by the physical lability of the virus, the complexity, caution, and cost involved in designing and evaluating vaccines designed for young infants, and the phenomenon of disease enhancement associated with killed or subunit vaccines. Furthermore, reduced immune responses early in life, and the inefficiency of protective immunity in the respiratory tract, will reduce the efficacy of an RSV vaccine. Even if an RSV vaccine is not highly effective in preventing infection, a substantial reduction in serious disease would be a major advance for human health. Live attenuated vaccines have been identified that are well tolerated in young children and infants, and their efficacy is presently being evaluated.

One of the notable features of RSV is its ability to infect very early in life despite the presence of maternal antibodies. Severe RSV disease early in life frequently is associated with abnormalities in pulmonary function that can last more than 10 years. However, the full impact of severe RSV infection in the context of the immature lung and immature immune system, and possible mechanisms of damage, remains to be elucidated. Whether there are long lasting effects on the immune response in some individuals remains unknown. The relative contributions of direct viral damage versus immune-mediated damage to RSV disease remains unclear and probably will vary among individuals, due in part to genetic differences that are now beginning to be identified. Studies also continue to identify the ways in which RSV inhibits or evades the immune system, and to elucidate the contribution of these effects to the ability of the virus to infect early in life and re-infect throughout life. The virus and host are engaged in a yearly contest whose parameters are still being unravelled.

Acknowledgements

The author was supported by the Intramural Research Programme of NIAID, NIH.

References

(2009). From the American Academy of Pediatrics: Policy statements–Modified recommendations for use of palivizumab for prevention of respiratory syncytial virus infections. Pediatrics *124,* 1694–1701.

Abu-Harb, M., Bell, F., Finn, A., Rao, W.H., Nixon, L., Shale, D., and Everard, M.L. (1999). IL-8 and neutrophil elastase levels in the respiratory tract of infants with RSV bronchiolitis. Eur Respir. J. *14,* 139–143.

Adkins, B., Leclerc, C., and Marshall-Clarke, S. (2004). Neonatal adaptive immunity comes of age. Nat. Rev. Immunol. *4,* 553–564.

Aherne, W., Bird, T., Court, S.D., Gardner, P.S., and McQuillin, J. (1970). Pathological changes in virus infections of the lower respiratory tract in children. J. Clin. Pathol. *23,* 7–18.

Ahmadian, G., Chambers, P., and Easton, A.J. (1999). Detection and characterization of proteins encoded by the second ORF of the M2 gene of *Pneumoviruses*. J. Gen. Virol. *80,* 2011–2016.

Alper, C.M., Winther, B., Mandel, E.M., Hendley, J.O., and Doyle, W.J. (2009). Rate of concurrent otitis media in upper respiratory tract infections with specific viruses. Arch. Otolaryngol. Head Neck Surg. *135,* 17–21.

Anderson, K., King, A.M., Lerch, R.A., and Wertz, G.W. (1992). Polylactosaminoglycan modification of the respiratory syncytial virus small hydrophobic (SH) protein: a conserved feature among human and bovine respiratory syncytial viruses. Virology *191,* 417–430.

Anderson, L.J., Hierholzer, J.C., Tsou, C., Hendry, R.M., Fernie, B.F., Stone, Y., and McIntosh, K. (1985). Antigenic characterization of respiratory syncytial virus strains with monoclonal antibodies. J. Infect. Dis. *151,* 626–633.

Asenjo, A., and Villanueva, N. (2000). Regulated but not constitutive human respiratory syncytial virus (HRSV) P protein phosphorylation is essential for oligomerization. FEBS Lett. *467,* 279–284.

Asenjo, A., Rodriguez, L., and Villanueva, N. (2005). Determination of phosphorylated residues from human respiratory syncytial virus P protein that are dynamically dephosphorylated by cellular phosphatases: a possible role for serine 54. J. Gen. Virol. *86,* 1109–1120.

Asenjo, A., Calvo, E., and Villanueva, N. (2006). Phosphorylation of human respiratory syncytial virus P protein at threonine 108 controls its interaction with the M2-1 protein in the viral RNA polymerase complex. J. Gen. Virol. *87,* 3637–3642.

Asenjo, A., Gonzalez-Armas, J.C., and Villanueva, N. (2008). Phosphorylation of human respiratory syncytial virus P protein at serine 54 regulates viral uncoating. Virology *380,* 26–33.

Atreya, P.L., Peeples, M.E., and Collins, P.L. (1998). The NS1 protein of human respiratory syncytial virus is a potent inhibitor of minigenome transcription and RNA replication. J. Virol. *72,* 1452–1461.

Ausar, S.F., Espina, M., Brock, J., Thyagarayapuran, N., Repetto, R., Khandke, L., and Middaugh, C.R. (2007). High-throughput screening of stabilizers for respiratory syncytial virus: identification of stabilizers and their effects on the conformational thermostability of viral particles. Hum. Vaccin. *3,* 94–103.

Bachi, T. (1988). Direct observation of the budding and fusion of an enveloped virus by video microscopy of viable cells. J. Cell Biol. *107,* 1689–1695.

Bachi, T., and Howe, C. (1973). Morphogenesis and ultrastructure of respiratory syncytial virus. J. Virol. *12,* 1173–1180.

Barbas, C.F.d., Crowe, J.E., Jr., Cababa, D., Jones, T.M., Zebedee, S.L., Murphy, B.R., Chanock, R.M., and Burton, D.R. (1992). Human monoclonal Fab fragments derived from a combinatorial library bind to respiratory syncytial virus F glycoprotein and neutralize infectivity. Proc. Natl. Acad. Sci. U.S.A. *89,* 10164–10168.

Barik, S. (1993). The structure of the 5′ terminal cap of the respiratory syncytial virus mRNA. J. Gen. Virol. *74,* 485–490.

Barr, J., Chambers, P., Pringle, C.R., and Easton, A.J. (1991). Sequence of the major nucleocapsid protein gene of pneumonia virus of mice: sequence comparisons suggest structural homology between nucleocapsid proteins of *Pneumoviruses*, paramyxoviruses, rhabdoviruses and filoviruses. J. Gen. Virol. *72,* 677–685.

Barr, J., Chambers, P., Harriott, P., Pringle, C.R., and Easton, A.J. (1994). Sequence of the phosphoprotein gene of pneumonia virus of mice: expression of multiple proteins from two overlapping reading frames. J. Virol. *68,* 5330–5334.

Barr, F.E., Pedigo, H., Johnson, T.R., and Shepherd, V.L. (2000). Surfactant protein-A enhances uptake of respiratory syncytial virus by monocytes and U937 macrophages. Am. J. Respir. Cell Mol. Biol. *23,* 586–592.

Bartlett, E.J., Amaro-Carambot, E., Surman, S.R., Collins, P.L., Murphy, B.R., and Skiadopoulos, M.H. (2006). Introducing point and deletion mutations into the P/C gene of human parainfluenza virus type 1 (HPIV1) by reverse genetics generates attenuated and efficacious vaccine candidates. Vaccine *24,* 2674–2684.

Baudoux, J.-P.B., Normand; Rheault, Patrick; Ruelle, Jean-Louis (2009). Recombinant RSV antigens (GlaxoSmithKline Biologicals).

Becker, S., and Soukup, J.M. (1999). Airway epithelial cell-induced activation of monocytes and eosinophils in respiratory syncytial viral infection. Immunobiology *201,* 88–106.

Beeler, J.A., and van Wyke Coelingh, K. (1989). Neutralization epitopes of the F glycoprotein of respiratory syncytial virus: effect of mutation upon fusion function. J. Virol. *63,* 2941–2950.

Behera, A.K., Matsuse, H., Kumar, M., Kong, X., Lockey, R.F., and Mohapatra, S.S. (2001). Blocking intercellular adhesion molecule-1 on human epithelial cells decreases respiratory syncytial virus infection. Biochem. Biophys. Res. Commun. *280,* 188–195.

Bendelja, K., Gagro, A., Bace, A., Lokar-Kolbas, R., Krsulovic-Hresic, V., Drazenovic, V., Mlinaric-Galinovic, G., and Rabatic, S. (2000). Predominant type-2 response in infants with respiratory syncytial virus (RSV) infection demonstrated by cytokine flow cytometry. Clin. Exp. Immunol. *121,* 332–338.

Bermingham, A., and Collins, P.L. (1999). The M2-2 protein of human respiratory syncytial virus is a regulatory factor involved in the balance between RNA replication and transcription. Proc. Natl. Acad. Sci. U.S.A. *96,* 11259–11264.

Bhella, D., Ralph, A., Murphy, L.B., and Yeo, R.P. (2002). Significant differences in nucleocapsid morphology within the *Paramyxoviridae*. J. Gen. Virol. *83,* 1831–1839.

Bitko, V., Shulyayeva, O., Mazumder, B., Musiyenko, A., Ramaswamy, M., Look, D.C., and Barik, S. (2007). Nonstructural proteins of respiratory syncytial virus suppress premature apoptosis by an NF-kappaB-dependent, interferon-independent mechanism and facilitate virus growth. J. Virol. *81,* 1786–1795.

Bonville, C.A., Rosenberg, H.F., and Domachowske, J.B. (1999). Macrophage inflammatory protein-1alpha and RANTES are present in nasal secretions during ongoing upper respiratory tract infection. Pediatr. Allergy Immunol. *10,* 39–44.

Borregaard, N., Sorensen, O.E., and Theilgaard-Monch, K. (2007). Neutrophil granules: a library of innate immunity proteins. Trends Immunol. *28,* 340–345.

Boukhvalova, M.S., Prince, G.A., Soroush, L., Harrigan, D.C., Vogel, S.N., and Blanco, J.C. (2006). The TLR4 agonist, monophosphoryl lipid A, attenuates the cytokine storm associated with respiratory syncytial virus vaccine-enhanced disease. Vaccine *24,* 5027–5035.

Boyce, T.G., Mellen, B.G., Mitchel, E.F., Jr., Wright, P.F., and Griffin, M.R. (2000). Rates of hospitalization for respiratory syncytial virus infection among children in medicaid. J. Pediatr. *137,* 865–870.

Brandenburg, A.H., Kleinjan, A., van Het Land, B., Moll, H.A., Timmerman, H.H., de Swart, R.L., Neijens, H.J., Fokkens, W., and Osterhaus, A.D. (2000a). Type 1-like immune response is found in children with respiratory syncytial virus infection regardless of clinical severity. J. Med. Virol. *62,* 267–277.

Brandenburg, A.H., van Beek, R., Moll, H.A., Osterhaus, A.D., and Claas, E.C. (2000b). G protein variation in respiratory syncytial virus group A does not correlate with clinical severity. J. Clin. Microbiol. *38,* 3849–3852.

Brock, S.C., Goldenring, J.R., and Crowe, J.E., Jr. (2003). Apical recycling systems regulate directional budding of respiratory syncytial virus from polarized epithelial cells. Proc. Natl. Acad. Sci. U.S.A. *100,* 15143–15148.

Buchholz, U.J., Finke, S., and Conzelmann, K.K. (1999). Generation of bovine respiratory syncytial virus (BRSV) from cDNA: BRSV NS2 is not essential for virus replication in tissue culture, and the human RSV leader region acts as a functional BRSV genome promoter. J. Virol. *73,* 251–259.

Buchholz, U.J., Granzow, H., Schuldt, K., Whitehead, S.S., Murphy, B.R., and Collins, P.L. (2000). Chimeric bovine respiratory syncytial virus with glycoprotein gene substitutions from human respiratory syncytial virus (HRSV): Effects on host range and evaluation as a live-attenuated HRSV vaccine. J. Virol. *74,* 1187–1199.

Bukreyev, A., Whitehead, S.S., Murphy, B.R., and Collins, P.L. (1997). Recombinant respiratory syncytial virus from which the entire SH gene has been deleted grows efficiently in cell culture and exhibits site-specific attenuation in the respiratory tract of the mouse. J. Virol. *71,* 8973–8982.

Bukreyev, A., Murphy, B.R., and Collins, P.L. (2000). Respiratory syncytial virus can tolerate an intergenic sequence of at least 160 nucleotides with little effect on transcription or replication *in vitro* and *in vivo*. J. Virol. *74,* 11017–11026.

Bukreyev, A., Yang, L., Fricke, J., Cheng, L., Ward, J.M., Murphy, B.R., and Collins, P.L. (2008). The secreted form of respiratory syncytial virus G glycoprotein helps the virus evade antibody-mediated restriction of replication by acting as an antigen decoy and through effects on Fc receptor-bearing leukocytes. J. Virol. *82,* 12191–12204.

Burke, E., Mahoney, N.M., Almo, S.C., and Barik, S. (2000). Profilin is required for optimal actin-dependent transcription of respiratory syncytial virus genome RNA. J. Virol. *74,* 669–675.

Buynak, E.B., Weibel, R.E., McLean, A.A., and Hilleman, M.R. (1978). Live respiratory syncytial virus vaccine administered parenterally. Proc. Soc. Exp. Biol. Med. *157,* 636–642.

Calder, L.J., Gonzalez-Reyes, L., Garcia-Barreno, B., Wharton, S.A., Skehel, J.J., Wiley, D.C., and Melero, J.A. (2000). Electron microscopy of the human respiratory syncytial virus fusion protein and complexes that it forms with monoclonal antibodies. Virology *271,* 122–131.

Cane, P.A. (1997). Analysis of linear epitopes recognised by the primary human antibody response to a variable region of the attachment (G) protein of respiratory syncytial virus. J. Med. Virol. *51,* 297–304.

Cane, P.A., and Pringle, C.R. (1995). Evolution of subgroup A respiratory syncytial virus: evidence for progressive accumulation of amino acid changes in the attachment protein. J. Virol. *69,* 2918–2925.

Cannon, M.J., Openshaw, P.J., and Askonas, B.A. (1988). Cytotoxic T cells clear virus but augment lung pathology in mice infected with respiratory syncytial virus. J. Exp. Med. *168,* 1163–1168.

Carbonell-Estrany, X., Simoes, E.A., Dagan, R., Hall, C.B., Harris, B., Hultquist, M., Connor, E.M., and Losonsky, G.A. (2010). Motavizumab for prophylaxis of respiratory syncytial virus in high-risk children: a noninferiority trial. Pediatrics *125,* e35–51.

Cardenas, S., Auais, A., and Piedimonte, G. (2005). Palivizumab in the prophylaxis of respiratory syncytial virus infection. Expert Rev. AntiInfect. Ther. *3,* 719–726.

Carter, S.D., Dent, K.C., Atkins, E., Foster, T.L., Verow, M., Gorny, P., Harris, M., Hiscox, J.A., Ranson, N.A., Griffin, S., *et al.* (2010). Direct visualization of the small hydrophobic protein of human respiratory syncytial virus reveals the structural basis for membrane permeability. FEBS Lett. *584,* 2786–2790.

Castagne, N., Barbier, A., Bernard, J., Rezaei, H., Huet, J.C., Henry, C., Da Costa, B., and Eleouet, J.F. (2004). Biochemical characterization of the respiratory syncytial virus P-P and P-N protein complexes and localization of the P protein oligomerization domain. J. Gen. Virol. *85,* 1643–1653.

Castro-Rodriguez, J.A., Holberg, C.J., Wright, A.L., Halonen, M., Taussig, L.M., Morgan, W.J., and Martinez, F.D. (1999). Association of radiologically ascertained pneumonia before age 3 yr with asthmalike symptoms and pulmonary function during childhood: a prospective study. Am. J. Respir. Crit. Care Med. *159,* 1891–1897.

Chanock, R.M., Roizman, B., and Myers, R. (1957). Recovery from infants with respiratory illness of a virus related to chimpanzee coryza agent. I. Isolation, properties and characterization. Am. J. Hygiene *66,* 281–290.

Chapman, J., Abbott, E., Alber, D.G., Baxter, R.C., Bithell, S.K., Henderson, E.A., Carter, M.C., Chambers, P., Chubb, A., Cockerill, G.S., *et al.* (2007). RSV604, a novel inhibitor of respiratory syncytial virus replication. Antimicrob. Agents Chemother. *51,* 3346–3353.

Cheng, X., Park, H., Zhou, H., and Jin, H. (2005). Overexpression of the M2-2 protein of respiratory syncytial virus inhibits viral replication. J. Virol. *79, 13,*943–13952.

Collarini, E.J., Lee, F.E., Foord, O., Park, M., Sperinde, G., Wu, H., Harriman, W.D., Carroll, S.F., Ellsworth, S.L., Anderson, L.J., *et al.* (2009). Potent high-affinity antibodies for treatment and prophylaxis of respiratory syncytial virus derived from B cells of infected patients. J. Immunol. *183,* 6338–6345.

Collins, P.L., and Mottet, G. (1992). Oligomerization and post-translational processing of glycoprotein G of human respiratory syncytial virus: altered O-glycosylation in the presence of brefeldin A. J. Gen. Virol. *73,* 849–863.

Collins, P.L., and Mottet, G. (1993). Membrane orientation and oligomerization of the small hydrophobic protein of human respiratory syncytial virus. J. Gen. Virol. *74,* 1445–1450.

Collins, P.L., and Murphy, B.R. (2005). New generation live vaccines against human respiratory syncytial virus designed by reverse genetics. Proc. Am. Thorac. Soc.

Collins, P.L., and Wertz, G.W. (1983). cDNA cloning and transcriptional mapping of nine polyadenylylated RNAs encoded by the genome of human respiratory syncytial virus. Proc. Natl. Acad. Sci. U.S.A. *80,* 3208–3212.

Collins, P.L., Olmsted, R.A., Spriggs, M.K., Johnson, P.R., and Buckler-White, A.J. (1987). Gene overlap and site-specific attenuation of transcription of the viral polymerase L gene of human respiratory syncytial virus. Proc. Natl. Acad. Sci. U.S.A. *84,* 5134–5138.

Collins, P.L., Hill, M.G., and Johnson, P.R. (1990a). The two open reading frames of the 22K mRNA of human respiratory syncytial virus: sequence comparison of antigenic subgroups A and B and expression *in vitro*. J. Gen. Virol. *71,* 3015–3020.

Collins, P.L., Purcell, R.H., London, W.T., Lawrence, L.A., Chanock, R.M., and Murphy, B.R. (1990b). Evaluation in chimpanzees of vaccinia virus recombinants that express the surface glycoproteins of human respiratory syncytial virus. Vaccine *8,* 164–168.

Collins, P.L., Mink, M.A., and Stec, D.S. (1991). Rescue of synthetic analogs of respiratory syncytial virus genomic RNA and effect of truncations and mutations on the expression of a foreign reporter gene. Proc. Natl. Acad. Sci. U.S.A. *88,* 9663–9667.

Collins, P.L., Hill, M.G., Camargo, E., Grosfeld, H., Chanock, R.M., and Murphy, B.R. (1995). Production of infectious human respiratory syncytial virus from cloned cDNA confirms an essential role for the transcription elongation factor from the 5′ proximal open reading frame of the M2 mRNA in gene expression and provides a capability for vaccine development. Proc. Natl. Acad. Sci. U.S.A. *92, 11*,563–11567.

Collins, P.L., Hill, M.G., Cristina, J., and Grosfeld, H. (1996). Transcription elongation factor of respiratory syncytial virus, a nonsegmented negative-strand RNA virus. Proc. Natl. Acad. Sci. U.S.A. *93,* 81–85.

Collins, P.L., Camargo, E., and Hill, M.G. (1999). Support plasmids and support proteins required for recovery of recombinant respiratory syncytial virus. Virology *259,* 251–255.

Collins, P.L., Bukreyev, A., and Murphy, B.R. (2008). What are the risks–hypothetical and observed–of recombination involving live vaccines and vaccine vectors based on nonsegmented negative-strain RNA viruses? J. Virol. *82,* 9805–9806.

Connors, M., Collins, P.L., Firestone, C.Y., and Murphy, B.R. (1991). Respiratory syncytial virus (RSV) F, G, M2 (22K), and N proteins each induce resistance to RSV challenge, but resistance induced by M2 and N proteins is relatively short-lived. J. Virol. *65,* 1634–1637.

Connors, M., Kulkarni, A.B., Firestone, C.Y., Holmes, K.L., Morse, H.C.d., Sotnikov, A.V., and Murphy, B.R. (1992). Pulmonary histopathology induced by respiratory syncytial virus (RSV) challenge of formalin-inactivated RSV-immunized BALB/c mice is abrogated by depletion of $CD4^+$ T cells. J. Virol. *66,* 7444–7451.

Cowton, V.M., and Fearns, R. (2005). Evidence that the respiratory syncytial virus polymerase is recruited to nucleotides 1 to 11 at the 3′ end of the nucleocapsid and can scan to access internal signals. J. Virol. *79, 11*,311–11322.

Cowton, V.M., McGivern, D.R., and Fearns, R. (2006). Unravelling the complexities of respiratory syncytial virus RNA synthesis. J. Gen. Virol. *87,* 1805–1821.

Crowe, J.E., Firestone, C.Y., Crim, R., Beeler, J.A., Coelingh, K.L., Barbas, C.F., Burton, D.R., Chanock, R.M., and Murphy, B.R. (1998). Monoclonal antibody-resistant mutants selected with a respiratory syncytial virus-neutralizing human antibody fab fragment (Fab 19) define a unique epitope on the fusion (F) glycoprotein. Virology *252,* 373–375.

Crowe, J.E., Jr. (2001). Influence of maternal antibodies on neonatal immunization against respiratory viruses. Clin. Infect. Dis. *33,* 1720–1727.

Crowe, J.E., Jr., Bui, P.T., Davis, A.R., Chanock, R.M., and Murphy, B.R. (1994). A further attenuated derivative of a cold-passaged temperature-sensitive mutant of human respiratory syncytial virus retains immunogenicity and protective efficacy against wild-type challenge in seronegative chimpanzees. Vaccine *12,* 783–790.

Crowe, J.E., Jr., Bui, P.T., Siber, G.R., Elkins, W.R., Chanock, R.M., and Murphy, B.R. (1995). Cold-passaged, temperature-sensitive mutants of human respiratory syncytial virus (RSV) are highly attenuated, immunogenic, and protective in seronegative chimpanzees, even when RSV antibodies are infused shortly before immunization. Vaccine *13,* 847–855.

Crowe, J.E., Jr., Firestone, C.Y., and Murphy, B.R. (2001). Passively acquired antibodies suppress humoral but not cell-mediated immunity in mice immunized with live attenuated respiratory syncytial virus vaccines. J. Immunol. *167,* 3910–3918.

Cuesta, I., Geng, X., Asenjo, A., and Villanueva, N. (2000). Structural phosphoprotein M2-1 of the human respiratory syncytial virus is an RNA binding protein. J. Virol. *74,* 9858–9867.

Culley, F.J., Pollott, J., and Openshaw, P.J. (2002). Age at first viral infection determines the pattern of T cell-mediated disease during reinfection in adulthood. J. Exp. Med. *196,* 1381–1386.

Dakhama, A., Vitalis, T.Z., and Hegele, R.G. (1997). Persistence of respiratory syncytial virus (RSV) infection and development of RSV-specific IgG1 response in a guinea-pig model of acute bronchiolitis. Eur Respir. J. *10,* 20–26.

Dall'Acqua, W.F., Kiener, P.A., and Wu, H. (2006). Properties of human IgG1s engineered for enhanced binding to the neonatal Fc receptor (FcRn). J. Biol. Chem. *281, 23*,514–23524.

Day, N.D., Branigan, P.J., Liu, C., Gutshall, L.L., Luo, J., Melero, J.A., Sarisky, R.T., and Del Vecchio, A.M. (2006). Contribution of cysteine residues in the extracellular domain of the F protein of human respiratory syncytial virus to its function. Virol. J. *3,* 34.

De Alarcon, A., Walsh, E.E., Carper, H.T., La Russa, J.B., Evans, B.A., Rakes, G.P., Platts-Mills, T.A., and Heymann, P.W. (2001). Detection of IgA and IgG but not IgE antibody to respiratory syncytial virus in nasal washes and sera from infants with wheezing. J. Pediatr. *138,* 311–317.

Delgado, M.F., Coviello, S., Monsalvo, A.C., Melendi, G.A., Hernandez, J.Z., Batalle, J.P., Diaz, L., Trento, A., Chang, H.Y., Mitzner, W., *et al.* (2009). Lack of antibody affinity maturation due to poor Toll-like receptor stimulation leads to enhanced respiratory syncytial virus disease. Nat. Med. *15,* 34–41.

DeVincenzo, J.P., El Saleeby, C.M., and Bush, A.J. (2005). Respiratory syncytial virus load predicts disease severity in previously healthy infants. J. Infect. Dis. *191,* 1861–1868.

DeVincenzo, J., Lambkin-Williams, R., Wilkinson, T., Cehelsky, J., Nochur, S., Walsh, E., Meyers, R., Gollob, J., and Vaishnaw, A. (2010a). A randomized, double-blind, placebo-controlled study of an RNAi-based therapy directed against respiratory syncytial virus. Proc. Natl. Acad. Sci. U.S.A. *107,* 8800–8805.

Devincenzo, J.P., Wilkinson, T., Vaishnaw, A., Cehelsky, J., Meyers, R., Nochur, S., Harrison, L., Meeking, P., Mann, A., Moane, E., *et al.* (2010b). Viral Load Drives Disease in Humans Experimentally Infected with Respiratory Syncytial Virus. Am. J. Respir. Crit. Care Med. *182,* 1305–1314.

Dickens, L.E., Collins, P.L., and Wertz, G.W. (1984). Transcriptional mapping of human respiratory syncytial virus. J. Virol. *52,* 364–369.

Donnelly, R.P., and Kotenko, S.V. (2010). Interferon-lambda: a new addition to an old family. J. Interferon Cytokine Res. *30,* 555–564.

Dupuy, L.C., Dobson, S., Bitko, V., and Barik, S. (1999). Casein kinase 2-mediated phosphorylation of respiratory syncytial virus phosphoprotein P is essential for the transcription elongation activity of the viral polymerase; phosphorylation by casein kinase 1 occurs mainly at Ser(215) and is without effect. J. Virol. *73,* 8384–8392.

Durbin, J.E., Johnson, T.R., Durbin, R.K., Mertz, S.E., Morotti, R.A., Peebles, R.S., and Graham, B.S. (2002). The role of IFN in respiratory syncytial virus pathogenesis. J. Immunol. *168,* 2944–2952.

Eisenhut, M. (2006). Extrapulmonary manifestations of severe respiratory syncytial virus infection–a systematic review. Crit. Care *10,* R107.

El Saleeby, C.M., Suzich, J., Conley, M.E., and DeVincenzo, J.P. (2004). Quantitative effects of palivizumab and donor-derived T cells on chronic respiratory syncytial virus infection, lung disease, and fusion glycoprotein amino acid sequences in a patient before and after bone marrow transplantation. Clin. Infect. Dis. *39,* e17–20.

El Saleeby, C.M., Li, R., Somes, G.W., Dahmer, M.K., Quasney, M.W., and DeVincenzo, J.P. (2010). Surfactant protein A2 polymorphisms and disease severity in a respiratory syncytial virus-infected population. J. Pediatr. *156,* 409–414.

Elliott, J., Lynch, O.T., Suessmuth, Y., Qian, P., Boyd, C.R., Burrows, J.F., Buick, R., Stevenson, N.J., Touzelet, O., Gadina, M., *et al.* (2007). Respiratory syncytial virus NS1 protein degrades STAT2 by using the Elongin-Cullin E3 ligase. J. Virol. *81,* 3428–3436.

Empey, K.M., Peebles, R.S., Jr., and Kolls, J.K. (2010). Pharmacologic advances in the treatment and prevention of respiratory syncytial virus. Clin. Infect. Dis. *50,* 1258–1267.

Ermers, M.J., Rovers, M.M., van Woensel, J.B., Kimpen, J.L., and Bont, L.J. (2009). The effect of high dose inhaled corticosteroids on wheeze in infants after respiratory syncytial virus infection: randomised double blind placebo controlled trial. BMJ *338,* b897.

Escribano-Romero, E., Rawling, J., Garcia-Barreno, B., and Melero, J.A. (2004). The soluble form of human respiratory syncytial virus attachment protein differs from the membrane-bound form in its oligomeric state but is still capable of binding to cell surface proteoglycans. J. Virol. *78,* 3524–3532.

Evans, J.E., Cane, P.A., and Pringle, C.R. (1996). Expression and characterisation of the NS1 and NS2 proteins of respiratory syncytial virus. Virus Res. *43,* 155–161.

Everard, M.L., Swarbrick, A., Wrightham, M., McIntyre, J., Dunkley, C., James, P.D., Sewell, H.F., and Milner, A.D. (1994). Analysis of cells obtained by bronchial lavage of infants with respiratory syncytial virus infection. Arch. Dis. Child. *71,* 428–432.

Falsey, A.R. (2007). Respiratory syncytial virus infection in adults. Semin. Respir. Crit. Care Med. *28,* 171–181.

Falsey, A.R., and Walsh, E.E. (1998). Relationship of serum antibody to risk of respiratory syncytial virus infection in elderly adults. J. Infect. Dis. *177,* 463–466.

Falsey, A.R., Hennessey, P.A., Formica, M.A., Cox, C., and Walsh, E.E. (2005). Respiratory syncytial virus infection in elderly and high-risk adults. N. Engl. J. Med. *352,* 1749–1759.

Falsey, A.R., Formica, M.A., Hennessey, P.A., Criddle, M.M., Sullender, W.M., and Walsh, E.E. (2006a). Detection of respiratory syncytial virus in adults with chronic obstructive pulmonary disease. Am. J. Respir. Crit. Care Med. *173,* 639–643.

Falsey, A.R., Singh, H.K., and Walsh, E.E. (2006b). Serum antibody decay in adults following natural respiratory syncytial virus infection. J. Med. Virol. *78,* 1493–1497.

Falsey, A.R., Walsh, E.E., Capellan, J., Gravenstein, S., Zambon, M., Yau, E., Gorse, G.J., Edelman, R., Hayden, F.G., McElhaney, J.E., *et al.* (2008). Comparison of the safety and immunogenicity of 2 respiratory syncytial virus (rsv) vaccines–nonadjuvanted vaccine or vaccine adjuvanted with alum–given concomitantly with influenza vaccine to high-risk elderly individuals. J. Infect. Dis. *198,* 1317–1326.

Fearns, R., and Collins, P.L. (1999a). Model for polymerase access to the overlapped L gene of respiratory syncytial virus. J. Virol. *73,* 388–397.

Fearns, R., and Collins, P.L. (1999b). Role of the M2-1 transcription antitermination protein of respiratory syncytial virus in sequential transcription. J. Virol. *73,* 5852–5864.

Fearns, R., Peeples, M.E., and Collins, P.L. (1997). Increased expression of the N protein of respiratory syncytial virus stimulates minigenome replication but does not alter the balance between the synthesis of mRNA and antigenome. Virology *236,* 188–201.

Fearns, R., Collins, P.L., and Peeples, M.E. (2000). Functional analysis of the genomic and antigenomic promoters of human respiratory syncytial virus. J. Virol. *74,* 6006–6014.

Fearns, R., Peeples, M.E., and Collins, P.L. (2002). Mapping the transcription and replication promoters of respiratory syncytial virus. J. Virol. *76,* 1663–1672.

Feldman, S.A., Hendry, R.M., and Beeler, J.A. (1999). Identification of a linear heparin binding domain for human respiratory syncytial virus attachment glycoprotein G. J. Virol. *73,* 6610–6617.

Fishaut, M., Tubergen, D., and McIntosh, K. (1980). Cellular response to respiratory viruses with particular reference to children with disorders of cell-mediated immunity. J. Pediatr. *96,* 179–186.

Fisher, R.G., Gruber, W.C., Edwards, K.M., Reed, G.W., Tollefson, S.J., Thompson, J.M., and Wright, P.F. (1997). Twenty years of outpatient respiratory syncytial virus infection: a framework for vaccine efficacy trials. Pediatrics *99,* E7.

Friedewald, W.T., Forsyth, B.R., Smith, C.B., Gharpure, M.A., and Chanock, R.M. (1968). Low-temperature-grown RS virus in adult volunteers. JAMA *203,* 690–694.

Fuentes, S., Tran, K.C., Luthra, P., Teng, M.N., and He, B. (2007). Function of the respiratory syncytial virus small hydrophobic protein. J. Virol. *81*, 8361–8366.

Furze, J.M., Roberts, S.R., Wertz, G.W., and Taylor, G. (1997). Antigenically distinct G glycoproteins of BRSV strains share a high degree of genetic homogeneity. Virology *231*, 48–58.

Gadomski, A.M., and Bhasale, A.L. (2006). Bronchodilators for bronchiolitis. Cochrane Database Syst. Rev. *3*, CD001266.

Gan, S.W., Ng, L., Lin, X., Gong, X., and Torres, J. (2008). Structure and ion channel activity of the human respiratory syncytial virus (hRSV) small hydrophobic protein transmembrane domain. Protein Sci. *17*, 813–820.

Gao, L., Yu, S., Chen, Q., Duan, Z., Zhou, J., Mao, C., Yu, D., Zhu, W., Nie, J., and Hou, Y. (2010). A randomized controlled trial of low-dose recombinant human interferons alpha-2b nasal spray to prevent acute viral respiratory infections in military recruits. Vaccine *28*, 4445–4451.

Garcia, J., Garcia-Barreno, B., Vivo, A., and Melero, J.A. (1993). Cytoplasmic inclusions of respiratory syncytial virus-infected cells: formation of inclusion bodies in transfected cells that coexpress the nucleoprotein, the phosphoprotein, and the 22K protein. Virology *195*, 243–247.

Gardner, P.S., McQuillin, J., and Court, S.D. (1970). Speculation on pathogenesis in death from respiratory syncytial virus infection. Br. Med. J. *1*, 327–330.

Garofalo, R., Kimpen, J.L., Welliver, R.C., and Ogra, P.L. (1992). Eosinophil degranulation in the respiratory tract during naturally acquired respiratory syncytial virus infection. J. Pediatr. *120*, 28–32.

Gern, J.E., Rosenthal, L.A., Sorkness, R.L., and Lemanske, R.F., Jr. (2005). Effects of viral respiratory infections on lung development and childhood asthma. J. Allergy Clin. Immunol. *115*, 668–674; quiz 675.

Ghildyal, R., Hartley, C., Varrasso, A., Meanger, J., Voelker, D.R., Anders, E.M., and Mills, J. (1999). Surfactant protein A binds to the fusion glycoprotein of respiratory syncytial virus and neutralizes virion infectivity. J. Infect. Dis. *180*, 2009–2013.

Ghildyal, R., Baulch-Brown, C., Mills, J., and Meanger, J. (2003). The matrix protein of Human respiratory syncytial virus localises to the nucleus of infected cells and inhibits transcription. Arch. Virol. *148*, 1419–1429.

Gill, M.A., and Welliver, R.C. (2009). Motavizumab for the prevention of respiratory syncytial virus infection in infants. Expert Opin. Biol. Ther. *9*, 1335–1345.

Ginocchio, C.C., Swierkosz, E., McAdam, A.J., Marcon, M., Storch, G.A., Valsamakis, A., Juretschko, S., Romero, J., and Yen-Lieberman, B. (2010). Multicenter study of clinical performance of the 3M Rapid Detection RSV test. J. Clin. Microbiol. *48*, 2337–2343.

Glasser, S.W., Witt, T.L., Senft, A.P., Baatz, J.E., Folger, D., Maxfield, M.D., Akinbi, H.T., Newton, D.A., Prows, D.R., and Korfhagen, T.R. (2009). Surfactant protein C-deficient mice are susceptible to respiratory syncytial virus infection. Am. J. Physiol. Lung Cell Mol. Physiol. *297*, L64–72.

Glezen, W.P., Paredes, A., Allison, J.E., Taber, L.H., and Frank, A.L. (1981). Risk of respiratory syncytial virus infection for infants from low-income families in relationship to age, sex, ethnic group, and maternal antibody level. J. Pediatr. *98*, 708–715.

Glezen, W.P., Taber, L.H., Frank, A.L., and Kasel, J.A. (1986). Risk of primary infection and reinfection with respiratory syncytial virus. Am. J. Dis. Child. *140*, 543–546.

Gomez, M., Mufson, M.A., Dubovsky, F., Knightly, C., Zeng, W., and Losonsky, G. (2009). Phase-I study MEDI-534, of a live, attenuated intranasal vaccine against respiratory syncytial virus and parainfluenza-3 virus in seropositive children. Pediatr. Infect. Dis. J. *28*, 655–658.

Gonzalez, I.M., Karron, R.A., Eichelberger, M., Walsh, E.E., Delagarza, V.W., Bennett, R., Chanock, R.M., Murphy, B.R., Clements-Mann, M.L., and Falsey, A.R. (2000). Evaluation of the live attenuated cpts 248/404 RSV vaccine in combination with a subunit RSV vaccine (PFP-2) in healthy young and older adults. Vaccine *18*, 1763–1772.

Gonzalez-Reyes, L., Ruiz-Arguello, M.B., Garcia-Barreno, B., Calder, L., Lopez, J.A., Albar, J.P., Skehel, J.J., Wiley, D.C., and Melero, J.A. (2001). Cleavage of the human respiratory syncytial virus fusion protein at two distinct sites is required for activation of membrane fusion. Proc. Natl. Acad. Sci. U.S.A. *98*, 9859–9864.

Gorman, J.J., Ferguson, B.L., Speelman, D., and Mills, J. (1997). Determination of the disulfide bond arrangement of human respiratory syncytial virus attachment (G) protein by matrix-assisted laser desorption/ionization time-of-flight mass spectrometry. Protein Sci. *6*, 1308–1315.

Gould, P.S., and Easton, A.J. (2005). Coupled translation of the respiratory syncytial virus m2 open reading frames requires upstream sequences. J. Biol. Chem. *280, 21*,972–21980.

Graham, B.S., Bunton, L.A., Rowland, J., Wright, P.F., and Karzon, D.T. (1991a). Respiratory syncytial virus infection in anti-mu-treated mice. J. Virol. *65*, 4936–4942.

Graham, B.S., Bunton, L.A., Wright, P.F., and Karzon, D.T. (1991b). Role of T lymphocyte subsets in the pathogenesis of primary infection and rechallenge with respiratory syncytial virus in mice. J. Clin. Invest *88*, 1026–1033.

Groskreutz, D.J., Babor, E.C., Monick, M.M., Varga, S.M., and Hunninghake, G.W. (2010). Respiratory syncytial virus limits alpha subunit of eukaryotic translation initiation factor 2 (eIF2alpha) phosphorylation to maintain translation and viral replication. J. Biol. Chem. *285*, 2,023–24031.

Hacking, D., Knight, J.C., Rockett, K., Brown, H., Frampton, J., Kwiatkowski, D.P., Hull, J., and Udalova, I.A. (2004). Increased *in vivo* transcription of an IL-8 haplotype associated with respiratory syncytial virus disease-susceptibility. Genes Immun. *5*, 274–282.

Hall, C.B. (1982). Respiratory syncytial virus: its transmission in the hospital environment. Yale J. Biol. Med. *55*, 219–223.

Hall, C.B. (2001). Respiratory syncytial virus and parainfluenza virus. N. Engl. J. Med. *344*, 1917–1928.

Hall, C.B. (2007). The spread of influenza and other respiratory viruses: complexities and conjectures. Clin. Infect. Dis. *45*, 353–359.

Hall, C.B. (2010). Respiratory syncytial virus in young children. Lancet *375*, 1500–1502.

Hall, C.B., Douglas, R.G., Jr., and Geiman, J.M. (1975). Quantitative shedding patterns of respiratory syncytial virus in infants. J. Infect. Dis. *132*, 151–156.

Hall, C.B., Douglas, R.G., Jr., and Geiman, J.M. (1976). Respiratory syncytial virus infections in infants: quantitation and duration of shedding. J. Pediatr. *89*, 11–15.

Hall, C.B., Douglas, R.G., Jr., Simons, R.L., and Geiman, J.M. (1978a). Interferon production in children with respiratory syncytial, influenza, and parainfluenza virus infections. J. Pediatr. *93*, 28–32.

Hall, W.J., Hall, C.B., and Speers, D.M. (1978b). Respiratory syncytial virus infection in adults: clinical, virologic, and serial pulmonary function studies. Ann. Intern. Med. *88*, 203–205.

Hall, C.B., Douglas, R.G., Jr., and Simons, R.L. (1981). Interferon production in adults with respiratory syncytial viral infection. Ann. Intern. Med. *94*, 53–55.

Hall, C.B., Hall, W.J., Gala, C.L., MaGill, F.B., and Leddy, J.P. (1984). Long-term prospective study in children after respiratory syncytial virus infection. J. Pediatr. *105*, 358–364.

Hall, C.B., Powell, K.R., MacDonald, N.E., Gala, C.L., Menegus, M.E., Suffin, S.C., and Cohen, H.J. (1986). Respiratory syncytial viral infection in children with compromised immune function. N. Engl. J. Med. *315*, 77–81.

Hall, C.B., Powell, K.R., Schnabel, K.C., Gala, C.L., and Pincus, P.H. (1988). Risk of secondary bacterial infection in infants hospitalized with respiratory syncytial viral infection. J. Pediatr. *113*, 266–271.

Hall, C.B., Walsh, E.E., Long, C.E., and Schnabel, K.C. (1991). Immunity to and frequency of reinfection with respiratory syncytial virus. J. Infect. Dis. *163*, 693–698.

Hall, C.B., Long, C.E., and Schnabel, K.C. (2001). Respiratory syncytial virus infections in previously healthy working adults. Clin. Infect. Dis. *33*, 792–796.

Hall, C.B., Weinberg, G.A., Iwane, M.K., Blumkin, A.K., Edwards, K.M., Staat, M.A., Auinger, P., Griffin, M.R., Poehling, K.A., Erdman, D., *et al.* (2009). The burden of respiratory syncytial virus infection in young children. N. Engl. J. Med. *360*, 588–598.

Hallak, L.K., Spillmann, D., Collins, P.L., and Peeples, M.E. (2000). Glycosaminoglycan sulfation requirements for respiratory syncytial virus infection. J. Virol. *74*, 10508–10513.

Halstead, D.C., and Jenkins, S.G. (1998). Continuous non-seasonal epidemic of respiratory syncytial virus infection in the southeast United States. South Med. J. *91*, 433–436.

Hanley, L.L., McGivern, D.R., Teng, M.N., Djang, R., Collins, P.L., and Fearns, R. (2010). Roles of the respiratory syncytial virus trailer region: Effects of mutations on genome production and stress granule formation. Virology *406*, 241–252.

Harcourt, J., Alvarez, R., Jones, L.P., Henderson, C., Anderson, L.J., and Tripp, R.A. (2006). Respiratory syncytial virus G protein and G protein CX3C motif adversely affect CX3CR1+ T-cell responses. J. Immunol. *176*, 1600–1608.

Hardy, R.W., and Wertz, G.W. (1998). The product of the respiratory syncytial virus M2 gene ORF1 enhances readthrough of intergenic junctions during viral transcription. J. Virol. *72*, 520–526.

Hardy, R.W., and Wertz, G.W. (2000). The Cys(3)-His(1) motif of the respiratory syncytial virus M2-1 protein is essential for protein function. J. Virol. *74*, 5880–5885.

Harmon, S.B., and Wertz, G.W. (2002). Transcriptional termination modulated by nucleotides outside the characterized gene end sequence of respiratory syncytial virus. Virology *300*, 304–315.

Harrison, A.M., Bonville, C.A., Rosenberg, H.F., and Domachowske, J.B. (1999). Respiratory syncytical virus-induced chemokine expression in the lower airways: eosinophil recruitment and degranulation. Am. J. Respir. Crit. Care Med. *159*, 1918–1924.

Hashem, M., and Hall, C.B. (2003). Respiratory syncytial virus in healthy adults: the cost of a cold. J. Clin. Virol. *27*, 14–21.

Heidema, J., Lukens, M.V., van Maren, W.W., van Dijk, M.E., Otten, H.G., van Vught, A.J., van der Werff, D.B., van Gestel, S.J., Semple, M.G., Smyth, R.L., *et al.* (2007). $CD8^+$ T-cell responses in bronchoalveolar lavage fluid and peripheral blood mononuclear cells of infants with severe primary respiratory syncytial virus infections. J. Immunol. *179*, 8410–8417.

Heidema, J., Rossen, J.W., Lukens, M.V., Ketel, M.S., Scheltens, E., Kranendonk, M.E., van Maren, W.W., van Loon, A.M., Otten, H.G., Kimpen, J.L., *et al.* (2008). Dynamics of human respiratory virus-specific $CD8^+$ T-cell responses in blood and airways during episodes of common cold. J. Immunol. *181*, 5551–5559.

Henderson, F.W., Collier, A.M., Clyde, W.A., Jr., and Denny, F.W. (1979). Respiratory-syncytial-virus infections, reinfections and immunity. A prospective, longitudinal study in young children. N. Engl. J. Med. *300*, 530–534.

Henderson, G., Murray, J., and Yeo, R.P. (2002). Sorting of the respiratory syncytial virus matrix protein into detergent-resistant structures is dependent on cell-surface expression of the glycoproteins. Virology *300*, 244–254.

Hendricks, D.A., McIntosh, K., and Patterson, J.L. (1988). Further characterization of the soluble form of the G glycoprotein of respiratory syncytial virus. J. Virol. *62*, 2228–2233.

Henrickson, K.J., Hoover, S., Kehl, K.S., and Hua, W. (2004). National disease burden of respiratory viruses detected in children by polymerase chain reaction. Pediatr. Infect. Dis. J. *23*, S11–18.

Hishiki, H., Ishiwada, N., Fukasawa, C., Abe, K., Hoshino, T., Aizawa, J., Ishikawa, N., and Kohno, Y. (2010). Incidence of bacterial coinfection with respiratory syncytial virus bronchopulmonary infection in pediatric inpatients. J. Infect. Chemother. *17*, 87–90.

Hogg, J.C., Williams, J., Richardson, J.B., Macklem, P.T., and Thurlbeck, W.M. (1970). Age as a factor in the distribution of lower-airway conductance and in the pathologic anatomy of obstructive lung disease. N. Engl. J. Med. *282*, 1283–1287.

Holberg, C.J., Wright, A.L., Martinez, F.D., Ray, C.G., Taussig, L.M., and Lebowitz, M.D. (1991). Risk factors for respiratory syncytial virus-associated lower respiratory illnesses in the first year of life. Am. J. Epidemiol. *133*, 1135–1151.

Horsfall, F., and R.G. Hahn (1939). A pneumonia virus of Swiss mice. Proc. Soc. Exp. Biol. Med. *40*, 684–686.

Hussell, T., and Openshaw, P.J. (1998). Intracellular IFN-gamma expression in natural killer cells precedes lung CD8$^+$ T cell recruitment during respiratory syncytial virus infection. J. Gen. Virol. *79*, 2593–2601.

Iversen, M.B., and Paludan, S.R. (2010). Mechanisms of type III interferon expression. J. Interferon Cytokine Res. *30*, 573–578.

Iwane, M.K., Edwards, K.M., Szilagyi, P.G., Walker, F.J., Griffin, M.R., Weinberg, G.A., Coulen, C., Poehling, K.A., Shone, L.P., Balter, S., *et al.* (2004). Population-based surveillance for hospitalizations associated with respiratory syncytial virus, influenza virus, and parainfluenza viruses among young children. Pediatrics *113*, 1758–1764.

Janssen, R., Bont, L., Siezen, C.L., Hodemaekers, H.M., Ermers, M.J., Doornbos, G., van 't Slot, R., Wijmenga, C., Goeman, J.J., Kimpen, J.L., *et al.* (2007). Genetic susceptibility to respiratory syncytial virus bronchiolitis is predominantly associated with innate immune genes. J. Infect. Dis. *196*, 826–834.

Jeffree, C.E., Rixon, H.W., Brown, G., Aitken, J., and Sugrue, R.J. (2003). Distribution of the attachment (G) glycoprotein and GM1 within the envelope of mature respiratory syncytial virus filaments revealed using field emission scanning electron microscopy. Virology *306*, 254–267.

Jeffree, C.E., Brown, G., Aitken, J., Su-Yin, D.Y., Tan, B.H., and Sugrue, R.J. (2007). Ultrastructural analysis of the interaction between F-actin and respiratory syncytial virus during virus assembly. Virology *369*, 309–323.

Jin, H., Cheng, X., Zhou, H.Z., Li, S., and Seddiqui, A. (2000). Respiratory syncytial virus that lacks open reading frame 2 of the M2 gene (M2-2) has altered growth characteristics and is attenuated in rodents. J. Virol. *74*, 74–82.

Johnson, J.E., Gonzales, R.A., Olson, S.J., Wright, P.F., and Graham, B.S. (2007). The histopathology of fatal untreated human respiratory syncytial virus infection. Mod. Pathol. *20*, 108–119.

Johnson, P.R., and Collins, P.L. (1988). The A and B subgroups of human respiratory syncytial virus: comparison of intergenic and gene-overlap sequences. J. Gen. Virol. *69*, 2901–2906.

Johnson, P.R., Jr., Olmsted, R.A., Prince, G.A., Murphy, B.R., Alling, D.W., Walsh, E.E., and Collins, P.L. (1987a). Antigenic relatedness between glycoproteins of human respiratory syncytial virus subgroups A and B: evaluation of the contributions of F and G glycoproteins to immunity. J. Virol. *61*, 3163–3166.

Johnson, P.R., Spriggs, M.K., Olmsted, R.A., and Collins, P.L. (1987b). The G glycoprotein of human respiratory syncytial viruses of subgroups A and B: extensive sequence divergence between antigenically related proteins. Proc. Natl. Acad. Sci. U.S.A. *84*, 5625–5629.

Johnson, R.A., Prince, G.A., Suffin, S.C., Horswood, R.L., and Chanock, R.M. (1982). Respiratory syncytial virus infection in cyclophosphamide-treated cotton rats. Infect. Immun. *37*, 369–373.

Johnston, S.L., and Siegel, C.S. (1990). Evaluation of direct immunofluorescence, enzyme immunoassay, centrifugation culture, and conventional culture for the detection of respiratory syncytial virus. J. Clin. Microbiol. *28*, 2394–2397.

Jones, B., Zhan, X., Mishin, V., Slobod, K.S., Surman, S., Russell, C.J., Portner, A., and Hurwitz, J.L. (2009). Human PIV-2 recombinant Sendai virus (rSeV) elicits durable immunity and combines with two additional rSeVs to protect against hPIV-1, hPIV-2, hPIV-3, and RSV. Vaccine *27*, 1848–1857.

Kahn, J.S., Schnell, M.J., Buonocore, L., and Rose, J.K. (1999). Recombinant vesicular stomatitis virus expressing respiratory syncytial virus (RSV) glycoproteins: RSV fusion protein can mediate infection and cell fusion. Virology *254*, 81–91.

Kalica, A.R., Wright, P.F., Hetrick, F.M., and Chanock, R.M. (1973). Electron microscopic studies of respiratory syncytial temperature-sensitive mutants. Arch. Gesamte Virusforsch *41*, 248–258.

Kallewaard, N.L., Bowen, A.L., and Crowe, J.E., Jr. (2005). Cooperativity of actin and microtubule elements during replication of respiratory syncytial virus. Virology *331*, 73–81.

Kapikian, A.Z., Bell, J.A., Mastrota, F.M., Johnson, K.M., Huebner, R.J., and Chanock, R.M. (1961). An outbreak of febrile illness and pneumonia associated with respiratory syncytial virus infection. American Journal of Hygiene *74*, 234–248.

Karron, R.A., Buonagurio, D.A., Georgiu, A.F., Whitehead, S.S., Adamus, J.E., Clements-Mann, M.L., Harris, D.O., Randolph, V.B., Udem, S.A., Murphy, B.R., *et al.* (1997a). Respiratory syncytial virus (RSV) SH and G proteins are not essential for viral replication *in vitro*: clinical evaluation and molecular characterization of a cold-passaged, attenuated RSV subgroup B mutant. Proc. Natl. Acad. Sci. U.S.A. *94*, 13961–13966.

Karron, R.A., Wright, P.F., Crowe, J.E., Jr., Clements, M.L., Thompson, J., Makhene, M., Casey, R., and Murphy, B.R. (1997b). Evaluation of two live, cold-passaged, temperature-sensitive respiratory syncytial virus (RSV) vaccines in chimpanzees, adults, infants and children. J. Infect. Dis. *176*, 1428–1436.

Karron, R.A., Singleton, R.J., Bulkow, L., Parkinson, A., Kruse, D., DeSmet, I., Indorf, C., Petersen, K.M., Leombruno, D., Hurlburt, D., *et al.* (1999). Severe respiratory syncytial virus disease in Alaska native children. RSV Alaska Study Group. J. Infect. Dis. *180*, 41–49.

Karron, R.A., Wright, P.F., Belshe, R.B., Thumar, B., Casey, R., Newman, F., Polack, F.P., Randolph, V.B., Deatly, A., Hackell, J., *et al.* (2005). Identification of a recombinant live attenuated respiratory syncytial virus vaccine candidate that is highly attenuated in infants. J. Infect. Dis. *191*, 1093–1104.

Kaul, T.N., Welliver, R.C., Wong, D.T., Udwadia, R.A., Riddlesberger, K., and Ogra, P.L. (1981). Secretory antibody response to respiratory syncytial virus infection. Am. J. Dis. Child. *135*, 1013–1016.

Kerr, M.H., and Paton, J.Y. (1999). Surfactant protein levels in severe respiratory syncytial virus infection. Am. J. Respir. Crit. Care Med. *159*, 1115–1118.

Khattar, S.K., Yunus, A.S., and Samal, S.K. (2001). Mapping the domains on the phosphoprotein of bovine respiratory syncytial virus required for N-P and P–L interactions using a minigenome system. J. Gen. Virol. *82*, 775–779.

Kim, H.W., Arrobio, J.O., Brandt, C.D., Jeffries, B.C., Pyles, G., Reid, J.L., Chanock, R.M., and Parrott, R.H. (1973). Epidemiology of respiratory syncytial virus infection in Washington, D.C. I. Importance of the virus in different respiratory tract disease syndromes and temporal distribution of infection. Am. J. Epidemiol. *98*, 216–225.

Kim, H.W., Leikin, S.L., Arrobio, J., Brandt, C.D., Chanock, R.M., and Parrott, R.H. (1976). Cell-mediated immunity to respiratory syncytial virus induced by inactivated vaccine or by infection. Pediatr. Res. *10*, 75–78.

Kim, C.K., Kim, S.W., Park, C.S., Kim, B.I., Kang, H., and Koh, Y.Y. (2003). Bronchoalveolar lavage cytokine profiles in acute asthma and acute bronchiolitis. J. Allergy Clin. Immunol. *112*, 64–71.

Klein Klouwenberg, P., Tan, L., Werkman, W., van Bleek, G.M., and Coenjaerts, F. (2009). The role of Toll-like receptors in regulating the immune response against respiratory syncytial virus. Crit. Rev. Immunol. *29*, 531–550.

Kochva, U., Leonov, H., and Arkin, I.T. (2003). Modeling the structure of the respiratory syncytial virus small hydrophobic protein by silent-mutation analysis of global searching molecular dynamics. Protein Sci. *12*, 2668–2674.

Kolakofsky, D., Roux, L., Garcin, D., and Ruigrok, R.W. (2005). Paramyxovirus mRNA editing, the 'rule of six' and error catastrophe: a hypothesis. J. Gen. Virol. *86*, 1869–1877.

Kolokoltsov, A.A., Deniger, D., Fleming, E.H., Roberts, N.J., Jr., Karpilow, J.M., and Davey, R.A. (2007). Small interfering RNA profiling reveals key role of clathrin-mediated endocytosis and early endosome formation for infection by respiratory syncytial virus. J. Virol. *81*, 7786–7800.

Kozak, M. (2005). Regulation of translation via mRNA structure in prokaryotes and eukaryotes. Gene *361*, 13–37.

Krempl, C., Murphy, B.R., and Collins, P.L. (2002). Recombinant respiratory syncytial virus with the G and F genes shifted to the promoter-proximal positions. J. Virol. *76*, 11931–11942.

Kristjansson, S., Bjarnarson, S.P., Wennergren, G., Palsdottir, A.H., Arnadottir, T., Haraldsson, A., and Jonsdottir, I. (2005). Respiratory syncytial virus and other respiratory viruses during the first 3 months of life promote a local TH2-like response. J. Allergy Clin. Immunol. *116*, 805–811.

Kulkarni, A.B., Connors, M., Firestone, C.Y., Morse, H.C.d., and Murphy, B.R. (1993). The cytolytic activity of pulmonary CD8$^+$ lymphocytes, induced by infection with a vaccinia virus recombinant expressing the M2 protein of respiratory syncytial virus (RSV), correlates with resistance to RSV infection in mice. J. Virol. *67*, 1044–1049.

Kulkarni, A.B., Collins, P.L., Bacik, I., Yewdell, J.W., Bennink, J.R., Crowe, J.E., Jr., and Murphy, B.R. (1995). Cytotoxic T cells specific for a single peptide on the M2 protein of respiratory syncytial virus are the sole mediators of resistance induced by immunization with M2 encoded by a recombinant vaccinia virus. J. Virol. *69*, 1261–1264.

Kunzelmann, K., Sun, J., Meanger, J., King, N.J., and Cook, D.I. (2007). Inhibition of airway Na+ transport by respiratory syncytial virus. J. Virol. *81*, 3714–3720.

Kuo, L., Fearns, R., and Collins, P.L. (1996a). The structurally diverse intergenic regions of respiratory syncytial virus do not modulate sequential transcription by a dicistronic minigenome. J. Virol. *70*, 6143–6150.

Kuo, L., Grosfeld, H., Cristina, J., Hill, M.G., and Collins, P.L. (1996b). Effect of mutations in the gene-start and gene-end sequence motifs on transcription of monocistronic and dicistronic minigenomes of respiratory syncytial virus. J. Virol. *70*, 6892–6901.

Kuo, L., Fearns, R., and Collins, P.L. (1997). Analysis of the gene start and gene end signals of human respiratory syncytial virus: quasi-templated initiation at position 1 of the encoded mRNA. J. Virol. *71*, 4944–4953.

Kurt-Jones, E.A., Popova, L., Kwinn, L., Haynes, L.M., Jones, L.P., Tripp, R.A., Walsh, E.E., Freeman, M.W., Golenbock, D.T., Anderson, L.J., *et al.* (2000). Pattern recognition receptors TLR4 and CD14 mediate response to respiratory syncytial virus. Nat. Immunol. *1*, 398–401.

Kusel, M.M., de Klerk, N.H., Holt, P.G., Kebadze, T., Johnston, S.L., and Sly, P.D. (2006). Role of respiratory viruses in acute upper and lower respiratory tract illness in the first year of life: a birth cohort study. Pediatr. Infect. Dis. J. *25*, 680–686.

Kwilas, S., Liesman, R.M., Zhang, L., Walsh, E., Pickles, R.J., and Peeples, M.E. (2009). Respiratory syncytial virus grown in Vero cells contains a truncated attachment protein that alters its infectivity and dependence on glycosaminoglycans. J. Virol. *83*, 10710–10718.

Kwilas, A.R., Yednak, M.A., Zhang, L., Liesman, R., Collins, P.L., Pickles, R.J., and Peeples, M.E. (2010). Respiratory syncytial virus engineered to express CFTR corrects the bioelectric phenotype of human cystic fibrosis airway epithelium *in vitro*. J. Virol. *84,* 7770–7781.

Lazar, I., Canaan, A., Weibel, C., and Kahn, J.S. (2006). Novel mutations in the respiratory syncytial virus G gene identified in viral isolates from a girl with severe combined immune deficiency treated with intravenous immune globulin. J. Clin. Virol. *37,* 168–173.

Le Nouen, C., Munir, S., Losq, S., Winter, C.C., McCarty, T., Stephany, D.A., Holmes, K.L., Bukreyev, A., Rabin, R.L., Collins, P.L., *et al.* (2009). Infection and maturation of monocyte-derived human dendritic cells by human respiratory syncytial virus, human *Metapneumovirus,* and human parainfluenza virus type 3. Virology *385,* 169–182.

Leaman, D.W., Longano, F.J., Okicki, J.R., Soike, K.F., Torrence, P.F., Silverman, R.H., and Cramer, H. (2002). Targeted therapy of respiratory syncytial virus in African green monkeys by intranasally administered 2–5A antisense. Virology *292,* 70–77.

Lee, F.E., Walsh, E.E., Falsey, A.R., Lumb, M.E., Okam, N.V., Liu, N., Divekar, A.A., Hall, C.B., and Mosmann, T.R. (2007). Human infant respiratory syncytial virus (RSV)-specific type 1 and 2 cytokine responses *ex vivo* during primary RSV infection. J. Infect. Dis. *195,* 1779–1788.

Legg, J.P., Hussain, I.R., Warner, J.A., Johnston, S.L., and Warner, J.O. (2003). Type 1 and type 2 cytokine imbalance in acute respiratory syncytial virus bronchiolitis. Am. J. Respir. Crit. Care Med. *168,* 633–639.

Levin, D.L., Garg, A., Hall, L.J., Slogic, S., Jarvis, J.D., and Leiter, J.C. (2008). A prospective randomized controlled blinded study of three bronchodilators in infants with respiratory syncytial virus bronchiolitis on mechanical ventilation. Pediatr. Crit. Care Med. *9,* 598–604.

LeVine, A.M., Gwozdz, J., Stark, J., Bruno, M., Whitsett, J., and Korfhagen, T. (1999). Surfactant protein-A enhances respiratory syncytial virus clearance *in vivo*. J. Clin. Invest *103,* 1015–1021.

LeVine, A.M., Elliott, J., Whitsett, J.A., Srikiatkhachorn, A., Crouch, E., DeSilva, N., and Korfhagen, T. (2004). Surfactant protein-d enhances phagocytosis and pulmonary clearance of respiratory syncytial virus. Am. J. Respir. Cell Mol. Biol. *31,* 193–199.

Levine, S., Peeples, M., and Hamilton, R. (1977). Effect of respiratory syncytial virus infection of HeLa-cell macromolecular synthesis. J. Gen. Virol. *37,* 53–63.

Levy, O. (2007). Innate immunity of the newborn: basic mechanisms and clinical correlates. Nat. Rev. Immunol. *7,* 379–390.

Liao, R.S., Tomalty, L.L., Majury, A., and Zoutman, D.E. (2009). Comparison of viral isolation and multiplex real-time reverse transcription-PCR for confirmation of respiratory syncytial virus and influenza virus detection by antigen immunoassays. J. Clin. Microbiol. *47,* 527–532.

Ling, Z., Tran, K.C., and Teng, M.N. (2009). Human respiratory syncytial virus nonstructural protein NS2 antagonizes the activation of beta interferon transcription by interacting with RIG-I. J. Virol. *83,* 3734–3742.

Liu, V., Dhillon, G.S., and Weill, D. (2010). A multidrug regimen for respiratory syncytial virus and parainfluenza virus infections in adult lung and heart-lung transplant recipients. Transpl. Infect. Dis. *12,* 38–44.

Liuzzi, M., Mason, S.W., Cartier, M., Lawetz, C., McCollum, R.S., Dansereau, N., Bolger, G., Lapeyre, N., Gaudette, Y., Lagace, L., *et al.* (2005). Inhibitors of respiratory syncytial virus replication target cotranscriptional mRNA guanylylation by viral RNA-dependent RNA polymerase. J. Virol. *79,* 13105–13115.

Llorente, M.T., Garcia-Barreno, B., Calero, M., Camafeita, E., Lopez, J.A., Longhi, S., Ferron, F., Varela, P.F., and Melero, J.A. (2006). Structural analysis of the human respiratory syncytial virus phosphoprotein: characterization of an alpha-helical domain involved in oligomerization. J. Gen. Virol. *87,* 159–169.

Lo, M.S., Brazas, R.M., and Holtzman, M.J. (2005). Respiratory syncytial virus nonstructural proteins NS1 and NS2 mediate inhibition of Stat2 expression and alpha/beta interferon responsiveness. J. Virol. *79,* 9315–9319.

Lofgren, J., Marttila, R., Renko, M., Ramet, M., and Hallman, M. (2010). Toll-like receptor 4 Asp299Gly polymorphism in respiratory syncytial virus epidemics. Pediatr. Pulmonol. *45,* 687–692.

Long, C.E., McBride, J.T., and Hall, C.B. (1995). Sequelae of respiratory syncytial virus infections. A role for intervention studies. Am. J. Respir. Crit. Care Med. *151,* 1678–1680; discussion 1680–1671.

Lopez, J.A., Bustos, R., Orvell, C., Berois, M., Arbiza, J., Garcia-Barreno, B., and Melero, J.A. (1998). Antigenic structure of human respiratory syncytial virus fusion glycoprotein. J. Virol. *72,* 6922–6928.

Lu, B., Ma, C.H., Brazas, R., and Jin, H. (2002). The major phosphorylation sites of the respiratory syncytial virus phosphoprotein are dispensable for virus replication *in vitro*. J. Virol. *76,* 10776–10784.

Lukens, M.V., van de Pol, A.C., Coenjaerts, F.E., Jansen, N.J., Kamp, V.M., Kimpen, J.L., Rossen, J.W., Ulfman, L.H., Tacke, C.E., Viveen, M.C., *et al.* (2010). A systemic neutrophil response precedes robust CD8(+) T-cell activation during natural respiratory syncytial virus infection in infants. J. Virol. *84,* 2374–2383.

Luongo, C., Yang, L., Winter, C.C., Spann, K.M., Murphy, B.R., Collins, P.L., and Buchholz, U.J. (2009). Codon stabilization analysis of the '248' temperature sensitive mutation for increased phenotypic stability of respiratory syncytial virus vaccine candidates. Vaccine *27,* 5667–5676.

McAuliffe, J.M., Surman, S.R., Newman, J.T., Riggs, J.M., Collins, P.L., Murphy, B.R., and Skiadopoulos, M.H. (2004). Codon substitution mutations at two positions in the L polymerase protein of human parainfluenza virus type 1 yield viruses with a spectrum of attenuation *in vivo* and increased phenotypic stability *in vitro*. J. Virol. *78,* 2029–2036.

McCurdy, L.H., and Graham, B.S. (2003). Role of plasma membrane lipid microdomains in respiratory syncytial virus filament formation. J. Virol. *77,* 1747–1756.

McDonald, T.P., Pitt, A.R., Brown, G., Rixon, H.W., and Sugrue, R.J. (2004). Evidence that the respiratory syncytial virus polymerase complex associates with lipid rafts in virus-infected cells: a proteomic analysis. Virology *330,* 147–157.

McGivern, D.R., Collins, P.L., and Fearns, R. (2005). Identification of internal sequences in the 3′ leader region of human respiratory syncytial virus that enhance transcription and confer replication processivity. J. Virol. *79,* 2449–2460.

McIntosh, K. (1978). Interferon in nasal secretions from infants with viral respiratory tract infections. J. Pediatr. *93,* 33–36.

McIntosh, K., Masters, H.B., Orr, I., Chao, R.K., and Barkin, R.M. (1978). The immunologic response to infection with respiratory syncytial virus in infants. J. Infect. Dis. *138,* 24–32.

McLellan, J.S., Chen, M., Kim, A., Yang, Y., Graham, B.S., and Kwong, P.D. (2010). Structural basis of respiratory syncytial virus neutralization by motavizumab. Nat Struct Mol. Biol. *17,* 248–250.

McNamara, P.S., Ritson, P., Selby, A., Hart, C.A., and Smyth, R.L. (2003). Bronchoalveolar lavage cellularity in infants with severe respiratory syncytial virus bronchiolitis. Arch. Dis. Child. *88,* 922–926.

McNamara, P.S., Flanagan, B.F., Baldwin, L.M., Newland, P., Hart, C.A., and Smyth, R.L. (2004). Interleukin 9 production in the lungs of infants with severe respiratory syncytial virus bronchiolitis. Lancet *363,* 1031–1037.

Madhi, S.A., and Klugman, K.P. (2004). A role for Streptococcus pneumoniae in virus-associated pneumonia. Nat. Med. *10,* 811–813.

Maffey, A.F., Barrero, P.R., Venialgo, C., Fernandez, F., Fuse, V.A., Saia, M., Villalba, A., Fermepin, M.R., Teper, A.M., and Mistchenko, A.S. (2010). Viruses and atypical bacteria associated with asthma exacerbations in hospitalized children. Pediatr. Pulmonol. *45,* 619–625.

Malhotra, R., Ward, M., Bright, H., Priest, R., Foster, M.R., Hurle, M., Blair, E., and Bird, M. (2003). Isolation and characterisation of potential respiratory syncytial virus receptor(s) on epithelial cells. Microbes Infect. *5,* 123–133.

Malley, R., DeVincenzo, J., Ramilo, O., Dennehy, P.H., Meissner, H.C., Gruber, W.C., Sanchez, P.J., Jafri, H., Balsley, J., Carlin, D., *et al.* (1998). Reduction of respiratory syncytial virus (RSV) in tracheal aspirates in intubated infants by use of humanized monoclonal antibody to RSV F protein. J. Infect. Dis. *178,* 1555–1561.

Mallipeddi, S.K., and Samal, S.K. (1993). Analysis of the ovine respiratory syncytial virus (RSV) G glycoprotein gene defines a subgroup of ungulate RSV. J. Gen. Virol. *74,* 2787–2791.

Manji, R., Lotlikar, M., Zhang, F., and Ginocchio, C.C. (2009). Clinical evaluation of NucliSENS magnetic extraction and NucliSENS analytical specific reagents for the real-time detection of respiratory syncytial virus (RSV) in paediatric respiratory specimens. J. Clin. Pathol. *62,* 998–1002.

Marsh, R., Connor, A., Gias, E., and Toms, G.L. (2007). Increased susceptibility of human respiratory syncytial virus to neutralization by anti-fusion protein antibodies on adaptation to replication in cell culture. J. Med. Virol. *79,* 829–837.

Martinello, R.A., Chen, M.D., Weibel, C., and Kahn, J.S. (2002). Correlation between respiratory syncytial virus genotype and severity of illness. J. Infect. Dis. *186,* 839–842.

Martinez, I., and Melero, J.A. (1998). Enhanced neutralization of human respiratory syncytial virus by mixtures of monoclonal antibodies to the attachment (G) glycoprotein. J. Gen. Virol. *79,* 2215–2220.

Martinez, I., Dopazo, J., and Melero, J.A. (1997). Antigenic structure of the human respiratory syncytial virus G glycoprotein and relevance of hypermutation events for the generation of antigenic variants. J. Gen. Virol. *78,* 2419–2429.

Mason, S.W., Aberg, E., Lawetz, C., DeLong, R., Whitehead, P., and Liuzzi, M. (2003). Interaction between human respiratory syncytial virus (RSV) M2-1 and P proteins is required for reconstitution of M2-1-dependent RSV minigenome activity. J. Virol. *77,* 10670–10676.

Masters, J.R. (2002). HeLa cells 50 years on: the good, the bad and the ugly. Nat. Rev. Cancer *2,* 315–319.

Matsuoka, T., Okamoto, Y., Matsuzaki, Z., Endo, S., Ito, E., Tsutsumi, H., Williamson, R.A., Sakurai, H., Burton, D.R., and Saito, I. (2002). Characteristics of immunity induced by viral antigen or conferred by antibody via different administration routes. Clin. Exp. Immunol. *130,* 386–392.

Meissner, H.C., and Long, S.S. (2003). Revised indications for the use of palivizumab and respiratory syncytial virus immune globulin intravenous for the prevention of respiratory syncytial virus infections. Pediatrics *112,* 1447–1452.

Mejias, A., and Ramilo, O. (2008). Review of palivizumab in the prophylaxis of respiratory syncytial virus (RSV) in high-risk infants. Biologics *2,* 433–439.

Melendi, G.A., Hoffman, S.J., Karron, R.A., Irusta, P.M., Laham, F.R., Humbles, A., Schofield, B., Pan, C.H., Rabold, R., Thumar, B., *et al.* (2007). C5 modulates airway hyperreactivity and pulmonary eosinophilia during enhanced respiratory syncytial virus disease by decreasing C3a receptor expression. J. Virol. *81,* 991–999.

Melero, J.A., Garcia-Barreno, B., Martinez, I., Pringle, C.R., and Cane, P.A. (1997). Antigenic structure, evolution and immunobiology of human respiratory syncytial virus attachment (G) protein. J. Gen. Virol. *78,* 2411–2418.

Mellow, T.E., Murphy, P.C., Carson, J.L., Noah, T.L., Zhang, L., and Pickles, R.J. (2004). The effect of respiratory synctial virus on chemokine release by differentiated airway epithelium. Exp. Lung Res. *30,* 43–57.

Miller, A.L., Bowlin, T.L., and Lukacs, N.W. (2004). Respiratory syncytial virus-induced chemokine production: linking viral replication to chemokine production *in vitro* and *in vivo*. J. Infect. Dis. *189,* 1419–1430.

Mills, J., Van Kirk, J.E., Wright, P.F., and Chanock, R.M. (1971). Experimental respiratory syncytial virus infection of adults. Possible mechanisms of resistance to infection and illness. J. Immunol. *107,* 123–130.

Miyairi, I., and DeVincenzo, J.P. (2008). Human genetic factors and respiratory syncytial virus disease severity. Clin. Microbiol. Rev. *21,* 686–703.

Mobbs, K.J., Smyth, R.L., O'Hea, U., Ashby, D., Ritson, P., and Hart, C.A. (2002). Cytokines in severe respiratory syncytial virus bronchiolitis. Pediatr. Pulmonol. *33,* 449–452.

Moghaddam, A., Olszewska, W., Wang, B., Tregoning, J.S., Helson, R., Sattentau, Q.J., and Openshaw, P.J. (2006). A potential molecular mechanism for hypersensitivity caused by formalin-inactivated vaccines. Nat. Med. *12,* 905–907.

Mohtasham, L., Auais, A., and Piedimonte, G. (2007). Nerve growth factor mediates steroid-resistant inflammation in respiratory syncytial virus infection. Pediatr. Pulmonol. *42,* 496–504.

Money, V.A., McPhee, H.K., Mosely, J.A., Sanderson, J.M., and Yeo, R.P. (2009). Surface features of a *Mononegavirales* matrix protein indicate sites of membrane interaction. Proc. Natl. Acad. Sci. U.S.A. *106,* 4441–4446.

Moore, M.L., Chi, M.H., Luongo, C., Lukacs, N.W., Polosukhin, V.V., Huckabee, M.M., Newcomb, D.C., Buchholz, U.J., Crowe, J.E., Jr., Goleniewska, K., *et al.* (2009). A chimeric A2 strain of respiratory syncytial virus (RSV) with the fusion protein of RSV strain line 19 exhibits enhanced viral load, mucus, and airway dysfunction. J. Virol. *83,* 4185–4194.

Moreno de Alboran, I., Martinez-alonso, C., Barbas, C.F., 3rd, Burton, D.R., and Ditzel, H.J. (1995). Human monoclonal Fab fragments specific for viral antigens from combinatorial IgA libraries. Immunotechnology *1,* 21–28.

Morris, J.A., Blount, R.E., and Savage, R.E. (1956). Recovery of cytopathic agent from chimpanzees with coryza. Proc. Soc. Exp. Biol. Med. *92,* 544–550.

Moudy, R.M., Harmon, S.B., Sullender, W.M., and Wertz, G.W. (2003). Variations in transcription termination signals of human respiratory syncytial virus clinical isolates affect gene expression. Virology *313,* 250–260.

Moudy, R.M., Sullender, W.M., and Wertz, G.W. (2004). Variations in intergenic region sequences of Human respiratory syncytial virus clinical isolates: analysis of effects on transcriptional regulation. Virology *327,* 121–133.

Mufson, M.A., Orvell, C., Rafnar, B., and Norrby, E. (1985). Two distinct subtypes of human respiratory syncytial virus. J. Gen. Virol. *66,* 2111–2124.

Munir, S., Le Nouen, C., Luongo, C., Buchholz, U.J., Collins, P.L., and Bukreyev, A. (2008). Nonstructural proteins 1 and 2 of respiratory syncytial virus suppress maturation of human dendritic cells. J. Virol. *82,* 8780–8796.

Munoz, F.M., Piedra, P.A., and Glezen, W.P. (2003). Safety and immunogenicity of respiratory syncytial virus purified fusion protein-2 vaccine in pregnant women. Vaccine *21,* 3465–3467.

Murata, Y., and Falsey, A.R. (2007). Respiratory syncytial virus infection in adults. Antivir. Ther. *12,* 659–670.

Murata, Y., Lightfoote, P.M., Falsey, A.R., and Walsh, E.E. (2010). Identification of and human serum reactogenicity to neutralizing epitopes within the central unglycosylated region of the respiratory syncytial virus attachment protein. Clin. Vaccine Immunol. *17,* 695–697.

Murawski, M.R., Bowen, G.N., Cerny, A.M., Anderson, L.J., Haynes, L.M., Tripp, R.A., Kurt-Jones, E.A., and Finberg, R.W. (2009). Respiratory syncytial virus activates innate immunity through Toll-like receptor 2. J. Virol. *83,* 1492–1500.

Murphy, B.R., Alling, D.W., Snyder, M.H., Walsh, E.E., Prince, G.A., Chanock, R.M., Hemming, V.G., Rodriguez, W.J., Kim, H.W., Graham, B.S., *et al.* (1986a). Effect of age and preexisting antibody on serum antibody response of infants and children to the F and G glycoproteins during respiratory syncytial virus infection. J. Clin. Microbiol. *24,* 894–898.

Murphy, B.R., Graham, B.S., Prince, G.A., Walsh, E.E., Chanock, R.M., Karzon, D.T., and Wright, P.F. (1986b). Serum and nasal-wash immunoglobulin G and A antibody response of infants and children to respiratory syncytial virus F and G glycoproteins following primary infection. J. Clin. Microbiol. *23,* 1009–1014.

Murphy, B.R., Olmsted, R.A., Collins, P.L., Chanock, R.M., and Prince, G.A. (1988a). Passive transfer of respiratory syncytial virus (RSV) antiserum suppresses the immune response to the RSV fusion (F) and large (G) glycoproteins expressed by recombinant vaccinia viruses. J. Virol. *62,* 3907–3910.

Murphy, B.R., Prince, G.A., Collins, P.L., Van Wyke Coelingh, K., Olmsted, R.A., Spriggs, M.K., Parrott, R.H., Kim, H.W., Brandt, C.D., and Chanock, R.M. (1988b). Current approaches to the development of vaccines effective against parainfluenza and respiratory syncytial viruses. Virus Res. *11,* 1–15.

Murphy, B.R., Sotnikov, A.V., Lawrence, L.A., Banks, S.M., and Prince, G.A. (1990). Enhanced pulmonary histopathology is observed in cotton rats immunized with formalin-inactivated respiratory syncytial virus (RSV) or purified F glycoprotein and challenged with RSV 3–6 months after immunization. Vaccine *8,* 497–502.

Murphy, B.R., Prince, G.A., Collins, P.L., Hildreth, S.W., and Paradiso, P.R. (1991). Effect of passive antibody on the immune response of cotton rats to purified F and G glycoproteins of respiratory syncytial virus (RSV). Vaccine *9,* 185–189.

Nair, H., Nokes, D.J., Gessner, B.D., Dherani, M., Madhi, S.A., Singleton, R.J., O'Brien, K.L., Roca, A., Wright, P.F., Bruce, N., *et al.* (2010). Global burden of acute lower respiratory infections due to respiratory syncytial virus in young children: a systematic review and meta-analysis. Lancet *375,* 1545–1555.

Navarro, J., Lopez-Otin, C., and Villanueva, N. (1991). Location of phosphorylated residues in human respiratory syncytial virus phosphoprotein. J. Gen. Virol. *72,* 1455–1459.

Newman, J.T., Riggs, J.M., Surman, S.R., McAuliffe, J.M., Mulaikal, T.A., Collins, P.L., Murphy, B.R., and Skiadopoulos, M.H. (2004). Generation of recombinant human parainfluenza virus type 1 vaccine candidates by importation of temperature-sensitive and attenuating mutations from heterologous paramyxoviruses. J. Virol. *78,* 2017–2028.

Noble, V., Murray, M., Webb, M.S., Alexander, J., Swarbrick, A.S., and Milner, A.D. (1997). Respiratory status and allergy nine to 10 years after acute bronchiolitis. Arch. Dis. Child. *76,* 315–319.

Norrby, E., Marusyk, H., and Orvell, C. (1970). Morphogenesis of respiratory syncytial virus in a green monkey kidney cell line (Vero). J. Virol. *6,* 237–242.

Noton, S.L., Cowton, V.M., Zack, C.R., McGivern, D.R., and Fearns, R. (2010). Evidence that the polymerase of respiratory syncytial virus initiates RNA replication in a nontemplated fashion. Proc. Natl. Acad. Sci. U.S.A. *107,* 10226–10231.

Olmsted, R.A., and Collins, P.L. (1989). The 1A protein of respiratory syncytial virus is an integral membrane protein present as multiple, structurally distinct species. J. Virol. *63,* 2019–2029.

Olmsted, R.A., Elango, N., Prince, G.A., Murphy, B.R., Johnson, P.R., Moss, B., Chanock, R.M., and Collins, P.L. (1986). Expression of the F glycoprotein of respiratory syncytial virus by a recombinant vaccinia virus: comparison of the individual contributions of the F and G glycoproteins to host immunity. Proc. Natl. Acad. Sci. U.S.A. *83,* 7462–7466.

Olson, M.R., and Varga, S.M. (2007). CD8 T cells inhibit respiratory syncytial virus (RSV) vaccine-enhanced disease. J. Immunol. *179,* 5415–5424.

Olson, M.R., Hartwig, S.M., and Varga, S.M. (2008). The number of respiratory syncytial virus (RSV)-specific memory CD8 T cells in the lung is critical for their ability to inhibit RSV vaccine-enhanced pulmonary eosinophilia. J. Immunol. *181,* 7958–7968.

Olszewska, W., and Openshaw, P. (2009). Emerging drugs for respiratory syncytial virus infection. Expert Opin. Emerg. Drugs *14,* 207–217.

Openshaw, P.J., Clarke, S.L., and Record, F.M. (1992). Pulmonary eosinophilic response to respiratory syncytial virus infection in mice sensitized to the major surface glycoprotein G. Int. Immunol. *4,* 493–500.

Ottolini, M.G., Porter, D.D., Hemming, V.G., Zimmerman, M.N., Schwab, N.M., and Prince, G.A. (1999). Effectiveness of RSVIG prophylaxis and therapy of respiratory syncytial virus in an immunosuppressed animal model. Bone Marrow Transplant *24,* 41–45.

Ottolini, M.G., Porter, D.D., Hemming, V.G., and Prince, G.A. (2000). Enhanced pulmonary pathology in cotton rats upon challenge after immunization with inactivated parainfluenza virus 3 vaccines. Viral Immunol. *13,* 231–236.

Paccaud, M.F., and Jacquier, C. (1970). A respiratory syncytial virus of bovine origin. Arch. Gesamte Virusforsch *30,* 327–342.

Pastey, M.K., Crowe, J.E., Jr., and Graham, B.S. (1999). RhoA interacts with the fusion glycoprotein of respiratory syncytial virus and facilitates virus-induced syncytium formation. J. Virol. *73,* 7262–7270.

Patel, H., Platt, R., Lozano, J.M., and Wang, E.E. (2004). Glucocorticoids for acute viral bronchiolitis in infants and young children. Cochrane Database Syst. Rev. CD004878.

Peret, T.C., Hall, C.B., Schnabel, K.C., Golub, J.A., and Anderson, L.J. (1998). Circulation patterns of genetically distinct group A and B strains of human respiratory syncytial virus in a community. J. Gen. Virol. *79,* 2221–2229.

Peret, T.C., Hall, C.B., Hammond, G.W., Piedra, P.A., Storch, G.A., Sullender, W.M., Tsou, C., and Anderson, L.J. (2000). Circulation patterns of group A and B human respiratory syncytial virus genotypes in 5 communities in North America. J. Infect. Dis. *181,* 1891–1896.

Perez, M., Garcia-Barreno, B., Melero, J.A., Carrasco, L., and Guinea, R. (1997). Membrane permeability changes induced in Escherichia coli by the SH protein of human respiratory syncytial virus. Virology *235,* 342–351.

Phipps, S., Lam, C.E., Mahalingam, S., Newhouse, M., Ramirez, R., Rosenberg, H.F., Foster, P.S., and Matthaei, K.I. (2007). Eosinophils contribute to innate antiviral immunity and promote clearance of respiratory syncytial virus. Blood *110,* 1578–1586.

Piedimonte, G., Hegele, R.G., and Auais, A. (2004). Persistent airway inflammation after resolution of respiratory syncytial virus infection in rats. Pediatr. Res. *55,* 657–665.

Piedra, P.A., Grace, S., Jewell, A., Spinelli, S., Bunting, D., Hogerman, D.A., Malinoski, F., and Hiatt, P.W. (1996). Purified fusion protein vaccine protects against lower respiratory tract illness during respiratory syncytial virus season in children with cystic fibrosis. Pediatr. Infect. Dis. J. *15,* 23–31.

Piedra, P.A., Cron, S.G., Jewell, A., Hamblett, N., McBride, R., Palacio, M.A., Ginsberg, R., Oermann, C.M., and Hiatt, P.W. (2003). Immunogenicity of a new purified fusion protein vaccine to respiratory syncytial virus: a multicenter trial in children with cystic fibrosis. Vaccine *21,* 2448–2460.

Polack, F.P. (2007). Atypical measles and enhanced respiratory syncytial virus disease (ERD) made simple. Pediatr. Res. *62,* 111–115.

Polack, F.P., Irusta, P.M., Hoffman, S.J., Schiatti, M.P., Melendi, G.A., Delgado, M.F., Laham, F.R., Thumar, B., Hendry, R.M., Melero, J.A., *et al.* (2005). The cysteine-rich region of respiratory syncytial virus attachment protein inhibits innate immunity elicited by the virus and endotoxin. Proc. Natl. Acad. Sci. U.S.A. *102,* 8996–9001.

Power, U.F., Nguyen, T.N., Rietveld, E., de Swart, R.L., Groen, J., Osterhaus, A.D., de Groot, R., Corvaia, N., Beck, A., Bouveret-Le-Cam, N., *et al.* (2001). Safety and immunogenicity of a novel recombinant subunit respiratory syncytial virus vaccine (BBG2Na) in healthy young adults. J. Infect. Dis. *184,* 1456–1460.

Prais, D., Schonfeld, T., and Amir, J. (2003). Admission to the intensive care unit for respiratory syncytial virus bronchiolitis: a national survey before palivizumab use. Pediatrics *112,* 548–552.

Prescott, W.A., Jr., Doloresco, F., Brown, J., and Paladino, J.A. (2010). Cost effectiveness of respiratory syncytial virus prophylaxis: a critical and systematic review. Pharmacoeconomics *28,* 279–293.

Prince, G.A., Potash, L., Horswood, R.L., Camargo, E., Suffin, S.C., Johnson, R.A., and Chanock, R.M. (1979). Intramuscular inoculation of live respiratory syncytial virus induces immunity in cotton rats. Infect. Immun. *23,* 723–728.

Prince, G.A., Horswood, R.L., and Chanock, R.M. (1985). Quantitative aspects of passive immunity to respiratory syncytial virus infection in infant cotton rats. J. Virol. *55,* 517–520.

Prince, G.A., Mathews, A., Curtis, S.J., and Porter, D.D. (2000). Treatment of respiratory syncytial virus bronchiolitis and pneumonia in a cotton rat model wit systemically administered monoclonal antibody (palivizumab) and glucocorticosteroid. J. Infect. Dis. *182,* 1326–1330.

Prince, G.A., Curtis, S.J., Yim, K.C., and Porter, D.D. (2001). Vaccine-enhanced respiratory syncytial virus disease in cotton rats following immunization with Lot 100 or a newly prepared reference vaccine. J. Gen. Virol. *82,* 2881–2888.

Puthothu, B., Forster, J., Heinze, J., Heinzmann, A., and Krueger, M. (2007). Surfactant protein B polymorphisms are associated with severe respiratory syncytial virus infection, but not with asthma. BMC Pulm. Med. *7,* 6.

Ramaswamy, M., Shi, L., Monick, M.M., Hunninghake, G.W., and Look, D.C. (2004). Specific inhibition of type I interferon signal transduction by respiratory syncytial virus. Am. J. Respir. Cell Mol. Biol. *30,* 893–900.

Ramaswamy, M., Shi, L., Varga, S.M., Barik, S., Behlke, M.A., and Look, D.C. (2006). Respiratory syncytial virus nonstructural protein 2 specifically inhibits type I interferon signal transduction. Virology *344,* 328–339.

Randolph, A.G., and Wang, E.E. (1996). Ribavirin for respiratory syncytial virus lower respiratory tract infection. A systematic overview. Arch. Pediatr. Adolesc. Med. *150,* 942–947.

Rimensberger, P.C., Burek-Kozlowska, A., Morell, A., Germann, D., Eigenmann, A.K., Steiner, F., Burger, R., Kuenzli, M., and Schaad, U.B. (1996). Aerosolized immunoglobulin treatment of respiratory syncytial virus infection in infants. Pediatr. Infect. Dis. J. *15,* 209–216.

Roberts, S.R., Lichtenstein, D., Ball, L.A., and Wertz, G.W. (1994). The membrane-associated and secreted forms of the respiratory syncytial virus attachment glycoprotein G are synthesized from alternative initiation codons. J. Virol. *68,* 4538–4546.

Roberts, S.R., Compans, R.W., and Wertz, G.W. (1995). Respiratory syncytial virus matures at the apical surfaces of polarized epithelial cells. J. Virol. *69,* 2667–2673.

Rodriguez, W.J., Gruber, W.C., Groothuis, J.R., Simoes, E.A., Rosas, A.J., Lepow, M., Kramer, A., and Hemming, V. (1997a). Respiratory syncytial virus immune globulin treatment of RSV lower respiratory tract infection in previously healthy children. Pediatrics *100,* 937–942.

Rodriguez, W.J., Gruber, W.C., Welliver, R.C., Groothuis, J.R., Simoes, E.A., Meissner, H.C., Hemming, V.G., Hall, C.B., Lepow, M.L., Rosas, A.J., *et al.* (1997b). Respiratory syncytial virus (RSV) immune globulin intravenous therapy for RSV lower respiratory tract infection in infants and young children at high risk for severe RSV infections: Respiratory Syncytial Virus Immune Globulin Study Group. Pediatrics *99,* 454–461.

Rodriguez, W.J., Arrobio, J., Fink, R., Kim, H.W., and Milburn, C. (1999). Prospective follow-up and pulmonary functions from a placebo-controlled randomized trial of ribavirin therapy in respiratory syncytial virus bronchiolitis. Ribavirin Study Group. Arch. Pediatr. Adolesc Med. *153,* 469–474.

Rohwedder, A., Keminer, O., Forster, J., Schneider, K., Schneider, E., and Werchau, H. (1998). Detection of respiratory syncytial virus RNA in blood of neonates by polymerase chain reaction. J. Med. Virol. *54,* 320–327.

Roman, M., Calhoun, W.J., Hinton, K.L., Avendano, L.F., Simon, V., Escobar, A.M., Gaggero, A., and Diaz, P.V. (1997). Respiratory syncytial virus infection in infants is associated with predominant Th-2-like response. Am. J. Respir. Crit. Care Med. *156,* 190–195.

Rosenberg, H.F., and Domachowske, J.B. (2001). Eosinophils, eosinophil ribonucleases, and their role in host defense against respiratory virus pathogens. J. Leukoc. Biol. *70,* 691–698.

Rothoeft, T., Fischer, K., Zawatzki, S., Schulz, V., Schauer, U., and Korner Rettberg, C. (2007). Differential response of human naive and memory/effector T cells to dendritic cells infected by respiratory syncytial virus. Clin. Exp. Immunol. *150,* 263–273.

Rudd, B.D., Burstein, E., Duckett, C.S., Li, X., and Lukacs, N.W. (2005). Differential role for TLR3 in respiratory syncytial virus-induced chemokine expression. J. Virol. *79,* 3350–3357.

Rudd, B.D., Smit, J.J., Flavell, R.A., Alexopoulou, L., Schaller, M.A., Gruber, A., Berlin, A.A., and Lukacs, N.W. (2006). Deletion of TLR3 alters the pulmonary immune environment and mucus production during respiratory syncytial virus infection. J. Immunol. *176,* 1937–1942.

Saitou, N., and Nei, M. (1987). The neighbor-joining method: a new method for reconstructing phylogenetic trees. Mol. Biol. Evol. *4,* 406–425.

Sakurai, H., Williamson, R.A., Crowe, J.E., Beeler, J.A., Poignard, P., Bastidas, R.B., Chanock, R.M., and Burton, D.R. (1999). Human antibody responses to mature and immature forms of viral envelope in respiratory syncytial virus infection: significance for subunit vaccines. J. Virol. *73,* 2956–2962.

Samal, S.K., and Collins, P.L. (1996). RNA replication by a respiratory syncytial virus RNA analog does not obey the rule of six and retains a nonviral trinucleotide extension at the leader end. J. Virol. *70,* 5075–5082.

Santangelo, P.J., and Bao, G. (2007). Dynamics of filamentous viral RNPs prior to egress. Nucleic Acids Res. *35,* 3602–3611.

Sastre, P., Oomens, A.G., and Wertz, G.W. (2007). The stability of human respiratory syncytial virus is enhanced by incorporation of the baculovirus GP64 protein. Vaccine *25,* 5025–5033.

Schlender, J., Bossert, B., Buchholz, U., and Conzelmann, K.K. (2000). Bovine respiratory syncytial virus nonstructural proteins NS1 and NS2 cooperatively antagonize alpha/beta interferon-induced antiviral response. J. Virol. *74,* 8234–8242.

Schlender, J., Hornung, V., Finke, S., Gunthner-Biller, M., Marozin, S., Brzozka, K., Moghim, S., Endres, S., Hartmann, G., and Conzelmann, K.K. (2005). Inhibition of toll-like receptor 7- and 9-mediated alpha/beta interferon production in human plasmacytoid dendritic cells by respiratory syncytial virus and measles virus. J. Virol. *79,* 5507–5515.

Schmidt, A.C., Wenzke, D.R., McAuliffe, J.M., St. Clair, M., Elkins, W.R., Murphy, B.R., and Collins, P.L. (2002). Mucosal immunization of rhesus monkeys against respiratory syncytial virus subgroups A and B and human parainfluenza virus type 3 using a live cDNA-derived vaccine based on a host range-attenuated bovine parainfluenza virus type 3 vector backbone. J. Virol. *76.*

Schuurhof, A., Bont, L., Siezen, C.L., Hodemaekers, H., van Houwelingen, H.C., Kimman, T.G., Hoebee, B., Kimpen, J.L., and Janssen, R. (2010). Interleukin-9 polymorphism in infants with respiratory syncytial virus infection: an opposite effect in boys and girls. Pediatr. Pulmonol. *45,* 608–613.

Schwarze, J., O'Donnell, D.R., Rohwedder, A., and Openshaw, P.J. (2004). Latency and persistence of respiratory syncytial virus despite T cell immunity. Am. J. Respir. Crit. Care Med. *169,* 801–805.

Shapiro, J.M., and Jean, R.E. (2001). Respiratory syncytial virus. N. Engl. J. Med. *345,* 1132–1133.

Shay, D.K., Holman, R.C., Newman, R.D., Liu, L.L., Stout, J.W., and Anderson, L.J. (1999). Bronchiolitis-associated hospitalizations among US children, 1980–1996. J. Am. Med. Assoc. *282,* 1440–1446.

Shinoff, J.J., O'Brien, K.L., Thumar, B., Shaw, J.B., Reid, R., Hua, W., Santosham, M., and Karron, R.A. (2008). Young infants can develop protective levels of neutralizing antibody after infection with respiratory syncytial virus. J. Infect. Dis. *198,* 1007–1015.

Sidwell, R.W., and Barnard, D.L. (2006). Respiratory syncytial virus infections: recent prospects for control. Antiviral Res. *71,* 379–390.

Siegrist, C.A. (2003). Mechanisms by which maternal antibodies influence infant vaccine responses: review of hypotheses and definition of main determinants. Vaccine *21,* 3406–3412.

Siezen, C.L., Bont, L., Hodemaekers, H.M., Ermers, M.J., Doornbos, G., Van't Slot, R., Wijmenga, C., Houwelingen, H.C., Kimpen, J.L., Kimman, T.G., *et al.* (2009). Genetic susceptibility to respiratory syncytial virus bronchiolitis in preterm children is associated with airway remodeling genes and innate immune genes. Pediatr. Infect. Dis. J. *28,* 333–335.

Sigurs, N., Gustafsson, P.M., Bjarnason, R., Lundberg, F., Schmidt, S., Sigurbergsson, F., and Kjellman, B. (2005). Severe respiratory syncytial virus bronchiolitis in infancy and asthma and allergy at age 13. Am. J. Respir. Crit. Care Med. *171,* 137–141.

Sigurs, N., Aljassim, F., Kjellman, B., Robinson, P.D., Sigurbergsson, F., Bjarnason, R., and Gustafsson, P.M. (2010). Asthma and allergy patterns over 18 years after severe RSV bronchiolitis in the first year of life. Thorax *65,* 1045–1052.

Sikkel, M.B., Quint, J.K., Mallia, P., Wedzicha, J.A., and Johnston, S.L. (2008). Respiratory syncytial virus persistence in chronic obstructive pulmonary disease. Pediatr. Infect. Dis. J. *27,* S63–70.

Simoes, E.A., Carbonell-Estrany, X., Rieger, C.H., Mitchell, I., Fredrick, L., and Groothuis, J.R. (2010). The effect of respiratory syncytial virus on subsequent recurrent wheezing in atopic and nonatopic children. J. Allergy Clin. Immunol. *126,* 256–262.

Skiadopoulos, M.H., Schmidt, A.C., Riggs, J.M., Surman, S.R., Elkins, W.R., St Claire, M., Collins, P.L., and Murphy, B.R. (2003). Determinants of the host range restriction of replication of bovine parainfluenza virus type 3 in rhesus monkeys are polygenic. J. Virol. *77,* 1141–1148.

Smith, B.J., Lawrence, M.C., and Colman, P.M. (2002). Modelling the structure of the fusion protein from human respiratory syncytial virus. Protein Eng. *15,* 365–371.

Smyth, R.L., Mobbs, K.J., O'Hea, U., Ashby, D., and Hart, C.A. (2002). Respiratory syncytial virus bronchiolitis: disease severity, interleukin-8, and virus genotype. Pediatr. Pulmonol. *33,* 339–346.

Snow, A., Dayyat, E., Montgomery-Downs, H.E., Kheirandish-Gozal, L., and Gozal, D. (2009). Pediatric obstructive sleep apnea: a potential late consequence of respiratory syncitial virus bronchiolitis. Pediatr. Pulmonol. *44,* 1186–1191.

Sonoguchi, T., Sakoh, M., Kunita, N., Satsuta, K., Noriki, H., and Fukumi, H. (1986). Reinfection with influenza A (H2N2, H3N2, and H1N1) viruses in soldiers and students in Japan. J. Infect. Dis. *153,* 33–40.

Spann, K.M., Collins, P.L., and Teng, M.N. (2003). Genetic recombination during coinfection of two mutants of human respiratory syncytial virus. J. Virol. *77,* 11201–11211.

Spann, K.M., Tran, K.C., Chi, B., Rabin, R.L., and Collins, P.L. (2004). Suppression of the induction of alpha, beta, and lambda interferons by the NS1 and NS2 proteins of human respiratory syncytial virus in human epithelial cells and macrophages [corrected]. J. Virol. *78,* 4363–4369.

Spann, K.M., Tran, K.C., and Collins, P.L. (2005). Effects of nonstructural proteins NS1 and NS2 of human respiratory syncytial virus on interferon regulatory factor 3, NF-kappaB, and proinflammatory cytokines. J. Virol. *79,* 5353–5362.

Srinivasakumar, N., Ogra, P.L., and Flanagan, T.D. (1991). Characteristics of fusion of respiratory syncytial virus with HEP-2 cells as measured by R18 fluorescence dequenching assay. J. Virol. *65,* 4063–4069.

Stark, J.M., Godding, V., Sedgwick, J.B., and Busse, W.W. (1996). Respiratory syncytial virus infection enhances neutrophil and eosinophil adhesion to cultured respiratory epithelial cells. Roles of CD18 and intercellular adhesion molecule-1. J. Immunol. *156,* 4774–4782.

Stec, D.S., Hill, M.G., and Collins, P.L. (1991). Sequence analysis of the polymerase L gene of human respiratory syncytial virus and predicted phylogeny of nonsegmented negative-strand viruses. Virology *183,* 273–287.

Stein, R.T., Sherrill, D., Morgan, W.J., Holberg, C.J., Halonen, M., Taussig, L.M., Wright, A.L., and Martinez, F.D. (1999). Respiratory syncytial virus in early life and risk of wheeze and allergy by age 13 years. Lancet *354,* 541–545.

Stevens, W.W., Sun, J., Castillo, J.P., and Braciale, T.J. (2009). Pulmonary eosinophilia is attenuated by early responding CD8(+) memory T cells in a murine model of RSV vaccine-enhanced disease. Viral Immunol. *22,* 243–251.

Sudo, K., Miyazaki, Y., Kojima, N., Kobayashi, M., Suzuki, H., Shintani, M., and Shimizu, Y. (2005). YM-53403, a unique anti-respiratory syncytial virus agent with a novel mechanism of action. Antiviral Res. *65,* 125–131.

Sullender, W.M., Mufson, M.A., Anderson, L.J., and Wertz, G.W. (1991). Genetic diversity of the attachment protein of subgroup B respiratory syncytial viruses. J. Virol. *65,* 5425–5434.

Sung, R.Y., Yin, J., Oppenheimer, S.J., Tam, J.S., and Lau, J. (1993). Treatment of respiratory syncytial virus infection with recombinant interferon alfa-2a. Arch. Dis. Child *69,* 440–442.

Sung, R.Y., Hui, S.H., Wong, C.K., Lam, C.W., and Yin, J. (2001). A comparison of cytokine responses in respiratory syncytial virus and influenza A infections in infants. Eur. J. Pediatr. *160,* 117–122.

Surman, S.R., Collins, P.L., Murphy, B.R., and Skiadopoulos, M.H. (2007). An improved method for the recovery of recombinant paramyxovirus vaccine candidates suitable for use in human clinical trials. J. Virol. Methods *141,* 30–33.

de Swart, R.L., van den Hoogen, B.G., Kuiken, T., Herfst, S., van Amerongen, G., Yuksel, S., Sprong, L., and Osterhaus, A.D. (2007). Immunization of macaques with formalin-inactivated human *Metapneumovirus* induces hypersensitivity to hMPV infection. Vaccine *25,* 8518–8528.

Swedan, S., Musiyenko, A., and Barik, S. (2009). Respiratory syncytial virus nonstructural proteins decrease levels of multiple members of the cellular interferon pathways. J. Virol. *83,* 9682–9693.

Tamura, K., Dudley, J., Nei, M., and Kumar, S. (2007). MEGA4: Molecular Evolutionary Genetics Analysis (MEGA) software version 4.0. Mol. Biol. Evol *24,* 1596–1599.

Tang, R.S., Spaete, R.R., Thompson, M.W., MacPhail, M., Guzzetta, J.M., Ryan, P.C., Reisinger, K., Chandler, P., Hilty, M., Walker, R.E., *et al.* (2008). Development of a PIV-vectored RSV vaccine: preclinical evaluation of safety, toxicity, and enhanced disease and initial clinical testing in healthy adults. Vaccine *26,* 6373–6382.

Tawar, R.G., Duquerroy, S., Vonrhein, C., Varela, P.F., Damier-Piolle, L., Castagne, N., MacLellan, K., Bedouelle, H., Bricogne, G., Bhella, D., *et al.* (2009). Crystal structure of a nucleocapsid-like nucleoprotein–RNA complex of respiratory syncytial virus. Science *326,* 1279–1283.

Techaarpornkul, S., Barretto, N., and Peeples, M.E. (2001). Functional analysis of recombinant respiratory syncytial virus deletion mutants lacking the small hydrophobic and/or attachment glycoprotein gene. J. Virol. *75,* 6825–6834.

Techaarpornkul, S., Collins, P.L., and Peeples, M.E. (2002). Respiratory syncytial virus with the fusion protein as its only viral glycoprotein is less dependent on cellular glycosaminoglycans for attachment than complete virus. Virology *294,* 296–304.

Teng, M.N., and Collins, P.L. (1998). Identification of the respiratory syncytial virus proteins required for formation and passage of helper-dependent infectious particles. J. Virol. *72,* 5707–5716.

Teng, M.N., and Collins, P.L. (1999). Altered growth characteristics of recombinant respiratory syncytial viruses which do not produce NS2 protein. J. Virol. *73,* 466–473.

Teng, M.N., and Collins, P.L. (2002). The central conserved cystine noose of the attachment G protein of human respiratory syncytial virus is not required for efficient viral infection *in vitro* or *in vivo*. J. Virol. *76,* 6164–6171.

Teng, M.N., Whitehead, S.S., Bermingham, A., Clair, M.S., Elkins, W.R., Murphy, B.R., and Collins, P.L. (2000). Recombinant respiratory syncytial virus that does not express the NS1 or M2-2 protein is highly attenuated and immunogenic in chimpanzees. J. Virol. *74,* 9317–9321.

Teng, M.N., Whitehead, S.S., and Collins, P.L. (2001). Contribution of the respiratory syncytial virus g glycoprotein and its secreted and membrane-bound forms to virus replication *in vitro* and *in vivo*. Virology *289,* 283–296.

Thompson, W.W., Shay, D.K., Weintraub, E., Brammer, L., Cox, N., Anderson, L.J., and Fukuda, K. (2003). Mortality associated with influenza and respiratory syncytial virus in the United States. JAMA *289,* 179–186.

Thomsen, S.F., Stensballe, L.G., Skytthe, A., Kyvik, K.O., Backer, V., and Bisgaard, H. (2008). Increased concordance of severe respiratory syncytial virus infection in identical twins. Pediatrics *121,* 493–496.

Thomsen, S.F., van der Sluis, S., Stensballe, L.G., Posthuma, D., Skytthe, A., Kyvik, K.O., Duffy, D.L., Backer, V., and Bisgaard, H. (2009). Exploring the association between severe respiratory syncytial virus infection and asthma: a registry-based twin study. Am. J. Respir. Crit. Care Med. *179,* 1091–1097.

Tortorolo, L., Langer, A., Polidori, G., Vento, G., Stampachiacchere, B., Aloe, L., and Piedimonte, G. (2005). Neurotrophin overexpression in lower airways of infants with respiratory syncytial virus infection. Am. J. Respir. Crit. Care Med. *172,* 233–237.

Tran, T.L., Castagne, N., Dubosclard, V., Noinville, S., Koch, E., Moudjou, M., Henry, C., Bernard, J., Yeo, R.P., and Eleouet, J.F. (2009). The respiratory syncytial virus M2-1 protein forms tetramers and interacts with RNA and P in a competitive manner. J. Virol. *83,* 6363–6374.

Trento, A., Casas, I., Calderon, A., Garcia-Garcia, M.L., Calvo, C., Perez-Brena, P., and Melero, J.A. (2010). Ten years of global evolution of the human respiratory syncytial virus BA genotype with a 60-nucleotide duplication in the G protein gene. J. Virol. *84,* 7500–7512.

Tripp, R.A., Jones, L.P., Haynes, L.M., Zheng, H., Murphy, P.M., and Anderson, L.J. (2001). CX3C chemokine mimicry by respiratory syncytial virus G glycoprotein. Nat. Immunol. *2,* 732–738.

Tristram, D.A., Welliver, R.C., Hogerman, D.A., Hildreth, S.W., and Paradiso, P. (1994). Second-year surveillance of recipients of a respiratory syncytial virus (RSV) F protein subunit vaccine, PFP-1: evaluation of antibody persistence and possible disease enhancement. Vaccine *12,* 551–556.

Tristram, D.A., Hicks, W., Jr., and Hard, R. (1998). Respiratory syncytial virus and human bronchial epithelium. Arch. Otolaryngol Head Neck Surg *124,* 777–783.

Utley, T.J., Ducharme, N.A., Varthakavi, V., Shepherd, B.E., Santangelo, P.J., Lindquist, M.E., Goldenring, J.R., and Crowe, J.E., Jr. (2008). Respiratory syncytial virus uses a Vps4-independent budding mechanism controlled by Rab11-FIP2. Proc. Natl. Acad. Sci. U.S.A. *105, 10,*209–10214.

Valarcher, J.F., Schelcher, F., and Bourhy, H. (2000). Evolution of bovine respiratory syncytial virus. J. Virol. *74,* 10714–10728.

Valarcher, J.F., Bourhy, H., Lavenu, A., Bourges-Abella, N., Roth, M., Andreoletti, O., Ave, P., and Schelcher, F. (2001). Persistent infection of B lymphocytes by bovine respiratory syncytial virus. Virology *291,* 55–67.

Valarcher, J.F., Furze, J., Wyld, S., Cook, R., Conzelmann, K.K., and Taylor, G. (2003). Role of alpha/beta interferons in the attenuation and immunogenicity of recombinant bovine respiratory syncytial viruses lacking NS proteins. J. Virol. *77,* 8426–8439.

Valarcher, J.F., Furze, J., Wyld, S.G., Cook, R., Zimmer, G., Herrler, G., and Taylor, G. (2006). Bovine respiratory syncytial virus lacking the virokinin or with a mutation in furin cleavage site RA(R/K)R109 induces less pulmonary inflammation without impeding the induction of protective immunity in calves. J. Gen. Virol. *87,* 1659–1667.

Valdovinos, M.R., and Gomez, B. (2003). Establishment of respiratory syncytial virus persistence in cell lines: association with defective interfering particles. Intervirology *46,* 190–198.

Varga, S.M., Wissinger, E.L., and Braciale, T.J. (2000). The attachment (G) glycoprotein of respiratory syncytial virus contains a single immunodominant epitope that elicits both Th1 and Th2 $CD4^+$ T-cell responses. J. Immunol. *165,* 6487–6495.

Ventre, K., and Randolph, A. (2004). Ribavirin for respiratory syncytial virus infection of the lower respiratory tract in infants and young children. Cochrane Database Syst. Rev. CD000181.

Villanueva, N., Hardy, R., Asenjo, A., Yu, Q., and Wertz, G. (2000). The bulk of the phosphorylation of human respiratory syncytial virus phosphoprotein is not essential but modulates viral RNA transcription and replication. J. Gen. Virol. *81,* 129–133.

de Waal, L., Power, U.F., Yuksel, S., van Amerongen, G., Nguyen, T.N., Niesters, H.G., de Swart, R.L., and Osterhaus, A.D. (2004). Evaluation of BBG2Na in infant macaques: specific immune responses after vaccination and RSV challenge. Vaccine *22,* 915–922.

Wagner, D.K., Clements, M.L., Reimer, C.B., Snyder, M., Nelson, D.L., and Murphy, B.R. (1987a). Analysis of immunoglobulin G antibody responses after administration of live and inactivated influenza A vaccine indicates that nasal wash immunoglobulin G is a transudate from serum. J. Clin. Microbiol. *25,* 559–562.

Wagner, D.K., Nelson, D.L., Walsh, E.E., Reimer, C.B., Henderson, F.W., and Murphy, B.R. (1987b). Differential immunoglobulin G subclass antibody titers to respiratory syncytial virus F and G glycoproteins in adults. J. Clin. Microbiol. *25,* 748–750.

Wagner, D.K., Muelenaer, P., Henderson, F.W., Snyder, M.H., Reimer, C.B., Walsh, E.E., Anderson, L.J., Nelson, D.L., and Murphy, B.R. (1989). Serum immunoglobulin G antibody subclass response to respiratory syncytial virus F and G glycoproteins after first, second, and third infections. J. Clin. Microbiol. *27,* 589–592.

Walsh, E.E., and Falsey, A.R. (2004a). Age related differences in humoral immune response to respiratory syncytial virus infection in adults. J. Med. Virol. *73,* 295–299.

Walsh, E.E., and Falsey, A.R. (2004b). Humoral and mucosal immunity in protection from natural respiratory syncytial virus infection in adults. J. Infect. Dis. *190,* 373–378.

Wang, S.Z., Xu, H., Wraith, A., Bowden, J.J., Alpers, J.H., and Forsyth, K.D. (1998). Neutrophils induce damage to respiratory epithelial cells infected with respiratory syncytial virus. Eur. Respir. J. *12,* 612–618.

Ward, K.A., Lambden, P.R., Ogilvie, M.M., and Watt, P.J. (1983). Antibodies to respiratory syncytial virus polypeptides and their significance in human infection. J. Gen. Virol. *64,* 1867–1876.

Waris, M. (1991). Pattern of respiratory syncytial virus epidemics in Finland: two-year cycles with alternating prevalence of groups A and B. J. Infect. Dis. *163,* 464–469.

Waris, M.E., Tsou, C., Erdman, D.D., Day, D.B., and Anderson, L.J. (1997). Priming with live respiratory syncytial virus (RSV) prevents the enhanced pulmonary inflammatory response seen after RSV challenge in BALB/c mice immunized with formalin-inactivated RSV. J. Virol. *71,* 6935–6939.

Wathen, M.W., Aeed, P.A., and Elhammer, A.P. (1991). Characterization of oligosaccharide structures on a chimeric respiratory syncytial virus protein expressed in insect cell line Sf9. Biochemistry *30,* 2863–2868.

Welliver, R.C., Kaul, T.N., Putnam, T.I., Sun, M., Riddlesberger, K., and Ogra, P.L. (1980). The antibody response to primary and secondary infection with respiratory syncytial virus: kinetics of class-specific responses. J. Pediatr. *96,* 808–813.

Welliver, R.C., Wong, D.T., Sun, M., Middleton, E., Jr., Vaughan, R.S., and Ogra, P.L. (1981). The development of respiratory syncytial virus-specific IgE and the release of histamine in nasopharyngeal secretions after infection. N. Engl. J. Med. *305,* 841–846.

Welliver, T.P., Garofalo, R.P., Hosakote, Y., Hintz, K.H., Avendano, L., Sanchez, K., Velozo, L., Jafri, H., Chavez-Bueno, S., Ogra, P.L., *et al.* (2007). Severe human lower respiratory tract illness caused by respiratory syncytial virus and influenza virus is characterized by the absence of pulmonary cytotoxic lymphocyte responses. J. Infect. Dis. *195,* 1126–1136.

Welliver, R.C., Checchia, P.A., Bauman, J.H., Fernandes, A.W., Mahadevia, P.J., and Hall, C.B. (2010). Fatality rates in published reports of RSV hospitalizations among high-risk and otherwise healthy children. Curr. Med. Res. Opin. *26,* 2175–2181.

Wertz, G.W., Krieger, M., and Ball, L.A. (1989). Structure and cell surface maturation of the attachment glycoprotein of human respiratory syncytial virus in a cell line deficient in O glycosylation. J. Virol. *63,* 4767–4776.

Whimbey, E., Champlin, R.E., Englund, J.A., Mirza, N.Q., Piedra, P.A., Goodrich, J.M., Przepiorka, D., Luna, M.A., Morice, R.C., Neumann, J.L., *et al.* (1995). Combination therapy with aerosolized ribavirin and intravenous immunoglobulin for respiratory syncytial virus disease in adult bone marrow transplant recipients. Bone Marrow Transplant *16,* 393–399.

Whimbey, E., Champlin, R.E., Couch, R.B., Englund, J.A., Goodrich, J.M., Raad, I., Przepiorka, D., Lewis, V.A., Mirza, N., Yousuf, H., *et al.* (1996). Community respiratory virus infections among hospitalized adult bone marrow transplant recipients. Clin. Infect. Dis. *22,* 778–782.

Whitehead, S.S., Firestone, C.Y., Collins, P.L., and Murphy, B.R. (1998a). A single nucleotide substitution in the transcription start signal of the M2 gene of respiratory syncytial virus vaccine candidate cpts248/404 is the major determinant of the temperature-sensitive and attenuation phenotypes. Virology *247,* 232–239.

Whitehead, S.S., Juhasz, K., Firestone, C.Y., Collins, P.L., and Murphy, B.R. (1998b). Recombinant respiratory syncytial virus (RSV) bearing a set of mutations from cold-passaged RSV is attenuated in chimpanzees. J. Virol. *72,* 4467–4471.

Whitehead, S.S., Bukreyev, A., Teng, M.N., Firestone, C.Y., St Claire, M., Elkins, W.R., Collins, P.L., and Murphy, B.R. (1999a). Recombinant respiratory syncytial virus bearing a deletion of either the NS2 or SH gene is attenuated in chimpanzees. J. Virol. *73,* 3438–3442.

Whitehead, S.S., Firestone, C.Y., Karron, R.A., Crowe, J.E., Jr., Elkins, W.R., Collins, P.L., and Murphy, B.R. (1999b). Addition of a missense mutation present in the L gene of respiratory syncytial virus (RSV) cpts530/1030 to RSV vaccine candidate cpts248/404 increases its attenuation and temperature sensitivity. J. Virol. *73,* 871–877.

Williams, J.V., Weitkamp, J.H., Blum, D.L., LaFleur, B.J., and Crowe, J.E., Jr. (2009). The human neonatal B cell response to respiratory syncytial virus uses a biased antibody variable gene repertoire that lacks somatic mutations. Mol. Immunol. *47,* 407–414.

Wilson, R.L., Fuentes, S.M., Wang, P., Taddeo, E.C., Klatt, A., Henderson, A.J., and He, B. (2006). Function of small hydrophobic proteins of paramyxovirus. J. Virol. *80,* 1700–1709.

Wong, D.T., Rosenband, M., Hovey, K., and Ogra, P.L. (1985). Respiratory syncytial virus infection in immunosuppressed animals: implications in human infection. J. Med. Virol. *17,* 359–370.

Wright, P.F., Karron, R.A., Belshe, R.B., Thompson, J., Crowe Jr, J.E., Boyce, T.G., Halburnt, L.L., Reed, G.W., Whitehead, S.S., Anderson, E.L., *et al.* (2000). Evaluation of a live, cold-passaged, temperature-sensitive, respiratory syncytial virus vaccine candidate in infancy. J. Infect. Dis. *182,* 1331–1342.

Wright, P.F., Gruber, W.C., Peters, M., Reed, G., Zhu, Y., Robinson, F., Coleman-Dockery, S., and Graham, B.S. (2002). Illness severity, viral shedding, and antibody responses in infants hospitalized with bronchiolitis caused by respiratory syncytial virus. J. Infect. Dis. *185,* 1011–1018.

Wright, P.F., Ikizler, M.R., Gonzales, R.A., Carroll, K.N., Johnson, J.E., and Werkhaven, J.A. (2005). Growth of respiratory syncytial virus in primary epithelial cells from the human respiratory tract. J. Virol. *79,* 8651–8654.

Wright, P.F., Karron, R.A., Madhi, S.A., Treanor, J.J., King, J.C., O'Shea, A., Ikizler, M.R., Zhu, Y., Collins, P.L., Cutland, C., *et al.* (2006). The interferon antagonist NS2 protein of respiratory syncytial virus is an important virulence determinant for humans. J. Infect. Dis. *193,* 573–581.

Wright, P.F., Karron, R.A., Belshe, R.B., Shi, J.R., Randolph, V.B., Collins, P.L., O'Shea, A.F., Gruber, W.C., and Murphy, B.R. (2007). The absence of enhanced disease with wild type respiratory syncytial virus infection occurring after receipt of live, attenuated, respiratory syncytial virus vaccines. Vaccine *25,* 7372–7378.

Wu, H., Pfarr, D.S., Johnson, S., Brewah, Y.A., Woods, R.M., Patel, N.K., White, W.I., Young, J.F., and Kiener, P.A. (2007). Development of motavizumab, an ultra-potent antibody for the prevention of respiratory syncytial virus infection in the upper and lower respiratory tract. J. Mol. Biol. *368,* 652–665.

Yano, H., Okitsu, N., Hori, T., Watanabe, O., Kisu, T., Hatagishi, E., Suzuki, A., Okamoto, M., Ohmi, A., Suetake, M., *et al.* (2009). Detection of respiratory viruses in nasopharyngeal secretions and middle ear fluid from children with acute otitis media. Acta Otolaryngol. *129,* 19–24.

Yeo, D.S., Chan, R., Brown, G., Ying, L., Sutejo, R., Aitken, J., Tan, B.H., Wenk, M.R., and Sugrue, R.J. (2009). Evidence that selective changes in the lipid composition of raft-membranes occur during respiratory syncytial virus infection. Virology *386,* 168–182.

Yoboua, F., Martel, A., Duval, A., Mukawera, E., and Grandvaux, N. (2010). Respiratory syncytial virus-mediated NF-kappa B p65 phosphorylation at serine 536 is dependent on RIG-I, TRAF6, and IKK beta. J. Virol. *84,* 7267–7277.

Young, S., O'Keeffe, P.T., Arnott, J., and Landau, L.I. (1995). Lung function, airway responsiveness, and respiratory symptoms before and after bronchiolitis. Arch. Dis. Child. *72,* 16–24.

Yunus, A.S., Khattar, S.K., Collins, P.L., and Samal, S.K. (2001). Rescue of bovine respiratory syncytial virus from cloned cDNA: entire genome sequence of BRSV strain A51908. Virus Genes *23,* 157–164.

Yusuf, S., Piedimonte, G., Auais, A., Demmler, G., Krishnan, S., Van Caeseele, P., Singleton, R., Broor, S., Parveen, S., Avendano, L., *et al.* (2007). The relationship of meteorological conditions to the epidemic activity of respiratory syncytial virus. Epidemiol. Infect. *135,* 1077–1090.

Zaman, K., Roy, E., Arifeen, S.E., Rahman, M., Raqib, R., Wilson, E., Omer, S.B., Shahid, N.S., Breiman, R.F., and Steinhoff, M.C. (2008). Effectiveness of maternal influenza immunization in mothers and infants. N. Engl. J. Med. *359,* 1555–1564.

Zhang, L., Peeples, M.E., Boucher, R.C., Collins, P.L., and Pickles, R.J. (2002). Respiratory syncytial virus infection of human airway epithelial cells is polarized, specific to ciliated cells, and without obvious cytopathology. J. Virol. *76,* 5654–5666.

Zheng, H., Storch, G.A., Zang, C., Peret, T.C., Park, C.S., and Anderson, L.J. (1999). Genetic variability in envelope-associated protein genes of closely related group A strains of respiratory syncytial virus. Virus Res. *59,* 89–99.

Zimmer, G., Trotz, I., and Herrler, G. (2001). N-glycans of F protein differentially affect fusion activity of human respiratory syncytial virus. J. Virol. *75,* 4744–4751.

Zimmer, G., Rohn, M., McGregor, G.P., Schemann, M., Conzelmann, K.K., and Herrler, G. (2003). Virokinin, a bioactive peptide of the tachykinin family, is released from the fusion protein of bovine respiratory syncytial virus. J. Biol. Chem. *278,* 46854–46861.

Zlateva, K.T., Lemey, P., Vandamme, A.M., and Van Ranst, M. (2004). Molecular evolution and circulation patterns of human respiratory syncytial virus subgroup a: positively selected sites in the attachment g glycoprotein. J. Virol. *78,* 4675–4683.

Zlateva, K.T., Lemey, P., Moes, E., Vandamme, A.M., and Van Ranst, M. (2005). Genetic variability and molecular evolution of the human respiratory syncytial virus subgroup B attachment G protein. J. Virol. *79,* 9157–9167.

Metapneumoviruses

James E. Crowe, Jr.* and John V. Williams

11

Abstract

The metapneumoviruses were discovered more recently than most paramyxoviruses, because of the difficulty in isolating live viruses in cell culture from clinical materials. Nevertheless, rapid progress has been made in identifying the salient features of the epidemiology and biology of these viruses. Remarkably, human *Metapneumovirus* appears to be one of the most common causes of severe lower respiratory tract illness in infants and children. The virus is also a major cause of illness in the elderly. Avian *Metapneumovirus* also was discovered relatively recently and is a major cause of disease in birds, especially in commercial poultry. Molecular biologists have defined the distinguishing features of the viral genomes, and animal models have been developed that allow detailed studies of pathogenesis, immune mechanisms, and protective efficacy of candidate vaccines.

History

Human *Metapneumovirus* (HMPV) was described first in the literature in 2001. Investigators used a molecular discovery approach to identify the genetic sequences of viruses isolated from infants and children experiencing disease that appeared to be similar to that caused by human RSV (HRSV) (van den Hoogen *et al.*, 2001). Although the discovery of the virus occurred relatively recently, the virus probably has been circulating in humans as long as most other human respiratory viruses. Examination of human antibodies in archived serum samples collected as many as four decades ago reveal reactivity with HMPV. Virus has been identified in archived nasal wash specimens from children dating back several decades (Williams *et al.*, 2004, 2006b). The reason that HMPV was discovered only recently, even though it is an exceedingly common human infection, has to do with its growth characteristics *in vitro*. The virus almost certainly had been isolated occasionally without specific identification, but inefficiently because of the slow growth of clinical isolates in cell culture and the requirement for trypsin activation. Since the original report, dozens of studies of the occurrence of HPMV in samples in diagnostic virology laboratories around the world have revealed that it is a ubiquitous pathogen especially in infants, children, and the elderly. The virus is found frequently in samples obtained from immunocompromised patients. The virus is often found to be the second most common viral pathogen in these populations, behind HRSV (Williams *et al.*, 2004).

Animal viruses

At the time of discovery of HMPV, investigators immediately realized from sequence alignments that this new human pathogen isolated directly from human respiratory secretions possessed a high level of homology with the avian virus designated avian *Metapneumovirus* (AMPV) (van den Hoogen *et al.*, 2001). The virus alternately had been termed turkey rhinotracheitis virus

*Corresponding author

(TRTV) or avian *Pneumovirus* (APV) in some reports in the past. This previously described avian pathogen was known to cause severe respiratory tract disease in turkeys in commercial settings, and investigators had shown that AMPV infects chickens and other birds (Njenga *et al.*, 2003). Interestingly, AMPV also was known to be difficult to isolate and grow in cell culture, but enough isolates had been obtained to lay the groundwork for early classification schemes, genetic sequence analysis, and genotyping. The genomic sequences of HMPV were found to have essentially the same organization as AMPV, and the viruses share nucleotide sequence identity at about two-thirds of positions. This level of homology raised the possibility that HMPV was an avian pathogen that had crossed over to humans recently. However, HPMV clinical strains did not replicate in turkeys or chickens, suggesting the virus was not of direct avian origin (van den Hoogen *et al.*, 2001).

AMPV was isolated first in Africa in 1978 and later was identified in birds in Europe and Asia (Njenga *et al.*, 2003). The first report of AMPV in the United States was in 1996. It is likely that the sequence of discovery of virus in these locations reflected emerging attention on a pathogen that was already circulating locally, rather than rapid transmission of emerging viruses worldwide. Despite the earlier identification of AMPV, there has now been less molecular epidemiology work reported on the prevalence and diversity for the avian virus than for HMPV. It does appear that there are major antigenic subgroups of AMPV, designated A, B, C and D (Njenga *et al.*, 2003), and viruses representing these subgroups exhibit differences in prevalence in differing geographic settings. The viruses circulating in the United States are of the antigenic subgroup C, while viruses from subgroups A, B and D are found in Old World settings. Interestingly, the genomic sequences of the viruses of subgroups A, B and D are closely related, while those of subgroup C are more distinct. Genotyping suggests that there is probably antigenic divergence between the A/B/D and C groups of viruses, but it is not clear yet the extent to which infection with virus of one subgroup would induce antibodies that recognize viruses of the other subgroups. When the genomes of AMPV were aligned with those of HMPV clinical isolates, the AMPV subgroup C viruses were found to be most highly related to HMPV. An AMPV subgroup C virus nucleoprotein (N) sequence shared 88% nucleotide identity with an HMPV sequence, while the identity of subgroup A or B AMPV sequences was closer to 70% (van den Hoogen *et al.*, 2001, 2002).

Infectious agent

Classification

HMPV and AMPV are non-segmented negative-strand RNA virus members of the order *Mononegavirales* and family *Paramyxoviridae*, like all viruses in this text. The family *Paramyxoviridae* has two subfamilies: (1) *Paramyxovirinae*, which includes a number of major human pathogens including measles, mumps, and the parainfluenza viruses, and (2) *Pneumovirinae*, which includes HRSV and animal RSVs and the MPVs. The subfamily *Pneumovirinae* organizes viruses in two genera: (1) *Pneumovirus*, which includes HRSV, ovine and bovine respiratory syncytial virus (BRSV) and pneumonia virus of mice (PVM) among others, and (2) *Metapneumovirus*, which includes HMPV and AMPV.

The classification of these viruses is based principally on distinct features of the virus genomes. All subfamily *Paramyxovirinae* possess nucleprotein (N), phosphoprotein (P), matrix (M), fusion (F) protein, and a large polymerase (L) protein. Sequence relatedness between genomes of viruses in the subfamilies *Pneumovirinae* and *Paramyxovirinae* are clear, especially in the N, L, and F genes, but overall the relatedness is modest. The short hydrophobic (SH) gene is found in *Pneumovirinae* and in some but not all *Paramyxovirinae.* The accessory genes V and C that are encoded by P genes and the haemagglutinin-neuramindase (HN) protein of subfamily *Paramyxovirinae* viruses are not found in *Pneumovirinae* such as HMPV or AMPV. Conversely, some proteins are found only in *Pneumovirinae,* such as the nonstructural genes NS1, NS2, the regulatory factors M2-1 and M2-2 and the highly glycosylated surface protein G. Consistent with the lack of an HA or HN protein, HMPV does not haemagglutinate red blood cells (van den Hoogen *et al.*, 2001). The pneumoviruses HMPV and HRSV possess the same structural proteins, but they are distinct because HPMV lacks the non-structural genes NS1 and NS2 (Biacchesi *et al.*, 2004a; van den Hoogen *et al.*, 2002).

HMPV has been classified based on genotype into two major subgroups, with a variety of smaller subgroups proposed (designated A1, A2a, A2b, B1, B2) that represent clades based on genetic relatedness. AMPV is classified into subgroups designated A, B, C and D (Njenga *et al.*, 2003).

Propagation

HMPV was isolated originally in what were essentially primary monkey kidney cells that had been passaged several times in culture (designated tertiary monkey kidney cells; van den Hoogen *et al.*, 2001). This approach was perhaps more feasible for these investigators because of their access to primate facilities with non-human primates without foamy virus infection. Typically, prolonged culture of primary monkey kidney cells from U.S. suppliers is not possible because foamy viruses in the original tissues cause marked cytopathic effect (CPE) in the cultured cells over time, obscuring the occurrence of CPE owing to a human virus like HMPV in the clinical sample tested. HMPV has been isolated and propagated in cell lines of primate origin that are easier to obtain, including Vero cells (African green monkey origin, lacking functional type I interferon response) or Vero E6 cells (Abiko *et al.*, 2007) and LLC-MK2 cells (rhesus monkey origin). The most permissive human airway epithelial cell line for HMPV is BEAS-2B (Tollefson *et al.*, 2010). Isolation from clinical specimens can be quite difficult, and often isolation requires blind serial passage in order to amplify the virus enough to detect CPE. Cell-culture adapted isolates may replicate independent of trypsin activation, but typically clinical isolates require added trypsin (Schickli *et al.*, 2005). This requirement also probably contributed to the delayed recognition of this major pathogen, since most diagnostic virology laboratories do not include trypsin activation in standard isolation protocols. When the growth of HPMV and HRSV is compared in cell culture monolayers, HMPV usually replicates more slowly and rarely achieves as high peak titres. AMPV can be isolated and propagated in Vero cells, quail fibroblasts, BGM, DF-1, or chicken embryo related cells (Tiwari *et al.*, 2006; Coswig *et al.*, 2010).

Morphology

The morphology of the infectious form of HMPV is not certain, though it is likely to be filamentous. When virus-infected cell culture supernatants are examined by electron microscopy, virion particles can be seen that are filamentous, but more spherical forms are also present (van den Hoogen *et al.*, 2001; Peret *et al.*, 2002). Particles identified as virions possess nucleocapsid diameters of 17 nm and envelope spike projections of about 13–17 nm, consistent with the predicted size of the envelope glycoproteins F and G. The spherical particles seen in electron microscopy had a varied diameter ranging from 150 to 600 nm (Peret *et al.*, 2002). The particles seen on electron microscopy are pleomorphic, but electron microscopy cannot distinguish between infectious and non-infectious particles.

Genome organization

HMPV uses a negative-sense non-segmented single-strand RNA genome (van den Hoogen *et al.*, 2002; Biacchesi *et al.*, 2003). Complete genomic sequences have been completed for representative strains of subgroup A (for example, the Netherlands strain NL/00/1 or Canadian strain CAN97–83) and subgroup B (for example, the Netherlands strain NL/1/99 or Canadian strain CAN98–75). Complete sequences of AMPV C virus are available, revealing this genome to be the longest of the metapneumoviruses (Govindarajan *et al.*, 2004; Govindarajan and Samal, 2005). Various genomes that have been sequenced exhibit minor variation in length, and the lengths are not restricted to the so-called 'rule of six' interval length of genomes found in some paramyxoviruses. The genome encodes nine proteins, with coding regions separated by intergenic regions that are in some cases longer than those found in pneumoviruses. Three is no obvious specificity of the intergenic regions other than to serve a spacer function. Overall, there is approximately 50% identity at the nucleotide level between HMPV and HRSV genomes in the genes that they share.

Since HMPV lacks the NS1 and NS2 genes seen in RSV strains, the genome is comparatively about two kilobases shorter. The order of some of the genes in the viral genome, SH, G, F and M2, is shuffled in HMPV. The HMPV gene order reading from the 3′ end is F-M2-SH-G, while HRSV encodes these genes in the SH-G-F-M2 order (Fig. 11.1). Single-stranded viruses like HMPV use gene-start and gene-end motifs in control of gene transcription to produce mRNA. The predicted motifs in HPMV are similar to those that have been studied well in HRSV. Within the transcribed viral mRNAs, the start codon is located in a canonical position at nucleotide14 in all but one of the predicted mRNAs. Unlike the paramyxovirus complexities of P/V/C expression, HMPV does not appear to use overlapping genes or reading frames in the P gene.

RNA and proteins

Of course, the virus must produce two types of RNAs during the life cycle, transcription of genes to produce mRNA for translation in order to produce newly synthesized viral proteins and replication of the full-length RNA genome to incorporate into new infectious virions.

Transcription

The virion particle brings polymerase L protein into the cell to initiate RNA transcription. The genome has nine genes that are transcribed sequentially in the 3′–5′ order into separate mRNAs. The beginning of each gene is marked by the gene-start signal that is highly conserved, and finishes with a gene-end signal that is generally conserved. The gene-end signal is thought to coordinate termination of the particular mRNA strand being transcribed and its polyadenylation upon completion. During replication of the genome, the gene-end sequences are ignored, but not surprisingly the termination of transcription at gene-end signals is not perfectly efficient. Therefore, read-through transcripts possessing more than one gene are expected in the RNA populations of infected cells; a special role for such transcripts has not been defined. There is a gradient of level of transcription of genes from the 3′ to 5′ end, as the polymerase falls off or is occupied with read-through transcripts.

Replication of genomic RNA

The genomic replication strategy is the conventional approach for all of the viruses in this text. During replication, the polymerase reads through all of the gene-start and gene-end signals to generate a full-length antigenomic RNA strand that is an exact positive-sense complementary copy of the negative-sense RNA genome. Subsequently, a complementary RNA strand copy of the antigenome is produced that becomes the newly replicated genomic RNA for new virion formation. Long RNAs such as this (over 13 kb) would be expected to be quite unstable in cells, however both antigenome and genome RNAs are encapsidated with N protein, which enhances the resistance to degradation. These long RNAs are not polyadenylated.

Replication of the full-length RNAs requires a mechanism of recognition of the non-coding ends of the genome. Both the 3′ and 5′ ends of the viral genome possess regulatory signals for the polymerase complex. The 3′ extragenic region comprises the leader sequence that serves as a promoter for transcription or for initiation of antigenomic RNA synthesis during full-length RNA replication. The 3′ and 5′ ends exhibit a

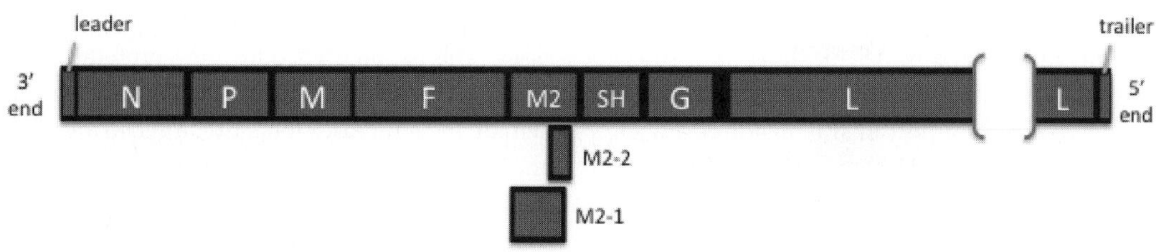

Figure 11.1 Organization of the single-strand negative-sense metapneumovirus genome, shown approximately to scale. The M2 gene has two overlapping reading frames.

high level of complementarity. The 5′ end serves as a promoter for the production of antigenomic RNA. Thus, functionally the two ends serve as similar types of promoters.

Proteins

Fusion F glycoprotein

HMPV F protein has a moderate level of sequence identity with HRSV at the amino acid level (33%), but only about 10% identity with *Paramyxovirinae* (van den Hoogen *et al.*, 2002; Biacchesi *et al.*, 2003). The HMPV F protein is a typical type I integral membrane protein with an N-terminal signal peptide sequence that directs protein entry into the endoplasmic reticulum/Golgi complex and trafficking through the secretory pathway (Fig. 11.2). The signal sequence is followed by a predicted signal peptide cleavage site. At the C terminus of the protein, a classical hydrophobic transmembrane domain mediates integration in the plasma cell membrane, serving as a membrane anchor near the C-terminus, followed by a short cytoplasmic tail. Presumably the tail domain mediates interaction with matrix protein or nucleocapsids during assembly and budding, but a direct interaction of the tail with other proteins has not been reported to date.

During infection, F protein traffics to the surface of cells via the secretory pathway, apparently independent of the presence of other viral proteins. F protein expressed in cell lines from plasmids encoding a cDNA copy of the F gene appears to distribute in the cell membrane. Like other related fusion proteins, the HMPV F protein is expressed as a F_0 precursor but possesses a cleavage activation site near the fusion peptide that facilitates cleavage to the F_1/F_2 form. The final active protein has two disulfide-linked subunits, in the order N terminus–F_2–F_1–C terminus. HMPV F protein only contains a single cleavage site, in contrast to HRSV, which has two. While the HRSV sites are canonical targets for furin cleavage, the residues at the HMPV F cleavage site (Arg-Gln-Ser-Arg) do not contain a consensus motif for furin-like proteases. Therefore, it is suspected that the activation of HMPV by cleavage requires an extracellular protease that is present in human airway secretions. The fact that trypsin activation is required for optimal growth of HMPV in cell culture is consistent with this prediction. Mutant viruses with decreased requirement for trypsin activation prior to growth in cell culture have been isolated by multiple passage in cell line monolayer cultures (Schickli *et al.*, 2005). These isolates typically possess a proline in place of serine in the cleavage site. Altering fusion protein cleavage sites in respiratory viruses often alters the tropism or pathogenesis of the mutant virus. HMPV isolates with this easily cleavable site, however, did not appear to be more virulent *in vivo*, at least when tested in hamsters (Schickli *et al.*, 2005). Additional synthetic sequences have been studied at the cleavage site that confer trypsin independence. The most cleavage-efficient mutant, with a sequence of Arg-Arg-Arg-Arg in the cleavage site, was tested in African green monkeys and these studies showed that there was no detectable change in the magnitude of replication in the upper or lower respiratory tract or in immunogenicity or protective efficacy against HMPV challenge (Biacchesi *et al.*, 2006).

Once at the surface of infected (or F gene cDNA-transfected) cells, the protein on the cell surface can mediate cell–cell fusion (so-called 'fusion from within'), resulting in the formation of multinucleated giant cells termed syncytia.

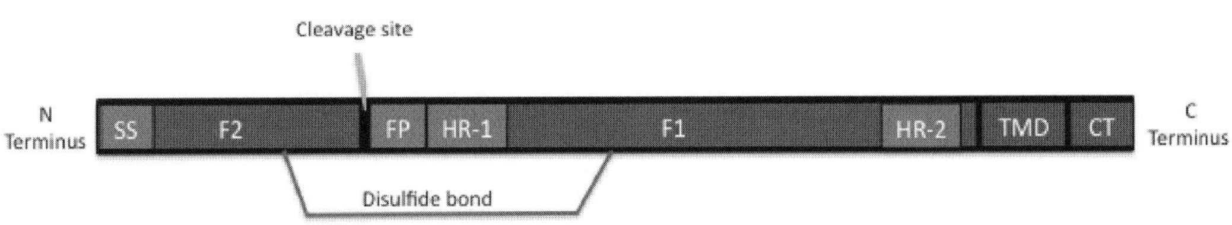

Figure 11.2 Organization of the metapneumovirus (MPV) fusion protein, shown to approximate scale. SS, signal sequence; F1/F2, F subunits 1 and 2; HR, heptad repeat; TMD, transmembrane domain; CT, cytoplasmic tail.

The syncytia are perhaps less dramatic than those caused by HRSV F protein, but they do occur in culture. It is less clear if HMPV causes syncytial formation in tissues during natural infection of humans. HMPV is similar to HRSV in the fact that F protein alone appears competent to fuse virus envelope to cell membranes or to accomplish cell–cell fusion. Most viruses in the family *Paramyxovirinae* require interaction of the attachment protein (typically HN) with the fusion protein for optimal fusogenicity of the particle. The F protein of AMPV-C appears to be more fusogenic when compared with the F protein of HMPV (de Graaf *et al.*, 2009). Fusion of membranes is accomplished by insertion of the fusion peptide at the N-terminus of the F_1 subunit that is created by cleavage. This peptide comprises a hydrophobic domain that inserts into the hydrophobic milieu of the target host cell plasma membrane. The structure of the full extracellular domain of HMPV F protein has not been determined yet, but sequence homologies with a wide variety of other viral fusion proteins suggests the mechanism of fusion, which involves two heptad repeat regions in the F_1 subunit. The N terminal heptad repeat (HR-1) is an alpha-helical domain that allows self-association in the trimeric F protein to form a trimeric coiled-coil structure. The C-terminal heptad repeat (HR-2) region forms an alpha-helical domain, but in the trimeric form of full-length F on particles, these helices are predicted to remain separate in the pre-fusion state. During fusion, by analogy with other F proteins, the metastable trimeric F protein undergoes conformation changes following cell attachment and insertion of the fusion peptide into cell membrane. The signals that trigger the conformational changes in F involved in membrane fusion presumably involve attachment events, but these signals have not been defined at a biochemical level. During fusion, the HR-2 helical domains associate with grooves on the outer surface of the HR-1 coiled-coil, forming a single stable hexameric structure of three HR-1 domains complexed with three HR-2 domains (Miller *et al.*, 2007). Synthetic peptides from HR-1 and HR-2 can be induced to form a thermostable structure with a melting temperature of approximately 90 °C, and helical content detected by circular dichroism consistent in mass with a hexameric coiled coil (Miller *et al.*, 2007). The physical movement of the HR-2 domains that are proximal to the viral lipid bilayer to a position where they can form a complex with the HR-1 domains is thought to bring the viral membrane in apposition with the cell membrane. This mechanism is common to a wide variety of viruses even beyond paramyxoviruses, and indeed even to non-viral membrane fusion mechanisms, such as those used by host cell vesicle or target soluble NSF attachment receptor (SNARE) proteins in yeast and mammalian cells. It is not clear the specific mechanism by which the outer and inner leaflets of the membranes fuse, or what drives the energy needed to accomplish membrane fusion. Interruption of the HMPV fusion mechanism has been explored using synthetic peptides. Inhibitory studies of HMPV HR-1 and HR-2 showed that a synthetic HR-1 peptide was a significant fusion inhibitor with a 50% inhibitory concentration and a 50% effective concentration of approximately 50 nM (Miller *et al.*, 2007). This is in contrast to studies with most paramyxoviruses in which HR-2 peptides typically exhibit a greater potency than those based on HR-1. Peptides derived from either of the HMPV HR regions do exhibit antiviral activity however (Miller *et al.*, 2007; Deffrasnes *et al.*, 2008).

Although the primary mechanism of attachment of HPMV has been suspected to be interactions of the G protein with a cellular receptor, clearly F protein plays a significant role in attachment. First, a recombinant HMPV strain that lacks the only other two envelope proteins, G and SH, replicates both in cell culture and in animal models (Biacchesi *et al.*, 2004b, 2005). Second, a specific interaction of HMPV F protein with a cell surface protein has been defined (Cseke *et al.*, 2009). All HMPV F sequences reported to date contain an Arg-Gly-Asp (RGD) sequence, which suggested that F might engage RGD-binding integrins on epithelial cell surfaces to mediate cell entry. Investigators showed that the divalent cation chelator EDTA, which disrupts heterodimeric integrin interactions, inhibits infectivity of hMPV. Function-blocking antibodies specific for $alpha_V beta_1$ integrin inhibited infectivity of hMPV. Transfection of non-permissive cells with $alpha_V$ or $beta_1$ cDNAs conferred hMPV infectivity, while in contrast reduction of $alpha_V$ and $beta_1$ integrin expression by siRNA inhibited hMPV infection. Recombinant HMPV F protein bound to cultured epithelial cells, but a mutant F protein in which the RGD motif had been altered to Arg-Gly-Glu (RGE) did not. These data suggested that HMPV F protein uses $alpha_V beta_1$ integrin as a functional receptor (Cseke *et al.*, 2009).

The F protein is a major target for neutralizing antibodies. Both murine and human monoclonal antibodies to F have been isolated (Ma *et al.*, 2005; Williams *et al.*, 2007; Hamelin *et al.*, 2008). Several neutralizing antigenic sites have been mapped for AMPV F (Brown *et al.*, 2009).

Attachment (G) glycoprotein

HMPV G is a type II transmembrane glycoprotein with a single hydrophobic region near the N terminus that functions as a signal peptide to direct movement through the endoplamic reticulum and Golgi complex, where the protein is highly modified post-translationally. Oligomers of G form in the endoplasmic reticulum, and a number of O-linked sugars are added during transit through the Golgi. On protein gels, G appears to run as a wide band with an apparent molecular weight of about 80 kDa, suggesting inhomogeneity of the glycosylation process (Luo *et al.*, 2010). The HMPV signal sequence remains uncleaved, allowing this domain also to serve as a membrane anchor. Thus, at the cell or virion surface, the bulk of the extracellular domain derives from the C terminus of the molecule. This ectodomain contains a high concentration of serine, threonine, and proline residues (van den Hoogen *et al.*, 2002; Biacchesi *et al.*, 2003). In contrast to the F protein, the HMPV G protein exhibits very little sequence identity with that of its counterpart in HRSV, and the HMPV G protein is shorter in length than that of HRSV. When G protein sequences of viruses from the major genotypes of HPMV A and B are examined, the sequence divergence within G does distinguish viruses of the subgroups. Further, these sequences appear more divergent than those of G proteins from RSV A and B subgroups. G genes of varying lengths are found in circulating strains, suggesting the protein is constantly evolving and is able to tolerate a high level of variability (Bastien *et al.*, 2004). AMPV strains in the field may be drifting genetically over time, such that vaccines based on 1980s sequences no longer induce protective antibodies against the G protein of current strains (Catelli *et al.*, 2010; Cecchinato *et al.*, 2010). A goose MPV has been reported that contained the largest G gene of any *Pneumovirus* or *Metapneumovirus*, with a predicted G protein of 585 amino acids, more than twice the sizes of G proteins from other AMPV C viruses or HMPV (Bennett *et al.*, 2005). Interestingly, this virus was avirulent but immunogenic, suggesting it could serve as an attenuated vaccine for poultry.

The function and structure of HMPV G protein are poorly understood. The protein is dispensable for growth in cell culture (Biacchesi *et al.*, 2004b, 2005). In animal models including hamsters and African green monkeys, a recombinant HMPV lacking the G protein or lacking both the G and SH proteins replicated, albeit to lower levels than the corresponding wild-type virus (Biacchesi *et al.*, 2004b, 2005). A recombinant AMPV lacking G replicated slightly better than did wild-type AMPV in Vero cells, but was attenuated in turkeys (Govindarajan *et al.*, 2010). There does not appear to be any obvious significant structural organization of the HMPV G ectodomain seen by sequence analysis, in contrast to the presence of a predicted cysteine noose structure found in the G protein of *Pneumoviruses* (Biacchesi *et al.*, 2003). The G protein genes of members of *Pneumovirinae* encode a secreted form of G whose translation begins at an alternative start codon located in the extracellular domain. In contrast, the HMPV G transcript does not appear to encode a secreted form.

Small hydrophobic (SH) protein

The HMPV SH protein is the third integral membrane protein, which is also anchored by a hydrophobic sequence near the N terminus that serves as both a signal sequence and a membrane anchor. The orientation of the protein is thus predicted to project the C-terminal domain as the extracellular region (van den Hoogen *et al.*, 2002; Biacchesi *et al.*, 2003). The function of the SH protein is poorly understood. The protein appears dispensable for replication in cultured cells and in experimentally inoculated hamsters or African green monkeys (Biacchesi *et al.*, 2004b; 2005). Interestingly, passage of HMPV in cell culture leads to frequent generation of virus subpopulations in which the SH gene has become truncated. It was speculated that truncation of the SH protein may provide a modest selective advantage to HMPV clinical isolates in cell culture, an unusual mode of culture adaptation (Biacchesi *et al.*, 2007). The SH protein of HMPV is about three times longer than the HRSV SH protein, which is predicted to have ion channel activity (van den Hoogen *et al.*, 2002; Biacchesi *et al.*, 2003).

Matrix (M) protein

The HMPV M protein is a 254 amino acid structural protein located inside the virion particle. In cells, M distributes in a diffuse but slightly reticulated cytoplasmic pattern, but the protein also appears slightly concentrated at or just below the plasmid membrane. It is suspected that the protein must interact with the cytoplasmic tails of the surface proteins, and with the nucleocapsid, however direct interactions have not been reported. Based on its homology with the M protein of other paramyxoviruses, the function of M is likely to inhibit RNA synthesis by binding to nucleocapsids and to facilitate virus assembly.

Replication complex proteins: nucleocapsid (N), phosphoprotein (P), polymerase (L) and M2-1

The HMPV N, P, L and M2-1 proteins associate with genomic RNA in a complex, and these proteins also serve as the polymerase complex. These proteins share a high level of sequence identity with the homologous HRSV proteins, and they are of a similar size (van den Hoogen *et al.*, 2002; Biacchesi *et al.*, 2003). We can infer the biology of these proteins in a general sense from studies of the HRSV proteins. The N protein probably associates in the cytoplasm with newly made genomic or antigenomic RNAs to form the helical nucleocapsid structure. AMPV may express an accessory in-frame truncated N2 ORF gene product (Alvarez and Seal, 2005). The P protein associates directly with N and maybe with viral RNA (Derdowski *et al.*, 2008). In cells, the N and P proteins are necessary and sufficient to form dense cytoplasmic granules that have been termed inclusion bodies (Derdowski *et al.*, 2008). Replication of RNA also requires the polymerase L. This very large protein is predicted to have at least six functional domains and enzymatic capabilities, including polymerase core motifs and ATP-binding motifs. Little work has been done to date to map the location of the functional domains by mutagenesis. These predicted domains are also present in AMPV (Govindarajan and Samal, 2004). Optimal transcription of HRSV genes requires the expression of M2-1, however, in cell culture the HMPV M2-1 gene can be altered with little to no effect on transcription (Buchholz *et al.*, 2005). An M2-1 deletion virus did not replicate in hamsters, suggesting that the effect on processive transcription mediated by M2-1 is greater *in vivo* than *in vitro*.

M2-2 protein

HMPV encodes a small M2-2 protein of 71 amino acids using an internal ORF in the mRNA of M2 that overlaps a 5′-end ORF encoding M2-1. It is suspected that the function of the M2-2 protein involves regulation of the switch from transcription to replication of genomic RNA. A recombinant HMPV virus in which the M2-2 gene was deleted exhibited a phenotype in which transcription was increased and RNA replication decreased, with a resulting reduction in virus replication and attenuation in non-human primates (Biacchesi *et al.*, 2005; Buchholz *et al.*, 2005). Others have reported that M2-2 inhibits both viral transcription and replication (Kitagawa *et al.*, 2010).

Replication

Attachment of HMPV to cells is facilitated by interaction with alpha$_V$beta$_1$ integrin, as described above. It is thought that the virus enters the cell after direct fusion of the viral membrane with the plasma cell membrane at the apical surface of airway cells at neutral pH. The virus replicates entirely in the cytoplasm, without apparent involvement of the cell nucleus. The viral particle brings its own polymerase into the cell to initiate transcription of viral genes. Transcription is sequential, in a 3′ to 5′ direction. Transcription of genes closer to the 3′ end of the genome is observed to be higher than those at the 5′ end, reflecting the falling off of the polymerase. RNA transcripts can be detected in infected cells within several hours. Since M2-1 appears to regulate the switch from transcription to replication, it is suspected that accumulation of sufficient amounts of M2-1 in infected cells reaches a threshold to mediate the switch such that the viral polymerase reads through the gene-start and gene-end signals to generate antigenome and subsequently genomic RNA. During replication, N protein appears to associate with both types of full-length RNAs. The N and P proteins are known to interact and they form inclusion bodies in infected cells; it is suspected that these bodies are the sites of RNA replication. The integral membrane proteins traffic through the secretory pathway to the cell surface; they are not present in the cytoplasmic inclusion bodies. It is presumed that the nucleocapsids must traffic to the cell surface where assembly and budding occur. The kinetic pattern of replication of HMPV in cultured cells appears to be significantly slower than that of most other paramyxoviruses in culture. It is not clear whether this stems from an inherently slow rate of replication of the virus, or the use of cell lines that are not optimally permissive for virus growth. Infection of epithelial cells alters the transcriptional profile of the cells markedly, affecting expression of over 1000 genes involved in cellular signalling, gene transcription and apoptosis (Bao *et al.*, 2008b).

Reverse genetics

Infectious recombinant HMPV can be generated in cells by reverse genetics following transfection with cloned cDNAs. Reverse genetic systems have been reported for both subgroups of HMPV, A and B, and for subgroup A and C AMPV (Biacchesi *et al.*, 2004a; Herfst *et al.*, 2004; Naylor *et al.*, 2004; Govindarajan *et al.*, 2006). The approach to HMPV rescue is similar to that used for most other paramyxoviruses (Buchholz *et al.*, 2006). A full-length cDNA encoding the HMPV RNA genome or antigenome in a plasmid is introduced into cells at the same time as separate plasmids encoding the N, P and L proteins. Expression of the proteins typically is controlled by T7 promoters in the plasmids, and T7 is provided by use of transfected cells such as *BHK-T7* cells that constitutively express the bacteriophage T7 RNA polymerase (Buchholz *et al.*, 1999). M2-1 protein expression is not required for rescue of infectious HMPV (Biacchesi *et al.*, 2004a; Buchholz *et al.*, 2005), in contrast to *Pneumovirinae* for which rescue is optimized when M2-1 is expressed. RNA recombination during co-infection of cells with two HMPV strains has not been reported, and probably is rare in nature if it occurs.

Pathogenesis

Host range

HMPV is a human virus, and accomplishes person-person spread efficiently. There does not appear to be an animal reservoir for HMPV, so humans are the only source of infection for the virus. A large number of animals have been studied following experimental inoculation, but all have been found to be semi-permissive for infection. Despite the high level of relatedness of HMPV to AMPV-C, HMPV does not productively infect birds such as turkeys or chickens (van den Hoogen *et al.*, 2001). Likewise, AMPV is a bird virus. AMPV infects birds in the wild and is a major pathogen in the commercial poultry industry. The F protein is the main determinant for host tropism. Mammalian Vero-118 cells supported replication of both HMPV and AMPV-C whereas only AMPV-C replicated to high titres in quail fibroblast (QT6) cells (de Graaf *et al.*, 2009). Recombinant HMPV viruses in which individual polymerase genes were exchanged with those of AMPV-C have been studied in minireplicon systems, *in vitro* replication kinetics, and growth in hamsters. Interestingly the chimeric viruses functioned well in the *in vitro* systems, but were revealed to be attenuated for growth *in vivo* (de Graaf *et al.*, 2008a).

Entry, spread, tissue tropism

HMPV is spread by contact with respiratory secretions from infected humans. Spread requires close contact with infected individuals or self-inoculation to the nasal mucosa by touching hands contaminated from contact with surfaces. Therefore, the major mode of spread appears to be by large droplets or through fomite contamination. Unlike measles virus, HMPV does not appear to spread by small-particle aerosols. AMPV is spread rapidly in commercial poultry settings where bird-bird contact is frequent. HMPV is probably one of the most infectious viruses of man, given that all subjects appear to be seropositive by the age of five years (van den Hoogen *et al.*, 2001), and most are infected in the first year or two of life. The incubation period for wild-type HMPV infection is not defined experimentally, but is thought to be about 4–5 days.

Since the virus is not spread by small-particle aerosol, virus replication must initiate in the nasopharynx. Virus can be recovered from nasal washes, aspirates or swabs obtained from ill children. It is more difficult to recover virus from the nasopharynx of adults during symptomatic naturally-acquired infection, even when these infections are documented by positive RT-PCR tests (Falsey, 2003, 2005a, 2006, 2008). Presumably this observation reflects a lower titre of virus shedding in nasal secretions in previously infected subjects who have partial immunity.

In a significant proportion of susceptible hosts (around one-third of infants), lower respiratory tract symptoms occur several days after the onset of nasal secretions, suggesting virus spreads rapidly in the respiratory tract. Viral spread from the upper to the lower respiratory tract (the airway below the vocal cords) probably involves aspiration of secretions or cell–cell fusion. The virus can be recovered from bronchoalveolar lavage fluids of patients supported by intubation and mechanical ventilation. There is a correlation between peak viral load in the nasopharynx or lower airway and the severity of disease (Bosis *et al.*, 2008). Resolution of severe disease and respiratory symptoms generally occurs about the same time as the resolution of detectable virus shedding. Infants and children probably shed infectious virus for a number of weeks. There is no evidence that HMPV infection causes a latent infection, although infection can be asymptomatic and persistent in the immunocompromised (Debiaggi *et al.*, 2006, 2007).

The virus is thought to replicate primarily in the most superficial layer of the respiratory epithelium and to shed from the apical surface of the infected cells into the airway lumen. Viraemia has not been described during infection of normal infants and children. There are reports of febrile seizures or even severe, fatal encephalitis temporally associated with HMPV infection, but neither HMPV nor antibodies to HMPV could be identified in the cerebrospinal fluid in these cases (Arnold *et al.*, 2009). The principal targets of infection in the airway are epithelial cells, but the airway contains a large number of monocyte/macrophages and pulmonary dendritic cells (DCs). HMPV does not appear to infect DCs efficiently (Guerrero-Plata *et al.*, 2006), but interaction of HMPV with DCs may alter the biology of these antigen-presenting cells (Le Nouen *et al.*, 2009).

At this time, there is very little histopathology data of HMPV infection in the lower airways of otherwise healthy infants or adults. One report suggested that the infection involves an unusual enlarged, darkly-staining pneumocyte ('smudge cell'; Sumino *et al.*, 2005). Histopathology studies of the lungs of HMPV-infected animals may be informative. HMPV-infected cotton rats showed the presence of viral antigens in respiratory epithelial cells lining the airway lumen throughout the respiratory tract (Williams *et al.*, 2005d). Histopathology studies of HMPV-infected cynomolgus macaques also showed virus antigens to be present in the airway, without submucosal spread (Kuiken *et al.*, 2004). The disease in young infants is predominantly one of wheezing, suggesting functional occlusion of small airways, especially the bronchioles. Typically, wheezing associated with respiratory virus infection of the bronchioles is associated with destruction of some of the bronchiolar epithelial cells by virus and the immune response, submucosal oedema, sloughed dead cells and marked secretions causing obstruction. Since infants have much smaller diameter bronchioles, their airway resistance is much higher than that of adults. Therefore, infants are at increased risk of wheezing during HMPV infection on the basis of their physiology. The occurrence of inspiratory crackles on exam and radiographic findings of pneumonia suggests the disease also can involve the lung parenchyma.

AMPV causes swollen head syndrome in broilers and broiler breeders, a condition marked by swelling of infraorbital sinuses (Aung *et al.*, 2008).

Animal models

A wide variety of small animals have been studied with HMPV, such as Syrian golden hamsters, guinea pigs, ferrets, cotton rats (*Sigmodon hispidus*), and various inbred strains of mice including C3H, CBA, SJL, C57Bl/10, BALB/c, 129, AKR, DBA/1, and DBA/2 (Alvarez *et al.*, 2004; MacPhail *et al.*, 2004; Hamelin *et al.*, 2005b; Williams *et al.*, 2005d). Infection of rodents results in a moderate level of virus replication that peaks on day 4–5 and is cleared quickly, in a pattern that is similar to that seen with HRSV. The most permissive small animal for infection in terms of virus peak titres is the cotton rat. Rodents do not exhibit clear clinical evidence of respiratory tract disease with HMPV, but pulmonary histopathological changes are evident and can be scored by a trained pathologist. The relatively low number of immunological reagents available for mechanistic studies in cotton rats has led many investigators to study immune responses to infection and vaccines in mice, for which there are obviously more strains and reagents. The BALB/c mouse is one of the more permissive strains, and has been used by many investigators because of its ready availability. Large animals also have been studied, especially cynomolgus macaques (*Macaca fascicularis*) and African green monkeys (*Cercopithecus aethiops*) (Kuiken *et al.*, 2004). These non-human primates shed virus with kinetics that are thought to be similar to those in children, with a peak on days 5–6 and resolution of shedding by about day 10 (van den Hoogen *et al.*, 2007). These animals mount a protective immune response following infection (van den Hoogen *et al.*, 2007). It is likely that cynomolgus macaques, African green monkeys and chimpanzees in captivity are infected by human strains from their handlers, since there is a high prevalence of seropositivity in these animals when screened in breeding facilities.

Immune responses

Human studies of HMPV immunity

Humoral immunity to HMPV in humans

Limited data exist describing mechanisms of immunity to HMPV in humans. In persons between the ages of six months and one year, the prevalence of seropositivity by immunofluorescent assay is 25%, increasing to 55% between one and two years, 70% between two and five years, and 100% over five years of age (van den Hoogen *et al.*, 2001). However, neutralizing titres are lower overall at 25% for 6–12 month olds, 31% for 1–2 years, 38% by 2–5 years, and 75–100% over 5 years of age. Thus, while all persons have been infected by age 5, a sizable proportion of those who are seropositive by immunofluorescent assay do not exhibit equivalent neutralizing titres. This study of HMPV seroepidemiology used a single virus isolate of the A1 subgroup; thus, it is possible that natural infection does not induce broadly cross-protective antibodies. Alternatively, neutralizing antibody titres may wane with increasing time since infection. One study using macaques showed waning immunity at 12 weeks and complete loss of protection at eight months following experimental infection (van den Hoogen *et al.*, 2007). Several other studies also have shown prevalent seropositivity, including one study that showed a rise in anti-HMPV ELISA titre following infection (Falsey *et al.*, 2003). One prospective study recruited adults and measured serum HMPV neutralizing titres in the fall and spring, while capturing acute HMPV infection by RT-PCR (Falsey *et al.*, 2005b). Mean serum binding and neutralizing antibody titres of HMPV-infected subjects were significantly lower compared with uninfected subjects. Seventy-one per cent of subjects with reciprocal $\log_2$ titres ≤ 10.5 were infected compared with 36% with reciprocal $\log_2$ titres >10.5. Further, this study found no difference in the mean acute antibody titres for patients with mild compared with severe illness. These limited data regarding protective antibody titres are provocative in light of the finding that reinfections with homotypic or heterotypic strains occur in both healthy and immunocompromised humans (Pelletier *et al.*, 2002; Ebihara *et al.*, 2004c; Williams *et al.*, 2004, 2006b; Madhi *et al.*, 2007). It is not known whether this phenomenon is due to limited cross-protective immunity or diminished responses to specific virus proteins.

Cellular immunity to HMPV in humans

Reports of severe and fatal infections in immunocompromised persons suggest that T cell immunity is important for virus clearance and resolution of illness (Pelletier *et al.*, 2002; Cane *et al.*, 2003; Larcher *et al.*, 2005; Martino *et al.*, 2005; Williams *et al.*, 2005b Englund *et al.*, 2006;). Major risk factors for severe HMPV disease in some of these reports were lymphopenia and cytotoxic therapy, further emphasizing a probable role for T cells in protection. It is not apparent whether HMPV illness is more severe in HIV-infected persons, leaving the role of CD4$^+$ T cells in HMPV immunity unclear (Madhi *et al.*, 2003, 2007; IJpma *et al.*, 2004; Garbino *et al.*, 2008). Putative HLA Class I-restricted CTL epitopes have been identified from HMPV-infected humans by ELISPOT (enzyme-linked immunosorbent spot) and chromium release assays (Herd *et al.*, 2008). However, the contribution of T cells during primary and secondary infection in humans is not known.

Animal studies of HMPV immunity

Cellular immunity to HMPV in mice

Few data exist regarding the role of T cells in HMPV immunity using animal models. Alvarez *et al.* infected BALB/c mice with HMPV and observed a biphasic pattern of virus shedding in the lungs, with an initial peak of 10^8 plaque forming units/gram tissue on day 7 post-infection, slight decline, and rebound to 10^7 plaque forming units/gram tissue on day 14 (Alvarez *et al.*, 2004). Virus remained detectable in the lungs of wild-type mice until day 60, despite a neutralizing antibody titre of 1:128, suggesting a role for T cells in controlling viral replication. Furthermore, mice depleted of T cells or NK cells by monoclonal antibody at day 28 or day 60 post infection had rebound of viral replication, with 10-fold higher titres of HMPV in lungs than non-depleted mice. However, other groups have failed to confirm this duration or biphasic pattern of shedding in mice (Kuiken *et al.*, 2004; MacPhail *et al.*, 2004; Darniot *et al.*, 2005; Guerrero-Plata *et al.*, 2005a,b; Hamelin *et al.*, 2005b, 2006a,b; Williams *et al.*, 2005d; Wyde *et al.*, 2005;Herd *et al.*, 2006; Huck *et al.*, 2007; Schildgen *et al.*, 2007; van den Hoogen *et al.*, 2007; Yim *et al.*, 2007; Kolli *et al.*, 2008). Thus, the finding in this model of viral persistence is interesting, but the meaning of these data is unclear.

Another group depleted CD4$^+$, CD8$^+$, or both subsets of T cells before HMPV inoculation (Kolli *et al.*, 2008). T cell-depleted animals exhibited diminished weight loss and lung histopathology. However, virus titres and lung pathology were evaluated only at one time point and virus titres in the lung were low in all groups including controls, making interpretation difficult. Secondary challenge was performed in continuously T cell-depleted mice and there were no differences in titre between any groups at the single time point examined, despite absence of antibody in the CD4$^+$ T cell-depleted mice. Challenge virus was inoculated 7 days after the last intraperitoneal injection of depleting monoclonal antibodies, and thus there may have been regeneration of T cell subsets. Therefore, these data, while intriguing, do not conclusively define the contribution of CD4$^+$ and CD8$^+$ T cells to protection or pathology.

Potential murine MHC class I-restricted epitopes for HMPV have been described (Herd *et al.*, 2006; Melendi *et al.*, 2007). One group used algorithms to identify candidate epitopes and tested predicted epitopes by peptide immunization (Herd *et al.*, 2006). They found that a few of these induced peptide-specific CTL responses against the same immunizing peptide and modest protection against virus challenge (reduction from 10^6 to 10^4 plaque forming units/gram tissue). However, the peptides were not screened against CTL induced by virus infection and thus it is not clear that these are true epitopes presented during natural infection. Melendi *et al.* used a similar bioinformatics approach to identify potential H-2^d-restricted epitopes (Melendi *et al.*, 2007). Predicted epitopes were screened as synthetic peptides in an ELISPOT assay using HMPV-immune splenocytes. These experiments identified H-2^d-restricted dominant M2$^{181-89}$ and subdominant N$^{307-315}$ epitopes. Primary T-cell lines generated by stimulation *ex vivo* with these peptides were CD8$^+$, IFN-γ secreting, and functional for lytic activity. Adoptive transfer of these cell lines into RAG$^{-/-}$ mice provided a modest

1.5-$\log_{10}$ reduction in virus titre. Lung pathology or signs of disease were not determined. These studies suggest that T cells contribute to protective immunity.

Innate immune responses to HMPV in humans and mice

Data on innate immune responses to HMPV are limited, but further suggest differences between HMPV and RSV. A Finnish study reported increased levels of interleukin (IL) 8 and decreased levels of RANTES (regulated upon activation, normal T cell expressed and secreted) in nasal secretions of HMPV-infected children compared with levels from RSV-infected children (Jartti *et al.*, 2002). A well-designed, prospective study found that infants infected with HMPV had significantly lower levels of IL-1β, IL-6, IL-8, IL-12, and TNF-α in nasal washes compared with RSV- or influenza virus-infected infants (Laham *et al.*, 2004). Experimental HMPV infection of human dendritic cells *ex vivo* also induced cytokine profiles that differed significantly from responses evoked by RSV (Guerrero-Plata *et al.*, 2006). BALB/c mice infected with HMPV produced significantly higher amounts of IFN-α compared with mice infected with RSV (Guerrero-Plata *et al.*, 2005a). Results of these studies are contradictory in some cases; nonetheless, they suggest that HMPV may induce more potent innate responses than RSV. These findings are interesting given the absence of known interferon-antagonist proteins encoded by HMPV. Two studies suggest that HMPV proteins possess immunomodulatory properties, but these have not been confirmed (Bao *et al.*, 2008a; Goutagny *et al*, 2010).

Clinical features

HMPV is associated with a variety of respiratory symptoms and diagnoses (Table 11.1). Children with HMPV infection typically present with upper respiratory symptoms such as rhinorrhoea, cough, and fever. Conjunctivitis, vomiting, diarrhoea, or rash are occasionally reported but are not prominent. The duration of symptoms prior to seeking medical attention is usually less than a week and limited data suggest that children may shed virus for 1–2 weeks (van den Hoogen *et al.*, 2003; Ebihara *et al.*, 2004a; Williams *et al.*, 2004). Only one study found evidence by RT-PCR of HMPV in patients' serum (Maggi *et al.*, 2003), suggesting that HMPV infection is limited to the respiratory tract. Studies in rodents and non-human primates have not detected HMPV in tissues outside the respiratory tract (Kuiken *et al.*, 2004; Alvarez *et al.*, 2004; Hamelin *et al.*, 2005b; Williams *et al.*, 2005d; Wyde *et al.*, 2005). The lower respiratory syndromes most frequently associated with HMPV are bronchiolitis, pneumonia, croup, and asthma exacerbation. Signs and symptoms of HMPV infection overlap sufficiently with those of other common respiratory viruses that reliable distinction cannot be made clinically. Reinfection with HMPV occurs, although repeated infections are more likely to be limited to the upper respiratory tract in otherwise healthy children (Williams *et al.*, 2004, 2006b). HMPV is associated with a substantial proportion of acute otitis media in children, and viral RNA has been detected in middle ear fluid (Schildgen and Simon, 2005; Suzuki *et al.*, 2005; Williams *et al.*, 2006a).

The link between HMPV infection and asthma is not clear. An Australian study of outpatient children with asthma did not find an association between HMPV and asthma exacerbations (Rawlinson *et al.*, 2003), while a large prospective study of outpatient children found a significant association between HMPV and the diagnosis of acute asthma exacerbation (Williams *et al.*, 2004). Studies of children hospitalized for wheezing and adults hospitalized for asthma exacerbations found a substantial number of these admissions were associated with HMPV infection (Williams *et al.*, 2005a,c). Measurements of asthma-associated cytokines in nasal washes of HMPV-infected infants have yielded conflicting results (Jartti *et al.*, 2002; Laham *et al.*, 2004). One retrospective study found a strong association between HMPV infection during infancy and a subsequent diagnosis of asthma (Garcia-Garcia *et al.*, 2007). All such studies are complicated by the difficulty of assigning the diagnosis of asthma during infancy, when acute wheezing is frequently associated with viral infections. However, the available data suggest an association between HMPV and asthma exacerbations.

Table 11.1 Presenting signs/symptoms and clinical diagnoses associated with human metapneumovirus infection of children

Sign/symptom	Bastien *et al.* (2003a)[1]	van den Hoogen *et al.* (2003)[1]	Esper *et al.* (2004)	Williams *et al.*, 2004)[2]	Williams *et al.* (2006b)[3]	Mullins *et al.* (2004)[1]	Dollner *et al.* (2004)[1]
Number of patients	12	25	53	49	118	26	50
Fever (%)	67	61	77	52	54	73	44
Rhinorrhoea (%)	92	80	64	88	82	77	90
Cough (%)	100	72	68	90	66	92	90
Wheezing (%)	83	24	51	52	NA	NR	56
Vomiting (%)	25	NR	NR	10	20	NR	36
Diarrhoea (%)	8	NR	NR	17	14	NR	14
Rash (%)	0	NR	NR	4	3	NR	2
Abnormal chest radiograph	NR	62	56	50	NA	85	67
Bronchiolitis (%)	67	NR	NR	59	NA	23	48
Pneumonia (%)	17	NR	NR	8	NA	23	34
Croup (%)	0	NR	NR	18	NA	NR	4
Asthma (%)	NR	NR	NR	14	NA	27	NR
Acute otitis media (%)	50	NR	NR	37	50	15	6

[1]Indicates studies that included only hospitalized children.

[2]Indicates study that included only children with lower respiratory tract illness.

[3]Indicates study included only children with upper respiratory tract illness.

NR, not reported. NA, not applicable.

HMPV is capable of causing severe infections in immunocompromised hosts, a phenomenon that has been well described for influenza virus, RSV and PIV. There are reports of fatal infection attributed to HMPV in cancer patients, and several studies suggest that HMPV is a relatively common cause of acute respiratory infection in children and adults with malignancy or haematopoietic stem cell transplants (Pelletier *et al.*, 2002; Cane *et al.*, 2003; Boeckh *et al.*, 2005; Franquet *et al.*, 2005; Williams *et al.*, 2005b). Further long-term prospective studies are needed to fully characterize disease due to HMPV in immunocompromised hosts.

Co-detection of other viruses with HMPV

All epidemiological studies of HMPV that have tested for other viruses have found co-detection rates of only 5–17%, usually with RSV, and most have not noted more severe disease in the patients with another virus detected (Jartti *et al.*, 2002; Bastien *et al.*, 2003b; Boivin *et al.*, 2003; Cane *et al.*, 2003; Cuevas *et al.*, 2003; Freymouth *et al.*, 2003; van den Hoogen *et al.*, 2003; Maggi *et al.*, 2003; Peiris *et al.*, 2003; Rawlinson *et al.*, 2003; Thanasugarn *et al.*, 2003; Viazov *et al.*, 2003; Dollner *et al.*, 2004; Ebihara *et al.*, 2004b; Esper *et al.*, 2004; Galiano *et al.*, 2004; Laham *et al.*, 2004; McAdam *et al.*, 2004; Mullins *et al.*, 2004; Williams *et al.*, 2004, 2006b; Bosis *et al.*, 2005). It is important to note that these studies used highly sensitive RT-PCR techniques that may detect viruses for several weeks after an acute infection. A few studies of hospitalized patients have noted much higher co-infection rates of 30–60% (Jartti *et al.*, 2002; Viazov *et al.*, 2003; Konig *et al.*, 2004;Garcia-Garcia *et al.*, 2006), raising the question whether HMPV infections are more severe if another virus is present. One British group addressed this question by using a nested RT-PCR assay to test bronchoalveolar lavage fluids from 30 intubated infants with RSV infection and detected HMPV in 21 of 30 (70%; Greensill *et al.*, 2003).

The authors subsequently used the same nested assay to test respiratory specimens from children admitted to the intensive care unit and the general wards. HMPV and RSV co-infection was detected in 18 of 25 (72%) intensive care patients and 15 of 171 (9%) general ward patients (Semple *et al.*, 2005). The authors concluded that dual infection with RSV and HMPV was associated with severe bronchiolitis. However, a Connecticut study of 46 inpatients with either mild or severe RSV disease found no co-infections with HMPV (Lazar *et al.*, 2004). A Dutch study also did not find any HMPV-RSV co-infections in children with severe RSV bronchiolitis (van Woensel *et al.*, 2006). Whether these conflicting findings are due to methodological differences or geographic variability in circulating virus strains is unknown. Further prospective studies are needed to clarify the nature of disease associated with co-detections; however, it is clear that the majority of HMPV infections are not associated with other viruses and that HMPV is a primary pathogen.

Epidemiology

HMPV is a major cause of acute respiratory tract disease in otherwise healthy infants and children worldwide, with a seasonal occurrence and spectrum of clinical illness similar to other common respiratory viruses (Boivin *et al.*, 2002, 2003; Jartti *et al.*, 2002; Bastien *et al.*, 2003b; Freymouth *et al.*, 2003; van den Hoogen *et al.*, 2003; Maggi *et al.*, 2003; Peiris *et al.*, 2003; Rawlinson *et al.*, 2003; Thanasugarn *et al.*, 2003; Viazov *et al.*, 2003; Dollner *et al.*, 2004; Esper *et al.*, 2004; McAdam *et al.*, 2004; Mullins *et al.*, 2004; Williams *et al.*, 2004, 2005c, 2006b; Principi *et al.*, 2006). Studies of children hospitalized for ARI in diverse regions have detected HMPV in 6% to 40% of ARI specimens. The average prevalence of HMPV is 5–10% overall in most paediatric studies of ARI, although it may be much higher during the peak months of HMPV circulation. Most studies have reported that HMPV is the second most common respiratory virus after RSV, with prevalence at least that of influenza virus and PIV (Jartti *et al.*, 2002; Boivin *et al.*, 2003; Cane *et al.*, 2003; Cuevas *et al.*, 2003; Freymouth *et al.*, 2003; Maggi *et al.*, 2003; Rawlinson *et al.*, 2003; Thanasugarn *et al.*, 2003; Viazov *et al.*, 2003; Dollner *et al.*, 2004; Ebihara *et al.*, 2004b; Esper *et al.*, 2004; Galiano *et al.*, 2004; Laham *et al.*, 2004; McAdam *et al.*, 2004; Mullins *et al.*, 2004; Williams *et al.*, 2004, 2006b; Bosis *et al.*, 2005).

HMPV infections occur in annual epidemics during the late winter-early spring months in temperate locations. The HMPV season overlaps considerably with those of RSV and influenza virus, but the peak incidence of HMPV is usually one to two months later than the peak of RSV. Prospective surveillance studies conducted over a 25-year period showed that HMPV was present in every season, though the prevalence varied from year to year (Williams *et al.*, 2004, 2006b). In most studies of patients with acute respiratory infection, the percentage of HMPV detection varies from 6% to 15%. HMPV infection in children is usually symptomatic; HMPV is rarely detected in nasal washes from asymptomatic patients (van den Hoogen *et al.*, 2003, 2004b; Osterhaus and Fouchier, 2003; Williams *et al.*, 2004, 2005c).

Hospitalization of children for HMPV infection occurs primarily in the first year of life, although a number of studies reported that the peak age of hospitalization for HMPV is from 6 to 12 months of age (Boivin *et al.*, 2003; van den Hoogen *et al.*, 2003; Maggi *et al.*, 2003; Peiris *et al.*, 2003; Thanasugarn *et al.*, 2003; Viazov *et al.*, 2003; Dollner *et al.*, 2004; Esper *et al.*, 2004; McAdam *et al.*, 2004; Mullins *et al.*, 2004; Bosis *et al.*, 2005; Klein *et al.*, 2006) and thus significantly later than the 2–3 month peak age of hospitalization for RSV. Whether this reflects a difference in the decline of maternal antibodies, later acquisition of HMPV infection or developmental airway physiology is not known. Males are at greater risk for lower respiratory tract illness with HMPV infection compared with females, similar to RSV and other common respiratory viruses (Jartti *et al.*, 2002; Boivin *et al.*, 2003; Cane *et al.*, 2003; Cuevas *et al.*, 2003; Freymouth *et al.*, 2003; Maggi *et al.*, 2003; Rawlinson *et al.*, 2003; Thanasugarn *et al.*, 2003; Viazov *et al.*, 2003; Dollner *et al.*, 2004; Ebihara *et al.*, 2004b; Esper *et al.*, 2004; Galiano *et al.*, 2004; Laham *et al.*, 2004; McAdam *et al.*, 2004; Mullins *et al.*, 2004; Bosis *et al.*, 2005).

HMPV may be more severe in patients with underlying medical conditions. Many HMPV studies are of hospitalized patients and thus subject to selection bias. Nonetheless, 30–85% of children hospitalized with HMPV have chronic conditions, such as asthma, chronic lung disease due to prematurity, congenital heart disease or cancer (Boivin *et al.*, 2003; Thanasugarn *et al.*, 2003; Dollner *et al.*, 2004; Esper *et al.*, 2004; van den Hoogen *et al.*, 2004b; McAdam *et al.*, 2004; Mullins *et al.*, 2004). Hospitalization of both adults and children for HMPV-associated lower respiratory tract illness is more likely in patients with underlying conditions such as asthma, chronic obstructive pulmonary disease, HIV infection, immune compromise, or prematurity (Jartti *et al.*, 2002; Pelletier *et al.*, 2002; Stockton *et al.*, 2002; Falsey *et al.*, 2003; Maggi *et al.*, 2003; Rawlinson *et al.*, 2003; Thanasugarn *et al.*, 2003; Viazov *et al.*, 2003; Dollner *et al.*, 2004; Esper *et al.*, 2004; Bosis *et al.*, 2005; Hamelin *et al.*, 2005a; Martino *et al.*, 2005; Williams *et al.*, 2005c). Fatal HMPV infections occur in cancer and transplant patients (Pelletier *et al.*, 2002; Cane *et al.*, 2003; Larcher *et al.*, 2005; Englund *et al.*, 2006). The basis for the increased severity of disease in these different groups is not known. For some, such as cancer patients and transplant recipients, extreme illness is probably related to poor control of viral replication.

HMPV is also a significant cause of acute respiratory disease in adults, particularly older adults >65 years of age, and adults with co-morbid illness such as chronic obstructive pulmonary disease, asthma, or cancer (Falsey *et al.*, 2003; Vicente *et al.*, 2004; Hamelin *et al.*, 2005a; Williams *et al.*, 2005a; Martinello *et al.*, 2006; Boivin *et al.*, 2007; van den Hoogen, 2007; Walsh *et al.*, 2008). HMPV infection is generally less frequent in otherwise healthy young adults than in children. In one prospective study, HMPV accounted for 11% of all hospitalizations for acute respiratory illnesses in older patients and 85% of these hospitalized adults had chronic heart or lung disease (Falsey *et al.*, 2003). Several studies have identified HMPV in hospitalized older adults at rates similar to those for influenza and RSV, highlighting the importance of HMPV in this vulnerable population (Falsey *et al.*, 2003; Walsh *et al.*, 2008).

Diagnosis

Traditional viral culture techniques are insensitive for the detection of HMPV. The virus grows slowly in a limited number of cell types and requires trypsin for replication in culture, conditions that are not used routinely in diagnostic virology laboratories (van den Hoogen *et al.*, 2001; Biacchesi *et al.*, 2004a; Deffrasnes *et al.*, 2005; Schickli *et al.*, 2005; Tollefson *et al.*, 2010). These are presumed to be the reasons why HMPV was not detected earlier. Even under ideal conditions, cytopathic effects produced by HMPV in cell culture are minimal and do not appear until several weeks of incubation or after repeated passages. There is no commercially available rapid antigen test for HMPV, although immunofluorescent assays have been reported and there are commercially available HMPV-specific monoclonal antibodies. In general, reliable diagnosis of HMPV currently depends on molecular techniques based on standard or real-time RT-PCR (Cote *et al.*, 2003; Mackay *et al.*, 2003; Maertzdorf *et al.*, 2004; Bellau-Pujol *et al.*, 2005; Kuypers *et al.*, 2005; Lopez-Huertas *et al.*, 2005; Gerna *et al.*, 2006).

Prevention and control

Vaccines

Several approaches to experimental HMPV vaccination of animals have been described. Live-attenuated reverse–engineered constructs with single gene deletions or HMPV F protein expressed in different virus backgrounds protect against viral replication in the lungs in mice, hamsters, or primates (Tang *et al.*, 2003, 2005; Biacchesi *et al.*, 2004b, 2005; Buchholz *et al.*, 2005, 2006; Skiadopoulos *et al.*, 2004, 2006). However, *in vitro* neutralization assays and heterotypic virus challenge show discrepant levels of cross-protection and cross-neutralization between subgroups (Skiadopoulos *et al.*, 2004; Tang *et al.*, 2005; Herfst *et al.*, 2007; van den Hoogen *et al.*, 2007). Limited data suggest that F protein immunization either as recombinant protein or as an alphavirus-vectored vaccine is protective and does not induce aberrant immune responses (Cseke *et al.*, 2007; Herfst *et al.*, 2007; Mok *et al.*, 2008). In contrast, formalin-inactivated (FI) HMPV, similar to FI-RSV vaccines, clearly is associated with enhanced disease in cotton rats and macaques (de Swart *et al.*, 2007; Yim *et al.*, 2007). Interestingly, in contrast to other human paramyxoviruses, the alternative glycoprotein G does not appear to contribute to antibody-mediated protection. Vaccine studies in mice using a chimeric PIV1 expressing HMPV G, alphavirus-vectored HMPV G, or recombinant HMPV G protein all failed to induce either neutralizing antibodies or protection (Skiadopoulos *et al.*, 2006; Mok *et al.*, 2008; Ryder *et al.*, 2010) A limitation of all these studies is the use of only one or two virus strains, despite the observation that field isolates demonstrate considerable diversity over time (van den Hoogen *et al.*, 2004a; Williams *et al.*, 2006b; de Graaf *et al.*, 2008b; Yang *et al.*, 2009). Nonetheless, several potential vaccine candidates are in development within nine years of the discovery of HMPV, a remarkable achievement.

Immunoprophylaxis

There are no licensed vaccines for HMPV or the related RSV. However, immunoprophylaxis for RSV with the monoclonal antibody palivizumab is safe, moderately effective, and recommended for high-risk infants (Pollack and Groothuis, 2002). Murine and human monoclonal antibodies directed against HMPV F have shown prophylactic and therapeutic benefit in animal models (Ulbrandt *et al.*, 2006; Williams *et al.*, 2007; Hamelin *et al.*, 2008).

Treatment

The majority of children infected with HMPV can be managed with supportive care. For infants and children who require hospitalization, the primary therapies are supplementary oxygen and intravenous hydration. Bronchodilators and corticosteroids have been used empirically, but there are no controlled trials of these medications for HMPV and no data to support or refute efficacy. One animal study suggested benefit with bronchodilator and corticosteroid treatment of experimental HMPV infection in cotton rats (Hamelin *et al.*, 2006a). The only currently licensed antiviral for the related RSV is ribavirin, a nucleoside analogue administered by aerosol. Both ribavirin and polyclonal human immunoglobulin exhibit *in vitro* neutralizing activity against HMPV equivalent to their activity against RSV (Wyde *et al.*, 2003). There are no published animal or human data for these interventions, although they may be worthy of consideration in severely immunocompromised hosts with lower respiratory tract illness due to HMPV. Ribavirin has been used in severely immunocompromised patients such as haematopoietic stem cell transplant recipients, often in conjunction with RSV-specific immunoglobulin, with some evidence for efficacy (DeVincenzo *et al.*, 2000).

Perspectives

Human *Metapneumovirus* is a leading cause of respiratory disease worldwide. All humans appear to experience primary HMPV infection during infancy and early childhood, and reinfections are common. Clinical manifestations of HMPV infection are similar to other respiratory viruses and include bronchiolitis, croup, pneumonia, and asthma exacerbation. Hospitalizations occur in otherwise healthy children, but are more common in children and adults with underlying conditions such as prematurity, asthma, immune compromise and cardiopulmonary disease. Numerous studies have defined much of the epidemiology of this virus, but little is known regarding pathogenesis and immunity. Despite the recent discovery of HMPV, candidate vaccines are already in development.

References

Abiko, C., Mizuta, K., Itagaki, T., Katsushima, N., Ito, S., Matsuzaki, Y., Okamoto, M., Nishimura, H., Aoki, Y., Murata, T., *et al.* (2007). Outbreak of human *Metapneumovirus* detected by use of the Vero E6 cell line in isolates collected in Yamagata, Japan, in 2004 and 2005. J. Clin. Microbiol. *45*, 1912–1919.

Alvarez, R., and Seal, B.S. (2005). Identification of a truncated nucleoprotein in avian *Metapneumovirus*-infected cells encoded by a second AUG, in-frame to the full-length gene. Virol. J. *2*, 31.

Alvarez, R., Harrod, K.S., Shieh, W.J., Zaki, S., and Tripp, R.A. (2004). Human *Metapneumovirus* persists in BALB/c mice despite the presence of neutralizing antibodies. J. Virol. *78*, 14003–14011.

Arnold, J.C., Singh, K.K., Milder, E., Spector, S.A., Sawyer, M.H., Gavali, S., and Glaser, C. (2009). Human *Metapneumovirus* associated with central nervous system infection in children. Pediatr. Infect. Dis. J. *28,* 1057–1060.

Aung, Y.H., Liman, M., Neumann, U., and Rautenschlein, S. (2008). Reproducibility of swollen sinuses in broilers by experimental infection with avian *Metapneumovirus* subtypes A and B of turkey origin and their comparative pathogenesis. Avian Pathol. *37,* 65–74.

Bao, X., Liu, T., Shan, Y., Li, K., Garofalo, R.P., and Casola, A. (2008a). Human *Metapneumovirus* glycoprotein G inhibits innate immune responses. PLoS Pathog. *4,* e1000077.

Bao, X., Sinha, M., Liu, T., Hong, C., Luxon, B.A., Garofalo, R.P., and Casola, A. (2008b). Identification of human *Metapneumovirus*-induced gene networks in airway epithelial cells by microarray analysis. Virology *374,* 114–127.

Bastien, N., Normand, S., Taylor, T., Ward, D., Peret, T.C., Boivin, G., Anderson, L.J., and Li, Y. (2003a). Sequence analysis of the N, P, M and F genes of Canadian human *Metapneumovirus* strains. Virus Res. *93,* 51–62.

Bastien, N., Ward, D., Van Caeseele, P., Brandt, K., Lee, S.H., McNabb, G., Klisko, B., Chan, E., and Li, Y. (2003b). Human *Metapneumovirus* infection in the Canadian population. J. Clin. Microbiol. *41,* 4642–4646.

Bastien, N., Liu, L., Ward, D., Taylor, T., and Li, Y. (2004). Genetic variability of the G glycoprotein gene of human *Metapneumovirus.* J. Clin. Microbiol. *42,* 3532–3537.

Bellau-Pujol, S., Vabret, A., Legrand, L., Dina, J., Gouarin, S., Petitjean-Lecherbonnier, J., Pozzetto, B., Ginevra, C., and Freymuth, F. (2005). Development of three multiplex RT-PCR assays for the detection of 12 respiratory RNA viruses. J. Virol. Methods *126,* 53–63.

Bennett, R.S., LaRue, R., Shaw, D., Yu, Q., Nagaraja, K.V., Halvorson, D.A., and Njenga, M.K. (2005). A wild goose *Metapneumovirus* containing a large attachment glycoprotein is avirulent but immunoprotective in domestic turkeys. J. Virol. *79,* 14834–14842.

Biacchesi, S., Skiadopoulos, M.H., Boivin, G., Hanson, C.T., Murphy, B.R., Collins, P.L., and Buchholz, U.J. (2003). Genetic diversity between human *Metapneumovirus* subgroups. Virology *315,* 1–9.

Biacchesi, S., Skiadopoulos, M.H., Tran, K.C., Murphy, B.R., Collins, P.L., and Buchholz, U.J. (2004a). Recovery of human *Metapneumovirus* from cDNA: optimization of growth *in vitro* and expression of additional genes. Virology *321,* 247–259.

Biacchesi, S., Skiadopoulos, M.H., Yang, L., Lamirande, E.W., Tran, K.C., Murphy, B.R., Collins, P.L., and Buchholz, U.J. (2004b). Recombinant human *Metapneumovirus* lacking the small hydrophobic SH and/or attachment G glycoprotein: deletion of G yields a promising vaccine candidate. J. Virol. *78,* 12877–12887.

Biacchesi, S., Pham, Q.N., Skiadopoulos, M.H., Murphy, B.R., Collins, P.L., and Buchholz, U.J. (2005). Infection of nonhuman primates with recombinant human *Metapneumovirus* lacking the SH, G, or M2-2 protein categorizes each as a nonessential accessory protein and identifies vaccine candidates. J. Virol. *79,* 12608–12613.

Biacchesi, S., Pham, Q.N., Skiadopoulos, M.H., Murphy, B.R., Collins, P.L., and Buchholz, U.J. (2006). Modification of the trypsin-dependent cleavage activation site of the human *Metapneumovirus* fusion protein to be trypsin independent does not increase replication or spread in rodents or nonhuman primates. J. Virol. *80,* 5798–5806.

Biacchesi, S., Murphy, B.R., Collins, P.L., and Buchholz, U.J. (2007). Frequent frameshift and point mutations in the SH gene of human *Metapneumovirus* passaged *in vitro*. J. Virol. *81,* 6057–6067.

Boeckh, M., Erard, V., Zerr, D., and Englund, J. (2005). Emerging viral infections after hematopoietic cell transplantation. Pediatr. Transplant *9* (*Suppl.* 7), 48–54.

Boivin, G., Abed, Y., Pelletier, G., Ruel, L., Moisan, D., Cote, S., Peret, T.C., Erdman, D.D., and Anderson, L.J. (2002). Virological features and clinical manifestations associated with human *Metapneumovirus*: a new paramyxovirus responsible for acute respiratory-tract infections in all age groups. J. Infect. Dis. *186,* 1330–1334.

Boivin, G., De Serres, G., Cote, S., Gilca, R., Abed, Y., Rochette, L., Bergeron, M.G., and Dery, P. (2003). Human *Metapneumovirus* infections in hospitalized children. Emerg. Infect. Dis. *9,* 634–640.

Boivin, G., De Serres, G., Hamelin, M.E., Cote, S., Argouin, M., Tremblay, G., Maranda-Aubut, R., Sauvageau, C., Ouakki, M., Boulianne, N., *et al.* (2007). An outbreak of severe respiratory tract infection due to human *Metapneumovirus* in a long-term care facility. Clin. Infect. Dis. *44,* 1152–1158.

Bosis, S., Esposito, S., Niesters, H.G., Crovari, P., Osterhaus, A.D., and Principi, N. (2005). Impact of human *Metapneumovirus* in childhood: comparison with respiratory syncytial virus and influenza viruses. J. Med. Virol. *75,* 101–104.

Bosis, S., Esposito, S., Osterhaus, A.D., Tremolati, E., Begliatti, E., Tagliabue, C., Corti, F., Principi, N., and Niesters, H.G. (2008). Association between high nasopharyngeal viral load and disease severity in children with human *Metapneumovirus* infection. J. Clin. Virol. *42,* 286–290.

Brown, P.A., Bonci, M., Ricchizzi, E., Jones, R.C., and Naylor, C.J. (2009). Identification of two regions within the subtype A avian *Metapneumovirus* fusion protein (amino acids 211–310 and 336–479) recognized by neutralizing antibodies. Virus Res. *146,* 13–18.

Buchholz, U.J., Finke, S., and Conzelmann, K.K. (1999). Generation of bovine respiratory syncytial virus (BRSV) from cDNA: BRSV NS2 is not essential for virus replication in tissue culture, and the human RSV leader region acts as a functional BRSV genome promoter. J. Virol. *73,* 251–259.

Buchholz, U.J., Biacchesi, S., Pham, Q.N., Tran, K.C., Yang, L., Luongo, C.L., Skiadopoulos, M.H., Murphy, B.R., and Collins, P.L. (2005). Deletion of M2 gene open reading frames 1 and 2 of human *Metapneumovirus*: effects on RNA synthesis, attenuation, and immunogenicity. J. Virol. *79,* 6588–6597.

Buchholz, U.J., Nagashima, K., Murphy, B.R., and Collins, P.L. (2006). Live vaccines for human *Metapneumovirus* designed by reverse genetics. Expert Rev Vaccines *5,* 695–706.

Cane, P.A., van den Hoogen, B.G., Chakrabarti, S., Fegan, C.D., and Osterhaus, A.D. (2003). Human *Metapneumovirus* in a haematopoietic stem cell transplant recipient with fatal lower respiratory tract disease. Bone Marrow Transplant *31,* 309–310.

Catelli, E., Lupini, C., Cecchinato, M., Ricchizzi, E., Brown, P., and Naylor, C.J. (2010). Field avian *Metapneumovirus* evolution avoiding vaccine induced immunity. Vaccine *28,* 916–921.

Cecchinato, M., Catelli, E., Lupini, C., Ricchizzi, E., Clubbe, J., Battilani, M., and Naylor, C.J. (2010). Avian *Metapneumovirus* (AMPV) attachment protein involvement in probable virus evolution concurrent with mass live vaccine introduction. Vet. Microbiol. *146,* 24–34.

Coswig, L.T., dos Santos, M.B., Hafez, H.M., Ferreira, H.L., and Arns, C.W. (2010). Propagation of avian *Metapneumovirus* subtypes A and B using chicken embryo related and other cell systems. J. Virol. Methods *167,* 1–4.

Cote, S., Abed, Y., and Boivin, G. (2003). Comparative evaluation of real-time PCR assays for detection of the human *Metapneumovirus*. J. Clin. Microbiol. *41,* 3631–3635.

Cseke, G., Wright, D.W., Tollefson, S.J., Johnson, J.E., Crowe, J.E., Jr., and Williams, J.V. (2007). Human *Metapneumovirus* fusion protein vaccines that are immunogenic and protective in cotton rats. J. Virol. *81,* 698–707.

Cseke, G., Maginnis, M.S., Cox, R.G., Tollefson, S.J., Podsiad, A.B., Wright, D.W., Dermody, T.S., and Williams, J.V. (2009). Integrin alphavbeta1 promotes infection by human *Metapneumovirus*. Proc. Natl. Acad. Sci. U.S.A. *106,* 1566–1571.

Cuevas, L.E., Nasser, A.M., Dove, W., Gurgel, R.Q., Greensill, J., and Hart, C.A. (2003). Human *Metapneumovirus* and respiratory syncytial virus, Brazil. Emerg. Infect. Dis. *9,* 1626–1628.

Darniot, M., Petrella, T., Aho, S., Pothier, P., and Manoha, C. (2005). Immune response and alteration of pulmonary function after primary human *Metapneumovirus* (hMPV) infection of BALB/c mice. Vaccine *23,* 4473–4480.

Debiaggi, M., Canducci, F., Sampaolo, M., Marinozzi, M.C., Parea, M., Terulla, C., Colombo, A.A., Alessandrino, E.P., Bragotti, L.Z., Arghittu, M., *et al.* (2006). Persistent symptomless human *Metapneumovirus* infection in hematopoietic stem cell transplant recipients. J. Infect. Dis. *194,* 474–478.

Debiaggi, M., Canducci, F., Terulla, C., Sampaolo, M., Marinozzi, M.C., Alessandrino, P.E., Colombo, A.A., Caldera, D., Bragotti, L.Z., Migliavacca, R., *et al.* (2007). Long-term study on symptomless human *Metapneumovirus* infection in hematopoietic stem cell transplant recipients. New Microbiol. *30,* 255–258.

Deffrasnes, C., Cote, S., and Boivin, G. (2005). Analysis of replication kinetics of the human *Metapneumovirus* in different cell lines by real-time PCR. J. Clin. Microbiol. *43,* 488–490.

Deffrasnes, C., Hamelin, M.E., Prince, G.A., and Boivin, G. (2008). Identification and evaluation of a highly effective fusion inhibitor for human *Metapneumovirus*. Antimicrob. Agents Chemother. *52,* 279–287.

Derdowski, A., Peters, T.R., Glover, N., Qian, R., Utley, T.J., Burnett, A., Williams, J.V., Spearman, P., and Crowe, J.E., Jr. (2008). Human *Metapneumovirus* nucleoprotein and phosphoprotein interact and provide the minimal requirements for inclusion body formation. J. Gen. Virol. *89,* 2698–2708.

DeVincenzo, J.P., Hirsch, R.L., Fuentes, R.J., and Top, F.H., Jr. (2000). Respiratory syncytial virus immune globulin treatment of lower respiratory tract infection in pediatric patients undergoing bone marrow transplantation – a compassionate use experience. Bone Marrow Transplant *25,* 161–165.

Dollner, H., Risnes, K., Radtke, A., and Nordbo, S.A. (2004). Outbreak of human *Metapneumovirus* infection in Norwegian children. Pediatr. Infect. Dis. J. *23,* 436–440.

Ebihara, T., Endo, R., Ishiguro, N., Nakayama, T., Sawada, H., and Kikuta, H. (2004a). Early reinfection with human *Metapneumovirus* in an infant. J. Clin. Microbiol. *42,* 5944–5946.

Ebihara, T., Endo, R., Kikuta, H., Ishiguro, N., Ishiko, H., Hara, M., Takahashi, Y., and Kobayashi, K. (2004b). Human *Metapneumovirus* infection in Japanese children. J. Clin. Microbiol. *42,* 126–132.

Ebihara, T., Endo, R., Kikuta, H., Ishiguro, N., Ishiko, H., and Kobayashi, K. (2004c). Comparison of the seroprevalence of human *Metapneumovirus* and human respiratory syncytial virus. J. Med. Virol. *72,* 304–306.

Englund, J.A., Boeckh, M., Kuypers, J., Nichols, W.G., Hackman, R.C., Morrow, R.A., Fredricks, D.N., and Corey, L. (2006). Brief communication: fatal human *Metapneumovirus* infection in stem-cell transplant recipients. Ann. Intern. Med. *144,* 344–349.

Esper, F., Martinello, R.A., Boucher, D., Weibel, C., Ferguson, D., Landry, M.L., and Kahn, J.S. (2004). A 1-year experience with human *Metapneumovirus* in children aged <5 years. J. Infect. Dis. *189,* 1388–1396.

Falsey, A.R. (2008). Human *Metapneumovirus* infection in adults. Pediatr. Infect. Dis. J. *27,* S80–83.

Falsey, A.R., Erdman, D., Anderson, L.J., and Walsh, E.E. (2003). Human *Metapneumovirus* infections in young and elderly adults. J. Infect. Dis. *187,* 785–790.

Falsey, A.R., Formica, M.A., Hennessey, P.A., Criddle, M.M., Sullender, W.M., and Walsh, E.E. (2005a). Detection of respiratory syncytial virus in adults with chronic obstructive pulmonary disease. Am. J. Respir. Crit. Care Med. *173,* 639–634.

Falsey, A.R., Hennessey, P.A., Formica, M.A., Cox, C., and Walsh, E.E. (2005b). Respiratory syncytial virus infection in elderly and high-risk adults. N. Engl. J. Med. *352,* 1749–1759.

Falsey, A.R., Criddle, M.C., and Walsh, E.E. (2006). Detection of respiratory syncytial virus and human *Metapneumovirus* by reverse transcription polymerase chain reaction in adults with and without respiratory illness. J. Clin. Virol. *35,* 46–50.

Franquet, T., Rodriguez, S., Martino, R., Salinas, T., Gimenez, A., and Hidalgo, A. (2005). Human *Metapneumovirus* infection in hematopoietic stem cell transplant recipients: high-resolution computed tomography findings. J. Comput. Assist. Tomogr. *29,* 223–227.

Freymouth, F., Vabret, A., Legrand, L., Eterradossi, N., Lafay-Delaire, F., Brouard, J., and Guillois, B. (2003). Presence of the new human *Metapneumovirus* in French children with bronchiolitis. Pediatr. Infect. Dis. J. *22,* 92–94.

Galiano, M., Videla, C., Puch, S.S., Martinez, A., Echavarria, M., and Carballal, G. (2004). Evidence of human *Metapneumovirus* in children in Argentina. J. Med. Virol. *72,* 299–303.

Garbino, J., Inoubli, S., Mossdorf, E., Weber, R., Tamm, M., Soccal, P., Aubert, J.D., Bridevaux, P.O., Tapparel, C., and Kaiser, L. (2008). Respiratory viruses in HIV-infected patients with suspected respiratory opportunistic infection. AIDS *22,* 701–705.

Garcia-Garcia, M.L., Calvo, C., Perez-Brena, P., De Cea, J.M., Acosta, B., and Casas, I. (2006). Prevalence and clinical characteristics of human *Metapneumovirus* infections in hospitalized infants in Spain. Pediatr. Pulmonol. *41,* 863–871.

Garcia-Garcia, M.L., Calvo, C., Casas, I., Bracamonte, T., Rellan, A., Gozalo, F., Tenorio, T., and Perez-Brena, P. (2007). Human *Metapneumovirus* bronchiolitis in infancy is an important risk factor for asthma at age 5. Pediatr. Pulmonol. *42,* 458–464.

Gerna, G., Sarasini, A., Percivalle, E., Genini, E., Campanini, G., and Grazia Revello, M. (2006). Simultaneous detection and typing of human *Metapneumovirus* strains in nasopharyngeal secretions and cell cultures by monoclonal antibodies. J. Clin. Virol. *35,* 113–116.

Goutagny, N., Jiang, Z., Tian, J., Parroche, P., Schickli, J., Monks, B.G., Ulbrandt, N., Ji, H., Kiener, P.A., Coyle, A.J., *et al.* (2010). Cell type-specific recognition of human metapneumoviruses (HMPVs) by retinoic acid-inducible gene I (RIG-I) and TLR7 and viral interference of RIG-I ligand recognition by HMPV-B1 phosphoprotein. J. Immunol. *184,* 1168–1179.

Govindarajan, D., and Samal, S.K. (2004). Sequence analysis of the large polymerase (L) protein of the US strain of avian *Metapneumovirus* indicates a close resemblance to that of the human *Metapneumovirus.* Virus Res. *105,* 59–66.

Govindarajan, D., and Samal, S.K. (2005). Analysis of the complete genome sequence of avian *Metapneumovirus* subgroup C indicates that it possesses the longest genome among metapneumoviruses. Virus Genes *30,* 331–333.

Govindarajan, D., Yunus, A.S., and Samal, S.K. (2004). Complete sequence of the G glycoprotein gene of avian *Metapneumovirus* subgroup C and identification of a divergent domain in the predicted protein. J. Gen. Virol. *85,* 3671–3675.

Govindarajan, D., Buchholz, U.J., and Samal, S.K. (2006). Recovery of avian *Metapneumovirus* subgroup C from cDNA: cross-recognition of avian and human *Metapneumovirus* support proteins. J. Virol. *80,* 5790–5797.

Govindarajan, D., Kim, S.H., and Samal, S.K. (2010). Contribution of the attachment G glycoprotein to pathogenicity and immunogenicity of avian *Metapneumovirus* subgroup C. Avian Dis. *54,* 59–66.

de Graaf, M., Herfst, S., Schrauwen, E.J., Choi, Y., van den Hoogen, B.G., Osterhaus, A.D., and Fouchier, R.A. (2008a). Specificity and functional interaction of the polymerase complex proteins of human and avian metapneumoviruses. J. Gen. Virol. *89,* 975–983.

de Graaf, M., Osterhaus, A.D., Fouchier, R.A., and Holmes, E.C. (2008b). Evolutionary dynamics of human and avian metapneumoviruses. J. Gen. Virol. *89,* 2933–2942.

de Graaf, M., Schrauwen, E.J., Herfst, S., van Amerongen, G., Osterhaus, A.D., and Fouchier, R.A. (2009). Fusion protein is the main determinant of *Metapneumovirus* host tropism. J. Gen. Virol. *90,* 1408–1416.

Greensill, J., McNamara, P.S., Dove, W., Flanagan, B., Smyth, R.L., and Hart, C.A. (2003). Human *Metapneumovirus* in severe respiratory syncytial virus bronchiolitis. Emerg. Infect. Dis. *9,* 372–375.

Guerrero-Plata, A., Baron, S., Poast, J.S., Adegboyega, P.A., Casola, A., and Garofalo, R.P. (2005a). Activity and regulation of alpha interferon in respiratory syncytial virus and human *Metapneumovirus* experimental infections. J. Virol. *79,* 10190–10199.

Guerrero-Plata, A., Casola, A., and Garofalo, R.P. (2005b). Human *Metapneumovirus* induces a profile of lung cytokines distinct from that of respiratory syncytial virus. J. Virol. *79,* 14992–14997.

Guerrero-Plata, A., Casola, A., Suarez, G., Yu, X., Spetch, L., Peeples, M.E., and Garofalo, R.P. (2006). Differential response of dendritic cells to human *Metapneumovirus* and respiratory syncytial virus. Am. J. Respir. Cell Mol. Biol. *34,* 320–329.

Hamelin, M.E., Cote, S., Laforge, J., Lampron, N., Bourbeau, J., Weiss, K., Gilca, R., DeSerres, G., and Boivin, G. (2005a). Human *Metapneumovirus* infection in adults with community-acquired pneumonia and exacerbation of chronic obstructive pulmonary disease. Clin. Infect. Dis. *41,* 498–502.

Hamelin, M.E., Yim, K., Kuhn, K.H., Cragin, R.P., Boukhvalova, M., Blanco, J.C., Prince, G.A., and Boivin, G. (2005b). Pathogenesis of human *Metapneumovirus* lung infection in BALB/c mice and cotton rats. J. Virol. *79,* 8894–8903.

Hamelin, M.E., Prince, G.A., and Boivin, G. (2006a). Effect of ribavirin and glucocorticoid treatment in a mouse model of human *Metapneumovirus* infection. Antimicrob Agents Chemother *50,* 774–777.

Hamelin, M.E., Prince, G.A., Gomez, A.M., Kinkead, R., and Boivin, G. (2006b). Human *Metapneumovirus* infection induces long-term pulmonary inflammation associated with airway obstruction and hyperresponsiveness in mice. J. Infect. Dis. *193,* 1634–1642.

Hamelin, M.E., Couture, C., Sackett, M., Kiener, P., Suzich, J., Ulbrandt, N., and Boivin, G. (2008). The prophylactic administration of a monoclonal antibody against human *Metapneumovirus* attenuates viral disease and airways hyperresponsiveness in mice. Antivir. Ther. *13,* 39–46.

Herd, K.A., Mahalingam, S., Mackay, I.M., Nissen, M., Sloots, T.P., and Tindle, R.W. (2006). Cytotoxic T-lymphocyte epitope vaccination protects against human *Metapneumovirus* infection and disease in mice. J. Virol. *80,* 2034–2044.

Herd, K.A., Nissen, M.D., Hopkins, P.M., Sloots, T.P., and Tindle, R.W. (2008). Major histocompatibility complex class I cytotoxic T lymphocyte immunity to human *Metapneumovirus* (hMPV) in individuals with previous hMPV infection and respiratory disease. J. Infect. Dis. *197,* 584–592.

Herfst, S., de Graaf, M., Schickli, J.H., Tang, R.S., Kaur, J., Yang, C.F., Spaete, R.R., Haller, A.A., van den Hoogen, B.G., Osterhaus, A.D., *et al.* (2004). Recovery of human *Metapneumovirus* genetic lineages a and B from cloned cDNA. J. Virol. *78,* 8264–8270.

Herfst, S., de Graaf, M., Schrauwen, E.J., Ulbrandt, N.D., Barnes, A.S., Senthil, K., Osterhaus, A.D., Fouchier, R.A., and van den Hoogen, B.G. (2007). Immunization of Syrian golden hamsters with F subunit vaccine of human *Metapneumovirus* induces protection against challenge with homologous or heterologous strains. J. Gen. Virol. *88,* 2702–2709.

van den Hoogen, B.G. (2007). Respiratory tract infection due to human *Metapneumovirus* among elderly patients. Clin. Infect. Dis. *44,* 1159–1160.

van den Hoogen, B.G., de Jong, J.C., Groen, J., Kuiken, T., de Groot, R., Fouchier, R.A., and Osterhaus, A.D. (2001). A newly discovered human *Pneumovirus* isolated from young children with respiratory tract disease. Nat. Med. *7,* 719–724.

van den Hoogen, B.G., Bestebroer, T.M., Osterhaus, A.D., and Fouchier, R.A. (2002). Analysis of the genomic sequence of a human *Metapneumovirus.* Virology *295,* 119–132.

van den Hoogen, B.G., van Doornum, G.J., Fockens, J.C., Cornelissen, J.J., Beyer, W.E., de Groot, R., Osterhaus, A.D., and Fouchier, R.A. (2003). Prevalence and clinical symptoms of human *Metapneumovirus* infection in hospitalized patients. J. Infect. Dis. *188,* 1571–1577.

van den Hoogen, B.G., Herfst, S., Sprong, L., Cane, P.A., Forleo-Neto, E., de Swart, R.L., Osterhaus, A.D., and Fouchier, R.A. (2004a). Antigenic and genetic variability of human metapneumoviruses. Emerg. Infect. Dis. *10,* 658–666.

van den Hoogen, B.G., Osterhaus, D.M., and Fouchier, R.A. (2004b). Clinical impact and diagnosis of human *Metapneumovirus* infection. Pediatr. Infect. Dis. J. *23,* S25–32.

van den Hoogen, B.G., Herfst, S., de Graaf, M., Sprong, L., van Lavieren, R., van Amerongen, G., Yuksel, S., Fouchier, R.A., Osterhaus, A.D., and de Swart, R.L. (2007). Experimental infection of macaques with human *Metapneumovirus* induces transient protective immunity. J. Gen. Virol. *88,* 1251–1259.

Huck, B., Neumann-Haefelin, D., Schmitt-Graeff, A., Weckmann, M., Mattes, J., Ehl, S., and Falcone, V. (2007). Human *Metapneumovirus* induces more severe disease and stronger innate immune response in BALB/c mice as compared with respiratory syncytial virus. Respir. Res. *8,* 6.

IJpma, F.F., Beekhuis, D., Cotton, M.F., Pieper, C.H., Kimpen, J.L., van den Hoogen, B.G., van Doornum, G.J., and Osterhaus, D.M. (2004). Human *Metapneumovirus* infection in hospital referred South African children. J. Med. Virol. *73,* 486–493.

Jartti, T., van den Hoogen, B., Garofalo, R.P., Osterhaus, A.D., and Ruuskanen, O. (2002). *Metapneumovirus* and acute wheezing in children. Lancet *360,* 1393–1394.

Kitagawa, Y., Zhou, M., Yamaguchi, M., Komatsu, T., Takeuchi, K., Itoh, M., and Gotoh, B. (2010). Human *Metapneumovirus* M2-2 protein inhibits viral transcription and replication. Microbes Infect. *12,* 135–145.

Klein, M.I., Coviello, S., Bauer, G., Benitez, A., Serra, M.E., Schiatti, M.P., Delgado, M.F., Melendi, G.A., Novalli, L., Pena, H.G., *et al.* (2006). The impact of infection with human *Metapneumovirus* and other respiratory viruses in young infants and children at high risk for severe pulmonary disease. J. Infect. Dis. *193,* 1544–1551.

Kolli, D., Bataki, E.L., Spetch, L., Guerrero-Plata, A., Jewell, A.M., Piedra, P.A., Milligan, G.N., Garofalo, R.P., and Casola, A. (2008). T lymphocytes contribute to antiviral immunity and pathogenesis in experimental human *Metapneumovirus* infection. J. Virol. *82,* 8560–8569.

Konig, B., Konig, W., Arnold, R., Werchau, H., Ihorst, G., and Forster, J. (2004). Prospective study of human *Metapneumovirus* infection in children less than 3 years of age. J. Clin. Microbiol. *42,* 4632–4635.

Kuiken, T., van den Hoogen, B.G., van Riel, D.A., Laman, J.D., van Amerongen, G., Sprong, L., Fouchier, R.A., and Osterhaus, A.D. (2004). Experimental human *Metapneumovirus* infection of cynomolgus macaques (Macaca fascicularis) results in virus replication in ciliated epithelial cells and pneumocytes with associated lesions throughout the respiratory tract. Am. J. Pathol. *164,* 1893–1900.

Kuypers, J., Wright, N., Corey, L., and Morrow, R. (2005). Detection and quantification of human *Metapneumovirus* in pediatric specimens by real-time RT-PCR. J. Clin. Virol. *33,* 299–305.

Laham, F.R., Israele, V., Casellas, J.M., Garcia, A.M., Lac Prugent, C.M., Hoffman, S.J., Hauer, D., Thumar, B., Name, M.I., Pascual, A., *et al.* (2004). Differential production of inflammatory cytokines in primary infection with human *Metapneumovirus* and with other common respiratory viruses of infancy. J. Infect. Dis. *189,* 2047–2056.

Larcher, C., Geltner, C., Fischer, H., Nachbaur, D., Muller, L.C., and Huemer, H.P. (2005). Human

Metapneumovirus infection in lung transplant recipients: clinical presentation and epidemiology. J. Heart Lung Transplant *24,* 1891–1901.

Lazar, I., Weibel, C., Dziura, J., Ferguson, D., Landry, M.L., and Kahn, J.S. (2004). Human *Metapneumovirus* and severity of respiratory syncytial virus disease. Emerg. Infect. Dis. *10,* 1318–1320.

Le Nouen, C., Munir, S., Losq, S., Winter, C.C., McCarty, T., Stephany, D.A., Holmes, K.L., Bukreyev, A., Rabin, R.L., Collins, P.L., *et al.* (2009). Infection and maturation of monocyte-derived human dendritic cells by human respiratory syncytial virus, human *Metapneumovirus,* and human parainfluenza virus type 3. Virology *385,* 169–182.

Lopez-Huertas, M.R., Casas, I., Acosta-Herrera, B., Garcia, M.L., Coiras, M.T., and Perez-Brena, P. (2005). Two RT-PCR based assays to detect human *Metapneumovirus* in nasopharyngeal aspirates. J. Virol. Methods *129,* 1–7.

Luo, L., Nishi, K., Macleod, E., Sabara, M.I., and Li, Y. (2010). Analysis of expression and glycosylation of avian *Metapneumovirus* attachment glycoprotein from recombinant baculoviruses. Virus Res. *153,* 244–249.

Ma, X., Endo, R., Ebihara, T., Ishiguro, N., Ishiko, H., and Kikuta, H. (2005). Production and characterization of neutralizing monoclonal antibodies against human *Metapneumovirus* F protein. Hybridoma (Larchmt) *24,* 201–205.

McAdam, A.J., Hasenbein, M.E., Feldman, H.A., Cole, S.E., Offermann, J.T., Riley, A.M., and Lieu, T.A. (2004). Human *Metapneumovirus* in children tested at a tertiary-care hospital. J. Infect. Dis. *190,* 20–26.

Mackay, I.M., Jacob, K.C., Woolhouse, D., Waller, K., Syrmis, M.W., Whiley, D.M., Siebert, D.J., Nissen, M., and Sloots, T.P. (2003). Molecular assays for detection of human *Metapneumovirus.* J. Clin. Microbiol. *41,* 100–105.

MacPhail, M., Schickli, J.H., Tang, R.S., Kaur, J., Robinson, C., Fouchier, R.A., Osterhaus, A.D., Spaete, R.R., and Haller, A.A. (2004). Identification of small-animal and primate models for evaluation of vaccine candidates for human *Metapneumovirus* (hMPV) and implications for hMPV vaccine design. J. Gen. Virol. *85,* 1655–1663.

Madhi, S.A., Ludewick, H., Abed, Y., Klugman, K.P., and Boivin, G. (2003). Human *Metapneumovirus*-associated lower respiratory tract infections among hospitalized human immunodeficiency virus type 1 (HIV-1)-infected and HIV-1-uninfected African infants. Clin. Infect. Dis. *37,* 1705–1710.

Madhi, S.A., Ludewick, H., Kuwanda, L., van Niekerk, N., Cutland, C., and Klugman, K.P. (2007). Seasonality, incidence, and repeat human *Metapneumovirus* lower respiratory tract infections in an area with a high prevalence of human immunodeficiency virus type-1 infection. Pediatr. Infect. Dis. J. *26,* 693–699.

Maertzdorf, J., Wang, C.K., Brown, J.B., Quinto, J.D., Chu, M., de Graaf, M., van den Hoogen, B.G., Spaete, R., Osterhaus, A.D., and Fouchier, R.A. (2004). Real-time reverse transcriptase PCR assay for detection of human metapneumoviruses from all known genetic lineages. J. Clin. Microbiol. *42,* 981–986.

Maggi, F., Pifferi, M., Vatteroni, M., Fornai, C., Tempestini, E., Anzilotti, S., Lanini, L., Andreoli, E., Ragazzo, V., Pistello, M., *et al.* (2003). Human *Metapneumovirus* associated with respiratory tract infections in a 3-year study of nasal swabs from infants in Italy. J. Clin. Microbiol. *41,* 2987–2991.

Martinello, R.A., Esper, F., Weibel, C., Ferguson, D., Landry, M.L., and Kahn, J.S. (2006). Human *Metapneumovirus* and exacerbations of chronic obstructive pulmonary disease. J. Infect. *53,* 248–254.

Martino, R., Porras, R.P., Rabella, N., Williams, J.V., Ramila, E., Margall, N., Labeaga, R., Crowe, J.E., Jr., Coll, P., and Sierra, J. (2005). Prospective study of the incidence, clinical features, and outcome of symptomatic upper and lower respiratory tract infections by respiratory viruses in adult recipients of hematopoietic stem cell transplants for hematologic malignancies. Biol. Blood Marrow Transplant *11,* 781–796.

Melendi, G.A., Zavala, F., Buchholz, U.J., Boivin, G., Collins, P.L., Kleeberger, S.R., and Polack, F.P. (2007). Mapping and characterization of the primary and anamnestic H-2(d)-restricted cytotoxic T-lymphocyte response in mice against human *Metapneumovirus.* J. Virol. *81,* 11461–11467.

Miller, S.A., Tollefson, S., Crowe, J.E., Jr., Williams, J.V., and Wright, D.W. (2007). Examination of a fusogenic hexameric core from human *Metapneumovirus* and identification of a potent synthetic peptide inhibitor from the heptad repeat 1 region. J. Virol. *81,* 141–149.

Mok, H., Tollefson, S.J., Podsiad, A.B., Shepherd, B.E., Polosukhin, V.V., Johnston, R.E., Williams, J.V., and Crowe, J.E., Jr. (2008). An alphavirus replicon-based human *Metapneumovirus* vaccine is immunogenic and protective in mice and cotton rats. J. Virol. *82,* 11410–11418.

Mullins, J.A., Erdman, D.D., Weinberg, G.A., Edwards, K., Hall, C.B., Walker, F.J., Iwane, M., and Anderson, L.J. (2004). Human *Metapneumovirus* infection among children hospitalized with acute respiratory illness. Emerg. Infect. Dis. *10,* 700–705.

Naylor, C.J., Brown, P.A., Edworthy, N., Ling, R., Jones, R.C., Savage, C.E., and Easton, A.J. (2004). Development of a reverse-genetics system for Avian *Pneumovirus* demonstrates that the small hydrophobic (SH) and attachment (G) genes are not essential for virus viability. J. Gen. Virol. *85,* 3219–3227.

Njenga, M.K., Lwamba, H.M., and Seal, B.S. (2003). Metapneumoviruses in birds and humans. Virus Res. *91,* 163–169.

Osterhaus, A., and Fouchier, R. (2003). Human *Metapneumovirus* in the community. Lancet *361,* 890–891.

Peiris, J.S., Tang, W.H., Chan, K.H., Khong, P.L., Guan, Y., Lau, Y.L., and Chiu, S.S. (2003). Children with respiratory disease associated with *Metapneumovirus* in Hong Kong. Emerg. Infect. Dis. *9,* 628–633.

Pelletier, G., Dery, P., Abed, Y., and Boivin, G. (2002). Respiratory tract reinfections by the new human

Metapneumovirus in an immunocompromised child. Emerg. Infect. Dis. *8,* 976–978.

Peret, T.C., Boivin, G., Li, Y., Couillard, M., Humphrey, C., Osterhaus, A.D., Erdman, D.D., and Anderson, L.J. (2002). Characterization of human *MetaPneumovirus*es isolated from patients in North America. J. Infect. Dis. *185,* 1660–1663.

Pollack, P., and Groothuis, J.R. (2002). Development and use of palivizumab (Synagis): a passive immunoprophylactic agent for RSV. J. Infect. Chemother. *8,* 201–206.

Principi, N., Bosis, S., and Esposito, S. (2006). Human *Metapneumovirus* in paediatric patients. Clin. Microbiol. Infect. *12,* 301–308.

Rawlinson, W.D., Waliuzzaman, Z., Carter, I.W., Belessis, Y.C., Gilbert, K.M., and Morton, J.R. (2003). Asthma exacerbations in children associated with rhinovirus but not human *Metapneumovirus* infection. J. Infect. Dis. *187,* 1314–1318.

Ryder, A.B., Tollefson, S.J., Podsiad, A.B., Johnson, J.E., and Williams, J.V. (2010). Soluble recombinant human *Metapneumovirus* G protein is immunogenic but not protective. Vaccine *28,* 4145–4152.

Schickli, J.H., Kaur, J., Ulbrandt, N., Spaete, R.R., and Tang, R.S. (2005). An S101P substitution in the putative cleavage motif of the human *Metapneumovirus* fusion protein is a major determinant for trypsin-independent growth in vero cells and does not alter tissue tropism in hamsters. J. Virol. *79, 10,*678–10689.

Schildgen, O., and Simon, A. (2005). Induction of acute otitis media by human *Metapneumovirus.* Pediatr. Infect. Dis. J. *24,* 1126.

Schildgen, O., Simon, A., and Williams, J. (2007). Animal models for human *Metapneumovirus* (HMPV) infections. Vet. Res. *38,* 117–126.

Semple, M.G., Cowell, A., Dove, W., Greensill, J., McNamara, P.S., Halfhide, C., Shears, P., Smyth, R.L., and Hart, C.A. (2005). Dual infection of infants by human *Metapneumovirus* and human respiratory syncytial virus is strongly associated with severe bronchiolitis. J. Infect. Dis. *191,* 382–386.

Skiadopoulos, M.H., Biacchesi, S., Buchholz, U.J., Riggs, J.M., Surman, S.R., Amaro-Carambot, E., McAuliffe, J.M., Elkins, W.R., St Claire, M., Collins, P.L., *et al.* (2004). The two major human *Metapneumovirus* genetic lineages are highly related antigenically, and the fusion (F) protein is a major contributor to this antigenic relatedness. J. Virol. *78,* 6927–6937.

Skiadopoulos, M.H., Biacchesi, S., Buchholz, U.J., Amaro-Carambot, E., Surman, S.R., Collins, P.L., and Murphy, B.R. (2006). Individual contributions of the human *Metapneumovirus* F, G, and SH surface glycoproteins to the induction of neutralizing antibodies and protective immunity. Virology *345,* 492–501.

Stockton, J., Stephenson, I., Fleming, D., and Zambon, M. (2002). Human *Metapneumovirus* as a cause of community-acquired respiratory illness. Emerg. Infect. Dis. *8,* 897–901.

Sumino, K.C., Agapov, E., Pierce, R.A., Trulock, E.P., Pfeifer, J.D., Ritter, J.H., Gaudreault-Keener, M., Storch, G.A., and Holtzman, M.J. (2005). Detection of severe human *Metapneumovirus* infection by real-time polymerase chain reaction and histopathological assessment. J. Infect. Dis. *192,* 1052–1060.

Suzuki, A., Watanabe, O., Okamoto, M., Endo, H., Yano, H., Suetake, M., and Nishimura, H. (2005). Detection of human *Metapneumovirus* from children with acute otitis media. Pediatr. Infect. Dis. J. *24,* 655–657.

de Swart, R.L., van den Hoogen, B.G., Kuiken, T., Herfst, S., van Amerongen, G., Yuksel, S., Sprong, L., and Osterhaus, A.D. (2007). Immunization of macaques with formalin-inactivated human *Metapneumovirus* induces hypersensitivity to hMPV infection. Vaccine *25,* 8518–8528.

Tang, R.S., Schickli, J.H., MacPhail, M., Fernandes, F., Bicha, L., Spaete, J., Fouchier, R.A., Osterhaus, A.D., Spaete, R., and Haller, A.A. (2003). Effects of human *Metapneumovirus* and respiratory syncytial virus antigen insertion in two 3′ proximal genome positions of bovine/human parainfluenza virus type 3 on virus replication and immunogenicity. J. Virol. *77,* 10819–10828.

Tang, R.S., Mahmood, K., Macphail, M., Guzzetta, J.M., Haller, A.A., Liu, H., Kaur, J., Lawlor, H.A., Stillman, E.A., Schickli, J.H., *et al.* (2005). A host-range restricted parainfluenza virus type 3 (PIV3) expressing the human *Metapneumovirus* (hMPV) fusion protein elicits protective immunity in African green monkeys. Vaccine *23,* 1657–1667.

Thanasugarn, W., Samransamruajkit, R., Vanapongtipagorn, P., Prapphal, N., Van den Hoogen, B., Osterhaus, A.D., and Poovorawan, Y. (2003). Human metapneumovirus infection in Thai children. Scand J. Infect. Dis. *35,* 754–756.

Tiwari, A., Patnayak, D.P., Chander, Y., and Goyal, S.M. (2006). Permissibility of different cell types for the growth of avian metapneumovirus. J. Virol. Methods *138,* 80–84.

Tollefson, S.J., Cox, R.G., and Williams, J.V. (2010). Studies of culture conditions and environmental stability of human metapneumovirus. Virus Res. *151,* 54–59.

Ulbrandt, N.D., Ji, H., Patel, N.K., Riggs, J.M., Brewah, Y.A., Ready, S., Donacki, N.E., Folliot, K., Barnes, A.S., Senthil, K., *et al.* (2006). Isolation and characterization of monoclonal antibodies which neutralize human metapneumovirus *in vitro* and *in vivo*. J. Virol. *80,* 7799–7806.

Viazov, S., Ratjen, F., Scheidhauer, R., Fiedler, M., and Roggendorf, M. (2003). High prevalence of human metapneumovirus infection in young children and genetic heterogeneity of the viral isolates. J. Clin. Microbiol. *41,* 3043–3045.

Vicente, D., Montes, M., Cilla, G., and Perez-Trallero, E. (2004). Human metapneumovirus and chronic obstructive pulmonary disease. Emerg. Infect. Dis. *10,* 1338–1339.

Walsh, E.E., Peterson, D.R., and Falsey, A.R. (2008). Human metapneumovirus infections in adults: another piece of the puzzle. Arch. Intern Med. *168,* 2489–2496.

Williams, J.V., Harris, P.A., Tollefson, S.J., Halburnt-Rush, L.L., Pingsterhaus, J.M., Edwards, K.M., Wright, P.F., and Crowe, J.E., Jr. (2004). Human metapneumovirus and lower respiratory tract disease in otherwise healthy infants and children. N. Engl. J. Med. *350,* 443–450.

Williams, J.V., Crowe, J.E., Jr., Enriquez, R., Minton, P., Peebles, R.S., Jr., Hamilton, R.G., Higgins, S., Griffin, M., and Hartert, T.V. (2005a). Human metapneumovirus infection plays an etiologic role in acute asthma exacerbations requiring hospitalization in adults. J. Infect. Dis. *192,* 1149–1153.

Williams, J.V., Martino, R., Rabella, N., Otegui, M., Parody, R., Heck, J.M., and Crowe, J.E., Jr. (2005b). A prospective study comparing human metapneumovirus with other respiratory viruses in adults with hematologic malignancies and respiratory tract infections. J. Infect. Dis. *192,* 1061–1065.

Williams, J.V., Tollefson, S.J., Heymann, P.W., Carper, H.T., Patrie, J., and Crowe, J.E. (2005c). Human metapneumovirus infection in children hospitalized for wheezing. J. Allergy Clin. Immunol. *115,* 1311–1312.

Williams, J.V., Tollefson, S.J., Johnson, J.E., and Crowe, J.E., Jr. (2005d). The cotton rat (Sigmodon hispidus) is a permissive small animal model of human metapneumovirus infection, pathogenesis, and protective immunity. J. Virol. *79,* 10944–10951.

Williams, J.V., Tollefson, S.J., Nair, S., and Chonmaitree, T. (2006a). Association of human metapneumovirus with acute otitis media. Int. J. Pediatr. Otorhinolaryngol. *70,* 1189–1193.

Williams, J.V., Wang, C.K., Yang, C.F., Tollefson, S.J., House, F.S., Heck, J.M., Chu, M., Brown, J.B., Lintao, L.D., Quinto, J.D., *et al.* (2006b). The role of human metapneumovirus in upper respiratory tract infections in children: a 20-year experience. J. Infect. Dis. *193,* 387–395.

Williams, J.V., Chen, Z., Cseke, G., Wright, D.W., Keefer, C.J., Tollefson, S.J., Hessell, A., Podsiad, A., Shepherd, B.E., Sanna, P.P., *et al.* (2007). A recombinant human monoclonal antibody to human metapneumovirus fusion protein that neutralizes virus *in vitro* and is effective therapeutically *in vivo*. J. Virol. *81,* 8315–8324.

van Woensel, J.B., Bos, A.P., Lutter, R., Rossen, J.W., and Schuurman, R. (2006). Absence of human metapneumovirus co-infection in cases of severe respiratory syncytial virus infection. Pediatr. Pulmonol. *41,* 872–874.

Wyde, P.R., Chetty, S.N., Jewell, A.M., Boivin, G., and Piedra, P.A. (2003). Comparison of the inhibition of human metapneumovirus and respiratory syncytial virus by ribavirin and immune serum globulin *in vitro*. Antiviral Res. *60,* 51–59.

Wyde, P.R., Chetty, S.N., Jewell, A.M., Schoonover, S.L., and Piedra, P.A. (2005). Development of a cotton rat-human metapneumovirus (hMPV) model for identifying and evaluating potential hMPV antivirals and vaccines. Antiviral Res. *66,* 57–66.

Yang, C.F., Wang, C.K., Tollefson, S.J., Piyaratna, R., Lintao, L.D., Chu, M., Liem, A., Mark, M., Spaete, R.R., Crowe, J.E., Jr., *et al.* (2009). Genetic diversity and evolution of human metapneumovirus fusion protein over twenty years. Virol. J. *6,* 138.

Yim, K.C., Cragin, R.P., Boukhvalova, M.S., Blanco, J.C., Hamlin, M.E., Boivin, G., Porter, D.D., and Prince, G.A. (2007). Human metapneumovirus: Enhanced pulmonary disease in cotton rats immunized with formalin-inactivated virus vaccine and challenged. Vaccine *25,* 5034–5040.

New and Emerging Paramyxoviruses 12

Danielle E. Anderson and Lin-Fa Wang*

Abstract

The family *Paramyxoviridae* is composed of a diverse group of viruses and is divided into two subfamilies, *Paramyxovirinae* and *Pneumovirinae*. This chapter will limit the discussion to viruses in *Paramyxovirinae* only. There are currently 40 virus species classified within this subfamily, but several remain unclassified at the genus level. In the past few decades, paramyxoviruses have emerged from terrestrial, volant and aquatic animals, demonstrating a vast host range and great viral genetic diversity. As molecular technology advances and viral surveillance programmes are implemented, it is expected that the opportunity for the discovery of new viruses will increase in the years to come. Here, we will provide a review of those new and emerging paramyxoviruses originating from non-human and mainly wildlife animals. The disease causing potential of many of these new viruses is not known, our focus will therefore be placed on the molecular characterization of these viruses and their host range and genetic relationship with other viruses in the same subfamily.

Introduction

Infectious diseases have played a significant role throughout human history. Investigation of disease dates back to ancient times and the quest for understanding through science has lead to the discovery of viruses and bacteria as the causative agents of various types of infection and illness. Pathogenicity of viruses and our vulnerability to the effects of infectious agents is constantly and sometimes spectacularly demonstrated through the emergence of new diseases, reappearance of pre-existing diseases and, unfortunately, bioterrorism. Emerging infectious viral diseases are those that have recently appeared in a population as a result of a new virus or the recognition of a previously undetected virus. Emergence of an infectious viral disease may occur as a result of the extension of the geographic or host range of the virus. Emerging infectious diseases are often zoonotic, hence having the ability to be transmitted between humans and animals either directly or via a vector. Recently, bats have been implicated as an important source of several emerging viruses. As new technology for detection of viruses becomes increasingly available and more powerful, more viruses are likely to be detected and enhanced molecular biology techniques will allow faster and more complete characterization of these viruses.

Currently, there are 40 viruses classified within the subfamily *Paramyxovirinae*. Over the last 50 years, many new paramyxoviruses have emerged and several remain unclassified at the genus level. The discovery of these viruses occurred not only as a result of disease outbreaks but also under circumstances where no disease correlation was evident. New paramyxoviruses have emerged from an extensive range of hosts, including terrestrial, volant and aquatic animals, demonstrating the vast host range and zoonotic potential of this family of viruses. These emergent viruses provide

*Corresponding author

virologists with a great opportunity to gain further insight into the evolution, genetic diversity, host specificity, epidemiology and pathogenesis of paramyxoviruses.

Taxonomy

The family *Paramyxoviridae* is divided into two subfamilies, the *Paramyxovirinae* (paramyxoviruses) and *Pneumovirinae* (pneumoviruses). The paramyxoviruses are currently grouped into five genera: *Morbillivirus, Respirovirus, Rubulavirus, Avulavirus* and *Henipavirus* (Lamb *et al.*, 2005; Table 12.1). There are currently nine virus species in the genus *Rubulavirus*, nine in *Avulavirus*, six in *Morbillivirus*, five in *Respirovirus* and two in *Henipavirus*. Several other new or emerging paramyxoviruses cannot be classified as part of any of the existing genera (Table 12.1). To date, the unclassified viruses within the *Paramyxoviridae* are *J paramyxovirus* (JPV), *Beilong paramyxovirus* (BeiPV), *Salem virus* (SalPV), *Fer-de-Lance virus* (FdlPV), *Tupaia paramyxovirus* (TupPV), *Nariva virus* (NarPV), *Mossman virus* (MosPV) and *Atlantic salmon paramyxovirus* (AsaPV). As indicated in Table 12.1, two new genera, *Ferlavirus* and *Aquaparamyxovirus*, have been proposed and are currently under consideration by the International Committee on Taxonomy of Viruses (ICTV).

Genome organization

Paramyxoviruses contain a negative-sense, non-segmented, single-stranded (NNS) RNA genome and viral genomes range from 15,178 (PorPV) to 19,212 (BeiPV) nucleotides in length. The genome structure is conserved with the nucleocapsid (N), phosphoprotein (P), matrix (M), fusion (F), attachment (HN/H/G) and large (L) genes present in all viruses. Although the overall genome organization is conserved across all known viruses in the subfamily, some species possess additional genes (Fig. 12.1). The term 'gene', when used in reference to paramyxoviruses, refers to the sequence encoding a single mRNA, even if more than one open reading frame (ORF) is present or more than one protein is encoded. Genes are flanked by untranslated regions (UTRs) at both 5′ and 3′ ends of the ORFs and separated by an intergenic region which is three nucleotides in length for viruses in the genera *Morbillivirus, Respirovirus* and *Henipavirus*, but can vary in size from 1 to 124 nucleotides for viruses in the genera *Rubulavirus* and *Avulavirus* (Lamb and Parks, 2007; Lau *et al.*, 2010). Conserved 3′ leader (55 nucleotides for all members of the subfamily) and 5′ trailer extragenic regions (variable size) frame the genome termini.

Virus structure

Paramyxovirus virions are pleomorphic and composed of two structural components; the helical ribonucleoprotein core and the lipoprotein bilayer membrane, acquired from the host cell, which constitutes the envelope. The NNS genomic RNA is always tightly encased by nucleocapsid proteins in a left-handed coil with distinct 'herringbone' morphology. The N protein, in conjunction with highly phosphorylated P proteins and multifunctional L protein that possesses catalytic polymerase activity, forms a complex that has RNA-dependent RNA polymerase activity. Projecting 8–12 nm from the lipid bilayer of the viral envelope are 2–4 glycoproteins (Lamb and Parks, 2007). All paramyxoviruses possess attachment (HN, H or G) proteins and F proteins that are activated by proteolytic cleavage (Conzelmann, 1998; Lamb and Parks, 2007). A small hydrophobic (SH) protein is present in some rubulaviruses (Hiebert *et al.*, 1985; Elango *et al.*, 1989). An SH protein is present in JPV and BeiPV, two unclassified rodent paramyxoviruses, and these viruses also contain a transmembrane (TM) protein (Jack *et al.*, 2005, 2008; Li *et al.*, 2006). The most abundant protein, M, is basic and slightly hydrophobic. The M protein plays a key role in viral morphogenesis by interacting with the cytoplasmic tails of the integral membrane proteins, the lipid bilayer and the nucleocapsids (Lamb and Parks, 2007).

Replication

The life cycle of a paramyxovirus begins with virus adsorption and entry into a host cell, followed by primary transcription and translation, and genome replication, and then secondary transcription and translation. When sufficient genomic RNA and viral proteins are produced in the infected cells, virus assembly occurs at the cell membrane followed by the release of progeny

Table 12.1 List of viruses in the subfamily *Paramyxovirinae*

Genus	Virus	Abbreviation
Rubulavirus	Mumps virus	MuV
	Human parainfluenza virus 2	HPIV-2
	Human parainfluenza virus 4	HPIV-4
	Porcine rubulavirus	PorPV
	Mapuera virus	MprPV
	Parainfluenza 5[1]	PIV-5
	Simian virus 41	SV-41
	Menangle virus[2]	MenPV
	Tioman virus[2]	TioPV
Avulavirus	Newcastle disease virus[3]	NDV
	Avian paramyxovirus 2	APMV-2
	Avian paramyxovirus 3	APMV-3
	Avian paramyxovirus 4	APMV-4
	Avian paramyxovirus 5	APMV-5
	Avian paramyxovirus 6	APMV-6
	Avian paramyxovirus 7	APMV-7
	Avian paramyxovirus 8	APMV-8
	Avian paramyxovirus 9	APMV-9
Morbillivirus	Measles virus	MeV
	Rinderpest virus	RPV
	Canine distemper virus	CDV
	Peste des petits ruminants virus	PPRV
	Phocine distemper virus	PDV
	Cetacean Morbillivirus virus	CeMV
Respirovirus	Sendai virus	SeV
	Human parainfluenza virus 1	HPIV-1
	Human parainfluenza virus 3	HPIV-3
	Bovine parainfluenza virus 3	BPIV-3
	Simian virus 10	SV-10
Henipavirus	Hendra virus	HeV
	Nipah virus	NiV
Ferlavirus[4]	Fer-de-Lance virus	FdlPV
Aquaparamyxovirus[4]	Atlantic salmon paramyxovirus	AsaPV
	Pacific salmon paramyxovirus	PsaPV
Unassigned	J paramyxovirus	JPV
	Beilong virus	BeiPV
	Salem virus	SalPV
	Tupaia paramyxovirus	TupPV
	Nariva virus	NarPV
	Mossman virus	MosPV

[1]Previously known as Simian virus 5 (SV-5).

[2]Tentative species in the genus.

[3]Also known as avian paramyxovirus 1 (APMV-1).

[4]New genera proposed to ICTV, yet to be formally approved.

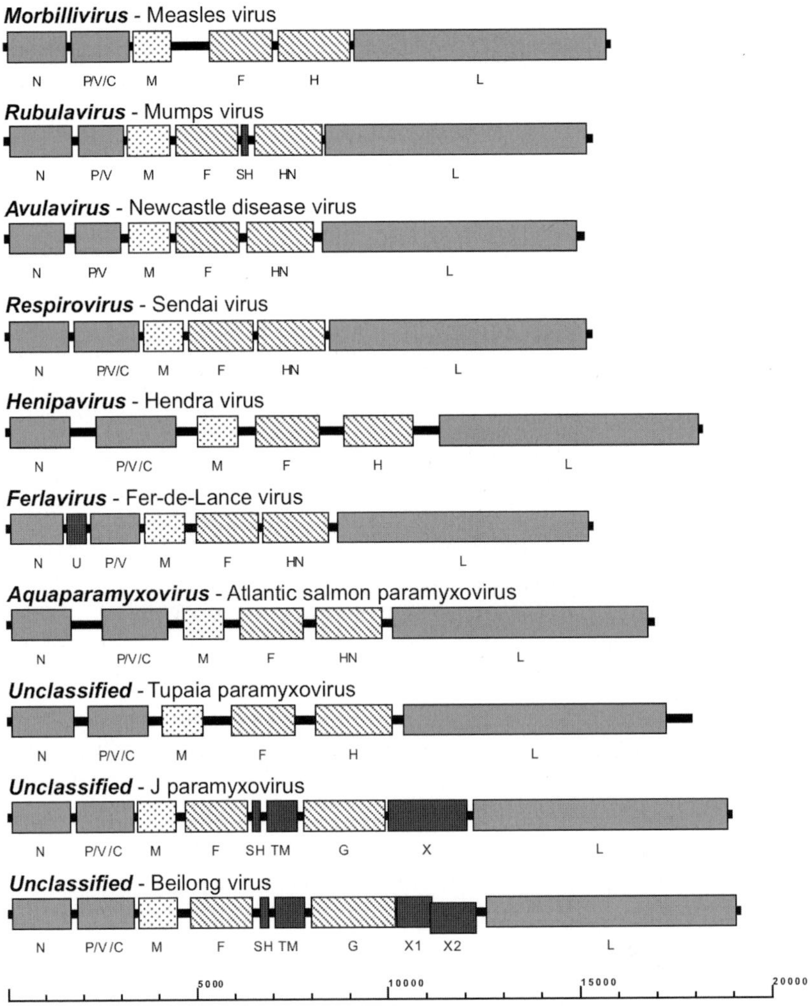

Figure 12.1 Representative genomes of paramyxoviruses from each genus. The genome, coding regions and untranslated regions are drawn to scale. The six conserved genes present in all paramyxovirus genomes are indicated as follows: light shaded = RNA polymerase and nucleocapsid genes (N, P and L); slanted = envelope membrane protein genes (F and attachment protein); dotted = matrix protein (M). The dark shaded boxes represent genes which are not commonly shared among members of the subfamily. The scale at the bottom represents genome size in nucleotides.

virions via a budding process. It is believed that genome replication of all paramyxoviruses takes place exclusively in the cytoplasm of an infected cell with no involvement of the cell nucleus (Lamb and Parks, 2007).

The Kingsbury–Kolakofsky model of transcription and replication dictates that a promoter sequence for polymerase entry exists on the 3′ end of both genome and antigenome, mRNA is produced via a polarity gradient of transcription, and relative amount of mRNA versus genomic RNA is modulated by a virus-encoded protein (Blumberg *et al.*, 1981; Vidal and Kolakofsky, 1989). Attempts to experimentally verify this model, however, have been unsuccessful (Chuang and Perrault, 1997; Fearns *et al.*, 1997; Whelan and Wertz, 2002; Qanungo *et al.*, 2004) and taking experimental findings into account, a consensus model for NNS RNA synthesis was proposed (Curran and Kolakofsky, 2008).

Transcriptional regulation

All paramyxovirus RNA synthesis begins with a single entry of the viral RNA-dependent RNA polymerase (vRdRp) at the 3′ end of the genome, with transcription of the 3′ leader beginning before mRNA synthesis. Initiation of mRNA synthesis is dependent on the distance between the genome 3′ end and gene start signal (Cordey and Roux, 2006). The strength of the replication promoter is important in determining initiation of mRNA synthesis (Vulliemoz *et al.*, 2005). A scanning model predicts that ribosomes initiate translation at the first AUG codon that is in favourable context according to the 'Kozak' sequence $5'\text{-GCC}^{\text{A}}\text{or}_{\text{G}}\text{CC}\underline{\text{AUG}}\text{G-3}'$ (Kozak, 1986). Monocistronic mRNA produced by the vRdRp contains a 5′ cap (Abraham *et al.*, 1975) and is polyadenylated by a stuttering mechanism that reiteratively copies a short run of 4–7 uridines leading to the termination and release of the mRNA. After releasing the mRNA, the vRdRp remains attached to the template before reinitiating transcription at the downstream gene. This stop/start mechanism of transcription is not perfect and failure of vRdRp to reinitiate transcription at a downstream site results in a gradient of mRNA production inversely proportional to the distance from the 3′ end of the genome (Lamb and Parks, 2007), and this transcription gradient is maintained throughout infection (Plumet *et al.*, 2005).

Rule of six

Paramyxovirus RNA is required to be bound by a helical nucleocapsid in order to be infectious. Electron microscopy analysis of SeV nucleocapsids revealed helices of left-handed symmetry, which suggested a stoichiometric relationship between six nucleotides of RNA and one N protein subunit (Egelman *et al.*, 1989). For efficient replication, SeV RNA must contain a total number of nucleotides that is a multiple of six, which was dubbed the 'rule of six' (Calain and Roux, 1993). The rule of six does not exclude other requirements for efficient replication, such as total minimum RNA length and *cis*-acting sequences. The RNA polymerase initiates more efficiently when *cis*-acting promoter sequences are found in the correct context with relation to N subunits, determined by the length of the entire genome (Hausmann *et al.*, 1996). Association of N monomers with hexameric genomic nucleotide sequences begins with the first nucleotide at the 5′ end of the genome and continues until the genome is precisely covered by N subunits to the 3′ end. Remarkably, all paramyxoviruses governed by the rule of six edit their P gene (Kolakofsky *et al.*, 1998, 2005).

Reverse genetics

The unique gene expression strategy utilized by NNS RNA viruses involves replication of the ribonucleoprotein complex and sequential synthesis of mRNA and neither the genomic, nor antigenomic RNA genome is capable of initiating an infectious cycle. Development of reverse genetic systems for RNA viruses within this replication category has provided a powerful tool for genetic manipulation and study of those viruses.

The establishment of a technique to introduce recombinant RNA into the genome of a segmented negative-sense, single-stranded RNA virus was achieved through the manipulation of influenza virus (Luytjes *et al.*, 1989). The first reverse genetic system to be established for a NNS RNA virus resulted from the manipulation of *Rabies virus*, a member of the family *Rhabdoviridae* (Schnell *et al.*, 1994) and the first paramyxovirus for which a reverse genetic system was developed was MeV (Sidhu *et al.*, 1995). Insertion of an additional ORF into the MeV genome was determined not to be detrimental, as the expression of the firefly luciferase reporter gene did not reduce MeV infectivity (Sidhu *et al.*, 1995).

In general, reverse genetics of paramyxoviruses implies the ability to recover infectious virus from a cDNA copy of the NNS genome. Establishment of a reverse genetic system involves the construction of a plasmid containing the viral antigenome, and support plasmids containing the N, P and L genes. Transfection of cells with protein expression plasmids occurs in conjunction with transfection of the plasmid containing the full-length viral genome, all plasmids being under the control of a T7 promoter (Neumann *et al.*, 2002). Reporter genes, such as GFP may be included as an additional gene (containing its own transcriptional start and stop signals) to visualize virus infection (von Messling *et al.*, 2004; Towner *et al.*, 2005).

Bat paramyxoviruses

Bats are arguably the most abundant, diverse and geographically dispersed vertebrates on earth. Bats have been shown to be the reservoir hosts of a variety of viruses responsible for severe disease outbreaks in humans and animals, including filoviruses, coronaviruses and paramyxoviruses (Calisher *et al.*, 2006; Wang *et al.*, 2006). In addition to the high profile Hendra and Nipah viruses, there are several other bat paramyxoviruses which were discovered during disease investigation or surveillance. A brief discussion will be provided at the end of this section highlighting the expectation that many more bat-borne paramyxoviruses are to be discovered in the near future thanks to the intensified surveillance activities around the world.

Menangle virus (MenPV)

In 1997, MenPV was isolated during an outbreak of reproductive disease in pigs at a piggery in New South Wales, Australia. The piggery experienced a decrease in farrowing rate from 82% to 62% and there were 27% less piglets born alive during the outbreak. Mummified piglets were born and stillborn piglets were frequently deformed, having severe degeneration of the brain and spinal cord (Philbey *et al.*, 1998). No symptoms of disease were observed in postnatal pigs. The outbreak began in April of 1997 and lasted for 21 weeks (Love *et al.*, 2001). Control of the MenPV outbreak and eradication of the virus from the piggery resulted following the administration of a serological testing and segregation programme (Kirkland *et al.*, 2001).

Virus isolation was achieved by inoculation of BHK-21 cells with heart, lung and brain tissue from affected piglets, where CPE and syncytia formation was observed. MenPV was initially identified as a paramyxovirus based on electron microscopic visualization of pleomorphic particles and 'herringbone' nucleocapsids. Subsequent electron microscopy studies have shown MenPV virions to be highly pleomorphic, and commonly filamentous (Yaiw *et al.*, 2008b). Serum samples collected from cattle, sheep, birds, rodents, cats and a dog were all negative for MenPV antibodies (Kirkland *et al.*, 2001). Two piggery workers became ill during the outbreak with an influenza-like illness in conjunction with a red, spotty rash. Other symptoms included malaise, chills, fever, sweating, headache, with weight loss observed after disease progression. Both of these workers had intense occupational exposure to infected pigs and serological analysis revealed the presence of MenPV antibodies in both cases (Chant *et al.*, 1998). Serum samples were obtained from 250 humans but the only MenPV positive samples were those corresponding to the two piggery workers (Kirkland *et al.*, 2001). The likelihood of MenPV infection of humans appears to be low without the prior amplification of the virus in pigs (Hooper, 2001).

A colony of fruit bats (*Pteropus poliocephalus*) was found to be roosting within 150 m of the piggery and these animals were speculated to be the potential source of MenPV infection in the pigs (Kirkland *et al.*, 2001). Serum was collected from bats and determined to be positive for MenPV antibodies (Philbey *et al.*, 1998; Kirkland *et al.*, 2001). Furthering these serological studies of fruit bats, antibodies against MenPV were found in four species of *Pteropus* spp. (*P. alecto*, *P. poliocephalus*, *P. conspicillatus* and *P. scapulatus*) from widespread areas of Australia (Philbey *et al.*, 2008). Neutralizing antibodies against MenPV are widespread in at least three species of Pteropus in Australia, but the virus has not been isolated from these animals, nor has disease been associated with natural exposure to MenPV (Philbey *et al.*, 2008), suggesting that bats are the reservoir host.

The 15,516 nucleotide genome of MenPV has been completely sequenced (Bowden *et al.*, 2001; Bowden and Boyle, 2005) and obeys the 'rule of six' (Calain and Roux, 1993). Complementarity between the genome termini of MenPV is perfect for 14 nucleotides (Bowden and Boyle, 2005). The 3′ leader sequence of MenPV is 55 nucleotides, matching the length of that of other paramyxovirus leader sequences (Bowden *et al.*, 2001), and the 5′ trailer sequence is 31 nucleotides, similar to the trailer region of rubulaviruses (Bowden and Boyle, 2005). MenPV genes are separated by intergenic regions ranging from 5 to 85 nucleotides in length (Bowden *et al.*, 2001; Bowden and Boyle, 2005). Based on the genome structure and gene sequence homology analysis, MenPV has been tentatively classified as a member of the genus *Rubulavirus* (Lamb *et al.*, 2005).

The MenPV N gene encodes a protein of 519 amino acid residues and has a predicted molecular mass of 58.9 kDa (Bowden *et al.*, 2001). The P gene is 1423 nucleotides long, and potentially encodes three proteins. Faithful transcription of the P gene encodes the V protein. The reading frame of the P protein is accessed via the insertion of two non-templated G residues in the RNA editing site. Insertion of one non-templated G residues potentially leads to the production of a W protein. The MenPV P gene does not encode a C protein and the M protein is predicted to be 373 amino acids in length (Bowden *et al.*, 2001).

The MenPV F protein is predicted to be a type I integral membrane protein. The conserved furin cleavage sequence R-X-K/R-R (Hosaka *et al.*, 1991) is present as a multibasic RRRR sequence (Bowden *et al.*, 2001). The HN protein of MenPV is 595 amino acids in length and is predicted to be a type II integral membrane protein (Bowden *et al.*, 2001). The conserved sequence of respiroviruses and rubulaviruses, NRKSCS, in the attachment protein forms part of the sialic acid binding site (Jorgensen *et al.*, 1987). Only two or three of these residues are present in MenPV, suggesting that MenPV does not possess the same degree of neuraminidase activity as detected in other rubulaviruses. This characteristic of the protein is consistent with experimental evidence that MenPV lacks haemagglutinin and neuraminidase activity (Philbey *et al.*, 1998). The L protein of MenPV is 2269 amino acids in length and contains the QGDNQ motif present in the majority of paramyxoviruses (Bowden and Boyle, 2005).

Tioman virus (TioPV)

In a search for the origin of NiV on the island of Tioman, Malaysia, in 1999, TioPV was isolated from the urine of fruit bats (*Pteropus hypomelanus*; Chua *et al.*, 2001). A total of 50 pooled urine samples were collected and inoculated onto Vero cells for virus isolation and formation of syncytia was observed. Electron microscopic examination of virus particles revealed pleomorphic particles that possessed an envelope and typical paramyxovirus nucleocapsids. The intracellular ultrastructure of TioPV differs from that of the closely related MenPV (Yaiw *et al.*, 2008b).

An animal study demonstrated that TioPV was 100% lethal in suckling mice 8–12 days post intracerebral inoculation. The post-mortem suckling mouse brain showed evidence of severe meningoencephalitis and TioPV could be readily reisolated. In contrast, adult mice infected with TioPV in a subsequent study did not succumb to the disease (Yaiw *et al.*, 2007b). Mice were inoculated either intracerebrally or intraperitoneally and only mice infected intracerebrally showed any signs of disease. Clinical signs included loss of balance, mobility and weight, but all mice survived the infection. Although a high mortality rate was not observed in the later study, TioPV was shown to be neurotropic, preferentially replicating in the neocortex and limbic system. Evidence of apoptosis was found in inflammatory cells, but not in neurons (Yaiw *et al.*, 2007b).

In an attempt to ascertain whether infection with TioPV has occurred in humans, sera samples were collected from 169 Tioman islanders from September 2001 to April 2002 (Yaiw *et al.*, 2007a). Of the 169 samples, five were positive for TioPV by ELISA. Of these five ELISA positive samples, three were positive in a serum neutralization assay. These results provide evidence of human exposure to TioPV, but based on the available clinical data it is unknown whether exposure to the virus resulted in clinical disease.

A serological survey of Old World fruit bats was conducted in Madagascar from May 2003 to July 2005 (Iehle *et al.*, 2007). Sera from 427 bats was analysed and TioPV antibodies were detected in three samples (two from *Pteropus rufus* and one from *Rousettus madascariensis*) indicating that pteropodid bats in Madagascar have been in contact with TioPV or a closely related bat *Rubulavirus*, even though no disease has been detected.

In light of the evidence that the paramyxoviruses MenPV and NiV originate from bats and involve pigs as intermediate or amplifying hosts (Kirkland *et al.*, 2002), a study to determine the susceptibility of pigs to TioPV infection was undertaken (Yaiw *et al.*, 2008a). TioPV was found to be capable of infecting and replicating in pigs and the main cellular targets were lymphocytes, thymic epithelioreticular cells, and the tonsillar epithelium. This study demonstrated that unsuspected mild TioPV infection could occur in naturally infected pigs and this may perhaps facilitate viral transmission to humans via contact with oral secretions.

To date, TioPV has not been reported to cause any natural disease in either humans or animals, which is in contrast to the closely related *Rubulavirus*, MenPV, which has caused a disease outbreak in pigs (Philbey *et al.*, 1998).

The TioPV genome has been completely sequenced and is 15,522 nucleotides in length (Chua *et al.*, 2002). At this length, the genome obeys the 'rule of six' (Calain and Roux, 1993). Complementarity between the TioPV genome termini is perfect for 17 nucleotides (Chua *et al.*, 2002). The 3′ leader sequence of TioPV is 55 nucleotides, matching the length of all other paramyxovirus leader sequences, and the 5′ trailer sequence is 23 nucleotides long. TioPV genes are separated by intergenic regions ranging in length from 8 to 70 nucleotides .

The TioPV N protein is 519 amino acids in length and has the highest sequence homology with the rubulaviruses (Chua *et al.*, 2001). The P gene is 1455 nucleotides long, and potentially encodes three proteins. Faithful transcription of the P gene encodes a 228 amino acid V protein. The P ORF is accessed via the insertion of two non-templated G residues in the RNA editing site and produces a 385 amino acid P protein. Insertion of one non-templated G residue potentially leads to the production of a large 211 amino acid W protein (Chua *et al.*, 2001). The TioPV P gene does not encode a C protein. The M protein is predicted to be 373 amino acids in length, the same size as the MenPV M protein (Chua *et al.*, 2002).

The TioPV F protein is predicted to be a type I integral membrane protein made up of 538 amino acid residues. The TioPV F protein has a multibasic cleavage site followed by a highly conserved fusion peptide (Chua *et al.*, 2002). The HN protein of TioPV is 593 amino acids in length and is predicted to be a type II integral membrane protein. Only two residues of the conserved sialic acid binding site sequence of the respiroviruses and rubulaviruses, NRKSCS, (Jorgensen *et al.*, 1987) are present in TioPV. TioPV is most related to MenPV, as based on sequence analysis and antigenic cross-reactivity, but the HN proteins of these viruses show insignificant homology with other rubulaviruses (Bowden *et al.*, 2001). This perhaps demonstrates a more divergent evolution of the attachment proteins when compared with other structural proteins. The L protein of TioPV is 2271 amino acids in length and contains the QGDNQ motif present in the majority of paramyxoviruses (Chua *et al.*, 2001).

Phylogenetic analysis of the N and P/V proteins concludes that TioPV is most closely related to MenPV (Chua *et al.* 2001, 2002). As for MenPV, TioPV has been tentatively classified in the genus *Rubulavirus* (Lamb *et al.*, 2005).

Mapuera virus (MprPV)

MprPV was isolated in 1979 from the salivary glands of an asymptomatic fruit bat (*Sturnira lilium*) captured in the Brazilian rainforest. Preliminary characterization using electron microscopy revealed virus particles of 250 nm and indicated that MprPV was a paramyxovirus (Zeller *et al.*, 1989). Further electron microscopy studies demonstrated that MprPV virions possessed surface projections of 10 nm in length, and characteristic 'herringbone' morphology of nucleocapsid filaments was also observed (Henderson *et al.*, 1995). Potent neuraminidase activity was assumed when hemadsorption of guinea pig erythrocytes was detected (Zeller *et al.*, 1989). To further classify the paramyxovirus, ELISA and immunofluorescence assays were performed. No reactivity was observed with HPIV-1–3 in the ELISA or with NarPV, JPV, MosPV and NDV using immunofluorescence (Zeller *et al.*, 1989). In initial experiments, MprPV was lethal in mice within 4 days of intracranial inoculation (Zeller *et al.*, 1989), but it is not known whether MprPV is capable of infecting and causing disease in other animal species.

The genome of MprPV has been completely sequenced and is 15,486 nucleotides in length (Wang *et al.*, 2007) and conforms to the 'rule of six'. Complementarity between the MprPV genome leader and trailer is perfect for 13 nucleotides.

The 3′ leader sequence of MprPV is the standard length of 55 nucleotides but the 5′ trailer, at 159 nucleotides, is significantly longer than that of other rubulaviruses. MprPV genes are separated by intergenic regions ranging in length from 6 to 58 nucleotides.

The MprPV N protein is 537 amino acids in length (Henderson *et al.*, 1995). The P gene is 1400 nucleotides long, and potentially encodes three proteins. Faithful transcription of the P gene encodes a 251 amino acid V protein (Wang *et al.*, 2007). The V protein of MprPV, unlike all other rubulaviruses, is able to sequester STAT1 and STAT2 in the cytoplasm, hence disrupting both IFN-α and IFN-β signalling (Hagmaier *et al.*, 2007). The P ORF is accessed via the insertion of two non-templated G residues in the RNA editing site and produces a 391 amino acid P protein. Insertion of one non-templated G residues potentially encodes a 201 amino acid W protein (Wang *et al.*, 2007). Although the MprPV P gene contains a potential predicted overlapping ORF, experimentally a C protein was not detected. The M protein is predicted to be 371 amino acids in length.

The MprPV F protein is predicted to be a type I integral membrane protein and consists of 537 amino acid residues. Paramyxoviruses contain the conserved sequence R-X-K/R-R (Hosaka *et al.*, 1991) that directs cleavage of the F protein by furin. The MprPV F protein contains the highly basic sequence RRKKR as the cleavage site, followed by a highly conserved fusion peptide (Wang *et al.*, 2007). The HN protein of MprPV is 582 amino acids in length and is predicted to be a type II integral membrane protein. All six residues of the conserved sialic acid binding site sequence of the respiroviruses and rubulaviruses, NRKSCS (Jorgensen *et al.*, 1987) are present in MprPV (Wang *et al.*, 2007). The L protein of MprPV is 2253 amino acids in length and contains the QGDNQ motif present in the majority of paramyxoviruses.

Phylogenetic analysis of the N, P, M, F, HN and L proteins demonstrate the close relationship between MprPV and PorPV (see below) and support the classification of both viruses as rubulaviruses (Wang *et al.*, 2007).

Porcine rubulavirus (PorPV)

Although the origin of PorPV has not been conclusively identified, we have included the discussion of PorPV in this section based on its close genetic relationship with MprPV and the preliminary surveillance data indicating the presence of PorPV-specific antibodies in at least one bat species in Mexico.

An outbreak of encephalitis occurred in piglets, in 1980, in farms around the town of La Piedad, district of Michoacan, Mexico. Young piglets displayed central nervous system manifestations and corneal opacity, while older pigs seem to be more resistant with only corneal opacity commonly observed. In pregnant sows the virus appeared to be responsible for reproductive disturbances such as stillbirth and mummification (Stephan *et al.*, 1988). The virus at the centre of the outbreak, La Piedad Michoacan virus (LPMV) or, blue-eye disease virus, is now classified as the *Rubulavirus* PorPV (Lamb *et al.*, 2005).

PorPV was isolated from the brain of a 5-day-old piglet with symptoms of nervous disorder, corneal opacity and pneumonia (Moreno-Lopez *et al.*, 1986). Following infection of pig kidney (PK-15) and bovine turbinate (BT) cells, CPE was evident in the form of syncytia and vacuolated cells. Electron microscopy analysis of PorPV revealed particles with the morphology of a paramyxovirus. Purified PorPV nucleocapsids also observed by electron microscopy revealed structures typical of paramyxovirus nucleocapsids (Sundqvist *et al.*, 1990).

In preliminary animal studies, PorPV was highly infectious and lethal in very young pigs. Infected piglets experienced fever, anorexia and conjunctivitis four days post infection, and by nine days post infection, the animals had developed neurological symptoms and corneal opacity. In the young piglets PorPV had a tropism for the lungs and the thalamus, midbrain and cerebral cortex in the brain (Moreno-Lopez *et al.*, 1986). In contrast, corneal opacity was the only sign of disease seen in older pigs, results that mimicked the disease outbreaks. Mice were also infected with PorPV during the initial virus characterization. PorPV was lethal for mice infected intracerebrally, but mice inoculated intraperitoneally all survived the infection.

A subsequent animal study was performed where pigs were inoculated intranasally to mimic natural infection (Allan *et al.*, 1996). Experimental infection of 3-day-old pigs induced neurological signs in all animals and the infection was lethal. Neurological signs were only observed in 30% of infected 17-day-old pigs and all other animals remained healthy for the duration of the experiment. PorPV infection did not produce any clinical signs in 30-day-old pigs. Results of this study confirm field observations that disease severity following PorPV infection is dependent on the age of the pig. Virus excretion experiments performed during the animal study demonstrated that excretion via the respiratory tract and urine were important methods of spread of PorPV.

Although not lethal, PorPV can replicate in tissues of seronegative pregnant pigs, cross the placenta and cause fetal death and mummification (Hernandez-Jauregui *et al.*, 2004). PorPV also has a tropism for epididymis in infected mature pigs (Ramirez-Mendoza *et al.*, 1997) and may be transmitted to the female reproductive tract in the semen (Solis *et al.*, 2007). Since the first discovery of PorPV in La Piedad, Michoacan State, in 1980, the virus has spread to other parts of central Mexico, but appears to be contained to that country (Wang *et al.*, 2007).

A serological survey was conducted on the subtropical Pacific coast of the state of Colima, Mexico, to investigate the role of bats in outbreaks of PorPV and rabies (Salas-Rojas *et al.*, 2004). In the study, 108 serum samples from 13 different bat species were analysed for antibodies against PorPV and one sample was positive. Alone, this survey is unable to indicate the origin of PorPV. More surveillance studies are needed to either prove or disprove the bat origin hypothesis.

The genome of PorPV has been completely sequenced and obeys the 'rule of six'. At 15,180 nucleotides, the PorPV genome is the smallest of all the paramyxoviruses that infect mammals (Wang *et al.*, 2007). Complementarity between the PorPV genome leader and trailer is imperfect with two mismatches in the terminal 19 nucleotides. The 3′ leader sequence of PorPV is the standard length of 55 nucleotides and the 5′ trailer is 43 nucleotides. PorPV genes are separated by intergenic regions ranging from 1–46 nucleotides in length.

The PorPV N protein is 545 amino acids in length and shares 56% sequence homology with MprPV (Wang *et al.*, 2007). The P gene is 1372 nucleotides long, potentially encoding three proteins. Transcription of the P gene encodes a 249 amino acid V protein (Berg *et al.*, 1992; Wang *et al.*, 2007). The 404 amino acid P protein is produced following the insertion of two non-templated G residues in the RNA editing site. Insertion of one non-templated G residue potentially encodes a 174 amino acid W protein (Wang *et al.*, 2007). Similar to MprPV, PorPV contains a putative C overlapping ORF within the P gene, potentially encoding a 126 amino acid C protein (Berg *et al.*, 1992), but the C protein was not detected experimentally (Wang *et al.*, 2007). The M protein is predicted to be 369 amino acids in length (Berg *et al.*, 1991).

The PorPV F protein is predicted to be a type I integral membrane protein containing 541 amino acid residues. The PorPV F protein contains the highly basic furin sequence, HRKKR, site followed by a highly conserved fusion peptide (Berg *et al.*, 1997; Wang *et al.*, 2007). The HN protein of PorPV is 576 amino acids in length and is predicted to be a type II integral membrane protein (Sundqvist *et al.*, 1992; Wang *et al.*, 2007). Molecular analysis of the HN protein correlated with haemagglutination and neuraminidase studies observed during the initial virus isolation (Moreno-Lopez *et al.*, 1986) and infectivity of PorPV depends on the presence of sialylated receptors in host cells (Reyes-Leyva *et al.*, 1993, 1997, 1999). The L protein of PorPV is 2251 amino acids in length and contains the QGDNQ motif present in the majority of paramyxoviruses (Svenda *et al.*, 1997).

Phylogenetic analysis of the N, P, M, F, HN and L proteins unequivocally indicate that PorPV is more closely related to MprPV than to any other member of the genus *Rubulavirus* (Wang *et al.*, 2007).

Other bat paramyxoviruses

Since the discovery of a large number of bat-borne coronaviruses, it has been hypothesized that bats are probably the natural hosts for all presently known coronavirus lineages and that all coronaviruses recognized in other species were derived from viruses residing in bats (Vijaykrishna *et al.*, 2007). Will the same apply to paramyxoviruses? The trend of bat paramyxovirus discovery suggests that this is possible.

PCR surveillance of bat faecal samples in Africa by a German group demonstrated the presence of novel paramyxovirus sequences which are closely related to henipaviruses, highlighting the genetic diversity of this group of viruses is not limited to bats in the Australasian region (Drexler *et al.*, 2009).

The genetic diversity of rubulaviruses has increased significantly owing to the discovery of MenPV and TioPV. More recently, a Hong Kong group has discovered three new bat rubulaviruses (Tuhokoviruses 1, 2 and 3) from a PCR surveillance study involving 489 bats of 11 different species (Lau *et al.*, 2010).

In another study conducted by US and Kenyan scientists, 11 paramyxoviruses were discovered in Kenyan bats (Tong *et al.*, 2010). Interestingly, some of these new viruses seem to be most closely related to JPV and BeiPV, raising the question whether these viruses could originate from bats before being established in rodents.

Rodent paramyxoviruses

Sendai virus (SeV), an extremely contagious respiratory virus, is one of the most important pathogens of rodents, and this *Respirovirus* has been indispensable in paramyxovirus research since isolation from laboratory rodents in the 1950s (Baker, 1998). At least four paramyxoviruses have been isolated from rodents since the 1960s.

J paramyxovirus (JPV)

The trapping of wild mice (*Mus musculus*) during a study investigating the pathology of feral rodents in northern Queensland, Australia, in 1972 (Mesina *et al.* 1974) lead to the isolation of JPV on four separate occasions (Jun *et al.*, 1977). The four moribund mice from which JPV was isolated all displayed extensive haemorrhagic lung lesions. Passaging of JPV in mouse kidney and lung primary cell lines, and continuous cell lines from mouse kidney (CSL217), human (MRC5 and HEp-2), pig kidney (PS), baby hamster kidney (BHK-21) and monkey kidney (Vero) resulted in CPE characterized by destruction of the cell monolayer and vacuolated syncytia formation. Cell lines where CPE was not observed were bovine kidney (MDBK), human (HeLa and Chang liver) and rat fibroblast (RT). CPE was associated with eosinophilic cytoplasmic inclusions, but no involvement of the nucleus was detected.

Electron microscopy exposed a viral nucleocapsid of 15 nm diameter, with 'herringbone' morphology, typical of paramyxoviruses. Evidence of soluble or virus-associated haemagglutinin was not detected and hemadsorption using standard techniques and erythrocytes from a variety of species was not observed (Jun *et al.*, 1977). Complement fixation assays were used in an attempt to detect a serological relationship between JPV and a range of paramyxoviruses all produced negative results.

Intranasal or subcutaneous inoculation of mice, pigs and rats with JPV did not produce clinical signs of disease in these animals, although neutralizing antibodies against the virus were detected. Viraemia, in conjunction with mild lethargy was observed in rats and mice inoculated with JPV, and varying degrees of haemorrhagic interstitial pneumonia was discovered in the animals upon autopsy, three weeks post-infection (Jun *et al.*, 1977). Serological surveillance by serum neutralization tests among Australian mammals detected JPV antibodies in 27/96 wild mice, 17/123 wild rats, 13/107 pigs, 1/157 cattle and 2/91 humans, but a similar survey using complement fixation experiments on American mammals failed to detect JPV reactive antibodies in any of 55 laboratory mice, 45 humans and 5 hamsters. Serum neutralization experiments, when taken into account with viral characterization and animal experiments, suggest JPV is a natural respiratory pathogen of feral *Mus musculus* in Queensland, Australia.

The entire genome of JPV has been sequenced and is 18,954 nucleotides in length (Jack *et al.*, 2005). The increase in genome size of JPV over that of other paramyxoviruses is due to the presence of two additional genes between the fusion and attachment genes, and an extensive second

ORF within the attachment protein gene. The genome organization of JPV is 3′-N-P/V/C-M-F-SH-TM-G-L-5′ (Fig. 12.1). At 18,954 nucleotides, the genome of the JPV is compatible with the 'rule of six'. The terminal end sequences of paramyxoviruses customarily display complementarity of 11–13 nucleotides (Lamb and Parks, 2007) but complementarity between the genome termini of JPV is imperfect, differing by two nucleotides.

The 3′ leader sequence of JPV is 55 nucleotides, matching the length of that of other paramyxovirus leader sequences, and the 5′ trailer sequence is 36 nucleotides, similar to the 33 nucleotide trailer region of the henipaviruses. JPV genes are separated by trinucleotide intergenic regions, with the bases GAA making up the nucleotide sequence in all gene junctions except between the 3′ leader and N gene where the sequence is AAA (Jack *et al.*, 2005). The genome of JPV consists of eight genes and may produce up to 12 proteins.

The JPV N protein contains 522 amino acids and has a predicted molecular mass of 58.37 kDa (Jack *et al.*, 2005). The P gene of JPV has the capacity to encode four putative proteins. The highly acidic P protein contains 499 amino acids, has a predicted molecular mass of 53.68 kDa, and is translated after faithful transcription of the P gene. There are a large number of potential sites for phosphorylation on the P protein. The C protein contains 152 amino acids, with a predicted molecular mass of 17.72 kDa, and is translated via an alternative ORF within the P gene. The V protein is produced as a result of the addition of one non-templated G residue into the RNA editing site of the P gene and contains 292 amino acids, with predicted molecular mass of 31.27 kDa. The W protein contains 310 amino acids, has a predicted molecular mass of 33.74 kDa, and is produced as a result of the addition of two non-templated G residues into the RNA editing site of the P gene.

The M protein of JPV contains 340 amino acids, has a predicted molecular mass of 37.69 kDa, and is structurally similar to M proteins of other known paramyxoviruses. The F protein is a type I glycoprotein containing 544 amino acids, with a predicted molecular mass of 59.74 kDa. The cleavage site of the F protein of JPV is monobasic, with the sequence GVPGVR (Jack *et al.*, 2005).

The SH protein, consisting of 69 amino acids, with a predicted molecular mass of 7.66 kDa, is translated from the novel SH gene of JPV (Jack *et al.*, 2005). SH proteins have been identified in PIV-5 (Hiebert *et al.*, 1985), MuV (Elango *et al.*, 1989), and APMV-6 (Chang *et al.*, 2001), with hydrophobic regions present in these proteins functioning as transmembrane anchors. Although the positioning of the JPV SH gene is analogous to that of the SH gene of PIV-5 and MuV, no sequence homology is observed between the JPV SH protein and any other paramyxovirus protein. Structurally, the JPV SH protein is similar to other SH proteins, having a predicted transmembrane domain at the N-terminus. Identification of the JPV SH protein in infected cells has been achieved by radio-immunoprecipitation, but the subcellular localization of the protein is yet to be determined. Dicistronic forms of SH mRNA, F-SH and SH-TM were also detected (Jack *et al.*, 2008).

The other novel gene within the JPV genome, TM, produces a type II transmembrane glycoprotein of 258 amino acids in length, with a predicted molecular mass of 29.09 kDa, and this protein lacks homology with any other proteins listed in the NCBI database. The TM protein, which is basic in nature, contains two potential sites for O-linked glycosylation and has a putative transmembrane domain. The TM protein was identified by western blot in JPV infected cells, and indirect immunofluorescence experiments supported the prediction that the TM protein is an integral membrane protein with the C-terminal region exposed at the cell surface (Jack *et al.*, 2008).

The attachment gene produces a protein containing 709 amino acids, with a predicted molecular mass of 78.13 kDa, and also contains an extraordinarily long untranslated region of 2218 nucleotides after the stop codon (Jack *et al.*, 2005). The untranslated region of the attachment gene consists of a 2115 nucleotide second ORF, which has been designated ORF-X, beginning directly after the stop codon. Paramyxoviruses with HN attachment proteins contain six of the seven conserved residues believed to be neuraminidase active site (Langedijk *et al.*, 1997). This conservation pattern is also present in the JPV G protein. The conserved hexapeptide sequence NRKSCS, thought to be the sialic acid binding site (Jorgensen *et al.*, 1987), is changed to NRRSCS in the JPV G protein. The conservation of the sialic acid binding site, along with the conservation of six out of seven resides of the neuraminidase active site suggests the attachment protein of JPV would have haemagglutination and neuraminidase activities. In contrast to predictions, experimental evidence suggested the absence of either haemagglutination or neuraminidase activities, and the JPV attachment protein was hence designated as a G protein (Jack *et al.*, 2005). The JPV G protein contains a putative hydrophobic transmembrane domain and five potential sites for the addition of N-linked carbohydrates. Analysis of mRNA transcripts by northern blot revealed monocistronic G mRNA transcripts. JPV may utilize an unconventional mechanism to translate a protein from ORF-X, either individually, or as part of a G–X fusion product of 1414 amino acids in length. No evidence of X synthesis or translation was detected by western blot, indirect immunofluorescence or radioimmunoprecipitation in JPV infected cells (Jack *et al.*, 2008).

The L gene of JPV produces a protein containing 2204 amino acids, with a predicted molecular mass of 254.46 kDa. The six conserved domains essential for polymerase activity are all present in JPV. JPV contains the conserved QGDNQ sequence (Jack *et al.*, 2005), not the altered QGDNE sequence observed in the other four paramyxoviruses with large genomes (Miller *et al.*, 2003). Mutation into the QGDNE sequence did not negatively affect replication of JPV in a minigenome assay system (Magoffin *et al.*, 2007).

Beilong paramyxovirus (BeiPV)

A routine search for sequences related to JPV using BLASTx on the NCBI server revealed two putative novel cDNAs, termed Angrem 104 (AF367870) and Angrem 52 (AY040225) (Liang *et al.*, 2003), from human mesangial cells with significant homology to JPV P, M and F genes (Jack *et al.*, 2005). The relationship between the Angrem cDNA sequences to other paramyxovirus was reported (Schomacker *et al.*, 2004; Basler *et al.*, 2005), but the close relationship to JPV was not ascertained by these groups as the JPV genome sequence had not yet been deposited in the NCBI database. Although non-retroviral RNA integration into a eukaryotic genome has been reported (Crochu *et al.*, 2004; Horie *et al.*, 2010), contaminating viral RNA in the mesangial cells was believed to be the more likely source of the homologous paramyxovirus sequence.

A virus was later isolated from the human mesangial cell line from which the cDNA sequences Angrem 104 and Angrem 52 were originally obtained. The isolated virus produced syncytia in Vero cells, and upon examination of cell cultures and supernatant by electron microscopy, pleomorphic virions were observed (Li *et al.*, 2006). The virions observed by negative contrast electron microscopy were enveloped, possessed surface projections, and contained ‘herringbone-like’ RNP, features characteristic of paramyxoviruses. The new paramyxovirus was named Beilong virus to represent the joint collaboration between the Beijing and Geelong researchers that lead to the isolation and characterization of the virus. Concurrently, isolation of this paramyxovirus from human mesangial cells, obtained from the American Type Culture Collection, was unsuccessfully attempted (Basler *et al.*, 2005). Later, it was discovered that a rat mesangial cell line, from the same laboratory in which the human mesangial cells were cultured, carried BeiPV, which is believed to be the original source of the virus (Li *et al.*, 2006).

The genome of BeiPV has been completely sequenced and at 19,212 nucleotides in length, the genome is the largest in the family *Paramyxoviridae* sequenced to date, which also is the largest in the order *Mononegavirales* (Li *et al.*, 2006). BeiPV, like JPV, contains the two novel genes, SH and TM, and an extended G gene (Fig. 12.1). The BeiPV genome length is compatible with the ‘rule of six’. Complementarity between the genome termini of BeiPV is imperfect, with differences at three nucleotides. The 3′ leader sequence of BeiPV is 55 nucleotides, matching the length of that of other paramyxovirus leader sequences, and the 5′ trailer sequence is 25 nucleotides, similar to the trailer region of the rubulaviruses. BeiPV genes are separated by conserved trinucleotide IGRs, with the bases GAA making up the nucleotide sequence in all gene junctions (Li *et al.*, 2006).

The BeiPV N protein contains 522 amino acids,

and has a predicted molecular mass of 58.09 kDa (Li *et al.*, 2006). The P gene of BeiPV has the capacity to encode four putative proteins, P, C, V and W. The P protein contains 496 amino acids, has a predicted molecular mass of 53.60 kDa, and is translated after faithful transcription of the P gene. The V protein is produced as a result of the addition of one non-templated G residue into the RNA editing site of the P gene and contains 292 amino acids, with predicted molecular mass of 31.90 kDa. The W protein contains 313 amino acids, has a predicted molecular mass of 34.46 kDa, and is produced as a result of the addition of two non-templated G residues into the RNA editing site of the P gene. The C protein contains 160 amino acids, with a predicted molecular mass of 18.44 kDa.

The M protein of BeiPV contains 340 amino acids, and has a predicted molecular mass of 37.43 kDa, and the F protein is a type I glycoprotein containing 544 amino acids, with a predicted molecular mass of 58.98 kDa. The cleavage site of the F protein of BeiPV is monobasic, with the sequence KLGNVK (Li *et al.*, 2006). The BeiPV F gene cleavage site is most similar to that of HeV and is not expected to be cleaved by furin. The L gene of BeiPV produces a protein containing 2172 amino acids with a predicted molecular mass of 249.97 kDa. BeiPV, like JPV, contains the conserved QGDNQ sequence.

The gene encoding the attachment protein of BeiPV, at 4527 nucleotides, is the largest of all paramyxovirus attachment protein genes identified to date. The attachment gene produces a protein containing 734 amino acids, with a predicted molecular mass of 81.43 kDa, and also contains a long untranslated region of 2284 nucleotides after the stop codon (Li *et al.*, 2006). The untranslated region of the attachment gene consists of two additional ORFs, which have been designated ORF-X1 and ORF-X2. ORF-X1 is present in the same reading frame as the G ORF and may encode a protein of 299 amino acids. ORF-X2, overlapping ORF-X1 by 31 nucleotides, is present in an alternative reading frame and may encode a protein of 394 amino acids (Fig. 12.1). Currently, there is no evidence to suggest that proteins are encoded by ORF-X1 and ORF-X2.

The BeiPV SH and TM genes are similar in size and location to those of JPV. The SH protein, consisting of 76 amino acids, encodes a protein with a predicted molecular mass of 9.05 kDa, and the TM protein, consisting of 254 amino acids, encodes a protein with a predicted molecular mass of 28.73 kDa (Li *et al.*, 2006). The function of these proteins is unknown.

Prior to the sequencing of the complete genome, a study based on partial sequence of the P, M and F genes suggested that BeiPV should be classified as a member of genus *Henipavirus* (Schomacker *et al.*, 2004). This initial proposal was not supported by the genome organization structure revealed by subsequent whole genome sequence analysis (Li *et al.*, 2006). Moreover, genetic analysis of BeiPV, JPV and NiV using a minireplicon system demonstrated that the genomic termini and proteins of the polymerase complex were exchangeable between BeiPV and JPV, but not NiV (Magoffin *et al.*, 2007). These studies determined that BeiPV and JPV are closely related, but both are distinct from the henipaviruses.

Mossman virus (MosPV)

MosPV was isolated in northern Queensland, Australia, on three occasions from 1970–1971. The virus was isolated from pooled organs of a female rat (*Rattus leucopus*), trapped near Mossman, Queensland, a pool of mites (*Laelaps echidninus*) and pooled organs from a male rat (*Rattus fuscipes*) (Campbell *et al.*, 1977). Infection of BHK-21 cells with MosPV produced CPE with syncytia formation, a hallmark of paramyxovirus infection. Electron microscopic analysis of MosPV displayed viral particles and free nucleocapsids, which when taken into account with cellular observations, lead to the conclusion that MosPV was a paramyxovirus (Campbell *et al.*, 1977).

Sequencing of MosPV revealed a genome of 16,650 nucleotides in length, with arrangement of genes consistent with that of the henipaviruses (Miller *et al.*, 2003). The MosPV genome conforms to the 'rule of six'. Characterization of the genome uncovered several features that may assist in classification of MosPV. The MosPV 3′ leader region of the genome was found to contain 55 nucleotides, consistent with all paramyxoviruses and the 5′ trailer region contains 51 nucleotides. The intergenic regions that separate the viral genes are all trinucleotide sequences.

The 528 amino acid N protein of MosPV shares most sequence similarity to that of NarPV, another rodent paramyxovirus (Lambeth *et al.*, 2009). The coding strategy determined for the P gene is similar to that of the henipaviruses, morbilliviruses and respiroviruses, where translation of unedited P gene mRNA produces the P protein and the V protein is produced by RNA editing (Miller *et al.*, 2003). The MosPV C protein is produced from an alternative ORF. The 340 amino acid MosPV M protein has the highest sequence identity with the NarPV M protein.

The F protein of MosPV is a type I integral membrane protein. The F genes of many paramyxoviruses contain the conserved sequence R-X-K/R-R (Hosaka *et al.*, 1991) that directs cleavage of the F protein by furin, but the MosPV sequence deviates from the conserved sequence and contains AGNKK as the predicted cleavage sequence (Miller *et al.*, 2003). The conserved sequence of respiroviruses and rubulaviruses, NRKSCS, in the attachment protein is thought to form part of the sialic acid binding site (Jorgensen *et al.*, 1987). Similar to henipaviruses and morbilliviruses, MosPV lacks this conserved site. The lack of conservation of the hexapeptide motif in the attachment protein, along with the failure to detect any neuraminidase activity, resulted in the attachment gene of MosPV to be designed G, a type II glycoprotein (Miller *et al.*, 2003). The conserved QGDNQ motif in the L gene was replaced with QGDNE (Miller *et al.*, 2003), an observation also made in HeV, NiV, TupPV and NarPV.

Nariva virus (NarPV)

NarPV was isolated from forest rodents (*Zygodontomys b. brevicauda*), trapped on Bush Bush Island in the Nariva swamp, Eastern Trinidad, in 1962 and 1963 (Tikasingh *et al.*, 1966). NarPV was isolated from mouse brains and when the virus infected Vero and BHK-21 cells, formation of syncytia was observed. Haemagglutination and cell binding studies indicate NarPV may not use sialic acid as a receptor for attachment (Karabatsos *et al.*, 1969). NarPV virus particles were visualized as 130–250 nm pleomorphic structures capable of haemagglutination of red blood cells (Walder, 1971). Intracytoplasmic inclusion bodies containing helical nucleocapsids engulfed within the matrix of lysosomes of brain cells were observed, and brain lesions were comparable to those of MeV infection (Walder, 1971).

The pathogenesis of NarPV was studied in hamsters, an animal closely related to the natural rodent host of the virus. Infection of suckling hamsters with NarPV resulted in acute necrotizing encephalitis with multinucleated cells observed in ependyma and neutrophils (Roos and Wollmann, 1979). Weanling hamsters infected with NarPV exhibited less cerebral parenchymal necrosis than suckling hamsters. NarPV produced non-productive infection in the weanlings, but despite the lack of infectious virus particles, the animals developed fatal encephalitis. The maturity of the host immune system appears to be important in NarPV pathogenesis, and age-related changes in susceptible neural cells, such as changes in cell surface receptors and cellular enzymes, may be critical in disease progression.

The genome of NarPV has been completely sequenced and is 15,276 nucleotides in length (Lambeth *et al.*, 2009). At this length, the genome of NarPV obeys the 'rule of six' and the genome organization is typical of most paramyxoviruses. Characterization of the genome revealed that NarPV is most closely related to MosPV. The NarPV 3′ leader region of the genome contains 55 nucleotides, consistent with all other paramyxoviruses and the gene boundaries all contain GAA intergenic triplets. The 5′ trailer of the genome is 42 nucleotides. The first and last 11–13 nucleotides of paramyxovirus genomes are highly conserved and complementary (Lamb and Parks, 2007), but the NarPV genome contains a C residue at the fifth position of both the leader and trailer, which is unique (Lambeth *et al.*, 2009).

The N protein of NarPV shares the most sequence identity with the MosPV N protein, and both viruses contain the sequence FAPGNYPLLWSYAMG in the conserved F-X4-Y-X3-Φ-S-Φ-A-M-G (X = any amino acid, Φ = aromatic amino acid) N-protein motif (Lambeth *et al.*, 2009). The coding strategy determined for the P gene was similar to that of most other paramyxoviruses where the faithfully transcribed P gene mRNA produces a P protein. Insertion of a single non-templated G residue at the RNA editing site produces the V protein. The coding capacity for a C protein exists in an alternative ORF. The M protein is 340 amino acids in length and is basic in nature.

The NarPV F protein is predicted to be a type I integral membrane protein. The conserved furin cleavage sequence R-X-K/R-R (Hosaka *et al.*, 1991) is not present in NarPV, but the virus instead contains SGRNK as the cleavage sequence (Lambeth *et al.*, 2009). The attachment protein, H, is predicted to be a type II integral membrane protein. The conserved NRKSCS sequence that forms the sialic acid binding site (Jorgensen *et al.*, 1987) in the attachment protein is absent and is replaced with the sequence AYDGCA. The conserved QGDNQ motif in the L gene was replaced with QGDNE, the motif also present in HeV, NiV, TupPV and MosPV.

Phylogenetically, NarPV is closely related to MosPV and both viruses are positioned between the *Morbillivirus* and *Henipavirus* genera, along with SalPV, JPV, BeiPV and TupPV (Lambeth *et al.*, 2009).

Aquatic paramyxoviruses

Disease outbreaks in wild and farmed fish occur sporadically, resulting in increased mortality and/or economic loss. Disease in aquatic animals has been linked to viruses such as herpesviruses and birnaviruses, and more recently, paramyxoviruses (Kvellestad *et al.*, 2003).

Atlantic and Pacific salmon paramyxovirus (AsaPV and PsaPV)

AsaPV was first isolated in 1995, in Norway, during an attempt to culture the causative agents of epitheliocystis in Atlantic salmon (*Salmo salar* L.) (Kvellestad *et al.*, 2003). Upon isolation, highly pleomorphic paramyxovirus virions with fine surface projections of 10 nm in length were observed by electron microscopy (Kvellestad *et al.*, 2003). AsaPV was analysed and found to have all three surface activities (haemagglutinin, neuraminidase and fusion) typical of most paramyxoviruses (Kvellestad *et al.*, 2003). Paramyxoviruses have also been isolated from Pacific salmon since 1982 (Batts *et al.*, 2008). PsaPV have mainly been isolated from Chinook salmon (*Oncorhynchus tshawytscha*) Batts *et al.*, 2008). Phylogenetic analyses revealed that PsaPV were most closely related to AsaPV, but were sufficiently distinct to suggest that they are not clearly members of any of the established genera in the family *Paramyxoviridae* (Batts *et al.*, 2008). The closest relatives of AsaPV and PsaPV are the respiroviruses (Fridell *et al.*, 2004).

The complete genome of AsaPV has been sequenced by two independent groups, although the length is conflicting between the two studies. The genome length was reported as 16,968 (Falk *et al.*, 2008) and 16,965 (Nylund *et al.*, 2008) nucleotides. Given that a genome length of 16,965 does not obey the 'rule of six', and there have not been any viruses classified within the subfamily *Paramyxovirinae* that do not conform to this rule, the genome length of 16,968 nucleotides is most likely the correct of the two.

The AsaPV 3′ leader region of the genome contains 55 nucleotides, consistent with all other paramyxoviruses and the gene boundaries all contain GAA intergenic triplets, except the F/HN junction, which contains a GUA triplet (Falk *et al.*, 2008; Nylund *et al.*, 2008.). The 5′ trailer of the genome is reported as either 51 (Falk *et al.*, 2008) or 54 (Nylund *et al.*, 2008) nucleotides. The first and last 11–13 nucleotides of paramyxovirus genomes are highly conserved and complementary (Lamb and Parks, 2007), and the first 12 (Falk *et al.*, 2008) to 13 (Nylund *et al.*, 2008) are complementary in AsaPV.

The AsaPV N protein is 350 amino acids in length and contains the sequence FAPGAYPLLWSYAMG in the conserved F-X4-Y-X3-Φ-S-Φ-A-M-G (X = any amino acid, Φ = aromatic amino acid) N-protein motif (Falk *et al.*, 2008; Nylund *et al.*, 2008). The coding strategy predicted for the P gene was similar to that of most other paramyxoviruses where the faithfully transcribed P gene mRNA produces a 575 amino acid P protein. Insertion of a single non-templated G residue at the RNA editing site produces the 328 amino acid V protein. A W protein is sometimes encoded from the P gene (Lamb and Parks, 2007) and the insertion of two non-templated G residues in the editing site may produce a W protein. Alternative ORFs within the P gene produce a C protein and possibly a hypothetical small basic (SB) protein (Falk *et al.*, 2008), similar to those present in the P genes of HeV (Yu *et al.*, 1998), TupPV (Tidona *et al.*, 1999) and FelPV (Kurath *et al.*, 2004). The M protein is 354 amino acids in length, is basic in nature and shares the highest amino acid identity with JPV (Nylund *et al.*, 2008).

Hydrophobicity and transmembrane analysis of the deduced 522 amino acid F protein demonstrate features of a type I integral transmembrane protein (Falk *et al.*, 2008; Nylund *et al.*, 2008). The conserved furin cleavage sequence R-X-K/R-R (Hosaka *et al.*, 1991) is not present in AsaPV, but the virus instead contains the monobasic cleavage site GTIVR. The AsaPV attachment protein, HN, is a type II integral membrane protein, is most similar to the respirovirus attachment proteins (Nylund *et al.*, 2008). The HN protein contains the sialic acid binding site motif, NRKSCS, which is conserved in respiroviruses and rubulaviruses (Jorgensen *et al.*, 1987). respiroviruses and rubulaviruses contain six of seven conserved neuraminidase active site residues (Langedijk *et al.*, 1997) and the globular head domain of AsaPV HN protein contains all seven of these residues (Falk *et al.*, 2008; Nylund *et al.*, 2008). These features support the finding of neuraminidase activity with AsaPV (Kvellestad *et al.*, 2003). The AsaPV L protein is predicted to be either 2209 (Falk *et al.*, 2008) or 2208 (Nylund *et al.*, 2008) amino acids in length and contains the conserved QGDNQ motif present in the majority of paramyxoviruses.

Phylogenetic trees (Fig. 12.2) place AsaPV as an independent branch, not closely related to other paramyxoviruses, but with the genus *Respirovirus* as the closest neighbour (Falk *et al.*, 2008; Nylund *et al.*, 2008). A proposal for the creation of a new genus was submitted to the International Committee on Taxonomy of Viruses (ICTV). The genus name proposed was *Aquaparamyxovirus*, with AsaPV the type species. Viral species to be included in this genus would be required to have fish hosts, low replication temperatures, and separate from paramyxoviruses in other known genera (G. Kurath, personal communication).

Reptilian paramyxoviruses

In reptiles, paramyxovirus infection is commonly associated with respiratory disease. Paramyxoviruses have been detected in snakes, lizards, iguanas and tortoises.

Fer-de-Lance virus (FdlPV)

FdlPV is a reptilian paramyxovirus that was isolated from a Fer-de-Lance viper (*Bothrops atrox*) in 1972, in Switzerland (Clark *et al.*, 1979). Replication of FdlPV requires the low body temperature of a cold-blooded reptile host. Reptile infection by a paramyxovirus is most commonly associated with respiratory disease and since 1979 FdlPV has been isolated from a variety of reptiles including snakes (Ahne *et al.*, 1999; Franke *et al.*, 2001; Kindermann *et al.*, 2001; Marschang *et al.*, 2009), lizards (Ahne *et al.*, 1999; Jacobson *et al.*, 2001; Marschang *et al.*, 2009) and tortoises (Marschang *et al.*, 2009). A study of 16 paramyxoviruses isolated from reptiles demonstrated strong antigenic relationships between viruses, irrespective of the virus origin (Ahne *et al.*, 1999). Sequence analysis of the L gene of FdlPV isolated from snakes, lizards and a tortoise phylogenetically places this virus between the respiroviruses and rubulaviruses (Marschang *et al.*, 2009).

The FdlPV genome has been completely sequenced and at 15,378 nucleotides in length (Kurath *et al.*, 2004) obeys the 'rule of six'. The genome contains all genes common to paramyxoviruses, along with an extra gene, designated U for unknown (Kurath *et al.*, 2004). The genome organization of FdlPV is 3′-N-U-P/V-M-F-HN-L-5′ (Fig. 12.1). Although containing an additional gene, the FdlPV genome is among the smallest of known paramyxoviruses. The 3′ leader region of the genome contains 55 nucleotides, consistent with all other paramyxoviruses and the gene boundaries all contain either GAA or GGA intergenic triplets. The 5′ trailer of the genome is 36 nucleotides. The first and last 18 nucleotides of the genome are complementary (Kurath *et al.*, 2004).

The FdlPV N protein contains 471 amino acids and has relatively low amino acid identity with other characterized paramyxoviruses (Kurath *et al.*, 2004). The primary transcript of the P gene encodes the V protein, and insertion of two

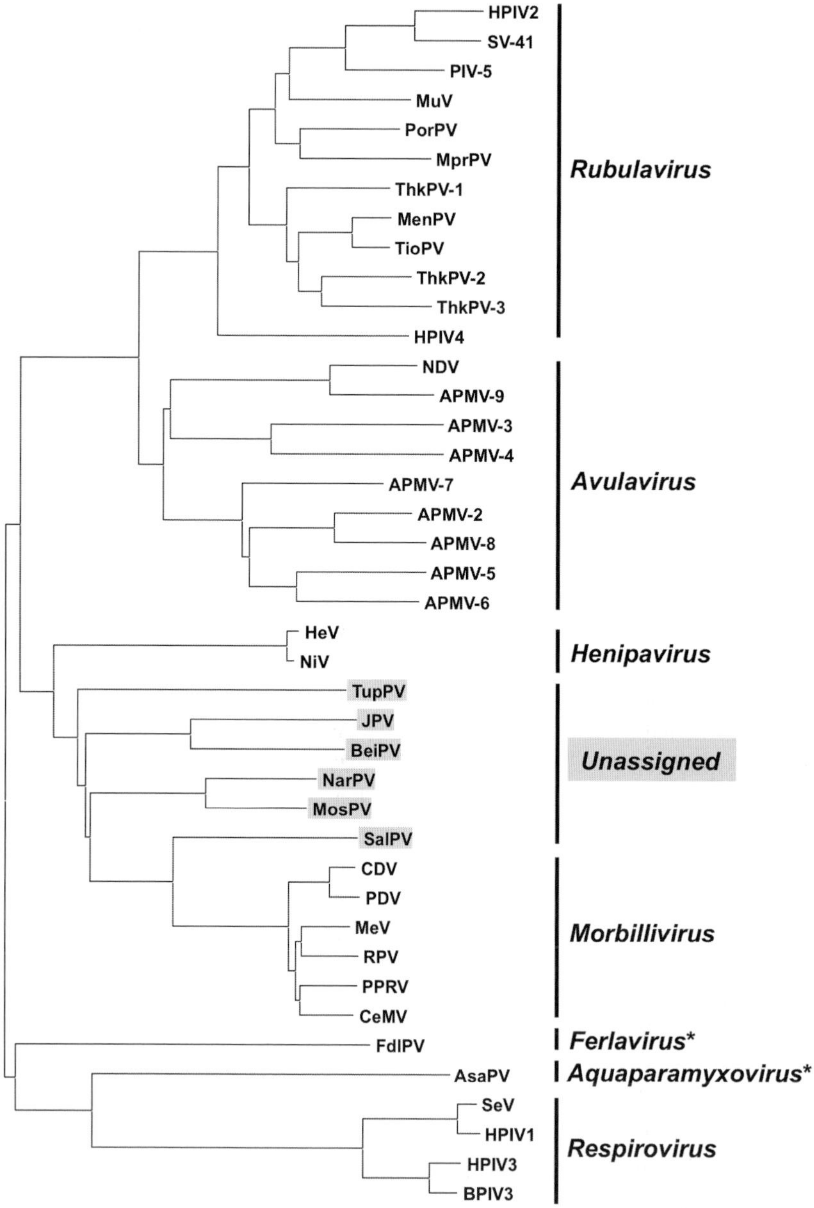

Figure 12.2 Phylogenetic tree of selected paramyxoviruses. Phylogenetic analysis using MEGA4.1 was performed on the amino acid sequence of N proteins from various members of the subfamily *Paramyxovirinae*. Asterisk indicates proposed new genus which is yet to be formally approved by ICTV. Unclassified viruses are highlighted by shading.

additional G residues by the RNA editing mechanism produces the P protein. This expression strategy is identical to that of the rubulaviruses. The P gene does not encode a C protein, a feature shared by most rubulaviruses (Wang *et al.*, 2007). The FdlPV M gene produces a protein predicted to be 351 amino acids in length.

The FdlPV F gene encodes a 545 amino acid type I integral membrane protein that is most similar to TupPV (Kurath *et al.*, 2004). Comparative analysis of the F gene revealed a multibasic consensus cleavage site typical of other paramyxoviruses, suggesting that the protein is cleaved by furin (Franke *et al.*, 2001, 2006). The FdlPV F protein has two potential N-linked glycosylation sites. The FdlPV attachment protein, HN, is a type II integral membrane protein, is most similar to the *Respirovirus* HN proteins and least similar to the morbillivirus H proteins (Kurath *et al.*, 2004). The HN protein contains the sialic acid binding site motif, NRKSCS, which is conserved in respiroviruses and rubulaviruses (Jorgensen *et al.*, 1987). respiroviruses and rubulaviruses contain six of seven conserved neuraminidase active site residues (Langedijk *et al.*, 1997) and the globular head domain of FdlPV HN protein also contains these residues (Kurath *et al.*, 2004). These features of the HN protein correlate with observed neuraminidase activity of FdlPV virion (Essbauer and Ahne, 2001). The FdlPV L protein is 2181 amino acids in length and contains the conserved QGDNQ motif present in the majority of paramyxoviruses.

The novel U gene present between the N and P genes is 604 nucleotides long and encodes a protein of either 167 or 143 amino acids, depending on which codon is the actual initiator of translation. Based on hydrophobicity analysis, the U protein could possibly be a small transmembrane protein, but this has not been experimentally confirmed (Kurath *et al.*, 2004). The U protein sequences from three different FdlPV isolates were analysed and found to be no more variable than any other snake paramyxovirus genes (Ahne *et al.*, 1999; Kurath *et al.*, 2004).

A proposal for the creation of a new genus was submitted to the ICTV. The genus name proposed was *Ferlavirus*, with FdlPV the type species. Demarcation criterion for this proposed genus thus include isolation from reptile hosts, ability to replicate at low temperatures, presence of a U gene between the N and P cistrons, and phylogenetic clustering with the type species FdlPV, separate from paramyxoviruses in other known genera (G. Kurath, personal communication).

Other unclassified paramyxoviruses

Tupaia paramyxovirus (TupPV)

TupPV was isolated from a South East Asian tree shrew (*Tupaia belangeri*), in the late 1970s (Tidona *et al.*, 1999). Primary kidney cells from the tree shrew demonstrated CPE after 7 days in culture. Supernatants were transferred to tree shrew fibroblast and kidney cell lines and the virus was purified by sucrose gradient ultracentrifugation. Electron microscopy revealed that the isolated virus was 300–1000 nm in diameter, pleomorphic and possessed an envelope with 8 nm long surface projections. TupPV growth was restricted to tree shrew cells and the virus did not replicate in any other mammalian cell line. Infection in newborn Syrian hamsters, C3H/HeJ mice and Sprague-Dawley rats could not be established using subcutaneous, intraperitoneal nor intracerebral routes of inoculation (Tidona *et al.*, 1999). Lack of a receptor on non-tree shrew cells is suggested as the limiting factor restricting TupPV host range and tropism (Springfeld *et al.*, 2005).

The TupPV genome has been completely sequenced, is 17,904 nucleotides in length (Springfeld *et al.*, 2005), and obeys the 'rule of six'. Analysis of the N and P genes revealed that TupPV is most closely related to the henipaviruses and morbilliviruses (Tidona *et al.*, 1999). The TupPV 3′ non-coding leader region of the genome contains 55 nucleotides, consistent with all other paramyxoviruses and genes are separated GAA intergenic triplets. The 5′ non-coding region of the genome is significantly longer than that of other paramyxoviruses, with a length of 590 nucleotides (Springfeld *et al.*, 2005) (Fig. 12.1).

The N protein of TupPV is 552 amino acids in length. Faithful transcription of the P gene leads to the production of the P protein and the incorporation of one non-templated G residue by RNA editing produces the V protein. TupPV produces a C protein from an alternative ORF within the P gene (Tidona *et al.*, 1999). The F protein of TupPV is 553 amino acids in length and is most similar to the F protein of MosPV (Miller *et al.*, 2003). In contrast to all other known F proteins, the TupPV F protein does not contain a monobasic or oligobasic cleavage sequence upstream of the signal peptide, suggesting cleavage of F by an unknown protease (Springfeld *et al.*, 2005). TupPV contains haemagglutinin but does not possess neuraminidase activity and the H attachment protein is glycosylated (Springfeld *et al.*, 2005). The L gene contains the QGDNE sequence present in HeV, NiV, MosPV and NarPV.

Salem virus (SalPV)

SalPV was isolated from a horse in 1992 during a disease outbreak that occurred simultaneously at three thoroughbred racetracks in New Hampshire and Massachusetts, USA (Renshaw *et al.*, 2000). Subsequent epidemiological analysis of this outbreak concluded that SalPV was not the aetiological agent of the disease, but the isolation of the virus was fortuitous (Glaser *et al.*, 2002).

SalPV was isolated from a co-culture of equine mononuclear cells and SV-40 transformed equine ovary cells. Serological classification was attempted by using immunofluorescence to screen SalPV infected cells against PorPV, HPIV-5, NDV, CDV, MeV and HeV sera, but was unsuccessful. Observation of pleomorphic virions by electron microscopy identified the virus as a paramyxovirus (Renshaw *et al.*, 2000). Intravenous inoculation of a 20 day old foal with SalPV resulted in productive infection in the animal, but no overt clinical signs of disease were observed throughout the duration of the experiment (Glaser *et al.*, 2002). This single experimental infection, along with serological evidence, suggests that horses may be the natural host of SalPV.

The complete genome of SalPV has been sequenced and is 16,698 nucleotides in length (Renshaw and Wang, unpublished data). The SalPV genome conforms to the 'rule of six'. The SalPV 3′ leader region of the genome contains the standard 55 nucleotides and the 5′ trailer region has a length of 35 nucleotides. The first and last 11 nucleotides of the genome are complementary.

The N protein of SalPV has a predicted length of 542 amino acids and, although larger, is most similar to the morbilliviruses (Renshaw *et al.*, 2000). In contrast to the morbilliviruses, the P gene of SalPV appears to encode a V protein, with the P protein produced through RNA editing. The P protein is produced from the P gene following the insertion of 6–13 non-templated G residues (Renshaw *et al.*, 2000). SalPV produces a C protein from an alternative ORF within the P gene. The SalPV L protein contains QGDNT motif in place of the QGDNQ/E, the only paramyxovirus to date that differs in this region (Renshaw and Wang, unpublished data).

Conclusion and perspectives

The rapid progress in paramyxovirus discovery during the last 15 years has challenged the notion of a 'typical' paramyxovirus. No longer can the paramyxovirus genome be considered a 15–16 kb strand of RNA consisting of six genes. The discovery and subsequent molecular characterization of HeV in 1994 lead to the finding that the virus had a larger genome than all other paramyxoviruses and also contained extended UTRs. Since this discovery, an increasing amount of novel features have been discovered within paramyxovirus genomes. Similar to HeV, the genome of AsaPV is long, with extended UTRs. Likewise, the TupPV genome contains long UTRs and an extended trailer region. These viruses differ in genome coding capacity, hence pushing the boundaries of the 'compact' paramyxovirus genome.

The SH gene, previously only known for some rubulaviruses, has since been found in JPV and BeiPV. The characterization of JPV and BeiPV genomes also uncovered another gene, TM, not previously seen in other paramyxoviruses. The presence of the X ORFs in the JPV and BeiPV genomes is intriguing and further investigation of these genes may lead to clues about the evolutionary histories of paramyxoviruses in general. The presence of the U gene in the FdlPV genome further demonstrates the 'elasticity' of the paramyxovirus genome by having an extra gene located at the upstream region of the transcription gradient. It is expected that as more paramyxoviruses are uncovered, more novel genomic features will be found.

The discovery and characterization of new and emerging paramyxoviruses broadens the knowledge of genome diversity and host range of this family of viruses. New and emerging viruses are potentially zoonotic and the possible disease outcome upon human infection is unknown. The broad host range of paramyxoviruses, ranging from humans, to snakes, to bats, demonstrates that this group of viruses is capable of infecting many species. This fact highlights the importance of studying paramyxovirus transmission and pathogenesis, with the goal to prevent future disease outbreaks. Upon infection, paramyxoviruses, such as the henipaviruses, can cause diseases with high mortality rates. On the other hand, paramyxovirus infection may also persist without evidence of disease. As there is currently no known molecular feature that distinguishes pathogenic from non-pathogenic paramyxoviruses, it is important to determine the best approaches to prevent or treat infections of individual paramyxoviruses.

Further understanding of viral transmission and pathogenesis can play an important role to develop prevention and treatment strategies. With the advancement in scientific techniques and rapid detection methods, it is likely new paramyxoviruses will continue to be discovered. Ongoing characterization and studies of new and emerging paramyxoviruses will undoubtedly lead to a better understanding of this diverse and important family of viruses.

References

Abraham, G., Rhodes, D.P., and Banerjee, A.K. (1975). The 5′ terminal structure of the methylated mRNA synthesized *in vitro* by vesicular stomatitis virus. Cell *5*, 51–58.

Ahne, W., Batts, W.N., Kurath, G., and Winton, J.R. (1999). Comparative sequence analyses of sixteen reptilian paramyxoviruses. Virus Res. *63*, 65–74.

Allan, G.M., McNeilly, F., Walker, I., Linne, T., Moreno-Lopez, J., Hernandez, P., Kennedy, S., Carroll, B.P., Herron, B., Foster, J.C., *et al.* (1996). A sequential study of experimental porcine paramyxovirus (LPMV) infection of pigs: immunostaining of cryostat sections and virus isolation. J. Vet. Diagn. Invest. *8*, 405–413.

Baker, D.G. (1998). Natural pathogens of laboratory mice, rats, and rabbits and their effects on research. Clin. Microbiol. Rev. *11*, 231–266.

Basler, C.F., Garcia-Sastre, A., and Palese, P. (2005). A novel paramyxovirus? Emerg. Infect. Dis. *11*, 108–112.

Batts, W.N., Falk, K., and Winton, J.R. (2008). Genetic Analysis of Paramyxovirus Isolates from Pacific Salmon Reveals Two Independently Co-circulating Lineages. J. Aquatic Anim. Health *20*, 215–224.

Berg, M., Sundqvist, A., Moreno-Lopez, J., and Linne, T. (1991). Identification of the porcine paramyxovirus LPMV matrix protein gene: comparative sequence analysis with other paramyxoviruses. J. Gen. Virol. *72*, 1045–1050.

Berg, M., Hjertner, B., Moreno-Lopez, J., and Linne, T. (1992). The P gene of the porcine paramyxovirus LPMV encodes three possible polypeptides P, V and C: the P protein mRNA is edited. J. Gen. Virol. *73*, 1195–1200.

Berg, M., Bergvall, A.C., Svenda, M., Sundqvist, A., Moreno-Lopez, J., and Linne, T. (1997). Analysis of the fusion protein gene of the porcine rubulavirus LPMV: comparative analysis of paramyxovirus F proteins. Virus Genes *14*, 55–61.

Blumberg, B.M., Leppert, M., and Kolakofsky, D. (1981). Interaction of VSV leader RNA and nucleocapsid protein may control VSV genome replication. Cell *23*, 837–845.

Bowden, T.R., and Boyle, D.B. (2005). Completion of the full-length genome sequence of Menangle virus: characterisation of the polymerase gene and genomic 5′ trailer region. Arch. Virol. *150*, 2125–2137.

Bowden, T.R., Westenberg, M., Wang, L.F., Eaton, B.T., and Boyle, D.B. (2001). Molecular characterization of Menangle virus, a novel paramyxovirus which infects pigs, fruit bats, and humans. Virology *283*, 358–373.

Calain, P., and Roux, L. (1993). The rule of six, a basic feature for efficient replication of Sendai virus defective interfering RNA. The J. Virol. *67*, 4822–4830.

Calisher, C.H., Childs, J.E., Field, H.E., Holmes, K.V., and Schountz, T. (2006). Bats: important reservoir hosts of emerging viruses. Clin. Microbiol. Rev. *19*, 531–545.

Campbell, R.W., Carley, J.G., Doherty, R.L., R.Domrow, Filippich, C., Gorman, B.M., and Karabatsos, N. (1977). Mossman virus, a paramyxovirus of rodents isolated in Queensland. SeArch. *8*, 435–436.

Chang, P.C., Hsieh, M.L., Shien, J.H., Graham, D.A., Lee, M.S., and Shieh, H.K. (2001). Complete nucleotide sequence of avian paramyxovirus type 6 isolated from ducks. J. Gen. Virol. *82*, 2157–2168.

Chant, K., Chan, R., Smith, M., Dwyer, D.E., and Kirkland, P. (1998). Probable human infection with a newly described virus in the family *Paramyxoviridae*. The NSW Expert Group. Emerg. Infect. Dis. *4*, 273–275.

Chua, K.B., Wang, L.F., Lam, S.K., Crameri, G., Yu, M., Wise, T., Boyle, D., Hyatt, A.D., and Eaton, B.T. (2001). Tioman virus, a novel paramyxovirus isolated from fruit bats in Malaysia. Virology *283*, 215–229.

Chua, K.B., Wang, L.F., Lam, S.K., and Eaton, B.T. (2002). Full length genome sequence of Tioman virus, a novel paramyxovirus in the genus *Rubulavirus* isolated from fruit bats in Malaysia. Arch. Virol. *147*, 1323–1348.

Chuang, J.L., and Perrault, J. (1997). Initiation of vesicular stomatitis virus mutant polR1 transcription internally at the N gene *in vitro*. J. Virol. *71*, 1466–1475.

Clark, H.F., Lief, F.S., Lunger, P.D., Waters, D., Leloup, P., Foelsch, D.W., and Wyler, R.W. (1979). Fer de Lance virus (FDLV): a probable paramyxovirus isolated from a reptile. J. Gen. Virol. *44*, 405–418.

Conzelmann, K.K. (1998). Nonsegmented negative-strand RNA viruses: genetics and manipulation of viral genomes. Annu. Rev. Genet. *32*, 123–162.

Cordey, S., and Roux, L. (2006). Transcribing paramyxovirus RNA polymerase engages the template at its 3′ extremity. J. Gen. Virol. *87*, 665–672.

Crochu, S., Cook, S., Attoui, H., Charrel, R.N., De Chesse, R., Belhouchet, M., Lemasson, J.J., de Micco, P., and de Lamballerie, X. (2004). Sequences of flavivirus-related RNA viruses persist in DNA form integrated in the genome of Aedes spp. mosquitoes. J. Gen. Virol. *85*, 1971–1980.

Curran, J., and Kolakofsky, D. (2008). Nonsegmented negative-strand RNA virus RNA synthesis *in vivo*. Virology *371,* 227–230.

Drexler, J.F., Corman, V.M., Gloza-Rausch, F., Seebens, A., Annan, A., Ipsen, A., Kruppa, T., Müller, M.A., Kalko, E.K.V., Adu-Sarkodie, Y., *et al.* (2009). Henipavirus RNA in African Bats. PLoS ONE *4,* e6367.

Egelman, E.H., Wu, S.S., Amrein, M., Portner, A., and Murti, G. (1989). The Sendai virus nucleocapsid exists in at least four different helical states. J. Virol. *63,* 2233–2243.

Elango, N., Kovamees, J., Varsanyi, T.M., and Norrby, E. (1989). mRNA sequence and deduced amino acid sequence of the mumps virus small hydrophobic protein gene. J. Virol. *63,* 1413–1415.

Essbauer, S., and Ahne, W. (2001). Viruses of lower vertebrates. J. Vet. Med. B Infect. Dis. Vet. Public Health *48,* 403–475.

Falk, K., Batts, W.N., Kvellestad, A., Kurath, G., Wiik-Nielsen, J., and Winton, J.R. (2008). Molecular characterisation of Atlantic salmon paramyxovirus (ASPV): a novel paramyxovirus associated with proliferative gill inflammation. Virus Res. *133,* 218–227.

Fearns, R., Peeples, M.E., and Collins, P.L. (1997). Increased expression of the N protein of respiratory syncytial virus stimulates minigenome replication but does not alter the balance between the synthesis of mRNA and antigenome. Virology *236,* 188–201.

Franke, J., Essbauer, S., Ahne, W., and Blahak, S. (2001). Identification and molecular characterization of 18 paramyxoviruses isolated from snakes. Virus Res. *80,* 67–74.

Franke, J., Batts, W.N., Ahne, W., Kurath, G., and Winton, J.R. (2006). Sequence motifs and prokaryotic expression of the reptilian paramyxovirus fusion protein. Arch. Virol. *151,* 449–464.

Fridell, F., Devold, M., and Nylund, A. (2004). Phylogenetic position of a paramyxovirus from Atlantic salmon Salmo salar. Dis. Aquat. Organ. *59,* 11–15.

Glaser, A.L., Renshaw, R.W., Trock, S.C., Brady, R.C., and Dubovi, E.J. (2002). Isolation of Salem virus, a novel equine paramyxovirus, and assessment of its etiologic role in a disease outbreak. Vet. Microbiol. *87,* 205–212.

Hagmaier, K., Stock, N., Precious, B., Childs, K., Wang, L.F., Goodbourn, S., and Randall, R.E. (2007). Mapuera virus, a rubulavirus that inhibits interferon signalling in a wide variety of mammalian cells without degrading STATs. J. Gen. Virol. *88,* 956–966.

Hausmann, S., Jacques, J.P., and Kolakofsky, D. (1996). Paramyxovirus RNA editing and the requirement for hexamer genome length. RNA *2,* 1033–1045.

Henderson, G.W., Laird, C., Dermott, E., and Rima, B.K. (1995). Characterization of Mapuera virus: structure, proteins and nucleotide sequence of the gene encoding the nucleocapsid protein. J. Gen. Virol. *76,* 2509–2518.

Hernandez-Jauregui, P., Ramirez Mendoza, H., Mercado Garcia, C., Moreno-Lopez, J., and Kennedy, S. (2004). Experimental porcine rubulavirus (La Piedad-Michoacan virus) infection in pregnant gilts. J. Comp Pathol. *130,* 1–6.

Hiebert, S.W., Paterson, R.G., and Lamb, R.A. (1985). Identification and predicted sequence of a previously unrecognized small hydrophobic protein, SH, of the paramyxovirus simian virus 5. J. Virol. *55,* 744–751.

Hooper, P.T. (2001). Identification and eradication of Menangle virus from pigs. Aust. Vet. J. *79,* 190–191.

Horie, M., Honda, T., Suzuki, Y., Kobayashi, Y., Daito, T., Oshida, T., Ikuta, K., Jern, P., Gojobori, T., Coffin, J.M., *et al.* (2010). Endogenous non-retroviral RNA virus elements in mammalian genomes. Nature *463,* 84–87.

Hosaka, M., Nagahama, M., Kim, W.S., Watanabe, T., Hatsuzawa, K., Ikemizu, J., Murakami, K., and Nakayama, K. (1991). Arg-X-Lys/Arg-Arg motif as a signal for precursor cleavage catalyzed by furin within the constitutive secretory pathway. J. Biol. Chem. *266, 12,*127–12130.

Iehle, C., Razafitrimo, G., Razainirina, J., Andriaholinirina, N., Goodman, S.M., Faure, C., Georges-Courbot, M.C., Rousset, D., and Reynes, J.M. (2007). Henipavirus and Tioman virus antibodies in pteropodid bats, Madagascar. Emerg. Infect. Dis. *13,* 159–161.

Jack, P.J., Boyle, D.B., Eaton, B.T., and Wang, L.F. (2005). The complete genome sequence of J. virus reveals a unique genome structure in the family *Paramyxoviridae*. J. Virol. *79,* 10690–10700.

Jack, P.J., Anderson, D.E., Bossart, K.N., Marsh, G.A., Yu, M., and Wang, L.F. (2008). Expression of novel genes encoded by the paramyxovirus J. virus. J. Gen. Virol. *89,* 1434–1441.

Jacobson, E., Origgi, F., Pessier, A., Lamirande, E., Walker, I., Whitaker, B., Stalis, I., Nordhausen, R., Owens, J., Nichols, D., *et al.* (2001). Paramyxovirus infection in caiman lizards (Draecena guianensis). J. Vet. Diagn Invest *13,* 143–151.

Jorgensen, E.D., Collins, P.L., and Lomedico, P.T. (1987). Cloning and nucleotide sequence of Newcastle disease virus hemagglutinin-neuraminidase mRNA: identification of a putative sialic acid binding site. Virology *156,* 12–24.

Jun, M.H., Karabatsos, N., and Johnson, R.H. (1977). A new mouse paramyxovirus (J virus). Aust. J. Exp. Biol. Med. *55,* 645–647.

Karabatsos, N., Buckley, S.M., and Ardoin, P. (1969). Nariva virus: further studies, with particular reference to its hemadsorption and hemagglutinating properties. Proc. Soc. Exp. Biol. Med. *130,* 888–892.

Kindermann, J., Kubber-Heiss, A., Kerschbaumer, P., and Nowotny, N. (2001). Phylogenetic analysis of the L and HN gene of ophidian paramyxoviruses. Brief report. Arch. Virol. *146,* 1021–1035.

Kirkland, P.D., Love, R.J., Philbey, A.W., Ross, A.D., Davis, R.J., and Hart, K.G. (2001). Epidemiology and control of Menangle virus in pigs. Aust. Vet. J. *79,* 199–206.

Kirkland, P.D., Daniels, P.W., Nor, M.N., Love, R.J., Philbey, A.W., and Ross, A.D. (2002). Menangle and Nipah virus infections of pigs. Vet. Clin. North Am. Food Anim Pract *18,* 557–571, ix.

Kolakofsky, D., Pelet, T., Garcin, D., Hausmann, S., Curran, J., and Roux, L. (1998). Paramyxovirus RNA synthesis and the requirement for hexamer genome length: the rule of six revisited. J. Virol. *72,* 891–899.

Kolakofsky, D., Roux, L., Garcin, D., and Ruigrok, R.W. (2005). Paramyxovirus mRNA editing, the 'rule of six' and error catastrophe: a hypothesis. J. Gen. Virol. *86,* 1869–1877.

Kozak, M. (1986). Point mutations define a sequence flanking the AUG initiator codon that modulates translation by eukaryotic ribosomes. Cell *44,* 283–292.

Kurath, G., Batts, W.N., Ahne, W., and Winton, J.R. (2004). Complete genome sequence of Fer-de-Lance virus reveals a novel gene in reptilian paramyxoviruses. J. Virol. *78,* 2045–2056.

Kvellestad, A., Dannevig, B.H., and Falk, K. (2003). Isolation and partial characterization of a novel paramyxovirus from the gills of diseased seawater-reared Atlantic salmon (Salmo salar L.). J. Gen. Virol. *84,* 2179–2189.

Lamb, R.A., and Parks, G.D. (2007). *Paramyxoviridae*: the viruses and their replication. In Fields Virology, Knipe, D.M., Howley, P.M., Griffin, D.E., Lamb, R.A., Martin, M.A., Roizman, B., and Straus, S.E., eds. (Lippincott Williams & Wilkins: Philadelphia, PA), pp. 1449–1496.

Lamb, R.A., Collins, P.L., Kolakofsky, D., Melero, J.A., Nagai, Y., Oldstone, M.B.A., Pringle, C.R., and Rima, B.K. (2005). *Paramyxoviridae.* In Virus Taxonomy: VIIIth Report of the International Committee on Taxonomy of Viruses, Fauquet, C.M., Mayo, M.A., Maniloff, J., Desselberger, U., and Ball, L.A., eds. (Elsevier Academic Press: London), pp. 655–668.

Lambeth, L.S., Yu, M., Anderson, D.E., Crameri, G., Eaton, B.T., and Wang, L.F. (2009). Complete genome sequence of Nariva virus, a rodent paramyxovirus. Arch. Virol. *154,* 199–207.

Langedijk, J.P., Daus, F.J., and van Oirschot, J.T. (1997). Sequence and structure alignment of *Paramyxoviridae* attachment proteins and discovery of enzymatic activity for a morbillivirus hemagglutinin. J. Virol. *71,* 6155–6167.

Lau, S.K., Woo, P.C., Wong, B.H., Wong, A.Y., Tsoi, H.W., Wang, M., Lee, P., Xu, H., Poon, R.W., Guo, R., *et al.* (2010). Identification and complete genome analysis of three novel paramyxoviruses, Tuhoko virus *1,* 2 and *3,* in fruit bats from China. Virology *404,* 106–116.

Li, Z., Yu, M., Zhang, H., Magoffin, D.E., Jack, P.J., Hyatt, A., Wang, H.Y., and Wang, L.F. (2006). Beilong virus, a novel paramyxovirus with the largest genome of non-segmented negative-stranded RNA viruses. Virology *346,* 219–228.

Liang, X., Zhang, H., Zhou, A., and Wang, H. (2003). AngRem104, an angiotensin II-induced novel upregulated gene in human mesangial cells, is potentially involved in the regulation of fibronectin expression. J. Am. Soc. Nephrol. *14,* 1443–1451.

Love, R.J., Philbey, A.W., Kirkland, P.D., Ross, A.D., Davis, R.J., Morrissey, C., and Daniels, P.W. (2001). Reproductive disease and congenital malformations caused by Menangle virus in pigs. Aust. Vet. J. *79,* 192–198.

Luytjes, W., Krystal, M., Enami, M., Parvin, J.D., and Palese, P. (1989). Amplification, expression, and packaging of foreign gene by influenza virus. Cell *59,* 1107–1113.

Magoffin, D.E., Mackenzie, J.S., and Wang, L.-F. (2007). Genetic analysis of J-virus and Beilong virus using minireplicons. Virology *364,* 103–111.

Marschang, R.E., Papp, T., and Frost, J.W. (2009). Comparison of paramyxovirus isolates from snakes, lizards and a tortoise. Virus Res. *144,* 272–279.

Mesina, J.E., Campbell, R.S., Glazebrook, J.S., Copeman, D.B., and Johnson, R.H. (1974). The pathology of feral rodents in North Queensland. Tropenmed. Parasitol. *25,* 116–127.

von Messling, V., Milosevic, D., and Cattaneo, R. (2004). Tropism illuminated: lymphocyte-based pathways blazed by lethal morbillivirus through the host immune system. Proc. Natl. Acad. Sci. U.S.A. *101,* 14216–14221.

Miller, P.J., Boyle, D.B., Eaton, B.T., and Wang, L.F. (2003). Full-length genome sequence of Mossman virus, a novel paramyxovirus isolated from rodents in Australia. Virology *317,* 330–344.

Moreno-Lopez, J., Correa-Giron, P., Martinez, A., and Ericsson, A. (1986). Characterization of a paramyxovirus isolated from the brain of a piglet in Mexico. Arch. Virol. *91,* 221–231.

Neumann, G., Whitt, M.A., and Kawaoka, Y. (2002). A decade after the generation of a negative-sense RNA virus from cloned cDNA – what have we learned? J. Gen. Virol. *83,* 2635–2662.

Nylund, S., Karlsen, M., and Nylund, A. (2008). The complete genome sequence of the Atlantic salmon paramyxovirus (ASPV). Virology *373,* 137–148.

Philbey, A., Kirkland, P., Ross, A., Field, H., Srivastava, M., Davis, R., and Love, R. (2008). Infection with Menangle virus in flying foxes (Pteropus spp.) in Australia. Aust. Vet. J. *86,* 449–454.

Philbey, A.W., Kirkland, P.D., Ross, A.D., Davis, R.J., Gleeson, A.B., Love, R.J., Daniels, P.W., Gould, A.R., and Hyatt, A.D. (1998). An apparently new virus (family *Paramyxoviridae*) infectious for pigs, humans, and fruit bats. Emerg. Infect. Dis. *4,* 269–271.

Plumet, S., Duprex, W.P., and Gerlier, D. (2005). Dynamics of viral RNA synthesis during measles virus infection. J. Virol. *79,* 6900–6908.

Qanungo, K.R., Shaji, D., Mathur, M., and Banerjee, A.K. (2004). Two RNA polymerase complexes from vesicular stomatitis virus-infected cells that carry out transcription and replication of genome RNA. Proc. Natl. Acad. Sci. U.S.A. *101,* 5952–5957.

Ramirez-Mendoza, H., Hernandez-Jauregui, P., Reyes-Leyva, J., Zenteno, E., Moreno-Lopez, J., and Kennedy, S. (1997). Lesions in the reproductive tract of boars experimentally infected with porcine rubulavirus. J. Comp. Pathol. *117,* 237–252.

Renshaw, R.W., Glaser, A.L., Van Campen, H., Weiland, F., and Dubovi, E.J. (2000). Identification and phylogenetic comparison of Salem virus, a novel paramyxovirus of horses. Virology *270,* 417–429.

Reyes-Leyva, J., Hernandez-Jauregui, P., Montano, L.F., and Zenteno, E. (1993). The porcine paramyxovirus LPM specifically recognizes sialyl (alpha 2,3) lactose-containing structures. Arch. Virol. *133,* 195–200.

Reyes-Leyva, J., Espinosa, B., Hernandez, J., Zenteno, R., Vallejo, V., Hernandez-Jauregui, P., and Zenteno, E. (1997). NeuAc alpha 2,3gal-glycoconjugate expression determines cell susceptibility to the porcine rubulavirus LPMV. Comp. Biochem. Physiol. B. Biochem. Mol. Biol. *118,* 327–332.

Reyes-Leyva, J., Espinosa, B., Santos, G., Zenteno, R., Hernández, J., Vallejo, V., and Zenteno, E. (1999). Purification and characterization of the Hemagglutinin-Neuraminidase of porcine rubulavirus LPMV. Glycoconjugate Journal *16,* 517–522.

Roos, R.P., and Wollmann, R. (1979). Non-productive paramyxovirus infection: Nariva virus infection in hamsters. Arch. Virol. *62,* 229–240.

Salas-Rojas, M., Sanchez-Hernandez, C., Romero-Almaraz Md Mde, L., Schnell, G.D., Schmid, R.K., and Aguilar-Setien, A. (2004). Prevalence of rabies and LPM paramyxovirus antibody in non-hematophagous bats captured in the Central Pacific coast of Mexico. Trans. R. Soc. Trop. Med. Hyg. *98,* 577–584.

Schnell, M.J., Mebatsion, T., and Conzelmann, K.K. (1994). Infectious rabies viruses from cloned cDNA. EMBO J. *13,* 4195–4203.

Schomacker, H., Collins, P.L., and Schmidt, A.C. (2004). In silico identification of a putative new paramyxovirus related to the *Henipavirus* genus. Virology *330,* 178–185.

Sidhu, M.S., Chan, J., Kaelin, K., Spielhofer, P., Radecke, F., Schneider, H., Masurekar, M., Dowling, P.C., Billeter, M.A., and Udem, S.A. (1995). Rescue of synthetic measles virus minireplicons: measles genomic termini direct efficient expression and propagation of a reporter gene. Virology *208,* 800–807.

Solis, M., Ramirez-Mendoza, H., Mercado, C., Espinosa, S., Vallejo, V., Reyes-Leyva, J., and Hernandez, J. (2007). Semen alterations in porcine rubulavirus-infected boars are related to viral excretion and have implications for artificial insemination. Res. Vet. Sci. *83,* 403–409.

Springfeld, C., von Messling, V., Tidona, C.A., Darai, G., and Cattaneo, R. (2005). Envelope targeting: hemagglutinin attachment specificity rather than fusion protein cleavage-activation restricts Tupaia paramyxovirus tropism. J. Virol. *79,* 10155–10163.

Stephan, H., Gay, G., and Ramirez, T. (1988). Encephalomyelitis, reproductive failure and corneal opacity (blue eye) in pigs, associated with a paramyxovirus infection. Vet. Rec. *122,* 6–10.

Sundqvist, A., Berg, M., Hernandez-Jauregui, P., Linne, T., and Moreno-Lopez, J. (1990). The structural proteins of a porcine paramyxovirus (LPMV). J. Gen. Virol. *71,* 609–613.

Sundqvist, A., Berg, M., Moreno-López, J., and Linné, T. (1992). The haemagglutinin–neuraminidase glycoprotein of the porcine paramyxovirus LPMV: comparison with other paramyxoviruses revealed the closest relationship to simian virus 5 and mumps virus. Arch. Virol. *122,* 331–340.

Svenda, M., Berg, M., Moreno-Lopez, J., and Linne, T. (1997). Analysis of the large (L) protein gene of the porcine rubulavirus LPMV: identification of possible functional domains. Virus Res. *48,* 57–70.

Tidona, C.A., Kurz, H.W., Gelderblom, H.R., and Darai, G. (1999). Isolation and molecular characterization of a novel cytopathogenic paramyxovirus from tree shrews. Virology *258,* 425–434.

Tikasingh, E.S., Jonkers, A.H., Spence, L., and Aitken, T.H. (1966). Nariva virus, a hitherto undescribed agent isolated from the Trinidadian rat, Zygodontomys b. brevicauda (J. A. Allen and Chapman). Am. J. Trop Med. Hyg *15,* 235–238.

Tong, S., Conrardy, C., Kuzmin, I.V., Agwanda, B., Neizgoda, M., Breiman, R.F., Rupprecht, C.E., and Anderson, L.J. (2010). Novel paramyxoviruses from Kenyan bats- a potential for emerging zoonotic diseases. Paper presented at: XIV International Conference on Negative Strand Viruses (Brugge, Belgium).

Towner, J.S., Paragas, J., Dover, J.E., Gupta, M., Goldsmith, C.S., Huggins, J.W., and Nichol, S.T. (2005). Generation of eGFP expressing recombinant Zaire ebolavirus for analysis of early pathogenesis events and high-throughput antiviral drug screening. Virology *332,* 20–27.

Vidal, S., and Kolakofsky, D. (1989). Modified model for the switch from Sendai virus transcription to replication. J. Virol. *63,* 1951–1958.

Vijaykrishna, D., Smith, G.J.D., Zhang, J.X., Peiris, J.S.M., Chen, H., and Guan, Y. (2007). Evolutionary Insights into the Ecology of Coronaviruses. J. Virol. *81,* 4012–4020.

Vulliemoz, D., Cordey, S., Mottet-Osman, G., and Roux, L. (2005). Nature of a paramyxovirus replication promoter influences a nearby transcription signal. J. Gen. Virol. *86,* 171–180.

Walder, R. (1971). Electron microscopic evidence of Nariva virus structure. J. Gen. Virol. *11,* 123–128.

Wang, L.-F., Shi, Z., Zhang, S., Field, H., Daszak, P., and Eaton, B.T. (2006). Review of bats and SARS. In Emerg. Infect. Dis. [serial on the Internet].

Wang, L.F., Hansson, E., Yu, M., Chua, K.B., Mathe, N., Crameri, G., Rima, B.K., Moreno-Lopez, J., and Eaton, B.T. (2007). Full-length genome sequence and genetic relationship of two paramyxoviruses isolated from bat and pigs in the Americas. Arch. Virol. *152,* 1259–1271.

Whelan, S.P., and Wertz, G.W. (2002). Transcription and replication initiate at separate sites on the vesicular stomatitis virus genome. Proc. Natl. Acad. Sci. U.S.A. *99,* 9178–9183.

Yaiw, K.C., Crameri, G., Wang, L., Chong, H.T., Chua, K.B., Tan, C.T., Goh, K.J., Shamala, D., and Wong, K.T. (2007a). Serological evidence of possible human infection with Tioman virus, a newly described paramyxovirus of bat origin. J. Infect. Dis. *196,* 884–886.

Yaiw, K.C., Ong, K.C., Chua, K.B., Bingham, J., Wang, L., Shamala, D., and Wong, K.T. (2007b). Tioman virus infection in experimentally infected mouse brain and its association with apoptosis. J. Virol. Methods *143*, 140–146.

Yaiw, K.C., Bingham, J., Crameri, G., Mungall, B., Hyatt, A., Yu, M., Eaton, B., Shamala, D., Wang, L.F., and Thong Wong, K. (2008a). Tioman virus, a paramyxovirus of bat origin, causes mild disease in pigs and has a predilection for lymphoid tissues. J. Virol. *82*, 565–568.

Yaiw, K.C., Hyatt, A., Vandriel, R., Crameri, S.G., Eaton, B., Wong, M.H., Wang, L.F., Ng, M.L., Bingham, J., Shamala, D., *et al.* (2008b). Viral morphogenesis and morphological changes in human neuronal cells following Tioman and Menangle virus infection. Arch. Virol. *153*, 865–875.

Yu, M., Hansson, E., Langedijk, J.P.M., Eaton, B.T., and Wang, L.-F. (1998). The attachment protein of Hendra virus has high structural similarity but limited primary sequence homology compared with viruses in the genus *Paramyxovirus*. Virology 251, 227–233.

Zeller, H.G., Karabatsos, N., Calisher, C.H., Digoutte, J.P., Murphy, F.A., and Shope, R.E. (1989). Electron microscopy and antigenic studies of uncharacterized viruses. I. Evidence suggesting the placement of viruses in families Arenaviridae, *Paramyxoviridae*, or Poxviridae. Arch. Virol. *108*, 191–209.

Other Books of Interest

Microbial Biofilms	2012
Bacterial Glycomics	2012
Non-coding RNAs and Epigenetic Regulation of Gene Expression	2012
Brucella	2012
Molecular Virology and Control of Flaviviruses	2012
Bacterial Pathogenesis	2012
Bunyaviridae	2011
Emerging Trends in Antibacterial Discovery	2011
Epigenetics	2011
Metagenomics	2011
Nitrogen Cycling in Bacteria	2011
Helicobacter pylori	2011
Microbial Bioremediation of Non-metals	2011
Lactic Acid Bacteria and Bifidobacteria	2011
Essentials of Veterinary Parasitology	2011
Hepatitis C: Antiviral Drug Discovery and Development	2011
Streptomyces: Molecular Biology and Biotechnology	2011
Alphaherpesviruses: Molecular Virology	2011
Recent Advances in Plant Virology	2011
Vaccine Design: Innovative Approaches and Novel Strategies	2011
Salmonella: From Genome to Function	2011
Insect Virology	2010
Environmental Microbiology: Current Technology and Water Applications	2010
Sensory Mechanisms in Bacteria: Molecular Aspects of Signal Recognition	2010
Bifidobacteria: Genomics and Molecular Aspects	2010
Molecular Phylogeny of Microorganisms	2010
Nanotechnology in Water Treatment Applications	2010
Iron Uptake and Homeostasis in Microorganisms	2010
Caliciviruses: Molecular and Cellular Virology	2010
Epstein-Barr Virus: Latency and Transformation	2010
Lentiviruses and Macrophages: Molecular and Cellular Interactions	2010
RNA Interference and Viruses: Current Innovations and Future Trends	2010
Retroviruses: Molecular Biology, Genomics and Pathogenesis	2010
Frontiers in Dengue Virus Research	2010
Metagenomics: Theory, Methods and Applications	2010
Microbial Population Genetics	2010
Borrelia: Molecular Biology, Host Interaction and Pathogenesis	2010
Aspergillus: Molecular Biology and Genomics	2010
Environmental Molecular Microbiology	2010
Neisseria: Molecular Mechanisms of Pathogenesis	2010
Lab-on-a-Chip Technology: Biomolecular Separation and Analysis	2009
Lab-on-a-Chip Technology: Fabrication and Microfluidics	2009

Index

Page numbers followed by an *f* indicate figures.

N

R

S

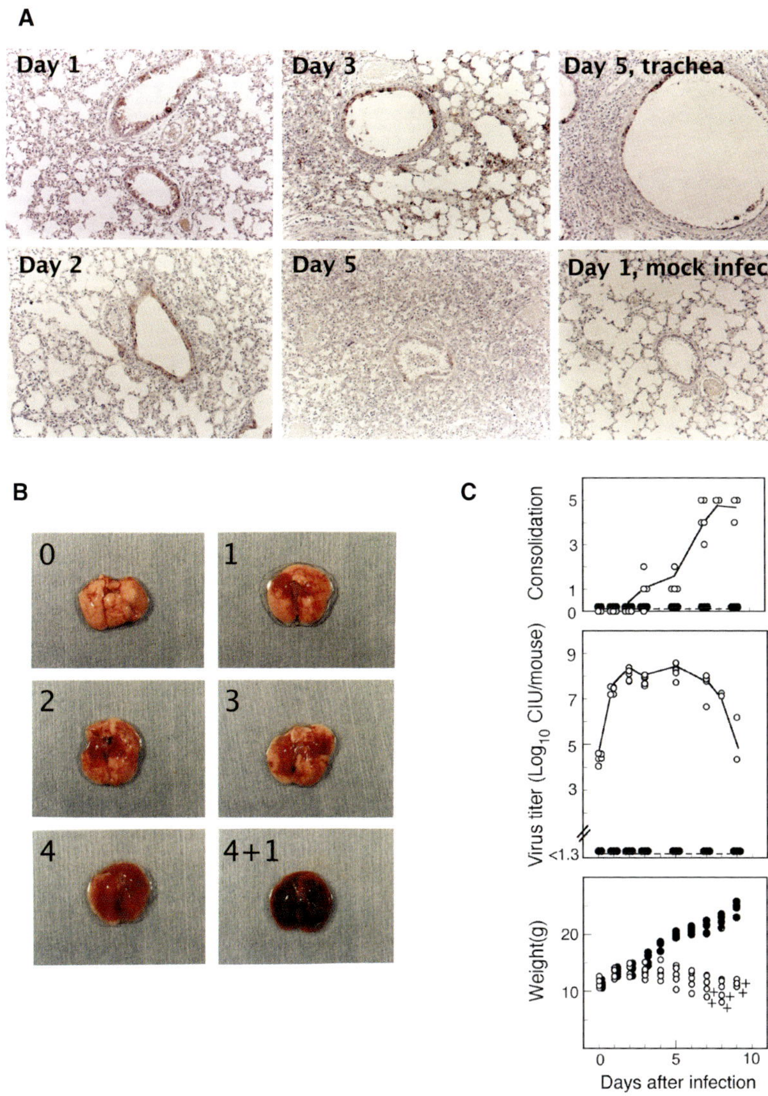

Figure 4.5 Progress of Sendai virus (SeV) infection in mice. 3-week-old ICR mice were intranasally inoculated with 10^7 CIU (cell infectious units) of SeV/mouse. (A) Immunohistochemical staining with mouse anti-SeV antibody and peroxidase conjugated anti-mouse IgG followed by haematoxylin-eosin staining. (B) Macroscopic view of the entire lung with consolidation scores. (C) Consolidation scores, virus growth and body weights of mice during the course of infection. Open circle, mouse infected; filled circle, uninfected; +, the mouse that died. For details see the text. Adapted from Kato *et al.* (1997a).

A

Case 201	Pre	1Mo	3Mo	5Mo
Peak walking distance (m)	74	138 (186%)	235 (318%)	>300m (complete)
Claudication onset distance (m)	43	118 (274%)	208 (483%)	>300m (complete)

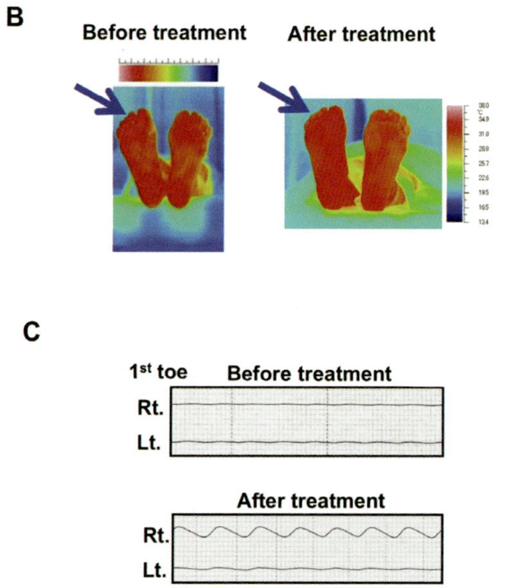

Figure 4.7 Improvement of efficacy parameters in a patient (Case No. 201) treated with Sendai virus (SeV) SeV/Δ*F*-hFGF2. This patient was a 65 year-old woman who had been suffering from rest pain of the right limb. Her left limb was also diseased, but did not show either rest pain or gangrene. (A) Time course of peak walking distance and claudication onset distance examined by treadmill test immediately before (pre) and 1, 3 and 5 months (Mo) after treatment. (B) Thermography assessing foot (sole) temperature. Both of the limbs before treatment showed similar temperature (left). In contrast, the gene transferred right limb, but not the untreated opposite limb, showed apparent increase of temperature by 3°C. Arrows indicate the right limbs before and after treatment. (C) Detection of flow wave on the first toe. At the time immediately before treatment, blood flow at the first toe was not detected in either the right (Rt) or left (Lt) limbs. Two months after gene transfer it became evident at the first toe of the treated Rt limb, but not the untreated Lt limb.